Lecture Notes in Computer Science 959

T0180884

Springer

Berlin
Heidelberg
New York
Barcelona
Budapest
Hong Kong
London
Milan
Paris
Tokyo

Ding-Zhu Du Ming Li (Eds.)

Computing
and Combinatorics

First Annual International Conference, COCOON '95
Xi'an, China, August 24-26, 1995
Proceedings

 Springer

Series Editors

Gerhard Goos, Karlsruhe University, Germany

Juris Hartmanis, Cornell University, NY, USA

Jan van Leeuwen, Utrecht University, The Netherlands

Volume Editors

Ding-Zhu Du
Department of Computer Science, University of Minnesota
Minneapolis, MN 55455, USA, and
Institute of Applied Mathematics, Chinese Academy of Sciences
Beijing 100080, China

Ming Li
Department of Computer Science, University of Waterloo
Waterloo, Ontario, Canada N2L 3G1

Cataloging-in-Publication data applied for

Die Deutsche Bibliothek - CIP-Einheitsaufnahme

Computing and combinatorics : first annual international
conference ; proceedings / COCOON '95, Xi'an, China, August
24 - 26, 1995. Ding-Zhu Du ; Ming Li (ed.). - Berlin ;
Heidelberg ; New York ; Barcelona ; Budapest ; Hong Kong ;
London ; Milan ; Paris ; Tokyo : Springer, 1995
  (Lecture notes in computer science ; Vol. 959)
  ISBN 3-540-60216-X
NE: Du, Ding-Zhu [Hrsg.]; COCOON <1, 1995, Xi'an>; GT

CR Subject Classification (1991): F.2, G.2.1-2, I.3.5, F.4.2

1991 Mathematics Subject Classification: 05Cxx, 68Q20, 68Q25, 68Q30,
68R05, 68R10

ISBN 3-540-60216-X Springer-Verlag Berlin Heidelberg New York

Typesetting: Camera-ready by author
SPIN 10486567   06/3142 – 5 4 3 2 1 0   Printed on acid-free paper

Preface

The papers in this volume were presented at the First Annual International Computing and Combinatorics Conference, held August 24-26, 1995, in Xi'an, China. The topics cover all aspects of theoretical computer science and combinatorics related to computing. The conference was sponsored by the National Natural Science Foundation of China, Xi'an Jiaotong University, and the Chinese Academy of Sciences.

These 52 regular and 22 short papers were selected from 120 submissions by a program committee consisting of Bonnie Berger, Avrim Blum, Shai Ben-David, Danny Z. Chen, Francis Y. Chin, Xiaotie Deng, Ding-Zhu Du, Shafi Goldwasser, Hiroshi Imai, Tao Jiang, Rao Kosaraju, Ming Li, Mike Paterson, Jyh-Jong Tsay, Vijay Vazirani, Osamu Watanabe, Derick Wood, F. Frances Yao, Andrew C.C. Yao, and Zezeng Zhang. These papers come from the following countries and regions: Australia, Canada, China (including Taiwan), Finland, France, Germany, Greece, Italy, Hong Kong, Japan, Korea, India, Israel, India, Norway, Russia, Singapore, Spain, Sweden, U.K., and USA. Every submitted paper was reviewed by three program committee members. In addition to the selected papers, three program committee members were invited to present plenary survey lectures on their research programs (Tao Jiang, Vijay Vazirani, and Derick Wood).

We wish to thank all who made this meeting possible: the authors for submitting papers, the program committee members for their excellent work in reviewing the papers, the sponsors, the local organizers, and Springer for their support and assistance. We would like to give a special thank-you to conference chairs Xiang-Sun Zhang and Zhao-Yong You who made this conference possible and -we hope- successful.

August 1995   Ding-Zhu Du
                                                                   Ming Li

Symposium Chairs:
Zhao-Yong You (Center for Applied Mathematics, Xi'an, China)
Xiang-Sun Zhang (Institute of Applied Mathematics, Chinese Academy of Sciences, China)

Program Committee Chairs:
Ding-Zhu Du (Institute of Applied Mathematics, Chinese Academy of Sciences, China and University of Minnesota, USA)
Ming Li (University of Waterloo, Canada)

Program Committee Members:
Bonnie Berger, MIT
Avrim Blum, Carnegie-Mellon University
Shai Ben-David, Technion
Danny Z. Chen, University of Notre Dame
Francis Y. Chin, Hong Kong University
Xiaotie Deng, York University
Shafi Goldwasser, MIT
Hiroshi Imai, University of Tokyo
Tao Jiang, McMaster University
Rao Kosaraju, Johns Hopkins University
Mike Paterson, University of Warwick
Jyh-Jong Tsay, National Chung Cheng University
Vijay Vazirani, Indian Institute of Technology
Osamu Watanabe, Tokyo Institute of Technology
Derick Wood, Hong Kong University of Science and Technology/UWO
F. Frances Yao, Xerox PARC
Andrew C.C. Yao, Princeton University
Zezeng Zhang, Xi'an University of Electronics

Organising Committee Members:
Ru-E Yang
Lin Cheng
Yong Li
Shu-Fen Qi

Contents

SESSION 11A: Scheduling

Plenary Survey Lectures:

The Complexity of Mean Payoff Games *

Uri Zwick[1] and Michael S. Paterson[2]

[1] Dept. of Computer Science, Tel Aviv University, Tel Aviv 69978, Israel
[2] Dept. of Computer Science, Univ. of Warwick, Coventry CV4 7AL, UK

Abstract. We study the complexity of finding the values and optimal strategies of *mean payoff games*, a family of perfect information games introduced by Ehrenfeucht and Mycielski. We describe a pseudo-polynomial time algorithm for the solution of such games, the decision problem for which is in NP ∩ co-NP. Finally, we describe a polynomial reduction from mean payoff games to the *simple stochastic games* studied by Condon. These games are also known to be in NP ∩ co-NP, but no polynomial or pseudo-polynomial time algorithm is known for them.

1 Introduction

Let $G = (V, E)$ be a finite directed graph in which each vertex has at least one edge going out of it. Let $w : E \to \{-W, \ldots, 0, \ldots, W\}$ be a function that assigns an integral weight to each edge of G. Ehrenfeucht and Mycielski [EM79] studied the following infinite two-person game played on such a graph. The game starts at a vertex $a_0 \in V$. The first player chooses an edge $e_1 = (a_0, a_1) \in E$. The second player then chooses an edge $e_2 = (a_1, a_2) \in E$, and so on indefinitely. The first player wants to maximise $\liminf_{n \to \infty} \frac{1}{n} \sum_{i=1}^{n} w(e_i)$. The second player wants to minimise $\limsup_{n \to \infty} \frac{1}{n} \sum_{i=1}^{n} w(e_i)$. Ehrenfeucht and Mycielski show that each such game has a value ν such that the first player has a strategy that ensures that $\liminf_{n \to \infty} \frac{1}{n} \sum_{i=1}^{n} w(e_i) \geq \nu$, while the second player has a strategy that ensures that $\limsup_{n \to \infty} \frac{1}{n} \sum_{i=1}^{n} w(e_i) \leq \nu$. Furthermore, they show that both players can achieve this value using a *positional strategy*, i.e., a strategy in which the next move depends only on the vertex from which the player is to move.

Without loss of generality, we may assume that the graph $G = (V, E)$ on which such a game is played is bipartite, with V_1 and V_2 being the partition of the vertices into the two 'sides' and with $E = E_1 \cup E_2$ such that $E_1 \subseteq V_1 \times V_2$ and $E_2 \subseteq V_2 \times V_1$. If the original graph is not bipartite, we simply duplicate the set of vertices.

To obtain their results for the infinite game, Ehrenfeucht and Mycielski [EM79] also consider the following finite version of the game. Again the game starts at a specific vertex of the graph, which is assumed to be bipartite. The

* Supported in part by the ESPRIT Basic Research Action Programme of the EC under contract No. 7141 (project ALCOM II). E-mail addresses of authors: zwick@math.tau.ac.il and Mike.Paterson@dcs.warwick.ac.uk .

players alternate in choosing successive edges that form a path, but the game ends as soon as a cycle is formed. The outcome of the game is then the mean weight of the edges on this cycle. The first player wants to maximise and the second player to minimise this outcome. This game is a finite perfect-information two-person game and so, by definition, has a value. Ehrenfeucht and Mycielski [EM79] show that the value ν of this finite game is also the value of the infinite game described above. Furthermore, they show, surprisingly perhaps, that both players have positional optimal strategies for the finite game. The positional optimal strategies of the finite game are also positional optimal strategies for the infinite game.

Ehrenfeucht and Mycielski [EM79] give no efficient algorithm for finding optimal strategies for the finite and infinite games. We complement their work by exhibiting an $O(|V|^3 \cdot |E| \cdot W)$ time algorithm for finding the values of the games played on a graph $G = (V, E)$. The algorithm finds the values of all the vertices of the graph. Games starting at different vertices may have different values, of course. We also give an $O(|V|^4 \cdot |E| \cdot \log(|E|/|V|) \cdot W)$ time algorithm for finding positional optimal strategies for both players. Our algorithm is polynomial in the size of the graph but only pseudo-polynomial in the weights. Our algorithm is polynomial if the weights are presented in unary notation. In particular, our algorithms work in polynomial time if the weights are taken from, say, $\{-1, 0, +1\}$. This is already a non-trivial case.

We also consider situations in which one player knows in advance the positional strategy to be used by the other player. Using a result of Karp [Kar78] we show that an optimal counter-strategy can be found in (strongly) polynomial time. This immediately implies that the decision problem associated with the game is in NP ∩ co-NP.

The decision problem corresponding to mean payoff games (MPG's) is thus in NP ∩ co-NP as well as in $\tilde{P}$ (pseudo-polynomial time), but is not yet known to be in P. This gives the MPG problem a rare status shared only by a few number-theoretic problems, such as *primality* [Pra75].

Mean payoff games have been considered independently by Gurvich, Karzanov and Khachiyan [GKK88]. They were not aware of the results of Ehrenfeucht and Mycielski and gave an alternative proof of the fact that both players in mean payoff games, or *cyclic games* as they call them, have positional optimal strategies. Gurvich *et al.* give an algorithm for finding such optimal strategies, but the worst-case complexity of their algorithm is exponential. Further generalisations and variants of mean payoff games have also been considered by Karzanov and Lebedev [KL93], who also point out that the decision problem corresponding to mean payoff games is in NP ∩ co-NP.

Condon [Con92] has recently studied the complexity of *simple stochastic games* (SSG's) introduced originally by Shapley [Sha53]. Condon shows that the decision problem corresponding to SSG's is also in NP ∩ co-NP. While MPG's are deterministic, SSG's are games of chance. We describe a simple reduction from MPG's to SSG's in two steps. We first describe a reduction from MPG's to *discounted payoff games* (DPG's), and then a reduction from DPG's to SSG's.

The reduction from MPG's to SSG's shows that SSG's are at least as hard as MPG's. We believe that the MPG problem is strictly easier then the SSG problem. As attempts to obtain polynomial time algorithms for SSG's have not yet borne fruit, it may be interesting to focus attention on the possibly easier problem of obtaining a polynomial time algorithm for MPG's.

2 Finding the values of a game

Let $G = (V, E)$ be a graph and let $w : E \to \{-W, \ldots, 0, \ldots, W\}$ be a weight function on its edges. Let $|V| = n$. We assume that the graph is bipartite with V_1 being the set of vertices from which player I is to play, and V_2 being the set of edges from which player II is to play.

Our first goal is to find, for each vertex $a \in V$, the value $\nu(a)$ of the finite and infinite games that start at a. Recall that the values of the finite and infinite games are equal. If $a \in V_1$ then player I (the maximiser) is to play first and if $a \in V_2$ then the second player (the minimiser) is to play first.

To reach this goal we consider a third version of the game. This time the two players play the game for exactly k steps constructing a path of length k, and the weight of this path is the outcome of the game. The length of the game is known is advance to both players. We let $\nu_k(a)$ be the value of this game started at $a \in V$, where player I or II plays first according to whether $a \in V_1$ or $a \in V_2$.

Theorem 1. *The values $\nu_k(a)$, for all $a \in V$, can be computed in $O(k \cdot |E|)$ time.*

Proof. The result follows easily from the following recursive relation

$$\nu_k(a) = \begin{cases} \max_{(a,b) \in E}\{w(a,b) + \nu_{k-1}(b)\} & \text{if } a \in V_1, \\ \min_{(a,b) \in E}\{w(a,b) + \nu_{k-1}(b)\} & \text{if } a \in V_2, \end{cases}$$

along with the initial condition, $\nu_0(a) = 0$ for every $a \in V$. □

It seems intuitively clear that $\lim_{k \to \infty} \nu_k(a)/k = \nu(a)$, where $\nu(a)$ is the value of the infinite game that starts at a. The next theorem shows that this is indeed the case. In the proof of this theorem we rely on the result, proved by Ehrenfeucht and Mycielski, that both players have positional optimal strategies. A *positional strategy for player I* is just a mapping $\pi_1 : V_1 \to V_2$ such that $(a_1, \pi_1(a_1)) \in E_1$ for every $a_1 \in V_1$. Similarly, a *positional strategy for player II* is a mapping $\pi_2 : V_2 \to V_1$ such that $(a_2, \pi_2(a_2)) \in E_2$ for every $a_2 \in V_2$.

Theorem 2. *For every $a \in V$ we have: $k \cdot \nu(a) - 2nW \leq \nu_k(a) \leq k \cdot \nu(a) + 2nW$.*

Proof. Let $\pi_1 : V_1 \to V_2$ be a positional optimal strategy for player I in the finite game starting at a. We show that if player I plays using the strategy π_1 then the outcome of a k-step game is at least $(k - n) \cdot \nu(a) - nW$. Consider a game in which player I plays according to π_1. Push (copies of) the edges played by the players onto a stack. Whenever a cycle is formed, it follows from the fact that π_1 is an optimal strategy for player I in the finite game, that the mean weight of the cycle formed is at least $\nu(a)$. The edges that participate in that cycle lie consecutively at the top of the stack. They are all removed and the process continues. Note that at each stage the stack contains at most n edges and the weight of each of them is at least $-W$. Player I can therefore ensure that the total weight of the edges encountered in a k-step game starting from a is at least $(k-n)\cdot\nu(a)-nW$. This is at least $k\cdot\nu(a)-2nW$ as $\nu(a)\leq W$.

Similarly, if player II plays according to a positional optimal strategy π_2 : $V_2 \to V_1$ of the finite game that starts at a, she can make sure that the mean weight of each cycle closed is at most $\nu(a)$. At most n edges are left on the stack and the weight of each of them is at most W. She can therefore ensure that the total weight of the edges encountered in a k-step game starting at a is at most $(k-n)\cdot\nu(a)+nW \leq k\cdot\nu(a)+2nW$. $\square$

We can now describe the algorithm for computing the exact values of the finite and infinite games.

Theorem 3. *Let $G = (V, E)$ be a directed graph and let $w : E \to \{-W,\ldots,0, \ldots,W\}$ be a weight function on its edges. The value $\nu(a)$, for every $a \in V$, corresponding to the infinite and finite games that start at all the vertices of V can be computed in $O(|V|^3\cdot|E|\cdot W)$ time.*

Proof. Compute the values $\nu_k(a)$, for every $a \in V$, for $k = 2n^3W$. This can be done, according to Theorem 1, in $O(|V|^3\cdot|E|\cdot W)$ time. For each vertex $a \in V$, compute the estimate $\nu'(a) = \nu_k(a)/k$. By Theorem 2, we get that

$$\nu'(a) - \frac{1}{n(n-2)} < \nu'(a) - \frac{2nW}{k} \leq \nu(a) \leq \nu'(a) + \frac{2nW}{k} < \nu'(a) + \frac{1}{n(n-2)}\ .$$

The value $\nu(a)$ is a rational number, with an even denominator whose size is at most n. The minimum distance between two possible values of $\nu(a)$ is at least $2/(n(n-2))$. The exact value of $\nu(a)$ is therefore the unique rational number with an even denominator of size at most n that lies in the interval $(\nu'(a) - \frac{1}{n(n-2)}, \nu'(a) + \frac{1}{n(n-2)})$. This number is easily found. $\square$

Slightly less accuracy is needed if we just want to know whether the value of each position is negative, zero or positive. This decision problem can therefore be decided more efficiently.

Theorem 4. *Let G and w be as in Theorem 3, and let T be an integer threshold. The decision whether $\nu(a) < T$, $\nu(a) = T$, or $\nu(a) > T$, for every $a \in V$, can be made in $O(|V|^2\cdot|E|\cdot W)$ time.*

Proof. The distance between T and the closest rational number with an even denominator of size at most n is at least $1/n$. It is therefore enough to compute the values $\nu_k(a)$ for $k = 4n^2W$, and this takes only $O(|V|^2\cdot|E|\cdot W)$ time. $\square$

3 Finding the optimal strategies

Given an algorithm for finding the value of any vertex of a graph, positional optimal strategies can be found using a simple method.

Theorem 5. *Let $G = (V, E)$ be a directed graph and let $w : E \to \{-W, \ldots, 0, \ldots, W\}$ be a weight function on its edges. Positional optimal strategies for both players for games played on G can be found in $O(|V|^4 \cdot |E| \cdot \log(|E|/|V|) \cdot W)$ time.*

Proof. Start by computing the values $\nu(a)$ for every $a \in V$. If all the vertices $a \in V_1$ have outdegree one, then player I has a unique strategy and this strategy is positional and optimal. Otherwise, consider any vertex $a \in V_1$ with outdegree $d > 1$. Remove any $\lceil d/2 \rceil$ of the edges leaving a, and recompute the value of a, $\nu'(a)$ say, for the resulting graph. If $\nu'(a) = \nu(a)$ then there is a positional optimal strategy for the player I which does not use any of the removed edges; if $\nu'(a) \neq \nu(a)$ then there is a positional optimal strategy for this player using one of the removed edges. Whichever is the case, we can now restrict attention to a subgraph G' with at least $\lfloor d/2 \rfloor$ fewer edges. Let $d(a)$ be the initial outdegree of a vertex $a \in V$. After $O(\sum_{a \in V_1} \log d(a))$ such experiments we are left with a positional optimal strategy for player I. A positional optimal strategy for player II is found in a similar way. As $\sum_{a \in V} \log d(a) \leq |V| \cdot \log(|E|/|V|)$, we get that the complexity of this algorithm is $O(|V|^4 \cdot |E| \cdot \log(|E|/|V|) \cdot W)$, as required. $\square$

An interesting open problem is whether finding positional optimal strategies is harder than just computing the values of a game. The algorithm we describe calls the full value-finding algorithm repeatedly, but uses only the value at a single vertex and ignores any information about the optimal moves of the players in the truncated games. Unfortunately, optimal moves in the truncated games may not conform to positional strategies. We think however that it should be possible to use the additional information gathered and improve our algorithm.

4 Playing against a known positional strategy

In this section we consider degenerate games in which there is only one edge out of each vertex for player II, say. This corresponds, for example, to cases in which player I knows in advance the positional strategy according to which player II is going to play. An $O(|V| \cdot |E|)$ algorithm of Karp [Kar78] (see also [CLR90], p. 548) for finding the maximum (or minimum) mean weight cycle of a weighted graph $G = (V, E)$ supplies, almost immediately, an efficient purely combinatorial algorithm for such special cases.

Theorem 6. *Let $G = (V, E)$ be a directed bipartite graph with a real weight function $w : E \to R$ on its edges, and assume that the outdegree of each vertex $v_2 \in V_2$ is exactly one. Then, the values of all the vertices and a positional optimal strategy $\pi_1 : V_1 \to V_2$ for player I can be found in $O(|V| \cdot |E|)$ time.*

Could Karp's algorithm be used to obtain a more efficient algorithm for the general case? The natural decision problem corresponding to MPG's is the following. Given an MPG G and a number ν, is the value of G at least ν? As a Corollary to Theorem 6 we get the following result.

Theorem 7. *The decision problem corresponding to MPG's is in NP $\cap$ co-NP.*

The simple observations of this section were first made by Gurvich, Karzanov and Khachiyan [GKK88], and by Karzanov and Lebedev [KL93]. They are included here for completeness.

5 Discounted payoff games

In this section we describe a *discounted* version of mean payoff games. This (fourth) variant, which is also interesting in its own right, will serve in the next section as a link between mean payoff games and simple stochastic games.

Let $0 < \lambda < 1$ be a real number. The weight of the ith edge, e_i, chosen by the players is now multiplied by $(1 - \lambda)\lambda^i$ and the outcome of the game is defined to be $(1 - \lambda) \sum_{i=0}^{\infty} \lambda^i w(e_i)$. The goal of the first player is again to maximise, and of the second player to minimise, this outcome. The number λ is called the *discounting factor* of the game.

Let $G = (V_1, V_2, E)$ be a directed bipartite graph, where $V_1 = \{u_1, \ldots, u_{n_1}\}$ and $V_2 = \{v_1, \ldots, v_{n_2}\}$. Let a_{ij} be the weight of the edge (u_i, v_j), or $-\infty$ if $(u_i, v_j) \notin E$. Similarly, let b_{ji} be the weight of the edge (v_j, u_i), or $+\infty$ if $(v_j, u_i) \notin E$. Let $x_i = x_i(\lambda)$ be the value of a discounted game started at u_i and let $y_j = y_j(\lambda)$ be the value of a discounted game started at v_j. The following theorem is easily verified. Its proof is omitted.

Theorem 8. *The value vectors $(x_1, \ldots, x_{n_1})$ and $(y_1, \ldots, y_{n_2})$ of the discounted games played on $G = (V_1, V_2, E)$ are the unique solutions of the equations:*

$$x_i = \max_{1 \le j \le n_2}\{(1 - \lambda)a_{ij} + \lambda y_j\} \text{ for } 1 \le i \le n_1 ,$$
$$y_j = \min_{1 \le i \le n_1}\{(1 - \lambda)b_{ji} + \lambda x_i\} \text{ for } 1 \le j \le n_2 .$$

It follows immediately from this theorem that both players of the discounted game again have positional optimal strategies. Let $\nu(\lambda)$ be the value of the discounted game with discounting factor λ. As λ tends to 1, we expect $\nu(\lambda)$ to tend to ν, the value of the non-discounted game. This follows from the next theorem whose proof is omitted due to lack of space.

Theorem 9. *Let $G = (V, E)$ be a graph on n vertices, let $w : E \to \{-W, \ldots, 0, \ldots, W\}$ be a weight function on its edges and let λ be a real number satisfying $0 < \lambda < 1$. Let $\nu(\lambda)$ and ν be the values of the discounted and mean payoff games played on the graph $G = (V, E)$ starting at $a \in V$. Then,*

$$\nu - 2n(1 - \lambda)W \ \le \ \nu(\lambda) \ \le \ \nu + 2n(1 - \lambda)W .$$

In particular, if we choose $\lambda = 1 - 1/(2n^3 W)$ then it is easy to verify that $|\nu(\lambda) - \nu| \le 1/(n(n - 2))$, and ν can be obtained from $\nu(\lambda)$ by rounding to the nearest rational with an even denominator of at most n, as was done in Section 2. We thus obtain a reduction from MPG's to discounted payoff games (DPG's).

6 Reduction to simple stochastic games

In this section we describe a simple polynomial reduction from discounted payoff games (DPG's) to simple stochastic games (SSG's). This reduction, combined with the reduction from MPG's to DPG's, shows that SSG's are at least as hard as MPG's. We believe that MPG's are in fact easier than SSG's.

A *simple stochastic game* is a two-person game played on a directed graph $G = (V, E)$ whose vertex set V is the union of three disjoint sets V_{max}, V_{min} and $V_{average}$. The graph also contains a special start vertex and two special vertices called the 0-sink and the 1-sink. Each edge emanating from an 'average' vertex has a rational probability attached to it. The probabilities attached to all the edges from each average vertex add up to 1.

A token is initially placed on the start vertex of the graph. At each step of the game the token is moved from a vertex to one of its neighbours, according to the following rules:

1. At a max vertex, player I chooses the edge along which the token is moved.
2. At a min vertex, player II chooses the edge.
3. At an average vertex, the edge along which the token is moved is chosen randomly according to the probabilities attached to the outgoing edges.

The game ends when the token reaches one of the sink vertices. Player I wins if the token reaches the 1-sink and player II wins otherwise, i.e., if the token reaches the 0-sink or if the game does not end. The *value* of such a game is the probability that player I wins the game when both players play optimally. As was the case for mean payoff games, the two players of a simple stochastic game have positional optimal strategies.

Simple stochastic games were first studied by Shapley [Sha53]. Many variants of them have been studied since then (see Peters and Vrieze [PV87] for a survey). Condon [Con92] was the first to study simple stochastic games from a complexity theory point of view. She showed that the natural decision problem corresponding to SSG's (i.e., given a game G and a rational number $0 < \alpha \leq 1$, is the value of G at least α?) is in NP $\cap$ co-NP. No polynomial time algorithm for SSG's is yet known. A subexponential randomised algorithm for SSG's was obtained by Ludwig [Lud95].

Condon [Con92] actually shows containment in NP $\cap$ co-NP of the decision problem that corresponds to SSG's of the following restricted form. The outdegree of each non-sink vertex is exactly two and the probability attached to each edge that emanates from an average vertex is $1/2$. She then describes a reduction from general SSG's to SSG's of this restricted form. Her reduction, however, is not polynomial. A general SSG on n vertices in which the denominators of all the (rational) probabilities are at most m is transformed into a restricted SSG of size polynomial in n and m, rather than in n and $\log m$. Her transformation can be easily modified however, as we show next, to yield a polynomial reduction.

It is easy to transform a SSG into an equivalent SSG in which the outdegree of each non-sink vertex is exactly two. Each vertex of fan-out k is simply replaced by a binary tree with k leaves. This increases the size of the graph

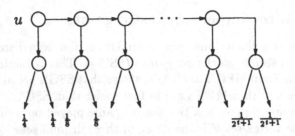

Fig. 1. Implementing an average vertex with arbitrary probabilities

(i.e., the number of vertices and edges) by only a constant factor. The remaining problem is therefore the simulation of binary average vertices with non-equal probabilities. Suppose we want to implement an average vertex u with two emanating edges (u, v_1) and (u, v_2), labeled respectively by the probabilities p/q and $(q - p)/q$, where p and q are integers and $2^{t-1} \leq q < 2^t$. Let $a_1 a_2 \ldots a_{t-1} a_t$ and $b_1 b_2 \ldots b_{t-1} b_t$ be the binary representations of p and $q - p$ respectively, where a_1 and b_1 are the most significant digits. We use the construct shown in Figure 1. All the vertices used are average vertices with equal probabilities. For every $2 \leq i \leq t + 1$, there are two emanating edges that are reached from u with probability 2^{-i}. If $a_i = 1$ then connect one of the edges which has probability $2^{-(i+1)}$ to v_1, and if $b_i = 1$ then connect one of these edges to v_2. All the unused edges are connected back to u. Is it easy to check that v_1 and v_2 are eventually reached with the appropriate probabilities. The number of vertices used in this construction is proportional to the number of bits needed to represent the transition probabilities. The reduction is therefore polynomial.

A simple stochastic game is said to *halt with probability 1* if, no matter how the players play, the game ends with probability 1. The proof of the following theorem can be found in Condon [Con92]. Note its similarity to Theorem 8.

Theorem 10. *Let $G = (V, E)$ be an SSG that halts with probability 1, and let $p(u, v)$ denote the probability attached to an edge (u, v) that emanates from an average vertex u. The values $\nu(v)$ of the vertices of G form the unique solution to the following set of equations:*

$$\nu(u) = \begin{cases} \max_{(u,v) \in E} \{\nu(v)\} & \textit{if } u \textit{ is a max vertex,} \\ \min_{(u,v) \in E} \{\nu(v)\} & \textit{if } u \textit{ is a min vertex,} \\ \sum_{(u,v) \in E} \{p(u, v) \cdot \nu(v)\} & \textit{if } u \textit{ is an average vertex,} \end{cases}$$

along with the conditions that $\nu(\text{0-sink}) = 0$ and $\nu(\text{1-sink}) = 1$.

We are finally in a position to describe a reduction from discounted payoff games (DPG's) to simple stochastic games (SSG's). Recall that we have already described a reduction from MPG's to DPG's.

Let $G = (V, E)$ be a DPG with discounting factor λ. If we add a constant c to all the weights of the game, the value of the game is increased by c. If we multiply all the weights of the game by a constant $c > 0$, the value of the game is multiplied by c. We can therefore scale the weights so that they will all be rational numbers in the interval $[0, 1]$. If the original weights were in the

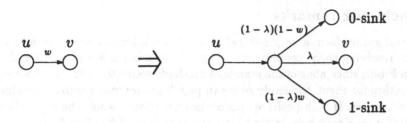

Fig. 2. Simulating a transition of a discounted payoff game.

range $\{-W, \ldots, 0, \ldots, W\}$, then the new weights will be rational numbers with denominators and numerators in the range $\{0, 1, \ldots, 2W\}$.

We construct an SSG $G' = (V', E')$, with the same value as the scaled DPG $G = (V, E)$ with discounting factor λ, in the following way. Each edge (u, v) with weight w in G is replaced by the construct shown in Figure 2. The simple stochastic game G' halts with probability 1, as in each transition there is a probability of $1 - \lambda$ of reaching a sink vertex. The values of the vertices of the discounted payoff game G satisfy the set of equations given in Theorem 8. The values of the vertices of the simple stochastic game G' satisfy the set of equations given in Theorem 10. These two sets of equations become *identical* once the intermediate variables, that correspond to the intermediate vertices introduced by the transformation described in Figure 2, are eliminated. As this set of equations has a unique solution, the values of the two games are equal. The transformation of G to G' can clearly be carried out in polynomial time. This completes the description of the reduction.

7 Some applications

In this section we briefly mention two applications of mean payoff games.

Consider a system that can be in one of n possible states. At each time unit, the system receives one of k possible requests. The system is allowed to change its state and then it has to serve the request. The transition from state i to state j costs a_{ij}, and serving a request of type t from state i costs b_{it}. What, in the worst case, is the average cost of serving a request?

Borodin, Linial and Saks [BLS92] performed a *competitive analysis* of such systems, which they call *on-line metrical task systems*. If we look at the worst case instead, we get a mean payoff game played between the system and an adversary that chooses the requests.

Consider finite-window on-line string matching algorithms (see [CHPZ95] for a definition). What, in the worst case, is the average number of comparisons that an optimal algorithm has to perform per text character? The problem can be formulated as a mean payoff game played between the designer of a string matching algorithm and an adversary that answers the queries made by an algorithm. The reward (the complement of cost) obtained by the algorithm at each stage is the amount by which it can shift its window. For each pattern string and window size we obtain a mean payoff game, the solution of which yields an optimal string matching algorithm for that pattern and window size.

8 Concluding remarks

Mean payoff games form a very natural class of full-information games and we think that resolving their complexity is an interesting issue. We conjecture that they lie in P but, since none of the standard methods seems to yield a polynomial time algorithm for them, the study of mean payoff games may require new algorithmic techniques. If such positive approaches are unsuccessful, the example of mean payoff games may help in exploring the structure of NP ∩ co-NP.

Acknowledgments

We would like to thank Sergiu Hart, Ehud Lehrer, Nimrod Megiddo, Moni Naor, Noam Nisan and Avi Wigderson for helpful discussions and suggestions, Alexander Karzanov for pointing out a flaw in the previous version of Theorem 5, and Thomas McCormick for bringing references [GKK88] and [KL93] to our attention.

References

[BLS92]   A. Borodin, N. Linial, and M.E. Saks. An optimal on-line algorithm for metrical task system. *Journal of the ACM*, 39(4):745–763, 1992.

[CHPZ95] R. Cole, R. Hariharan, M. Paterson, and U. Zwick. Tighter lower bounds on the exact complexity of string matching. *SIAM Journal on Computing*, 24:30–45, 1995.

[CLR90]   T.H. Cormen, C.E. Leiserson, and R.L. Rivest. *Introduction to algorithms.* The MIT Press, 1990.

[Con92]   A. Condon. The complexity of stochastic games. *Information and Computation*, 96(2):203–224, February 1992.

[EM79]   A. Ehrenfeucht and J. Mycielski. Positional strategies for mean payoff games. *International Journal of Game Theory*, 8:109–113, 1979.

[GKK88]   V.A. Gurvich, A.V. Karzanov, and L.G. Khachiyan. Cyclic games and an algorithm to find minimax cycle means in directed graphs. *USSR Computational Mathematics and Mathematical Physics*, 28:85–91, 1988.

[Kar78]   R.M. Karp. A characterization of the minimum cycle mean in a digraph. *Discrete Mathematics*, 23:309–311, 1978.

[KL93]   A.V. Karzanov and V.N. Lebedev. Cyclical games with prohibitions. *Mathematical Programming*, 60:277–293, 1993.

[Lud95]   W. Ludwig. A subexponential randomized algorithm for the simple stochastic game problem. *Information and Computation*, 117:151–155, 1995.

[Pra75]   V.R. Pratt. Every prime has a succinct certificate. *SIAM Journal on Computing*, 4(3):214–220, 1975.

[PV87]   H.J.M. Peters and O.J. Vrieze. Surveys in game theory and related topics. CWI Tract 39, Centrum voor Wiskunde en Informatica, Amsterdam, 1987.

[Sha53]   L.S. Shapley. Stochastic games. *Proc. Nat. Acad. Sci. U.S.A.*, 39:1095–1100, 1953.

Approximation of coNP Sets
by NP-complete Sets

Kazuo IWAMA* and Shuichi MIYAZAKI**

Department of Computer Science and Communication Engineering
Kyushu University, Hakozaki, Higashi-ku, Fukuoka 812, Japan

Abstract. It is said that a set L_1 in a class C_1 *approximates* a set L_2 in a class C_2 if L_1 is a subset of L_2. Approximation L_1 is said to be optimal if there is no approximation L_1' such that $L_1' \supset L_1$ and $L_1' - L_1$ is infinite. When C_1=P and C_2=NP, it is known that there is no optimal approximation under a quite general condition unless P=NP. In this paper we discuss the case where C_1=the class of NP-complete sets and C_2=coNP. A similar result as above that shows the difficulty of the optimal approximation is obtained. Approximating coNP sets by NP-complete sets play an important role in the efficient generation of test instances for combinatorial algorithms.

1 Introduction

It is said that a set L_1 in a class C_1 *approximates* a set L_2 in a class C_2 if L_1 is a subset of L_2. When both A and B approximate L, A is said to be *better* than B if $A \supset B$ and $A - B$ is infinite. In many occasions of algorithm development for hard combinatorial problems, developing a *better algorithm* is equivalent to seeking a *better approximation* of an easier class. Consider, for example, the problem of obtaining a satisfying assignment of a CNF formula. One can see that to claim, for example, "our backtracking algorithm can solve a wide variety of formulas within practical time" is the same as to claim "we found a large set in P that approximates the NP set of all satisfiable CNF formulas".

A similar situation occurs when we generate benchmark instances for testing the performance of combinatorial algorithms. A recently popular approach is to generate instances by randomized algorithms which guarantee the answer of generated instances. In the case of satisfiability testing of CNF formulas, the generator consists of independent two randomized algorithms, one of which generates instances in SAT, the set of satisfiable formulas, and the other generates instances in $UNSAT$, the set of unsatisfiable ones. Generation algorithms are basically nondeterministic Turing Machines. So, it is not surprising that NP languages can be generated in polynomial time; i.e., generating SAT is relatively easy. However, generating coNP languages (e.g., $UNSAT$) efficiently is

* This research was supported by Scientific Research Grant, Ministry of Education, Japan, No. 04650318, and Engineering Adventure Group Linkage Program (EAGL), Japan.
** Research Fellow of the Japan Society for the Promotion of Science. This research was supported by Scientific Research Grant, Ministry of Education, Japan, No. 2273

hard. It turns out that coNP sets cannot be generated in polynomial time unless NP=coNP. Thus we probably have to depend on approximation, namely, trying to generate some NP-subset of the coNP language.

Also in this case, better approximation is always desirable. However, that is not enough this time: Since the approximation set, say L_1, is used to test the performance of algorithms, L_1 must not be too easy. For example, suppose that we would use the following approach to generate unsatisfiable formulas: (i) First generate an arbitrary formula f. (ii) Next select two variables x_i and x_j at random and add four clauses $(x_i + x_j)$, $(x_i + \overline{x_j})$, $(\overline{x_i} + x_j)$ and $(\overline{x_i} + \overline{x_j})$ to f. This algorithm surely generates unsatisfiable formulas. However, as one can see easily, this approximation is in P and is clearly irrelevant for benchmark instances.

Recently, several generation algorithms for coNP languages based on this approximation idea have been proposed including for unsatisfiable CNF formulas [AIM95, IM95b], non-Hamiltonian graphs[IM95a], non-k-colorable graphs[PU92], and others [San]. Among others, the AIM generator for the satisfiability testing introduced by Asahiro, Iwama and Miyano [AIM95] has already been used to provide benchmarks for DIMACS Special Year of combinatorial optimization (93-94). The feature of this generator is shown in [IM95b]: There are an infinite sequence of sets $L_0, L_1, \cdots, L_i, L_{i+1}, \cdots$ such that (i) each L_i approximates UN-SAT, (ii) each L_i is NP-complete, namely, L_i is not too easy and (iii) $L_i \subset L_{i+1}$ and $L_{i+1} - L_i$ is D^P-complete; i.e., L_{i+1} is essentially better than L_i. Intuitively, the AIM generator can generate "unlimitedly" better approximation of $UNSAT$.

Then a natural question is whether or not there is some *optimal approximation*, namely, the set L such that no better approximation than L exists in that class. A negative result, so-called P-levelability, already exists to this question in the case of approximating NP sets by P sets: [ORS86] shows that if an NP set L is honestly paddable then L has no optimal approximation in P unless P=NP.

The purpose of this paper is to obtain a similar result in the case of approximating coNP sets by NP-complete sets. Its motivation was mentioned above. Unfortunately our result is weaker than that of P-levelability; it is shown that there exists a considerably general condition for an NP-complete-set B and a coNP-set A such that B is not an optimal approximation of A. (The P-levelability says that a similar condition is needed only for an NP-set A such that no P-set B is an optimal approximation of A.) To demonstrate the generality, we shall show that the condition is satisfied for three combinations of coNP set A and its (probably the best) known NP-complete-approximation B. Finally we shall briefly discuss a certain relaxation of the condition to further increase the generality.

2 Main Results

Before presenting our main theorem, we shall first review the P-levelability result, where so-called padding functions play a key roll.

Proposition 1. Let A be a set of strings that is not in P and is closed under some mapping ψ which is length-increasing and can be computed in polynomial time. Then any P-approximation B of A is not optimal.

Proof [BDG89]. Suppose that B is optimal. A set C is defined as follows:

$$C = \{f \mid f \notin B \text{ but } \psi(f) \in B\}$$

As B is in P, and ψ is computable in polynomial time, C is also in P. Now consider, for each f in A, the following sequence: $f, \psi(f), \psi^2(f) = \psi(\psi(f)), \cdots$. Since A is closed under ψ, each value is in A. Now the set

$$D_f = \{f, \psi(f), \psi^2(f), \cdots, \psi^n(f), \cdots\}$$

is infinite and is in P since ψ is makes the size strictly larger. Because B was assumed to be optimal, D_f must eventually be contained in B; it must "cross" the boundary denoted by the set B.

As A is not in P, $A - B$ is infinite. For each f in $A - B$, the sequence D_f reaches some member g of C and the size of g is larger than that of f. Thus only finitely many f may reach any fixed g, and hence there must be infinitely many such g. Thus C is infinite. Since $C \cap B = \emptyset$ by the definition of C, $C \cup B$ contains infinitely more elements than B and is still in P since both C and B are in P, which contradicts the assumption that B is optimal. $\square$

It turns out that the length-increasing condition for ψ can be relaxed: The mapping ψ is said to be *monotone* if the following two conditions are met: (i) For any string f, $|\psi^{-1}(f)|$ is finite, where $\psi^{-k}(f) = \{g \mid f = \psi^k(g)\}$. (ii) There exists a polynomial $p(x)$ such that for any string f, $\psi^{-p(|f|)}(f) = \phi$. Now suppose that ψ is monotone. Then it is not hard to see that:

(a) The set D_f in the above proof is infinite and is in P.

(b) $A - B$ is also infinite since finitely many f may reach any fixed g.

Thus Proposition 1 holds for any monotone ψ. Obviously, if ψ is length-increasing then ψ is monotone.

Proposition 1 is extended to NP-approximation of non NP sets:

Proposition 2. Let A be a set of strings that is not in NP and is closed under some mapping ψ which is polynomial-time computable and monotone. Then any NP-approximation B of A is not optimal.

Proof. Almost the same as above but we define C as

$$C = \{f \mid \psi(f) \in B\}$$

Then C is in NP and so is $C \cup B$ as well. (If we would add the condition $f \notin B$ then we could no longer claim that C is in NP.) Although C and B are no longer disjoint, we can show that $C \cup B$ is infinitely larger than B by a similar argument as above. $\square$

Thus it is straightforward to extend the difficulty of obtaining optimal P-approximation into the difficulty of obtaining optimal NP-approximation. Then what would happen if NP-approximation is replaced by NPC-approximation, i.e., approximation by NP-complete sets? Suppose that B is an NP-complete set which approximates a set A (see Fig. 1). Since B is NP-complete, there are another NP-complete set X and a polynomial time reduction α such that $B \supseteq \alpha(X)$ and $\overline{B} \supseteq \alpha(\overline{X})$. Therefore, even if $A - B$ is infinite, most (except finite elements) of that portion might be occupied by $\alpha(\overline{X})$. If that were the case, extension of B is obviously not possible unless we modify the set $\alpha(\overline{X})$. Unfortunately, the fact that B is NP-complete does not give us, by itself, any

14

specific information about this $\alpha(\overline{X})$ or we have to prove the difficulty of NPC-approximation without any particular assumption about $\alpha(\overline{X})$. This is a major obstacle in extending Proposition 2 into the case of NPC-approximation.

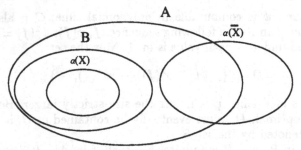

Fig. 1. An NPC-approximation

Now we present our main theorem. We need two mappings ψ and δ: ψ is the same as before, (being polynomial-time computable and monotone). δ also needs to be polynomial-time computable and monotone. Furthermore, δ has to satisfy the following three conditions. (For particular examples of δ and ψ, see Sec.3.)

(i) δ is one-to-one.

(ii) δ^{-1} can be computed in deterministic polynomial time. (Actually, this condition is too strong; the following condition is enough: There is an nondeterministic polynomial time algorithm for deciding, for given f in A, whether f cannot be expressed as $\delta(f')$ for some f', i.e., $\delta^{-1}(f) = \Lambda$.)

(iii) δ and ψ are disjoint, i.e., for any sets S_1 and S_2, $\delta(S_1) \cap \psi(S_2) = \emptyset$.

Theorem 1. Suppose that A is a coNP set and B is an NPC-approximation of A such that both A and $\overline{A}$ are closed under ψ and δ and $\overline{B}$ is closed under δ, where, ψ and δ satisfy the condition described above. Then if NP $\neq$ coNP, B is not optimal.

Remark. (i) Note that B must be in NP but its NP-completeness does not have to be already known. This theorem says that if we assume that B is NP-complete then we can make an infinitely larger NP-complete set under the described conditions. (ii) The condition that $\overline{B}$ is closed under δ will be relaxed in Sec.4. (iii) A may not be coNP but may be any class $\neq$NP.

Proof. Since B is NP-complete, there is another NP-complete set X and a polynomial-time reduction α such that $B \supseteq \alpha(X)$ and $\overline{B} \supseteq \alpha(\overline{X})$. The first case to be considered is that there are such X and α that $\alpha(\overline{X})$ is in $\overline{A}$. This case is easy to treat since we can apply the same expansion of B as Proposition 2. Clearly we do not have to care about an overlap of the extended B with $\alpha(\overline{X})$ because the latter is outside A. (In this case, two sets B and $\overline{A}$ are P-inseparable [ESY84, GS88]. For example, it is known that if P$\neq$NP then there exist two NP sets which are P-inseparable, but no particular P-inseparable sets are known.)

Hence, from now on, we can assume that there are no such α and X or that any pair of the set X and the reduction α satisfies that $\alpha(\overline{X}) \cap A \neq \emptyset$. Let $C = \alpha(\overline{X}) \cap A$. See Fig. 2.

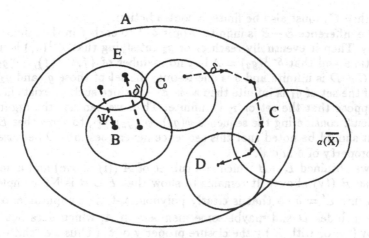

Fig. 2. Proof of Theorem 1

Now we shall show the following important property about this set C: Let

$$C_0 = \{\delta^{-\infty}(f) \mid f \in C\}$$

where $g = \delta^{-\infty}(f)$ iff $\delta^{-1}(g) = \Lambda$ and for some $i \geq 0$, $\delta^i(g) = f$. Then C_0 is infinite by the following reason: Suppose hypothetically that C_0 is finite. Then given f in $\alpha(X \cup \overline{X})$, we can decide whether f is in C or not in polynomial time. (One can apply δ repeatedly to each element in the finite set C_0. Since δ is monotone, it is easy to see whether that repeated application can get to f. If so, f is in C.) Now we can modify the reduction α into the following α': If $\alpha(x)$ is in C then $\alpha'(x) = \alpha(x)$. Otherwise $\alpha'(x)$ is some trivial element in $\overline{A}$. One can easily see that α' is also polynomial-time computable and it can act as a reduction just like α. But this time $\alpha'(\overline{X}) \subseteq \overline{A}$ or $\alpha'(\overline{X}) \cap A = \emptyset$, which contradicts our current assumption on the reduction α.

Now let $D = \delta(C)$ and

$$E = \{f \mid \text{(a) } \delta^{-1}(f) \in B, \text{ or (b) } \delta^{-1}(f) = \Lambda \text{ and } \psi(f) \in B\}$$

(i) $E \subseteq A$ by the closure property of δ and ψ.

(ii) Since B is in NP and δ^{-1} and ψ are polynomially computable functions, E is also in NP. (To check (a), the calculation of δ^{-1} is not necessary. To check (b), we need to check whether $\delta^{-1}(f) = \Lambda$ in nondeterministic polynomial time; for this reason, we need the condition (ii) previously given for the mapping δ.)

(iii) $E \cap D = \emptyset$: Suppose that f is in D. Then $\delta^{-1}(f)$ must be in C by the definition of D and consequently above condition (b) on E does not hold. E's elements f can be written as $f = \delta(g)$ for some $g \in B$. Thus, f cannot be in D since if f were in D then it cannot be written as $f = \delta(g)$ for $g \in C$. B and C are disjoint.

(iv) The difference $C - D$ is infinite: Suppose that we apply δ to some element f_0 in C_0 many times. Then f_0 changes to $f_1 = \delta(f_0)$, then $f_2 = \delta(f_1)$, $f_3 = \delta(f_2)$, and so on (see Fig. 2). A similar sequence also starts from a different element, say, g_0, in C_0. Note that these two sequences never intersect since δ is one-to-one. Now it turns out that if C includes f_i, f_{i+1}, $\cdots$, f_j, then D includes from f_{i+1} through f_{j+1}; namely $C - D$ includes f_i. Now one can see that if $C - D$ would be finite then C_0 must also be finite, a contradiction.

(v) The difference $E - B$ is infinite: Apply δ^{-1} to each f in the infinite $C - D$ repeatedly. Then it eventually reach g_1 or g_2 satisfying that $\delta^{-1}(g_1)$ is in B for the first time and that $\delta^{-1}(g_2) = \Lambda$ but no member of $\{f, \delta^{-1}(f), \cdots, g_2\}$ is in B. Since $C - D$ is infinite and δ is one-to-one, the set of those g_1 and g_2 is also infinite. If the set of g_1 is infinite then so is $E - B$ since such g_1 exists in $E - B$. Hence suppose that the set of g_2 is infinite. Then we can use the argument of Proposition 2 considering the sequence $\psi(g_2), \psi^2(g_2), \cdots$, to show that $E - B$ is infinite. It should be noted that this sequence never goes into D because of the disjoint property of δ and ψ.

Now we obtained $E \cup B$ which is a subset of A ((i) above) and is infinitely larger than B ((v) above). It remains to show that $E \cup B$ is NP-complete: We modify α into $\alpha' = \delta \cdot \alpha$ that is clearly polynomial-time computable. $\alpha'(\overline{X})(= \delta(\alpha(\overline{X})))$ includes D and maybe some members in $\overline{A}$, which does not overlap with E by (iii) or with B by the closure property of δ. (Thus we "shifted" C to D by δ. Note that E may intersect with C but not with D.) Also $\alpha'(X)$ must be contained in $E \cup B$ since $\alpha(X)$ is in B and E contains such f that for some g in B, $\delta(g) = f$. $\square$

3 Application of Theorem 1

The objective of this section is to claim the generality of Theorem 1 by applying it to three particular NPC-approximations, those for $UNSAT$, non-Hamiltonian graphs and non-3-colorable graphs. (As for the last one, its approximation we will be discussing is not known to be NP-complete. This does not cause any problem as mentioned before.) As is shown below, it is not hard to find the mapping δ and ψ that satisfy the conditions in each application. Recall that the conditions are: (i) δ and ψ are polynomial-time computable and monotone. (ii) δ is one-to-one. (iii) δ^{-1} is computable in polynomial time. (iv) δ and ψ are disjoint. (v) A and $\overline{A}$ are closed under δ and ψ, and $\overline{B}$ is closed under δ where A denotes the coNP set and B its NPC-approximation.

3.1 Unsatisfiable CNF formulas

We take the following approximation, called the set of *formulas provable with* $d(n)$ *duplications*, denoted by $PRV(d(n))$ [IM93]: A CNF formula f of n variables is in $PRV(d(n))$ if for some variable x in f, $x\overline{x}$ is implied from f by applying the following operations repeatedly, but operation (3) can be used at most $d(n)$ times:

(1) An arbitrary literal is added to an arbitrary clause A.

(2) Two clauses $(A + x)(A + \overline{x})$ are merged into a single clause A.

(3) An arbitrary clause A is duplicated to AA.

It is conjectured that $PRV(d(n))$ is NP-complete if $d(n)$ is a polynomial[IM93] and is actually NP-complete if $d(n)$ is constant[IM95b]. Suppose that f is a CNF formula using n variables, $x_1, \cdots, x_n$. Then, $\delta(f)$ and $\psi(f)$ are defined as follows:

$\delta(f)$: For each clause C in f, if there is x_n in C, then add x_n to C (i.e., C contains two or more x_n's). Furthermore, if there is $\overline{x_n}$ in C, then add $\overline{x_n}$ to C also.

ψ: Add a single clause (x_{n+1}) to f.

Theorem 2. δ and ψ meet the above conditions (i)–(v).

Proof. Recall that x_n is the variable of the largest index in f. Clearly, δ and ψ meet the condition (i). If $\delta^{-1}(f)$ exists, then there is no clause in f which includes exactly one x_n (or $\overline{x_n}$). Then, by removing a single x_n (and $\overline{x_n}$) from each clause which has two or more x_n (or $\overline{x_n}$), we can obtain $\delta^{-1}(f)$ uniquely. This can be done in polynomial time. Thus δ meets the conditions (ii) and (iii). If $\delta^{-1}(f)$ exists, then (x_n) cannot be its clause. However, if $\psi^{-1}(f)$ exists, there must be a clause (x_n) in f. Thus the condition (iv) is satisfied. Clearly, δ and ψ map a satisfiable (unsatisfiable) formula to a satisfiable (unsatisfiable) one. Then both A and $\overline{A}$ are closed under δ and ψ. To show that $\overline{B}$ is closed under δ, it is enough to show that if $\delta(f)$ is proved by using the operation (3) i times, f can also be proved by using (3) at most i times. This is almost obvious because to prove f, we once use operation (1) to get $\delta(f)$ and then can simulate the proof for $\delta(f)$, which needs operation (1) just i times. Thus δ and ψ meet the condition (v) as well.□

3.2 Non-Hamiltonian Graphs

Our approximation is called *Non-sub-2-factor* graphs [Chv85], denoted by *NS2F*. *NS2F* is the set of all graphs which satisfy the following condition: there exists a partition of V into disjoint subsets R, S and T (i.e., $T = G - R - S$) such that $T \neq V$, which satisfies $w(T) \leq |S| + \sum_{Q_i} \left\lfloor \frac{edge(Q_i,T)}{2} \right\rfloor$, where $w(T)$ denotes the number of connected components of T, Q_i denotes each subset of vertices connected in R. $edge(Q_i, T)$ denotes the number of edges which have one endpoint in Q_i and the other in T. [IM95a] proved that NS2F is NP-complete and better than another well-known NP-complete approximation, the set of graphs which are non-one-tough [BHS90].

δ and ψ are defined as follows:

$$\delta(f) = \begin{cases} T_1(f) \text{ if } n \text{ is even} \\ T_2(f) \text{ if } n \text{ is odd} \end{cases}$$

$$\psi(f) = \begin{cases} T_1(f) \text{ if } n \text{ is odd} \\ T_2(f) \text{ if } n \text{ is even} \end{cases}$$

where T_1 and T_2 are as shown in Fig. 3.

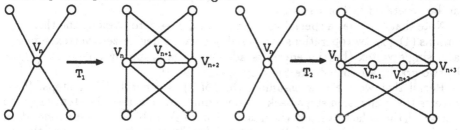

Fig. 3. Function T_1 and T_2

Formally, when $G = (V, E)$ is given, $T_1(G) = (V', E')$, where $V' = V \cup \{v_{n+1}, v_{n+2}\}$, $E' = E \cup \{(v_n, v_{n+1}), (v_{n+1}, v_{n+2})\} \cup \{(v_{n+2}, v_i) | (v_n, v_i) \in E\}$. $T_2(G) =$

(V'', E'') where $V'' = V \cup \{v_{n+1}, v_{n+2}, v_{n+3}\}, E'' = E \cup \{(v_n, v_{n+1}), (v_{n+1}, v_{n+2}),$ $(v_{n+2}, v_{n+3})\} \cup \{(v_{n+3}, v_i) | (v_n, v_i) \in E\}$. (Here, n is the number of vertices in f.)

Theorem 3. δ and ψ meet the conditions (i)–(v). (Proof is omitted.)

3.3 Non-3-Colorable Graphs

Hajós calculus is a popular system for generating the set HC of non-k-colorable graphs (k is a constant≥ 3) [PU92]. We only consider the case where $k = 3$ in this paper. It is known that HC contains all the non-k-colorable graphs but whether the generation only needs polynomial time for all such graphs is open. (Probably this is not true since if so NP=coNP.) Here we consider $HC(d(m))$ that is the set of non-3-colorable graphs of m edges which can be generated by at most $d(m)$ applications of Hajós calculus rules. So, $HC(d(m))$ is in NP if $d(m)$ is a polynomial. It is not known whether it is NP-complete. Hajós calculus starts with a set of K_4's and consists of the following three rules:

(1) Add (any number of) vertices and edges.

(2) Let G_1 and G_2 be disjoint graphs, a_1 and b_1 adjacent vertices in G_1, and a_2 and b_2 adjacent vertices in G_2. Construct the graph G_3 from $G_1 \cup G_2$ as follows. First, remove edges (a_1, b_1) and (a_2, b_2); then add an edge (b_1, b_2); lastly, contract vertices a_1 and a_2 into a single vertex.

(3) Contract two nonadjacent vertices into a single vertex, and remove any resulting duplicated edges; the single vertex can be either of two original vertices.

δ and ψ are defined as shown in Fig.4. Formally, for a given $G = (V, E)$, $\delta(G) = (V', E)$ and $\psi(G) = (V', E')$, where $V' = V \cup \{v_{n+1}\}$, $E' = E \cup \{(v_n, v_{n+1})\}$.

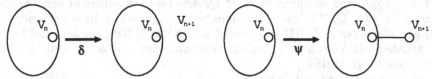

Fig. 4. Function δ and ψ

Theorem 4. δ and ψ meet the conditions (i)–(v).

Proof. We will show that $\overline{B}$ (=the set of non-3-colorable graphs that is not in $HC(d(m))$) is closed under δ. (It is easy to see that other conditions are satisfied.) To do so, we shall prove that if $\delta(G)$ is generated in t steps, then G can be generated in t or less steps.

Note that $\delta(G)$ has a special vertex v_{n+1} of degree zero. Among the three operations (1)–(3), two operations can possibly create a degree-zero vertex; namely, (a) a single vertex (but no edges) is added to the graph by operation (1), or (b) two graphs are joined like Fig.5 by operation (2).

Recall that we are now assuming that $\delta(G)$ is generated in t steps. If we traverse this generation steps backward, we must come across the step, say, step i, in which the isolated vertex was created either (a) or (b) above. Our generation of G is the same as that of $\delta(G)$ except for this ith step. We will change the ith step as follows: If v_{n+1} is generated by (a), we will eliminate only the addition of v_{n+1}. If v_{n+1} becomes degree-zero by (b), we use the operation (1) instead of (2) to obtain G'_3 by adding vertices and edges to G_1 (see Fig.6). Thus, we can obtain G within t steps.□

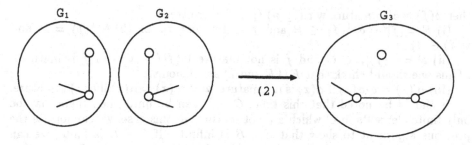

Fig. 5. Proof of Thorem 4

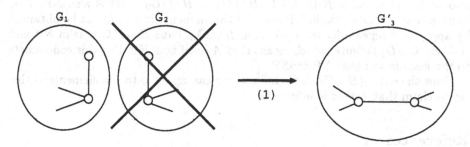

Fig. 6. Proof of Theorem 4

4 Weaker Conditions for δ

Recall that $\overline{B}$ must be closed under δ in Theorem 1. This condition may be too strong for the following reason: Consider, for example, that δ is length-increasing (and is hence monotone). Then one can see that if $|f| = n$ then $|\delta^{2^n}(f)| > 2^n$. Now suppose that $f = \delta^{2^n}(g)$ for some g. Since A and $\overline{A}$ are closed under δ, one can test whether f is in A or $\overline{A}$ not directly but by testing whether g is in A or $\overline{A}$. Then it follows that we can test whether f is in A or not in deterministic polynomial time if the same test can be done for g in exponential time. (Note that the number of possible g's is polynomial in $|f|$. The number of maximal repetitions of applying δ is also polynomial.) Thus it is quite natural that B contains such f, or it is not natural that $\delta^i(g)$ for $g \in A - B$ is not in B when i is exponentially large.

It is said that a set S is closed under a mapping δ *with respect to mature function* $p(n)$ if the following condition is met: For any f in S such that f cannot be written as $f = \delta^i(g)$ for any $i \geq p(|g|)$, $\delta(f) \in S$. f is said to be *mature with respect to* $p(n)$ if it can be written as $f = \delta^i(g)$ for some $i \geq p(|g|)$ and some $g \in S$. Thus the new closure condition only requires that if f is not mature then $\delta(f)$ must stay within the set.

Theorem 5. Theorem 1 is true if $\overline{B}$ is closed under δ with respect to the mature function which may be any polynomial $p(n) > 0$.

Proof. Since $p(n)$ may be any polynomial, it may be 1 (a constant function); we shall prove the theorem for this case. Note that if f can be written as $f = \delta(g)$ then f is mature and $\delta(f)$ may be in B. If $f \in \overline{B}$ and f cannot be written as $f = \delta(g)$, i.e., $\delta^{-1}(f) = \Lambda$, then $\delta(f)$ must be in $\overline{B}$. The proof is almost the same as theorem 1 except for the construction of the polynomial time reduction α', D and E as follows. Also an important property is that if f is mature w.r.t. $p(n)$

then $\delta(f)$ is also mature w.r.t. $p(n)$ (proof is omitted).

(i) $E = \{f|$(a) $\delta^{-1}(f) \in B$ and f is not mature, or (b) $\delta^{-1}(f) = \Lambda$ and $\psi(f) \in B\}$.

(ii) $D = \{\delta(f)|f \in C$ and f is not mature$\} \cup \{f|f \in C$ and f is mature$\}$. (Thus one should check that $B \cup E$ and D are disjoint.)

(iii) $\alpha'(x) = \delta \cdot \alpha(x)$ if $\alpha(x)$ is not mature, and $\alpha'(x) = \alpha(x)$ if $\alpha(x)$ is mature.

It should be noted that this time, $C - D$ can be finite, i.e., there may be only finite elements in C which are not mature. In this case, we cannot use the previous argument to show that $E - B$ is infinite. If $C - D$ is finite, we can claim that $A - (B \cup C \cup D)$ is infinite by the following reason: Suppose that $A - (B \cup C \cup D)$ is finite. Then since $C_0 - B \subseteq A - (B \cup C \cup D)$, $C_0 - B$ is also finite. Then since B is in NP, $B \cup C_0 = B \cup (C_0 - B)$ is also NP. This implies that $C \cup D$ is also in NP since all the elements in $C \cup D$ can be obtained by applying δ repeatedly to elements in $B \cup C_0$. Thus $B \cup C \cup D$ is in NP and $A - (B \cup C \cup D)$ is finite, which means that A itself is in NP. But this contradicts to the assumption that NP$\neq$coNP.

Now since $A - (B \cup C \cup D)$ is infinite, we can apply ψ to the elements in this set to claim that $E - B$ is infinite.$\Box$

References

[AIM95] Y. Asahiro, K. Iwama and E. Miyano, " Random generation of test instances with controlled attributes," In *Proc. Second DIMACS Challenge Workshop*, (M. Trick and D. Johnson, Ed.) American Mathematical Society, 1995.

[BDG89] J. Balcazar and J. Diaz and J. Gabarro, "Structural Complexity II," Springer, 1989.

[BHS90] D. Bauer, S.L. Hakimi and E. Schmeichel, "Recognizing tough graphs is NP-hard," *Discrete Applied Mathematics*, 28, pp. 191-195, 1990.

[Chv85] V. Chvátal, "Hamiltonian cycles", In *The traveling salesman problem* (John Wiley and Sons Ltd.), pp. 403-429, 1985.

[ESY84] S. Even and A. Selman and Y. Yacobi, "The complexity of promise problems with application to public-key cryptography," *Information and Control*, 61, 1984.

[GS88] J. Grollmann and A.L. Selman, "Complexity measures for public-key cryptosystems," *SIAM J. Comput.*, 17, pp. 309-334, 1988.

[IM93] K. Iwama and E. Miyano, "Security of test-case generation with known answers," In *Proc. AAAI Spring Symposium Series*, 1993.

[IM95a] K. Iwama and E. Miyano, "Better approximations of non-Hamiltonian graphs," *manuscript*, 1995.

[IM95b] K. Iwama and E. Miyano, "Intractability of read-once resolution," In *Proc. 10th IEEE Conference on Structure in Complexity Theory*, 1995, to appear.

[ORS86] P. Orponen, D. Russo and U. Schoning, "Optimal approximations and polynomially levelable sets," *SIAM J. Comput.*, 15, pp.399-408, 1986.

[San] L. Sanchis, "Generating hard and diverse test sets for NP-hard graph problems," *Discrete Applied Mathematics*, to appear.

[PU92] T. Pitassi and A. Urquhart, "The complexity of Hajós calculus," In *Proc. 33rd IEEE Symp. on Foundations of Computer Science*, pp.187-196, 1992.

How to Draw a Planar Clustered Graph

Qing-Wen Feng   Robert F. Cohen   Peter Eades

Department of Computer Science, University of Newcastle
University Drive, Callaghan NSW 2308, AUSTRALIA
Email: qwfeng, rfc, eades@cs.newcastle.edu.au

Abstract. In this paper, we introduce and show how to draw a practical graph structure known as *clustered graphs*. We present an algorithm which produces planar, straight-line, convex drawings of clustered graphs in $O(n^{2.5})$ time. We also demonstrate an area lower bound and an angle upper bound for straight-line convex drawings of C-planar graphs. We show that such drawings require $\Omega(2^n)$ area and the smallest angle is $O(1/n)$. Our bounds are unlike the area and angle bounds of classical graph drawing conventions in which area bound is $\Omega(n^2)$ and angle bounds are functions of the maximum degree of the graph. Our results indicate important tradeoff between line straightness and area, and between region convexity and area.

1   Introduction

Many systems, particularly those which present relational information, include a graph drawing function. Examples include CASE tools, reverse engineering systems, software design systems, etc. Such systems have motivated a great deal of research on algorithms for drawing graphs; the survey [1] contains over 250 references.

As the amount of information that we want to visualize becomes larger and more complicated, classical graph models tend to be insufficient. Some more powerful graph formalisms for representing information have been introduced, e.g. hypergraphs [3], compound digraphs [18], cigraphs [12] and higraphs [8]. Some new graphic systems, such as TomSawyer [17] and D-ABDUCTOR [15], have included such graph models. Fig. 1 shows the drawing of a compound digraph produced in the idea organizing system D-ABDUCTOR. However, algorithms for automatically drawing these types of graphs appears difficult. To date only heuristic algorithms have been presented and implemented [18, 16], and the planar drawing of these graphs has not been studied. In this paper, we introduce a practical and simple graph model called *clustered graphs*. Clustered graphs are similar to compound graphs but quite simple. We believe that results of clustered graphs can be easily extended to compound graphs and also to some other graph models.

A clustered graph consists of a graph G and a recursive partitioning of the vertices of G (see Fig. 2). Each partition is known as a *cluster* of a subset of the vertices of G. Clustering appears in the diagrams produced in a wide number

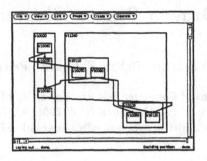

Fig. 1. Drawing of a Compound Digraph from D-ABDUCTOR

of applications areas, such as software engineering [20], knowledge representation [9], idea organization [11], and VLSI design [8].

We extend the definitions of connectivity and planarity to clustered graphs. A clustered graph is a *connected* clustered graph if the vertices in each cluster induce a connected subgraph in the underlying graph. In a drawing of a clustered graph, vertices and edges are drawn as points and curves as usual. Clusters are drawn as simple closed curves that define a closed region of the plane. The region for each cluster contains the drawing of the subgraph induced by its vertices and no other vertices. In other words, a cluster contains the regions for all its subclusters and does not intersect the region for any other cluster. A clustered graph is *C-planar* if it admits a drawing that has no crossings between both the drawings of distinct edges and the drawing of an edge and a region boundary. Note that the planarity of the underlying graph does not imply the existence of a C-planar drawing of a clustered graph (see Fig. 3). An efficient algorithm for testing C-planarity in connected clustered graphs and a general characterization of C-planar clustered graphs are presented in [5].

Our main results in this paper are summarized as follows:

− We present an algorithm for drawing connected C-planar clustered graphs such that edges are drawn as straight line segments and clusters are drawn

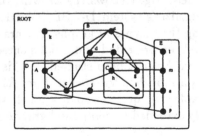

Fig. 2. An Example of a Clustered Graph

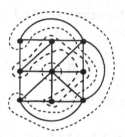

Fig. 3. A Non C-planar Clustered Graph with Planar Underlying Graph

as convex[1] regions. The algorithm works in $O(n^{2.5})$ time.

- We present an area lower bound and an angle upper bound for straight-line convex drawings of C-planar graphs. We show that such drawings require $\Omega(2^n)$ area and the smallest angle is $O(1/n)$. This area bound is unlike the area bound for classical graphs which is $\Omega(n^2)$ (see [4]). Our angle bound is also different from any angle bound of classical graph drawing conventions in which angle bounds are functions of the degree of the graph (see [14, 6]). Our results indicate important tradeoffs between line straightness and area, and between region convexity and area.

The rest of the paper is organized as follows. In Section 2, we present some terminology and some basic results. In Section 3, we describe a straight-line convex drawing algorithm for clustered graphs. Section 4 gives an area lower bound and an angle upper bound for straight-line convex drawing of clustered graphs. We conclude in Section 5 with some interesting open problems.

2 Preliminaries

A *clustered graph* $C = (G, T)$ consists of an undirected graph G and a rooted tree T such that the leaves of T are exactly the vertices of G. Each node ν of T represents a *cluster* $V(\nu)$ of the vertices of G that are leaves of the subtree rooted at ν. Note that tree T describes an inclusion relation between clusters. T is the *inclusion tree* of C, and G is the *underlying graph* of C. We let $T(\nu)$ denote the subtree of T rooted at node ν and $G(\nu)$ denote the subgraph of G induced by the cluster associated with node ν. We define $C(\nu) = (G(\nu), T(\nu))$ to be the *sub-clustered graph* associated with node ν.

A *drawing* of a clustered graph $C = (G, T)$ is a representation of the clustered graph in the plane. Each vertex of G is represented by a point. Each edge of G is represented by a simple curve between the drawings of its endpoints. For each

---

[1] There is a notion of convexity for drawings of classical planar graphs: a drawing is "convex" if every face is a convex polygon. Our algorithm does output drawings in which the underlying graph is convex in this classical sense, but our notion is stronger, and requires that the clusters are represented by convex regions.

node ν of T, the cluster $V(\nu)$ is drawn as simple closed region R that contains the drawing of $G(\nu)$, such that:

- the regions for all sub-clusters of R are completely contained in the interior of R;
- the regions for all other clusters are completely contained in the exterior of R;
- if there is an edge e between two vertices of $V(\nu)$ then the drawing of e is completely contained in R.

Given a drawing $\mathcal{D}$ of $C = (G, T)$, we produce a *consistent drawing* $\mathcal{D}'$ of G by removing the boundary curves from $\mathcal{D}$.

We say that the drawing of edge e and region R have an *edge-region crossing* if the drawing of e crosses the boundary of R more than once. A drawing of a clustered graph is *C-planar* if there are no edge crossings or edge-region crossings. If a clustered graph C has a C-planar drawing then we say that it is *C-planar* (see Fig. 2). Note that a C-planar drawing also contains a planar drawing of the underlying graph.

An edge is said to be *incident* to a cluster $V(\nu)$ if one end of the edge is a vertex of that cluster but the other endpoint is not in $V(\nu)$. An *embedding* of C includes an embedding of G plus the circular ordering of edges crossing the boundary of the region of each non trivial cluster (a cluster which is not a single vertex). In other words, an embedding of a clustered graph consists of the circular ordering of edges around each cluster which are incident to that cluster. A clustered graph $C = (G, T)$ is a *connected clustered graph* if each cluster induces a connected subgraph of G.

Suppose that $C_1 = (G_1, T_1)$ and $C_2 = (G_2, T_2)$ are two clustered graphs such that T_1 is a subtree of T_2, and for each node ν of T_1, $G_1(\nu)$ is a subgraph of $G_2(\nu)$. Then we say C_1 is a *sub-clustered graph* of C_2, and C_2 is a *super-clustered graph* of C_1.

The following results from [5] characterize C-planarity in a way which can be exploited by our drawing algorithms.

Theorem 1. *[5] A connected clustered graph* $C = (G, T)$ *is C-planar if and only if graph G is planar and there exists a planar drawing $\mathcal{D}$ of G, such that for each node ν of T, all the vertices and edges of $G - G(\nu)$ are in the outer face of the drawing of $G(\nu)$.*

Theorem 2. *[5] A clustered graph* $C = (G, T)$ *is C-planar if and only if it is a sub-clustered graph of a connected and C-planar clustered graph.*

Using the above theorems, we can assume that we are given a connected clustered graph when drawing a C-planar clustered graph; and the embedding of the underlying graph G uniquely determines the embedding of the clustered graph.

3 Straight-line Convex Drawing

In this section, we present an algorithm which produces a C-planar drawing for
a certain kind of clustered graph. The output drawings are:

- *convex* in that the region representing each cluster is convex, and
- all edges are drawn as *straight lines*.

Our algorithm applies a well known algorithm of Tutte (see [19]). The input
to Tutte's algorithm is a triconnected planar graph G, a face F of G, and a
convex polygon P with the same number of vertices as F. The algorithm creates
a straight-line planar drawing of G in which every face is a convex polygon.

   We need some connectivity conditions for the algorithm. Suppose that $C =
(G, T)$ is a clustered graph; for each node ν of T, the *skeleton* $\Gamma(\nu)$ is the subgraph
of $G(\nu)$ consisting of the vertices and edges on the outer faces of the child clusters
of ν. A clustered graph and its skeleton are shown in Fig. 4. We say that C is
skeleton-triconnected if $\Gamma(\nu)$ is triconnected for every ν in T. Intuitively, our
algorithm represents a child μ of a node ν of T by the outer face of $G(\mu)$ in the
skeleton $\Gamma(\nu)$. The input to the algorithm below must be skeleton-triconnected.
Further, we require that for every ν, $G(\nu)$ is biconnected; by Theorem 1, this
ensures that the vertices of $G(\mu)$ which have connections with $G - G(\nu)$ are on
the boundary of the outer face of $G(\mu)$.

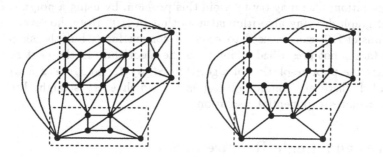

Fig. 4. A Clustered Graph and its Skeleton

Algorithm 1 Straight_Line_Convex_Draw
   Input: a skeleton-triconnected clustered graph $C =
      (G, T)$ such that for each node ν of T, $G(\nu)$ is
      biconnected; a C-planar embedding of C; a node
      ν of T; a polygon P with the same number of ver-
      tices as the outside face of $G(\nu)$
   Output: a C-planar straight-line convex drawing of C.

1. Compute the skeleton $\Gamma(\nu)$.
2. Apply Tutte's algorithm to find a layout of $\Gamma(\nu)$ with the outer face drawn as P. For each child μ of ν, denote the drawing of the face of $\Gamma(\nu)$ corresponding to μ by $P(\mu)$. If μ is not a leaf, then draw a closed curve bounding $P(\mu)$ and just outside $P(\mu)$ to represent the cluster $V(\mu)$.
3. For each child μ of ν, apply Straight_Line_Convex_Draw recursively, with node μ of T and polygon $P(\mu)$.

□

Tutte's algorithm requires solving a sparse system of $O(n)$ linear equations; this can be done in $O(n^{1.5})$ time using the sparse Gaussian elimination method of [13]. Thus the complexity of *Straight_Line_Convex_Draw* is bounded above by $\sum_{\nu \in T} |V(\nu)|^{1.5}$. We assume that each node ν in T has at least two children except for leaf nodes. If T has n leaves, then T has less then $2n$ nodes. It follows that *Straight_Line_Convex_Draw* requires $O(n^{2.5})$ time, where n is the number of vertices in G. The following theorem summarizes the performance of algorithm *Straight_Line_Convex_Draw*.

Theorem 3. *Suppose that $C = (G, T)$ is a skeleton-triconnected clustered graph such that for each node ν of T, $G(\nu)$ is biconnected. Then Algorithm* Straight_Line _Convex_Draw *computes a straight line convex drawing of C in time $O(n^{2.5})$.*

Tutte's algorithm has poor resolution properties: nodes may be exponentially close together [7]; in other words, the output may have exponential area for a given resolution. One may try to avoid this problem by using a polynomial area classical graph drawing algorithm such as that in [10]. Note, however, that the algorithms of [10] can be used to produce convex faces, but the shape of the outside face is not prescribed as input to the algorithm. In the next section we show that the poor resolution of algorithm *Straight_Line_Convex_Draw* cannot be avoided: there are compound graphs for which every straight line convex drawing requires exponential resolution.

4 Area and Angle Requirements

In this section, we present an area lower bound and an angle upper bound for C-planar straight-line convex drawings of clustered graphs. Our bounds are unlike the bounds for classical graphs. Planar drawing of classical graphs requires $\Omega(n^2)$ area [4] and with angle resolution dependent on the degree of the graph [14, 6]. We present a class of clustered graphs which, in any C-planar straight-line convex drawing, require exponential area and an angle inversely proportional to the size of the clustered graph. This class of clustered graph is an extension of the graphs defined in [2] (see Fig. 5).

For each $n \geq 0$ we define a clustered graph $C_n = (G_n, T_n)$ as follows. The first graph G_0 consists of two vertices a_0 and b_0, and the second graph G_1 consists of G_0 plus vertices a_1 and b_1 and edges (a_0, b_1), (b_0, a_1), (a_0, a_1) and (b_0, b_1). For each $n > 1$, G_n is defined from G_{n-1} by adding vertices a_n and b_n, and edges (a_n, b_{n-1}), (b_n, a_{n-1}), (a_n, a_{n-1}), (b_n, b_{n-1}), (a_n, b_{n-2}) and (b_n, a_{n-2}). The tree

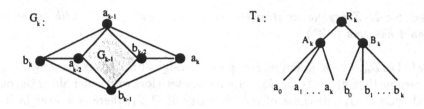

Fig. 5. Recursive Definition of $C_n = (G_n, T_n)$

T_n is composed of a root node which has two children A_n and B_n; A_n has children $a_0, a_1, \ldots, a_n$, and B_n has children $b_0, b_1, \ldots, b_n$. Note that for $n > 0$, $C_n = (G_n, T_n)$ is a C-planar clustered graph with $2n + 2$ vertices and $6n - 2$ edges. For $n > 2$, G_n is 3-connected and thus has has a unique planar embedding. Further, C_n is a connected clustered graph, and the outer face of each cluster is bounded by a single cycle. Therefore C_n has a unique C-planar embedding.

We need to show that the choice of an outside face in a drawing does not effect the result of area requirement. In this embedding the vertices a_{k+1} and b_{k+1} are embedded in the face f_k of G_k bounded by the cycle $(a_{k-1}, a_k, b_{k-1}, b_k, a_{k-1})$ for $k = 2, 3, \ldots, n - 1$. Consider a planar drawing $\mathcal{D}'_n$ of G_n in which f_k is not drawn as the outer face of G_k. By relabeling a_k as a_{n-k}, and b_k as b_{n-k}, $\mathcal{D}'_n$ becomes a planar drawing of G_n in which each f_k is drawn as the outer face of G_k. Thus we can assume that face f_k of G_k is drawn as the outer face of G_k. In particular, a_{k+1} and b_{k+1} are drawn in the outer face of G_k.

The clustered graphs C_n is undirected. However, for the purposes of the proofs of the next two theorems, we shall adopt a direction for each edge. For $k > 1$ we direct (a_{k-1}, a_k) from a_{k-1} to a_k, (b_{k-1}, b_k) from b_k to b_{k-1}, and (a_i, b_j) from b_j to a_i. The *vertex resolution* of a drawing is the minimum distance between a pair of vertices.

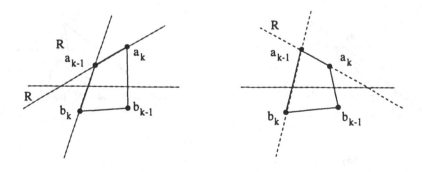

Fig. 6. Illustration of the Proof of Theorem 4

Theorem 4. *Every planar straight-line convex drawing of C_n with vertex resolution 1 has area $\Omega(2^n)$.*

Proof. Let $\mathcal{D}_{n+1}$ be a minimum area planar straight-line convex drawing of C_{n+1}. We assume that $a_k, a_{k-1}, b_k, b_{k-1}$ are in counter-clockwise order along the outer facial cycle of G_k . Because of the convexity of $\mathcal{D}_{n+1}$, there is a straight line l that can separate $a_0, \ldots, a_n, a_{n+1}$ from $b_0, \ldots, b_n, b_{n+1}$. We can assume that line l is the x-axis. Further assume that the vertices $a_0 \ldots a_n, a_{n+1}$ are above line l, and that $b_0, \ldots, b_n, b_{n+1}$ are below line l.

Next, we show that for each $1 \leq k \leq n$, a_k is *above* a_{k-1}. Note that a_k, a_{k-1}, and b_k are *visible* from b_{k+1}, and b_{k+1} is drawn in the outer face of G_k Therefore, b_{k+1} lies in the intersection of these two half planes (shown as region R in Fig. 6). Suppose that a_k has y-coordinate less than or equal to that of a_{k-1}, and we know that b_k is also below a_{k-1}; then region R is above the horizontal line that goes through a_{k-1}. Consequentially, b_{k+1} is above the x-axis. This contradicts with the assumption that b_{k+1} is below the x-axis. Thus we can deduce that a_k is above a_{k-1}. Similarly, we can show that b_k is drawn below b_{k-1}.

It follows that all arcs in the drawing $\mathcal{D}_{n+1}$ are directed upward; in other words, $\mathcal{D}_{n+1}$ contains a planar straight-line *upward* drawing of G_n. It is shown in [2] that planar straight-line *upward* drawing of digraph G_n requires area $\Omega(2^n)$; hence a C-planar straight-line convex drawing of C_n requires area $\Omega(2^n)$. □

Theorem 5. *In every planar straight-line convex drawing of C_n, the smallest angle between two edges incident to a vertex is $O(1/n)$.*

Proof. Again, let $\mathcal{D}_{n+1}$ be a planar straight-line convex drawing of C_{n+1}. Suppose that ω is the smallest angle between two edges which are incident to the same vertex in drawing $\mathcal{D}_{n+1}$. For each $1 \leq k \leq n$, let α_k be the angle formed by the arc (a_{k-1}, a_k) with the x-axis, let β_k be the angle formed by the arc (b_k, b_{k-1})

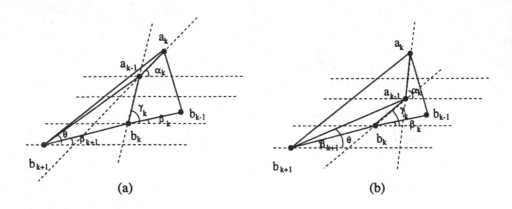

(a)   (b)

Fig. 7. Illustration of the Proof of Theorem 5

with the x-axis, and let γ_k be the angle formed by the arc (b_k, a_{k-1}) with the x-axis. Let θ be the angle between arc (b_{k+1}, a_{k-1}) and the x-axis. Consider the two cases shown in Fig. 7:

1. $\alpha_k \leq \gamma_k$. Because a_k, a_{k-1} and b_k are all *visible* from b_{k+1}, and $\alpha_k \leq \gamma_k$, b_{k+1} lies on the left of the line which goes through the arc (a_{k-1}, a_k), and below the horizontal line which goes through b_k. Then $\theta < \alpha_k$.
2. $\gamma_k \leq \alpha_k$. Because a_k, a_{k-1} and b_k are all *visible* from b_{k+1}, and $\gamma_k \leq \alpha_k$, b_{k+1} lies on the left of the line which goes through the arc (b_k, a_{k-1}), and below the horizontal line which goes through b_k. Then $\theta < \gamma_k \leq \alpha_k$.

Since b_k is below a_{k-1}, we have $\theta = \beta_{k+1} + \angle a_{k-1} b_{k+1} b_k$. It follows that

$$\beta_{k+1} = \theta - \angle a_{k-1} b_{k+1} b_k \leq \theta - \omega \leq \alpha_k - \omega.$$

Similarly, by a simple reflection of upside-down and leftside-right, we get

$$\alpha_{k+1} \leq \beta_k - \omega.$$

So, for each $1 < k < n + 1$, we have:

$$0 \leq \alpha_k \leq \beta_{k-1} - \omega$$

and

$$0 \leq \beta_k \leq \alpha_{k-1} - \omega.$$

Therefore,

$$0 \leq \alpha_n + \beta_n \leq \ldots \leq \alpha_1 + \beta_1 - 2(n-1)\omega \leq 4\pi - 2(n-1)\omega.$$

It follows that ω is $O(1/n)$. $\qquad\square$

5   Conclusion and Open Problems

This paper (with [5]) represents the first formal attempt to investigate drawing algorithms for *clustered graphs*. We are particularly concerned about planarity concepts for these structures. We have presented an efficient planar layout algorithm for clustered graphs; and we have demonstrated an area lower bound and an angle upper bound for planar straight-line convex drawings.

Nevertheless, there are several interesting open problems on planar layout of clustered graphs, such as the following.

- Does every C-planar clustered graph admits a straight-line convex drawing? If 'yes', then can we find a polynomial time algorithm to construct a straight-line convex drawing of clustered graphs (including non skeleton-triconnected clustered graphs), or show that the problem is NP-hard?
- Can we find some functions which relate line straightness (number of bends and total edge length) and regularity of regions to area requirements?

References

1. G. Di Battista, P. Eades, and R. Tamassia. Algorithms for drawing graphs: an annotated bibliography. Technical report, Department of Computer Science, Brown University, 1993. To appear in Computational Geometry and Applications.
2. G. Di Battista, R. Tamassia, and I. G. Tollis. Area requirement and symmetry disdplay of planar upward drawings. *Discrete & Computational Geometry*, 7:381–401, 1992.
3. Claude Berge. *Hypergraphs*. North-Holland, 1989.
4. H. de Fraysseiz, J. Pach, and R. Pollack. Small sets supporting fary embeddings of planar graphs. In *Proc. 20th ACM Symposium on Theory of Computing*, pages 426–433, 1988.
5. Qing-Wen Feng, Peter Eades, and Robert F. Cohen. Clustered graphs and c-planarity. Technical Report 04, Department of Computer Science, The University of Newcastle, Australia, 1995.
6. M. Formann, T. Hagerup, J. Haralambides, M. Kaufmann, F.T. Leighton, A. Simvonis, E. Welzl, and G. Woeginger. Drawing graphs in the plane with high resolution. In *Proc. 31th IEEE Symp. on Foundations of Computer Science*, pages 86–95, 1990.
7. Patrick L. Garvan. Drawing 3-connected planar graphs as convex polyhedra. Technical Report 02, Department of Computer Science, The University of Newcastle, Australia, 1995.
8. D. Harel. On visual formalisms. *Communications of the ACM*, 31(5):514–530, 1988.
9. T. Kamada. *Visualizing Abstract Objects and Relations*. World Scientific Series in Computer Science, 1989.
10. G. Kant. Drawing planar graphs using the *lmc*-ordering. In *Proc. 33th IEEE Symp. on Foundations of Computer Science*, pages 101–110, 1992.
11. J. Kawakita. The KJ method – a scientific approach to problem solving. Technical report, Kawakita Research Institute, Tokyo, 1975.
12. Wei Lai. *Building Interactive Digram Applications*. PhD thesis, Department of Computer Science, University of Newcastle, Callaghan, New South Wales, Australia, 2308, June 1993.
13. R. J. Lipton, D. J. Rose, and R. E. Tarjan. Generalized nested dissection. *SIAM J. Numer. Anal.*, 16(2):346–258, 1979.
14. S. Malitz and A. Papakostas. On the angular resolution of planar graphs. In *Proc. 24th ACM Symp. on Theory of Computing*, pages 527–538, 1992.
15. K. Misue and K. Sugiyama. An overview of diagram based idea organizer: D-abductor. Technical Report IIAS-RR-93-3E, ISIS, Fujitsu Laboratories, 1993.
16. S. C. North. Drawing ranked digraphs with recursive clusters. preprint, 1993. Software Systems and Research Center, AT & T Laboratories.
17. Tom Sawyer Software. Graph layout toolkit. available from bmadden@TomSawyer.COM.
18. K. Sugiyama and K. Misue. Visualization of structural information: Automatic drawing of compound digraphs. *IEEE Transactions on Systems, Man and Cybernetics*, 21(4):876–892, 1991.
19. W. T. Tutte. How to draw a graph. *Proceedings of the London Mathematical Society*, 3(13):743–768, 1963.
20. Rebecca Wirfs-Brock, Brian Wilkerson, and Lauren Wiener. *Designing Object-Oriented Software*. P T R Prentics Hall, Englewood Cliffs, NJ 07632, 1990.

An Efficient Orthogonal Grid Drawing Algorithm For Cubic Graphs

Tiziana Calamoneri and Rossella Petreschi

University of Rome "La Sapienza", Dept. of Computer Science,
Via Salaria 113, 00198 Roma, Italy

Abstract. In this paper we present a new algorithm that constructs an orthogonal drawing of a graph G with degree at most three. Even if we do not require any limitations neither to planar nor to biconnected graphs, we reach the best known results in the literarture: each edge has at most 1 bend, the total number of bends is $\leq \frac{n}{2} + 1$, and the area is $\leq (\frac{n}{2} - 1)^2$.

1 Introduction

The orthogonal grid drawing of a graph $G = (V, E)$ is a drawing such that the edges are polynomial chains consisting of horizontal and vertical segments and the vertices and the bends of the edges have integer coordinates. The graphs that admit such a drawing must have maximum degree 4. Inside this set of graphs, cubic graphs (i.e. regular graphs of degree 3), constitute an interesting and complex class of graphs, despite their apparent semplicity. Moreover cubic graphs and at most cubic graphs (i.e. graphs with maximum degree 3) seem to be a natural model for a lot of practical problems in different fields like Chemistry, Phisics, Computer Science, etc. [3, 6, 10].

Generally speaking, an orthogonal graph drawing algorithm reads as input a combinatorial description of a graph and produces a graphical representation of the graph that optimizes some cost functions (area, number of bends, etc.) or satisfies some other constraint.

Several results regarding orthogonal drawings of cubic graphs have appeared in the literature just in the last years [2, 5, 7, 8, 9]. Table 1 summarizes these results.

In this paper we present a new algorithm that constructs an orthogonal drawing of a graph G with degree at most three in which each edge has at most 1 bend, the total number of bends is $\leq \frac{n}{2} + 1$, and the area is $\leq (\frac{n}{2} - 1)^2$.

So, we reach the best known results in the leterature, even if we do not require any limitations neither to planar nor to biconnected graphs. Infact, at the best of our knowledge the only paper dealing with so general graphs is [2], whose upper bounds are worse than ours.

Troughout this paper we follow the standard graph-theoretic terminology of [1] and [4], and we consider only finite, simple, loopless and connected graphs.

	Input	Output	Time	Gridsize	Total number of bends	Max numb. of bends per edge	Planarity test
DLV93 [5]	3-connected planar at most cubic graph	orthogonal drawing with the minimum number of bends	$O(n^4 m \log n)$	not investigated	minimum	not investigated	no
LMP94 [8]	2-connected at most cubic graph	orthogonal planar drawing if G is planar, nothing if G is not planar	$O(n)$ ammortized	$O(n^2)$	$n/2+1$	1	yes
BK94 [2]	connected at most cubic graph and a layout of the graph if it is planar	orthogonal drawing (planar if G is planar)	$O(n)$	$(n-1) \times (n-1)$	$2n-1$	2	no
PT94 [9]	2-connected at most cubic graph	orthogonal drawing (not necessarily planar)	$O(n)$	$(n/2+1) \times n/2$	$n/2+3$	1 except one edge bending twice	no
K94 [7]	planar at most cubic graph	orthogonal planar drawing	$O(n)$	$\lceil n/2 \rceil \times \lceil n/2 \rceil$	$\lfloor n/2 \rfloor + 1$	1	no

Table 1. Known results

2 Drawing biconnected graphs

In this section we will present an algorithm to obtain single bend drawings for general biconnected graphs of maximum degree 3. Without loss of generality, we assume that the input graph is cubic: if it is not, a vertex of degree 2 is a dot along a single edge. Observe that a biconnected graph cannot have any vertex of degree 1.

The basic idea of the algorithm consists in adding to the drawing, one at a time, the vertices of the graph, ordered according to an st-numbering. In a st-numbered graph the vertices 1 and n (corresponding to the vertices s and t, respectively) are connected and for each vertex i, $2 \leq i \leq n - 1$, two adjacents j and k exist such that $j < i < k$.

Let $G_k = (V_k, E_k)$ be the subgraph induced by the first k vertices in the st-numbering ($V_k = \{1, 2, \ldots, k\}$) and D_k a drawing for G_k. During the k-th step, the vertex k and the edges (i, k), with $i < k$, are added to D_{k-1}. So, at each step k, where $1 < k < n$, by st-numbering's properties, it is possible to draw at least one and at most two edges, together with vertex k. Only vertex n has three adjacents in D_{n-1} and it is drawn as a *comb graph*, shown in Fig. 1. Namely, given a biconnected cubic graph G and an st-numbering on it, the number of vertices v having two adjacents i and j such that $i, j < v$, is $\frac{n}{2} - 1$.

We build the drawing in such a way that at least one of the two possible edges is connected in a directed way (i.e. it does not introduce any bend).

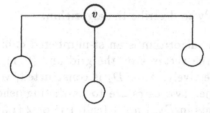

Fig. 1. A comb graph

Now on the four directions on the grid respect to each vertex are distinguished by labels from the set $\{Left, Right, Up, Down\}$. We call *free* a direction respect to a vertex if no edge is present on it.

Before describing the algorithm, we have to consider two operations that will be useful in the following.

In Fig. 2 a *rotation* is shown. The advantage of this operation is the changing of the free directions respect to the rotated vertex, even if there is an increase of the number of bends.

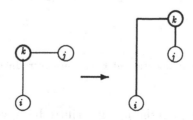

Fig. 2. Rotation of the vertex k

In Fig. 3 a *bend's movement* is represented. In this case the changing of the labels does not require any increase of the number of bends.

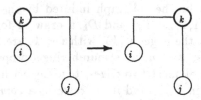

Fig. 3. Movement of the vertex k on a bend

Now we are ready to describe the algorithm.

The input of the algorithm is an st-numbered cubic graph. Let (x_k, y_k) be the coordinates of the vertex k on the grid and $(0,1), (0,2)$ the coordinates in D_2 of 1 and 2 respectively. After D_2 is constructed, vertices from 3 to n-1 are added one at a time. Two cases are to be distinguished according to the fact that the k's adjacents in D_{k-1} are 1 (call it i) or 2 (i and j).

In the first case, the edge $\{i, k\}$ is added to D_{k-1} as a straight line. One k's coordinate coincides with the corresponding i's while the other needs the increment either of a new row or of a new column of the current gridsize.

Not so easy is the case when $\{i, k\}$ and $\{j, k\}$ are to be added to D_{k-1}. In order to limit the number of bends, the algorithm try to put in a straight line way both the edges; this is possible only in the situation of fig. 4 in which neither new rows, nor new columns are added.

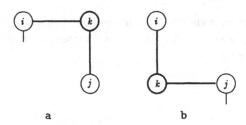

a   b

Fig. 4. The insertion of k does not introduce any bend

In all the other situations the algorithm introduces only a new bend: two cases are to be distinguished according to the existence of a common free direction for i and j, or not.

In fig. 5 and in fig. 6 are shown the two different cases in which it is possible to see that either a new row or a new column is added to the drawing: in the first case on the grid the graph is extended from the border while in the second case we have an expansion from the inner part of the drawing.

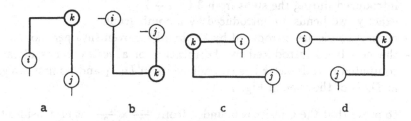

a   b   c   d

Fig. 5. The two k's adjacents have a common free direction

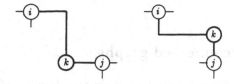

Fig. 6. The two k's adjacents have different free directions

The anlysis of all the cases of the algorithm is done either on the current drawing or on the drawing modified by some bend's movements. The operation of rotating a vertex is done only if no one of the previous case is verified.

The last step of the algorithm put on the grid the vertex n according to the comb graph design, after a possible operation of movement of bend and/or a rotation.

Theorem 1. *There is a linear time and space algorithm to draw a 2-connected at most cubic graph on an $\frac{n-1}{2} \times \frac{n-1}{2}$ grid with at most $\frac{n}{2} + 1$ bends and every edge bends at most once.*

Proof. The algorithm works correctly; infact it is easy to see that it analyzes all the possible combinations of vertices' free directions, either directly or after a rotation and/or a bend's movement.

Moreover, it is always possible to draw an edge without going over a vertex, infact every new vertex put on the grid lays either on a new row or on a new column, in such a way that its free directions do not cross any vertex. The only case in which neither a new row nor a new column is introduced is the case of

Fig. 4, but in that case the free directions are not stopped with further vertices because they were the free directions of vertices i and j.

The limitation of the total number of bends is the consequence of the following three facts:

a - at most $\frac{n}{2} - 1$ vertices have two adjacents then at most $\frac{n}{2} - 1$ bends are introduced during the steps from 3 to $n - 1$
b - exactly two bends are introduced by a comb graph
c - no new bends are introduced by the bend's movement operation
d - the new bend introduced by the rotation of a vertex k has already been computed in a. Infact k has two adjacents in D_{k-1} and its first assignement in D_k is of the type in Fig. 4.

To prove that the gridsize is bounded from $\frac{n-1}{2} \times \frac{n-1}{2}$ let us consider that the vertices 1 and 2 do not introduce any area. An insertion of a new row or a new column, but not both, is required by each vertex from 3 to $n-1$. If exactly h rows are introduced, then at most $n-3-h$ columns are introduced. The vertex n can introduce at most both one row and one column. Then $A \leq (h+1) \cdot (n-2-h)$, that is bounded by $\frac{n-1}{2} \times \frac{n-1}{2}$.

No further details are needed to prove the linearity in time and space of the algorithm. □

3   Drawing connected graphs

In this section we point the attention on the drawing of general at most cubic graphs. The approach to the problem is through the biconnected components, hence the algorithm of Section 2 will play a fundamental role also in this case.

Let us call $B_1 = (V_1, E_1), \ldots, B_K = (V_K, E_K)$ the biconnected components of $G = (V, E)$ and let $a_1, \ldots, a_r$ be the articulation points ($|V| = |V_1| + \ldots + |V_K| + r$).

Our idea consists in drawing on the grid separately each B_i and then in connecting the drawings through points a_i. The edges incident to vertices a_i are drawn taking into account both the limitation of the total number of bends and the total area of the drawing of G.

We will prove that, in this way, the upper bounds of the area and of the total number of bends remain the same of biconnected case. To show that, it is not restrictive to suppose that only one separating vertex a exists. Infact, if there is more than one, by deleting a from G some simply connected components are obtained, and it is possible to inductively repeat the same argument. So, suppose that when a is deleted from G, only biconnected components remain.

Observe that the adjacents of a: v_1, v_2 and v_3 have at most two adjacents on the drawing of their biconnected component, therefore they have at least two free directions and the algorithm mantains empty the row (or the column) from all their free directions to the boundary. So, it is not restrictive to think (just for semplicity of explanation) that v_1, v_2 and v_3 lay on the boundary of their biconnected components drawings.

We have to distinguish two different cases, according to the fact that the deletion of a disconnects the graph in two or in three biconnected components (see Fig. 7a and b).

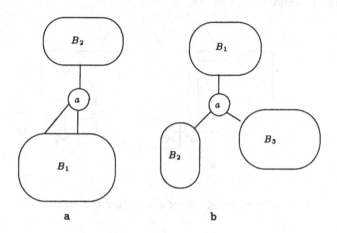

Fig. 7. Two different types of articulation points

1 - There are two biconnected components B_1 and B_2. Let us suppose $v_1, v_2 \in B_1$ and $v_3 \in B_2$. There are different cases according to the mutual position of v_1 and v_2.

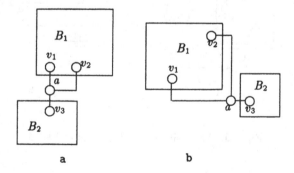

Fig. 8. Two ways of connecting two components

In Fig. 8a and 8b the way to connect a to v_1, v_2 and v_3 is shown both when v_1 and v_2 have a free direction on the same side of the drawing of B_1 and when they have it on two different but consecutive sides of the drawing of B_1. In both these cases at most 2 bends are introduced.

In Fig. 9 the case in which the free directions of v_1 and v_2 are on two different and opposite sides of the drawing of B_1 is shown. It is possible to see that this case is reduced to one of the previous ones. Infact v_1 and v_2 have at least two free directions, therefore another available row (or column) going to the boundary must exists.

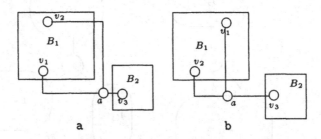

Fig. 9. Third way of connecting two components

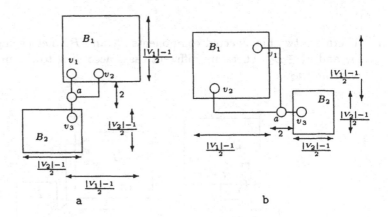

Fig. 10. Worst situations for cases of Fig. 8

For what concerns the gridsize, in case of Fig. 8a the worst case happens when v_1 lays on the lower-left corner of the embedding of B_1 and v_3 is on the higher-right corner of B_2 (see Fig. 10a). In case of Fig. 8b the worst case occurs when v_3 is exactly in the middle of the side on which it lays (see Fig. 10b). Therefore the values of the area in the two cases are $A \leq (\frac{|V_1|-1}{2} + \frac{|V_2|-1}{2} + 2) \times (\frac{|V_1|-1}{2} + \frac{|V_2|-1}{2} - 1)$ and $A \leq (\frac{|V_1|-1}{2} + \frac{|V_2|-1}{4} + 1) \times (\frac{|V_1|-1}{2} + \frac{|V_2|-1}{2} + 2)$ respectively, that are both less than $\frac{n-1}{2} \times \frac{n-1}{2}$.

2 - There are three biconnected components B_1, B_2 and B_3. It is always possible to put in the grid the embeddings of the components in such a way that the two smallest components are put beside, while the biggest one is put on the opposite side (see Fig. 11a). In order to compute the gridsize, observe that the situation in which v_1 lays on the lower-left corner of the embedding of B_1 and v_3 on the higher-right corner of B_3 is the worst case; it does not matter where v_2 lays (see Fig. 11b). It is not restrictive to think that $|V_2| \geq |V_1|$; therefore the grid has its width not greater than $\frac{|V_3|-1}{2} + 2 + \frac{|V_2|-1}{2}$ and its height not greater than $max(\frac{|V_3|-1}{2}, \frac{|V_1|-1}{2} + \frac{|V_2|-1}{2} + 1, \frac{|V_3|-1}{2} + \frac{|V_1|-1}{2} - 1)$. In all cases, the area is less than $\frac{n-1}{2} \times \frac{n-1}{2}$.

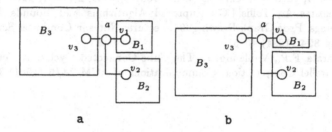

a   b

Fig. 11. Way of connecting three components and worst case for the gridsize

From the previous observations, it is possible to state the following theorem.

Theorem 2. *There is a linear time and space algorithm to draw a one-bend drawing of a connected at most cubic graph on an $\frac{n-1}{2} \times \frac{n-1}{2}$ grid with at most $\frac{n}{2} + 1$ bends.*

Proof. It is easy to see that the procedure inserting a separating vertex in the drawing runs in constant time, so the whole algorithm runs in linear time. Let us prove that the maximum number of bends is $\frac{n}{2} + 1$. The drawing of a and its connections introduce at most two bends in the drawing of G. Each biconnected component B_i has at most $\frac{|V_i|}{2}$ bends because its t node is adjacent to a, and in B_i it has at most two adjacents. Then , the increasing of number of bends introduced by inserting a is balanced by the decreasing in each biconnected component.

Bounds for the gridsize immediately follows from the previous details. ☐

References

1. Aho, A., Hopcroft, J. K. Ullman, J. D.: The design and analysis of computer algorithms. Addison Wesley, Reading, MA, (1973)

2. Biedl, T., Kant, G.: A Better Euristic for Orthogonal Graph Drawings. Algorithms - ESA '94. Proceedings, Lectures Notes in Computer Science, Springer-Verlag **855** (1994) 24–35
3. Bjorken, J.D., Drell, S.D.: Quantum Electrodynamics. Mc-Graw Hill
4. Cormen, T.H., Leiserson, C.E., Rivest, R.L.: *Introduction to algorithms*. the MIT Press, Cambridge,MA, (1990)
5. Di Battista, G. ,Liotta, G., Vargiu, F.: Spirality of Orthogonal Representations and Optimal Drawings of Series-Parallel Graphs and 3-Planar Graphs. Lectures Notes in Computer Science, Springer-Verlag **709** (1993) 151–162
6. Greenlaw, R. ,Petreschi, R.: Cubic graphs, Tech. Rep. University of New Hampshire, Durham,N.H.,USA **15** (1993)
7. Kant, G.: Drawing Planar Graphs Using the canonical ordering. Algorithmica - Special Issue on Graph Drawing (to appear)
8. Liu, Y. Marchioro, P., Petreschi, R.: At most single bend embedding of cubic graphs. Applied Mathematics (Chin. Journ.) **9/B/2** (1994) 127–142
9. Papakostas, A., Tollis, I.G.: Improved Algorithms and Bounds for Orthogonal Drawings. Proc.Graph Drawing '94, Lectures Notes in Computer Science, Springer-Verlag **894** (1994) 40–51
10. Preparata, F.P., Vuillemin, J.: The Cube-Connected Cycles: A Versatile Network for Parallel Computation. Communications of ACM **24/5** (1981) 300–309

Constrained Independence System and Triangulations of Planar Point Sets*

Siu-Wing Cheng[1] and Yin-Feng Xu[2]

[1] Department of Computer Science, The Hong Kong University of Science & Technology, Clear Water Bay, Hong Kong. E-mail : scheng@cs.ust.hk
[2] School of Management, Xi'an Jiaotong University, Xi'an, Shaanxi, 710049, PRC.

Abstract. We propose and study a new constrained independence system. We obtain a sequence of results, including a matching theorem for bases of the system and introducing a set of light elements which give a lower bound for the objective function of a minimization problem in the system. We then demonstrate that the set of triangulations of a planar point set can be modeled as constrained independence systems. The corresponding minimization problem in the system is the well-known minimum weight triangulation problem. Thus, we obtain two matching theorems for triangulations and a set of light edges (or light triangles) that give a lower bound for the minimum weight triangulation. We also prove directly a third matching theorem for triangulations. We show that the set of light edges is a superset of some subsets of edges of a minimum weight triangulation that were studied before.

1 Introduction

Independence system plays a very important role in the study of discrete optimization problems and matroid is a well-known example [10]. In this paper, we propose a *constrained independence system* and study some of its combinatorial properties. In particular, we prove a matching theorem for bases of such a system. We define a minimization problem for such a system and presents two heuristics that are based on a greedy approach and a local search method, respectively. We also study how to obtain a lower bound to the minimization problem.

The constrained independence problem can be used to model the set of all possible triangulations of a given planar point set. Thus, our result implies a matching between edges/triangles of different triangulations of the same point set. This is a new structural property of triangulations that were not known before. The minimization problem for a constrained independence system then translates to the open problem of finding a minimum weight triangulation. We present two lower bounds to this problem.

* Research of the first author is supported partially by RGC grant HKUST 190/93E. Research of the second author is supported partially by RGC grants HKUST 190/93E and HKUST 181/93E. The work is done while the second author visits Department of Computer Science, The Hong Kong University of Science and Technology.

The definition of constrained independence system is very similar to that of matroid. However, the minimization for a constrained independence system can be shown to be NP-hard [9], while the optimization problem for matroid can be solved efficiently by a greedy strategy. This implies that if the minimum weight triangulation problem can be solved in polynomial time, then properties (perhaps geometrical) other than those captured by the constrained independence system have to be discovered.

The rest of this paper is organized as follows. In Section 2, we define the constrained independence system and prove several properties, including the matching theorem and a lower bound for the minimization problem. In Section 3, we model the triangulations with constrained independence systems. In Section 3.1, we first present two matching theorems for triangulations that follow from the result in Section 2. We also prove directly another different matching theorem for triangulations. In Section 3.2, we use the result in Section 2 to define *light edges* and *light triangles*, which will give lower bounds to the minimum weight triangulation problem. The minimum weight triangulation problem is to find a triangulation of a given point set such that the sum of the Euclidean lengths of the edges used is minimum. The β-skeleton for $\beta = \sqrt{2}$ studied by triangulations studied by Keil [6] and the stable line segments studied by Xu [11] are two distinct subgraphs of a minimum weight triangulation. We prove that the set of light edges is a superset of the $\sqrt{2}$-skeleton and the set of stable line segments. Thus, the length sum of the light edges gives us a better lower bound on the weight of the minimum weight triangulation. However, the set of light edges is not necessarily a subgraph of a minimum weight triangulation. Recently, Cheng and Xu [3] prove that for $\beta > 1/\sin(\tan^{-1}(3/\sqrt{2\sqrt{3}}) \approx \pi/3.1$, the β-skeleton is a subgraph of a minimum weight triangulation. This is almost the smallest β possible because there is a counter-example for $\beta < \pi/3$ [6].

One of our matching theorems Theorem 1 is also obtained independently by Aichholzer *et al* [1]. Light edges and triangles are also defined in [1]. Theorem 1 can easily be extended [1] to give an injective mapping from a triangulation to the set of all possible $n(n-1)/2$ edges. By minimizing the sum of edge lengths in the image, a lower bound to the minimum weight triangulation can be obtained. We can apply this result to any triangulation containing the set of light edges. Then a lower bound stronger than the sum of lengths of all light edges can be obtained. It is also described in [1] how to compute the set of light edges in $O(n^2 \log n)$ time. At the time of writing, our results and the results in [1] are being combined in one full paper.

2 Constrained independence system

Given E a finite set of elements and $\mathcal{J}$ a nonempty collection of subsets of E, $(E, \mathcal{J})$ is an independence system if $Y \in \mathcal{J}$ whenever $X \in \mathcal{J}$ and $Y \subseteq X$. The elements of $\mathcal{J}$ are called *independent sets*. A maximal independent set is called a *base*. Any subset of E that does not belong to $\mathcal{J}$ is called a *dependent set*. A minimal dependent set is called a *circuit*. A lot of combinatorial optimization problems can be formulated as optimization problems on different independence systems by associating appropriate non-negative weight $w(e)$ with each element e of E.

We define a *constrained independence system* as follows. $(E, \mathcal{J})$ is a constrained independence system if it is an independence system and it satisfies the following conditions:

Condition(1). If $X, Y \in \mathcal{J}$ with $|Y| = p$ and $|X| = p + 1$, then there is an element $e \in E - Y$ such that $Y \cup \{e\} \in \mathcal{J}$.

Condition(2). The cardinality of every circuit is two.

Fact. Every base of an independence system that satisfies condition(1) has the same cardinality. $\square$

The definition of constrained independence system is very similar to that of matroid. A matroid is an independence system such that if $X, Y \in \mathcal{J}$ with $|Y| = p$ and $|X| = p+1$, then there is an element $e \in X - Y$ such that $Y \cup \{e\} \in \mathcal{J}$. This is more restrictive than condition(1). Also, in general, a matroid may not satisfy condition(2).

2.1 A matching theorem

We prove a matching theorem for bases of a constrained independence system. Let $\mathcal{B}$ be the set of all bases.

Theorem 1. *Given $B, B' \in \mathcal{B}$, there exists a perfect matching between B and B', such that if $e \in B$ and $e' \in B'$ are matched, then either $e = e'$ or $\{e, e'\}$ is a circuit.*

Proof. Construct a bipartite graph with vertex set B and B'. There is an edge between $e \in B$ and $e' \in B'$ if $e = e'$ or $\{e, e'\}$ is a circuit. For any subset $F \subseteq B$, let $N(F)$ be the set of vertices in B' that are adjacent to some vertex in F. We now show that $|N(F)| \geq |F|$ and thus, by the König-Hall matching theorem [2], there is a perfect matching between B and B'.

We claim that $F \cup (B' - N(F)) \in \mathcal{J}$. Otherwise, $F \cup (B' - N(F))$ contains some circuit $\{e, e'\}$ (which contains exactly two elements by condition(2)) such that $e \in F$ and $e' \in B' - N(F)$. But this implies that $e' \in N(F)$, a contradiction. Since $F \cup (B' - N(F)) \in \mathcal{J}$ and all bases have the same cardinality, $|F \cup (B' - N(F))| \leq |B'|$. Moreover, $|F \cup (B' - N(F))| = |F| + |B'| - |N(F)|$. So we conclude that $|N(F)| \geq |F|$. This completes the proof. $\square$

2.2 The minimization problem

In this section, we consider optimization problems for a constrained independence system which can be formulated as

$$\min_{B \in \mathcal{B}} \sum_{e \in B} w(e),$$

where $w(e)$ is the (non-negative) weight of e. In general, it is very difficult to find an optimal solution for the objective function. However, local optimum can be computed efficiently using a greedy approach.

A base B is a *local optimum* if for every $e^* \in B$ and $e' \in E - B$ such that $\{e'\} \cup (B - \{e^*\})$ is a base, we have $\sum_{e \in B} w(e) \le \sum_{e \in \{e'\} \cup (B - \{e^*\})} w(e)$.

```
Algorithm Greedy /* Input : (E, J); Output : Bg */
Bg := ∅;
Q := E;
while Q ≠ ∅ do
    find e ∈ Q such that w(e) is minimized;
    Q := Q - {e};
    if Bg ∪ {e} ∈ J then Bg := Bg ∪ {e}
end while
```

Theorem 2. *The output of algorithm* Greedy, B_g, *is a base and is a local optimum.*

Proof. It is clear that $B_g \in \mathcal{J}$. Assume to the contrary that B_g is not maximal. Then there exists $e \in E - B_g$ such that $B_g \cup \{e\} \in \mathcal{J}$, which implies that $\{e\} \cup A \in \mathcal{J}$ for all $A \subseteq B_g$. Thus, when e was examined by algorithm Greedy, e should have been included in B_g, a contradiction. Hence, B_g is a base.

Assume to the contrary that B_g is not a local optimum. Then there exists $e \in B_g$ and $e' \in E - B_g$ such that $\{e'\} \cup (B_g - \{e\})$ is a base and $w(e') < w(e)$. Therefore, algorithm Greedy should examine e' before e. Since $\{e'\} \cup (B_g - \{e\}) \in \mathcal{J}$, $\{e'\} \cup A \in \mathcal{J}$ for all $A \subseteq B_g - \{e\}$. Thus, e' will be included and so e will not be included later, which contradicts our assumption. □

Remark. Since the proof of Theorem 2 does not make use of condition(1) and condition(2), the result is true for all independence systems. □

The greedy solution can be viewed as an approximate solution to the minimization problem. Sometimes, we may be given a base B and we are asked to obtain an improved solution. We can handle this problem with algorithm ESA which uses repeated *element substitution* to obtain an improved solution. A base B' is obtained from another base B by element substitution, if there exists $e \in B$ and $e' \in B' - B$ such that $B' = \{e'\} \cup (B - \{e\})$.

```
Algorithm ESA /* Input : A base B; Output : A base B* */
B* := B;
while B* is not a local optimum do
    find e ∈ B*, e' ∈ E - B* s.t. {e'} ∪ (B* - {e}) is a base and w(e') < w(e)
    B* := (B* - {e}) ∪ {e'}
end while
```

From the working of algorithm ESA, it is clear that the following is true.

Theorem 3. *Let B and B* be the input and output of algorithm* ESA, *respectively. Then B* is a local optimum and* $\sum_{e \in B^*} w(e) \leq \sum_{e \in B} w(e)$. *If B is not a local optimum, then the inequality holds strictly.*

We do not know of any polynomial-time algorithm that can solve the general minimization problem. Thus, it may be helpful to derive lower bound to the objective function, which can then guide the running of some heuristics or exhaustive search algorithms. We describe one such lower bound in the following. An element $e \in E$ is *light* if $w(e) < w(e')$ for all $e' \in E$ such that $\{e, e'\} \notin \mathcal{J}$. Let $L(E)$ be the set of all light elements.

Theorem 4. $L(E) \in \mathcal{J}$ *and* $\sum_{e \in L(E)} w(e) \leq \min_{B \in \mathcal{B}} \sum_{e \in B} w(e)$.

Proof. Assume to the contrary that $L(E) \notin \mathcal{J}$. Then there exists a circuit in $L(E)$. By condition(2), this circuit is $\{e_1, e_2\} \notin \mathcal{J}$ for some $e_1, e_2 \in L(E)$. But this implies that $w(e_1) < w(e_2)$ and $w(e_2) < w(e_1)$, a contradiction. Let B^* be a base such that $\sum_{e \in B^*} w(e) = \min_{B \in \mathcal{B}} \sum_{e \in B} w(e)$. Since $L(E) \in \mathcal{J}$, by condition(1), there is a base B such that $L(E) \subseteq B$. We apply Theorem 1 to B and B^*. Thus, every $e_1 \in L(E)$ is matched with an $e_2 \in B^*$ such that either $e_1 = e_2$ or $\{e_1, e_2\} \notin \mathcal{J}$. If $\{e_1, e_2\} \notin \mathcal{J}$, then $w(e_1) < w(e_2)$ as $e_1 \in L(E)$. So we conclude that $w(e_1) \leq w(e_2)$ and hence $\sum_{e \in L(E)} w(e) \leq \sum_{e \in B^*} w(e)$. $\square$

So $\sum_{e \in L(E)} w(e)$ is a lower bound to $\min_{B \in \mathcal{B}} \sum_{e \in B} w(e)$. In the proof above, we mention that there exists a base B such that $L(E) \subseteq B$. We can identify one such possible B to be B_g, where B_g is the output of algorithm Greedy.

Theorem 5. $L(E) \subseteq B_g$.

Proof. Let $e \in L(E)$. For all $e' \in E$, if $\{e, e'\}$ is a circuit, then $w(e) < w(e')$ as $e \in L(E)$. Therefore, no such e' can be examined before e in algorithm Greedy, which implies that e should be included in the greedy solution when it is examined. This completes the proof. $\square$

3 Applications in triangulations

Let S be a finite planar point set. To simplify our exposition, we assume that no three points in S are collinear. Let $E(S)$ denote the set of edges with endpoints in S. Let $\Delta(S)$ denote the set of empty triangles with vertices in S. (In this paper, we treat a triangle as the closed region bounded by its boundary edges.) A triangle is empty if its interior does not contain any point in S. Given an edge e (resp. a triangle Δ), we use $\text{int}e$ (resp. $\text{int}\Delta$) to denote the interior of e (resp. Δ).

A triangulation $T_e(S)$ is a maximal subset of $E(S)$ such that $\text{int}e_1 \cap \text{int}e_2 = \emptyset$ for any two distinct edges $e_1, e_2 \in T_e(S)$. Let n denote the cardinality of S. From Euler's formula, we have [4]

$$|T_e(S)| = 3n - 3 - |CH(S)|, \tag{1}$$

where $CH(S)$ denotes the set of boundary edges of the convex hull of S. We can equivalently view a triangulation as a maximal subset $T_t(S)$ of $\Delta(S)$ such that $\text{int}\Delta_1 \cap \text{int}\Delta_2 = \emptyset$ for any two distinct triangles $\Delta_1, \Delta_2 \in T_t(S)$. The following can also derived from Euler's formula.

$$|T_t(S)| = 2n - 2 - |CH(S)|. \tag{2}$$

3.1 Matching theorems for triangulations

Given a point set S of size n, define $\mathcal{J}_e(S)$ to be the collection of subsets of $E(S)$ such that for all $X \in \mathcal{J}_e(S)$ and $e_1, e_2 \in X$, $\text{int}e_1 \cap \text{int}e_2 = \emptyset$. We claim that $(E(S), \mathcal{J}_e(S))$ is a constrained independence system. By Equation 1, every base in $\mathcal{J}_e(S)$ has the same cardinality and hence condition(1) is satisfied. Condition(2) is also satisfied since the smallest dependent set must be an intersecting pair of edges in $E(S)$. By Theorem 1, we have the following corollary.

Corollary 6. *Given a finite point set S and two triangulations $T_1, T_2 \in \mathcal{J}_e(S)$, there exists a perfect matching between T_1 and T_2 such that if $e_1 \in T_1$ and $e_2 \in T_2$ are matched, then $\text{int}e_1 \cap \text{int}e_2 \neq \emptyset$.*

We can also consider the constrained independence system $(\Delta(S), \mathcal{J}_t(S))$, where $\mathcal{J}_t(S)$ is the collection of subsets of $\Delta(S)$ such that for all $X \in \mathcal{J}_t(S)$ and $\Delta_1, \Delta_2 \in X$, $\text{int}\Delta_1 \cap \text{int}\Delta_2 = \emptyset$. We then obtain another corollary from Theorem 1.

Corollary 7. *Given a finite point set S and two triangulations $T_1, T_2 \in \mathcal{J}_t(S)$, there exists a perfect matching between T_1 and T_2 such that if $\Delta_1 \in T_1$ and $\Delta_2 \in T_2$ are matched, then $\text{int}\Delta_1 \cap \text{int}\Delta_2 \neq \emptyset$.*

Corollaries 6 and 7 are also obtained independently by [1]. We can actually obtain a matching theorem stronger than Corollary 7 by applying the König-Hall Theorem directly.

Theorem 8. *Given a finite point set S and two triangulations $T_1, T_2 \in \mathcal{J}_t(S)$, there exists a perfect matching between T_1 and T_2 such that if $\Delta_1 \in T_1$ and $\Delta_2 \in T_2$ are matched, then*

1. *$\text{int}\Delta_1 \cap \text{int}\Delta_2 \neq \emptyset$*
2. *$V(\Delta_1) \cap V(\Delta_2) \neq \emptyset$, where $V(\Delta_i)$ denotes the vertex set of Δ_i, $i = 1$ to 2.*

Proof. Construct a bipartite graph with one vertex set corresponding to triangles in T_1 and the other vertex set corresponding to triangles in T_2. For convenience, we use $\Delta_i \in T_i$, $i = 1$ or 2, to denote both a triangle and the corresponding vertex in the bipartite graph. There is an edge between $\Delta_1 \in T_1$ and $\Delta_2 \in T_2$ if $\text{int}\Delta_1 \cap \text{int}\Delta_2 \neq \emptyset$ and $V(\Delta_1) \cap V(\Delta_2) \neq \emptyset$. For any subset $F \subseteq T_1$, let $N(F) \subseteq T_2$ be the set of triangles that are neighbors of some triangle in F. To complete the proof, it suffices to show that $|N(F)| \geq |F|$. Let x be a vertex of a triangle Δ_1 in F. There is a set of triangles in $N(F)$ that share the vertex x with

Δ_1. Moreover, the union of the angles of these triangles at x must contain the angle of Δ_1 at x. Thus, if we sum up the angles of all triangles in $N(F)$ and F, we obtain the inequality $\pi|N(F)| \geq \pi|F|$, which implies that $|N(F)| \geq |F|$. □

By Theorem 8, two matched triangles are either identical or they are in one of the four configurations shown in Figure 1.

Fig. 1.

3.2 Light edges and triangles

Given a finite point set S in the plane, the minimum weight triangulation problem is to triangulate the point set so that the sum of edge lengths is minimized. The complexity of the problem has not been resolved: neither is it known to be NP-hard nor is a polynomial-time algorithm known to exist [5]. Also, no approximation algorithm is known that achieves a constant approximation ratio. The best approximation algorithm is due to Plaisted and Hong [8] which achieves an approximation ratio of $O(\log n)$, where n is the size of the given point set. In this section, we make use of Theorem 4 to give a lower bound on the sum of edge lengths in a minimum weight triangulation.

Given an edge $e \in E(S)$, define $w(e)$ to be its length. An edge e is light if $w(e') > w(e)$ for all edge $e' \in L(S)$ such that $\text{int} e \cap \text{int} e' \neq \emptyset$. We use $L(E(S))$ to denote the set of light edges.

We can define light triangles similarly. Given a triangle $\Delta \in \Delta(S)$, define $w(\Delta)$ to be its perimeter. A triangle Δ is light if $w(\Delta') > w(\Delta)$ for all triangle $\Delta' \in \Delta(S)$ such that $\text{int}\Delta \cap \text{int}\Delta' \neq \emptyset$. We use $L(\Delta(S))$ to denote the set of light triangles.

In [7], a β-skeleton of a planar point set is introduced and it is proved in [6] that a $\sqrt{2}$-skeleton is a subgraph of a minimum weight triangulation. We describe the definition of β-skeleton below. For $\beta > 1$, the forbidden neighborhood for points x and y is defined to be the union of the two disks of radius $\beta \cdot d(x, y)/2$ that pass through both x and y (see Figure 2), where $d(x, y)$ denotes the distance between x and y. The edge xy is in the β-skeleton if the forbidden neighborhood for x and y does not contain any point of the given point set. (There is actually another variant of β-skeleton, the definition of which is based on lune-like neighborhood provided in [7]. The above results refer to the disk-based forbidden neighborhood described.) We use $\beta(E(S))$ to denote the β-skeleton, for $\beta > 1$.

We now prove that $\beta(E(S)) \subseteq L(E(S))$ for $\beta \geq 2/\sqrt{3}$ which implies that the β-skeletons obtained in [3, 6] are subsets of $L(E(S))$.

Theorem 9. $\beta(E(S)) \subseteq L(E(S))$ for $\beta \geq 2/\sqrt{3}$.

Proof. Refer to Figure 2. Let the union of the two disks shown be the forbidden neighborhood of x and y. Suppose that the forbidden neighborhood is empty. Note that the disk centers v and w are on opposite sides of xy. It suffices to show that every edge that intersects xy is longer than xy.

Suppose that an edge ab intersects xy at a point z and the disk centers are on opposite sides of ab. By geometry, $|az| > |yz|$ and $|bz| > |xz|$. Thus, $|ab| > |xy|$.

Suppose that an edge bc intersects xy and v and w lie on the same side of bc. Draw a line segment from c through x to a point d on the boundary of the other disk. By geometry, $|bc| > |cd|$. Thus, it suffices to show that $|cd| \geq |xy|$ for $\beta \geq 2/\sqrt{3}$. Imagine that cd is actually an elastic rod that passes through x and connects two points on the boundary of the two disks. If we move the endpoints of the elastic rod to rotate it about x, its length will change. Let $\theta = \angle cxy$ and $\alpha = \angle vxy$. Let r be the radius of the two disks. $|cx| = 2r\cos(\theta - \alpha)$. Similarly, $|dx| = 2r\cos(\pi - \theta - \alpha) = 2r|\cos(\theta + \alpha)|$. Thus, both $|cx|$ and $|dx|$ are convex functions in θ, which implies that $|cd|$ is also a convex function in θ.

Thus, $|cd|$ is minimum when c coincides with x and d moves to a point e such that ex is tangent to the disk with center v. Refer to Figure 2. Let r be the radius of the two disks. Then in order that xy is light, we require that $2r\cos(\pi/2 - 2\alpha) \geq |xy|$. Since $|xy|/2r = \cos\alpha$, this is equivalent to $\sin 2\alpha \geq \cos\alpha$ or $\cos\theta \leq \sqrt{3}/2$. Since $\beta = 2r/|xy| = 1/\cos\alpha$, this is equivalent to requiring that $\beta \geq 2/\sqrt{3}$. □

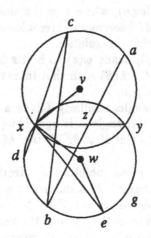

Fig. 2.

Another kind of subgraph of a minimum triangulation is the intersection of the edge set of all triangulations. This common intersection is called the *stable line segments* in [11]. Since each stable line segment is not intersected by any other edge in $E(S)$, $L(E(S))$ contains all the stable line segments by definition.

By considering the independence systems $(E(S), \mathcal{J}_e(S))$ and $(\Delta(S), \mathcal{J}_t(S))$ and applying Theorem 4, we obtain the following lower bounds for a minimum weight triangulation.

Corollary 10. *Let $MWT_e(S)$ and $MWT_t(S)$ be the set of edges and triangles, respectively, of a minimum weight triangulation.*

1. $\sum_{e \in L(E(S))} w(e) \leq \sum_{e \in MWT_e(S)} w(e)$
2. $\sum_{\Delta \in L(\Delta(S))} w(\Delta) \leq \sum_{\Delta \in MWT_t(S)} w(\Delta)$.

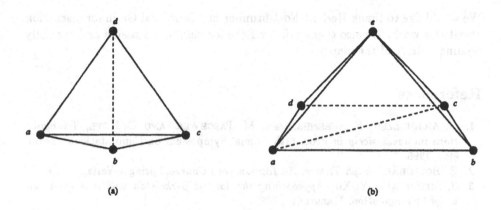

Fig. 3.

The two formulations $(E(S), \mathcal{J}_e(S))$ and $(E(S), \mathcal{J}_t(S))$ are not equivalent. Refer to Figure 3(a). By construction, ac is longer than bd, ad and bc are longer than bd, and ab and bc are the shortest edges. So the set of light edges $\{ab, bc, bd, ad, cd\}$ is exactly the minimum weight triangulation. The triangle abc has the smallest perimeter and it is the only light triangle. Thus, in this example, running Greedy on edges will give us a better triangulation (the optimal one) than running Greedy on triangles. Refer to Figure 3(b). By construction, ad and bc are shortest and cd is shorter than ae and be. Also, the sum of lengths of ae and be is less than the sum of lengths of cd and ac. So the set of light edges is $\{ad, ab, bc, cd, de, ce\}$. The set of light triangles is $\{ade, bce\}$. Thus, in this example, running Greedy on triangles will give us a better triangulation (the optimal one) than running Greedy on edges.

Theorem 5 implies that $L(E(S))$ is a subset of the greedy triangulation. Given any triangulation, we can use algorithm ESA to try to obtain an improved triangulation. The repeated element substitution in algorithm ESA translates to repeated swapping of diagonals of convex quadrilaterals formed by two neighboring triangles.

4   Conclusion and related work

We introduce the constrained independence system and prove some properties which lead to some structural characterization about triangulations of a planar point set. Whether there exists stronger matching theorems for constrained independence system or triangulations is open. It is also interesting to study the application of the matching theorem to the minimum weight triangulation problem.

Acknowledgement

We would like to thank Herbert Edelsbrunner and Mordecai Golin for discussion about this work. We also thank Günter Rote for helpful discussion and carefully reading a draft of this paper.

References

1. O. AICHHOLZER, F. AURENHAMMER, M. TASCHWER, AND G. ROTE, *Triangulations intersect nicely*, in Proc. 11th Annual Symposium on Computational Geometry, 1995.
2. B. BOLLOBÁS, *Graph Theory. An Introductory Course*, Springer-Verlag, 1979.
3. S. CHENG AND Y. XU, *Approaching the largest β-skeleton within a minimum weight triangulation*. Manuscript, 1995.
4. H. EDELSBRUNNER, *Algorithms in Combinatorial Geometry*, Springer-Verlag, 1987.
5. M. GAREY AND D. JOHNSON, *Computers and Intractability. A guide to the Theory of NP-completeness*, Freeman, 1979.
6. M. KEIL, *Computing a subgraph of the minimum weight triangulation*, Computational Geometry: Theory and Applications, 4 (1994), pp. 13–26.
7. D. KIRKPATRICK AND J. RADKE, *A framework for computational morphology*, in Computational Geometry, G. Toussaint, ed., Elsevier, Amsterdam, 1985, pp. 217–248.
8. D. PLAISTED AND J. HONG, *A heuristic triangulation algorithm*, Journal of Algorithms, 8 (1987), pp. 405–437.
9. G. ROTE. personal communication.
10. D. WELSH, *Matroid Theory*, Academic Press, New York, 1976.
11. Y. XU, *Minimum weight triangulation problem of a planar point set*, PhD thesis, Institute of Applied Mathematics, Academia Sinica, Beijing, 1992.

Three Dimensional Weak Visibility: Complexity and Applications

Caoan Wang[1] and Binhai Zhu[2]

[1] Dept. of Computer Science, Memorial University of Newfoundland, CANADA
[2] Group C-3, MS M986, Los Alamos National Laboratory, Los Alamos, NM USA

Abstract. In this paper, we study the complexity of 3D weak visibility. We obtain an $O(n^8)$ time and $\Theta(n^6)$ space algorithm to compute the weakly visible region of a triangle F from another triangle G among general scenes, which are a set of n disjoint triangles. We also consider the cases when the scenes are rectilinear objects and polyhedral terrains. We show that in these special situations the weakly visible regions can be computed much faster in $O(n^6)$ time and $O(n^4)$ space. With these results, we obtain the first known polynomial time algorithm to decide whether or not a simple polyhedron is weakly (internally or externally) visible.

1 Introduction

In 1981, Avis and Toussaint first introduced the *weak visibility* of a simple polygon [AT81]. Since then much research has been conducted in this direction [AGT86, GMP+90, BMT91, GM91, Che92, Che93]. More recently, some cases of the weak visibility in 3D has also been studied [CS89, Pla92, BDEG94]. 3D weak visibility has many applications in practice, especially in computer vision and computer graphics. An example is the 3D image rendering. Current rending techniques use radiosity methods to trace lights in order to provide the most realistic rendering of a scene [CW93]. In the inner loop of these methods the computation of form factors is involved. Informally, a form factor is defined as a pair of atomic triangles as the fraction of one object visible to the other through obstacles in the scene. The computation of 3D weakly visibility is closely related to the computation of a form factor. Because of the importance of 3D weak visibility, there has been some work in computer vision on computing the aspect graph of polyhedral objects and in computer graphics on computing the global illumination. In fact the fundamental concept of EEE events, which is used in this paper, is first proposed in computing aspect graphs [GM90, GCS91]. In [Tel92], a special case of 3D weak visibility is studied when an area light source illuminates through a bunch of convex "holes".

In this paper we study the general weak visibility in 3D. Instead of computing *critical points* at which the topology of the viewed scene change [CS89, BDEG94], we are more interested in computing the whole *weakly visible region*. We show that given n disjoint triangles in 3D there is a polynomial ($O(n^8)$) time algorithm to compute the weakly visible region of a triangle F from a triangle G.

It should be noted that this general case contains polyhedral objects as special cases. Although the time and space complexity for this general case is really high we show that if the scenes are rectilinear or polyhedral terrains then the weakly visible regions can be computed much faster. Moreover, no previous work has been done on the weak visibility for these two important special cases. In Section 3, we show some theoretical applications of these results. In practice, to treat such a complex algorithm one can compute only the visible parts which are most important with respect to a specific application [TH93].

2 Computing Weakly Visible Regions

2.1 Preliminary

We begin by giving some elementary definitions: Given a set of n disjoint triangles in 3D, a point v of F is said to be *weakly visible* to another triangle G if and only if there exists a point $u \in G$ such that u, v are visible, i.e., the open line segment $\overline{uv}$ does not intersect with any triangle. The union of all the points of F which are weakly visible from G is called the *weakly visible region* from G. F is weakly visible to G if and only if every $v \in F$ is weakly visible to G, in other words, the weakly visible region of F from G is the whole F.

Now we follow [GM90, GCS91] to give a characterization of the situations when some part (i.e., some points) of F is weakly visible from G. We are especially interested in a weakly visible *event*, which is defined as a continuous set of points t's on F which are weakly visible from G and some part of the boundary of the ϵ-disk centered at t is not weakly visible from G. Now let us consider the *line of sights* through these event points on F. There are three situations for the line of sights: (1) $\overline{uv}$ is determined by a vertex of G and a point on the boundary e_1 of a triangle F_i; (2) $\overline{uv}$ is determined by a point on the boundary e_1 of a triangle F_i and an edge e_2 of F_j; and (3) $\overline{uv}$ is determined by three edges e_1, e_2, e_3 of some triangles F_i, F_j, F_k respectively. The line of sights of the first case define a series of *opaque EV* events, which are a set of halfspaces on F. Let $e_1 = p_1 + xa_1$, $e_2 = p_2 + ya_2$ and let $e_3 = p_3 + za_3$. (Note that x, y, z are variables.) The direction $\mathbf{d}$ of a line of sight of the third case, which defines a series of *opaque EEE* events, is determined by the follow equation

$$\mathbf{d} = [(\mathbf{p_1} + x\mathbf{a_1} - \mathbf{p_2}) \times \mathbf{a_2}] \times [(\mathbf{p_1} + x\mathbf{a_1} - \mathbf{p_3}) \times \mathbf{a_3}]. \qquad (1)$$

In other words, the direction of the line of light passing through x, e_2 and e_3 is the intersection of the plane through x, e_2 (whose norm is $(\mathbf{p_1} + x\mathbf{a_1} - \mathbf{p_2}) \times \mathbf{a_2}$) and the plane through x, e_3 (whose norm is $(\mathbf{p_1} + x\mathbf{a_1} - \mathbf{p_3}) \times \mathbf{a_3}$). This implies that $\mathbf{d}$ is a quadratic function of x and the locus of viewpoints of an opaque EEE event is not a plane but a special quadratic surface such that for any point v' on the surface there exists another point u' and the line through u' and v' is also on the surface. Furthermore, since e_1, e_2, e_3 are line segments instead of lines, the wrapped surface defined by e_1, e_2, e_3 is a quadratic surface concatenated with two linear surfaces. (The line of sights of the second case define a series of *opaque*

EE events, which are special cases of the opaque EV events and the opaque EEE events.) It is clear that the intersection of this surface with F (and symmetrically, G) is a "hyperspace" defined by a piece of quadratic curve followed with two rays (see Figure 1, case (1)). Consequently, the weakly visible region of F from G is the union of all these (linear) halfspaces and (quadratic) hyperspaces. In the next three subsections, we study the problem of computing weakly visible regions when the scenes are general (i.e., a set of disjoint triangles), terrain, and rectilinear, respectively.

2.2 General Scenes

In this subsection we consider the situation when the scene is a set of n disjoint triangles. We mainly discuss the EEE events, since EV and EE events can be dealt with similarly. It should be noted that it is not always the case that three edges define an opaque EEE event; furthermore, it seems very difficult to characterize an opaque EEE event. Consequently we consider a superset of opaque EEE events which we define as follows. Suppose that a triple of edges of three triangles are in general position (i.e., no two edges are coplanar). If we assume all triangles are transparent, then three arbitrary edges in general position define a *transparent* EEE event. Clearly, since there are $\Theta(n^3)$ number of transparent EEE events between F and G, the arrangement of these $\Theta(n^3)$ (linear) halfspaces and (quadratic) hyperspaces has a combinatorial complexity of $O(n^6)$ and can be computed in $O(n^6 2^{\alpha(n)})$ time, where $\alpha(n)$ is the inverse of the Ackermann's function [EGP+92]. From now on, all the notations of events and line of sights are in regard with *transparent* ones.

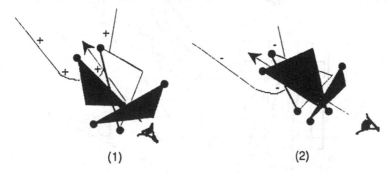

(1) (2)

Fig. 1. The hyperspaces defined by transparent EEE events.

As we have just mentioned, the weakly visible region of F from G is defined as the union of all opaque EV, EE and EEE events. Since we do not have a nice characterization of an opaque event, we use an algorithm which involves computing all transparent events. It is clear that a face C in the arrangement of the $O(n^3)$ halfspaces and hyperspaces is either weakly visible or invisible from G. Moreover, C is weakly visible to G if and only if for every point $p \in C$, p is weakly visible to G, in other words, not all of the line of sights reaching p

are blocked (intersected) by some triangles. However, since there are an infinite number of points in C, this property does not immediately imply an efficient algorithm. The following lemma makes a polynomial time algorithm possible.

Lemma 1. *A face C in the arrangement is weakly visible to G if and only if for every vertex v of C, v is weakly visible from G.*

Consequently, to decide whether C is weakly visible to G we just need to decide whether every vertex of C is weakly visible from G, i.e., whether all the line of sights reaching each vertex of C are blocked. To solve this problem, we reverse the line of sights by considering all the line of sights from vertex v of C to G. For each triangle F_i between F and G we draw a pyramid cone through F_i and starting at v. Each pyramid cone defines a triangle $T_{<v,F_i>}$ on the plane through G. Consequently, all the line of sights from G reaching v are blocked if and only if the union of the triangle $T_{<v,F_i>}$'s contains G. This latter decision problem is unlikely to be solvable in $o(n^2)$ time at this moment [GO93]. However, an $O(n^2)$ upper bound can be obtained without too much difficulty. In the worst case, it takes $\Theta(n^2)$ time and space to compute the union of $O(n)$ triangles [MMP+91]. Consequently, it takes $O(n^2)$ time to test whether the union of these $O(n)$ triangles contains G (i.e., whether v is weakly visible to G). Therefore, testing whether C is weakly visible from G takes $O(|C|n^2)$ time and computing the weakly visible region of F from G takes $O(n^8)$ time.

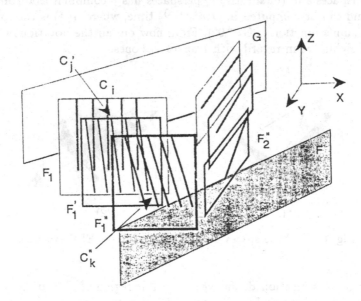

Fig. 2. The combinatorial complexity of the weakly visible region between two triangles among a general scene is $\Theta(n^6)$.

Theorem 2. *Given a set of n disjoint triangles in 3D, the weakly visible region of a triangle F from another triangle G can be computed in $O(n^8)$ time and $\Theta(n^6)$ space.*

The following example shows that in the worst case the weakly visible region can have $\Omega(n^6)$ number of vertices. First, we erect two parallel, vertical walls F, G which intersects XZ- and YZ-plane at an angle of $\pi/4$. Second, we erect two vertical walls F_1'', F_2'' parallel to the XZ- and YZ-plane respectively such that they almost touch each other (Figure 2). Then we erect two vertical walls F_1', F_1 which are parallel to F_1'' and two vertical walls F_2', F_2 which are parallel to F_2'' respectively. Finally we construct n vertical *cusp*'s on F_1. On F_1' we construct n cusps $C_j' (1 \le j \le n)$ from left to right such that the two long edges of a cusp have a degree of $\epsilon_j' (1 \le j \le n)$ with the YZ-plane and moreover ϵ_j''s are close to zero and $\epsilon_j' < \epsilon_{j+1}'$ for $1 \le j \le n - 1$. Similarly, on F_1'' we construct n cusps C_k'''s such that they have the same properties as those of C_j''s. The lengths of the cusps can be made very large while the width of the cusps on the three levels, i.e., the distance between the leftmost and rightmost cusp, can be made very small with respect the lengths of the cusps. Therefore, a triple of cusps, C_i of F_1, C_j' of F_1' and C_k'' of F_1'', for any $1 \le i, j, k \le n$, determine a vertical weakly visible strip from G to F (Figure 3). In this example there are $O(n^3)$ disjoint vertical weakly visible strips, which are all very long. Symmetrically, we construct horizontal cusps on F_2, F_2' and F_2'' so that they determine $O(n^3)$ disjoint horizontal weakly visible strips from G to F. Clearly, the union of these vertical and horizontal strips has a combinatorial complexity of $\Omega(n^6)$.

2.3 Terrains

A polyhedral terrain (terrain) T is a connected piecewise linear function over (x, y). In other words, it is a connected polyhedral surface in 3D such that any vertical line intersects it at most once and the orthogonal projection of a terrain on the XY-plane is a bounded planar subdivision. Polyhedral terrains have found applications in GIS, spatial databases, computer graphics and computational geometry (for a complete list of references, see [Zhu94]).

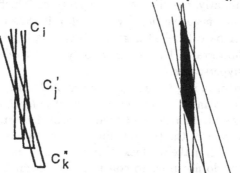

Fig. 3. A vertical weakly visible stripe.

In this subsection we consider the special case when the scene is a polyhedral terrain. In other words, we want to compute the weakly visible region between two faces F and G of a given polyhedral terrain. It turns out that given a general terrain T, if there is no other constraint about visibility then the weakly visible region between two faces F and G could also have a combinatorial complexity of $\Omega(n^6)$ (by modifying the example in Figure 1). The reason is that if the scene is a general terrain T and there is no restriction on visibility then a line of sight could be beneath some faces of T. However, the standard definition of visibility regarding terrains does not allow any line of sight to be beneath T. In other words, two points a, b on T are said to be *visible* if the line segment $\overline{ab}$ does not intersect any point strictly below T. Under this visibility model, we show that the weakly visible region between two faces can be computed in $O(n^6)$ time and $O(n^4)$ space, which is significantly lower than the general case. We first show the following lemma.

Lemma 3. *Let F, G be two faces of a terrain. The weakly visible region of F from G is the same as the weakly visible region of F from the boundary edges of G.*

Proof: Let x be any point in the weakly visible region of F from G. Furthermore, let $\overline{ux}$ be a line of sight from G to F where $u \in G$. Erect a vertical plane H through $\overline{ux}$. Let the highest intersection of H and G be u'. Because no line of sight could be beneath the terrain $\triangle uxu'$ is empty. Clearly $\overline{u'x}$ is also a line of sight and u' lies on the boundary of G. $\square$

With this lemma, we can see that in general no EEE event exists when the scene is a terrain (except when one of the three edges defining an EEE event is an edge of G). Therefore, we only need to consider the EE events, whose number clearly has a combinatorial complexity of $\Theta(n^2)$. However, we must make sure that any EE event, which define a hyperspace on F, do not allow line of sights beneath the terrain.

We first label the local visibility property of a hyperspace defined by a transparent EE events, i.e., we make the two triangles containing the corresponding two edges opaque and keep all other triangles transparent. This can be done in constant time with respect to the two triangles containing the two edges. The side of the hyperspace which is not blocked by the two triangles is labelled positive (locally visible); similarly, the side of the hyperspace which is blocked by the two triangles is labelled negative (locally invisible). It should be noted that a line of sight of a hyperspace labelled negative is always beneath the terrain and always intersects the terrain. Therefore, we only need to deal with line of sights of locally visible hyperspaces.

We start with more definitions. A *legal* hyperspace on F (from G) is a locally visible hyperspace such that any line of sight reaching it is *legal*, i.e., is above T. An illegal line of sight is said to be *restricted* by an edge of the terrain if it touches that edge and any movement of the (illegal) line of sight by a small amount of ϵ along the $+Z$ direction will make it blocked by some triangles. Clearly under our model of visibility we do not need to consider any illegal hyperspace. The following lemma shows the special structure of a legal hyperspace.

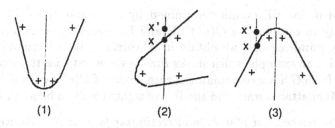

Fig. 4. Legal and illegal hyperspaces among a terrain.

Lemma 4. *A hyperspace S on F is legal only if there exists a plane H which is parallel to the Z-axis and H cuts S into two infinite pieces which are both labelled positive in the $+Z$ direction (see Figure 4, case (1)).*

Proof: Suppose it is not the case, then we know that at least one of the two rays of S, which results from one of the two edges defining the hyperspace and a vertex u of G, restricts line of sights $\overline{ux}$, where x is an arbitrary point of the upper ray of S (see Figure 4, case (2) and (3)). We claim that $\overline{ux}$'s are all illegal, because if they were legal then following Lemma 3 all $\overline{ux'}$'s, where x' is an arbitrary point on the intersection of F and the vertical plane H through $\overline{ux}$, would be all legal. This contradicts the fact that $\overline{ux}$ is restricted. $\square$

With the above lemma, we can identify and discard all illegal hyperspaces. There are still $O(n^2)$ legal hyperspaces and $O(n^2)$ locally invisible hyperspaces left, the arrangement of these hyperspaces clearly has a combinatorial complexity of $O(n^4)$. Now we simply run the algorithm in Theorem 2 to compute the weakly visible region from G to F. The result is summarized by the following theorem.

Theorem 5. *Given a polyhedral terrain, the weakly visible region of a face F from another face G can be computed in $O(n^6)$ time and $O(n^4)$ space.*

2.4 Rectilinear Scenes

In this section we consider the case when the scene is rectilinear. A line segment in 3D is called *rectilinear* if it is perpendicular to either the XY-, XZ- or the YZ-plane. A polyhedral object A is called *rectilinear* if and only if every edge of A is rectilinear. A *Manhattan terrain* M with n vertices is a connected 3D rectilinear polyhedral surface such that the intersection of any vertical line with M is either empty, a point, or a vertical line segment. We show that among a rectilinear scene, the weakly visible region between two faces F and G of the given scene can be computed with an $O(n^6)$ time and $\Theta(n^4)$ space algorithm, which is again significantly lower than the general case. The reason in this case is simple: we do not have any EEE event. In fact, assume e_1 are horizontal and e_2, e_3 are vertical, all with respect to F, then there is a line of sight touches a point x on e_1, e_2 and e_3 if and only if x, e_2, e_3 are coplanar. This line of sight is contained in the EE events determined by e_1, e_2 and e_1, e_3. Consequently, we need only to consider the $O(n^2)$ EE and EV events. The algorithm can be obtained by mimicing that we obtain in Theorem 2. In the full version of the paper, we give an example which shows that in the worst case the weakly visible region of F from G has a combinatorial complexity of $\Omega(n^4)$, even if F, G are the faces of a Manhattan terrain and the line of sights could be beneath the terrain.

Theorem 6. *Given a set of n disjoint rectilinear faces in 3D, the weakly visible region of a face F from another face G can be computed in $O(n^6)$ time and $\Theta(n^4)$ space.*

3 Applications

3.1 Determining if a polyhedron is weakly internally visible

In [AT81] an $O(n^2)$ time algorithm is given to decide whether or not a simple polyhedron is weakly internally visible from one of its edges. This bound is improved to linear later by Sack and Suri [SS86]. A question arises naturally: how fast can one test whether or not a 3D simple polyhedron is weakly visible from one of its faces? Up to this writing no algorithm is known to answer this question. With our algorithm this question can be answered in $O(n^{10})$ time. What we do is to fix a face G of the given polyhedron P and compute from G all the weakly visible regions of all the non-adjacent faces of G. If there is a face which is not weakly visible from G then P is not weakly visible from G. This clearly takes $O(n \times n^8) = O(n^9)$ time. There are $O(n)$ candidate faces hence the whole algorithm takes $O(n^{10})$ time. Therefore, we have the following result.

Corollary 7. *Given a simple polyhedron with n vertices, we can decide if it is weakly internally visible in $O(n^{10})$ time and $O(n^6)$ space. If the answer is positive we can return all the faces from which the polyhedron is weakly internally visible within the same time.*

3.2 Determining if a polyhedron is weakly externally visible

In [BMT91] a linear time algorithm is given to decide whether or not a simple polygon is weakly externally visible from a line segment in space. As in the above subsection, we are interested in determining whether a 3D simple polyhedron is weakly externally visible from a triangle in space. We show that this question can also be answered in $O(n^{10})$ time. We first compute an enclosing tetrahedron T of P and then we show the following lemma.

Lemma 8. *P is weakly externally visible if and only if it is weakly externally visible from at most three faces of T.*

With the above lemma, it is easy to obtain an algorithm to test whether P is weakly externally visible. After obtaining the enclosing tetrahedron, we simple test whether P is weakly externally visible by one, two or three faces of T. The time bound is clearly $O(n^{10})$.

Theorem 9. *Given a simple polyhedron with n vertices, we can decide if it is weakly externally visible in $O(n^{10})$ time and $O(n^6)$ space.*

3.3 Face guarding a polyhedral terrain

In [CS89], it is shown that the problem of computing the minimum number of vertex guards of a terrain is NP-complete. It is shown, by modifying the proof of [CS89], that the problem of computing the minimum number of edge guards is also NP-complete [Zhu94]. It is not difficult to prove, again by modifying the proof in [CS89], that computing the minimum number of face guards is NP-hard. However, it is not straightforward that this problem belongs to NP. With our algorithm of computing weakly visible region, given $k = O(n)$ face guards we can decide in $O(n^9)$ time if an arbitrary face of the terrain are covered by the k face guards. The algorithm is similar to that in Theorem 2 except that we have the arrangement of $O(kn^2) = O(n^3)$ hyperspace which is of combinatorial complexity of $O(n^6)$. Computing the weakly visible region from all the k face guards to that face takes $O(n^6 \times n^2 \times k) = O(n^9)$ time. In total, whether the whole terrain can be guarded by the k given face guards can be decided in $O(n \times n^9) = O(n^{10})$ time. Consequent this decision problem belongs to NP.

Corollary 10. *The problem of computing the minimum number of face guards of a polyhedral terrain is NP-complete.*

4 Concluding Remarks

One of the most outstanding open problems related to this paper is whether we could compute the weakly visible regions via opaque events. Although the number of opaque events is the same order as that of transparent events in the worst case, in practice the former is much smaller. In [Tel92], Teller did characterize opaque events for a special case of the problem; however, for our general problem Teller's method, i.e., the Plueker coordinates representation of lines, might not work. The other interesting problem is to reduce the gaps between the time and space complexity in computing the weakly visible regions in this paper.

References

[AGT86]   D. Avis, T. Gum, and G. Toussaint. Visibility between two edges of a simple polygon. *Visual Comput.*, 2(6):342–357, December 1986.

[AT81]   D. Avis and G. T. Toussaint. An optimal algorithm for determining the visibility of a polygon from an edge. *IEEE Trans. Comput.*, C-30:910–1014, 1981.

[BDEG94]   M. Bern, D. Dobkin, D. Eppstein, and R. Grossman. Visibility with a moving point of view. *Algorithmica*, 11:360–378, 1994.

[BMT91]   B. K. Bhattacharya, A. Mukhopadhyay, and G. T. Toussaint. A linear time algorithm for computing the shortest line segment from which a polygon is weakly visible. In *Proc. 2nd Workshop Algorithms Data Struct.*, volume 519 of *Lecture Notes in Computer Science*, pages 412–424. Springer-Verlag, 1991.

[Che92]   D. Z. Chen. An optimal parallel algorithm for detecting weak visibility of a simple polygon. In *Proc. 8th Annu. ACM Sympos. Comput. Geom.*, pages 63–72, 1992.

[Che93]   D. Z. Chen. Optimally computing the shortest weakly visible subedge of a simple polygon. In *Proc. 4th Annu. Internat. Sympos. Algorithms Comput. (ISAAC 93)*, volume 762 of *Lecture Notes in Computer Science*, pages 323–332. Springer-Verlag, 1993.

[CS89]   R. Cole and M. Sharir. Visibility problems for polyhedral terrains. *J. Symbolic Computation*, 7:11–30, 1989.

[CW93]   M. Cohen and J. Wallace. *Radiosity and Realistic Image Synthesis*. Academic Press, 1993.

[EGP$^+$92] H. Edelsbrunner, L. Guibas, J. Pach, R. Pollack, R. Seidel, and M. Sharir. Arrangements of curves in the plane: topology, combinatorics, and algorithms. *Theoret. Comput. Sci.*, 92:319–336, 1992.

[GCS91]   Z. Gigus, J. Canny, and R. Seidel. Efficiently computing and representing aspect graphs of polyhedral objects. *IEEE Trans. PAMI*, 13(6):542–551, 1991.

[GM90]   Z. Gigus and J. Malik. Computing the aspect graphs for line drawings of polyhedral objects. *IEEE Trans. PAMI*, 12(2):113–122, 1990.

[GM91]   S. K. Ghosh and D. M. Mount. An output-sensitive algorithm for computing visibility graphs. *SIAM J. Comput.*, 20:888–910, 1991.

[GMP$^+$90] S. K. Ghosh, A. Maheshwari, S. P. Pal, S. Saluja, and C. E. Veni Madhavan. Characterizing weak visibility polygons and related problems. In *Proc. 2nd Canad. Conf. Comput. Geom.*, pages 93–97, 1990.

[GO93]   A. Gajentaan and M. Overmars. On a class of $O(n^2)$ problems in computational geometry. Report 1993-15, Department of Comput. Sci., Utrecht Univ., 1993.

[MMP$^+$91] J. Matoušek, N. Miller, J. Pach, M. Sharir, S. Sifrony, and E. Welzl. Fat triangles determine linearly many holes. In *Proc. 32nd Annu. IEEE Sympos. Found. Comput. Sci.*, pages 49–58, 1991.

[Pla92]   H. Plantinga. An algorithm for finding the weakly visible faces from a polygon in 3-d. In *Proc. 4th Canad. Conf. Comput. Geom.*, pages 45–51, 1992.

[SS86]   J.-R. Sack and S. Suri. An optimal algorithm for detecting weak visibility of a polygon. Report SCS-TR-114, School Comput. Sci., Carleton Univ., Ottawa, ON, 1986.

[Tel92]   S. J. Teller. Computing the antipenumbra of an area light source. *Comput. Graph.*, 26(4):139–148, July 1992.

[TH93]   S. Teller and P. Hanrahan. Global visibility algorithms for illumination computations. In *Proc. SIGGRAPH '93*, pages 239–246, 1993.

[Zhu94]   B. Zhu. *Computational Geometry in Two and a Half Dimensions*. PhD thesis, School of Computer Science, McGill University, Montreal, Canada, 1994.

Rectangulating Rectilinear Polygons in Parallel

Sung Kwon KIM

Department of Computer Science
Kyungsung University
Pusan 608-736
Korea
ksk@csd.kyungsung.ac.kr.

Abstract: In NC, one can decompose an n-vertex rectilinear polygon into $O(n)$ rectangles, allowing Steiner points, so that any horizontal or vertical segment inside the polygon intersects $O(\log n)$ rectangles.

1 Introduction

Let $\mathcal{P}$ be a rectilinear polygon. A decomposition into rectangles, or a *rectangulation*, of $\mathcal{P}$, denoted by $\mathcal{DR}$, is a decomposition of $\mathcal{P}$ into non-overlapping rectangles, allowing Steiner points. The *stabbing number* of $\mathcal{DR}$ is the maximum number of rectangles intersected by a rectilinear segment lying entirely inside $\mathcal{P}$. The *size* of $\mathcal{DR}$ is its number of rectangles.

In [3], de Berg and van Kreveld have shown that for any rectilinear polygon $\mathcal{P}$ with n vertices a rectangulation with size $O(n)$ and stabbing number $O(\log n)$ is realizable. Their algorithm works as follows: Choose a horizontal edge e of $\mathcal{P}$ and compute the region $\mathcal{R}$ that is vertically visible from e. Removing $\mathcal{R}$ from $\mathcal{P}$ leaves several polygons $\mathcal{P}_1, \mathcal{P}_2, \ldots$, each containing an edge e_i which is part of the boundary of $\mathcal{R}$. Compute the region $\mathcal{R}_i$ in $\mathcal{P}_i$ that is horizontally visible from e_i. Removing $\mathcal{R}_i$ from $\mathcal{P}_i$ leaves several polygons, each containing an edge which is part of the boundary of $\mathcal{R}_i$. Repeat this procedure, alternating between the horizontal and vertical direction, until the entire region of $\mathcal{P}$ is removed. This will decompose $\mathcal{P}$ into $O(n)$ histograms so that any rectilinear segment inside $\mathcal{P}$ intersects at most three histograms. (A *histogram* is a rectilinear polygon having an edge from which its interior is visible either horizontally or vertically.) Rectangulate each histogram with m vertices so that its rectangulation has size $O(m)$ and stabbing number $O(\log m)$ as described in [3] to obtain a rectangulation of the entire polygon $\mathcal{P}$.

This algorithm works in a very sequential way and seems hard to parallelize. In this paper we will show that a rectangulation of $\mathcal{P}$ with size $O(n)$ and stabbing number $O(\log n)$ can also be obtained in NC, i.e., in time $O(\log^{c_1} n)$ using n^{c_2} processors for some constants $c_1, c_2 > 0$ in the PRAM (parallel random access machine). Our parallel algorithm works in a totally different way from de Berg and van Kreveld's.

A similar problem of decomposing a (non-rectilinear) polygon into non-overlapping triangles with Steiner points allowed has been studied. Chazelle et al. [2] have shown that any polygon $\mathcal{P}$ with n vertices can be decomposed into non-overlapping Steiner triangles such that any (arbitrarily oriented) segment inside P intersects only $O(\log^2 n)$ triangles. Hershberger and Suri [4] have improved the bound to $O(\log n)$.

In its overall strategy, our parallel algorithm takes after those of Chazelle et al. and Hershberger and Suri, but its details are quite different.

2 Preliminaries

In the rest of this paper, every geometric object (a polygon, a line, a segment, etc.) is rectilinear, unless otherwise specified. So, we will omit *rectilinear* hereafter if no confusion occurs. Let P be a polygon. An edge of the boundary of P is associated with two internal angles at its end vertices formed by it and its adjacent edges. Such an angle is either 90 or 270 degrees. An edge is said to be *concave* if both of its angles are 270 degrees and *convex* if both are 90 degrees. A horizontal concave edge is said to be *downward (upward)* if the interior of P is below (above) it. A vertical concave edge is said to be *rightward (leftward)* if the interior of P is to the right of (to the left of) it. In Figure 1(a), a polygon has two concave edges numbered 1 and 2. Edge 1 is rightward and edge 2 is upward.

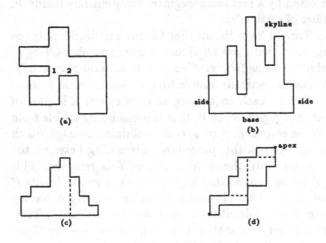

Fig. 1.

A *histogram* is a polygon having an edge, called its *base*, from which its interior is (either vertically or horizontally) visible. Two edges of a histogram that are adjacent to its base are called its *sides*. The boundary of a histogram, excluding its base and sides, is called its *skyline*. See Figure 1(b). A histogram is called a *staircase* if its skyline is either monotone increasing or decreasing. A staircase with increasing skyline is shown in Figure 2(a). A histogram is called a *pyramid* if its skyline initially monotone increases and then monotone decreases.

See Figure 1(c). A polygon is called a *zigzag* if its boundary contains two vertices, called the *apexes*, and consists of two monotone chains of horizontal and vertical edges connecting these apexes. See Figure 1(d). A staircase is a zigzag.

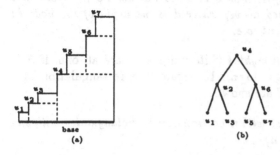

base

(a)

(b)

Fig. 2.

2.1 Rectangulation of staircases

A staircase is a histogram whose skyline is either monotone increasing or decreasing. Let S be a staircase whose base is horizontal and whose skyline is monotone increasing. Let $(u'_1, u'_2, \ldots, u'_{m'})$ be the vertices of the skyline of S where u'_1 has the minimum x-coordinate and $u'_{m'}$ has the maximum x-coordinate. Delete all even-indexed vertices and rename the remaining ones $(u_1, u_2, \ldots, u_m)$. That is, $u_i = u'_{2i-1}$ for $1 \leq i \leq m$. Note that each u_i with its adjacent edges forms an internal angle of 90 degrees. See Figure 2(a).

Suppose each u_i is associated with a non-negative integer weight w_i. We will explain later in Section 3.3 what these weights are. Let $\sigma = \sum_{i=1}^{m} w_i$. S will be partitioned into m rectangles so that each rectangle has a u_i as its upper left corner as follows: Choose the unique j such that $\sum_{i=1}^{j-1} w_i \leq \frac{1}{2}\sigma < \sum_{i=1}^{j} w_i$. Draw from u_j a vertical line to the base and a horizontal line to the right side of S. This partitions S into two staircases S_1 and S_2 with skylines $(u_1, \ldots, u_{j-1})$ and $(u_{j+1}, \ldots, u_m)$, respectively, and a rectangle having u_j as its upper left corner. Recursively, repeat partitioning S_1 and S_2. This partitioning clearly produces a rectangulation of S into m rectangles and each of these rectangles has a u_i as its upper left corner.

This recursive partitioning can be modeled by a binary tree, called the *staircase rectangulation tree (SRT)*. The root of *SRT* is associated with the u_j chosen from S and the rectangle having the u_j as its upper left corner and it has two children which are associated with two vertices chosen from S_1 and S_2, respectively, and two rectangles having each of these two vertices as its upper left corner. In this fashion, each node[1] of *SRT* is associated with a vertex u_i and a rectangle having this u_i as its upper left corner.

---

[1] We use *node* for graph theoretical purpose and *vertex* for geometric purpose.

Lemma 1. *A node associated with u_i in SRT is at depth[2] at most $\log \frac{\sigma}{w_i}$.*

So, the height of SRT is $\log \sigma$.

Lemma 2. *If a vertical segment s intersecting the base of S also intersects a horizontal edge e in the skyline of S, then s intersects at most $d+1$ rectangles in the rectangulation of S obtained above, where d is the depth of the node in SRT associated with the u_i adjacent to e.*

We say that a staircase is *unweighted* if its weights w_i are all one. If S is unweighted, then SRT is of height $\log m$. In Figure 2, a rectangulation of a unweighted staircase and its SRT are shown.

Lemma 3. *A unweighted staircase with n vertices can be rectangulated with size $O(n)$ and stabbing number $O(\log n)$.*

2.2 Decomposition of pyramids and zigzags into staircases

A pyramid is a histogram whose skyline monotone increases and then monotone decreases. A pyramid with horizontal base can be partitioned into two staircases by drawing a vertical line from an endpoint of the topmost edge of its skyline as in Figure 1(c).

A zigzag is a polygon bounded by two monotone chains, namely, the *upper* and *lower* chains, connecting two apexes. See Figure 1(d) for a zigzag consisting of two monotone increasing chains, which can be decomposed into a number of staircases by considering a rightward ray containing the horizontal edge of minimum y-coordinate in the upper chain. The ray reflects upward (rightward) if it hits a vertical (horizontal) edge. The locus of this ray partitions the zigzag into a number of staircases. Any other zigzag can be decomposed into staircases by a similar method.

The following lemmas are easy to prove.

Lemma 4. *A pyramid or a zigzag can be decomposed into staircases so that any segment inside it intersects at most two staircases.*

Lemma 5. *A pyramid or a zigzag with n vertices can be rectangulated with size $O(n)$ and stabbing number $O(\log n)$.*

3 Rectangulation of polygons

3.1 Decomposition of polygons into kites

Let $\mathcal{P}$ be a polygon. A path between two points inside $\mathcal{P}$ is a chain of vertical and horizontal segments connecting them. A shortest path between two points inside $\mathcal{P}$ is a path of minimum length between them. The length of a path is the

---

[2] The root is at depth zero.

sum of the lengths of its constituent segments. Note that there may be an infinite number of (rectilinear) shortest paths between two points, while there is only one geodesic shortest path between them. A vertical segment and a horizontal segment are said to *cross* if they intersect. Two horizontal (or vertical) segments are said to *overlap* if they intersect. Two paths cross if there is a vertical segment in a path crossing a horizontal one in the other. Two paths overlap if there are two vertical (or horizontal) segments, one in each path, which overlap.

Let us define a *kite*, which is an important geometric object for our algorithm. Let a, b and c be three vertices of $\mathcal{P}$, appearing in clockwise order. Define a kite $K(a, b, c)$ to be the area-maximal region bounded by three non-crossing shortest paths among a, b and c, namely, $P(a, b)$, $P(b, c)$, and $P(a, c)$, inside $\mathcal{P}$. A kite consists of a number of polygons, called *bodies*, which are connected by paths. There is only one body, called the *major body*, that is bounded by all three shortest paths; each of the others, called the *minor bodies*, is bounded by two of them only. Two shortest paths may overlap and this overlap may consist of several paths; each of these paths is called a *link*. Endpoints of links are called *junctions*. A major body contains three junctions and a minor body does two.

Remark on Figure 3(a). $K(a, b, c)$, shaded, consists of a major body, a minor body, and a link (f, g). The major body contains three junctions, a, b and f, and the minor body does two, c and g. Though a, b and c are not endpoints of links, they are also treated as junctions.

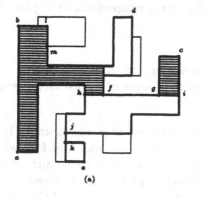

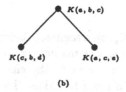

Fig. 3.

Our decomposition procedure of $\mathcal{P}$ into kites works recursively as follows: Let $p_1, \ldots, p_n$ be the vertices of $\mathcal{P}$, listing in clockwise order. Let $a = p_n$, $b = p_1$ and $c = p_{\lfloor n/2 \rfloor}$ and compute $K(a, b, c)$. This divides up the boundary of $\mathcal{P}$ excluding the edge (a, b) into two chains, one from b to c and the other from c to a. At the second stage, find the middle vertex of each chain and compute the shortest paths from it to the end vertices of the chain. This draws $K(p, q, r)$ for $(p, q, r) \in$

$\{(c, b, p_{\lfloor n/4 \rfloor}), (a, c, p_{\lfloor 3n/4 \rfloor})\}$. Only $P(p, r)$ and $P(q, r)$ are newly computed, and $P(p, q)$ is not but instead the one computed at the previous stage is used. In general, at stage i of the procedure, $i \geq 2$, for each pair of vertices (p, q) such that a shortest path between them was computed at stage $i - 1$, find the middle vertex r of the chain between p and q and compute $K(p, q, r)$, where $P(p, q)$ is the one computed at stage $i - 1$ and $P(p, r)$ and $P(q, r)$ are newly computed at stage i. The procedure terminates when each chain consists of a single edge.

Let $\mathcal{DK}$ denote this decomposition of $\mathcal{P}$ into kites. Decomposition of a polygon after the second stage of the procedure is shown Figure 3(a). The *root kite* is the one formed at the initial stage, i.e., $K(p_n, p_1, p_{\lfloor n/2 \rfloor})$.

Our decomposition procedure can be modeled by a binary tree of height $\log n$, which is called the *decomposition tree (DT)*. Each node of DT at depth i, $i \geq 0$, corresponds to a distinct kite produced at stage $i + 1$. Specifically, the root corresponds to the root kite and its two children correspond to the two kites computed at the second stage; and if $K(p, q, r)$ is computed at stage $i > 0$ and $K(p, r, r_1)$ and $K(r, q, r_2)$ are computed at stage $i + 1$, then the node corresponding to $K(p, q, r)$ is the parent of the nodes corresponding to $K(p, r, r_1)$ and $K(r, q, r_2)$. $K(p, q, r)$ is the *parent kite* of $K(p, r, r_1)$ and $K(r, q, r_2)$, which are its *child kites*. Of three boundary chains of $K(p, q, r)$, $P(p, r)$ and $P(q, r)$ are called the *child boundary chains* for they bound the kite and its child kites, and $P(p, q)$, the *parent boundary chain* for it bounds the kite and its parent kite. For the root kite, the edge (p_1, p_n) is assumed to be its parent boundary chain. We will use $K(p, q, r)$ to denote a kite whose parent boundary chain is $P(p, q)$ and whose child boundary chains are $P(p, r)$ and $P(q, r)$.

The root and its children in DT for the decomposition in Figure 3(a) are depicted in Figure 3(b).

Lemma 6. *Let s be a segment inside $\mathcal{P}$.*
(i) The number of kites in $\mathcal{DK}$ that intersect s is at most $2 \log n$.
(ii) The nodes corresponding to the kites that intersect s constitute a single path in DT.

In $\mathcal{DK}$, some segments belong to more than two kites. For example, in Figure 3(a), segment (l, m) belong to $K(a, b, c)$ and $K(c, b, d)$ and $K(d, b, r)$ where r is the middle vertex of the chain from b to d. So, $\mathcal{DK}$ may require superlinear storage to store full kites. To reduce storage, a kite will be stored as a list of bodies (each body is stored as a list of its boundary edges); links will not be stored. Thus, each edge belonging to a body boundary is stored at most twice. To distinguish from $\mathcal{DK}$, we will denote by $\mathcal{DB}$ this decomposition of $\mathcal{P}$ into bodies. No essential difference exists between them; the interior of $\mathcal{P}$ is thought to be partitioned into kites in $\mathcal{DK}$ and into bodies in $\mathcal{DB}$.

3.2 Decomposition of kites into staircases

As mentioned before, a kite is formed by three shortest paths, consisting of a major body, a number of minor bodies and links, and a minor body is bounded by parts of two shortest paths and contains two junctions. The boundary of a minor body can be divided into two chains, both starting at one junction and ending at the other. Since both chains are shortest paths between the junctions and a minor body is area-maximal (as a kite is area-maximal), both are monotone. Hence, a minor body is a zigzag having junctions as its apexes.

So, we can state the following lemma.

Lemma 7. *A minor body can be rectangulated with size $O(n)$ and stabbing number $O(\log n)$, where n is its number of vertices.*

A major body is bounded by parts of three shortest paths, and contains three junctions. In the following, we will show how major bodies are decomposed into simpler objects.

Let $K = K(p, q, r)$ be a kite and K' its parent kite. Note that $P(p, q)$, bounding K and K', is as *outward* as possible with respect to K' in the sense that it was drawn to make K' as large as possible. If K is the root kite, then we assume that $K' = \emptyset$ and outwardness holds.

Let B be the major body of K with junctions j_1, j_2 and j_3 where j_1 is on the overlap between $P(p, q)$ and $P(p, r)$, j_2 between $P(p, q)$ and $P(q, r)$, and j_3 between $P(p, r)$ and $P(q, r)$. In other words, B is bounded by three chains C_1, C_2, and C_3, where C_i is the chain of B from j_i to $j_{(i \bmod 3)+1}$. There is no minor body of K bounded by the parent boundary chain and a child boundary chain due to outwardness of the parent boundary chain. That is, the part of K between p and j_1 is a link. Similarly, the part between q and j_2 is a link.

Remark on Figure 3(a): $K(a, c, e)$ is a kite consisting of a major body, a minor body, and three links (one between a and h, one between c and i, and one between j and k). The minor body is bounded by $P(a, e)$ and $P(c, e)$ and contains two junctions e and k. C_1 consists of a single segment (h, i), C_2 is the chain from i to j and C_3 is the chain from j to h.

C_1 is called the *window* of K, denoted by W. A window is the part of the boundary of a major body which is shared by the body B and its parent kite K'. Significance of W is that only through it, one can go from B to K'. If a segment intersecting K' also intersects a rectangle in K, then the rectangle must be contained in B and thus the segment must intersect W. No minor body of K intersects $P(p, q)$ and thus W.

We claim that each window W consists of either one or two edges. Suppose W consists of more than two edges, $e_1, e_2, e_3, \ldots$. Assume that e_2 is horizontal. Then e_1 and e_3 must be vertical. Wlog, assume that e_1 is above the line ℓ containing e_2. If e_3 is below ℓ, then we have a contradiction that $P(p, q)$ is not as outward as possible because e_2 can be translated to make K' larger without increasing the length of $P(p, q)$. (e_2 does not touch the boundary of K.) If e_3 is above ℓ, then e_2 is a convex edge in K, which is a contradiction to that W is part of a shortest path. So, W consists of either one (called a *straight* window) or two edges (called an *L-shaped* window). For the root kite, the edge (p_1, p_n) is a straight window.

To decompose B into simpler objects, we consider two cases according to the shape of W.

Case (1): W is a straight window.

Wlog, assume that W is horizontal and B is below W. We claim that B contains at most one concave edge, i.e., that C_2 and C_3 together contain at most one concave edge. We will first show that C_2 contains at most one concave edge. Suppose C_2 contains more than one concave edge, $e_1, e_2, \ldots$, appearing in this order when walking C_2 from j_2 to j_3. Remember that C_2 is a shortest path between j_2 to j_3. Since W is horizontal and B is below W, e_1 must be leftward. Since the subchain between e_1 and e_2 is monotone, e_2 must be downward. Thus, j_3 is above the horizontal line containing e_2. Then, C_2 and C_3 overlap at e_2, which contradicts to the definition of B. So, C_2 (and similarly, C_3) cannot contain more than one concave edge.

Next, we will show that if C_2 contains a concave edge, then C_3 contains none, and vice versa. Assume that C_2 contains a concave edge e. Then e must be leftward. Let ℓ be the line containing e inside B as in Figure 4(c). Clearly, j_2 is to the right of ℓ as e is leftward. (Otherwise, C_2 and C_3 would overlap at e.) So, ℓ intersects W at a. Construct a monotone path from j_1 to j_3: Start at j_1, go straight rightward to a, go straight downward to e and take C_2 to j_3. This path is monotone and thus a shortest path between j_1 and j_3. Thus, C_3, which is also a shortest path, must be monotone (and has no concave edge). The claim is proved.

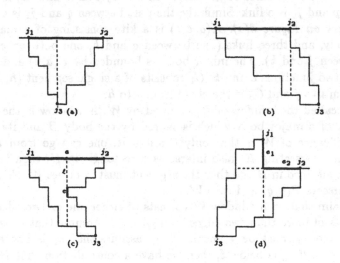

Fig. 4.

Case (1.1): B contains no concave edge.

Let x_i be the the x-coordinate of j_i for $1 \leq i \leq 3$. If $x_3 < x_1$ ($x_3 = x_1$, $x_1 < x_3 < x_2$, $x_3 = x_2$, $x_2 < x_3$), then B is a zigzag (a staircase, a pyramid, a staircase, a zigzag). For the case of a pyramid, B can be decomposed into two staircases by drawing a vertical line inside B containing j_3 as in Figure 4(a). For the case of a zigzag, B can be decomposed into a staircase and a smaller zigzag by drawing a vertical line inside B containing j_1 (j_2) if $x_3 < x_1$ (if $x_2 < x_3$). In Figure 4(b), a line containing j_2 is drawn.

Case (1.2): B contains one concave edge.

B can be divided into two staircases and a zigzag by drawing a vertical line inside B containing the concave edge as in Figure 4(c).

Case (2): W is an L-shaped window.

Let W consist of a vertical segment e_1 and a horizontal segment e_2. Wlog, assume that their intersection is the bottommost point of e_1 and the the leftmost point of e_2 as in Figure 4(d). Then, both C_2 and C_3 are monotone. Drawing two lines, one containing e_1 and the other e_2, decomposes B into two staircases, one having e_1 and the other e_2 as base, and one zigzag as in Figure 4(d).

Lemma 8. *(i) A major body can be decomposed into at most two staircases and at most one zigzag.*
(ii) Each of these staircases has part of the window as its base.
(iii) A segment crossing the window intersects exactly one of these staircases.

By Lemma 8 together with the fact that a minor body is a zigzag, we have decomposed each body in $\mathcal{DB}$ into a number of zigzags and staircases. Since each zigzag can be decomposed into staircases, we can say that $\mathcal{P}$ has been decomposed into a number of staircases. We denote by $\mathcal{DS}$ this decomposition of $\mathcal{P}$ into staircases.

Lemma 9. *In $\mathcal{DS}$, any segment inside $\mathcal{P}$ intersects $O(\log n)$ staircases.*

3.3 Assigning weights to staircases

Assume that a decomposition of $\mathcal{P}$ into staircases, $\mathcal{DS}$, has been obtained. We distinguish staircases into two types. Staircases whose bases are part of a window are called *window* staircases, and the others *non-window* ones. In Section 3.2, a window was defined to be the part of the boundary of a major body, which is shared by it and its parent kite. So, a window staircase is part of a major body.

If all (window or non-window) staircases are unweighted and if each staircase is rectangulated by Lemma 3, then we have a rectangulation of $\mathcal{P}$ with stabbing number $O(\log^2 n)$ by Lemma 9.

To reduce the stabbing number, all window staircases will be weighted as follows: Let S be a window staircase and B, the major body containing it. See Figure 5, which redraws Figure 4(b). Define $(u_1, u_2, \ldots, u_m)$ as in Section 2.1. Let v_i and v_{i+1}, $1 \le i \le m$, be the vertices in B adjacent to u_i. Then, all v_i's are in the boundary of $\mathcal{P}$ because the child boundary chains of B are as outward as possible. Each u_i, $1 \le i \le m$, will be assigned a non-negative integer weight w_i, the number of edges in the bay associated with u_i. The *bay* associated with u_i is the region of $\mathcal{P}$ separated from B by the edges (u_i, v_i) and (u_i, v_{i+1}). In Figure 5, four bays, one per u_i, are shown. Window staircases corresponding to Figure 4(a), (c) and (d) can be weighted in a similar fashion.

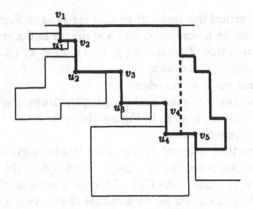

Fig. 5.

Rectangulate all weighted and unweighted staircases by the method in Section 2.1. Let $\mathcal{DR}$ denote the resulting rectangulation. In the following, we will show that $\mathcal{DR}$ has stabbing number $O(\log n)$ and consists of $O(n)$ rectangles, and that it can be computed in NC.

Theorem 10. $\mathcal{DR}$ *has stabbing number* $O(\log n)$.

Theorem 11. *The size of* $\mathcal{DR}$ *is* $O(n)$.

Theorem 12. $\mathcal{DR}$ *can be found in* NC.

By Theorems 10, 11, and 12, we have proved the theorem.

Theorem 13. *Given a rectilinear polygon* $\mathcal{P}$ *with* n *vertices, a rectangulation of* $\mathcal{P}$ *with size* $O(n)$ *and stabbing number* $O(\log n)$ *can be computed in* NC.

References

1. M.J. Atallah, R. Cole, and M.T. Goodrich, Cascading divide-and-conquer: A technique for designing parallel algorithms, *SIAM J. Comput.*, 18 (1989) 499–532.
2. B. Chazelle, H. Edelsbrunner, M. Grigni, L. Guibas, J. Hershberger, M. Sharir, and J. Snoeyink, Ray shooting in polygons using geodesic triangulations, *Proc. 18th ICALP*, Lecture Notes in Computer Science, vol. 510, pp. 661–673, 1991.
3. M. de Berg and M. van Kreveld, Rectilinear decompositions with low stabbing number, *Information Processing Letters*, 52 (1994) 215–221.
4. J. Hershberger and S. Suri, A pedestrian approach to ray shooting: Shoot a ray, take a walk, *Proc. 4th SODA*, pp. 54–63, 1993.

Efficient Randomized Incremental Algorithm For The Closest Pair Problem Using Leafary Trees

V. Kamakoti, Kamala Krithivasan and C. Pandu Rangan

Department of Computer Science and Engineering, Indian Institute of Technology, Madras - 600 036, Tamilnadu, India. Email: rangan@iitm.ernet.in

Abstract. We present a new data structure, the *Leafary tree*, for designing an efficient randomized algorithm for the *Closest Pair Problem*. Using this data structure, we show that the Closest Pair of n points in D-dimensional space, where, $D \geq 2$, is a fixed constant, can be found in $O(n \log n / \log \log n)$ expected time. The algorithm does not employ hashing.

Key words : Closest pair, Computational Geometry, Randomized Algorithms.

1 Introduction

The *Closest Pair Problem (CPP)* is to find a closest pair in a given set of n points. It is well known that this problem requires $\Omega(n \log n)$ time in the algebraic computation tree model [6] and optimal algorithms already exist. However, if the model of computation is changed then $\Omega(n \log n)$ is no longer a lower bound. We summarize the major results and the corresponding models.

1. $O(n)$ expected time randomized algorithms are presented in [4, 5, 7].
   They use hashing and assumes the floor function as a unit cost operation.
2. An $O(n \log \log n)$ time deterministic algorithm is presented in [3] which uses hashing and assumes floor function as a unit cost operation.
3. In [1] an $O(n)$ expected time randomized algorithm using the real-RAM model is presented. This algorithm assumes that the input coordinates are in a range $[0..U-1]$ for some constant U. In a restricted real-RAM model where LOG and EXP are not allowed they have shown that the running time of the algorithm is $O(n + \log \log U)$ if the input coordinates are integers and $O(n + \log \log(\delta_{max}/\delta_{min}))$ if the input coordinates are reals, where δ_{max} is the largest and δ_{min} is the shortest distance between the points. They further assume that all the numbers manipulated by these algorithms contain $O(\log n + \log U)$ bits.

1.1 Motivation

As seen above, the result in [3], when compared to that in [7] implies that the major part of the reduction in run time, from $O(n \log n)$ in the algebraic computation tree model to $O(n \log \log n)$ is due to hashing and floor function. Randomization has helped cutting the remaining $O(\log \log n)$ factor. It is proved that even the randomized lower bound for the CPP is $\Omega(n \log n)$ [8].

In [4] a simple incremental $O(n)$ expected time randomized algorithm is presented. The algorithm assumes that the floor function takes unit cost and employs the Dynamic perfect hashing [2]. In fact they have shown that the algorithm completes with very high probability (i.e, $1 - O(n^{-s})$, for some constant $s > 0$) only in $O(n \log n / \log \log n)$ time. Further the use of dynamic perfect hashing necessitates the following additional assumptions:

1. $\lfloor x_{max}/\delta_{min} \rfloor$ and $\lfloor y_{max}/\delta_{min} \rfloor$ are integers from a bounded universe where x_{max} and y_{max} are the maximum abscissa and ordinate respectively of all the input points and δ_{min} is the minimum closest pair distance. Normally the bound on the universe is such that it is less than some *polynomial* in n [10]. The above assumption is extended for any dimension.
2. Time required to compute a prime number greater than an integer function of the coordinates of the input points is less; and,
3. An integer $I \geq \lfloor x_{max}/\delta_{min} \rfloor$ and $\lfloor y_{max}/\delta_{min} \rfloor$ should be known in advance, i.e., before the execution of the algorithm.

The challenging open problem in this context is to devise an algorithm for the CPP running in less than $O(n \log n)$ time without employing hashing.

1.2 Contribution of this paper

In this paper we propose a simple and elegant data structure called *Leafary tree* and use the same to arrive at an $O(n \log n / \log \log n)$ expected time implementation of the algorithm presented in [4] with slight modifications. We also prove a high confidence bound for an $O(n \log n / \log \log \log n)$ run time. We assume that the floor function takes unit time and ONLY the assumption 1 mentioned above. Strictly speaking, the assumption 3 mentioned above is impractical because δ_{min} cannot be known in advance.

2 The Leafary tree - A New Data Structure

Recall that a *k-ary tree* is a tree in which every vertex has at most k children. We define a Leafary tree as a k-ary tree L, where k is a function f_L, of the number of leaves l of the tree. We can number the subtrees of any internal node of the Leafary tree as $0,1,2,...,f_L - 1$, from left to right.

2.1 The Leafary Search Tree

A Leafary Search Tree (LST) is a leafary tree L, used for searching integers in the range $[0..H]$, for some integer H. We assume henceforth that,

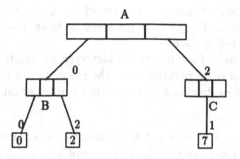

Fig. 1. Leafary Tree

1. H will denote the maximum number of leaves of the LST L and f_L will be the upper bound on the number of children of any internal node in L.
2. All the leaves of the LST L are in the level $\lceil log(H)/log(f_L) \rceil$ which we denote by *Maxlevel*. The level of a vertex i in L is the number of edges in the path between i and the root of L.
3. The leaves are numbered $0,1,2,...,H$ in order from left to right.

2.2 Operations on LST

Insertion: Suppose we have to search an integer in the integer set $I = \{ i_1, i_2, ..., i_n \}$, where, the integers are in the range $[0..H]$, we construct a LST L for I as follows. Let leaf i correspond to the integer i. Let us consider the following example, where, $H = 8$, $f_L = \lfloor log_2(H) \rfloor = 3$, and $I = \{0, 2, 7\}$. The LST constructed will be as shown in Figure 1. Note that all the nodes are not generated but only the paths between the root and the leaves corresponding to the inserted integers are generated. We will now describe a simple method for locating a leaf, say leaf 7, in the LST L shown in Figure 1. Since $f_L = 3$ for the LST, we find the 3-ary representation for 7 which is $(21)_3$. It is interesting to see that if we traverse the node A which is the second child of the root and then to the first child of A we reach the leaf 7. Generalizing the above concept, to locate any leaf q in a LST L we first find its f_L-ary representation. Let $q = (b_r, b_{r-1}, ..., b_1)$ in f_L-ary representation, where, $r = \lceil log(H)/log(f_L) \rceil$. Then, we traverse the path which comprises of the b_r^{th} child of the root and then the b_{r-1}^{th} child of the b_r^{th} child of the root and so on. Since the label of the maximum leaf is H, the f_L-ary representation has $\lceil log_{f_L}(H) \rceil = \lceil log(H)/log(f_L) \rceil$ digits.

For every non-leaf node i in the LST L, we associate a variable $ENTRY(i)$ which is defined as follows: Let $R(i)$ be the set of integers such that, $j \in R(i)$ if and only if $0 \le j \le f_L - 1$ and the jth subtree of node i is already inserted in L. Let $ENTRY(i) = \sum_{j \in R(i)} 2^j$. Later we see that $ENTRY(i) \le n$, for all nodes i of the LSTs built during the execution of the algorithm. Initially, the LST will be the root node with no integers inserted and the ENTRY value of the root is set to 0. While an integer is inserted into the LST L, the pointers corresponding to every subtree of a newly created (if one is created) internal node A should be initialized to *Null* and modified whenever a subtree of the node A is created. But we will be spending an extra $O(f_L)$ time for every node created, which is undesirable. This can be avoided as follows:

We shall now define a Boolean function called Isentry(r, q) which is **true** if the rth subtree of that internal node pointed to by the pointer q, in the LST is inserted, else it is **false** and then discuss a *constant time* implementation of the same.

Lemma 1. *Let $R = \{ k_1, k_2, ..., k_r \}$ be a set of r distinct positive integers such that, $k_i < k_{i+1}$, for all i, $1 \leq i < r$, and let $k_{r+1} = \infty$ and $k = \sum_{i=1}^{r} 2^{k_i}$. Then for any integer $q \geq 0$, such that, $k_j < q \leq k_{j+1}$, $\lfloor log_2(k \bmod 2^q) \rfloor = k_j$.*

Let $T(k) = 2^k$, for all k, $0 \leq k \leq f_L$. Let $LOG(i) = \lfloor log_2(i) \rfloor$, for $1 \leq i \leq f_L$. Let $f_L \leq 2^b$. We can easily see that for all i, $0 \leq i \leq b-1$, $LOG(j) = i$, for all $T(i) \leq j \leq T(i+1) - 1$. Hence, the LOG and the T arrays can be computed in $O(f_L)$ time. Note that for any two integers a and b, $a \bmod b = a - b\lfloor a/b \rfloor$. Lemma 1 implies the following $O(1)$ time implementation of the function ISENTRY:

Function ISENTRY(j, node) : Boolean;

begin
    If $(LOG(ENTRY(node) \bmod T(j+1)) = j)$ then return true
        else return false;
end;(Function)

Theorem 2. *Function Isentry can be implemented in $O(1)$ time but with $O(f_L)$ time of preprocessing in a real-PRAM model in which LOG and EXP are NOT unit cost operations.*

Henceforth, we will assume that the function Isentry is implemented for a LST L by performing the $O(f_L)$ time preprocessing while constructing L. The procedure shown in Figure 2 and Theorem 2 imply the following Theorem.

Theorem 3. *Procedure INSERTL takes $O(log_2(H)/log_2(f_L))$ time.*

Building an LST can be done by repeated Insertions. Searching an LST for an integer p is similar to that of Insertion. Hence the following theorem.

Theorem 4. *Building an LST L takes $O(f_L + \frac{|I| log_2(H)}{log_2(f_L)})$ time, where, I is the set of integers to be inserted into the LST L and searching a LST L takes $O(log_2(H)/log_2(f_L))$ time.*

2.3 Leafary Tree and Grids Storing Point Sets

Grids: A two-dimensional grid G consists of $h \times v$ square boxes whose each side measure δ and arranged as a matrix of h columns and v rows. Each box can be uniquely assigned a index (i, j) which is a two-tuple, such that $0 \leq i \leq h - 1$

Procedure INSERTL(p,L,c)

/* This procedure inserts the integer p along with the data c associated with p, into a
LST L, which could hold a maximum of H leaves. */
/* The function Root(L) returns a pointer to the root node of LST L */

1. **begin**
2. Let $(p_r, p_{r-1}, \ldots, p_1)$ be the f_L-ary representation of p, where, $r =$ Maxlevel;
3. Currentnode := Root(L);
4. **for** $i = r$ **to** 2 **do**
   (a) **begin**
   (b) **if** NOT(Isentry(p_i,Currentnode)) **then**
      i. getnode(Q);
      ii. Make Q the p_i^{th} child of Currentnode;
      iii. ENTRY(Currentnode) := ENTRY(Currentnode) + T(p_i);
      iv. ENTRY(Q) := 0;
   (c) Currentnode := p_i^{th} child of Currentnode;
   (d) End; /* At this stage Currentnode = parent of leaf p. */
5. **if** NOT(Isentry(p_1,Currentnode)) **then**
   (a) make the leaf p as the p_1^{th} child of Currentnode;
   (b) ENTRY(Currentnode) := ENTRY(Currentnode) + T(p_1);
6. Store c in the p_1^{th} child of Currentnode;
7. return L;
8. end;(Procedure)

Fig. 2. The Procedure to Insert into a Leafary tree

and $0 \leq j \leq v - 1$. We say that the dimension of G is $h \times v$. We call h and v as the *horizontal* and *vertical* dimensions of G respectively. We also assume that the bottom left corner of the grid is the origin of the coordinate system. The boxes (i, j) of G can be numbered uniquely by an integer $f(i, j) = i * v + j$ from the set $\{0, \ldots, ((h \times v) - 1)\}$ using the standard *row major* numbering of arrays. The notion of grids can be easily extended to general D dimensions.

Storing Point Sets on Grids and Leafary Trees: We initially consider storing a two-dimensional point set P on a grid G. Henceforth we assume without loss of generality that the points are in the first quadrant else we can make a translation of the points in linear time. To store P in G with box size d, the dimension of $G = (\lfloor x_m/d \rfloor + 1) \times (\lfloor y_m/d \rfloor + 1)$, where $x_m = $ maximum$(x \mid (x, y) \in P)$ and $y_m = $ maximum$(y \mid (x, y) \in P)$. Each point $p = (x, y)$ of P will be mapped onto the box with index $(\lfloor x/d \rfloor, \lfloor y/d \rfloor)$ in G. A box b in G is said to be *non-empty* if and only if at least one point of P is mapped onto it. As discussed in the previous section each of the non-empty boxes will be assigned an unique integer from the set $\{ 0, 1, \ldots, ((\lfloor x_m/d \rfloor + 1) \times (\lfloor y_m/d \rfloor + 1) - 1) \}$.

To maintain G in a Leafary Search Tree LST, we number the boxes of G in *row major* ordering and store the *non-empty* boxes in a LST such that the non-empty box with number i of G is stored in the leaf numbered i of the LST.

The LST can search integers in the range $[0..((\lfloor x_m/d \rfloor + 1) \times (\lfloor y_m/d \rfloor + 1) - 1)]$.
We define the following operations on Grids.

1. $Build(P, d)$: Returns a grid G with mesh size d that contains the points in P along with the horizontal (M_x) and vertical (M_y) dimensions of G.
2. $Insert(G, p)$: Inserts point p into grid G.
3. $Report(G, b)$: Returns all points in grid box b.

We use the Leafary Trees to store the grid G. Now, Theorems 3 and 4 imply the following Theorem.

Theorem 5. *If a grid G storing the point set P is stored on a Leafary tree, then the operations Insert and Report take $O(\log_2(H)/\log_2(|\ f_L\ |))$ time and the operation Build takes $O(|\ P\ | \log_2(H)/\log_2(|\ f_L\ |))$ time, where, H is the maximum number of boxes in G.*

3 The Closest Pair Problem

Let $S = \{\ p_1, p_2, ..., p_n\ \}$ be the given set of n points. We discuss the algorithm for the two-dimensional case ($D = 2$). Let $S_i = \{\ p_1, p_2, ..., p_i\ \}$ and $d(p_i, p_j)$ denote the distance between the points p_i and p_j in L_t - metric, $1 \le t \le \infty$. The *closest pair distance* in S: $\delta(S) = \min(d(p_i, p_j)\ |\ i \ne j)$. The *Closest Pair* problem is to find a pair of points $p, q \in S$, such that $d(p, q) = \delta(S)$.

For sake of completeness, we briefly state the closest pair algorithm presented in [4]. We find $\delta(S_2), \delta(S_3)$ and so on till $\delta(S_n) = \delta(S)$ which is the required result. Note that, $\delta(S_2), \delta(S_3), ..., \delta(S_n)(= \delta(S))$ is a *decreasing* sequence and $\delta(S_{i+1}) < \delta(S_i)$ if and only if $\exists p \in S_i$ such that $d(p, p_{i+1}) < \delta(S_i)$. Assume that we store the points of S_i on a *grid* G consisting of *square boxes or meshes* of size $\delta(S_i)$. Let b be the grid box in G in which the new point $p_{i+1} = (x, y)$ is located, then the index of b is the integer pair $(\lfloor x/\delta(S_i) \rfloor, \lfloor y/\delta(S_i) \rfloor)$. Every point in S_i that is within a distance $\delta(S_i)$ of p_{i+1} must be located in one of the 9 grid boxes that are *adjacent* to b. We call these 9 boxes the neighbors of b. Figure 3 shows the nine neighbors of box 5 that are numbered 1 to 9

Fact 1 *Each grid box in G can contain at most 4 points from S_i, else some pair of them would be less than $\delta(S_i)$ apart, contradicting the definition of $\delta(S_i)$. If G is a D-dimensional grid, $D > 2$, then it contains at most $(D + 1)^D$ points of S_i [8].*

Fact 1 implies that p_{i+1} will be tested with at most 36 points that lie on the 9 grid boxes adjacent to b. Hence $\delta(S_{i+1})$ is found. If $\delta(S_{i+1}) < \delta(S_i)$ then we construct a new grid with mesh size $\delta(S_{i+1})$ and insert all the points in the set S_{i+1} into it. This we do until we find $\delta(S_n)\ (= \delta(S))$.

Let us assume that at stage $i + 1$ the grid G storing the points in S_i is of horizontal dimension M_x and vertical dimension M_y. As seen earlier, the point $p_{i+1} = (x, y)$ maps on to a box b given by the index $(\lfloor x/\delta(S_i) \rfloor, \lfloor y/\delta(S_i) \rfloor)$. Now,

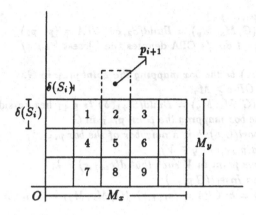

Fig. 3. p_{i+1} lies outside the grid G

to use the existing *LST* for finding the nearest neighbors of b, it is necessary that $\lfloor x/\delta(S_i) \rfloor < M_x$ and $\lfloor y/\delta(S_i) \rfloor < M_y$. In other words p_{i+1} should lie within the grid G. If this is not the case then we cannot use the *LST* storing G for finding the nearest neighbor of b because the *row major* ordering changes (see Figure 3) and hence the grid should be reconstructed for the new dimensions.

Suppose S is a collection of D-dimensional points, $D \geq 2$ then any box in a D-dimensional grid has 3^D nearest neighbors. This and Fact 1 imply a straightforward extension of the algorithm for the D-dimensional case.

Lemma 6. *Let $p_1, p_2, ..., p_n$ be a random permutation of the points in the two-dimensional point set S, and G be the grid storing the point set S_{i-1} then $Pr[p_i$ lies outside $G] \leq 2/i$.*

Proof. We use Seidel's Backwards Analysis [9]. Let $x_{max} = \text{maximum}(x \mid (x, y) \in S_i)$ and $y_{max} = \text{maximum}(y \mid (x, y) \in S_i)$. Let $A = \{ p \mid p = (x, y)$ and $x = x_{max} \}$ and $B = \{ p \mid p = (x, y)$ and $y = y_{max} \}$

We can easily see that if p_i lies outside G then *at least one* of the following conditions is true.

(1) $p_i \in A$ and $| A | = 1$
(2) $p_i \in B$ and $| B | = 1$
(3) Both (1) and (2).

We have just shown that, regardless of the composition of S_i, there are at most 2 possible choices of p_i which will permit p_i to lie outside G. Since $p_1, p_2, ..., p_n$ is a random permutation, the point p_i is a random point from S_i. Therefore, the probability that p_i lies outside G is at most $2/i$.

In a similar fashion we can prove the following Lemma.

Algorithm CP($p_1, p_2, ..., p_n$)
(1) $\delta = d(p_1, p_2)$; $(G, M_x, M_y) = Build(S_2, \delta)$; $CPA = (p_1, p_2)$;
(2) For $i = 2$ to $n - 1$ do / CPA denotes the Closest Pair */*
(3) begin
(4)   Let $b = (d, e)$ be the box mapping the point p_{i+1} in G.
(5)   if $d \geq M_x$ OR $e \geq M_y$
(6)   then $(G, M_x, M_y) = Build(S_{i+1}, \delta)$ / p_{i+1} lies outside G */*
(7)   Let b be the box mapping the point p_{i+1} in G.
(8) $V = \{$ Report(G,b): b is a neighbor of the box p_{i+1} $\}$;
(9) $k = min(d(p_{i+1}, q) \mid q \in V)$;
(10)   Let r be some point in V such that $d(p_{i+1}, r) = k$;
(11)   if $k \geq \delta$ then Insert(G,p_{i+1})
(12)   else $\delta = k$; $CPA = (p_{i+1}, r)$; $(G, M_x, M_y) = Build(S_{i+1}, \delta)$;
(13) end;
(14) return(δ, CPA).

Fig. 4. The Algorithm for CPP in D-Dimensions

Lemma 7. *Let $p_1, p_2, ..., p_n$ be a random permutation of the points in the D-dimensional point set S, and G be the grid storing the point set S_{i-1} then $Pr[p_i$ lies outside $G] \leq D/i$.*

Lemma 8 [4]. *Let $p_1, p_2, ..., p_n$ be a random permutation of the points in S, then $Pr[\delta(S_i) < \delta(S_{i-1})] \leq 2/i$.*

Figure 4 gives the Algorithm for CPP in D-dimensions.

3.1 Time Complexity

Let h_i be the maximum number of boxes of the grid G during iteration i. We can easily see that $h_1, h_2, ..., h_n$ is a *non decreasing* sequence. Henceforth, let us assume that $H = h_n$. We set $f_L = \log_2(n)$ for all the LSTs L, built during the execution of Algorithm CP. This implies that $ENTRY(i) \leq n$ for each internal node i of the $LSTs$ built at any stage of the execution of algorithm CP. Recall that, $D \geq 2$ is a fixed constant. Lemmas 7 and 8 implies that in iteration i, the steps 6 and 12 of Algorithm CP will be executed with probability D/i and $2/i$ respectively. This and Theorem 3 imply that steps 6 and 12 take $O(\log_2(H)/\log\log(n))$ expected time. From assumption 1 in Section 1.1, $H = O(n^k)$ for some constant $k > 0$. The following Theorem can be easily proved.

Theorem 9. *Given a point set S in the D-dimensional space, D a constant (≥ 2) the Algorithm CP takes $O(n \log n / \log\log n)$ expected time.*

4 Confidence bound

We first solve the following Lemma.

Lemma 10. *Let X_i, $1 \leq i \leq n$ be n independent random variables defined as follows:*

$$X_i = \begin{cases} O(i \ln n / \ln \ln n) & \text{with Probability } K/i \\ 0 & \text{with Probability } 1 - K/i \end{cases}$$

where $K > 0$ is a constant and let $X = \sum_{i=1}^{n} X_i$ and $\mu = E(X)$, the expectation of X. Then,
a). $Pr(X \geq eKsn \ln n / \ln \ln n) \leq O(s^{-Kes})$, where, $s > 0$ is a constant.
b). $Pr(X \geq 2eKsn \ln \ln n / \ln \ln \ln n) \leq O((\log n)^{-Kes})$, where, $s > 0$ is a constant and e is the base of the natural logarithm.

Proof. It is clear that the expectation of $E(X) = \mu = O(Kn \ln n / \ln \ln n)$. Using *Markov's inequality* we know that

$Pr(X \geq c\mu) = Pr(e^{tX} \geq e^{tc\mu}) \leq E(e^{tX})/e^{tc\mu}$ where, $c > 1$ and $t > 0$.
Using the above inequality in our context we get

$$Pr(X \geq cKn \ln n / \ln \ln n) \leq E(e^{t \sum_{i=1}^{n} X_i})/e^{tcKn \ln n / \ln \ln n}$$

Since X_i's are independent we get,
$$Pr(X \geq cKn \ln n / \ln \ln n) \leq \Pi_{i=1}^{n} E(e^{tX_i})/e^{tcKn \ln n / \ln \ln n} \qquad ...(1)$$
From definition of X_i we see that
$E(e^{tX_i}) = (1 + (K/i)(e^{ti \ln n / \ln \ln n} - 1)) < e^{(K/i)(e^{ti \ln n / \ln \ln i} - 1)}$ as $1 + x < e^x$.
Substituting in (1) we get
$Pr(X \geq cKn \ln n / \ln \ln n) \leq \Pi_{i=1}^{n} e^{(K/i)(e^{ti \ln n / \ln \ln i} - 1)}/e^{tcKn \ln n / \ln \ln n} \qquad ...(2)$
$$\leq e^{\sum_{i=1}^{n} (K/i)(e^{ti \ln n / \ln \ln n} - 1)}/e^{tcKn \ln n / \ln \ln n}$$
Since $(K/i)(e^{ti \ln n / \ln \ln n} - 1) \leq (K/n)(e^{tn \ln n / \ln \ln n} - 1)$ for each $1 \leq i \leq n$ and $t > 0$, we get

$Pr(X \geq cKn \ln n / \ln \ln n) \leq e^{K(e^{tn \ln n / \ln \ln n} - 1)}/e^{tKcn \ln n / \ln \ln n}$

Substituting $t = \ln c(\ln \ln n / n \ln n)$ in the above equation we get
$Pr(X \geq cKn \ln n / \ln \ln n) \leq O(e^{Kc - Kc \ln c}) \qquad ...(3)$
Substituting $c = es$ ($c = 2es \ln \ln n / \ln \ln \ln n$) in (3) we prove case a (case b) of the lemma.

Let X_i' and X_i'' be the cost of steps 6 and 11 of Algorithm CP in iteration i. Clearly, each of X_i' and X_i'', $1 \leq i \leq n$ is a *random variable*. Since the points $p_1, p_2, ..., p_n$ is a random permutation, the random variables X_i', $1 \leq i \leq n$ are *independent*. Similarly, the random variables X_i'', $1 \leq i \leq n$ are also *independent*. From the assumption that $H < \text{Polynomial}(n)$ [10] we see that X_i' satisfies the definition of X_i of Lemma 5 with $K = D$ and X_i'' satisfies the definition of X_i of Lemma 5 with $K = 2$. Since D is a constant, we get the following Theorem.

Theorem 11. *The implementation of the closest pair algorithm to find the closest pair in D dimensions in a given set of n points, that uses a Leafary tree runs in $O(n \log n / \log \log n)$ time with probability $1 - O(s^{-k})$ and in $O(n \log n / \log \log \log n)$ time with probability $1 - O((\log n)^{-k})$, for every constant $s, k \ (k > s > 0)$.*

5 Conclusion

In [4], a simple incremental randomized algorithm was presented which uses hashing and has a run time of $O(n \log n / \log \log n)$ with high probability. In this paper we have presented an algorithm which does not employ hashing. However, the reliability bounds, while very good for all the practical purposes, it is a bit weaker than that of the hashing based method. This is probably the trade off we have done for sparing the hashing. Thus we leave the following question as open: *Is there a tree based implementation that matches hashing based methods both in expected time and reliability bounds ?*

Acknowledgement: This work was partially supported by the Indo-German Project on Computational Geometry and its Applications, sponsored jointly by MHRD, India and KFA, Germany.

References

1. Dietzfelbinger, M., Hagerup, T., Katajainen, J., Penttonen, M.: A reliable randomized algorithm for the Closest-pair problem, Technical report, Personal Communication.
2. Dietzfelbinger, M., Meyer auf der Heide, F.: A new universal class of hash functions and dynamic hashing in real time, Proc. ICALP 90, Lecture Notes in Computer Science, Vol. **443**, Springer-Verlag, Berlin, 1990, 6–19.
3. Fortune, S., Hopcroft, J.: A note on Rabin's Nearest-Neighbor algorithm, Information Processing Letters, Vol **8**, No 1, (1979), 20–23.
4. Golin, M., Raman, R., Schwarz, C., Smid, M.: Simple Randomized Algorithms for Closest Pair Problems, Technical Report, Max-Planck-Institut Für Informatik, Saarbrucken, Germany, 1992.
5. Khuller, S., Matias, Y.: A simple randomized sieve algorithm for the closest pair problem, Proc. Third Canadian Conference on Computational Geometry, (1991), 130–134.
6. Preparata, F.P., Shamos, M.I.: Computational Geometry - an Introduction, Springer-Verlag, New York, 1985.
7. Rabin, M.: Probabilistic algorithms in Algorithms and Complexity: New directions and Recent results (J.F. Traub ed.), (1976), pp 21–39.
8. Schwarz, C.: Data Structures and Algorithms for the Dynamic Closest Pair Problem, Thesis, Max Planck Institut, Saarbrücken, Germany.
9. Seidel, R.: Backwards Analysis of Randomized Geometric Algorithms, Report TR-92-014, Department of Computer Scince, University of California Berkeley, Berkeley, CA, (1992).
10. Willard, D.E.: Application of Fusion tree method to computational Geometry and searching, Proceedings of the SODA, 1992, 286–295.

Testing Containment of Object-Oriented Conjunctive Queries is $\prod_2^p$-hard

Edward P.F. Chan[*]
Department of Computer Science
University of Waterloo
Waterloo, Ontario, Canada N2L 3G1

Ron van der Meyden
Information Science Laboratory
NTT Basic Research Laboratories
Atsugi-shi, Kanagawa 243-01, Japan

May 27, 1995

Abstract

We study the complexity of testing containment for a class of object-oriented conjunctive queries. We show that the containment problem is Π_2^p-hard. Together with a previous result, the containment problem is complete in Π_2^p.

1 Introduction

The problem of query optimization is difficult in an object-oriented database system (OODB). A natural first step is to use the typing constraints imposed by the schema to transform a query into an equivalent one that logically accesses a minimal set of objects. In an OODB, classes are named collections of similar objects. A class C could be refined into subclasses. Conversely, the class C is said to be a superclass of its subclasses. Subclasses are specialization of their superclasses. Consequently, objects in a subclass are also objects in its superclasses. Specialization of a class is often achieved by refining and/or adding properties to its superclasses. Since properties of a superclass are also properties of its subclasses, a subclass is said to inherit the properties of its superclasses. Class-subclass relationships form an acyclic directed graph called inheritance or generalization hierarchy.

Inheritance is a powerful modeling tool, because it gives a better structured and more concise description of the schema and it helps in factoring out shared implementations in applications. Objects belonging to the same class share some common properties. Properties are attributes or methods defined on types; they are applicable only to the instance of the types. In effect, therefore, types are constraints imposed on objects in the classes. Properties are formally denoted as attribute-type pairs in this paper. A natural first step in query optimization is to use the typing constraints implied by the schema to minimize the search space for variables involved in the query.

Example 1.1 *The following is a schema for a vehicle rental database. It keeps track of all rental transactions for vehicles in the company. In this application, Auto, Trailer and*

---

[*]Part of this work was done while the author visited NTT Basic Research Laboratories in Tokyo.

Truck are subclasses of the superclass Vehicle. There are clients, called discount customers, who are known to the company and receive special treatments. Discount customers receive a special rate and are not required to have deposit on the vehicles rented. However, discount customers only allow to rent automobiles, and not other types of vehicles. Let us assume further that all superclasses are partitioned by their respective subclasses.

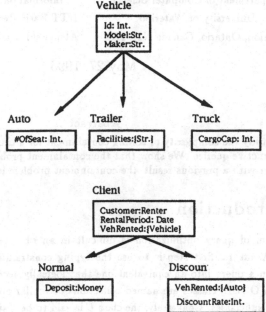

Suppose we want to find out all those vehicles that have been rented to a discount client. Express in a calculus-like language, the query looks like:

Q_1: { x | $\exists y$ ($x \in$ Vehicle & $y \in$ Discount & $x \in y$. VehRented)}.

Since discount clients are allowed to rent Auto only, the above query is equivalent to the following query:

Q_2: { x | $\exists y$ ($x \in$ Auto & $y \in$ Discount & $x \in y$. VehRented)}.

Q_2 *is considered to be more optimal since the number of variables as well as their search spaces are minimal, given the typing constraints implied by the schema.* □

In this paper, we shall define a class of queries called conjunctive queries for an object-oriented database. In general, variables in an object-oriented query range over heterogeneous sets of objects. This constitutes a significant divergence from its relational counterpart. As shown in [3], to solve the equivalence and optimization problems, we need first to resolve the satisfiability and containment problems. Let us consider the following example to see how the containment problem in our case is different from the containment problem in the relational model.

Example 1.2 *The following schema records the employer-employee relationships among a group of people. The Employee attribute indicates the set of employees hired by a person.*

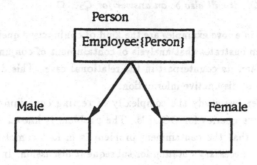

Consider the following two queries defined on the above inheritance hierarchy. Q_1 retrieves all persons x who hires a person u and a male v such that u is also an employee of v and u hires a female employee w. Q_2 finds all those persons x who hires a male employee y who in term hires a female employee z. Expressed in our language, they are as follows.

Q_1: { x | $\exists u \, \exists v \, \exists w$ *(x∈ Person & u∈ Person & v∈ Male & w∈ Female & u∈x.Employee & v∈x.Employee & u∈v.Employee & w∈u.Employee)*}.

Q_2: { x | $\exists y \, \exists z$ *(x∈ Person & y∈ Male & z∈ Female & y∈x.Employee & z∈y.Employee)*}.

The above two queries are best visualized as the following two graphs.

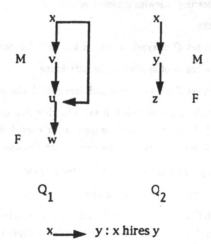

We claim that Q_2 contains Q_1, meaning that whenever there is an answer for Q_1, it will also be the answer for Q_2. The person u is either a male or a female. If u is a male, *then y and z in Q_2 can be mapped to u and w, respectively. Similarly if u is a female, variables y and z in Q_2 can be mapped to v and u, respectively. Thus, whenever there is an answer for Q_1, it will also be an answer for Q_2.* □

The queries in above examples are the kind of conjunctive queries we are interested in. Example 1.2 demonstrates that analysis of containment of conjunctive queries is perhaps more difficult than its counterpart in the relational case. This dues to the fact it might involve analysis of disjunctive information.

In this paper, we study the complexity of testing containment of a class of queries called conjunctive queries for OODB's. The problem is known to be in Π_2^p [2]. In this paper, we show that the containment problem is in fact complete in Π_2^p. In Section 2, we introduce the necessary notation for subsequent discussion. In Section 3, we prove our complexity result.

2 Definitions and Notation

In this section, we introduce notation that is necessary for the rest of the discussion. To make the lower bound result as strong as possible, and isolate those features responsible for the complexity, the hardness proof should make the smallest possible set of semantic assumptions, and use as few features of the query language as possible. Hence, the definitions given here will be a simplification of those of [2].

2.1 Types, Classes and Schemas

We suppose given the following pairwise disjoint sets:

1. The set Z of integers.

2. A countably infinite set O of symbols which are called *object identifiers*.

3. A countably infinite set A of symbols, the *attributes*.

4. A countably infinite set C of symbols which are called *classes*.

The set Z is said to be the set of *atomic values*. The elements of A are used as attribute names in tuple types, and the elements of C as names for user-defined classes.

A *type* over a set $C \subseteq C$ of class names is an expression defined as follows:

1. The expression *Integer* is a type, called the *atomic type*.

2. Every element of C is a type, called a *class*.

3. If $a_1, \ldots, a_n$ are distinct attributes in A and $t_1, \ldots, t_n$ are atomic types or classes, where $n \geq 0$, then $[a_1:t_1, \ldots, a_n:t_n]$ is a type, called a *tuple* type. The type t_i is said to be the *type* of the attribute a_i, for each $i = 1 \ldots n$.

A *schema* S is a triple (C, σ, $\prec$), where C is a finite subset of C, σ is a function to be described below, and $\prec$ is a partial order on C, called the *subclass* relationship. In practice, the subclass relationship represents the user-defined *inheritance hierarchy*, but because the result we prove is independent of inheritance, we give a simplified account of typing in class hierarchies that factors out the semantics of inheritance.

A class $A \in \mathbb{C}$ is said to be *terminal* if there is no other class $B \neq A$ such that $B \prec A$. Otherwise A is *non-terminal*. We write *terminal*(S) for the set of terminal classes in S. We write *type-expr*(C) for the set of all types that only involve classes in C. The function σ maps each terminal class in *terminal*(S) to a tuple type in *type-expr*(C) that describes its structure. Attributes in $\sigma(C)$ for $C \in \mathbb{C}$ are called the *attributes* of C. The *type of C.A*, denote *type*(C.A), is the type t of A in $\sigma(C)$.

2.2 States, Domains and Objects

Let $S = (\mathbb{C}, \sigma, \prec)$ be a schema. Let O be a finite subset of $\mathcal{O}$ and I_c be a function mapping objects in O to *terminal*(S). Intuitively, $I_c(o)$ represents the terminal class to which the object o belongs.

Given O and I_c, each type expression T in *type-expr*(C) is interpreted as a set of possible values, called the *domain* of T, denoted as *dom*(T). In order to represent *null values* and inapplicable attributes, we introduce a new symbol Λ. The *domain* of a type with respect to (w.r.t.) O and I_c is defined as follows:

1. $dom(Integer) = \mathcal{Z} \cup \{\Lambda\}$.

2. For each class $D \in \mathbb{C}$, $dom(D) = \{ o \in O \mid I_c(o) \prec D\} \cup \{\Lambda\}$.

3. For each tuple type $[a_1{:}t_1, \ldots, a_n{:}t_n]$, $dom([\, a_1{:}t_1, \ldots, a_n{:}t_n]) = \{[a_1{:}v_1, \ldots, a_n{:}v_n] \mid v_i \in dom(t_i) \text{ for all } i=1 \ldots n\}$.

A *state* s on a schema $S = (\mathbb{C}, \sigma, \prec)$ is a triple (O, I_c, I_u), where O is a finite subset of $\mathcal{O}$, I_c is a function from O to *terminal*(S), and I_v is a function from O to tuple values in domains of types with respect to O. The function I_v is required to satisfy the following condition:

$$\forall o \in O \; I_v(o) \in dom(\sigma(I_c)).$$

That is, the tuple value of an object must be of the type associated with the terminal class to which that object belongs.

The *domain* of a state s, denoted by *dom*(s), is the set $\mathcal{Z} \cup O \cup \{\Lambda\}$.

2.3 A Class of Object-Preserving Conjunctive Queries

Queries are constructed from a set of variables, symbols from the set of atomic values, equality operator '=', membership operator '$\in$', logical operator '&' and the existential quantifier '$\exists$'.

First, we define the concept of *term*. Terms will denote objects or atomic values. A term τ is an expression of the following form: c, x or $x.A$, where $c \in \mathcal{Z}$ is an atomic value, x is a variable and A is an attribute. A term of the form x or $x.A$ is called a *variable term*.

An *atom* or an *atomic formula* is one of the following:

1. A *range atom*, of the form $x \in C$, where C is a class or atomic type. Such an atom asserts that x denotes an object or value in the class or atomic type C.

2. An *equality atom*, of the form $\tau_1 = \tau_2$, where τ_1 and τ_2 are terms. Such an atom asserts that these terms denote the same object or atomic value.

A *conjuctive formula* Φ is constructed from atomic formulas using only the logical operator '&' and existential quantification '∃'. We will write $\exists x \in C[\Phi]$ for $\exists x[x \in C \& \Phi]$. *Bound* and *free* variables are defined in the usual manner. A *conjunctive query* is an expression of the form $\{ s_0 \mid \Phi(s_0) \}$, where $\Phi(s_0)$ is a conjunctive formula with only the variable s_0 free. A *union* query is an expression $Q_1 \cup \ldots \cup Q_n$ where the Q_i are conjunctive queries.

2.4 The Answer of a Query

To account for the possible presence of null values in instances, queries are evaluated in a slightly nonstandard way, following [2]. We remark that the definitions below may be shown to correspond to a number of different interpretations of the occurence of Λ as the value of some attribute of an object. The simplest of these interprets this to mean that the attribute is inapplicable to the object. The other interprets this to mean that either the attribute is inapplicable, or it is applicable but the object or atomic value denoted is unknown.

Given a state $s = (O, I_c, I_v)$ on a schema $S = (C, \sigma, \prec)$, we first define the denotation of terms. Let α be an *assignment* mapping variables occurring in the query Q to $dom(s)$. The denotation $[\tau]$ of a term τ of Q with respect to α is defined as follows. If c is an atomic value then $[c] = c$. If x is a variable of Q then $[x] = \alpha(x)$. If x is a variable with $\alpha(x)$ equal to an object o and $I_v(o) = [a_1 : v_1, \ldots, a_n : v_n]$ then $[x.a_i] = v_i$ for each $i = 1 \ldots n$. In all other cases $[x.A] = \Lambda$.

The answer to a conjunctive query $Q = \{x_0 \mid \Phi(x_0)\}$ in an instance $s = (O, I_c, I_v)$ may now be defined as follows. By renaming variables, we may assume that each variable is quantified at most once in Φ. The formula Φ is said to be *true* in s under the assignment α to the variables of Φ if no term in Φ has denotation Λ with respect to α, and for all atoms At of Φ:

1. if At is $x \in C$ then $[x] \in dom(C)$, and

2. if At is $\tau_1 = \tau_2$ then $[\tau_1] = [\tau_2]$.

If α is an assignment under which the formula Φ is true in s, then the object or atomic value $\alpha(x_0)$ is called an *answer* to Q in s. We write $Q(s)$ for the set of answers of Q in s. The answer $Q(s)$ of a union query $Q = Q_1 \cup \ldots \cup Q_n$ is the set $Q = Q_1(s) \cup \ldots \cup Q_n(s)$.

A query Q is *satisfiable* if there is a state s such that $Q(s)$ is non-empty. Unless otherwise stated, we assume queries from now on are satisfiable. Given two queries F and G (on a schema S), F is said to *contain* G, denoted $F \supseteq G$, if $F(s) \supseteq G(s)$, for all states s on S. Two queries F and G are said to be *equivalent*, denoted $F \equiv G$, if they contain each other.

2.5 Terminal Conjunctive Queries

A conjunctive query in *normal form* is a conjunctive query of the form $\{x_0 \mid x_0 \in C_0 \ \& \ \exists x_1 \in C_1 \ldots \exists x_n \in C_n \ \Phi(x_0, \ldots, x_n, c_1, \ldots, c_n)\}$ where Φ is a conjunction of atoms containing only the variables $x_0, \ldots, x_n$ and atomic values $c_1, \ldots, c_n$, such that

1. all atoms are of the form $x_i.A = \tau$, where τ is either a variable x_j or an atomic value c_j, and

2. for each variable x_i and attribute A, the term $x_i.A$ occurs in at most one atom.

Such a query is said to be *terminal* if the C_i are all terminal classes. Terminal conjunctive queries play an important role in the analysis of conjunctive queries in general, as is illustrated by the following proposition.

Proposition 2.1 *Every satisfiable conjunctive query is equivalent to a union of terminal conjunctive queries in normal form.*

For the proof, see [1]. Since the queries we will write in this paper satisfy the structural condition on Φ, the main step of concern to us is the following. Suppose that Q is a conjunctive query of the form $\{x_0 \mid x_0 \in C_0 \ \& \ \exists x_1 \in C_1 \ldots \exists x_n \in C_n \ \Phi\}$. A satisfiable query of the form $\{x_0 \mid x_0 \in C_0 \ \& \ \exists x_1 \in D_1 \ldots \exists x_n \in D_n \ \Phi\}$, where D_i is a terminal subclass of C_i for each i, will be called an *expansion* of Q. It is straightforward to show that every conjunctive query is equivalent to the union of its expansions.

2.6 Containment of Conjunctive Queries

We now state a condition characterizing containment for terminal conjunctive queries in normal form. Let $Q_1 = \{x_0 \mid x_0 \in C_0 \ \& \ \exists x_1 \in C_1 \ldots \exists x_n \in C_n \ \Phi(x_0, \ldots, x_n, c_1, \ldots, c_n)\}$ and $Q_2 = \{y_0 \mid y_0 \in C_0 \ \& \ \exists y_1 \in D_1 \ldots \exists y_m \in D_m \ \Psi(y_0, \ldots, y_m, d_1, \ldots, d_m)\}$ be terminal conjunctive queries in normal form. A *variable mapping* from Q_2 to Q_1 is a a function μ that maps the variables $y_1 \ldots y_m$ of Q_2 to the set $\{x_0, \ldots, x_n, c_1, \ldots, c_n\}$ of variables and constants of Q_2. This mapping may be extended to the atomic values by putting $\mu(c) = c$ for all atomic values c.

We say that μ is a *containment mapping* from Q_2 to Q_1 if $\mu(y_0) = x_0$, for each atom $y_i.A = \tau$ of Q_2 we have that $\mu(y_i).A = \mu(\tau)$ is an atom of Q_1, and for each atom $y_i \in D_i$ of Q_2 we have that $\mu(y_i) \in D_i$ is an atom of Q_1.

Theorem 2.2 *If Q_1 and Q_2 are terminal conjunctive queries in normal form then $Q_1 \subseteq Q_2$ if and only if there is a containment mapping μ from Q_2 to Q_1.*

[Proof]: See [3]. □

From Theorem 2.2 and Proposition 2.1, a characterization of containment of conjunctive queries can be found.

Theorem 2.3 *Let $M = Q_1 \cup \cdots \cup Q_s$ and $N = P_1 \cup \cdots \cup P_t$ be two unions of terminal conjunctive queries. Then $M \subseteq N$ if and only if for each Q_i in M, there is a P_j in N such that $Q_i \subseteq P_j$.*

[Proof]: [3]. □

Theorem 2.4 *The problem of determining containment for conjunctive queries is in* Π_2^p.

[Proof]: See [2]. □

3 The Main Result

In this Section, we show that the containment problem is Π_2^p- hard.

A Π_2 formula of quantified propositional logic is an expression

$$\varphi = \forall p_1 \ldots p_n \exists p_{n+1} \ldots p_{n+m} [\alpha]$$

where α is a formula of propositional logic containing only the propositional variables $p_1 \ldots p_{n+m}$. Such an expression is true if for every assignment of boolean truth value to the variables $p_1 \ldots p_n$, there exists an assignment of truth values to the variables $p_{n+1} \ldots p_{n+m}$ under which the formula α is true. The set Π_2-SAT is the set of all true Π_2 formulae. This is a generalization of the problem of satisfiability to the polynomial hierarchy. It is known that the set Π_2-SAT is complete for the level Π_2^p of this hierarchy [4, 5].

Theorem 3.1 *There exists a fixed schema* S *such that problem of deciding, given queries* Q_1 *and* Q_2, *whether* $Q_1 \subseteq Q_2$ *with respect to* S *is* Π_2^p-*hard.*

[Proof]: By reduction from Π_2-SAT. We show that for every Π_2 formula φ, there is a pair of conjunctive queries Q_1, Q_2 such that φ is true iff Q_1 is contained in Q_2.

Define the schema S to contain classes C, R, G, V, AND and NOT. The subclass relationships between these classes are given by $R \prec C$ and $G \prec C$. Thus, the terminal classes are R, G, V, AND and NOT. The tuple type for both R and G is $[a : C, b : INT, c : V]$, for V is the empty tuple type $[]$, for AND is $[in_1 : V, in_2 : V, out : V]$, and for NOT is $[in : V, out : V]$.

We first describe the query Q_1. Part of this query will encode the truth tables for conjunction and negation. Intuitively, there are (four) variables ranging over AND which represent the lines of the truth table for '$\wedge$', and there are (two) variables ranging over NOT which represent the lines of the truth table of '$\neg$', and there are variables t and f in V which denote the truth values *true* and *false*, respectively. We write $TT(t, f)$ for the following formula:

$$\exists u_1 \in AND \ [u_1.in_1 = t \ \& \ u_1.in_2 = t \ \& \ u_1.out = t] \ \& $$
$$\exists u_2 \in AND \ [u_2.in_1 = t \ \& \ u_2.in_2 = f \ \& \ u_2.out = f] \ \& $$
$$\exists u_3 \in AND \ [u_3.in_1 = f \ \& \ u_3.in_2 = t \ \& \ u_3.out = f] \ \& $$
$$\exists u_4 \in AND \ [u_4.in_1 = f \ \& \ u_4.in_2 = f \ \& \ u_4.out = f] \ \& $$
$$\exists v_1 \in NOT \ [v_1.in = t \ \& \ v_1.out = f] \ \& $$
$$\exists v_2 \in NOT \ [v_2.in = f \ \& \ v_2.out = t].$$

Next, suppose φ has n universally quantified variables. Part of the query Q_1 will have the function of assigning a truth value to each of these variables. We construct for each

$i = 1 \ldots n$ the query $ASGN_i(t, f)$ given by

$$\exists w_{i1} \in R \; \exists w_{i2} \in C \; \exists w_{i3} \in G \; [w_{i1}.a = w_{i2} \; \& \quad w_{i2}.a = w_{i3} \quad \&$$
$$w_{i2}.b = i \qquad \& \; w_{i3}.b = i \; \&$$
$$w_{i2}.c = t \qquad \& \; w_{i3}.c = f].$$

We now define the query Q_1 to be

$$\{t \mid t \in V \; \& \; \exists f \in V \; [TT(t, f) \; \& \; ASGN_1(t, f) \; \& \ldots \& \; ASGN_n(t, f)]\}.$$

After moving the quantifiers to the front, it is clear that this is a query in normal form.

Let α be a formula of propositional logic in the propositional constants $p_1 \ldots p_{n+m}$. We define inductively the formula Φ_α with free variables amongst $\mathbf{x} = x_{p_n} \ldots x_{p_{n+m}}$ and x_α. If α is the propositional constant p_i, where $i = 1 \ldots n$, then

$$\Phi_\alpha = \exists z_{i1} \in R \exists z_{i2} \in G[z_{i1}.a = z_{i2} \; \& \; z_{i2}.b = i \; \& \; z_{i2}.c = x_{p_i}].$$

If α is the propositional constant p_i, where $i = n+1 \ldots n+m$, then Φ_α is the null formula true. (Note that the variable x_α is x_{p_i} in these cases.) If $\alpha = \beta \& \gamma$ then $\Phi_\alpha(\mathbf{x}, x_\alpha)$ is

$$\exists y_\beta \in AND \; \exists x_\beta x_\gamma \in V \left[\begin{array}{l} y_\beta.in_1 = x_\beta \; \& \; y_\beta.in_2 = x_\gamma \; \& \; y_\beta.out = x_\alpha \; \& \\ \Phi_\beta(\mathbf{x}, x_\beta) \; \& \; \Phi_\gamma(\mathbf{x}, x_\gamma) \end{array} \right]$$

If $\alpha = \neg \beta$ then $\Phi_\alpha(\mathbf{x}, x_\alpha)$ is

$$\exists y_\beta \in NOT \; \exists x_\beta \in V \; [x.in = x_\beta \& y_\beta.out = x_\alpha \& \Phi_\beta(\mathbf{x}, x_\beta)].$$

We define Q_2 to be the query

$$\{x_\alpha \mid x_\alpha \in V \; \& \; \exists x_{p_1} \ldots x_{p_n} \in V \; [\Phi_\alpha(\mathbf{x}, x_\alpha)]\}$$

Observe that the class C does not occur in Q_2. Thus, after moving the quantifiers to the front (strictly, we first need to rename the repeated quantified variables arising from repeated subformulae), this query is a terminal conjunctive query in normal form.

Note that expansions of Q_1 are obtained by replacing each of the n range atoms $w_{i2} \in C$ by either $w_{i2} \in R$ or $w_{i2} \in G$. There are therefore 2^n such expansions. We first show that each of these expansions uniquely determines an assignment of truth values to the propositional constants $p_1 \ldots p_n$.

Suppose $1 \leq i \leq n$. Because the constant i has only two occurrences in Q_1, if μ is a mapping from the variables z_{i1}, z_{i2}, x_{p_i} to the variables of an expansion E of Q_1 that preserves the atom $z_{i2}.b = i$ of the formula Φ_{p_i}, then we must have $\mu(z_{i2}) = w_{i2}$ or $\mu(z_{i2}) = w_{i3}$. In case the atom $w_{i2} \in C$ of $ASGN_i$ is expanded as $w_{i2} \in G$, the mapping $z_{i1} \mapsto w_{i1}, z_{i2} \mapsto w_{i2}, x_{p_i} \mapsto t$, is the only mapping that preserves all the equality and range atoms of Φ_{p_i}. This determines the assignment of $true$ to p_i. Similarly, in case $w_{i2} \in C$ is expanded as $w_{i2} \in R$ the mapping $z_{i1} \mapsto w_{i2}, z_{i2} \mapsto w_{i3}, x_{p_i} \mapsto f$ is the only such mapping. This determines the assignment of $false$ to p_i.

Next, suppose that μ is mapping from the variables of the formulae Φ_{p_i} for $i = 1 \ldots n$ and the variables $x_{p_{n+1}} \ldots x_{p_{n+m}}$ to the variables of an expansion E of Q_1, such that the atoms of Q_2 containing these variables are preserved. As noted above, this implies that the variables $x_{p_1} \ldots x_{p_n}$ are mapped to either t or f. The same holds for the variables $x_{p_{n+1}} \ldots x_{p_{n+m}}$, because of the range atoms $x_{p_i} \in V$ in Q_2. Let θ be the truth value assignment that assigns the constant p_i to be *true* if and only if $\mu(x_{p_i}) = t$. Under these conditions, a straightforward induction on the complexity of α shows that there exists a unique extension of the mapping μ to a mapping from the variables of Q_2 to the variables of Q_1 that preserves all the atoms of Q_2. Furthermore, we have $\mu(x_\alpha) = t$ iff the formula α is true with respect to the assignment θ. Note also that this mapping μ is a containment mapping from Q_2 to the expansion E if and only if $\mu(x_\alpha) = t$.

It now follows from the observations above that there exists a containment mapping from Q_2 to E for each expansion E of Q_1 if and only if the quantified formula φ is true. □

References

[1] Chan, E.P.F., "Testing Satisfiability of a Class of Object-Oriented Conjunctive Queries," *Theoretical Computer Science (134)*, pp. 287-309, 1994.

[2] Chan, E.P.F., "Complexity of Testing Containment, Equivalence and Minimization of Object-Preserving Conjunctive Queries," *TR-93-36*, Department of Computer Science, University of Waterloo, 1993.

[3] Chan, E.P.F., "Containment and Optimization of Object-Preserving Conjunctive Queries," *TR-94-07*, Department of Computer Science, University of Waterloo, 1994. An extended abstract appeared in *Proceedings of ACM PODS 1992*, pp. 192-202.

[4] Stockmeyer, L.J. and Meyer, A.R., "Word Problems requiring exponential time," *Proc. of the 5th STOC*, pp.1-9, 1973.

[5] Wrathall, C, "Complete Sets and the Polynomial-time Hierarchy," *Theoretical Computer Science (3)*, pp. 23-33, 1976.

Computing Infinite Relations Using Finite Expressions: A New Approach To The Safety Issue In Relational Databases

Ruogu Zhang

Massachusetts Institute of Technology, Cambridge MA 02139, USA

Abstract. Commonly used methods to deal with the safety issue in relational databases are based on "placing constraint of finiteness" on relational calculus. This paper describes how safety can be achieved using a novel approach. A new representation is created to denote the relations defined by relational calculus. An algorithm which can symbolically handle infinite relations is designed to efficiently compute all these relations. Therefore the constraint of finiteness on relational calculus can be removed.

1 Introduction

We are in an age of information generation, collection, and processing. Computer database technology has been rapidly developed to meet today's challenges and needs. There is a continuing demand for researchers to improve solutions for many important problems. The safety issue in relational databases is one of them.[1, 2, 4, 5, 3, 7, 6].

In the relational model, sets of data are represented by relations. A form of logic called "relational calculus" is commonly used for expressing operations on relations. A limitation we face in relational calculus is the finiteness of relations. If relations range over an infinite set, the computer database may crash since the memory and time are both limited. Therefore, the safety issue, or how to avoid the fatal infiniteness of computation in a database system, has arisen. Much research has been devoted to this problem. All current approaches are based on "placing the constraint of finiteness" on relational calculus[1, 2, 4, 5, 3, 7, 6]. In other words, the relation created at any operation step must be guaranteed to be finite. The constraint of finiteness introduces many difficulties into the definition of relational languages such as disallowing certain operations. A restricted form of relational calculus, such as safe domain relational calculus (DRC) and safe tuple relational calculus (TRC), has to be defined[7, 4, 2]. This leads to tedious definitions and complicated rules.

This paper proposes a completely new approach for dealing with the safety issue. This approach doesn't impose any restriction on relational calculus. Instead, a new form of representation to denote a relation (either finite or infinite) defined by a formula of relational calculus, called a +_relation_, is created. It can be proven that the set of +_relations_ associated with formulas of relational calculus has a closure property. In other words, a formula of relational calculus, recursively defined (by using logical operators and quantifiers) from formulas which have $+$_relations, also has its associated $+$_relation. An advanced algorithm that can efficiently compute all these $+$_relations is designed. The designed algorithm can symbolically handle infinite relations, therefore, the constraint of finiteness on relational calculus can be removed. In the process of obtaining answers to a query, the algorithm can find the intermediate and final results, whether they are finite or infinite relations. In the case that the final answer is an infinite relation, the algorithm can find it and represent it by a $+$_relation. The answer denoted by the $+$_relation is explicit in the sense that it indicates desired tuples in a straightforward way.

2 A New Representation of Relations in the Relational Model

The data model of a database system is a mathematical formalism with two parts: a notation for describing data, and a set of operations used to manipulate that data.

The mathematical concept underlying the relational model is the set-theoretic relation. A relation is any subset of the Cartesian product of a list of domains. A domain is a set of values. The Cartesian product of domains $D_1, D_2, ..., D_k$, is the set of all k-tuples $(v_1, v_2, ..., v_k)$ such that v_1 is in D_1, v_2 is in D_2, and so on. The domains are often given names, called attributes.

Operations on the database require data manipulation language or query language. In relational calculus, queries are expressed by writing logical formulas that the tuples in the answer must satisfy. The process of obtaining answers to a query is one of proving that certain facts follow logically from the facts in the relational database. Two forms of relational calculus, called "domain relational calculus" and "tuple relational calculus", form the basis for most commercial query languages in relational database systems.

In relational calculus, each formula defines a relation whose attributes correspond to the free variables of the formula. $F(x_1, ..., x_n)$ implies that formula F has free variables $x_1, ..., x_n$ and no others. The query, or expression, denoted by F is

$$\{x_1, ..., x_n | F(x_1, ..., x_n)\}$$

The answer is the set of tuples $a_1, ..., a_n$ such that when we substitute a_i for x_i, $1 \leq i \leq n$, the formula $f(a_1, ..., a_n)$ becomes true. The query language consisting of expressions in the above form is called domain relational calculus (DRC).

It should be observed that negation is a source of infiniteness. The relations defined by DRC expressions need not be finite. For example, $\{x, y | \neg p(x, y)\}$ is a legal.DRC expression defining the set of infinite pairs (x, y) that are not in the relation of predicate p. To avoid such expressions, as mentioned earlier, a subset of DRC formulas called "safe formulas" is introduced. The definition of safe DRC formulas is complicated. A number of rules are applied: there are no uses of the ∀ quantifier; whenever an OR operator is used, the two formulas connected must have the same set of free variables; all variables in the body of a rule should be limited, and so on[6, 1, 2, 3, 4, 5, 7].

The approach presented in this paper will remove restrictions on relational calculus. A new representation of relations is introduced for dealing with the safety problem. We number all the members in a domain of a relational model, every tuple in a relation will correspond to a tuple of positive integers. Then we can easily map a relation to one which consists of tuples of positive integers by using Hashing methods. Without loss of generality, the relations discussed here are subsets of the set of positive integers.

Definition 1. Assume M is a positive integer. Then an $M+\_tuple$ is a tuple consisting of n elements, each element is either a number less than or equal to M, or a + symbol. When M is understood, we may omit the number M.

Definition 2. If $t = (a_1, a_2, ..., a_n)$, where all $a_1, a_2, ..., a_n$ are positive integers, is a tuple, then the $M+\_projection$ of t, denoted by t^+, is defined as $t^+ = (b_1, b_2, ..., b_n)$, where $b_j = a_j$ if $a_j \leq M$ and $b_j = +$ if $a_j > M$, for all $j = 1, 2, ..., n$. When M is understood, we may omit the number M.

Definition 3. An $M+\_relation$ is a set of different $M+\_tuples$. We say an $M+\_relation$ R^+ represents a relation R, if for an arbitrary n-tuple t it holds that $t \in R \Longleftrightarrow t^+ \in R^+$. When M is understood, we may omit the number M.

Intuitively, the symbol + means an arbitrary integer greater than M. In the case $M=10$ and a relation has two attributes. $R^+ = \{(1,2), (9,+), (+,+)\}$ is a $+\_relation$ consisting of three $+\_tuples$ $(1,2)$, $(9,+)$ and $(+,+)$. Let us determine all pairs in the relation R represented by this $+\_relation$. Since for every tuple $t = (9, b)$ with $b > 10$, $t^+ = (9, +)$, and $(9, +) \in R^+$, we know all pairs $(9, b)$ with $b > 10$ belong to R. It is not difficult to see that R consists of $(1,2)$, all $(9, b)$ with $b > 10$, and all (a, b) with $a > 10$, $b > 10$. R is an infinite relation, and it can be represented by the finite relation R^+.

When $M=10$, for a relation of two attributes, there are $M+1=11$ choices for each argument of a pair in the set R^+. Therefore, there are at most $(10+1)^2$ pairs in the set R^+. Generally, if there are n attributes, there are at most $(M+1)^n$ $+\_tuples$ in R^+. In conclusion, R^+ is always finite! It is the finiteness of R^+ which guarantees "safety" under the operation of the relational calculus.

In brief, formulas are expressions that denote relations in relational calculus. The formulas of relational calculus are recursively defined by using logical operators and quantifiers. We use R_F to denote the relation defined by a formula F of relational calculus and call R_F the relation associated with formula F. If we write $F(x_1, ..., x_n)$ to mean that formula F has free variables $x_1, ..., x_n$ and if $(a_1, ..., a_n)$ is a tuple, then $F(a_1, ..., a_n)$ is true if and only if the tuple $(a_1, ..., a_n)$ belongs to R_F.

Facts are stored as predicates (atomic formulas) and their finite associated relations in a relational database. Assume there is a finite number of predicates and each one is associated with a finite relation R_p. We can recursively define the formulas of relational calculus[6] and their associated relations as follows:

1. Every literal $p(x_1, ..., x_n)$, where p is a predicate symbol and $x_1, ..., x_n$ are variables or constants, is a formula. All occurrences of variables among $x_1, ..., x_n$ are free. These formulas are called atomic formulas. For atomic formula $p(x_1, x_2, ..., x_n)$, where p is a predicate symbol and $x_1, x_2, ..., x_n$ are different free variables, R_p is its associated relation. For atomic formula $p(x_1, x_2, ..., x_n)$, where p is a predicate symbol and $x_1, x_2, ..., x_n$ are free variables or constants, if there are m different variables $z_1, z_2, ..., z_m$ among $x_1, x_2, ..., x_n$, its associated relation consists of all tuples $(z_1, z_2, ..., z_m)$ such that $(x_1, x_2, ..., x_n)$ belongs to R_p.

2. Every arithmetic comparison $x \# y$ is a formula, where x and y are variables or constants, and $\#$ is one of the six arithmetic comparison operators, $=$, $>$, and so on. It is required that such formulas are attached by logical $\wedge$ to another formula that defines a finite relation. These formulas are also called atomic formulas. If g is a formula associated with a finite relation R_g, then the relation associated with $g \wedge (x \# y)$ or $(x \# y) \wedge g$ consists of all tuples in R_g satisfying $x \# y$.

3. If formula $F(x_1, x_2, ..., x_n)$ is associated with relation R_F, then $\neg F(x_1, x_2, ..., x_n)$ is also a formula associated with relation $R_{\neg F}$, where $R_{\neg F}$ consists of all n-tuples $(x_1, x_2, ..., x_n)$ which do not belong to R_F.

4. If formula $F(x_1, ..., x_i, y_1, .., y_j)$, where $x_1, ..., x_i, y_1, .., y_j$ are $i + j$ free variables, is associated with relation $R_F(x_1, ..., x_i, y_1, .., y_j)$, formula $G(y_1, ..., y_j, z_1, .., z_k)$ where $y_1, ..., y_j, z_1, .., z_k$ are $j + k$ free variables, is associated with relation $R_G(y_1, ..., y_j, z_1, .., z_k)$, F and G have j common free variables $(y_1, .., y_j)$, then $F(x_1, ..., x_i, y_1, .., y_j) \wedge G(y_1, ..., y_j, z_1, .., z_k)$, where $x_1, ..., x_i, y_1, .., y_j, z_1, .., z_k$ are $i + j + k$ free variables, is also a formula associated with relation $R_{(F \wedge G)}$, where $R_{(F \wedge G)}$ consists of all $(i + j + k)$-tuples $(x_1, ..., x_i, y_1, .., y_j, z_1, .., z_k)$ such that $(x_1, ..., x_i, y_1, .., y_j)$ belongs to R_F and $(y_1, ..., y_j, z_1, .., z_k)$ belongs to R_G.

5. If formula $F(x_1, ..., x_i, y_1, .., y_j)$, where $x_1, ..., x_i, y_1, .., y_j$ are $i + j$ free variables, is associated with relation $R_F(x_1, ..., x_i, y_1, .., y_j)$, formula $G(y_1, ..., y_j, z_1, .., z_k)$, where $y_1, ..., y_j, z_1, .., z_k$ are $j + k$ free variables, is associated with relation $R_G(y_1, ..., y_j, z_1, .., z_k)$, F and G have j common free variables $(y_1, .., y_j)$, then $F(x_1, ..., x_i, y_1, .., y_j) \vee G(y_1, ..., y_j, z_1, .., z_k)$, where $x_1, ..., x_i, y_1, .., y_j, z_1, .., z_k$ are $i + j + k$ free variables, is also a formula associated with relation $R_{(F \vee G)}$, where $R_{(F \vee G)}$ consists of all $(i + j + k)$-tuples $(x_1, ..., x_i, y_1, .., y_j, z_1, .., z_k)$ such that either $(x_1, ..., x_i, y_1, .., y_j)$ belongs to R_F or $(y_1, ..., y_j, z_1, .., z_k)$ belongs to R_G.

6. If formula $F(x_1, ..., x_i, ..., x_j)$, where $x_1, ..., x_i, ..., x_j$ are j free variables, is associated with relation $R_F(x_1, ..., x_i, ..., x_j)$, then $\forall x_i F(x_1, ..., x_i, ..., x_j)$, where $(x_1, ..., x_{i-1}, x_{i+1}, ..., x_j)$ are $i - 1$ free variables, is also a formula associated with relation $R_{\forall x_i F}$, where $R_{\forall x_i F}$ consists of all $(i-1)$-tuples $(x_1, ..., x_{i-1}, x_{i+1}, ..., x_j)$ such that for all x_i, $(x_1, ..., x_i, ..., x_j)$ belongs to R_F.

7. If formula $F(x_1, ..., x_i, ..., x_j)$, where $x_1, ..., x_i, ..., x_j$ are j free variables, is associated with relation $R_F(x_1, ..., x_i, ..., x_j)$, then $\exists x_i F(x_1, ..., x_i, ..., x_j)$ is

also a formula associated with relation $R_{\exists x_i F}$, where $R_{\exists x_i F}$ consists of all $(i-1)$-tuples $(x_1, ..., x_{i-1}, x_{i+1}, ..., x_j)$ such that there is at least one x_i such that $(x_1, ..., x_i, ..., x_j)$ belongs to R_F.

If $F(x_1, ..., x_n)$ is a formula with free variables $x_1, ..., x_n$, and $a_1, ..., a_n$ are constants, then the value of $F(a_1, ..., a_n)$ is defined as either *true* or *false*, according to whether the tuple $(a_1, ..., a_n)$ is in R_F or not.

3 Construction of +_relations for Formulas of Relational Calculus

To obtain answers to a query from a relational database, we need to find the set of tuples which make the query formula true, i.e., to find the relation associated with the query formula. By our new approach, for any query, we need to find a +_relation which represents the relation associated with the query formula.

It can be proven that every formula of relational calculus F has a M+_relation that represents its associated relation R_F. M, the maximum of the domain sizes (the number of members in a domain) in the database system can be omitted.

Theorem 4. *There is an effective way to construct an M+_relation R_F^+ that represents the relation R_F associated with any formula F of relational calculus.*

Proof. The prove is by mathematical induction. Proof of the basis is trivial because every predicate p is associated with a finite relation R_p, and for every finite relation there is a straightforward +_relation to represent it.

Two cases have to be considered for induction: first, assuming that the new formula F is defined by $\wedge$ operation from an atomic formula $x\#y$ with another formula G and formula G is associated with a finite relation, a +_relation R_F^+ that represents the relation associated with F should be constructed. This is true because R_G is finite, the +_relation R_G^+ is already constructed and is a set of ordinary tuples, i.e., there is no + symbol at all. The +_relation R_F^+ can be constructed by selecting from R_G^+ those tuples that satisfy $x\#y$.

Second, assume that the new formula F is defined by $\neg$, $\wedge$, $\vee$, $\exists$, and $\forall$ operations from formulas whose associated +_expressions have been already constructed, and show how to construct a new +_relation R_F^+ that represents the relation associated with F. Since $\forall_x F$ is logically equivalent to $\neg(\exists_x(\neg F))$ and $F \vee G$ is equivalent to $\neg(\neg F \wedge \neg G)$, the induction can be reduced to three parts: $\neg$, $\wedge$ and $\exists$. Each part includes a Rule and a Lemma.

3.1 Proof on $\neg$ Operation.

Assume that $F(x_1, ..., x_n)$ is a formula associated with relation $R_F(x_1, ..., x_n)$, and there is a +_relation R_F^+ representing R_F. As mentioned before, R_F^+ consists of some +_tuples. Each tuple has n elements. If the maximum integer that appears is M, each element is either an integer between 1 and M, or a + symbol. The complete set of these kinds of tuples is of size $(M+1)^n$. Let t be an arbitrary tuple $t = (x_1, ..., x_n)$ and t^+ be its $+$_projection_. Rule 1 below is used to construct another $+$_relation_ $R^+_{\neg F}$:

Rule 1 *Choose and add t^+ to $R^+_{\neg F}$ if and only if t^+ is in the complete set but not in R^+_F.*

Lemma 5. $R^+_{\neg F}$ *represents the relation associated with* $\neg F$.

Proof. Let t be an arbitrary n-tuple, $t = (x_1, ..., x_n)$, and t^+ be its $+$_projection_. t belongs to the relation associated with $\neg F$ if and only if it does not belong to the relation associated with F, by the definition of $R_{\neg F}$. The later is true if and only if t^+ is not in R^+_F, since R^+_F represents R_F, the relation associated with F. By Rule 1, t^+ is not in R^+_F if and only if t^+ is in $R^+_{\neg F}$. Combining the above facts, t belongs to the relation associated with $\neg F$ if and only if t^+ is in $R^+_{\neg F}$. In other words, $R^+_{\neg F}$ is the $+$_relations_ representing the relation associated with $\neg F$. □

3.2 Proof on $\wedge$ Operation

Assume $F(x_1, .., x_i, y_1, .., y_j)$ is a formula associated with relation $R_F(x_1, .., x_i, y_1, .., y_j)$ represented by a $+$_relation_ R^+_F, and $G(y_1, ..., y_j, z_1, .., z_k)$ is a formula associated with relation $R_G(y_1, ..., y_j, z_1, .., z_k)$ represented by a $+$_relation_ R^+_G. F and G have j common free variables $(y_1, .., y_j)$. Then $F(x_1, .., x_i, y_1, .., y_j) \wedge G(y_1, ..., y_j, z_1, .., z_k)$ is a formula with $i + j + k$ variables associated with relation $R_{F \wedge G}(x_1, .., x_i, y_1, .., y_j, z_1, .., z_k)$, where $R_{F \wedge G}$ consists of all $(i+j+k)$-tuples $(x_1, .., x_i, y_1, .., y_j, z_1, .., z_k)$ such that $(x_1, .., x_i, y_1, .., y_j)$ belongs to R_F and $(y_1, ..., y_j, z_1, .., z_k)$ belongs to R_G. Let t be an arbitrary tuple $t = (x_1, .., x_i, y_1, .., y_j, z_1, .., z_k)$ and $f = (x_1, .., x_i, y_1, .., y_j)$, $g = (y_1, ..., y_j, z_1, .., z_k)$. Let t^+, f^+ and g^+ be their $+$_projections_ respectively. A $+$_relation_ $R^+_{F \wedge G}$ is constructed from R^+_F and R^+_G according to the following Rule 2.

Rule 2 *Choose and add a $+$_relation_ t^+ into $R^+_{F \wedge G}$ if and only if f^+ is in R^+_F and g^+ is in R^+_G.*

Lemma 6. $R^+_{F \wedge G}$ *is the $+$_relation_ representing the relation associated with* $F \wedge G$

Proof. t belongs to $R_{F \wedge G}$ if and only if f belongs to R_F and g belongs to R_G, by the definition of $R_{F \wedge G}$. The later is true if and only if f^+ belongs to R^+_F and g^+ belongs to R^+_G, because R^+_F and R^+_G represent R_F and R_G respectively. According to Rule 2, f^+ belongs to R^+_F and g^+ belongs to R^+_G if and only if t^+ belongs to $R^+_{F \wedge G}$. Combining these ideas, t belongs to $R_{F \wedge G}$ if and only if t^+ belongs to $R^+_{F \wedge G}$. In other words, $R^+_{F \wedge G}$ is the $+$_relation_ that represents the relation associated with $F \wedge G$. □

3.3 Proof on ∃ Operation

Assume $F(x_1, ..., x_i, ..., x_j)$ is a formula with free variables $x_1, ..., x_i, ..., x_j$ associated with relation $R_F(x_1, ..., x_i, ..., x_j)$ represented by a $+\_relation$ R_F^+. Then $\exists x_i F(x_1, ..., x_i, ..., x_j)$ is a formula with free variables $x_1, ..., x_{i-1}, x_{i+1}, ..., x_j$. Let $R_{\exists x_i F}$ be the relation associated with $\exists x_i F$. Then $R_{\exists x_i F}$ consists of all $(i-1)$-tuples $(x_1, ..., x_{i-1}, x_{i+1}, ..., x_j)$ such that there is some x_i such that $(x_1, ..., x_i, ..., x_j)$ belongs to R_F. Let t be an arbitrary j-tuple $t = (x_1, ..., x_i, ..., x_j)$, and u be a $(j-1)$-tuple obtained from t with the i-th component removed. Let t^+ and u^+ be the $+\_projections$ of t and u respectively. A $+\_relation$ $R_{\exists x_i F}^+$ is constructed according to the following Rule 3.

Rule 3 *Choose and add u^+ into $R_{\exists x_i F}^+$ if and only if there exists some x_i such that t^+ belongs to R_F^+.*

Lemma 7. $R_{\exists x_i F}^+$ *thus constructed represents the relation associated with $\exists x_i F$.*

Proof. u belongs to $R_{\exists x_i F}$ if and only if there exists some x_i such that $t = (x_1, ..., x_i, ..., x_j)$ belongs to R_F, by the definition of $R_{\exists x_i F}$. The later is true if and only if there exists some x_i such that t^+ belongs to R_F^+, since R_F^+ represents the relation associated with F. According to Rule 3, thisis equivalent to the fact that u^+ is in $R_{\exists x_i F}^+$. Putting these facts together, u belongs to $R_{\exists x_i F}$ if and only if u^+ is in $R_{\exists x_i F}^+$. In other words, $R_{\exists x_i F}^+$ is the $+\_relation$ representing the relation associated with $\exists x_i F$. □

In the meantime, it has also been proven that the set of $+\_relation$ associated to formulas of relational calculus has a closure property. Every formula of relational calculus has its associated $+\_relation$.

4 The Construction Algorithm and an Example

According to the proof of Theorem 1, it is not hard to design a construction algorithm that transfers an arbitrary formula f in DRC to R_f^+, given R_p for every predicate p in the database system.

Construction Algorithm: First, rewrite the formula so that $\forall$ and $\vee$ operators are all eliminated. Then repeatedly perform the following operations: Find one of the innermost subformulas g whose $+\_relations$ have not been calculated, among those which are not in the form $x\#y$, where x, y are two variables or constants and $\#$ is an arithmetic comparator. If there is no such a subformula, then R_f^+ has already been calculated, the computation stops. Otherwise, calculate R_g^+ according to the following different situations:

1. g is an atomic formula, $g = p(x_1, ..., x_n)$, where p is a predicate, $x_1, ..., x_n$ are variables or constants. Assume among $x_1, ..., x_n$, there are altogether m different variables $z_1, ..., z_m$. Then create R_g^+ by selecting all tuples $(z_1, ..., z_m)$ such that $(x_1, ..., x_n)$ belongs to R_p, the relation associated with p.

2. g is of the form $h \wedge (x\#y)$ or $(x\#y) \wedge h$, where x and y are some variables or constants, $\#$ is an arithematic comparator, h is a subformula whose $+\_relation$

R_h^+ has been already calculated. Then create R_g^+ by selecting from R_h^+ all the tuples which satisfy $x\#y$.

3. g is of the form $\neg h$, $h_1 \wedge h_2$ or $\exists_x h$, where h,h_1,h_2 are subformulas whose $+\_relations$ have been already constructed. Then create the $+\_relation$ R_g^+ from R_h^+, $R_{h_1}^+, R_{h_2}^+$, by using Rule 1, Rule 2 or Rule 3, respectively.

We consider an example where a predicate PDR is involved. A triple (p, d, r) is in the relation associated with PDR if and only if a person p had a test on disease d and got a result r. We may also assume that each person had at most one test on each disease.

We may define that the person who takes all disease tests and has a negative result for each test, is *healthy*. We also consider that the person who has a positive test result on any of the diseases is *sick*. A person is "undecided" if and only if he missed some tests, but for all tests taken, had negative results.

Let us try write the formula for $Undecided(p)$. In order to do so, we need to define some subformulas as follows:

$$Disease(d) = \exists_p \exists_r PDR(p, d, r) \tag{1}$$

$$OK(p, d) = PDR(p, d, negative) \tag{2}$$

$$\neg Healthy(p) = \exists_d (Disease(d) \wedge \neg OK(p, d)) \tag{3}$$

$$sick(p) = \exists_d \exists_r (PDR(p, d, r) \wedge r = positive) \tag{4}$$

$$Undecided(p) = \neg Healthy(p) \wedge \neg sick(p) \tag{5}$$

The above are, respectively: all diseases that appear in PDR, person p has a negative result on disease d, person p is not OK on some disease d, person p has a positive result on any disease d.

The relation $PDR(p, d, r)$ stored in the database is given as follows: (we use 1,2,3... to express people and diseases, and use 1 to express positive results, 2 to express negative results, e.g. the triple $(2,1,1)$ means person 2 had a positive result on the test for disease 1.).

$$R_{PDR(p,d,r)} = \{ (2,1,1), (2,3,2), (2,4,2), (3,1,2), (3,2,2), (3,3,2),$$
$$(3,4,2), (5,2,2), (5,3,2), (7,1,2), (7,2,1), (7,4,1)\} \tag{6}$$

We should note that all given relations such as the one shown above are actually a special kind of $+\_relation$ (without $+$ symbol).

To reduce the complexity, by $+$ we mean all elements that are greater than the maximum integer which appears as a value of a special attribute.

In order to determine the $+\_relation$ for $Undecided(p)$, we will calculate $+\_relations$ for all its subformulas, starting from the innermost ones according to the construction algorithm.

1. By using Rule 3 twice on (6), we select all values of the second attributes in (6) and obtain:

$$R_{Disease(d)}^+ = \{(1), (2), (3), (4)\} \tag{7}$$

2. According to situation 1 in the algorithm, since $OK(p, d) = PDR(p, d, 2)$, we select all pairs (p, d) such that $(p, d, 2)$ appears in (6), and have:

$$R^+_{OK(p,d)} = \{(2,3), (2,4), (3,1), (3,2), (3,3), (3,4), (5,2), (5,3), (7,1)\} \quad (8)$$

3. By Rule 1, $R^+_{\neg OK(p,d)}$ can be found as the complement of (8):

$$R^+_{\neg OK(p,d)} = \{\ (1,1), (1,2), (1,3), (1,4), (1,+), (2,1), (2,2), (2,+),$$
$$(3,+), (4,1), (4,2), (4,3), (4,4), (4,+), (5,1), (5,4),$$
$$(5,+), (6,1), (6,2), (6,3), (6,4), (6,+), (7,2), (7,3),$$
$$(7,4), (7,+), (+,1), (+,2), (+,3), (+,4), (+,+)\} \quad (9)$$

By Rule 2, removing from (9) all pairs whose second elements do not belong to the disease list (7), we have

$$R^+_{Disease(d) \wedge \neg OK(p,d)} = \{\ (1,1), (1,2), (1,3), (1,4), (2,1), (2,2),$$
$$(4,1), (4,2), (4,3), (4,4), (5,1), (5,4),$$
$$(6,1), (6,2), (6,3), (6,4), (7,2), (7,3),$$
$$(7,4), (+,1), (+,2), (+,3), (+,4)\} \quad (10)$$

Since $\neg Healthy(p) = \exists_d(Disease(d) \wedge \neg OK(p, d))$, by Rule 3, $R^+_{\neg Healthy(p)}$ can be found as the set of values of the first attribute in (10).

$$R^+_{\neg Healthy(p)} = \{(1), (2), (4), (5), (6), (7), (+)\} \quad (11)$$

4. According to situation 2 in the algorithm, we select from (6) those tuples whose third element is 1 and have:

$$R^+_{PDR(p,d,r) \wedge r=positive} = \{(2,1,1), (7,2,1), (7,4,1)\} \quad (12)$$

Then by using Rule 3 twice, from (12) we select all values of the first attribute in (12) and have:

$$R^+_{Sick(p)} = R^+_{\exists_d \exists_r (PDR(p,d,r) \wedge r=positive)} = \{(2), (7)\} \quad (13)$$

5. By using Rule 1, $R^+_{\neg Sick(p)}$ is obtained as the complement of (13).

$$R^+_{\neg Sick(p)} = \{(1), (3), (4), (5), (6), (+)\} \quad (14)$$

The final result, $R^+_{Undecided(p)}$, is obtained by taking the intersection of (11) and (13) according to Rule 2.

$$R^+_{Undecided(p)} = \{(1), (4), (5), (6), (+)\} \quad (15)$$

That is, persons 1, 4, 5, 6 and all others who do not appear in our database. From our answer, we can tell whether a given person is in it, which satisfies what we required in the introduction - it is explicit.

In our example, many intermediate results and the final result are infinite relations. In the related literature, all kinds of methods are proposed to prevent such "unsafe" formulas, since there is no effective way to compute them in conventional database systems as of now.

By using our approach in relational database systems, we do not have to worry about "safety". No matter what the formula is, we simply compute all intermediate results and the final result effectively. The method is always safe!

5 Conclusion

The relational database model is very popular because it supports powerful, yet simple and declarative data manipulation languages. Relational calculus forms the basis for most commercial data manipulation languages in relational systems. However, formulas of relational calculus are not always safe, since negation can create infinite relations from finite ones. To deal with this important problem, current systems take a restrictive approach. They avoid any infinite relation by a wide variety of means. Various restrictions are introduced to guarantee safety. When the final result is an infinite relation, there is no way to obtain it.

This paper presents a totally different approach to solve the safety issue. The original idea is to symbolically represent and compute infinite relations, while working under finite real-world constraints. The approach outlined in this paper achieves these goals. The popular data manipulation language in relational systems can be freely used without restrictions.

References

1. C. Beeri, S. Naqvi, R. Ramakrishnan, O. shmueli, and S. Tsur: Sets and negation in a logic database language(LDL1). Proc. Sixth ACM Symp. on Principles of Database Systems, pp. 21-37, 1987
2. A. V. Gelder and W. R. Topor: Safety and correct translation of relational calculus formulas. Proc. sixth ACM Symp. on Principles of Database Systems, pp.313-327, 1987
3. G. M. Kuper G. M.: Logic programming with sets. Proc. sixth ACM Symp. on Principles of Database Systems, pp.11-20, 1987
4. R. Ramakrishnan, F. Bancilhon, and A. Sillberschatz: Safety of recursive Horn clauses with infinite relations. Proc. sixth ACM Symp. on Principles of Database Systems, pp. 328-339, 1987
5. O. Shmuel: Decidability and Expressiveness aspects of logic queries. Proc. sixth ACM Symp. on Principles of Database Systems, pp. 237-249
6. J. D. Ullmman: Database and Knowledge-base Systems vol. 1, Classical Database Systems, Computer-Science Press, 1988
7. C. Zaniolo: Safety and Compilation of Nonrecursive Horn ClausesProc. First Intl. Conf. on Expert Database Systems, BenjaminCummings, Menlo Park, CA, 1986

Set-Term Unification in a Logic Database Language

Seung Jin Lim and Yiu-Kai Ng

Computer Science Department, Brigham Young University, Provo, UT 84602, U.S.A.

Abstract. Deterministic, parallel set-term unification algorithms for higher-order logic-based database languages, of which set terms have the commutative and idempotent properties, are lacking. As a result, an efficient, deterministic inference mechanism that can be used to determine answers to queries of these database languages is non-existent. To overcome these shortcomings, we propose a set-term unification approach for LDL/NR, a higher-order logic database language. Our approach not only computes all *generalized ground unifiers* of two given set terms in LDL/NR without *duplicates*, but also takes advantage of existing multiple processors for (potentially) computing all these unifiers in *parallel*.

1 Introduction

In a logic database system, unification is a process of assigning terms to variables in a database query Q for retrieving data from the database in order to compute the answers to Q. Unification algorithms for functions with the associativity, commutativity, and idempotency properties [Sie89] in first-order logic database languages have been proposed [LC88]; however, these algorithms cannot be adopted to handle set-term unification in higher-order logic database languages since functions and set terms have different semantics (See [STZ92], P. 93). [DOPR91, Sto93] proposed set-term unification algorithms that compute unifiers for a given pair of set terms with the commutative and idempotent properties; however, these algorithms have several problems. One of these problems is that *duplicated* unifiers, which are generated for a given pair of set terms, must be 'filtered' in order to obtain the desired result. However, redundant unifiers are undesirable. Another problem is that these algorithms are non-deterministic and hence are more difficult to implement and are less efficient in execution when compared with the deterministic set-term unification algorithm that we propose.

In this paper, we examine the set-term unification problem of LDL/NR, a (higher-order) logic database language for nested relations. LDL/NR is of interest because it captures the constraints of nested relations precisely. Furthermore, since for each complex-object type there is a nested-relation type with the same "information capacity", LDL/NR can handle complex-data type queries. In Section 3, we present a set-term unification approach for LDL/NR which generates in *parallel unique* generalized ground unifiers (defined in Section 2) for any two given set terms with the commutative and idempotent properties in LDL/NR. In Section 4, we give the concluding remarks.

2 LDL/NR

LDL/NR consists of constants and variables, which are of *atomic type*. There are two other types: *tuple type* and *set type*, defined in LDL/NR. Set type and tuple type can be used alternatively to form complex data types.

Definition 1. A *type* is recursively defined as follows: (i) an *atomic type* is a type, (ii) a *set type* $s\{r\}$ is a type, where r is either an atomic type or a tuple type, and (iii) a *tuple type* $p(s_1, \ldots, s_n)$ is a type, where each s_i, $1 \leq i \leq n$, is either an atomic type or a set type. $\Box$

Definition 2. A *term* is recursively defined as follows: (i) a constant is a term, (ii) a variable is a term, (iii) for a set-type $s\{r\}$, $s\{t_1, \ldots, t_m\}$ is a term called *set term*, where each t_i is of type r, and (iv) for a tuple-type $p(s_1, \ldots, s_n)$, $p(t_1, \ldots, t_n)$ is a term called *tuple term*, where t_i is of type s_i. $\Box$

Definition 3. A set term S in LDL/NR satisfies the *commutative* property if given any two elements e_1 and e_2, *concat* $(e_1, concat\ (e_2, S)) = concat\ (e_2, concat\ (e_1, S))$. A set term S in LDL/NR satisfies the *idempotent* property if given any element e, $concat(e, concat(e, S)) = concat(e, S)$. $\Box$

Definition 4. Let S_1 and S_2 be two set terms in LDL/NR. A substitution θ, which is a finite set of bindings $\{V_1/t_1, \ldots, V_m/t_m\}$, where V_i is a variable and t_j is a term, is a *unifier* for S_1 and S_2 if $S_1 \circ \theta = S_2 \circ \theta$. If either S_1 or S_2 is a set of ground terms, then θ is a *matcher* for S_1 and S_2. Further, θ is a *generalized ground unifier (ggu)* of S_1 and S_2 if composing θ with another substitution yields a *ground unifier*, i.e., unifier whose terms include no variable, of S_1 and S_2. $\Box$

3 An Approach for Solving the Set-Term Unification Problem of LDL/NR

In this section, we present an approach for solving the set-term unification problem of LDL/NR, based on which our set-term unification algorithm in [Lim94] is constructed. The proposed solution for set-term unification is constructed by exploiting general combinatorics and permutation theories. It is assumed that any two given set terms S_1 and S_2 in LDL/NR, which form an input of the proposed algorithm, satisfy the *commutative* and *idempotent* properties and the *unification condition*.

Definition 5. (*Unification Condition*) Given any two set terms S_1 and S_2 (in LDL/NR), S_1 and S_2 are of the following form:

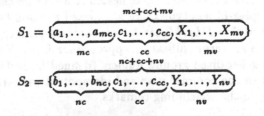

where each a_i $(1 \leq i \leq mc)$, b_j $(1 \leq j \leq nc)$ and c_k $(1 \leq k \leq cc)$ denotes a constant, and each X_m $(1 \leq m \leq mv)$ and Y_n $(1 \leq n \leq nv)$ denotes a variable. $a_1, \ldots, a_{mc}$ are called m-*compliants*, $b_1, \ldots, b_{nc}$ are called n-*compliants*, and $c_1, \ldots, c_{cc}$ are common constants. It is further required that $|S_1| \geq |S_2| > 0$, $mv+nv > 0$, $mv \geq nc$, $nv \geq mc$, and $mc, nc, cc, mv, nv \geq 0$. □

Our approach for solving the set-term unification problem involves computing bindings for all the compliants in S_1 and S_2, generating bindings for the remaining unbound variables in S_2 and S_1, and permuting the computed bindings. Before proceeding further, we give two definitions that are used throughout the process of computing all *ggu$_s$* of S_1 and S_2.

Definition 6. If the cardinality of a set, subset, multiset, or permutation Δ is l, then Δ is called an l-set, l-subset, l-multiset, or l-permutation, respectively. □

Definition 7. Let S_1 and S_2 be two set terms, let $\kappa = \{a_1, \ldots, a_{mc}\}$ be the set of m-compliants in S_1, and let $\theta = \{V_1/t_1, \ldots, V_n/t_n\}$ be a substitution. A subset (multiset) Δ of S_1 is an m-*idempotent* subset (multiset) or we say that Δ satisfies m-*idempotency* if $\kappa \subseteq \Delta$. If $\kappa \subseteq \{t_1, \ldots, t_n\}$, then θ satisfies m-*idempotency*. N-*idempotent* subset (multiset) and n-*idempotency* are defined accordingly for the set of n-compliants in S_2. □

Given two set terms S_1 and S_2, there exist a number of *ggu$_s$* of S_1 and S_2. The process of computing all *ggu$_s$* of S_1 and S_2 consists of the following four consecutive steps as shown in Figure 1, and the detailed description of each step is given in each of the following subsections. We take advantage of existing multiple processors to compute in *parallel* all the possible sets of bindings for m-compliants, subsets and multisets, and permutations in Steps 1, 3, and 4, respectively.

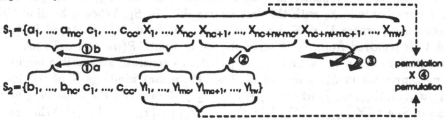

Fig. 1. Steps for computing all *ggu$_s$* of S_1 and S_2

Step 1. Assign all n-compliants $b_1, \ldots, b_{nc}$ in S_2 to the first nc variables in S_1 to yield the initial set of bindings in buf_1. Then, assign all m-compliants $a_1, \ldots, a_{mc}$ in S_1 to mc variables, denoted $Y_{i_1}, \ldots, Y_{i_{mc}}$ $(1 \leq i_j \leq nv, 1 \leq j \leq mc)$ as shown in Figure 1, that are chosen from the existing nv variables in S_2 to yield the set of bindings in buf_2. Indeed, there are $P(nv, mc)$ possible sets of bindings of m-compliants that are computed in parallel, where $P(nv, mc)$ denotes the mc-permutations of nv variables in S_2. Hence, there are $P(nv, mc)$ different versions of buf_2. The (initial) set of bindings in buf_1

and in buf_2 satisfy n- and m-idempotency, respectively. (From now on whenever we mention buf_1, buf_2, S_1, and S_2, we refer to a particular version of buf_1, buf_2, S_1, and S_2 unless stated otherwise.)

Step 2. Assign the remaining unbound variables $Y_{i_{mc+1}}, \ldots, Y_{i_{nv}}$ ($1 \leq i_j \leq nv$, $mc + 1 \leq j \leq nv$) in S_2 to $nv\text{-}mc$ remaining unbound variables $X_{nc+1}, \ldots,$ $X_{nc+nv-mc}$ in S_1 and append the computed bindings to buf_1.

Step 3. Assign an element in S_2 to one of the remaining unbound variables $X_{nc+nv-mc+1}, \ldots, X_{mv}$ in S_1 and generate all possible combinations of these bindings. This is accomplished by first computing different l-subsets of S_2, $1 \leq l \leq min(mv\text{-}nc\text{-}nv\text{+}mc, nc\text{+}cc\text{+}nv)$, and then generating all $(mv\text{-}nc\text{-}nv\text{+}mc)$-multisets $\{e_1, \ldots, e_{mv-nc-nv+mc}\}$ from the l-subsets followed by variable substitution of X_i ($nc\text{+}nv\text{-}mc\text{+}1 \leq i \leq mv$) for e_i ($1 \leq i \leq mv\text{-}nc\text{-}nv\text{+}mc$) in each multiset. Each set of these bindings is appended individually to buf_1 to yield a complete set of bindings for variables in S_1.

Step 4. Permute the set of bindings in buf_1 computed in Step 3 and merge each of these permutations with the set of bindings in buf_2 computed in Step 1 to generate a number of ggu_s of S_1 and S_2. (All the ggu_s of S_1 and S_2 are constructed by using different versions of buf_1 and buf_2 as computed in Steps 1 - 4.)

3.1 Generating Bindings for the Compliants

Before generating bindings for the compliants in S_1 and S_2, we first remove common constants $c_1, \ldots, c_{cc}$ from S_1 since they do not play any role in computing the desired ggu_s of S_1 and S_2. Hereafter, we generate bindings for the m- and n-compliants in S_1 and S_2, respectively. These computations are performed to ensure that each unifier of S_1 and S_2 to be constructed satisfies m- and n-idempotency. Since eventually the set of bindings for variables in S_1 is permuted to generate all possible ggu_s of S_1 and S_2, we simply bind the n-compliants in S_2 with the first nc variables in S_1. When a binding V/b is computed, where $V \in S_1$ and b is an n-compliant, V/b is stored in buf_1 and V is removed from S_1. Afterwards, we compute $P(nv, mc)$ different mc-permutations of nv variables in S_2 in parallel to generate all possible bindings of m-compliants in S_1. Each set of these $P(nv, mc)$ bindings is of the form $\{Y_{i_1}/a_1, \ldots, Y_{i_{mc}}/a_{mc}\}$, where $Y_{i_j} \in S_2$, $1 \leq j \leq mc$, and the remaining variables in S_2 are $\{Y_{i_{mc+1}}, \ldots, Y_{i_{nv}}\}$ ($= \{Y_1, \ldots, Y_{nv}\} - \{Y_{i_1}, \ldots, Y_{i_{mc}}\}$). When each binding U/a is computed for each m-compliant a, where $U \in S_2$, we replace U in S_2 by a and remove a from S_1. After computing the bindings for all compliants, $buf_1 = \{X_1/b_1, \ldots, X_{nc}/b_{nc}\}$, where X_i ($1 \leq i \leq nc$) is the ith variable in S_1 and b_i ($1 \leq i \leq nc$) is the ith n-compliant in S_2, $buf_2 = \{Y_{i_1}/a_1, \ldots, Y_{i_{mc}}/a_{mc}\}$, where Y_{i_j} ($1 \leq j \leq mc$) is a chosen variable from $\{Y_1, \ldots, Y_{nv}\}$ in S_2 and a_i ($1 \leq i \leq mc$) is the ith m-compliant in S_1, $S_1 = \{X_{nc+1}, \ldots, X_{mv}\}$, and $S_2 = \{b_1, \ldots, b_{nc}, c_1, \ldots, c_{cc}, a_1, \ldots, a_{mc}, Y_{i_{mc+1}}, \ldots, Y_{i_{nv}}\}$.

Example 1. Suppose that $S_1 = \{a, b, V, W, X, Y, Z\}$ and $S_2 = \{b, c, A, B\}$. The existence of the n-compliant (resp. m-compliant) c (resp. a) indicates that $V \in$

S_1 (resp. A or $B \in S_2$) must be bound with c (resp. a) in order to compute a ggu of S_1 and S_2. Hence, after all the compliants have been bound, $buf_1 = \{V/c\}$, $S_1 = \{W, X, Y, Z\}$, $buf_2 = \{A/a\}$ and $S_2 = \{b, c, a, B\}$ (resp. $buf_2 = \{B/a\}$ and $S_2 = \{b, c, a, A\}$). $\square$

3.2 Generating Bindings for the Remaining Variables in S_2

After all the n- and m-compliants are assigned to variables in S_1 and S_2, respectively, the remaining variables $Y_{i_{mc+1}}, \ldots, Y_{i_{nv}}$ in S_2 are assigned to $nv\text{-}mc$ variables (out of $mv\text{-}nc$ variables) in S_1. Again, since eventually the computed set of bindings for all variables in S_1 is permuted for generating all ggu, of S_1 and S_2, we simply choose the first $nv\text{-}mc$ remaining variables in S_1 and bind them with the remaining variables in S_2. (Since by Definition 5, $|S_1| \geq |S_2|$, $mc+mv \geq nc+nv$, and hence $mv-nc \geq nv-mc$. There are sufficient number of remaining variables in S_1 that can be bound with the remaining variables in S_2.) Hereafter, we append the set of bindings $\{X_{nc+1}/Y_{i_{mc+1}}, \ldots, X_{nc+nv-mc}/Y_{i_{nv}}\}$ to buf_1 and remove $X_{nc+1}, \ldots, X_{nc+nv-mc}$ from S_1, where X_i ($nc+1 \leq i \leq nc+nv-mc$) is the ith variable in the original S_1, and Y_{i_j} ($mc+1 \leq j \leq nv$) is one of the remaining unbound variables in S_2. Hence, after these bindings are generated, $buf_1 = \{X_1/b_1, \ldots, X_{nc}/b_{nc}, X_{nc+1}/Y_{i_{mc+1}}, \ldots, X_{nc+nv-mc}/Y_{i_{nv}}\}$, $S_1 = \{X_{nc+nv-mc+1}, \ldots, X_{mv}\}$, and buf_2 and S_2 remain unchanged.

Example 2. Consider the result of Example 1 where there is a variable B (resp. A) in S_2 after the bindings of all the compliants have been generated. B (resp. A) is assigned to the first remaining unbound variable, i.e., W, in S_1. Hereafter, $buf_1 = \{V/c, W/B\}$ (resp. $buf_1 = \{V/c, W/A\}$) and $S_1 = \{X, Y, Z\}$. $\square$

Proposition 8. *The total number of bindings generated for all the compliants in S_1 and S_2 and for assigning all the remaining variables in S_2 to variables in S_1 is $nc+nv$.*

Proof. Trivial.

3.3 Generating Multisets in Parallel

After all the compliants have been bound and the remaining variables in S_2 are assigned to variables in S_1, there are $mv\text{-}nc\text{-}nv+mc$ remaining variables $X_{nc+nv-mc+1}, \ldots, X_{mv}$ in S_1 to be bound with any element in S_2. Hence, in order to generate all possible bindings for the remaining variables in S_1, we first compute all subset groups S^l, $1 \leq l \leq min(mv\text{-}nc\text{-}nv+mc, nc+cc+nv)$, such that each $S^l \subseteq S_2$, and then generate all ($mv\text{-}nc\text{-}nv+mc$)-multisets from each S^l. Eventually, each element in an ($mv\text{-}nc\text{-}nv+mc$)-multiset M is assigned to one of the remaining, distinct variables in S_1, and the bindings are appended to buf_1 (generated in Section 3.2) to yield one of the completed sets of bindings for variables in S_1. Before proceeding further, we need the definition of n-ary *left-leaning tree* which is used to generate all subsets, multisets, and permutations in the following subsections.

Definition 9. An n-ary *left-leaning tree* T of height h satisfies the following conditions:

1. The root node of T has n children.
2. Node p at level i in T has $k - j + 1$ children, where $1 \leq i \leq h - 1$, $1 \leq j \leq n$, k is the number of children of p's parent, and j denotes that p is the jth child of its parent.

There are $\binom{n + i - 1}{i}$ nodes at level i of T, $0 \leq i \leq h$. $\square$

Generating Subset Groups in Parallel To generate all possible bindings $\{X_{nc+nv-mc+1}/e_1, \ldots, X_{mv}/e_{mv-nc-nv+mc}\}$ for the remaining variables $X_{nc+nv-mc+1}, \ldots, X_{mv}$ in S_1, where $e_i \in S_2$, $1 \leq i \leq mv\text{-}nc\text{-}nv+mc$, we first create each subset group $S^l = \{\Delta_1^l, \ldots, \Delta_{C(nc+cc+nv,l)}^l\}$, $1 \leq l \leq min(mv\text{-}nc\text{-}nv+mc, nc+cc+nv)$. To generate all subset groups, we choose each element from S_2 to form a subset in subset group S^1, and choose any two distinct elements from S_2 to form a subset in subset group S^2, and so on. (Hereafter, each subset in each subset group S^l is expanded to the cardinality of $mv\text{-}nc\text{-}nv+mc$ to yield an $(mv\text{-}nc\text{-}nv+mc)$-multiset.) Indeed, we can construct a *forest* F to generate each subset $\Delta^l = \{e_1, \ldots, e_l\}$ in each subset group S^l in parallel. This is accomplished by constructing a collection of $(|S_2| - l + 1)$-ary, i.e., $(nc + cc + nv - l + 1)$-ary, left-leaning trees of height l, $1 \leq l \leq min(mv\text{-}nc\text{-}nv+mc, nc+cc+nv)$, in F. Each tree T_l in F represents the process of generating S^l, and different nodes at level i $(0 \leq i \leq l - 1)$ in T_l denote different partial l-subsets $SB_i = \{e_1, \ldots, e_l\}^1$ in S^l. It is assumed that each of the components $e_1, \ldots, e_{i-1}$ in SB_i at level i has already been assigned an element in S_2. A leaf node in T_l denotes a fully constructed l-subset Δ^l of S_2, and the leaf nodes of T_l yield all the l-subsets in S^l. We construct F as follows:

Step 1. Construct the root node of each T_l in F, $1 \leq l \leq min(mv\text{-}nc\text{-}nv+mc, nc+cc+nv)$, in parallel.

Step 2. For each node N at level i $(0 \leq i \leq l - 1)$ in T_l, create k $(1 \leq k \leq nc+cc+nv\text{-}l+1)$ children of N in parallel. At a node N at level i $(1 \leq i \leq l)$, e_i in SB_i (which is the partial subset associated with N) is set to the $(k_1 + k_2)$-th element in S_2, where e_{i-1} is the k_1-th element in S_2, SB_i is the k_2-th child of its parent, and $k_1 = 0$ when $i = 1$.

Step 3. Terminate the construction of each T_l when all the leaf nodes of T_l have been constructed.

Our approach guarantees that the computed subsets in different subset groups are distinct and can be generated in parallel since they are generated according to the lexicographical order of indices of the elements in S_2. (The

---

[1] A partial l-subset SB is a subset of S_2 being constructed such that there exists at least one component of SB that has not been assigned an element of S_2. If every component of SB is assigned an element of S_2, then SB is called a *fully constructed* (or simply an) l-subset of S_2.

same approach is applied to generate multisets and permutations in the following subsections.) The first two columns of Table 1 show the subset groups S^ls, $1 \leq l \leq min(mv\text{-}nc\text{-}nv+mc,\ nc+cc+nv)$ and the number of possible subsets in each subset group S^l, respectively.

Subset group	Number of possible subsets in the subset group	Number of multisets generated from each subset in the subset group
S^1	$\binom{nc+cc+nv}{1}$	$\binom{mv-nc-nv+mc-1}{1-1}$
S^2	$\binom{nc+cc+nv}{2}$	$\binom{mv-nc-nv+mc-1}{2-1}$
...	...	...
$S^{MIN\dagger-1}$	$\binom{nc+cc+nv}{MIN-1}$	$\binom{mv-nc-nv+mc-1}{MIN-2}$
S^{MIN}	$\binom{nc+cc+nv}{MIN}$	$\binom{mv-nc-nv+mc-1}{MIN-1}$
Total	$\sum_{i=1}^{MIN}\binom{nc+cc+nv}{i}$	$\sum_{i=1}^{MIN}\binom{nc+cc+nv}{i}$ $\times \binom{mv-nc-nv+mc-1}{i-1}$

† MIN is $min(mv\text{-}nc\text{-}nv+mc,\ nc+cc+nv)$.

Table 1. Number of subsets and $(mv\text{-}nc\text{-}nv+mc)$-multisets in each subset group

Example 3. Consider the result of Example 2 where there are three remaining unbound variables in S_1 that can be bound with any element b, c, a, and B (assume $A/a \in buf_2$) in S_2. Hence, $S^1 = \{\{b\}, \{c\}, \{a\}, \{B\}\}$, $S^2 = \{\{b, c\}, \{b, a\}, \{b, B\}, \{c, a\}, \{c, B\}, \{a, B\}\}$, and $S^3 = \{\{b, c, a\}, \{b, c, B\}, \{b, a, B\}, \{c, a, B\}\}$. □

Proposition 10. *The number of subsets in all subset groups that can be generated from* $S_1 = \{X_{nc+nv-mc+1}, \ldots, X_{mv}\}$ *and* $S_2 = \{b_1, \ldots, b_{nc}, c_1, \ldots, c_{cc}, a_1, \ldots, a_{mc}, Y_{i_{mc+1}}, \ldots, Y_{i_{nv}}\}$ *is*

$$\tau_{num\text{-}subsets} = \sum_{i=1}^{min(mv-nc-nv+mc,nc+cc+nv)}\binom{nc+cc+nv}{i}.$$

Proof. Since each subset group S^i, $1 \leq i \leq min(mv\text{-}nc\text{-}nv+mc,\ nc+cc+nv)$, is computed by an unordered selection of i elements of $nc+cc+nv$ elements in S_2, the construction of each subset group S^i is an ordinary combination of i elements from $nc+cc+nv$ elements, and there are $\binom{nc+cc+nv}{i}$ subsets in S^i. Hence, the total number of distinct i-subsets of S_2 in all subset groups is the sum of the combinations, which is $\tau_{num\text{-}subsets}$ given above and is shown in column 2 of Table 1. □

Generating Multiset Groups in Parallel A subset Δ^l in subset group S^l, $1 \leq l \leq min(mv\text{-}nc\text{-}nv+mc,\ nc+cc+nv)$, is used to generate a number of $(mv\text{-}nc\text{-}nv+mc)$-multisets M^l, where for each $e \in M^l$, $e \in \Delta^l$. Since $mv\text{-}nc\text{-}nv+mc \geq l$, an element in Δ^l may appear more than once in M^l. Indeed, an element of Δ^l

appears at least once and at most $mv\text{-}nc\text{-}nv\text{+}mc$ times in M^l. The third column of Table 1 shows the number of possible $(mv\text{-}nc\text{-}nv\text{+}mc)$-multisets that can be generated from each subset in a subset group. Since subsets Δ's in all subset groups S's are distinct, and since we adopt the same strategy as used to compute subsets and subset groups to generate multisets, all the multisets in different multiset groups which are generated from Δ's are distinct. We construct each l-ary left-leaning tree T of height $mv\text{-}nc\text{-}nv\text{+}mc\text{-}l$ for computing each $(mv\text{-}nc\text{-}nv\text{+}mc\text{-}l)$-multiset M from a given subset Δ^l in parallel, and then concatenate M and Δ^l to yield one of the $(mv\text{-}nc\text{-}nv\text{+}mc)$-multisets M^l. We generate M by using a partial $(mv\text{-}nc\text{-}nv\text{+}mc\text{-}l)$-multiset PM with elements chosen from Δ^l. Different nodes N at level i $(1 \leq i \leq mv\text{-}nc\text{-}nv\text{+}mc\text{-}l\text{-}1)$ in T denote different partial $(mv\text{-}nc\text{-}nv\text{+}mc\text{-}l)$-multisets $PM_i = \{e_1, \ldots, e_{mv-nc-nv+mc-l}\}$. e_i in each PM_i at level i is assigned a different element in Δ^l. A leaf node in T denotes a fully constructed M^l. We construct T as follows:

Step 1. Construct the root node of T.

Step 2. For each node N at level i $(0 \leq i \leq mv\text{-}nc\text{-}nv\text{+}mc\text{-}l\text{-}1)$, create k $(1 \leq k \leq l)$ children of N in parallel. At a node N at level i $(1 \leq i \leq mv\text{-}nc\text{-}nv\text{+}mc\text{-}l)$, e_i in PM_i (which is the partial multiset associated with N) is set to the $(k_1 + k_2 - 1)$-th element in Δ^l, where e_{i-1} is the k_1-th element in Δ^l, PM_i is the k_2-th child of its parent, and $k_1 = 1$ when $i = 1$.

Step 3. Terminate the construction of T when all the leaf nodes of T have been created. Append each fully constructed $(mv\text{-}nc\text{-}nv\text{+}mc\text{-}l)$-multiset M, which is associated with a leaf node of T, to Δ^l to yield an M^l of Δ^l.

Example 4. Consider the result of Example 3. If Δ^2 is the subset $\{b, c\}$ of S_2 and we want to expand Δ^2 to yield a number of 3-multisets, then we need to replicate one of the 'b' and 'c' in Δ^2. We can choose 'b' and 'c' as the third element of the 3-multisets $\{b, c, b\}$ and $\{b, c, c\}$, respectively in parallel. On the other hand, if S_1 consists of four variables and we are supposed to construct 4-multisets from the subset $\Delta^2 = \{b, c\}$, the three 4-multisets, $\{b, c, b, b\}$, $\{b, c, b, c\}$, and $\{b, c, c, c\}$, can be generated from Δ^2 in parallel. $\square$

Proposition 11. *Given* $S_1 = \{X_{nc+nv-mc+1}, \ldots, X_{mv}\}$ *and* $S_2 = \{e_1, \ldots, e_{nc+cc+nv}\}$, *the number of* $(mv\text{-}nc\text{-}nv\text{+}mc)$-*multisets generated from* S_1 *and* S_2 *is*

$$\tau_{num\_multisets} = \sum_{i=1}^{min(mv-nc-nv+mc,nc+cc+nv)} \binom{nc + cc + nv}{i} \times \binom{mv - nc - nv + mc - 1}{i - 1}$$

Proof. The number of $(mv\text{-}nc\text{-}nv\text{+}mc)$-multisets that are generated from a subset $\Delta^l = \{e_1, \ldots, e_l\}$, $1 \leq l \leq min(mv\text{-}nc\text{-}nv\text{+}mc, nc\text{+}cc\text{+}nv)$, is the number of all positive integer solutions of the equation $q_1 + \ldots + q_l = mv\text{-}nc\text{-}nv\text{+}mc$, where $1 \leq q_i$, $1 \leq i \leq l$, and q_i denotes the number of $e_i \in \Delta^l$ that has been chosen. Thus, the number of $(mv\text{-}nc\text{-}nv\text{+}mc)$-multisets generated from Δ^l is $\binom{mv - nc - nv + mc - 1}{l - 1}$. Furthermore, since the number of subsets in S^i is $\binom{nc + cc + nv}{i}$, $1 \leq i \leq min(mv\text{-}nc\text{-}nv\text{+}mc, nc\text{+}cc\text{+}nv)$, as shown in Proposition 10, the total number of multisets generated from all the subset groups is $\tau_{num\_multisets}$ and is shown in the third column of Table 1. $\square$

109

3.4 Generating Permutations of a Set of Bindings in Parallel

When a multiset $M = \{e_1, \ldots, e_{mv-nc-nv+mc}\}$ is generated, each $e_i \in M$ ($1 \leq i \leq mv\text{-}nc\text{-}nv\text{+}mc$) is used to compute a binding X_i/e_i, where X_i is the ith variable of $\{X_{nc+nv-mc+1}, \ldots, X_{mv}\}$ in S_1. Hence, M and $\{X_{nc+nv-mc+1}, \ldots, X_{mv}\}$ yield the set of bindings $\{X_{nc+nv-mc+1}/e_1, \ldots, X_{mv}/e_{mv-nc-nv+mc}\}$ which is concatenated with buf_1. For example, in Example 4, $\Delta^2 = \{b, c\}$ yields two 3-multisets $\{b, c, b\}$ and $\{b, c, c\}$ which in turn yield the sets of bindings $\theta_1 = \{X/b, Y/c, Z/b\}$ and $\theta_2 = \{X/b, Y/c, Z/c\}$, respectively after variable substitution, and concatenating θ_1 and θ_2 with buf_1 yield $buf_1 = \{V/c, W/B, X/b, Y/c, Z/b\}$ and $buf_1 = \{V/c, W/B, X/b, Y/c, Z/c\}$, respectively, assuming that $buf_1 = \{V/c, W/B\}$.

Since a computed binding for any variable in S_1 or S_2 is stored in its corresponding buffer, i.e., buf_1 or buf_2, after each variable in S_1 (resp. S_2) has been bound with either a variable or a constant in S_2 (resp. S_1), buf_1 and buf_2 contain all the bindings for variables in S_1 and S_2, respectively. Hence, $buf_1 \cup buf_2$ yields a substitution θ, which is a ggu of S_1 and S_2. (The formal proof of the correctness of our approach is shown in [Lim94].) In fact, other ggu_s of S_1 and S_2 can be generated from θ. This is accomplished by first computing all the possible permutations of buf_1, and then concatenating each permutation of buf_1 with buf_2. A permutation of $buf_1 = \{V_1/t_1, V_2/t_2, \ldots, V_n/t_n\}$ is an ordered selection of t_i ($1 \leq i \leq n$) for V_j ($1 \leq j \leq n$).

The permutations of buf_1 are generated by using an **inventory vector** inv $= < v_1, v_2, \ldots, v_m >$ to avoid computing duplicated permutations, where m denotes the number of distinct terms in buf_1. (Given a binding V/t in buf_1, t is the term in V/t.) In inv, v_i ($1 \leq i \leq m$) denotes the *number* of the ith distinct term in buf_1 that can be used in a permutation. Thus, $\sum_{i=1}^{m} v_i = |buf_1|$. Each v_i, $1 \leq i \leq m$, is decremented by one when the corresponding ith term is chosen as a component of the permutation being constructed. In order to generate each permutation $P = \{V_1/e_1, \ldots, V_n/e_n\}$ from $buf_1 = \{V_1/t_1, \ldots, V_n/t_n\}$ in parallel, we construct an m-ary tree T of height $n = |buf_1|$. Each node N at level i ($1 \leq i \leq n-1$) in T denotes a different partial permutation $P_i = \{V_1/p_1, \ldots, V_n/p_n\}$. p_j ($1 \leq j \leq n$) in different P_i at level i is assigned a different term in buf_1. A leaf node in T denotes a fully constructed P. We construct T as follows:

Step 1. Construct the root node of T.

Step 2. For each node N at level i ($0 \leq i \leq n-1$), create m children of N in parallel, where m is the number of non-zero components in inv. p_j ($1 \leq j \leq n$) in each partial permutation P_i (which is associated with a node M at level i, $1 \leq i \leq n$) is set to the k_1-th distinct usable[2] term in buf_1, where P_i is the k_1-th child of its parent. Decrement v_{k_1} by 1 when the k_1-th distinct usable term in buf_1 is chosen.

Step 3. Terminate the construction of T when each permutation of buf_1, which is associated with a leaf node in T, has been generated.

---

[2] A term in buf_1 is the kth usable term if its corresponding vector component v_k in inv is non-zero.

One of the advantages of our permutation approach is that duplicated permutations caused by duplicated terms in buf_1 are eliminated. In addition, each permutation can be generated in parallel and each one yields a different set of bindings for the variables in buf_1. The concatenation of each permutation of buf_1 with buf_2 yields a ggu of S_1 and S_2. (Recall that there are different versions of buf_1 and buf_2. Each version of buf_1 and each version of buf_2 together yield a subset of all the possible ggu, of S_1 and S_2.)

Proposition 12. *The number of permutations that can be generated from a set of bindings buf_1 is $\tau_{num\_perm} = P(m; q_1, q_2, \ldots, q_t) = C(m, q_1)\, C(m - q_1, q_2) \cdots C(m - q_1 - \ldots - q_{t-1}, q_t)$, where $q_1 + \ldots + q_t = m = |buf_1|$, t is the number of distinct terms in buf_1, and q_i $(1 \le i \le t)$ denotes the number of the ith distinct term in buf_1 that has been chosen.*

Proof. τ_{num_perm} is the number of different ways to enumerate t distinct elements over m positions in a permutation. Since the number of m-permutations with constrained repetition[3] such that each element is used at least once is known to be $P(m; q_1, q_2, \ldots, q_t) = C(m, q_1)C(m - q_1, q_2)\cdots C(m - q_1 - \ldots - q_{t-1}, q_t)$ [MKB86], τ_{num_perm} is defined as above. □

4 Concluding Remarks

In this paper, we present an approach for set-term unification in LDL/NR which computes generalized ground unifiers for a given pair of set terms in LDL/NR in parallel. This approach can be used to determine answers to LDL/NR queries involving set terms with the properties of idempotency and commutativity.

References

[DOPR91] A. Dovier, E. Omodeo, E. Pontelli, and G. Rossi. {log}: A logic programming language with finite sets. In *Proceedings of 8th International Conference on Logic Programming*, pages 111–124. The MIT Press, June 1991.

[LC88] P. Lincoln and J. Christian. Adventures in Associative-Commutative Unification. In *Proceedings of 9th International Conference on Automated Deduction*, pages 359–367, 1988.

[Lim94] S. J. Lim. Set-Term Unification in a Logic Database language. Master's thesis, Brigham Young University, December 1994.

[MKB86] J. L. Mott, A. Kandel, and T. P. Baker. *Discrete Mathematics for Computer Scientists and Mathematicians*. Prentice-Hall, New Jersey, 1986.

[Sie89] J. H. Siekmann. Unification Theory. *Journal of Symbolic Computation*, 7:207–274, 1989.

[Sto93] F. Stolzenburg. An Algorithm for General Set Unificiation and its Complexity. In *Proceedings of the Workshop on Logic Programming with Sets*, June 1993.

[STZ92] O. Shmueli, S. Tsur, and C. Zaniolo. Compilation of Set Terms in the Logic Data Language (LDL). *Journal of Logic Programming*, 12(1):89–119, 1992.

---

[3] Constrained repetition means that $q_1 + \ldots + q_t = n$ and $q_i \ge 0$, $1 \le i \le t$.

Computations with Finite Closure Systems and Implications

Marcel Wild

THD, Fachbereich Mathematik

64289 Darmstadt, Germany

wild@mathematik.th-darmstadt.de

Abstract. *Closure systems* $\mathbf{C} \subseteq 2^M$ *on a finite set M arise in many areas of discrete mathematics. They are conveniently encoded by either the family $Irr(\mathbf{C})$ of meet irreducible closed sets, or by implicational bases Σ. We significantly improve six (partly little known) algorithms in order to settle the problems (a),(b),...(g) listed below. In particular, the algorithm LINCLOSURE, which is well known in relational database theory, is enhanced. It appears as a subroutine in three of our algorithms. Applications in database theory, matroid theory and algebra will be pointed out.*

1 Introduction

Recall that a *closure system* on a set M is a family $\mathbf{C} \subseteq 2^M$ which is closed under intersections. Without further mention, in this paper M will always be *finite*. The sets $X \in \mathbf{C}$ shall be called *closed* sets or *flats* (in analogy to matroid theory). Often not all flats $X \in \mathbf{C}$ are explicitely given, e.g. due to the sheer size of $\mathbf{C}$. But there are two important devices which uniquely determine $\mathbf{C}$. The first device is the subfamily $Irr(\mathbf{C}) \subseteq \mathbf{C}$ of all meet irreducible flats. Hereby $X \in \mathbf{C}$ is called *meet irreducible* if $X \neq \bigcap\{Y \in \mathbf{C} : X \subset Y\}$. Note that $Irr(\mathbf{C})$ is the *unique* minimal $\cap$-generating family of $\mathbf{C}$. For example $Irr(\mathbf{C})$ is the family of all hyperplanes in case of a matroidal closure system $\mathbf{C}$.

The second device is slightly more sophisticated. An *implication* on the set M is an ordered pair (A, B) of subsets, written as $A \to B$. Here A is the *premise* and B the *conclusion* of the implication. Let $\Sigma = \{A_1 \to B_1, \ldots, A_n \to B_n\}$ be a family of implications. Call a subset $X \subseteq M$ Σ-*closed* if $A_i \subseteq X$ entails $B_i \subseteq X$ for all $1 \leq i \leq n$. The family Σ is an *(implicational) base* of a closure system $\mathbf{C} \subseteq 2^M$ if the closed sets coincide with the Σ-closed sets. So called *direct* implicational bases shall also be important. For instance, a direct implicational base Σ of a matroid is obtained by taking all implications $(C - \{x\}) \to C$, where C is a circuit and $x \in C$. The following tasks arise in various areas of discrete mathematics:

(a) compute the *closure* of a set using a given implicational base;

(b) compute a *minimum base* of C from an arbitrary base;

(c) compute *all flats* of C from a generating family of C;

(d) compute *all flats* of C from a base of C;

(e) compute a *minimum (direct) base* of C from a generating family of C;

(f) compute $Irr(\mathbf{C})$ from a generating family of C;

(g) compute $Irr(\mathbf{C})$ from a base of C;

Altogether seven algorithms will be presented to solve problems (a) to (g). Algorithms 1 to 6 improve methods from [G], [MR], [S], and the Algorithm NEXTCLOSURE is taken over from [G]. Algorithm 1 settles (a), Algorithm 2 settles (b), NEXTCLOSURE manages both (c) and (d), Algorithm 3 and 4 settle the two subproblems of (e), Algorithm 5 and Algorithm 6 solve (f) respectively (g).

In Section 2 we shall discuss Algorithm 1 in detail. It improves upon the algorithm LINCLOSURE, which is frequently used in relational database theory. Algorithm 6 is treated in Section 3. It enhances a method of Mannila and Räihä to compute $Irr(\mathbf{C})$ from a base of C. In Section 4 an outline of Algorithms 2,3,4,5 and NEXTCLOSURE is given. Details can be found in [W4] which extends the present article. Throughout, applications in relational database theory [MR],[M], Combinatorics, Algebra and Formal Concept Analysis [Wi] will be pointed out.

2 Computing closures of sets using implicational bases

Let $\Sigma := \{A_1 \rightarrow B_1, \dots, A_n \rightarrow B_n\}$ be a family of implications on the set M. Denote by X^Σ the Σ-*closure* of $X \subseteq M$, i.e. the smallest Σ-closed set containing X. Define

(1)
$$X^\bullet := X \cup \bigcup \{B_i \mid 1 \le i \le n, \ A_i \subseteq X\}$$

It is readily verified that $X^\Sigma = X^{\bullet \cdots \bullet}$. ($M$ being finite, the iteration eventually becomes stationary). This prompts the next simple algorithm.

Algorithm 0 *folklore*
Input: A family $\Sigma = \{A_1 \rightarrow B_1, \dots, A_n \rightarrow B_n\}$
            of implications on M and a nonempty subset $X \subseteq M$.
Output: The Σ-closure X^Σ.

```
oldclosure := ∅:
newclosure := X:
while newclosure ≠ oldclosure do
    oldclosure := newclosure:
    for all i from 1 to n do
        if Aᵢ ⊆ newclosure then newclosure := newclosure ∪ Bᵢ:    fi:
    od:
od:
Xᴿ := newclosure:
```

The while-loop in Algorithm 0 is executed at most $min(|\Sigma|, |M|)$ times, which implies that the complexity of Algorithm 0 is $O(|\Sigma||M|min(|\Sigma|,|M|)) = O(|\Sigma||M|^2)$. Trivial examples [M,p.65] show that the while-loop may indeed be executed as often as $|M| - 1$ times, and each scanning of Σ may yield *only one* implication $A_i \to B_i$ with $A_i \subseteq$ *newclosure*. In order to improve that shortcoming, the following devices appear in the algorithm LINCLOSURE [M,p.65],[MR,p.150] which is frequently used in relational database theory. First, for each $x \in M$ a list *contain*$[x]$ of all indices i with $x \in A_i$ is set up. Second, for each index i a dynamic variable *count*(i) counts how many elements of A_i are outside the current *newclosure*. The kernel of LINCLOSURE works like this. A (yet untreated) element x of *newclosure* is picked. Then for all i in *contain*$[x]$ the variable *count*(i) is decreased by 1. If *count*$(i) = 0$ then B_i is added to *newclosure*. The algorithm LINCLOSURE has complexity $O(s(\Sigma))$, where $s(\Sigma) := |A_1| + \cdots + |A_n| + |B_1| + \cdots + |B_n|$ is the <u>size</u> of Σ (see [M,p.68], [MR,p.152]). It improves upon Algorithm 0 whenever the while-loop of Algorithm 0 is transversed many times (as in the example [M,p.65]). But note that the lists *contain*$[x] \subseteq \{1,\ldots,n\}$ can be long, and each $i \in$ *contain*$[x]$ gives rise to one arithmetical and one logical operation. Moreover, processing *contain*$[x]$ might not even yield a new conclusion B_i to be added to *newclosure*. In fact these manipulations render LINCLOSURE slower than Algorithm 0 whenever its while-loop is only passed a few times! Here the important thing is rather to determine *at once* all i with $A_i \subseteq$ *newclosure*. Algorithm 1 below achieves this goal by dropping the variables *count*(i) and by *combining* the lists *contain*$[x]$. Namely, it is easy to see that $\Sigma - (\bigcup_{x \in T}$ *contain*$[x])$ $(T := M -$ *newclosure*$)$ comprises precisely the indices i with $A_i \subseteq$ *newclosure*.

Algorithm 1 *improving LINCLOSURE*
Input: A family $\Sigma := \{A_1 \to B_1, \ldots, A_n \to B_n\}$ of implications on M
            and a nonempty subset $X \subseteq M$.
Output: The Σ-closure X^Σ.

```
for all x ∈ M do
    contain[x] := ∅:
    for i from 1 to n do
        if x ∈ Aᵢ then contain[x] := contain[x] ∪ {i}:   fi:
```

```
        od:
od:
oldimpl := ∅
oldclosure := ∅:
newclosure := X:
while newclosure ≠ oldclosure do
        oldclosure := newclosure:
        T := M − newclosure:
        applyingimpl := Σ − (⋃ₓ∈T contain[x]):
    newimpl := applyingimpl − oldimpl:
    oldimpl := applyingimpl:
    for all i ∈ newimpl do
        newclosure := newclosure ∪ {Bᵢ} :
        od:
od:
X^Σ := newclosure:
```

What is the complexity of Algorithm 1? Initializing the $|M|$ lists $contain[x]$ costs $O(|\Sigma||M|^2)$. The "easy" task of computing one set $applyingimpl := \Sigma - (\bigcup_{x \in T} contain[x])$ has legally complexity $O(|\Sigma||M|)$. Since the while-loop is executed at most $|M|$ times, it follows that Algorithm 1 has complexity $O(|\Sigma||M|^2)$, which is actually the same as the complexity of Algorithm 0. Yet in practice Algorithm 1 takes a fraction of the time of Algorithm 0 and also of LINCLOSURE. Philosophy: Doing few set operations with big sets[1] is better than doing many set operations with small sets.

Applications. In discrete mathematics structures "generated" by a finite set of elements arise frequently. Often this fits the setting of Algorithm 1. For instance consider subalgebras of universal algebras, such as semigroups, groups, rings, lattices, etc. An implicational base is immediately derived from the "multiplication table" of the algebra. Problem (b) of how to get *smaller* implicational bases is mentioned in Section 4. Algorithm 1 is also an important subroutine in Algorithm 2, 3, 4.

---

[1] The most time-consuming part of Alg. 1 is the *initialization* of the lists $contain[x]$. Assuming that Σ is constant, and only $X \subseteq M$ varies, this needs to be done only once. For comparison, in Algorithm 4 we used all three algorithms as a subroutine. For a family **G** of 15 random 5-element subsets $X_k \subseteq M := \{1, \ldots, 10\}$ we had $|\Sigma_C| = 43$ and $|C| = 90$ (these terms are irrelevant here but defined in Section 4). The times of Algorithm 4, using respectively subroutines Algorithm 1, Algorithm 0, LINCLOSURE, were 8sec, 16sec, 32sec. For bigger tasks also the proportions between the times increase: Taking 19 random 13-element subsets $X_i \subseteq \{1, \ldots, 20\}$ we had $|\Sigma_C| = 210$ and $|C| = 901$. The corresponding times were 1130sec, 4438sec, 12661sec. The rankings of LINCLOSURE and Algorithm 0 might be switched for other types of Σ (as occuring in database theory?), but very likely Algorithm 1 stays best.

3 Computing $Irr(C)$ from a base of C

More precisely, the setting is this: Let $\Sigma := \{A_1 \to B_1, \ldots, A_n \to B_n\}$ be a given family of implications on M. Denote by $C(\Sigma)$ the (unknown) closure system of all Σ-closed subsets of M. How can we compute $Irr(C(\Sigma))$ directly from Σ? For this purpose, put $\Sigma_i := \{A_1 \to B_1, \ldots, A_i \to B_i\}$ and $C_i := C(\Sigma_i)$. Furthermore, for $x \in M$, let $max(x, i)$ be the set of all $X \in C_i$ maximal with the property that $x \notin X$. For families $G, H \subseteq 2^M$ we put $G * H := \{X \cap Y : X \in G, Y \in H\}$. Finally, for all $x \in M$ and $1 \leq i \leq n$ set

$$(2) \qquad S(x, i) := max(x, i-1) \cup \bigcup_{a \in A_i} max(x, i-1) * max(a, i-1).$$

Lemma 1 [MR, Thm. 13.6]: With the notation as above one has for all $x \in M$ and all $1 \leq i \leq n$ that $max(x, i) \subseteq S(x, i)$.

This Lemma was used in [MR,Alg.13.2] as follows to compute $max(x, i)$ from the already computed families $max(w, i-1)$. For each "candidate" $Z \in S(x, i)$ check if $Z^{\Sigma_i} = Z$ and if $x \in (Z \cup \{y\})^{\Sigma_i}$ for all $y \in M - (Z \cup \{x\})$ (whereby LINCLOSURE is used to compute the Σ_i-closures). Obviously $Z \in max(x, i)$ iff it passes both tests. We shall extend Lemma 1 in a way that allows us to do away with the above Σ_i-closures. For this purpose, denote by $max'(x, i-1)$ and $max''(x, i-1)$ the subfamilies of $max(x, i-1)$ where $A_i \to B_i$ holds, respectively fails. Formally,

$$(3) \qquad max'(x, i-1) := \{Z \in max(x, i-1) : A_i \nsubseteq Z \text{ or } B_i \subseteq Z\},$$

$$(4) \qquad max''(x, i-1) := max(x, i-1) - max'(x, i-1).$$

Moreover set

$$(5) \qquad T(x, i) := max'(x, i-1) \cup \bigcup_{a \in A_i} max''(x, i-1) * max(a, i-1).$$

Lemma 2:
(a) For all $x \in M$ and all $1 \leq i \leq n$ one has $max'(x, i-1) \subseteq max(x, i) \subseteq T(x, i)$.
(b) Furthermore $max(x, i)$ coincides with the family of all inclusion-maximal members of $T(x, i)$.

Proof. (a) Consider $Z \in max'(x, i-1)$. By definition Z is Σ_i-closed with $x \notin Z$. Assume there was a Σ_i-closed $Z_0 \supset Z$ with $x \notin Z_0$. Since Z_0 is a fortiori Σ_{i-1}-closed, this contradicts $Z \in max(x, i-1)$. Hence there is no such Z_0, i.e $Z \in max(x, i)$. Concerning the second inclusion in Lemma 2 (a), note that the sets from $max''(x, i-1)$ are not even Σ_i-closed. Thus, it follows from Lemma 1 that

$$(6) \qquad max(x, i) \subseteq max'(x, i-1) \cup \bigcup_{a \in A_i} max(x, i-1) * max(a, i-1).$$

Consider any $Z = X \cap Y \subset X$ with $X \in max'(x, i-1)$ and $Y \in max(a, i-1)$ $(a \in A_i)$. Then $Z \notin max(x, i)$ since $Z \subset X$ and $X \in max(x, i)$. Hence the righthand side of (6) can be further shrinked to $T(x, i)$.

(b) Each $Z \in T(x, i)$ is Σ_i-closed, since either $Z \in max'(x, i-1)$ or $A_i \not\subseteq Z$. Trivially $x \notin Z$ for all $Z \in T(x, i)$. Consider any $Z \in max(x, i) \subseteq T(x, i)$. By definition Z is a maximal set of C_i with $x \notin Z$. A fortiori it is a maximal set of $T(x, i) \subseteq C_i$. Conversely, let Z be any maximal set in $T(x, i)$. Since $Z \in C_i$ with $x \notin Z$, there is some $Z_0 \supseteq Z$ with $Z_0 \in max(x, i) \subseteq T(x, i)$. But $Z_0 \supset Z$ is impossible since Z is maximal in $T(x, i)$. So $Z = Z_0 \in max(x, i)$. •

```
Algorithm 6 *improving [MR,Alg.13.2]*
Input:   A family Σ := {A₁ → B₁,...,Aₙ → Bₙ} of implications on a set M.
Output:  The collection Irr(C) of meet irreducible flats
         of the closure system C := C(Σ).

for all x ∈ M do max(x) := {M - {x}}:  od:
for i from 1 to n do
    for all x ∈ M do
        max'(x) := {Z ∈ max(x) :  Aᵢ ⊈ Z or Bᵢ ⊆ Z}:
        max''(x) := max(x) - max'(x):
        T(x) := max'(x):
        for all a ∈ Aᵢ do
            for all X ∈ max''(x) and all Y ∈ max(a) do
                T(x) := T(x) ∪ {X ∩ Y}:
            od:
        od:
        max(x) := {Z ∈ T(x)| Z is inclusion-maximal}:
    od:
od:
Irr(C) := ⋃ₓ∈M max(x):
```

In the main loop 'for i from 1 to n do' the new families $max(x, i)$ are computed from the old families $max(x) = max(x, i-1)$ and $max(a) = max(a, i-1)$ $(a \in A_i)$ according to Lemma 2. That is, the new $max(x) = max(x, i)$ is the family of all maximal members of $T(x) = T(x, i)$. A straightforward method[2] for obtaining $max(x)$ from $T(x)$ goes like this: Start with $max(x) := \emptyset$. For each $Z_0 \in T(x)$ kick all sets Z from the current family $max(x)$ with $Z \subseteq Z_0$, and add Z_0 to $max(x)$ at the end. If however $Z_0 \subseteq Z$ occurs at some point, then leave $max(x)$ unchanged and pick the next $Z_0 \in T(x)$.

Algorithm 6 improves upon [MR,Alg.13.2] (with subroutine LINCLOSURE) quite a bit. For instance, we measured 3sec, respectively 127sec, for some Σ with $|\Sigma| = 10$ and $|Irr(C(\Sigma))| = 45$. For Σ consisting of 100 random implications $A_i \to B_i$ on $M := \{1, \ldots, 15\}$ with $|A_i|, |B_i| \leq 4$, Algorithm 6 found $|Irr(C(\Sigma))| = 167$ in 1291 sec ([MR,Alg.13.2] was cancelled after ≈ 15000sec). Both algorithms have exponential worst case complexity in the size of their output, but see [MR,ch.13.4] for a discussion of the average case.

Applications. So called "Armstrong relations" are important tools in the design of relational databases. Knowing $Irr(C(\Sigma))$ immediately yields an Armstrong relation for Σ with $|Irr(C(\Sigma))| + 1$ records [MR,Thm.14.4]. Various other problems, which are exponential in the worst case, are easily solved once $Irr(C(\Sigma))$ is known. This includes deciding Primality, Boyce-Codd-Normal Form, or Third Normal Form [MR,ch.13]. Concerning applications in Algebra, some base Σ of 976 implications derived from the multiplication table of the symmetric group S_5 yielded 53 meet irreducible subgroups of S_5. Using the base Σ of all 1351 implications $\{x, y\} \to \{x \wedge y, x \vee y\}$ with $x, y \in 2^6$ incomparable, Algorithm 6 found that the lattice of all sublattices of the Boolean lattice 2^6 has 42 meet irreducibles. Equally important as the meet irreducible subalgebras are the meet irreducible congruence relations of an algebra. They correspond to its subdirectly irreducible factors (see also [HW]).

4 Outline of the other algorithms

We sketch the main features of our Algorithms 2,3,4,5, and NEXTCLOSURE, which take care of the problems (b) to (f) defined in the Introduction. Details[3] can be found in [W4].

As to (b), a base Σ of $C \subseteq 2^M$ is *redundant* if some $(A \to B) \in \Sigma$ is a *consequence* of the other implications, i.e. $B \subseteq A^{\Sigma - \{A \to B\}}$. Shock [S] claims to have an algorithm which computes a minimum cardinality base from an arbitrary base of C. It is based on the fact that a nonredundant base of C with *full* implications $A \to B$ (i.e. B is the closure of A) has automatically minimum cardinality. However, dispite Shocks assertion, it turns out that his final base is generally redundant. Having corrected that flaw and having

---

[2] A more sophisticated method, which pays off for big $T(x)$ is obtained by a simplification of Algorithm 5 in Section 4.

[3] The author is glad to email a postscript file of [W4] upon request.

fine-tuned some other details, our method (Algorithm 2) has complexity $O(|\Sigma|^2|M|^2)$ and was successfully applied to rather large families Σ. *Applications.* Besides applications in relational database theory [MR], e.g. the implicational base derived from the "multiplication table" of an algebraic structure can be minimized. For instance, some obvious 2278-element base for the closure system of all subgroups of the symmetric group S_5 boils down to a 976-element minimum base.

The algorithm NEXTCLOSURE from [G] takes care of problems (c) and (d) (see also [GR]). It computes all closed sets in some "lectic" linear ordering of 2^M. Its complexity is necessarily exponential with respect to the input, but quadratic with respect to the size of its output. It is interesting to compare it with two other methods [C] and [W2] for solving (c) respectively (d). Despite an interesting observation, Chase's method does not compete with NEXTCLOSURE, but [W2] is superior to NEXTCLOSURE whenever $|\Sigma|$ is small, say ≤ 70, and $|C(\Sigma)|/|\Sigma|$ is big. In fact, there is no upper bound on $|C(\Sigma)|$! For instance, [W2] was used to compute the cardinalities of all 318 distributive lattices $FD(P)$ freely generated by a poset $(P, \leq)$ of cardinality 6, and it would yield $|FD(P)|$ for "most" $(P, \leq)$ with $|P| = 7$, e.g. $|FD(P)| = 11147992$ for $(P, \leq) = 2 + 2 + 1 + 1 + 1$. Note that [W2] more generally applies to the enumeration of all semantic models of a family Σ of propositional formulae. It not only counts, but also generates them in a compact manner (e.g. only 145321 "entities" were used for the 11147992 elements above). *Applications.* The original place of NEXTCLOSURE was in *Formal Concept Analysis* [Wi], which can be roughly described as "data analysis through lattices". Namely, let $\mathbf{K} = (G, M, I)$ be a *context*. Here G and M are finite sets of "objects" and "attributes" respectively. Further, $I \subseteq G \times M$ is a binary relation with $(g, m) \in I$ being interpreted as "object g has attribute m". The associated *concept lattice* $\mathbf{L}(\mathbf{K})$ is dually isomorphic to the closure system $\mathbf{C} \subseteq 2^M$ generated by the family $\mathbf{G} = \{g' \subseteq M : g \in G\}$, where $g' := \{m \in M : (g, m) \in I\}$. NEXTCLOSURE computes $\mathbf{C}$ from $\mathbf{G}$. Concerning another application, note that in the worst case NEXTCLOSURE has to compute $|M|$ closures in order to find the lectic successor of a closed set X. For *matroidal* closure systems $\mathbf{C} \subseteq 2^M$ the upper bound $|M|$ drastically reduces to the number of upper covers of X within $\mathbf{C}$. Several interesting invariants of a matroid can be read off its flat lattice [Wh,ch.6].

As to (e), a family Σ of implications on a set M is *direct* if $X^\Sigma = X^*$ for all $X \subseteq M$ (cf. (1)). Algorithm 3 is an alternative to Algorithm 15.2 in [MR], and computes the *canonic direct base* Σ^C of a closure system $\mathbf{C}$ from a generating family $\mathbf{G} = \{X_1, \ldots, X_p\}$ of flats. It inductively updates the canonic direct bases Σ_k of the closure systems $\mathbf{C}_k$

generated by $\{X_1, \ldots, X_k\}$. Algorithm 3, like [MR,Alg.15.2], does not have polynomial complexity with respect to its output; the intermediate families Σ might be bigger than the final family Σ^C (but seldom are). An asset of Algorithm 3 is Algorithm 1 as a fast subroutine, together with some further tricks reducing the number of set operations involving the auxiliary sets $contain[x]$. Unfortunately, the inductive approach of Algorithm 3 does not lend itself to a computation of the *canonic minimum base* Σ_C since the subproblem of calculating bases of "projection" closure spaces is NP-complete. Still, there are methods which apparently work well in practice (see [MR,p.155], [Go]). It would be interesting to see how Algorithm 3 provided with such a subroutine performs in calculating Σ_C. We persued another idea to get a minimum base Σ of C from a generating family G of flats: Compute Σ^C with Algorithm 3 and then reduce it to a minimum base Σ using Algorithm 2. How efficient is this combination of Algorithm 3 and Algorithm 2? The competition is a method of Ganter [G] which is based on NEXTCLOSURE and which yields Σ_C. Algorithm 4 only differs from that method in the use of Algorithm 1 instead of Algorithm 0 as a fast subroutine (cf. footnote 1). The drawback of Algorithm 4 is that *all* closed sets $X \in C$ are produced as a "side product". Indeed, it turns out that (Algorithm 3 + Algorithm 2) prevails for small G and big C, whereas Algorithm 4 prevails for big G and small C. *Applications.* Algorithms 3 and 4 e.g. apply to the closure systems C_R arising in relational database theory: A *database relation* on M can be thought of as a set $R := \{r^1, \ldots, r^n\}$ of *records* $r^i := (r^i_m : m \in M)$ with entries indexed by M. An implication (called *functional dependency*) $A \to B$ holds in R if any two records r^i, r^j agreeing on A also agree on B, i.e. $(\forall m \in A)r^i_m = r^j_m \Rightarrow (\forall m \in B)r^i_m = r^j_m$. Let Σ_R be the (unknown) family of all implications holding in R. According to [MR,p.256] a generating set of $C_R := C(\Sigma_R)$ is easily obtained as $G := \{ag(i,j) : 1 \le i < j \le n\}$, where $ag(i,j) := \{m \in M : r^i_m = r^j_m\}$ is the *agree set* of r^i and r^j.

There is an obvious way to settle problem (f). A speed up, Algorithm 5, is obtained with a trick similar to the one that made the difference between Algorithm 0 and Algorithm 1. Although Algorithm 5 has the same complexity $O(|M||G|^2)$ as the naive approach, practise speaks another language: For an input G of 2000 random 10-element sets $X_i \subseteq M := \{1, 2, \ldots, 20\}$ it took 6246sec doing it the naive way, whereas Algorithm 5 did the job in 360sec. *Applications.* In Formal Concept Analysis one has to reduce contexts $K = (G, M, I)$, which amounts to filter out $Irr(C)$ from a generating set G of a closure system C. The more $|G| > |M|$, or dually $|M| > |G|$, Algorithm 5 prevails over the naive approach. In Combinatorics often the inclusion-maximal (or minimal) members of a family $G \subseteq 2^M$ need to be determined. A simplification of Algorithm 5 does the job (see also footnote 2).

References

[C] P.J. Chase, Efficient subsemilattice generation, J. Comb. Theory 10 (1971) 181-182.

[G] B. Ganter, Algorithmen zur Formalen Begriffsanalyse, p.241-254 in: B. Ganter, R. Wille, K. E. Wolff (ed.): Beitrage zur Begriffsanalyse, B.I. Mannheim/Wien/Zürich.

[GR] B. Ganter, K. Reuter, Finding all closed sets: A general approach, Order 8 (1991) 283-290.

[Go] G. Gottlob, Computing covers for embedded functional dependencies, in: Proceedings of the Sixth ACM SIGACT-SIGMOD-SIGART Symposium on Principles of Database Systems, p.58-69 (1987).

[HW] C. Herrmann, M. Wild, A polynomial algorithm for testing congruence modularity, submitted.

[M] D. Maier, The Theory of Relational Databases, Computer Science Press, 1983.

[MR] H. Mannila, K.J. Räihä, The design of relational databases, Addison Wesley 1992.

[S] R. C. Shock, Computing the minimum cover of functional dependencies, Inf. Proc. Letters 22 (1986) 157-159.

[Wh] N. White (ed.), Matroid Applications, Enc. Math. and its Appl., Vol. 40 (1992), Cambridge University Press.

[W1] M. Wild, A theory of finite closure spaces based on implications, Advances in Mathematics 108 (1994) 118-139.

[W2] M. Wild, An enumerative principle of exclusion, submitted.

[W3] M. Wild, Cover preserving order embeddings into Boolean lattices, Order 9 (1992) 209-232.

[W4] M. Wild, Computations with finite closure systems and implications, Report Nr. 1708, Technische Hochschule Darmstadt 1994 (21 pages).

[Wi] R. Wille, Concept lattices and conceptual knowledge systems, Computers Math. Applic. Vol. 23 (1992) 493-515.

Maximum Tree-Packing in Time $O(n^{5/2})$

Andrzej Lingas *
Lund University

Abstract

The problem of determining the maximum number of node-disjoint subtrees of a tree T on n_t nodes isomorphic to a tree S on n_s nodes is shown to be solvable in time $O(n_s^{3/2} n_t)$. The same asymptotic bounds are observed for the corresponding problems where topological imbedding and subgraph homeomorphism are respectively substituted for subgraph isomorphism.

1 Introduction

A *matching* of a graph H is a set of disjoint edges of H. The disjoint edges can be seen as node-disjoint copies of the graph K_2. This observation led Hell and Kirkpatrick to natural generalization of matching where an arbitrary graph G is substituted for K_2 [6]. Following [6], a *G-packing* of a graph H is a set of node-disjoint subgraphs H_1, H_2, ...,H_l of H such that each H_i is isomorphic to G. In analogy to the problem of maximum matching in H, the problem of *maximum G-packing in H* is to determine the maximum cardinality of G-packing of H. Hell and Kirkpatrick proved this problem to be NP-complete in general [10], and Berman *et al.* showed it to be NP-complete even when H is a planar graph [2].

In this paper we consider the maximum G-packing restricted to trees. Both the guest graph G and the host graph H are assumed to be arbitrary (connected) trees. In particular, the guest tree is of an arbitrary size. Further, we shall denote the guest tree by S and the host tree by T. The assumption on connectivity of S is important since even the problem of determining whether a forest is isomorphic to a subgraph of a tree is known to be NP-complete [3].

The problem of maximum S-packing in T can be seen as a natural generalization of the problem of determining whether S is isomorphic to a subtree of T. The latter problem is known in the literature as the *subtree isomorphism problem* [14]. Edmonds and Matula independently provided polynomial time solutions to the subtree isomorphism problem by reduction to a collection of maximum bipartite matching problems [14] (In fact, Edmonds considered a more general problem of determining the largest subtree of T isomorphic to a subtree of S and therefore used maximum weighted bipartite matching [14]). Presently the best known upper time-bound for subtree isomorphism is due to Matula who showed the subtree isomorphism problem

---

*Department of Comptuter Science, Lund University, Box 118, S-221 00 Lund, Sweden, email: Andrzej.Lingas@dna.lth.se

to be solvable in time $O(n^{5/2})$ (Reynolds has independently obtained this bound for the restricted case of rooted trees in [16, 17]).

In this paper, we generalize Matula's method for subtree isomorphism to include the maximum tree-packing problem. In effect, for trees S, T on respectively n_s and n_t nodes, we can solve the problem of maximum S-packing in T in time needed to solve the maximum matching problem in a bipartite graph on $n_s + n_t$ nodes, i.e., in time $O(n_s^{3/2} n_t)$ [9]*.

By a slight modification of our algorithm for maximum S-packing in T, we obtain the same asymptotic bound for the corresponding problem of determining the maximum number of node-disjoint subtrees of T in which S can be topologically imbedded [4]. Consequently, we also obtain the same bound for the problem of determining the maximum number of node-disjoint subtrees of T homeomorphic to S.

Our results have potential applications in editing, image clustering, chemical structure analysis and computational biology.

2 Preliminaries

We shall adhere to a standard notation for undirected graphs [5]. In particular, a graph is isomorphic to another graph if there is one-to-one correspondence between nodes of the two graphs preserving adjacency. Such a correspondence is called an *isomorphism* between the two graphs. A graph can be *imbedded* in another graph if it is isomorphic to a subgraph of the other graph. The isomorphism between the graph and a subgraph of the other graph is also called an *imbedding* of the graph in the other graph.

For a graph G, we shall denote the set of nodes of G by $V(G)$ and the set of edges of G by $E(G)$.

Following [14], for two adjacent nodes u, w, of a tree T, the limb $T[u, w]$ is the maximal subtree of T containing the edge (u, w) and having u as a leaf. Clearly, a tree on n nodes has $2n - 2$ limbs. A child limb of $T[u, w]$ is a limb $T[w, w']$ of T where $w' \neq u$. The edge (u, w) is called the *root-edge* of the limb $T[u, w]$. A limb is *trivial* if it has only two nodes, i.e., it is equivalent to its root-edge. A limb of T is *complete* if it covers the whole tree T.

Let $S[v, s]$ be a limb of a tree S. The limb $S[v, s]$ can be *root-imbedded* in the limb $T[u, w]$ if there is an isomorphism between $S[v, s]$ and a subtree of $T[u, w]$ that maps v on u and s on w, respectively.

For limbs A, B of trees S, T, respectively, by $G_{A,B}$ we shall denote the bipartite graph with nodes corresponding to the child limbs of A and B respectively and edges connecting the node pairs whenever the corresponding child limb of A can be root-imbedded in the corresponding child limb of B.

The following lemma lies in the heart of all known polynomial-time solutions to the subtree isomorphism problem.

---

*The author of [14] seems to forget in Theorem 3.1 that the maximum matching problem for a bipartite graph on respectively p and q nodes can be solved in time $O(p^{3/2} q)$ even when $p \leq q$, see p. 231 in [9]. Therefore, the weaker bound $O(n_s n_t^{3/2})$ for subtree isomorphism is reported in [14].

Lemma 2.1 *([13]) Let A and B be limbs of two trees respectively. The limb A can be root-imbedded in the limb B if and only if the cardinality of maximum matching of $G_{A,B}$ is equal to the number of child limbs of A.*

Let the *height* of a limb denote the maximum length of a path including its root edge. The above lemma enables us to determine in polynomial-time whether there is a root-imbedding between each pair of limbs of two trees S, T, by applying maximum bipartite matching in the order of non-decreasing height. The tree S is isomorphic to the tree T if and only if a complete limb of S can be root-imbedded in a limb of T.

3   Maximum tree-packing in polynomial time

By refining the idea of the polynomial-time solution to the subtree isomorphism problem we can also solve the problem of maximum S-packing in T in polynomial time. To start with we need the following notation.

For a limb B of T, let $N(B)$ be the cardinality of maximum S-packing in B, i.e., the maximum number of node-disjoint subtrees of B isomorphic to S. Next, let $L(B)$ be a list of all limbs A of S that can be root-imbedded in B so the remaining part of B still contains $N(B)$ subtrees isomorphic to S. In case $L(B)$ is empty and each set of $N(B)$ disjoint subtrees of B isomorphic to S uses the root edge of B, $L(B)$ is called *blocked*. Further, let $L_{A,B}$ denote the bipartite graph with nodes corresponding to the child limbs of A and B respectively and edges connecting the node pairs whenever none of the L lists of the child limbs of B is blocked and the corresponding child limb of A is in the L list of the corresponding child limb of B. Finally, let $L_{A,B}^*$ denote the bipartite graph obtained from $L_{A,B}$ by adding an additional node corresponding to the root edge of A to the nodes corresponding to the child limbs of A and creating an edge between this node and each node corresponding to a child limb of B whose L list is not empty.

The following lemmata enable us to recursively determine $N(B)$ and $L(B)$ in the order of non-decreasing height of limbs B of T. The first of them is an obvious generalization of Lemma 2.1 and trivially follows from the definitions of the graphs $L_{A,B}$ and $L_{A,B}^*$.

Lemma 3.1 *Let $A = S[v,s]$ and $B = T[u,w]$ be limbs of the trees S and T respectively.*
The limb A can be root-imbedded in the limb B such that each child limb C of B admits an S-packing which has cardinality $N(C)$ and is node-disjoint from the imdedding of A if and only if the cardinality of maximum matching of $L_{A,B}$ is equal to the number of child limbs of A.
Analogously, there is an imbedding of A in B such that s is mapped onto w, u is not an image of any node of A, and each child limb C of B admits an S-packing which has cardinality $N(C)$ and is node-disjoint from the imdedding of A if and only if the cardinality of maximum matching of $L_{A,B}^*$ is equal to one plus the number of child limbs of A.

The next lemma *de facto* implies a polynomial-time algorithm for determining $N(B)$ and $L(B)$ for all limbs of the host tree T.

Lemma 3.2 *Let B be a non-trivial limb of the tree T. Let l be the number of child limbs C of B for which $L(C)$ is blocked.*

(1) If $l > 0$ then $N(B) = 1 - l + \sum_{child\ limb\ C\ of\ B} N(C)$, and the list $L(B)$ is empty and not blocked.

(2) If $l = 0$ and there is a complete limb A of S such that the cardinality of maximum matching of $L_{A,B}^*$ is equal to one plus the number of child limbs of A then $N(B) = 1 + \sum_{child\ limb\ C\ of\ B} N(C)$ and the list $L(B)$ is empty and not blocked.

(3) If $l = 0$ and (2) doesn't hold, and there is a complete limb A of S such that the cardinality of maximum matching of $L_{A,B}$ is equal to the number of child limbs of A then $N(B) = 1 + \sum_{child\ limb\ C\ of\ B} N(C)$, and the list $L(B)$ is empty and blocked.

(4) If (1,2,3) don't hold then $N(B) = \sum_{child\ limb\ C\ of\ B} N(C)$, the list $L(B)$ is not empty and contains a limb A of S if and only if the cardinality of maximum matching of $L_{A,B}$ is equal to the number of child limbs of A.

Proof: Let $B = T[u, w]$.

(1) Note that there are at most $-l + \sum_{child\ limb\ C\ of\ B} N(C)$ node-disjoint subtrees of B isomorphic to S and not including w. Also, the node w can belong to at most one of the $N(B)$ node-disjoint subtrees of T isomorphic to S. Hence, we obtain the inequality $N(B) \leq 1 - l + \sum_{child\ limb\ C\ of\ B} N(C)$. To obtain the reverse inequality we can proceed without node conflicts as follows. Pack $N(C)$ copies of S in each child limb C of B with not blocked list, $N(C)$ copies of S in a single child limb C of B with a blocked list, and $N(C) - 1$ copies of S in each of the remaining $l - 1$ limbs C of B with blocked lists.

By the above argumentation, the node w is covered by any maximum S-packing of B. Hence, the list $L(B)$ is empty. On the other hand, our packing achieving the maximum $N(B)$ doesn't use the node u. Consequently the list $L(B)$ is not blocked.

(2) Since $l = 0$ and at most one of the $N(B)$ subtrees isomorphic to S may include w, we have $N(B) \leq 1 + \sum_{child\ limb\ C\ of\ B} N(C)$. To obtain the reverse inequality it is sufficient to apply the second part of Lemma 3.1 to the complete limb A of S whose existence is assumed in (2). Since again the node w has to be covered by any maximum S-packing in B and the edge (u, w) is not used in the imbedding of A, the list $L(B)$ is empty and not blocked.

(3) The proof here is analogous to that for (2) and relies on Lemma 3.1. The only difference is that the complete limb A can be only root-imbedded in the limb B since the cardinality of maximum matching of $L_{A,B}$ is equal to the number of child limbs of A, and the cardinality of maximum matching of $L_{A,B}^*$ is not greater. Consequently, the edge (u, w) is used and the list $L(B)$ is blocked.

(4) Since (2,3) don't hold, it is impossible to pack an additional complete limb of S into B. Hence, we have $N(B) = \sum_{child\ limb\ C\ of\ B} N(C)$. The remaining part of the thesis here follows from Lemma 3.1. $\qquad\square$

The above lemma combined with a polynomial time algorithm for maximum bipartite matching [9] and applied to limbs of T in non-decreasing height order yields a polynomial-time solution to the problem of maximum S-packing in all limbs of T. In Section 5, we shall present and analyze a more refined solution to the tree-packing problem using the idea of the so called *rooted bipartite matching* [14].

4 Rooted bipartite matching

In [14], Matula has shown that all the maximum bipartite problems necessary to determine whether the limb $S[v, s]$ can be root-imbedded in a limb of the form $T[\ , w]$ can be simultaneously solved in time proportional to that needed to solve such a single matching problem. The same idea can be used to solve all the root-imbedding problems between the limbs of the form $S[\ , s]$ and the limb $T[u, w]$ simultaneously. The groups of related maximum bipartite matchings problems can be specified in terms of rooted bipartite matching [14].

Let $G = (V_1 \cup V_2, E)$ be a bipartite graph. For $b \in V_2$, the b-rooted matching problem is to determine the set $R_b(G)$ of all nodes a in V_1 such that there is a maximum matching of G including the edge (a, b) and $\#V_1 - 1$ other edges.

Matula has shown that the rooted bipartite matching problem can be solved in time proportional to that taken by the maximum bipartite matching problem (Theorem 3.4 in [14]).

For positive integers p, q, let $B_{p,q}$ denote the class of bipartite graphs $(V_1 \cup V_2, E)$ where $\#V_1 = p$ and $\#V_2 = q$. Next, let $mb(p, q)$ denote the worst-case time in the form $O(p^\alpha q^\beta)$, where $\alpha \geq 1$ and $\beta \geq 1$, needed to find a maximum matching in a graph in $B_{p,q}$. In terms of our notation we can express the fact due to Matula as follows.

Lemma 4.1 *([14]) For $G = (V_1 \cup V_2, E)$ in $B_{p,q}$ and $b \in V_2$, the b-rooted matching problem can be solved in time $mb(p, q)$.*

For a node s of the tree S and a limb B of T, let $L_{s,B} = (V_1 \cup V_2, E)$ denote the bipartite graph with nodes in V_1 corresponding to the limbs of the form $S[s, \ ]$, and nodes in V_2 corresponding to the child limbs of B and the root edge of B, and edges connecting $a \in V_1$ with $b \in V_2$ whenever none of the L lists of the child limbs of B is blocked and the limb of the form $S[s, \ ]$ corresponding to a is in the L list of the child limb of B corresponding to b or b corresponds to the root edge of B. Next, let $L_{s,B}^*$ denote the bipartite graph obtained from $L_{s,B}$ by adding an additional node (corresponding to an edge of the form $(s, \ )$) to V_1 and creating an edge between this node and each node corresponding to a child limb of B whose L list is not empty. Finally, we shall skip the index b in $R_b(L_{s,B})$ and $R_b(L_{s,B}^*)$ whenever b is the root edge of B.

By the above definitions and Lemma 3.1, we obtain the following useful lemma.

Lemma 4.2 *Let $A = S[v, s]$ and $B = T[u, w]$ be limbs of the trees S and T respectively.*
The limb A can be root-imbedded in the limb B such that each child limb C of B admits an S-packing which has cardinality $N(C)$ and is node-disjoint from the imbedding of A if and only if the node of the graph $L_{s,B}$ corresponding to $S[s, v]$ (observe the reversed order of v, s) is in the set $R(L_{s,B})$.
Analogously, there is an imbedding of A in B such that s is mapped onto w, u is not an image of any node of A, and each child limb C of B admits an S-packing which has cardinality $N(C)$ and is node-disjoint from the imdedding of A if and only if the node of $L_{s,B}^*$ corresponding to $S[s, v]$ is in the set $R(L_{s,B}^*)$.

Proof: To prove the first part suppose first that there is such a root-imbedding of A in B. In $L_{s,B}$, we can match each node corresponding to a child limb A' of A

with the node corresponding to the child limb B' of B where A' is root-imbedded into B' by the imbedding. It remains to match the node corresponding to the limb $S[s, v]$ with the node corresponding to the root edge of B to obtain a matching of cardinality equal to the number of limbs of the form $S[s, \ ]$. We conclude that $S[s, v]$ is in $R(L_{s,B})$. Conversely, if $S[s, v]$ is in $R(L_{s,B})$ then there is a matching of $L_{s,B}$ incident in particular to all nodes corresponding to the child limbs of A and including the edge corresponding to the pair ($S[s, v]$, the root edge (u, w) of B). Thus, we can map the root edge (v, s) of A onto the root edge (u, w) of B, and root-imbed each child limb A' of A in the child limb B' of B where the node corresponding to A' is matched with the node corresponding to B'. By the definition of the graph $L_{s,B}$, the resulting root-imbedding satisfies the required properties.

The proof of the second part is analogous. The set of child limbs of A can be regarded as extended by the edge (s, v) which has to be root-imbedded in one of the child limbs of B. This corresponds to the fact that the node corresponding to the edge of the form $(s, \ )$ has to be matched with a node corresponding to a child limb of B in any matching of $L_{s,B}$ qualifying the node corresponding to $S[s, v]$ to $R(L_{s,B})$. $\square$

5 The algorithm for tree packing

To specify our algorithm for tree-packing we neeed the following definition.

For a limb $T[x, y]$ of T, a *proper* limb of $T[x, y]$ is a limb of $T[x, y]$ that do not contain the root edge (x, y) of $T[x, y]$. Equivalently, a limb of $T[x, y]$ is proper if it is a child limb of either $T[x, y]$ or a proper limb of $T[x, y]$.

Our algorithm recursively determines the value $N(B)$ and the list $L(B)$ for all proper limbs B of a complete limb of T in the order of non-decreasing height.

1. Pick a leaf x of T, determine the limb $T[x, y]$ and order the proper limbs of $T[x, y]$ by non-decreasing height;

2. For each proper limb B of $T[x, y]$ set $N(B)$ to 0 and $L(B)$ to an empty set;

3. For each limb A of S of height 1 and each proper limb B of $T[x, y]$ insert A into the list $L(B)$;

4. **for** $h = 1, ..., height(T)$ **do**

    for each proper limb $B = T[u, w]$ of $T[x, y]$ of height h **do**
      begin
        Compute the number l of the lists of child limbs of B which are "blocked";
        Set $N(B)$, $N'(B)$ to $\min(0, 1 - l) + \sum_{child \ limb \ C \ of \ B} N(C)$;
        if $l > 0$ then set $L(B)$ to an empty list and go to E;
        for each node s of S **do**
          begin
            Compute the sets $R(L^*_{s,B})$, $R(L_{s,B})$;
            if $R(L^*_{s,B})$ contains a node corresponding to a limb $S[s, v]$ where $S[v, s]$ is a complete limb **then** set $N(B)$ to $N'(B) + 1$, set $L(B)$ to an empty list and go to E;

for each limb $S[s,v]$ corresponding to a node in $R(L_{s,B})$ **do**

    if $S[v,s]$ is not complete **then** insert $S[v,s]$ into $L(B)$ **else** set
        $N(B)$ to $N'(B)+1$, set $L(B)$ to an empty "blocked" list and go
        to E

  end

  E:**end**

Theorem 5.1 *For a tree T on n_t nodes and a tree S on n_s nodes, the problem of maximum S-packing in T can be solved in time $O(mb(n_s, n_t))$, hence, in time $O((n_s)^{3/2}n_t)$.*

Proof: By Lemma 3.2 and Lemma 4.2, our algorithm for computing $N(B)$ and $L(B)$ for all proper limbs of a complete limb $T[x,y]$ of T is correct. The value $N(T[x,y])$ yields the maximum number of node-disjoint subtrees of T isomorphic to S.

It remains to analyze the time complexity of the algorithm. We may assume w.l.o.g. $n_s \leq n_t$. To make possible constant time insertions in the lists $L(B)$, we can represent them by linear integer tables of lenght $O(n_s)$. The first entry of such a table can code one of the three actual states of the table (i.e., empty, empty and blocked, and non-empty, respectively). The other entries one-to-one correspond to the limbs of S and are set to one if the corresponding limb is in $L(B)$. To be able to form the tables, we can number the edges of S, e.g, by depth-first search, and then number the limbs of S by using the numbers of their root edges (each edge yields two limbs). Clearly, this preprocessing can be done in time $O(n_s n_t)$ [1]. Also, we can easily determine the height of proper limbs of $T[x,y]$ and order them by non-decreasing height in time $O(n_s n_t)$ [1]. The second and third step of the algorithm can trivially be implemented in time $O(n_s n_t)$.

Let us estimate the time-performance of the block under the loop for B. It is dominated by the cost of the loop for s. The latter is in turn dominated by the cost of computing the sets $R(L_{s,B}^*)$, $R(L_{s,B})$ and the cost of the loop for $S[s,v]$. Recall that $B = T[u,w]$. The graphs $L_{s,B}^*$, $L_{s,B}$ can be computed in time proportional to their size, i.e., in time $O(deg(s)deg(w))$. Then, the sets $R(L_{s,B}^*)$, $R(L_{s,B})$ can be determined in time $O(mb(deg(s), deg(w)))$ by Lemma 4.1. Since $\#R(L_{s,B}) = O(deg(s))$, the loop for $S[s,v]$ can be performed in time $O(deg(s))$. We conclude that the block under the loop for B takes time $O(\sum_{s \in V(S)} mb(deg(s), deg(w)))$. Each node of T can occur at most once as the second node of the root edge of a proper limb of $T[x,y]$, i.e., as w. Consequently, the fourth step and the whole algorithm take time $O(\sum_{w \in T(S)} \sum_{s \in V(S)} mb(deg(s), deg(w)))$. Since $\sum_{s \in V(S)} deg(s) = 2\#E(S) = 2n_s - 2$, we have $O(\sum_{s \in V(S)} mb(deg(s), deg(w))) = O(mb(n_s, deg(w)))$. Analogously, we have $\sum_{w \in V(T)} mb(n_s, deg(w)) = O(mb(n_s, n_t))$ which completes the proof. $\quad\square$

6 Extensions

We can easily extend our results on maximum tree-packing by replacing the concept of subgraph isomorphism with the more general concepts of topological imbedding [4] and subgraph homeomorphism [5].

A graph G can be *topologically imbedded* in a graph H if after contracting some paths in H with internal nodes of degree two the resulting graph contains a subgraph isomorphic to G. The corresponding partial $1-1$ mapping of $V(G)$ into $V(H)$ is called a *topological imbedding* of G into H. A *topological G-packing* of a graph H is a set of node-disjoint subgraphs H_1, H_2, ..., H_l of H such that G can be topologically imbedded in each H_i. The problem of *maximum topological G-packing* in H is to determine the cardinality of maximum topological G-packing in H.

To solve the problem of maximum topological S-packing in T for trees S and T, we need only to slightly modify our algorithm for maximum S-packing in T. For this purpose, we replace the notion of root-imbedding for limbs with that of *topological root-imbedding*. A limb $S[v,s]$ can be *topologically root-imbedded* in a limb $T[u,w]$ if there is a topological imbedding of $S[v,s]$ in $T[u,w]$ which maps v onto u.

Now to obtain a lemma corresponding to Lemma 3.1 it is sufficient to extend the equivalent condition in the first part of Lemma 3.1 as follows:

... if and only if the cardinality of maximum matching of $L_{A,B}$ is equal to the number of child limbs of A or A is in the L list of a child limb of B.

The extension by the alternative condition corresponds to the fact that the root edge of A may be mapped onto a path in a topological root-imbedding of A in B. Analogously, in Lemma 3.2 we need only to to add this condition in (4).

(4') If (1,2,3) don't hold then $N(B) = \sum_{child\ limb\ C\ of\ B} N(C)$, the list $L(B)$ is not empty and contains a limb A of S if and only if the cardinality of maximum matching of $L_{A,B}$ is equal to the number of child limbs of A or A is in the L list of a child limb of B.

Finally, in Lemma 4.2 we need to modify the first part analogously.

...if and only if the node of the graph $L_{s,B}$ corresponding to $S[s,v]$ is in the set $R(L_{s,B})$ or $S[v,s]$ is in the L list of a child limb of B.

These slight lemma modifications cause a corresponding modification of our algorithm for S-packing in the condition of the inner-most loop.

for each edge (v,s) of S where $S[s,v]$ corresponds to a node in $R(L_{s,B})$ or $S[v,s]$ is in the L list of a child limb of B do

Checking this modified condition takes time $O(deg(s) + deg(w))$. Hence, the whole loop for $S[s,v]$ will take time $O(deg(s)deg(w))$. Consequently, the time performance of the block under the loop for B will be dominated by the time needed to solve the rooted matching problems. Hence, we obtain the following extension of Theorem 5.1.

Theorem 6.1 *For a tree T on n_t nodes and a tree S on n_s nodes, the problem of maximum topological S-packing in T can be solved in time $O(mb(n_s, n_t))$, hence, in time $O((n_s)^{3/2} n_t)$.*

Recall that two graphs G, H are *homeomorphic* to each other if and only if after contracting the paths in G and H with internal nodes of degree two to single edges, the resulting graphs are isomorphic.

Note that if we preprocess S by replacing its paths with internal nodes of degree two by single edges, our modified algorithm will compute the maximum number of node-disjoint subtrees of T homeomorphic to S. The preprocessing can be easily done in linear time. Thus, we also have:

Theorem 6.2 *For a tree T on n_t nodes and a tree S on n_s nodes, the maximum number of node-disjoint subtrees of T homeomorphic to S can be computed in time $O(mb(n_s, n_t))$, hence, in time $O((n_s)^{3/2}n_t)$.*

7 Final remarks

1) Our algorithm for maximum (maximum topological or maximum homeomorphic, respectively) S-packing in T can be also extended to produce the maximum number of node-disjoint subtrees of T isomorphic (homeomorphic, respectively) to S by retracing appropriate bipartite matchings in top-down fashion (see also page 98 in [14]).

2) The related problems of maximum tree-packing where the subtrees are required to be only edge-disjoint seem much harder. Polynomial-time solutions seem obvious only when the guest tree is of constant size.

3) It is an interesting problem whether maximum tree-packing and its extensions admit randomized NC algorithms like subtree isomorphism [12].

4) Since the method for subtree-isomorphism can be generalized to include subgraph isomorphism for k-connected or bounded-degree partial k-trees [4, 15], it is likely that our method for maximum tree-packing could be also extended to include similarly restricted classes of partial k-trees.

Acknowledgements

Thanks go to Anders Dessmark for valuable comments.

References

[1] A.V. Aho, J.E. Hopcroft and J.D. Ullman. The Design and Analysis of Computer Algorithms. Addison-Wesley, Reading, MA, 1974.

[2] F. Berman, D. Johnson, T. Leighton, P.W. Shor, and L. Snyder. Generalized planar matching. Journal of Algorithms, 11 (1990), pp. 153-184.

[3] M.R. Garey, D.S. Johnson. Computers and Intractability. A Guide to the Theory of NP-completeness. Freeman, San Francisco, 1979.

[4] A. Gupta, N. Nishimura. Sequential and Parallel Algorithms for Embedding Problems on Classes of Partial k-Trees. Proc. SWAT'94, LNCS 824, pp. 172-182.

[5] F. Harary. Graph Theory. Addison-Wesley, Reading, Massachusetts, 1979.

[6] P. Hell and D.G. Kirkpatrick. On generalized matching problems. Information Processing Letters 12 (1981), pp. 33-35.

[7] P. Hell and D.G. Kirkpatrick. Packing by cliques and by finite families of graphs. Discrete Mathematics 49 (1984), pp. 45-59.

[8] P. Hell and D.G. Kirkpatrick. Packings by complete bipartite graphs. SIAM Journal on Algebraic and Discrete Methods 7 (1986), pp. 199-209.

[9] J. Hopcroft and R. Karp. *An $n^{5/2}$ algorithm for maximum matching in bipartite graphs.* SIAM J.Comput., 4:4, 1975.

[10] D.G. Kirkpatrick and P. Hell. Proc. 10th Annual ACM Symposium on Theory of Computing, 1978, pp. 240-245.

[11] D.G. Kirkpatrick and P. Hell. On the complexity of general factor problems. SIAM Journal on Computing 12 (1983), pp. 601-609.

[12] A. Lingas and M. Karpinski. Subtree Isomorphism is NC reducible to Bipartite Perfect Matching. Information Processing Letters 30(1989), pp. 27-32.

[13] D. Matula An algorithm for subtree identification (abstract). SIAM Rev. 10 (1968), pp. 273-274.

[14] D. Matula. Subtree isomorphism in $O(n^{5/2})$. Annals of Discrete Mathematics 2 (1978), pp. 91-106.

[15] J. Matousek and R. Thomas. On the complexity of finding iso- and other morphisms for partial k-trees. Discrete Mathematics 108 (1992), pp. 343-364.

[16] S. W. Reyner. An analysis of a good algorithm for the subtree problem. SIAM J. Comput. 6 (1977), 730-732.

[17] R.M. Verma and S. W. Reyner. An analysis of a good algorithm for the subtree problem, corrected. SIAM J. Comput. 18,5 (1989), pp. 906-908.

Optimal Algorithms for Finding Connected Components of an Unknown Graph

Weiping Shi[1] and Douglas B. West[2]

[1] Department of Computer Science
University of North Texas, Denton, TX 76203, USA
[2] Department of Mathematics
University of Illinois at Urbana-Champaign
Urbana, IL 61801, USA

Abstract. We want to find the connected components of an unknown graph G with a known vertex set V. We learn about G by sending an oracle a query set $S \subseteq V$, and the oracle tells us the vertices connected to S. We want to use the minimum number of queries, adaptively, to find the components. The problem is also known as interconnect diagnosis of wiring networks in VLSI. The graph has n vertices and k components, but k is not part of the input. We present a deterministic algorithm using $O(\min\{k, \log n\})$ queries and a randomized algorithm using expected $O(\min\{k, \log k + \log\log n\})$ queries. We also prove matching lower bounds.

1 Introduction

In this paper, we study how to find connected components of an unknown undirected graph $G = (V, E)$. Vertices u and v are *connected* if there is a path between them. The connection relation is an equivalence relation on the vertex set V. The *components* of G are its maximal connected subgraphs, and the vertex sets of the components are the equivalence classes of the connection relation. In our search problem, we are given V but not the set E of edges. Also we do not know the number of components or their sizes. The only operation we may use to obtain information about G is to query an oracle. For any query set $S \subseteq V$, the oracle will tell us $Q(S)$, the set of vertices connected to vertices of S:

$$Q(S) = \bigcup_{u \in S} \{v \in V : u \text{ and } v \text{ are connected.}\}$$

Our objective is to find the components, adaptively, using the minimum number of queries.

This problem comes from the interconnect diagnosis of wiring networks of logic circuits. It has applications to design and testing of very large scale integration (VLSI), multi-chip module (MCM) and printed circuit board (PCB) systems [2, 5, 7, 9]. A wiring network consists of a set of nets, each having one driver and one receiver. The logic value of a good net is specified by its driver and observed by its receiver. When some nets are involved in a short fault, their receivers all receive the logical OR of the values of their drivers. To diagnose a wiring network, a test engineer sends a "test vector" of logical 0's and 1's from the drivers and observes the outputs. The task of interconnect diagnosis is to adaptively apply test vectors to the nets to identify all the faults. Diagnosing a wiring network is the same as finding the connected components of the graph of short faults; applying test vectors is the same as querying the oracle.

Kautz [7] studied the problem for the special case of testing $G = \overline{K}_n$. Garey, Johnson and So [4] observed that with the complication of partial information about G, minimizing the number of queries to test $G = \overline{K}_n$ is NP-complete (reduction from chromatic number). For our problem of identifying all components, Jarwala and Yau [5] provided an approach using $\log n + (n - k)$ queries. Cheng et al. [2] studied non-adaptive versions of the problem, where the inputs of all queries are decided before asking the oracle any question. Shi and Fuchs [9] proved $n - 1$ queries are necessary and sufficient to find all components nonadaptively. They also presented a recursive version of the deterministic algorithm of Section 2, for the special case of the interconnect diagnosis problem. Chen and Hwang [1] used a different model, called "group testing", where the inputs of each query are two sets S and T, and the oracle answers "yes" or "no" depending on whether some vertex in S is connected to some vertex in T.

Table 1 summaries our results, where n is the number of vertices and k is the number of components, which is not part of the input. We measure the query complexity in terms of both the input size n and the output size k. We make no assumption on k other than $1 \leq k \leq n$. We present algorithms achieving the upper bounds and prove matching lower bounds. Note that randomization may permit an exponential reduction in the number of queries in terms of the input size. This effect was first observed by Fiat et al. in studying layered graph traversal [3]. Recently, Kavraki et al., [8] also studied the problem of finding connected components of an unknown graph, but their oracle takes only one vertex in each query.

Table 1. Number of queries required to find connected components.

Deterministic	$\Theta(\min\{k, \log n\})$
Randomized	$\Theta(\min\{k, \log k + \log \log n\})$
Nondeterministic	$\Theta(\log k)$

2 Deterministic Algorithm

There is a straightforward deterministic algorithm [5] to find the components in k queries: Iteratively pick a vertex v, use it as a query, record $Q(\{v\})$ as a component, and delete $Q(\{v\})$. Each query finds one new component. The upper bound of k can be improved when $n < 2^k$; we present a divide-and-conquer algorithm that uses $\log n$ queries to find all components. The algorithm iteratively maintains a *component structure* $P = \{(S_j, R_j) : j = 1, 2, \ldots, t\}$, where $\{S_j\}$ is a collection of subsets of vertices that form a partition of V, and $\{R_j\}$ is a collection of "chosen" subsets of vertices such that $R_j \subseteq S_j$ and $Q(R_j) = S_j$. This property of the chosen subsets implies that each S_j is a union of components. The aim is to refine the partition into components and to reduce the chosen subsets to single vertices. the algorithm terminates when each R_j is a single vertex in a component S_j.

Formally, Algorithm 1 initializes with $P \leftarrow \{(V, V)\}$, and then pereforms $\log n$ iterations of $P \leftarrow \mathcal{D}(P)$.

Procedure $\mathcal{D}(P)$.
Input: A component structure $P = \{(S_j, R_j) : j = 1, 2, \ldots, t\}$.
Output: A refined component structure $P' = \{(S'_j, R'_j) : j = 1, 2, \ldots, t'\}$.
1: For each j such that $|R_j| > 1$,
2:      Arbitrarily pick $R'_j \subset R_j$ such that $|R'_j| = \lceil |R_j|/2 \rceil$.
3: Perform the single query $\cup R'_j$, with result $S' \leftarrow Q(\cup R'_j)$.
4: $P' \leftarrow \emptyset$.
5: For each j such that $|R_j| > 1$,
6:      Let $T_j = S_j \cap S'$.
7:      $P' \leftarrow P' \cup \{(T_j, R'_j)\}$.
8:      If $T_j \neq S_j$ then $P' \leftarrow P' \cup \{(S_j - T_j, R_j - T_j)\}$.
9: Return P'.
End of Procedure.

For example, suppose G has two components G_1 and G_2. Suppose in the first call of Procedure $\mathcal{D}(P)$, $P = \{(V, V)\}$, R'_1 has vertices from both components. Then $S' = V$, and $P' = \{(V, R'_1)\}$. In other words, the new partition still has all vertices in a single block, but the chosen subset is smaller. Applying the procedure again and suppose this time R'_1 has vertices only from G_1. Then T_1 will be $V(G_1)$ and $S_1 - T_1$ will be $V(G_2)$. The corresponding chosen subsets for $S_1 - T_1$ will contain vertices only in G_2. Maintaining the partition that we have discovered enables us to disassemble the information in steps 5-8 from the single combined query in step 3.

Theorem 1. *Given a component structure in which the largest chosen subset has m vertices, $\log m$ iterations of Procedure $\mathcal{D}$ (one query each) finds the components of the graph as the blocks of the final component structure.*

Proof. With each query, we divide by 2 the maximum size of the chosen subsets. Hence there are at most $\log m$ queries. To prove that the algorithm works it suffices to show that the property $Q(R_j) = S_j$ is maintained by Procedure $\mathcal{D}$. If so, then when all $|R_j| = 1$, all vertices in S_j are connected to the chosen vertex, and S_j is the component containing it.

When some $|R_j| > 1$, the procedure performs a query and splits S_j into T_j and $S_j - T_j$. Because $Q(R_j) = S_j$, the query $Q(\cup R'_j)$ tells us that $Q(R'_j) = T_j$. By definition of connection, there cannot be any path between T_j and $S_j - T_j$. If $S_j - T_j$ is nonempty, then it has additional components. Since all of S_j was connected to R_j, we now have $Q(R_j - T_j) = S_j - T_j$. $\qquad\square$

Note that the number of queries depends only on the size of the largest chosen subset; we will use this property in the randomized algorithm. Note also that an algorithm using $2(\min\{k, \log n\})$ queries can be obtained by alternating between the k-query algorithm and Algorithm 1.

3 Randomized Algorithm

In this section, we present a randomized algorithm that can sometimes reduce the number of queries exponentially. The algorithm first calls the randomized Procedure $\mathcal{R}$ for $\log\log n$ iterations, and then it calls the deterministic Procedure $\mathcal{D}$ to complete the refinement of the partition.

Algorithm 2.
Input: A set of vertices V, $|V| = n$.
Output: All connected components.
1: $P \leftarrow \{(V, V)\}$.
2: **For** $i \leftarrow 0$ to $\log\log n$ **do**
3:      $P \leftarrow \mathcal{R}(P)$.
4: **While** there exists $(S_j, R_j) \in P$ such that $|R_j| > 1$
5:      $P \leftarrow \mathcal{D}(P)$.
6: **For** each $(S_j, R_j) \in P$
7:      Report S_j as one component.
End of Algorithm.

Procedure $\mathcal{R}(P)$.
Input: A component structure $P = \{(S_j, R_j) : j = 1, 2, \ldots, t\}$.
Output: A refined component structure $P' = \{(S'_j, R'_j) : j = 1, 2, \ldots, t'\}$.
1: **For** each j such that $|R_j| > 1$,
2:      Randomly pick $R'_j \subset R_j$ such that $|R'_j| = \left\lceil \sqrt{|R_j|} \right\rceil$.
3:      Perform the single query $\cup R'_j$, with result $S' \leftarrow Q(\cup R'_j)$.
4: $P' \leftarrow \emptyset$.
5: **For** each j such that $|R_j| > 1$,
6:      Let $T_j = S_j \cap S'$.
7:      $P' \leftarrow P' \cup \{(T_j, R'_j)\}$.
8:      **If** $T_j \neq S_j$ then $P' \leftarrow P' \cup \{(S_j - T_j, R_j - T_j)\}$.
9: **Return** P'.
End of Procedure.

Again we maintain a partition and chosen subsets. The only difference between Procedure $\mathcal{R}$ and Procedure $\mathcal{D}$ is in line 2, where we randomly pick $\sqrt{|R|}$ vertices in $\mathcal{R}$, instead of arbitrarily pick $|R|/2$ vertices in $\mathcal{D}$. This difference leads to the rapid reduction in the size of the query set. The idea is still to cut S_j into S'_j connected to R'_j and $S_j - S'$ connected to $R_j - S'$. We will prove that after $\log\log n$ iterations of $\mathcal{R}$, the maximum size of the chosen subsets will be at most $k^4 \log n$ with probability at least $1 - 1/\log n$. Therefore $4\log k + \log\log n$ iterations of Procedure $\mathcal{D}$ will complete the search.

The correctness of Algorithm 2 follows as in Theorem 1, but the complexity analysis is more delicate. Let $T(r, i)$ be the maximum of the expected number of queries used after iteration i, where the maximum is taken over all graphs and all component structures on n vertices in which r is the maximum size of a chosen subset and i iterations have been performed. We have the following probabilistic recurrence relation:

$$T(r, i) \le \begin{cases} 0 & \text{if } r = 1, \\ \log r & \text{if } r > 1,\ i = \log\log n, \\ 1 + \max\{T(\sqrt{r}, i+1), T(h(r), i+1)\} & \text{if } r > 1,\ i < \log\log n. \end{cases}$$

where $h()$ is a random variable representing the maximum number of vertices in the chosen subsets of the output P'. In the rest of this section, we will use martingales to estimate $E[T(n, 0)]$. Karp [6] developed some general methods that apply to solutions of order $\Omega(\log n)$.

Lemma 2. *If $0 \le \alpha \le 1$ and m is a positive integer, then $\alpha(1 - \alpha)^m < \frac{1}{em}$.*

Proof. An application of calculus, omitted due to space limitation. □

Lemma 3. *If f is a concave function and X is an integer-valued random variable, then $E[f(X)] \le f(E[X])$.*

Proof. Easy from the definitions, omitted due to space limitation. □

Given n and k, define a sequence of random variables $X_0, X_1, \ldots, X_{\log\log n}$ by setting $X_0 = n$ and $X_i = \max\{\sqrt{X_{i-1}}, h(X_{i-1})\}$ for $i > 0$. Intuitively, X_i is a bound on the size of the maximum chosen subset after i iterations.

Lemma 4. *For the sequence of random variables defined above, the conditional expectation $E[X_i \mid X_{i-1}]$ is bounded by*

$$E[X_i \mid X_{i-1}] < \sqrt{X_{i-1}} \cdot k^2.$$

Proof.

$$\begin{aligned} E[X_i \mid X_{i-1}] &= E[\max\{\sqrt{X_{i-1}}, h(X_{i-1})\} \mid X_{i-1}] \\ &\le E[\sqrt{X_{i-1}} + h(X_{i-1}) \mid X_{i-1}] \\ &= E[\sqrt{X_{i-1}} \mid X_{i-1}] + E[h(X_{i-1}) \mid X_{i-1}] \\ &= \sqrt{X_{i-1}} + E[h(X_{i-1}) \mid X_{i-1}]. \end{aligned}$$

Now we examine $h(X_{i-1})$. Let $P = \{(S_j, R_j) \mid j = 1, 2, \ldots, t\}$ be an arbitrary component structure in which the largest of the chosen subsets is R_j of size X_{i-1}. Suppose that G has components $G_1, \ldots, G_k$, and their contributions to R_j have sizes $n_1, \ldots, n_k$, with $n_l = |V(G_l) \cap R_j|$. When we apply $\mathcal{R}(P)$, the chosen subset R_j is split into a chosen subset of size $m = \sqrt{X_{i-1}}$, and possibly a new block $S_j - S'$ has chosen subset $R_j - S'$ consisting of the vertices of R_j not in $Q(R'_j)$. These leftover vertices belong to the components not hit by the vertices of R'_j. Each time we select a vertex for R'_j, it has probability $n_l/|R_j|$ of belonging to G_l. Hence the probability of missing G_l completely is $(1 - n_l/|R_j|)^m$. When we do miss G_l, it contributes n_l vertices to $R_j - S'$. Hence the expected size of the leftover set is $\sum_{l=1}^{k} n_l(1 - n_l/|R_j|)^m$. This expression grows with $|R_j|$. Furthermore, there are never more than k blocks in the partition. We conclude that

$$E[h(X_{i-1}) \mid X_{i-1}] = E\left[\max_{1 \le j \le t} |R_j - T_j| \mid X_{i-1}\right] \le E\left[\sum_{j=1}^{t} |R_j - T_j| \mid X_{i-1}\right]$$

$$\le k \max_{1 \le j \le t} E[|R_j - T_j| \mid X_{i-1}] = k \sum_{l=1}^{k} n_l \left(1 - \frac{n_l}{X_{i-1}}\right)^m$$

$$= k X_{i-1} \sum_{l=1}^{k} \frac{n_l}{X_{i-1}} \left(1 - \frac{n_l}{X_{i-1}}\right)^m.$$

Letting $\alpha_l = n_l/X_{i-1}$, we have $\sum_{l=1}^{k} \alpha_l = 1$ and $0 \le \alpha_l \le 1$ for $l = 1, 2, \ldots, k$. Using Lemma 2 and $m = \sqrt{X_{i-1}}$, we obtain

$$E[h(X_{i-1}) \mid X_{i-1}] \le k X_{i-1} \sum_{l=1}^{k} \alpha_l(1 - \alpha_l)^m \le k X_{i-1} k \max_{0 \le \alpha \le 1} \alpha(1 - \alpha)^m$$

$$< \frac{k^2 X_{i-1}}{em} = \frac{k^2 \sqrt{X_{i-1}}}{e}.$$

For $k \ge 2$, we conclude that $E[X_i \mid X_{i-1}] < \sqrt{X_{i-1}} + \frac{k^2\sqrt{X_{i-1}}}{e} < \sqrt{X_{i-1}} \cdot k^2$. $\quad\square$

Theorem 5. *For graphs with n vertices and k components, the expected number of queries used by Algorithm 2 to find all components is $O(\log k + \log\log n)$.*

Proof. Consider the maximum size of the chosen subsets after i iterations. From Lemma 3 and Lemma 4, we have

$$E[X_i] < E[\sqrt{X_{i-1}} \cdot k^2] \le \sqrt{E[X_{i-1}]} \cdot k^2 < \cdots < E[X_0]^{\frac{1}{2^i}} k^4 = n^{\frac{1}{2^i}} k^4.$$

With $i = \log\log n$, we obtain $E[X_{\log\log n}] < k^4$. From Markov's inequality,

$$\Pr[X_{\log\log n} \ge k^4 \log n] < \frac{k^4}{k^4 \log n} = \frac{1}{\log n}.$$

With probability at most $1/\log n$, we may be left with huge chosen subsets after $\log\log n$ iterations; Procedure $\mathcal{D}$ can resolve these instances with at most $\log n$ further iterations. Since we always call Procedure $\mathcal{D}$ after $\log\log n$ iterations of Procedure $\mathcal{R}$, we have

$$E[T(n,0)] \le \log\log n + \log(k^4 \log n) + \frac{1}{\log n} \log n \le 2 \log\log n + 4 \log k + 1.$$

$$\square$$

4 Lower Bounds

Theorem 6. *On a graph with k components, every algorithm to find the components uses at least $\log k$ queries.*

Proof. Since the response to a query $Q(S)$ cuts each subset $U \subseteq V$ that is known to be a union of components into at most two disconnected subsets $U \cap Q(S)$ and $U - Q(S)$, we need at least $\log k$ queries to separate the set of vertices into k components. □

Theorem 6 implies a lower bound for any nondeterministic algorithm. On the other hand, a nondeterministic algorithm can guess the components and use $\lceil \log k \rceil + 1$ queries to verify them.

Theorem 7. *On an n-vertex graph with k components, every deterministic algorithm uses at least $\min\{k, \log n\}$ queries.*

Proof. Consider the following adversary. If we pick a set S of vertices to make the first query, then our adversary makes S one component if $|S| \leq n/2$, or makes $V - S$ one component otherwise. Therefore after the first query, we are left with a sub-problem of $k - 1$ components and at least $n/2$ vertices. □

The randomized lower bound is more difficult to derive. We will use Yao's corollary [10] of von Neumann's minimax principle. Consider a matrix game in which the strategies of the row player are deterministic algorithms, the strategies of the column player are input instances, and the payoff is the cost of the deterministic algorithm on the input instance. The row player seeks to minimize cost by randomizing over deterministic algorithms; this is a randomized algorithm. The column player is the adversary. Yao applied the von Neumann minimax theorem to this game, obtaining the result that the expected performance of the "optimal" randomized algorithm on the worst input instance for it equals the expected cost of the worst input distribution against the best deterministic algorithm for it. More precisely, if P denotes a distribution over deterministic algorithms A and Q denotes a distribution over input instances G, we have

$$\min_{P}\{\max_{G} E_P(c(A, G))\} = \max_{Q}\{\min_{A} E_Q(c(A, G))\}.$$

Hence to provide a lower bound for a randomized algorithm, it suffices to prove a lower bound for the expectation of every deterministinc algorithm against a particularly bad input distribution.

We will apply this to n-vertex graphs generated at random as follows. There are k components $C_1, \ldots, C_k$ that are cliques of sizes $n^{\epsilon_1}, n^{\epsilon_2}, \ldots, n^{\epsilon_k}$, respectively, except for a slight adjustment. We define $\epsilon_1 = 0$, $\epsilon_2 = 1$ and $\epsilon_3 = 1/2$. For $i \geq 4$, ϵ_i is chosen with the following distribution:

$$\Pr\left[\epsilon_i = \epsilon_{i-1} + \left(\frac{1}{2}\right)^{i-2}\right] = \Pr\left[\epsilon_i = \epsilon_{i-1} - \left(\frac{1}{2}\right)^{i-2}\right] = \frac{1}{2}. \qquad (1)$$

The sizes of the first three cliques are $1, n$ and $\sqrt{n}$. The total number of vertices generated is $\sum_{i=1}^{k} n^{\epsilon_i} = n + o(n)$. To obtain an n-vertex graph, we delete the excess $o(n)$ vertices from the giant clique that initially has order n. We are left with an isolated vertex, a giant component, and $k - 2$ intermediate components. The assignment of vertices to components is made randomly.

Fig. 1 is an example distribution of the exponents. Since it becomes crowded quickly, we only provide labels up to ϵ_7.

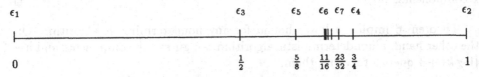

Fig. 1. Example distribution of the exponents.

Let π be a permutation of $\{1, 2, \ldots, k\}$ such that $\epsilon_{\pi(1)} < \epsilon_{\pi(2)} < \cdots < \epsilon_{\pi(k)}$. We call each interval $[\epsilon_{\pi(i)}, \epsilon_{\pi(i+1)})$ an ϵ-interval. The length of $[\epsilon_{\pi(i)}, \epsilon_{\pi(i+1)})$ is $\epsilon_{\pi(i+1)} - \epsilon_{\pi(i)}$. The sequence of real values $\epsilon_{\pi(1)}, \ldots, \epsilon_{\pi(k)}$ partition $[0, 1)$ into $k - 1$ disjoint ϵ-intervals.

Lemma 8. *Let $k \geq 3$ and $x \in [0, 1)$ be a real number. Then the probability that x is in an ϵ-interval of length 2^{-i} is 2^{-i}, for any $1 \leq i \leq k - 3$.*

Proof. During the selection of $\{\epsilon_i\}$, let the *active interval* be the interval of values that are candidates for later choices. When ϵ_i is chosen, the active interval is $(\epsilon_i - (1/2)^{i-2}, \epsilon_i + (1/2)^{i-2})$, which has length 2^{-i+3}. If x lies in an ϵ-interval of length 2^{-i} after ϵ_{i+3} is chosen, then x must remain in the active interval until ϵ_{i+2} is chosen, at which point the active interval has length 2^{-i+1}. Then ϵ_{i+3} must be chosen in the half of the active interval not containing x. Altogether, this requires i choices (from ϵ_4 through ϵ_{i+3}) to be made in a certain way, and the probability of this is 2^{-i}. $\qquad\square$

Lemma 9. *Given positive real numbers $x_1, \ldots, x_k$ and an integer n, let $P_1 = P(x_1, \ldots, x_k)$ be the distribution over graphs consisting of disjoint cliques of orders $n^{x_1}, \ldots, n^{x_k}$ that is obtained by assigning vertices to components at random. If δ and ϵ_i are positive real numbers, then the expected number of queries used by a deterministic algorithm against $P_2 = P(\epsilon_1 + \delta, \ldots, \epsilon_k + \delta)$ is at least the minimum expected number of queries used by a deterministic algorithm against $P(\epsilon_1, \ldots, \epsilon_k)$.*

Proof. (Sketch) Let A be an algorithm against P_2; we use A to specify an algorithm against P_1. Given G_1 drawn from P_1, we construct a corresponding G_2 drawn from P_2. We use the algorithm A on G_2, governed by queries to the oracle for G_1, to find the components of G_1. This establishes an algorithm against P_1 number of queries used by A against P_2. $\qquad\square$

Lemma 10. *Let $T(n,k)$ be the number of queries used by a deterministic algorithm A on input distribution (1). Then $E[T(n,k)] = \Omega(\min\{k, \log\log n\})$.*

Proof. Let S be the set of vertices the deterministic algorithm picks to make the first query. Assume without loss of generality that $S \neq V$. Let $x = \log_n \frac{n}{|S|} \in [0,1)$. The expected number of vertices of component i that are in S will be

$$\frac{n^{\epsilon_i}}{n} \cdot |S| = \frac{n^{\epsilon_i}}{n^x}.$$

It is easy to show that if $\epsilon_i > x$, then component i will be in $Q(S)$ with high probability, and if $\epsilon_i < x$, then component i will be in $V - Q(S)$ with high probability. Since $x \in [0,1)$, there must exist an ϵ-interval $[\epsilon_{\pi(i)}, \epsilon_{\pi(i+1)})$ that contains x. Therefore after the first query, we are left with two disjoint sets of vertices $Q(S)$ and $V - Q(S)$, where $Q(S)$ contains large components $C_{\pi(i+1)}, C_{\pi(i+2)}, \ldots, C_{\pi(k)}$, and $V - Q(S)$ contains small components $C_{\pi(1)}, C_{\pi(2)}, \ldots, C_{\pi(i)}$.

Let the length of the ϵ-interval $[\epsilon_{\pi(i)}, \epsilon_{\pi(i+1)})$ be 2^{-j}, then we must have $\pi(i) = j + 2$ or $\pi(i+1) = j + 2$. If $\pi(i) = j + 2$, then

$$\epsilon_{j+2} - \frac{1}{2^j} < \epsilon_{j+3}, \epsilon_{j+4}, \ldots, \epsilon_k < \epsilon_{j+2}.$$

Therefore $V - Q(S)$ has i components of sizes $n^{\epsilon_{\pi(1)}}, n^{\epsilon_{\pi(2)}}, \ldots, n^{\epsilon_{\pi(i)}}$. Among the i components, the following $k - j - 1$ components form a small instance of the same input distribution (1):

$$n^{\epsilon_{j+2}}, n^{\epsilon_{j+3}}, \ldots, n^{\epsilon_k}. \tag{2}$$

On the other hand, if $\pi(i+1) = j + 2$, then

$$\epsilon_{j+2} < \epsilon_{j+3}, \epsilon_{j+4}, \ldots, \epsilon_k < \epsilon_{j+2} + \frac{1}{2^j}.$$

Therefore $Q(S)$ will contain $k - i$ components of sizes $n^{\epsilon_{\pi(i+1)}}, n^{\epsilon_{\pi(i+2)}}, \ldots, n^{\epsilon_{\pi(k)}}$ and S contains the same $k - i$ components but the sizes are smaller: $n^{\epsilon_{\pi(i+1)} - x}$, $n^{\epsilon_{\pi(i+2)} - x}, \ldots, n^{\epsilon_{\pi(k)} - x}$. Among the $k - i$ components, the following $k - j - 1$ components form a small instance of the input distribution (1):

$$n^{\epsilon_{j+2} - x}, n^{\epsilon_{j+3} - x}, \ldots, n^{\epsilon_k - x}. \tag{3}$$

According to Lemma 9, finding the components of $Q(S)$ is as hard as finding the components of the graph induced by S, whose components have sizes given by (3).

The first query leaves us with one of two sub-problems of the types described above By Lemma 8, the probability is 2^{-j} that one of these contains a problem with $k - j - 1$ components as described in (2) or (3). These are smaller instances of problem (1). The numbers of vertices in (2) and (3) are approximately n^{-2^j}. Therefore we have the following recurrence relation:

$$E[T(n,k)] \geq 1 + \sum_{j=1}^{k} \frac{1}{2^j} E[T(n^{\frac{1}{2^j}}, k - j - 1)] \tag{4}$$

Solving (4) by induction, the Lemma is proved. □

Since $\log k$ is always a lower bound, we have

Theorem 11. *For every randomized algorithm finding the components of graphs with n vertices and k components, there is a graph in that class for which the expected number of queries used by the algorithm is $\Omega(\min\{k, \log k + \log\log n\})$.*

Acknowledgments Weiping Shi was supported by NSF grant MIP-9309120. Douglas West was supported by NSA/MSP grant MDA904-90-H-4011. The authors wish to thank Neal Brand, Edward Reingold and Steve Tate for discussions and three anonymous program committee members for pointing out errors in an earlier version of the paper and for improving the presentation.

References

1. C. C. Chen and F. Hwang. Detecting and locating electrical shorts using group testing. *IEEE Trans. on Circuits and Systems* 36 (8), pp. 1113–1116, Aug. 1989.
2. W.-T. Cheng, J. L. Lewandowski and E. Wu. Optimal diagnostic methods for wiring interconnects. *IEEE Trans. on Computer-Aided Design* 11 (9), pp. 1161–1166, Sept. 1992.
3. A. Fiat, D. P. Foster, H. Karloff, Y. Rabani, Y. Ravid, and S. Vishwanathan. Competitive algorithms for layered graph traversal. *FOCS*, pp. 288–297, 1991.
4. M. Garey, D. Johnson, and H. So. An application of graph coloring to printed circuit testing. *IEEE Trans. Circuits and Systems* 23 (10), pp. 591–599, Oct. 1976.
5. N. Jarwala and C. W. Yau. A new framework for analyzing test generation and diagnosis algorithms for wiring interconnect. *Proc. International Testing Conference*, pp. 63–70, 1989.
6. R. M. Karp. Probabilistic recurrence relations. *J. of ACM* 41 (6), Nov. 1994, pp. 1136–1150.
7. W. H. Kautz. Testing for faults in wiring networks. *IEEE Trans. on Computers*, 23 (4), pp. 358–363, April 1973.
8. L. Kavraki, J.-C. Latombe, R. Motwani and P. Raghavan. Randomized query processing in robot motion planning. *STOC*, 1995.
9. W. Shi and W. K. Fuchs. Optimal interconnect diagnosis of wiring networks. To appear *IEEE Trans. on VLSI Systems*, Sept. 1995.
10. A. Yao, Probabilistic Computations: Towards a Unified Measure of Complexity. *Proc. FOCS*, pp. 222–227, 1977.

The Multi-Weighted Spanning Tree Problem (Extended Abstract)

Joseph L. Ganley[1]   Mordecai J. Golin[2]   Jeffrey S. Salowe[3]

[1] Cadence Design Systems, Inc., 555 River Oaks Parkway, Building 2, MS 2A2,
San Jose, California 95134, ganley@cadence.com
[2] Department of Computer Science, Hong Kong UST,
Clear Water Bay, Kowloon, Hong Kong, golin@cs.ust.hk
[3] QuesTech, Inc., 7600A Leesburg Pike,
Falls Church, Virginia 22043, jsalowe@nvl.army.mil

Abstract. Consider a graph in which each edge is associated with q weights. In this paper we discuss different aspects of the problem of minimizing the minimum-spanning-tree cost simultaneously with respect to the different weights.

1 Introduction

Consider an n-vertex, m-edge graph $G = (V, E)$ in which each edge e is associated with q weights $w_i(e)$ for $1 \leq i \leq q$. Multiple-weight graph problems arise if there are two or more objective functions to optimize. For instance, if a road network is to be built between a set of cities, then the material cost and construction time may be independent considerations. VLSI applications of multi-weighted graphs to FPGA routing appear in Alexander and Robins [1]. Another example is the multicast problem in networks, which seeks a tree with minimum cost satisfying a delay bound [19]; the delay and the cost represent distinct edge weights. For the multicast problem in networks, it appears that practitioners use very naïve methods such as minimizing solely delay. Our investigations are aimed at developing a framework for good approximation strategies.

The problem of finding a tree that is simultaneously short with respect to each edge weight is NP-complete (the proof is not difficult, but since it appears to be part of the folklore we do not present it here). We therefore consider an approximation strategy, the first one that we know of with nontrivial performance bounds on a multi-weighted graph problem. Our approximation strategy is based on the following idea. Each tree T corresponds to a point $p(T) = (w_1(T), \ldots, w_q(T))$. Let P be the set of all $p(T)$ for all distinct spanning trees T. Rather than optimizing with respect to P, we optimize with respect to $extr(P)$, the set of extreme points of P. We give bounds on the quality of this approximation (the triangle inequality need not hold).

To search in $extr(P)$, we consider convex combinations of the edge weights; that is, the new weight of an edge e is $w_\Lambda(e) = \sum_{i=1}^{q} \lambda_i w_i(e)$, given that $\sum_{i=1}^{q} \lambda_i = 1$. Here, $\Lambda = (\lambda_1, \lambda_2, \ldots, \lambda_k)$. Such a weight function is called a *combined weight function*, and a minimum spanning tree with respect to a combined weight function is called a *combined-weight minimum spanning tree*. We use parametric search and cutting techniques to find an optimal value for Λ. The cutting techniques are tied to results on concave functions in a way that may be applicable to other problems. We note that there are more practical ways to obtain an optimal value for Λ, but ours have the best asymptotic time bounds.

In addition, we obtain results on the shape of the convex hull of P when the edge weights obey the triangle inequality. That is, we can guarantee that extreme points lie in certain regions. The shape of the convex hull is important in obtaining performance bounds on the approximation when the triangle inequality holds.

There are several related literatures on approximation algorithms and multiple weight functions in graphs, but none deal with the problem that we consider here. (As cited above, there are some applications-oriented papers that attempt to deal with multiple-weight problems.) Regarding approximation algorithms, several results indeed deal with multiple objective functions. Ravi et al. [26] attempt to minimize the cost of a network and the maximum degree of any node in the network and Khuller et al. [18] attempt to minimize the length of a spanning tree and the source-to-vertex distance in that tree. Goemans and Williamson [10] give a general approximation technique for a wide variety of graph problems. It is not clear how our problem fits into the framework of Goemans and Williamson; this is an avenue for further study. We use a different approach based on extreme points in convex hulls and geometric searching techniques.

Regarding multiple weight functions in graphs, there are at least two separate literatures. One literature focuses on "parametric combinatorial optimization," and the other centers on "hierarchical network design." The hierarchical network design problem is to find a shortest network that connects two distinguished vertices by a path whose edge lengths are determined by a primary weight measure, and the remaining vertices are connected by edges whose weights are determined by the secondary weight measure. The literature on hierarchical network design [5, 7, 8] deals with heuristics since the problem is known to be NP-complete [7]. It is quite different from the problems considered in this paper. (There is one other paper by Kozyrev [20], written in Russian, that may contain related results. The English abstract states that the paper presents an algorithm to find a shortest tree with respect to weight class 2 from among all minimum spanning trees with respect to weight class 1.)

The work on "parametric combinatorial optimization" is more relevant. It began with the study of minimum-ratio problems, such as finding minimum-ratio spanning trees [2, 24, 25] and computing a minimum mean cycle of a digraph [6, 21, 16]. The motivation for studying minimum ratio problems stems from the interpretation of one set of weights as cost and the other set of weights as time, in which case the minimum-ratio solution minimizes the cost-to-time ratio. The work on minimum-ratio problems focuses on the parametric problem where the edge weight is $w_1(e) - \lambda w_2(e)$. For instance, in the minimum-ratio spanning tree problem, the goal is to compute a tree T that minimizes the ratio $w_1(T)/w_2(T)$, and the objective is to find a value of λ that makes the weight of the resulting combined-weight minimum spanning tree T zero. The tree T is then a minimum-ratio spanning tree (see [2, 24, 25]).

A general framework for parametric combinatorial optimization appears in Gusfield [14] and Hassin and Tamir [15], among others. Gusfield deals with graphs whose edge weights are linear in the two weights, and Hassin and Tamir deal with matroids whose elements are a convex function of two weights.

Our work lies in the interface between approximation algorithms and parametric combinatorial optimization. We obtain the following results.

1. We obtain performance guarantees on the length of a minimum spanning tree with respect to edge weight $w_A(e)$ when edge weight function obeys the triangle inequality. These bounds indicate the length of a combined-weight minimum spanning tree compared to a single-weight minimum spanning tree. Equivalently, these bounds imply that extreme points of the convex hull of P lie in certain regions, particularly certain halfspaces. We then relate this performance bound to the multi-weighted spanning tree problem.

2. We use parametric search to find a combined-weight minimum spanning tree T'_δ that minimizes $w_2(T'_\delta)$ subject to $w_1(T'_\delta) \leq \delta$. The latter algorithm approximates a solution to the multi-weighted spanning tree problem.

3. In situations where parametric search does not easily apply, we use the theory of cuttings (see Matoušek [23]) and some results on concave functions to devise a general strategy to minimize a concave objective function among $extr(P)$. As a concrete example, we show how to find a longest combined-weight minimum spanning tree when $q \geq 2$. This methodology should be useful in other parametric combinatorial optimization problems.

Henceforth we use the notation that if $T = (V, E')$, then $w(T) = \sum_{e \in E'} w(e)$, where w is the weighting function.

2 Properties of the Extreme Points of P

In this section, we bound the number of points in $extr(P)$. We also specify their locations by providing bounds on $w_A(M_A)$, where M_A is a minimum spanning tree with respect to $w_A(e)$. These latter bounds only apply when each set of weights obeys the triangle inequality.

2.1 The Size of $extr(P)$

Define the quantity $C(m, n, q)$ to be the maximum number of distinct combined-weight minimum spanning trees over all values of Λ. Note that $C(m, n, q)$ also provides a bound on the size of $extr(P)$.

The general idea is originally due to Chandrasekaran [2], who considered the following set of line segments in the plane, one segment for each edge $e \in E$: $y = (w_1(e) - w_2(e))x + w_2(e)$, for $0 \leq x \leq 1$. These segments form an arrangement A of m line segments. Chandrasekaran's basic observation is the following. Consider a particular value $x = \lambda$. The weight of edge e is $\lambda w_1(e) + (1 - \lambda)w_2(e)$, which is the y-coordinate of the intersection of the segment for e with the line $x = \lambda$.

As λ varies from 0 to 1, the topology of a combined-weight minimum spanning tree changes, but only at the vertices of the arrangement A. Further, it must be the case that the two segments that cross correspond to edges on a cycle, in the following way. If e_1 has greater weight than e_2 at $x = \lambda - \epsilon$ and e_1 has the same weight as e_2 at $x = \lambda$, then e_1 replaces e_2 in a minimum spanning tree at $x = \lambda + \epsilon$ if and only if the cycle in $M(\lambda - \epsilon) \cup e_2$ contains e_1. As a consequence, there are only $O(m^2)$ distinct minimum spanning tree topologies as λ varies between 0 and 1, provided there is no value of λ such that three or more weights are identical.

To count $C(m, n, q)$, we define a lexicographic ordering among spanning trees. Arbitrarily number the edges of E from 1 to m, and let T_1 precede T_2 in the lexicographic ordering if the characteristic vector for T_1 is greater in numerical value than the characteristic vector for T_2. For a particular value of Λ, consider only the lexicographically first minimum spanning tree. Then $C(n, m, q)$ is the maximum number of distinct combined-weight minimum spanning trees over all n-vertex, m-edge, q-weighted graphs. Chandrasekaran's argument implies that $C(n, m, 2) = O(m^2)$; Gusfield [13] decreased the $O(m^2)$ upper bound by a more careful counting strategy: he showed that $C(n, m, 2) = O(m\sqrt{n})$. (Katoh et al. [17] bound the number of transitions in L_1 and L_∞ minimum spanning trees for points moving in d-dimensional space. Their bound is $O(n^{5/2}\alpha(n))$. Additional work by Gusfield [14] implies that when $q = 2$, all of the combined-weight minimum spanning trees can be constructed in $O(f(n, m)C(n, m, 2))$ time.)

Suppose there are q weights, with $q > 2$. The arrangement of line segments is replaced by a q-dimensional arrangement of m $(q-1)$-dimensional hyperplanes, and the interval between 0 and 1 is replaced by a $(q-1)$-dimensional hypercube (the ith coordinate corresponds to λ_i, and λ_q is dependent on the other values).

Theorem 1. $C(n, m, q) = O(m^{2q-2})$.

Proof. Each pair of hyperplanes h_i and h_j intersects in a $(q-2)$-dimensional affine variety $h_i \cap h_j$. Consider $h_i \cap h_j$ and the vertical hyperplane h_{ij} that passes through $h_i \cap h_j$. This hyperplane intersects the surface $x_q = 0$ in a $(q-2)$-dimensional affine variety h'_{ij}. The set of all $(q-2)$-dimensional affine varieties h'_{ij} forms a $(q-1)$-dimensional arrangement in $x_q = 0$. This arrangement will be called the *projected arrangement*. It can be shown that there is a one-to-one correspondence between distinct (lexicographically first) combined-weight minimum spanning trees and $(q-1)$-dimensional regions in the projected arrangement. The number of such regions is $m^{2(q-1)}$. $\quad\square$

2.2 Location of the Extreme Points of P

In this subsection, we obtain bounds on $w_A(M_A)$ when all weights obey the triangle inequality. One can use these bounds to say that there is a point in P, and therefore in $extr(P)$, in the halfspace $\{x = (x_1, x_2, \ldots, x_q) : (\lambda_1, \lambda_2, \ldots \lambda_q) \cdot x \le D\}$, for a value D given in Theorem 3. Implications to the multi-weighted spanning tree problem appear at the end of this section. (Note that if the triangle inequality does not hold, then the bounds on the maximum length of a combined-weight minimum spanning tree are arbitrarily bad [27].)

Let M_i be (the edges in) a minimum spanning tree for G with respect to the ith set of weights, and let M_A be a minimum spanning tree with respect to weights $w_A(e)$. We assume that each set of weights $w_i(e)$ obeys the triangle inequality. If the triangle inequality does not hold, then it is easy to show that there is no upper bound on the weight of $w_A(M_A)$ in terms of Λ and $w_i(M_i)$.

It is clear that $w_A(M_A) \geq \sum_{i=1}^{q} \lambda_i w_i(M_i)$, regardless of the triangle inequality, and that this bound is tight when the minimum spanning trees are identical for each set of weights. Since the triangle inequality implies that each edge must have weight less than or equal to the length of a minimum spanning tree, $w_A(M_A) \leq (n-1) \sum_{i=1}^{q} \lambda_i w_i(M_i)$. Henceforth, let $J_1 = \sum_{i=1}^{q} \lambda_i w_i(M_i)$, and let $J_2 = \prod_{i=1}^{q} \lambda_i w_i(M_i)$.

Theorem 2. *There is a set of edge weights, each edge weight satisfying the triangle inequality, such that $w_A(M_A) \geq J_2 n^{1-1/q}$.*

Proof. Omitted in this extended abstract.

Theorem 2 shows that J_2, as well as J_1, is an important quantity in the maximum length of a combined-weight minimum spanning tree. There is a set of edge weights whose combined-weight minimum spanning tree is at least $\max\{J_1, J_2 n^{1-1/q}\}$.

On the other hand, it is possible to obtain an upper bound on the maximum length of a combined-weight minimum spanning tree involving J_1, J_2, and n.

Theorem 3. *For all sets of edge weights with each weight set satisfying the triangle inequality,*

$$w_A(M_A) = O\left((J_1 + J_2 n^{1-1/q}) \log^{1+\frac{q-1}{2}-\frac{1}{q}} n \right).$$

Proof. Omitted in this extended abstract.

Suppose we construct a combined-weight minimum spanning tree M_A, and we are interested in comparing $w_i(M_A)$ to $w_i(M_i)$, which is a lower bound on the ith weight measure. Theorem 3 shows that $w_i(M_A)$ is not arbitrarily longer than $w_i(M_i)$. This result is summarized in the corollary below.

Corollary 4. *Let $w_i(e)$, $1 \leq i \leq q$, satisfy the triangle inequality. For all i,*

$$\frac{w_i(M_A)}{w_i(M_i)} \leq \frac{w_A(M_A)}{\lambda_i w_i(M_i)},$$

where the bound on $w_A(M_A)$ is given in Theorem 3.

Theorem 2 has a direct implication on the multi-weighted spanning tree problem when every weight set satisfies the triangle inequality. For ease of exposition, let $q = 2$. Let T be a tree that satisfies $w_1(T) \leq B_1$ and minimizes $w_2(T)$, and let $\Lambda = (\lambda, 1 - \lambda)$. For the weights satisfying Theorem 2, $w_\Lambda(M_\Lambda) \leq w_\Lambda(T) \leq \lambda B_1 + (1 - \lambda)w_2(T)$. Algebra yields

$$w_2(T) \geq \frac{\sqrt{n\lambda(1-\lambda)w_1(M_1)w_2(M_2)} - \lambda B_1}{1 - \lambda}.$$

The equation is maximized when $\lambda = \frac{nw_1(M_1)w_2(M_2)}{nw_1(M_1)w_2(M_2)+4B_1^2}$. The resulting inequality is

$$w_2(T) \geq \frac{nw_1(M_1)w_2(M_2)}{4B_1}.$$

As a consequence, there is a weight set for which a feasible solution requires that $B_2 \geq nw_1(M_1)w_2(M_2)/4B_1$.

Similar statements can be derived if $q > 2$.

3 Algorithms

We now turn to algorithmic questions on combined-weight minimum spanning trees. We show that $extr(P)$ can be searched efficiently for a variety of properties. The search techniques focus on piecewise linear surfaces defined over regions of a high-dimensional arrangement. We use known results on cuttings and some results on concave functions to devise a fast searching algorithm. We then present an approximation algorithm to the multi-weight spanning tree problem and bound its behavior.

3.1 Searching Extreme Points

Say that a tree $T(v)$ is optimal if point $v \in extr(P)$ minimizes function $f : Z^q \to \Re$. That is, the combined-weight minimum spanning tree $T(v)$ corresponding to v has minimum f value. We concentrate primarily on functions that are monotone or unimodal (i.e., convex or concave).

Using either parametric search or slope selection results, one can find optimal trees for monotone or unimodal functions when $q = 2$. The following theorem is a minor improvement over results appearing in the literature.

Theorem 5. *Let G be a graph with two edge weights, and let f be a monotone or unimodal function. An optimal tree with respect to f can be found in $O(m \log n + f(n, m) \log n)$ time, where $f(n, m) = \Omega(n + m)$ is the amount of time needed to compute a minimum spanning tree on a graph with n nodes and m edges.*

Proof. Omitted in this extended abstract.

It is possible to generalize Theorem 5 to the case $q > 2$, though it appears difficult to use parametric search to do so.

Theorem 6. *In a graph with $q \geq 2$ edge weights, an optimal tree with respect to a unimodal function can be found in $O((q-1)!m^2+(q-1)!f(m,n)\log^{q-1}n)$ time, where $f(n,m) = O(m\log\beta(m,n))$ is the amount of time needed to compute a minimum spanning tree on a graph with n nodes and m edges. Here, $\beta(m,n) = \min\{i : \lg^{(i)} n \leq m/n\}$.*

To prove this theorem, we need some intermediate results. Let a concave function $\hat{g} : \Re^d \rightarrow \Re$ be called *good* if, for any k-dimensional affine variety t, where $k \leq d$, the unique maximum value of $\hat{g}$ along t is at a vertex of $A \cap T$. (There is a similar definition for good convex functions.) A property is good if the underlying function is good. (All results are proven for concave functions only; similar results hold for convex functions.)

Lemma 7. *Consider a good concave function $\hat{g}$, let A be an arrangement of hyperplanes H in $\Re^d$, and let t be a hyperplane in $\Re^d$. If the maximum value along t is known, then it is possible to decide if the overall maximum is above, on, or below t in at most two function evaluations.*

Proof. Consider any line ℓ through the vertex v realizing the maximum value of $\hat{g}$ on t that does not lie in t, and consider a point v^* lying on ℓ with the property that no hyperplanes in H intersect ℓ between v and v^*. If the value of $\hat{g}$ at v^* is less than the value of $\hat{g}$ at v, then the maximum value of $\hat{g}$ cannot lie on the same side of t as v^*. This is because a basis can be established for all points on that side of t, and with respect to the value of $\hat{g}$ at v, all values are decreasing in the directions defined by the basis. The lemma follows from the observation that the function $\hat{g}$ is a good convex function. □

Lemma 8. *Consider a good concave function $\hat{g}$ with respect to arrangement A of m hyperplanes. Then for any k-simplex over A, the maximum value corresponding to that simplex can be found in $O(k!r^{\binom{k}{2}-k}m+k!r^{\binom{k}{2}}g(m)\log^k m)$ time, where $g(m)$ is the time required to evaluate $\hat{g}$ and $r > 1$ is a constant.*

Proof. We use the theory of cuttings (see Matoušek [23]) to subdivide the problem. The theory of cuttings states that, given an arrangement A of m hyperplanes in $\Re^k$ and a parameter r, it is possible to subdivide $\Re^k$ into $O(r^k)$ k-simplexes, each intersecting at most m/r of the hyperplanes in A. This subdivision is called a $1/r$-cutting. If r is a constant, a $1/r$-cutting can be computed in $O(m)$ time.

The proof of the lemma is by induction on the dimension k. Let H be the set of m hyperplanes in A. If $k = 1$, the simplex is a line segment ℓ. Since the function $\hat{g}$ is good and the maximum along ℓ occurs at $\ell \cap H$, which is a sequence of points along ℓ, the maximum can be found by enumerating these points and performing a binary search. The time complexity is $O(m + g(m)\log m)$.

For convenience, let $g = g(m)$; the time complexity of a function evaluation will not necessarily change for recursive calls. Let $T(m,k)$ be the time to find the maximum value within the k-dimensional simplex for an arrangement of m hyperplanes in $\Re^k$. Consider a k-simplex that is known to contain the maximum. Each k-simplex is composed of k $(k-1)$-dimensional facets. By the inductive hypothesis, the maximum in each of the facets can be found in $T(m, k-1)$ time. Using a minor variant of Lemma 7, k additional probes can be used to determine whether the maximum lies on one of these facets or inside the simplex. In the latter case, the problem is restricted to the hyperplanes intersecting the simplex. We compute a cutting and use a variant of Lemma 7 to decide in which simplex the maximum lies. There are kr^k such calls, and each takes $T(m/r, k-1)$ time; because we have a cutting, the number of hyperplanes intersecting any simplex is at most m/r. We then consider the problem recursively on the simplex that contains the maximum, costing $T(m/r, k)$ time. The total time is:

$$T(m, k) \leq kT(m, k-1) + km + kr^k T(m/r, k-1) + T(m/r, k) + O(m)$$
$$T(m, 1) = O(m + g \log m)$$
$$T(k, k) = O(1)$$

It can be shown that $T(m, k) = O(k! r^{\binom{k}{2-k}} m + k! r^{\binom{k}{2}} g(m) \log^k m)$, where $r > 1$ is a small constant. □

Proof of Theorem 6. Follows directly from Lemma 8, noting that the current fastest deterministic minimum spanning tree algorithm is due to Gabow et al. [9] and takes time $O(m \log \beta(m, n))$, where $\beta(m, n) = \min\{i : \lg^{(i)} n \leq m/n\}$. □

It appears that the problem of finding a combined-weight minimum spanning tree with maximum length fits into the framework of Theorems 5 and 6.

Lemma 9. *The length of a combined-weight minimum spanning tree as a function of Λ is a good concave function.*

Proof. Consider a particular minimum spanning tree topology T. Since the edge weights vary as a function of the first $q - 1$ terms in Λ, the length of T defines a hyperplane in q-dimensional space. A minimum spanning tree for a particular value of Λ must correspond to a point in the closed halfspace beneath this hyperplane. Therefore the points that correspond to minimum spanning trees form the bounding surface of a convex polyhedron.

Goodness follows because the length of a combined-weight minimum spanning tree is a piecewise linear function whose normal vector only changes at faces of the projected arrangement. □

As a consequence, we have:

Theorem 10. *Let G be a graph with $q \geq 2$ edge weights. If $q = 2$, a maximum-length combined-weight minimum spanning tree can be found in $O(f(n, m) \log n)$ time, and if $q > 2$, a maximum length combined-weight minimum spanning tree can be found in $O(m^2)$ time. Here $f(n, m) = O(m \log \beta(m, n))$.*

3.2 An Approximation Algorithm

To approximate the multi-weighted spanning tree problem, consider the variant where one wants to find a tree T_δ with minimum $w_2(T_\delta)$ that satisfies $w_1(T_\delta) \leq$

149

δ. Such a problem would arise if one of the parameters, such as total cost, is constrained by a particular bound. We consider finding a combined-weight minimum spanning tree T'_δ that satisfies $w_1(T_\delta) \leq \delta$ and has minimum $w_2(T'_\delta)$ value. We then provide bounds on $w_2(T'_\delta)$ in terms of $w_2(T_\delta)$. The intuition behind the bound stems from the observation that the point corresponding to T_δ dominates no point and T'_δ is an extreme point.

Lemma 11. *The function $w_1(M_\Lambda)$ is a monotonically nondecreasing function of λ_1, and the function $w_2(M_\Lambda)$ is a monotonically nonincreasing function of λ_1.*

We can therefore find tree T'_δ by finding the largest value λ_1^* of λ_1 such that $w_1(M_\Lambda) \leq \delta$. Once λ_1^* is found, $w_2(T'_\delta)$ can be bounded in terms of the optimal value $w_2(T_\delta)$, λ_1^*, and $w_1(T'_\delta)$.

Theorem 12. $w_2(T_\delta) \leq w_2(T'_\delta) \leq w_2(T_\delta) + \frac{\lambda_1^*}{1-\lambda_1^*}(\delta - w_1(T'_\delta))$

Proof. T_δ was selected so that $w_2(T_\delta) \leq w_2(T)$, for all T such that $w_1(T) \leq \delta$, yielding the first inequality. The second inequality stems from the observation that $\lambda_1^* w_1(T_\delta) + \lambda_2^* w_2(T_\delta) \geq \lambda_1^* w_1(T'_\delta) + \lambda_2^* w_2(T'_\delta)$. This inequality becomes

$$w_2(T'_\delta) \leq w_2(T_\delta) + \frac{\lambda_1^*}{\lambda_2^*}(w_1(T_\delta) - w_1(T'_\delta)).$$

Note that $w_1(T'_\delta) \leq w_1(T_\delta) \leq \delta$, yielding the second inequality. ☐

Finally, we mention that many of the techniques in this section work for all matroids if we replace $f(n, m)$ by the amount of time it takes to find a basis of size n in a matroid that has m elements. We are also currently trying to apply our methodology to some of the applications mentioned in the introduction.

Acknowledgements. We thank G. Robins for proposing the problem. J. Ganley's and J. Salowe's research was supported by NSF Grants CCR-9224789 and MIP-9107717; M. Golin's was supported by HK RGC CRG grant HKUST 181/93E. Some of this work was performed while M. Golin and J. Salowe were visiting the Max-Planck-Institut für Informatik in Saarbrücken and while J. Ganley and J. Salowe were working at the University of Virginia in Charlottesville, Virginia.

References

1. M. J. Alexander and G. Robins, "A new approach to FPGA routing based on multi-weighted graphs," *ACM/SIGDA International Workshop on Field-Programmable Gate Arrays* (1994).
2. R. Chandrasekaran, "Minimum ratio spanning trees," *Networks* 7 (1977) 335–342.
3. R. Cole, J. S. Salowe, W. L. Steiger, and E. Szemerédi, "An optimal-time algorithm for slope selection," *SIAM J. Comput.* 18(4) (1989) 792–810.
4. J. Cong, A. Kahng, G. Robins, M. Sarrafzadeh, and C. K. Wong, "Provably good performance-driven global routing," *IEEE Trans. on CAD* 11(6), (1992), 739–752.

5. J. R. Current, C. S. Revelle, and J. L. Cohon, "The hierarchical network design problem," *Eur. J. Oper. Res.* 27 (1986) 55-67.

6. G. B. Dantzig, W. O. Blattner, and M. R. Rao, "Finding a cycle in a graph with minimum cost to time ratio with application to a ship routing problem", *Theory of Graphs, International Symposium*, Gordon and Breach (1967) 77-84.

7. C. Duin and T. Volgenant, "Reducing the hierarchical network design problem," *Eur. J. Oper. Res.* 39 (1989) 332-344.

8. C. Duin and T. Volgenant, "The multi-weighted Steiner tree problem," *Ann. Oper. Res.* 33 (1991) 451-469.

9. H. H. Gabow, Z. Galil, T. Spencer, and R. E. Tarjan, "Efficient algorithms for finding minimum spanning trees in undirected and directed graphs," *Combinatorica* 6(2) (1986) 109-122.

10. M. X. Goemans and D. P. Williamson, "A general approximation technique for constrained forest problems," *Proc. 3rd SODA* (1992) 307-316.

11. M. J. Golin, C. Schwarz, and M. Smid, "Further Dynamic Computational Geometry", *Proc. Fourth Canad. Conf. Comp. Geom.* (1992) 154-159.

12. M. R. Garey and D. S. Johnson, Computers and Intractability, W. H. Freeman (1979).

13. D. Gusfield, "Bounds for the parametric minimum spanning tree problem," *Proc. West Coast Conf. Comb., Graph Theory and Comput.*, 173-181, 1980.

14. D. Gusfield, "Parametric combinatorial computing and a problem of program module distribution," *J. ACM* 30(3) (1983) 551-563.

15. R. Hassin and A. Tamir, "Maximizing classes of two-parameter objectives over matroids," *Math. Oper. Res.* 14(2) (1989) 362-375.

16. R. M. Karp and J. B. Orlin, "Parametric shortest path algorithms with an application to cyclic staffing," *Disc. Appl. Math.* 3 (1981) 37-45.

17. N. Katoh, T. Tokuyama, and K. Iwano, "On minimum and maximum spanning trees of linearly moving points", *Proc. 33rd IEEE Symp. Found. Comput. Sci.* (1992) 396-405.

18. S. Khuller, B. Raghavachari, N. Young, "Balancing minimum spanning and shortest path trees", *Third Symposium on Discrete Algorithms*, 1993.

19. V. P. Kompella, J. C. Pasquale, and G. C. Polyzos, "Multicasting for Multimedia Applications," *Proc. of IEEE INFOCOM '92* (1992).

20. V. P. Kozyrev, "Finding spanning trees that are optimal with respect to several ranked criteria," *Combinatorial Analysis, No. 7* (1986) 66-73, 163-164.

21. E. L. Lawler, "Optimal cycles in doubly weighted directed linear graphs", *Theory of Graphs, International Symposium*, Gordon and Breach (1967) 209-214.

22. H.-P. Lenhof, J. S. Salowe, and D. E. Wrege, "Two methods to mix shortest-path and minimum spanning trees," (1993), manuscript.

23. J. Matoušek, "Approximations and optimal geometric divide-and-conquer," *Proc. Twenty-third ACM Symp. Theory Comput.* (1991) 505-511.

24. N. Megiddo, "Combinatorial optimization with rational objective functions," *Math. Oper. Res.* 4(4) (1979) 414-424.

25. N. Megiddo, "Applying parallel computation algorithms in the design of serial algorithms," *J. ACM* 30(4) (1983) 852-865.

26. R. Ravi, M. V. Marathe, S. S. Ravi, D. J. Rosenkrantz, H. B. Hunt III, "Many birds with one stone: Multi-objective approximation algorithms," *Proc. 25th Symp. Theory Comput.* (1993) 438-447.

27. G. Robins, personal communication.

Algorithmic Graph Embeddings
(Extended Abstract)

Jianer Chen *

Department of Computer Science, Texas A&M University
College Station TX 77843-3112, USA

Abstract. The complexity of embedding a graph into a variety of topological surfaces is investigated. A new data structure for graph embeddings is introduced and shown to be superior to the previously known data structures. In particular, the new data structure efficiently supports all on-line operations for general graph embeddings. Based on this new data structure, very efficient algorithms are developed to solve the problem "given a graph G and an integer k, construct a genus k embedding for the graph G" for a large range of the integers k and for a large class of graphs.

1 Introduction

Graph embedding is a fundamental yet difficult problem, and it has many applications in diverse problem domains. The most studied is graph planar embeddings. The well-known Hopcroft-Tarjan algorithm shows that a planar embedding of a planar graph can be constructed in linear time [19]. Graph planar embeddings have been studied extensively in a variety of areas such as Computational Geometry [22] and Graph Drawing [9].

On the other hand, the computational complexity of constructing embeddings of a graph into non-planar surfaces is not well-understood yet. The complexity of graph minimum genus problem remained as a basic open problem in Garey and Johnson's list [16] until recently Thomassen proved that the problem is *NP*-complete [24]. Algorithms have also been developed for embedding a graph into a variety of surfaces. It is demonstrated by Furst, Gross, and McGeoch [15] that a maximum genus embedding of a graph can be constructed in time $O(n^4 \log^6 n)$. Filotti described an $O(n^6)$ time algorithm for embedding cubic graphs of minimum genus 1 into the torus [11]. Filotti, Miller, and Reif [12] derived an $O(n^{O(k)})$ time algorithm that embeds graphs of minimum genus $\leq k$ into a surface of genus k, which was improved recently by Djidjev and Reif [10] who developed an algorithm of time $O(2^{O(k)!}n^{O(1)})$. Frederickson and others have considered the complexity of graph embeddings that give the minimum "hammock number" or minimum "face cover", and studied their applications to other computational graph problems [13, 14]. Very recently, Mohar described a linear time algorithm for embedding graphs of crosscap number 1 into the projective plane [21], and Chen described a linear time algorithm for embedding graphs of bounded average genus [2]. Chen, Kanchi, and Kanevsky have also studied approximation algorithms for graph minimum genus embeddings [8].

---

* Supported in part by the United States National Science Foundation grant CCR-9110824 and by a P. R. China HTP-863 grant.

An open problem posed by Furst, Gross, and McGeoch [15] is to determine the complexity of graph embeddings into a general surface. There are several theoretical and practical reasons why this problem should be studied. First, the distribution of graph embeddings into topological surfaces provides a very useful isomorphism heuristic [4, 5, 6, 17]. Secondly, embedding a graph into a certain surface helps efficiently solving other computational graph problems [2, 3, 7, 13, 14, 15, 22]. Finally, study of graph embeddings has direct applications in the areas such as circuit layout and VLSI design.

In this paper, we will develop a number of efficient algorithms that embed a graph into a variety of surfaces. We start by introducing a new data structure for graph embeddings. We demonstrate that the new data structure is superior to the previously known data structures for graph embeddings. In particular, the new data structure efficiently supports *all* on-line operations for general graph embeddings. Based on this new data structure, we present an $O(m \log n)$ time greedy algorithm that constructs an embedding of genus at least $\beta(G)/8$ for a graph G, where n, m, and $\beta(G)$ are the number of vertices, the number of edges, and the cycle rank of the graph G, respectively. Then we show how we can continuously move in time $O(m \log n)$ for a graph from one embedding to another embedding. This implies, in particular, that there is an $O(m \log n)$ time algorithm that, given a planar graph G and an integer $k \leq \beta(G)/8$, constructs an embedding of genus k for G. We will also study the complexity of embedding a graph into a surface of "average genus". We present an $O(m^2)$ time randomized algorithm that, given a graph G and an integer k, either reports with a very small error probability that k is not equal to the average genus of G, or produces an embedding of genus k for G. These computational results demonstrate a very rich and interesting structure for computational complexity of graph embeddings.

We briefly review the fundamentals of the theory of graph embeddings. For further description, see Gross and Tucker [18].

Unless stated explicitly, all graphs in our discussion are supposed to be connected simple graphs in which each vertex is of degree at least 3. An *embedding* must have the "cellularity property" that the interior of every face is simply connected.

A *rotation* at a vertex v is a cyclic permutation of the edge-ends incident on v. A list of rotations, one for each vertex of the graph, is called a *rotation system*.

An embedding of a graph G in an orientable surface induces a rotation system, as follows: the rotation at vertex v is the cyclic permutation corresponding to the order in which the edge-ends are traversed in an orientation-preserving tour around v. Conversely, by the Heffter-Edmonds principle, every rotation system induces a unique embedding of G into an orientable surface. This bijectivity enables us to study graph embeddings based on graph rotation systems, which is a more combinatorial structure. We will interchangeably use the phrases "an embedding of a graph" and "a rotation system of a graph".

Suppose that we insert a new edge $e = (u, v)$ into an embedding $\Pi(G)$, where u and v are vertices of G. There are two possible cases.

If the edge-ends u and v of e are inserted between two corners of the same face f, then the new edge e splits the face f into two faces. In this case, the embedding genus remains the same. If the edge-ends u and v of e are inserted between corners of two different faces f_1 and f_2, then both these faces are merged by e into one larger face. In this case, the embedding genus is increased by 1.

Edge deletion is the inverse operation of edge insertion. Let e be an edge to be deleted from the embedding $\Pi(G)$. If e is on the boundary of two different faces, then deleting the edge e will merge these two faces into a larger face and keep the same embedding genus. If the two sides of e are on the boundary of the same face, then deleting the edge e will result in two new faces and decrease the embedding genus by 1.

2 A new data structure

We will evaluate the performance of a data structure for graph embeddings based on the following graph embedding operations:

- FACE-TRACE(f). output a boundary walk of the face f.
- VERTEX-TRACE(v). output the edges incident on the vertex v in the (circular) ordering of the rotation at v.
- QUERY(c_1, c_2). Return *true* if the two face corners c_1 and c_2 belong to the same face of the current embedding and *false* otherwise.
- INSERT(c_1, c_2, e). Insert the new edge e between the face corners c_1 and c_2.
- DELETE(e). Delete the edge e from the current embedding.
- GENUS. Report the genus of the current embedding.

A number of data structures have been proposed. Most of these data structures are only valid for planar embeddings of a graph. Tamassia [23] proposed a data structure that supports a special case of DELETE operation and each of the other operations in $O(\log n)$ time on planar embedded graphs. Very recently, Italiano, La Poutre, and Rauch [20] proposed a new data structure that supports each of the above operations in $O(\log^2 n)$ time on planar embedded graphs.

Not much has been done for general graph embeddings. Frequently, a rotation system of a graph is represented in the *edge-list form*, which for each vertex v contains the list of its incident edges, arranged in the order according to the rotation at v. It is not difficult to see that this representation does not efficiently support many embedding operations. Another data structure, *the doubly-connected-edge-list* (DCEL) that has been widely used in computational geometry [22], can be directly used for representing a rotation system of a graph. The DCEL structure efficiently supports the operations FACE-TRACE and VERTEX-TRACE, and can be converted in linear time to and from the edge-list form of a graph rotation system [22]. Unfortunately, the DCEL structure does not seem to support the operations INSERT and DELETE efficiently. In the worst case, a single INSERT or DELETE operation may take time as large as $O(n^2)$.

We introduce a new data structure and show that the new data structure efficiently supports all the operations listed at the beginning of this section.

Each face is given by a sequence of vertices and edges that corresponds to a boundary traversing of the face. The vertex appearances and the edge appearances in the sequence will be called *vertex nodes* and *edge nodes*, respectively. The sequence is represented by a concatenable data structure [1].

Definition 1. Let $\Pi(G)$ be an embedding of a graph $G = (V, E)$ with face set F. A *doubly-linked-face-list* (DLFL) for the embedding $\Pi(G)$ is a 4-tuple $L = \langle \mathcal{F}, \mathcal{V}, \mathcal{E}, g \rangle$, where the *face list* $\mathcal{F}$ consists of a set of $|F|$ sequences, each corresponds to a face in the embedding $\Pi(G)$. Moreover, the roots of the sequences are connected by a circular doubly linked list. The *vertex array* $\mathcal{V}$ has $|V|$ items, each $\mathcal{V}[i]$ is a linked list of pointers to the vertex nodes of the sequences in $\mathcal{F}$ that are labeled by the corresponding vertex. The *edge array* $\mathcal{E}$ has $|E|$ items, each $\mathcal{E}[i]$ consists of two pointers to the two edge nodes of the sequences in $\mathcal{F}$ that are labeled by the corresponding edge. The integer g is the genus of the embedding $\Pi(G)$.

The following theorem shows the relationship between DLFL and DCEL.

Theorem 2. *The DLFL structure and the DCEL structure of an embedding of a graph can be converted from one to the other in linear time.*

Corollary 3. *The DLFL structure for an embedding of a graph can be built in linear time and requires linear storage.*

For those operations that are supported by the DCEL structure efficiently, the DLFL structure has equally-good performance.

Theorem 4. *Based on the DLFL structure, the operation FACE-TRACE(f) can be done in time linear to the size of the face f, and the operation VERTEX-TRACE(v) can be done in time linear to the degree of the vertex v.*

It is obvious that the operation GENUS can be done in constant time based on the DLFL structure because the DLFL structure keeps the value of the genus for the embedding $\Pi(G)$. Furthermore, the DLFL structure also supports the other embedding operations efficiently, as given by the following theorem.

Theorem 5. *The DLFL structure supports the operations QUERY, INSERT, and DELETE in logarithmic time.*

3   Embeddings on high genus surfaces

In this section, we present an efficient algorithm that, given a graph G of n vertices and m edges, constructs an embedding of G whose genus is at least $\beta(G)/8$, where $\beta(G) = m - n + 1$ is the *cycle rank* of the graph G.

Let v be a vertex of degree d in a graph G. We say that we *split* the vertex v into two vertices v_1 and v_2 of degree $d' + 1$ and $d - d' + 1$, respectively, if we replace the vertex v in the graph G by two adjacent new vertices v_1 and v_2 such that the vertex v_1 is of degree $d' + 1$ and adjacent to v_2 and d' of the original neighbors of v and the vertex v_2 is of degree $d - d' + 1$ and adjacent to v_1 and the rest $d - d'$ original neighbors of v. Note that a splitting operation on a vertex of a graph does not change the cycle rank of the graph. The inverse operation of vertex splitting is *edge contraction* that deletes an edge $e = \{u, v\}$ as well as the two vertices u and v and adds a new vertex w that is adjacent to all original neighbors of u and v.

The edge contraction operation can be extended to an embedding of the graph G. Let $\Pi(G)$ be an embedding of the graph G such that the rotations at the two ends u and v of the edge e are

$$u: \quad v, u_1, u_2, \cdots, u_s \qquad \text{and} \qquad v: \quad u, v_1, v_2, \cdots, v_t$$

respectively. The contraction of the edge e in the embedding $\Pi(G)$ is an embedding of the graph from the graph G by contracting the edge e in which every vertex has the same rotation as in $\Pi(G)$ except the new added vertex w, whose rotation is $w: \quad u_1, \cdots, u_s, v_1, \cdots, v_t$. It is easy to see that edge contraction operation does not change the embedding genus.

The following definitions apply as well to disconnected graphs in which vertex degree may be less than 3. An *adjacency matching* in a graph H is a pairing in a subset of the edges of H such that two paired edges share a common endpoint. A *maximum adjacency matching* in H is an adjacency matching that maximizes the number of pairs. A *maximal suspended chain* C in H is a simple path in H such that all interior vertices of C have degree 2 in H and both end-vertices of C have degree not equal to 2.

Let G be a 3-regular graph and H a spanning subgraph of G. For each degree 2 vertex v in H, the unique edge in $G - H$ that has v as one of its endpoint will be called "the edge associated with v".

Consider the algorithm given in Fig. 1.

Lemma 6. *The graph H_0 in Step 4 has a one-face embedding $\Pi(H_0)$ whose genus is equal to the number of pairs in the maximum adjacency matching $\mathcal{M}$.*

Lemma 7. *The graph H_h constructed by Step 6 in Algorithm 1 is a spanning subgraph of H in which no vertex has degree 1. Moreover, for each degree 2 vertex in H_h, at most one of its neighbors is of degree 2.*

We need to verify the validity of case 2 in Step 6. Let e_v, e_u, and e_w be the three edges associated with the three vertices v, u, and w, respectively, in case 2 of Step 6 in Algorithm 1.

Lemma 8. *In case 2 of Step 6 in Algorithm 1, either a proper insertion of the edge e_u increases the embedding genus, or proper insertions of the edge e_u and one of the edges e_v and e_w increase the embedding genus.*

Proof. (Sketch) Since C_i is a maximal suspended chain in the graph H_{i-1}, we can talk about the two "sides" of the chain C_i in the embedding $\Pi(H_{i-1})$. If the two sides of the chain C_i belong to different faces in the embedding $\Pi(H_{i-1})$, we can insert the edge e_u so that the embedding genus increases. On the other hand, if the two sides of the chain C_i belong to the same face f of the embedding $\Pi(H_{i-1})$, then inserting the edge e_u will split one side of the chain C_i into two subwalks, each contains one of the vertices v and w, such that the two subwalks belong to different faces in the new embedding, while leave the entire other side of the chain C_i belonging to the same face in the new embedding. Therefore, for at least one of the vertices v and w, its two corners belong to different faces in the new embedding. We can thus insert the associated edge properly to increase the embedding genus. $\qquad\square$

Algorithm 1.

Input: A graph G of n vertices and m edges.

Output: An embedding of G of genus at least $\beta(G)/8$.

1. Use vertex splitting to convert the graph G into a 3-regular graph H;
2. Construct a spanning tree T of H;
3. Construct a maximum adjacency matching $\mathcal{M}$ in the co-tree $H - T$.
4. Construct a one-face embedding $\Pi(H_0)$ for the graph $H_0 = T + \mathcal{M}$;
5. Let $\mathcal{L} = \{C_1, C_2, \ldots, C_h\}$ be the list of all maximal suspended chains in H_0;
6. for each maximal suspended chain C_i in the list $\mathcal{L}$

    case 1. C_i has at most two interior vertices.

    If there is a way to insert the edges associated with the interior vertices of C_i and increase the embedding genus, then insert the edges and increase the embedding genus. Let H_i and $\Pi(H_i)$ be the resulting graph and embedding, respectively

    Otherwise, let $H_i = H_{i-1}$ and $\Pi(H_i) = \Pi(H_{i-1})$.

    case 2. C_i has more than two interior vertices.

    Pick the first three interior vertices v, u, and w on the chain C_i in that order. Insert the edge associated with u. If inserting u does not increase the embedding genus, insert one of the edges associated with v and w to increase the embedding genus. Let H_i and $\Pi(H_i)$ be the resulting graph and embedding, respectively.

    If any of the above edge insertion breaks a maximal suspended chain C_j, $i < j$, in the list $\mathcal{L}$, then replace the chain C_j in the list $\mathcal{L}$ by the two shorter chains and increase h by 1.

7. Arbitrarily insert the edges of H that are not used in Step 6. Let the resulting embedding of the graph H be $\Pi(H)$.
8. In the embedding $\Pi(H)$, contract those edges resulted from the vertex splitting operations in Step 1. The result is an embedding for the graph G.

Fig. 1. Constructing an embedding on a high genus surface

Lemma 9. *The embedding $\Pi(H)$ of graph H constructed by Step 7 of Algorithm 1 has genus at least $\beta(H)/8$.*

Proof. We count how many edges are not inserted in Step 6. By Lemma 7, the graph H_h is a spanning subgraph of H with no degree 1 vertices.

Since the graph H is 3-regular, every vertex in the graph H_h has degree either 2 or 3. Let n_2 and n_3 be the number of vertices of degree 2 and 3, respectively, of the graph H_h. Then $n_2 + n_3 = n$. We say that a degree 2 vertex v in H_h is *covered* by a degree 3 vertex w if v and w are adjacent in H_h. By Lemma 7, in the graph H_h, each degree 2 vertex is adjacent to at most one degree 2 vertex. Thus, every degree 2 vertex in H_h is covered by at least one degree 3 vertex. Since each degree 3 vertex in the graph H_h can cover at most three degree 2 vertices, we have $n_3 \geq n_2/3$. Thus, $n \geq n_2 + n_2/3$. This gives $3n/4 \geq n_2$. Since each edge in $H - H_h$ is associated with exactly two degree 2 vertices in the graph H_h, we conclude that the number of edges in $H - H_h$ is bounded by $3n/8$. In other words, the cycle rank $\beta(H_h)$ of the graph H_h is at least $\beta(H) - 3n/8$.

By Lemma 6, the genus of the embedding $\Pi(H_0)$ is equal to the number of pairs

157

in the maximum adjacency matching $\mathcal{M}$. Moreover, according to Step 6 of Algorithm 1, the embedding genus is increased by at least 1 by inserting at most two edges in $H - H_0$. Therefore, the genus of the embedding $\Pi(H_h)$ is at least half of cycle rank $\beta(H_h)$, which is at least $\beta(H) - 3n/8$.

Since the embedding $\Pi(H)$ constructed by Step 7 of Algorithm 1 has genus at least as larger as that of the embedding $\Pi(H_h)$, and the graph H is 3-regular thus $\beta(H) = \frac{n}{2} + 1$, we conclude that the embedding $\Pi(H)$ has genus at least $\frac{n}{16} + \frac{1}{2}$, which is larger than $\beta(H)/8$. $\square$

Now we can conclude with the following theorem.

Theorem 10. *Algorithm 1 runs in time $O(m \log n)$ and constructs for a given graph G an embedding of genus at least $\beta(G)/8$.*

4 Embeddings on the surface of genus k

The problem "given a graph G and an integer k, construct an embedding of genus k for the graph" is NP-hard in its general form [24]. In this section, we will show, in contrast with this general result, that for a large range of the integers k and for a large class of graphs, the problem can be solved efficiently.

We start with a very useful algorithm that efficiently moves continuously from one embedding of a graph to another embedding of the same graph. This algorithm will be crucial in many cases for finding graph embeddings on a surface of a given genus.

Theorem 11. *There is an $O(m \log n)$ time algorithm such that given two embeddings $\Pi_1(G)$ and $\Pi_2(G)$ of a graph G and an integer γ such that γ is between the genera of $\Pi_1(G)$ and $\Pi_2(G)$, the algorithm constructs an embedding of genus γ for the graph G.*

Proof. Let the genera of the embeddings $\Pi_1(G)$ and $\Pi_2(G)$ be γ_1 and γ_2, respectively. Consider the algorithm given in Fig. 2.

Since each edge deletion on an embedding can decrease the embedding genus by at most 1 and never increase the embedding genus, and since each edge insertion to an embedding can increase the embedding genus by at most 1 and never decrease the embedding genus, each execution of the body of the inner for loop in Algorithm 2 can change the embedding genus by at most 1. Therefore, the algorithm must stop at some point and end up with an embedding $\Pi(G)$ of genus exact γ.

Algorithm 2 consists of traversing the rotation at each vertex in the embedding $\Pi_2(G)$ and rearranging the rotation at each vertex in the embedding $\Pi_1(G)$. Each edge in the embedding $\Pi_1(G)$ is rearranged by an edge deletion followed by an edge insertion. Therefore, the number of edge insertions and edge deletions performed on each vertex of G is proportioned to the degree of the vertex. Thus, the total number of edge insertions and edge deletions performed by Algorithm 2 is proportional to the number m of edges of the graph G. In conclusion, if we use the data structure DLFL introduced in Section 2, Algorithm 2 runs in time $O(m \log n)$. $\square$

158

Algorithm 2.
Input: Two embeddings $\Pi_1(G)$ and $\Pi_2(G)$ and an integer γ between γ_1 and γ_2.
Output: An embedding $\Pi(G)$ of G of genus γ.

1. If $\gamma = \gamma_1$ then Stop
   Otherwise, let $\Pi(G) = \Pi_1(G)$;
2. for each vertex v of G
   Let the rotation at v in the embedding $\Pi_2(G)$ be $v: u_1, u_2, \ldots, u_r$.
   for $i = 2$ to r do
      Delete the edge $e_i = \{v, u_i\}$ from the embedding $\Pi(G)$, then
      reinsert the edge e_i into the embedding so that the edge e_i
      follows the edge e_{i-1} in the rotation at the vertex v, and the
      position of the other endpoint of e_i is unchanged.
      If γ equals the genus of the current embedding $\Pi(G)$, Stop.

Fig. 2. Moving from one embedding to another embedding

Corollary 12. *There is an algorithm such that given two embeddings $\Pi_1(G)$ and $\Pi_2(G)$ of genus γ_1 and γ_2, respectively, for a graph G, $\gamma_1 \leq \gamma_2$, the algorithm constructs in time $O(\max\{(\gamma_2 - \gamma_1)m, m\log n\})$ $\gamma_2 - \gamma_1 + 1$ embeddings $\Pi_{\gamma_1}(G)$, $\Pi_{\gamma_1+1}(G), \ldots, \Pi_{\gamma_2}(G)$ for the graph G, such that the genus of the embedding $\Pi_i(G)$ is i, for $\gamma_1 \leq i \leq \gamma_2$,*

The algorithm in Corollary 12 is optimal when $\gamma_2 - \gamma_1 = \Omega(\log n)$ since constructing any embedding of the graph G takes time at least $\Omega(m)$.

Theorem 13. *There is an $O(n\log n)$ time algorithm such that given a planar graph G and an integer $k \leq \beta(G)/8$, the algorithm constructs an embedding of genus k for the graph G.*

Proof. The algorithm first constructs a planar embedding for the graph G in linear time [19], then constructs an embedding of genus at least $\beta(G)/8$ for the graph G in time $O(m\log n) = O(n\log n)$, according to Theorem 10. Now the theorem follows directly from Theorem 11. □

Theorem 14. *There is a polynomial time algorithm such that given a planar graph G and an integer k, the algorithm either reports that there is no genus k embedding for the graph G, or constructs an embedding of genus k for G.*

Theorem 14 can be extended to any class of graphs for which the minimum genus embedding problem is solvable in polynomial time, such as the graph classes considered in [10, 11, 12, 21].

Now we study the graph embeddings that are related to the "average genus" of the graphs. Each graph G is associated with a sequence of integers $g_0, g_1, g_2, \ldots$, called the *genus distribution* of G, where g_i is the number of embeddings of genus i for the graph G. The *average genus* of G is defined to be the value

$$\gamma_{avg}(G) = \frac{\sum_{i=0} i \cdot g_i}{\sum_{i=0} g_i}$$

The average genus of a graph plays an important role in the recent study of topological invariants of graphs [2, 4, 5, 6, 17].

Denote by W_G the total number of embeddings of the graph G.

Lemma 15. *For any real number $\epsilon > 0$, there are at least $\epsilon W_G/(1 + \epsilon)$ embeddings of the graph G that are of genus $\leq (1 + \epsilon)\gamma_{avg}(G)$.*

Lemma 16. *For any real number $\epsilon > 0$, there are at least $\epsilon W_G/3$ embeddings of the graph G that are of genus $\geq (1 - \epsilon)\gamma_{avg}(G)$.*

For any real number r, denote by *round(r)* the unique integer in the interval $(r - 0.5, r + 0.5]$.

Theorem 17. *For any real number $\delta > 0$, there is a time $O(m^2)$ randomized algorithm such that given a graph G and an integer k, the algorithm either reports $k \neq round(\gamma_{avg}(G))$ with error probability less than δ, or constructs an embedding of genus k for the graph G.*

Proof. (Sketch) Randomly pick $2m$ embeddings of the graph G. According to Lemmas 15 and 16, if either all these embeddings have genus larger than k or all these embeddings have genus smaller than k, then with a very high probability we have $k \neq round(\gamma_{avg}(G))$. Otherwise, pick among these embeddings one embedding of genus less than or equal to k and one embedding of genus larger than or equal to k. Then apply Theorem 11. □

5 Final remarks

We have developed a new data structure for graph embeddings. The data structure is superior to the previously known data structures and efficiently supports all online operations for graph embeddings. We have shown a number of applications of the new data structure, including an $O(m \log n)$ time algorithm for constructing embeddings of graphs on surfaces of high genus, and an $O(m \log n)$ time algorithm for continuously moving from one embedding of a graph to another embedding of the same graph. We have presented efficient algorithms for constructing all kinds of embeddings for planar graphs and for other classes of graphs. An efficient randomized algorithm has also been developed to construct graph embeddings on surfaces of genus equal to the average genus of the graph. Our computational results indicate that there is a very interesting and rich computational structure related to graph embeddings, which definitely deserves further investigation.

Our results show that construction of an embedding of a graph on high genus surfaces is in general computationally feasible. A closely related problem is the complexity of constructing low genus embeddings for a graph. It is unknown whether there is a polynomial time algorithm that approximates the minimum genus embedding of a graph to a constant ratio. A recent result by Chen, Kanchi, and Kanevsky [8] partially hints the difficulty of approximating graph minimum genus embeddings: for any real number ϵ, $0 \leq \epsilon < 1$, the problem of embedding a graph G of n vertices into a surface of genus $\gamma_{\min}(G) + n^\epsilon$ is NP-hard.

References

1. Aho, A. V., Hopcroft, J. E., and Ullman, J. D.: *The Design and Analysis of Computer Algorithms*. Addison-Wesley, (1974)
2. Chen, J.: A linear time algorithm for isomorphism of graphs of bounded average genus. *SIAM J. on Discrete Mathematics* 7 (1994) 614-631
3. Chen, J., Archdeacon, D., and Gross, J. L.: Maximum genus and connectivity. *Discrete Mathematics* (1995) to appear
4. Chen, J., Gross, J. L.: Limit points for average genus I. 3-connected and 2-connected simplicial graphs. *J. Comb. Theory Ser. B* 55 (1992) 83-103
5. Chen, J., Gross, J. L.: Limit points for average genus II. 2-connected non-simplicial graphs. *J. Comb. Theory Ser. B* 56 (1992) 108-129
6. Chen, J., Gross, J. L.: Kuratowski-type theorems for average genus. *J. Comb. Theory Ser. B* 57 (1993) 100-121
7. Chen, J. and Kanchi, S. P.: Graph embeddings and graph ear decompositions. *Lecture Notes in Computer Science* 790 (1994) 376-387
8. Chen, J., Kanchi, S. P., and Kanevsky, A.: On the complexity of graph embeddings. *Lecture Notes in Computer Science* 709 (1993) 234-245
9. Di Battista, G., Eades, P., and Tamassia, R.: Algorithms for automatic graph drawing: an annotated bibliography. *Tech. Report* (1993) Dept. Computer Science, Brown University
10. Djidjev, H. and Reif, J.: An efficient algorithm for the genus problem with explicit construction of forbidden subgraphs. *Proc. 23rd Annual ACM Symposium on Theory of Computing* (1991) 337-347
11. Filotti, I. S.: An algorithm for imbedding cubic graphs in the torus. *Journal of Computer and System Sciences* 20 (1980) 255-276
12. Filotti, I. S., Miller, G. L., and Reif, J. H.: On determining the genus of a graph in $O(v^{O(g)})$ steps. *Proc. 11th Annual ACM Symp. on Theory of Comput.* (1979) 27-37
13. Frederickson, G. N.: Using cellular graph embeddings in solving all pairs shortest paths problems. *Proc. 30th IEEE Symp. on Foundations of Computer Science* (1989) 448-453
14. Frederickson, G. N. and Janardan, R.: Designing networks with compact routing tables. *Algorithmica* 3 (1988) 171-190
15. Furst, M. L., Gross, J. L., and McGeoch, L. A.: Finding a maximum-genus graph imbedding. *Journal of ACM* 35-3 (1988) 523-534
16. Garey, M. R. and Johnson, D. S.: *Computers and Intractability: A Guide to the Theory of NP-Completeness*. W. H. Freeman, 1979
17. Gross, J. L. and Furst, M. L.: Hierarchy for imbedding-distribution invariants of a graph. *J. Graph Theory* 11 (1987) 205-220
18. Gross, J. L., Tucker, T. W.: *Topological Graph Theory*. Wiley-Interscience, New York (1987)
19. Hopcroft, J. E. and Tarjan, R. E.: Efficient planarity testing. *Journal of the ACM* 21 (1974) 549-568.
20. Italiano, G. F., La Poutre, J. A., and Rauch, M. H.: Fully dynamic planarity testing in planar embedded graphs. *Lecture Notes in Computer Science* 726 (1993) 576-590.
21. Mohar, B.: Projective planarity in linear time. *Journal of Algorithms* 15 (1993) 482-502
22. Preparata, F. P. and Shamos, M. I.: *Computational Geometry: An Introduction*. Springer-Verlag 1985
23. Tamassia, R.: A dynamic data structure for planar graph embedding. *Lecture Notes in Computer Science* 317 (1988) 576-590.
24. Thomassen, C.: The graph genus problem is *NP*-complete. *Journal of Algorithms* 10 (1989) 568-576

Analysis of Quorum-Based Protocols for Distributed (k+1)-Exclusion *

Divyakant Agrawal, Ömer Eğecioğlu, and Amr El Abbadi

Department of Computer Science, University of California
Santa Barbara, CA 93106, USA

Abstract. A generalization of the majority quorum for the solution of the distributed $(k + 1)$-exclusion problem is proposed. This scheme produces a family of quorums of varying sizes and availabilities indexed by integral divisors r of k. The cases $r = 1$ and $r = k$ correspond to known majority based quorum generation algorithms MAJ and DIV, whereas intermediate values of r interpolate between these two extremes. A cost and availability analysis of the proposed methods is also presented.

1 Introduction

The problem of distributed mutual exclusion has been extensively studied and many interesting protocols for its solution have been proposed. Most of these protocols attempt to provide high performance by reducing the number of messages involved or by improving the degree of fault-tolerance and hence improving the chances of achieving mutual exclusion in the presence of site and communication failures. A generalization of the mutual exclusion problem is the *k-mutual exclusion problem*, where no more than k processes are allowed to enter the critical section simultaneously. Since $k + 1$ or more processes are excluded, this problem is also referred to as the $(k + 1)$-*exclusion problem*.

The distributed $(k + 1)$-exclusion problem was first solved by Raymond [12], who provided a simple extension to the Ricart and Agrawala's mutual exclusion algorithm [13]. Srimani and Reddy [15] improved on this protocol by using the notion of *privilege* of Suzuki and Kasami [16]. This solution reduces the number of required messages to achieve mutual exclusion. Recently, there has been a significant interest in fault-tolerant methods to solve the $(k + 1)$-exclusion problem based on the notion of quorums [6]. Fujita et al. [4] discuss some simple techniques and then propose a scheme with small quorum sizes. Huang et al. [7] propose an alternative method with small quorums and high availability in the presence of failures.

The majority quorum algorithm [6] for mutual exclusion has been widely used to develop various quorum-based protocols for mutual exclusion as well as for $(k + 1)$-exclusion by a suitable partitioning of the sites [1, 11, 9, 4]. In this

---

* Supported in part by the NSF under grant number IRI94-11330, by the NASA under grant number NAGW-3888.

paper we generalize the majority quorum algorithm for constructing quorums for distributed $(k+1)$-exclusion and analyze the resulting methods. We propose a sequence of algorithms MAJ_r with varying costs and availabilities, indexed by integral divisors r of k. When $r = 1$, the quorums correspond to those produced by MAJ (which partitions the sites into a single class), and when $r = k$, the quorums correspond to the quorums produced by DIV (which partitions the sites into k classes) [4]. For intermediate values of r, a sliding tradeoff between the number of quorum sets and quorum size is achieved.

Our analysis of availability is based on estimates for truncated binomial sums [2, 11], and the Vandermonde convolution identity. Recently, Kakugawa et al. [8] analyzed MAJ versus a centralized site solution adapted for $(k + 1)$-exclusion called SGL, and showed that for networks with low probability of site failure, MAJ provides optimal availability performance over *all* quorum-based mechanisms, whereas for sites with a high probability of failures SGL gives higher availability than MAJ. Our analysis is similar except that our quorums generalize majority, and instead of the two extremes, we consider a whole spectrum of protocols. The analytical results we present are of two kinds:

1. For p close to zero (where p is the probability that a site is up in a system of n sites), there is a nonzero p_n such that for $p < p_n$, DIV has better availability than MAJ. Furthermore the limit of p_n as n gets large is nonzero.
2. For p close to 1, the increase in availability achieved by MAJ for large n is insignificant compared to that attained by MAJ_r for $r > 1$.

Note that 1 is mostly of theoretical interest, whereas 2 has practical implications for systems with low probability of site failures. This is because the size of the quorums produced by MAJ_r for $r > 1$ are as small as a half of those produced by MAJ, and consequently the communication costs of acquiring such quorums is less than the cost of acquiring a MAJ quorum.

2 The Problem Statement

A distributed system consists of a set of sites that communicate with each other by sending messages over a communication network. We assume that every site has the capability to send a message to any other site when there is a communication path between them. The sites are either *fail-stop* or may fail to send or receive messages. Communication links may fail by crashing, or by failing to deliver messages. Although quorum-based protocols are resilient both site and communication failures, our analysis assumes site failures only [11, 8].

Distributed mutual exclusion is a classical technique for providing access to shared resources. We postulate the existence of a resource in the network, which may be accessed by a single process at a time. To access the resource, a process (site) p_i is required to receive permission from a set of sites S_i. If all sites in S_i grant permission to p_i, then it is allowed to access the resource. To ensure mutual exclusion the sets S_i are required to satisfy the *intersection property*: For any $i \neq j$, $S_i \cap S_j \neq \phi$. These and related concepts were formalized and analyzed in terms of the notions of *quorums* and *coteries* [6, 10, 5]. In $(k + 1)$-exclusion, up to k processes are allowed to access the resource simultaneously. Thus, if we consider $k+1$ sets of sites that grant permission to access the resource then there must exist at least two among these $k + 1$ sets with a non-empty intersection. The $(k + 1)$-exclusion problem can now be stated in terms of the requirements **(I)** and **(II)** below:

(I) The (k+1)-Intersection Property. For any $k + 1$ sets $S_1, S_2, \cdots, S_{k+1}$, there exist two distinct sets S_i and S_j such that $S_i \cap S_j \neq \phi$.

Note that quorums constructed to ensure the traditional mutual exclusion condition also ensure the above property. Hence, in order to eliminate trivial solutions to the $(k+1)$-exclusion problem we add an additional restriction [4, 7].

(II) The k-Non-intersection Property. There exist k sets $S_1, S_2, \cdots, S_k$ such that for any two distinct sets S_i and S_j, $S_i \cap S_j = \phi$.

3 A General Paradigm for $(k + 1)$-Exclusion

One of the simplest approaches to ensure mutual exclusion in a distributed system is to use majority quorums [6] of size $\lfloor \frac{n}{2} \rfloor + 1$. For 3-exclusion, we can reduce the size of the permission sets to $\lfloor \frac{n}{3} \rfloor + 1$. Clearly, the 3-intersection property holds since any three sets of sites with size $\lfloor \frac{n}{3} \rfloor + 1$ chosen from n sites will always have two sets with non-empty intersection. Similarly, the 2-non-intersection property also holds for $n > 5$, since it is possible to construct two disjoint sets when only one-third the number of sites from n are used for each. For $(k + 1)$-exclusion, it suffices to take the quorum size to be $\lfloor \frac{n}{k+1} \rfloor + 1$. This majority based construction for the $(k + 1)$-exclusion problem is referred to as MAJ [4].

One possibility to achieve $(k + 1)$-exclusion is to consider k instances of any mutual exclusion solution. A process wishing $(k + 1)$ exclusive access to a resource acquires permission from any of the k instances. This ensures the $(k + 1)$-intersection property since any $k + 1$ quorums chosen will consist of at least two quorums in the same instance of the mutual exclusion solution, and hence must have a non-empty intersection. Similarly, the k-non-intersection property is satisfied if each of the k processes choose a quorum from different instances of the mutual exclusion solution. This construction for the $(k + 1)$-exclusion problem is referred to as DIV [4].

The two generalizations MAJ and DIV of majority quorums for the solution of the $(k + 1)$-exclusion problem are at the opposite ends of a spectrum. In MAJ, the original mutual exclusion problem is generalized whereas in DIV the sites in the network are partitioned into k classes with each class using any traditional approach to enforce mutual exclusion. We explore the possibility of enforcing $(k + 1)$-exclusion by varying the number r of classes from 1 to k, and define a quorum generation method MAJ_r for any r dividing k. To simplify the presentation, we assume we are given a set of n sites where $n = kN$ for some

$N \geq 1$. In MAJ_r, the $(k + 1)$-exclusion problem is solved by partitioning the sites into r disjoint classes where $r = k/i$, for some integer i. Within each class, we choose the quorums of size q_r such that at least two sets from any collection of $i+1$ sets within the class intersect. More precisely MAJ_r denotes the method in which

1. The sites $1, 2, \ldots, n$ are partitioned into r classes of size $n/r = iN$ each.
2. From each class, all subsets of size

$$q_r = \left\lfloor \frac{iN}{i+1} \right\rfloor + 1 = \left\lfloor \frac{n}{k+r} \right\rfloor + 1 \qquad (1)$$

are taken as quorums. Here $r = k/i$.

It should be clear that MAJ_r produces sets that satisfy both (I) and (II), aside from the trivial cases of small parameters for which $\lfloor \frac{n}{k+r} \rfloor = \lfloor \frac{n}{k+r-1} \rfloor$ (see [17]). Furthermore, $MAJ_1 = MAJ$, and $MAJ_k = DIV$.

4 Availability Measures

The communication cost associated with obtaining mutual exclusion by using the quorum approach is directly proportional to the quorum size. The availability and the fault-tolerance characteristics of a particular method are determined by the number of ways in which a quorum can be constructed from a given number of sites in the network. The total number of sets produced by MAJ_r is

$$r \binom{iN}{q_r}, \qquad (2)$$

where q_r is as given in (1). Let $q_1 = q_{MAJ}$ be the quorum size and T_{MAJ} be the number of sets in MAJ. Then

$$q_{MAJ} = \left\lfloor \frac{n}{k+1} \right\rfloor + 1 , \qquad T_{MAJ} = \binom{n}{q_{MAJ}} . \qquad (3)$$

The other extreme of the partitioning approach arises when $r = k$. The quorum size $q_k = q_{DIV}$ and the total number of quorums T_{DIV} for DIV are given by the formulas

$$q_{DIV} = \left\lfloor \frac{n}{2k} \right\rfloor + 1 , \qquad T_{DIV} = k \binom{\frac{n}{k}}{q_{DIV}} . \qquad (4)$$

When MAJ and DIV are evaluated purely in terms of the communication costs incurred to enforce $(k + 1)$-exclusion, DIV is preferable due to its smaller sized quorums. However, if the evaluation criterion includes the number of quorum sets produced, the outcome is not so trivial. Fujita et al. [4] conjectured that the partitioned approach restricts the number of ways a quorum can be selected and hence will provide inferior availability. First we explore this issue in the context of these two approaches and isolate the instances in which DIV actually performs better than MAJ.

Suppose p is the probability that a site is up. If $q = q_r$ is the quorum size given by the formula (1), then the probability $C_r(p)$ that a quorum set is available in a given class for the method MAJ_r is

$$C_r(p) = \binom{iN}{q} p^q (1-p)^{iN-q} + \binom{iN}{q+1} p^{q+1} (1-p)^{iN-q-1} + \cdots + \binom{iN}{iN} p^{iN} . \quad (5)$$

Let $AV_r(p)$ denote the probability that a quorum set is available when the method MAJ_r is used. For the extreme cases of MAJ and DIV, we also use the notation $AV_{MAJ}(p)$ for $AV_1(p)$, and $AV_{DIV}(p)$ for $AV_k(p)$. The probability that none of the r classes of the partition has a quorum set available is $(1-C_r(p))^r$, and therefore

$$AV_r(p) = 1 - (1 - C_r(p))^r . \quad (6)$$

Example 1. Suppose $k = 2$ and $n = 10$. Then for $i = 2$ we get $r = 1$ and $MAJ_1 = MAJ$. From (5),

$$C_{MAJ}(p) = 210p^4(1-p)^6 + 252p^5(1-p)^5 + 210p^6(1-p)^4 + 120p^7(1-p)^3 +$$
$$+ 45p^8(1-p)^2 + 10p^9(1-p) + p^{10}.$$

For $i = 1$, $r = 2$ and $MAJ_2 = DIV$. In this case $C_{DIV}(p) = 10p^3(1-p)^2 + 5p^4(1-p) + p^5$. Therefore, from (6)

$$AV_{MAJ}(p) = C_{MAJ}(p),$$
$$AV_{DIV}(p) = 20p^3 - 30p^4 + 12p^5 - 100p^6 + 300p^7 - 345p^8 + 180p^9 - 36p^{10} .$$

Since $AV_{DIV}(p) - AV_{MAJ}(p) = 20p^3(1-p)^6(1-6p)$, setting $AV_{DIV}(p) > AV_{MAJ}(p)$ and solving for p, we find that whenever $p < 1/6$, DIV provides better quorum availability than MAJ.

5 Availability of *MAJ* and *DIV* for 3-Exclusion

We now generalize the above example by deriving asymptotic results for the availabilities of MAJ_r for $r = 1$ (MAJ) and $r = k$ (DIV). For simplicity of exposition, here we present our proofs for the case $k = 2$ (3-exclusion), although a similar analysis can be carried out for arbitrary k. We start out by studying the polynomials $AV_{MAJ}(p)$ and $AV_{DIV}(p)$ for large n. Since in this case

$$q_1 = q_{MAJ} = \left\lfloor \frac{n}{3} \right\rfloor + 1 , \quad q_2 = q_{DIV} = \left\lfloor \frac{n}{4} \right\rfloor + 1$$

from formula (1), it is convenient to assume that $n = 12m$ for some $m \geq 1$. Then $q_{MAJ} = 4m + 1$, $q_{DIV} = 3m + 1$ with

$$AV_{MAJ}(p) = \sum_{j=4m+1}^{12m} \binom{12m}{j} p^j (1-p)^{12m-j} , \quad (7)$$

$$AV_{DIV}(p) = 1 - \left[1 - \sum_{j=3m+1}^{6m} \binom{6m}{j} p^j (1-p)^{6m-j} \right]^2 . \quad (8)$$

The values of p in $0 < p < p_n$ for which $AV_{DIV}(p) > AV_{MAJ}(p)$ for small values of $n = 12m$ are tabulated in Table 1 (computed with the aid of MACSYMA). Even though the numbers p_n are decreasing, the limit of p_n as n gets large is nonzero, as shown in Theorem 1.

n	12	24	36	48	60	72	84	96
p_n	0.0461	0.0423	0.0408	0.0399	0.0394	0.0391	0.0388	0.0386

Table 1. Range of values $0 < p < p_n$ for which $AV_{DIV}(p) > AV_{MAJ}(p)$.

Theorem 1. *For 3-exclusion, DIV provides better availability for large n than MAJ when the probability p of a site being up satisfies $p < 0.0299$.*

Proof. We set $AV_{DIV}(p) > AV_{MAJ}(p)$ and find the values of p that satisfy this inequality as n gets large. From (7) and (8), $AV_{DIV}(p) > AV_{MAJ}(p)$ if and only if

$$\left[\sum_{j=0}^{3m} \binom{6m}{j} p^j (1-p)^{6m-j} \right]^2 < \sum_{j=0}^{4m} \binom{12m}{j} p^j (1-p)^{12m-j} \tag{9}$$

Let $q = p/(1-p)$. Factoring out and cancelling $(1-p)^{12m}$ from both sides, inequality (9) can be written in terms of q as

$$\left[\sum_{j=0}^{3m} \binom{6m}{j} q^j \right]^2 < \sum_{j=0}^{4m} \binom{12m}{j} q^j . \tag{10}$$

Write

$$\left[\sum_{j=0}^{3m} \binom{6m}{j} q^j \right]^2 = \sum_{j=0}^{6m} c_j q^j .$$

For $j \leq 3m$,

$$c_j = \sum_{i=0}^{j} \binom{6m}{i} \binom{6m}{j-i} = \binom{12m}{j}$$

by the Vandermonde convolution identity [3, 14]. This means that the terms involving powers q^j up to $j = 3m$ on both sides of (10) are identical. Cancelling these terms, (10) can be written as

$$\sum_{j=4m+1}^{6m} c_j q^j < \sum_{j=3m+1}^{4m} \left[\binom{12m}{j} - c_j \right] q^j .$$

For large m and q close to zero, we can discard the effect of all but the lowest power of q on both sides. Thus for large n, in order for the inequality $AV_{DIV}(p) > AV_{MAJ}(p)$ to hold, the parameter q should satisfy

$$c_{4m+1}q^{4m+1} < \left[\binom{12m}{3m+1} - c_{3m+1}\right]q^{3m+1}.$$

However,

$$c_{3m+1} = \sum_{i=1}^{3m}\binom{6m}{i}\binom{6m}{3m+1-i} = \binom{12m}{3m+1} - 2\binom{6m}{3m+1}$$

by the Vandermonde identity again. Similarly,

$$c_{4m+1} < \binom{12m}{4m+1}.$$

Therefore, in order to guarantee that $AV_{DIV}(p) > AV_{MAJ}(p)$, it suffices to have

$$q^m < 2\binom{6m}{3m+1}/\binom{12m}{4m+1}.$$

Using Stirling's approximation,

$$\binom{12m}{4m+1} \sim \frac{3^{12m+1}}{2^{8m+1}\sqrt{3\pi m}} \quad \text{and} \quad \binom{6m}{3m+1} \sim \frac{2^{6m}}{\sqrt{3\pi m}}.$$

It follows that

$$q^m < (\frac{2^{6m+1}}{\sqrt{3\pi m}})/(\frac{3^{12m+1}}{2^{8m+1}\sqrt{3\pi m}}) = \frac{2^{14m}}{3^{12m+1}}.$$

Taking m-th roots and letting m go to infinity, we find that the condition $q < 2^{14}/3^{12}$ guarantees that $AV_{DIV}(p) \geq AV_{MAJ}(p)$. In terms of p this bound becomes

$$p < \frac{2^{14}/3^{12}}{1 + 2^{14}/3^{12}} = 0.029907...$$

which is the content of the Theorem.

The above result for MAJ and DIV for 3-exclusion can be generalized in two directions, both valid for large n. First of all, it is possible to analytically compare the availability of MAJ and DIV for arbitrary k and find the values of p for which DIV performs better than MAJ as a function of k. Alternately, given k, it is possible to compute the range of values of p for which MAJ_r performs better than MAJ. In the former setting there is a constant $b_k = O(1/k^2)$ such that whenever $0 < p < b_k$, $AV_k(p) > AV_{MAJ}(p)$ for large n (i.e. in this range of p, DIV provides better availability than MAJ for $(k+1)$-exclusion). For the latter generalization, there is a function $B(k,r)$ such that for $p < B(k,r)$, $AV_r(p) > AV_{MAJ}(p)$ for large n. (i.e. in this range of p, MAJ_r provides better availability than MAJ for $(k+1)$-exclusion). These generalizations are of theoretical interest as they are valid only for unrealistically small values of p, and therefore are not given here.

6   Availability of MAJ_r Quorums

In general MAJ has better availability than DIV as well as all the other MAJ_r, except for systems with high site failure probability. On the other hand, from the formula in (1), MAJ quorums have twice the size of DIV, and are always larger than MAJ_r quorums for $r > 1$. In this section, we show that for large n, and p close to 1, the increase in availability provided by MAJ itself as compared to MAJ_r is actually quite small. This is significant since in most current systems we expect sites to have a low probability of failure. Hence in these cases we can use the smaller sized quorums of MAJ_r without losing much on availability, while reducing the communication overhead by up to a factor of 2.

Let $D_r(p) = |AV_{MAJ}(p) - AV_r(p)|$ denote the magnitude of the absolute error in availability made when MAJ_r instead of MAJ itself is used. Then

$$D_r(p) = \left| \sum_{j=0}^{q_{MAJ}-1} \binom{n}{j} p^j (1-p)^{n-j} - \left[\sum_{j=0}^{q_r-1} \binom{n/r}{j} p^j (1-p)^{n/r-j} \right]^r \right|$$

Factoring $(1-p)^n$ and letting $q = p/(1-p)$, we can write

$$D_r(p) = (1-p)^n \left| \sum_{j=0}^{q_{MAJ}-1} \binom{n}{j} q^j - \left[\sum_{j=0}^{q_r-1} \binom{n/r}{j} q^j \right]^r \right|$$

Consider the coefficients c_j in the expansion

$$\left[\sum_{j=0}^{q_r-1} \binom{n/r}{j} q^j \right]^r = \sum_{j=0}^{r(q_r-1)} c_j q^j \ .$$

By the Vandermonde convolution identity [3, 14], $c_j \leq \binom{n}{j}$, with equality in the range $0 \leq j \leq q_r - 1$. Therefore

$$D_r(p) = (1-p)^n \left| \sum_{j=q_r}^{q_{MAJ}-1} \binom{n}{j} q^j - \sum_{j=q_r}^{r(q_r-1)} c_j q^j \right| \leq (1-p)^n \sum_{j=q_r}^{r(q_r-1)} \binom{n}{j} q^j \ .$$

However, q^j is an increasing function of j whenever $p \geq 0.5$. Thus for $p \geq 0.5$

$$D_r(p) \leq p^{r(q_r-1)}(1-p)^{n-r(q_r-1)} \sum_{j=q_r}^{r(q_r-1)} \binom{n}{j} \ .$$

But since $1 \leq r \leq k$, $r(q_r - 1) = (r/(k+r))\,n \leq n/2$, and therefore

$$\sum_{j=q_r}^{r(q_r-1)} \binom{n}{j} \leq 2^{n-1} \ .$$

Furthermore, $n - (r/(k+r))n = (k/(k+r))n$, and 1 is an upper bound for $p^{r(q_r-1)}$. Therefore

$$D_r(p) \leq \frac{1}{2}[2(1-p)^{\frac{k}{k+r}}]^n .$$

If p is close enough to 1 so that $2(1-p)^{\frac{k}{k+r}} < 1$, or equivalently when

$$p > 1 - (\frac{1}{2})^{\frac{k+r}{k}} , \tag{11}$$

the error $D_r(p)$ goes to zero as n gets large. To summarize, we have proved

Theorem 2. *If p is in the range given by (11), then*

$$|AV_{MAJ}(p) - AV_r(p)| \leq \frac{1}{2}[2(1-p)^{\frac{k}{k+r}}]^n , \tag{12}$$

and the right hand side of (12) goes to zero as n gets large.

To get a rough idea of the magnitude of the difference in availability which will hold for all algorithms MAJ_r at once, we note that $(k+r)/k \leq 2$ since r is a divisor of k. Therefore whenever $p > 1 - (1/2)^2 = 0.75$, the error in availability satisfies

$$|AV_{MAJ}(p) - AV_r(p)| \leq \frac{1}{2}(2\sqrt{1-p})^n , \tag{13}$$

regardless of the value of r.

Example 2. Consider a system with $n = 100$ sites in which the probability of a site being up is $p = 0.9$. From (13) the difference between the availabilities of MAJ and DIV for any $(k+1)$-exclusion is about 6.3×10^{-21}, a negligible amount. On the other hand when $k = 9$, MAJ requires quorums of size $q_{MAJ} = 11$, while DIV requires quorums of size only $q_{DIV} = 6$.

7 Discussion

In this paper we proposed a family of quorum-based protocols MAJ_r that generalize majority quorums for distributed $(k+1)$-exclusion. These protocols are indexed by integral divisors r of k, with $MAJ_1 = MAJ$ and $MAJ_k = DIV$. In addition, we analyzed the whole spectrum of resulting protocols with respect to availability and communication cost. We first showed that each MAJ_r, $r > 1$, provides higher availability than MAJ in a large network where the probability of a site being up is sufficiently small. Next, we concentrated on the other end of the probability spectrum in which the network has a low probability of site failure. In this case, we showed that the increase in availability achieved by MAJ over any member of the family MAJ_r, $r > 1$ decreases rapidly as the number of sites gets large. Since the communication overhead of solving the $(k+1)$-exclusion problem is directly proportional to the quorum sizes, reductions up to a factor of two in communication overhead can be achieved without significant sacrifice in availability.

References

1. D. Agrawal and A. El Abbadi. Exploiting Logical Structures of Replicated Databases. *IPL*, 33(5):255–260, Jan. 1990.
2. M. Ahamad and M. H. Ammar. Performance Characterization of Quorum-Consensus Algorithms for Replicated Data. *IEEE Trans. on Software Engineering*, 15(4):492–495, Apr. 1989.
3. D. Cohen. *Basic Techniques of Combinatorial Theory*. John Wiley & Sons, 1978.
4. S. Fujita, M. Yamashita, and T. Ae. Distributed k-mutual exclusion problem and k-coteries. In *Proc. on the Symp. on Algorithms*, pp. 22–31, 1991.
5. H. Garcia-Molina and D. Barbara. How to assign votes in a distributed system. *J. of the ACM*, 32(4):841–860, Oct. 1985.
6. D. K. Gifford. Weighted Voting for Replicated Data. In *Proc. of 7th ACM Symp. on Operating Systems Principles*, pp. 150–159, Dec. 1979.
7. S. Huang, J. Jiang, and Y. Kuo. K-coteries for fault-tolerant k entries to a critical section. In *Proc. of 13th Int. Conf. on Dist. Comp. Sys.*, pp. 74–81, June 1993.
8. H. Kakugawa, S. Fujita, M. Yamashita, and T. Ae. Availability of k-Coteries. *IEEE Trans. on Computers*, 42(5):553–558, May 1993.
9. A. Kumar. Hierarchical Quorum Consensus - A New Algorithm for Managing Replicated Data. *IEEE Trans. on Computers*, 39(9):996–1004, Sep. 1991.
10. M. Maekawa. A $\sqrt{n}$ algorithm for mutual exclusion in decentralized systems. *ACM Trans. on Computer Systems*, 3(2):145–159, May 1985.
11. S. Ranagarajan and S. Tripathi. Robust Distributed Mutual Exclusion Algorithms. In *Proc. of 5th Int. Workshop on Distributed Algorithms*, pp. 295–308, Oct. 1991. Lecture Notes in Computer Science, 579, Springer-Verlag.
12. K. Raymond. A distributed algorithm for multiple entries to a critical section. *IPL*, 30(4):189–193, Feb. 1989.
13. G. Ricart and A. K. Agrawala. An optimal algorithm for mutual exclusion in computer networks. *Communications of the ACM*, 24(1):9–17, January 1981.
14. J. Riordan. *An Introduction to Combinatorial Analysis*. Princeton Univ. Press, 1978.
15. P. K. Srimani and R. L. N. Reddy. Another distributed algorithm for multiple entries to a critical section. *IPL*, 41(1):51–57, Jan. 1992.
16. I. Suzuki and T. Kasami. A Distributed Mutual Exclusion Algorithm. *ACM Trans. on Computer Systems*, 3(4):344–349, Nov. 1985.
17. S.-M. Yuan and H.-K. Chang. Comments on "Availability of k-Coterie". *IEEE Trans. on Computers*, 43(12):1457, Dec. 1994.

A Highly Fault-Tolerant Quorum Consensus Method for Managing Replicated Data

Xuemin Lin[1] and Maria E. Orlowska[2]

[1] Department of Computer Science, University of Western Australia
Nedlands, WA 6907, Australia. e-mail: lxue@cs.uwa.oz.au.
[2] Department of Computer Science, The University of Queensland
QLD 4072, Australia. email: maria@cs.uq.oz.au.

Abstract. The main objective of data replication is to provide high availability of data for processing transactions. Quorum consensus (QC) methods are frequently applied to managing replicated data. In this paper, we present a new QC method. The proposed QC approach has a low message overhead: 1) In the best case, each transaction operation process needs only to communicate with $O(\sqrt{n}\log^{1-\frac{1}{2\log_3 2}} n)$ ($\approx O(\sqrt{n}\log^{0.208} n)$) remote sites ($n$ is the number of sites storing the manipulating data item). 2) In the worst case, each transaction operation process may be forced to communicate with $O(\sqrt{n}\log^{\frac{1}{2\log_3 2}} n)$ ($\approx O(\sqrt{n}\log^{0.792} n)$) remote sites. Further, we can show that the proposed QC method is highly fault-tolerant. The proposed approach is also fully distributed, that is, each site in a distributed system bears equal responsibility.

Key words: concurrency control, distributed computing, fault tolerance, quorum consensus method.

1 Introduction

Through the replication of data, distributed database system reliability can be increased if an effective approach is found for managing replicated data. Meanwhile mutual consistency among the replicated copies of data should be maintained by synchronizing transactions at different sites, so that a global serialization order can be ensured. Thus, the problem of replicated data management involves a compromise between two conflicting goals: maximizing data availability and maintaining the consistency of data. *Quorum consensus* (QC) methods [2,3] are widely used in managing replicated data.

Using a QC method, an operation of a transaction issued at a site in a distributed database system can proceed only if permission is granted by a group of other sites storing the replicas of the manipulating data. A general protocol of a QC method for processing a transaction operation can be described as follows.

Given a data item (object), at each site s_i, the regulations for forming a read quorum group S_i^r and a write quorum group S_i^w are assigned, where S_i^r and S_i^w are both in terms of a subset of the set of the sites storing replicated copies of the object, so that the intersection of a read quorum group and a write quorum group is not empty, neither is the intersection of two write quorum groups. These two *intersection invariants* can be formally defined as:

for each pair of sites s_i and s_j, and any formed S_i^r, S_i^w, S_j^r and S_j^w, $S_i^w \cap S_j^w \neq \emptyset$, $S_i^r \cap S_j^w \neq \emptyset$, and $S_j^r \cap S_i^w \neq \emptyset$.

A read (write) operation should get permission from each site in S_i^r (S_i^w) for processing it.

If a 2-phase locking mechanism [1] is applied, a QC method will force, through the intersection invariants, the situation that a write and a read cannot take place simultaneously on different copies of the same object, and similarly, neither can two writes. Thus, mutual consistency can be maintained.

Recent trends in developing new QC methods include coupling high data availability with a low communication cost for processing transactions. The achievements of a low communication cost are either

1. through the minimization of the number of remote sites which a transaction operation process has to communicate with [4,8,9], or
2. through the minimization of the total communication cost for processing a given set of transactions [5,6].

In this paper, we restrict our interests to 1. Interested readers may refer to [5,6] for a detailed discussion about minimizing the total communication cost.

To maximize the utilization of a distributed system, and then to enhance the overall performance of transaction processes, a *fully distributed* QC approach is investigated in [4,8,9]. In a fully distributed QC approach, each site bears almost equal responsibility in performing a QC approach. The lower bound of a quorum group size in a fully distributed QC method has been shown as $\sqrt{n}$ [8], where n is the number of sites storing a given object. This lower bound has been achieved by the QC method in [8].

A rigorous analysis [9] shows that by the QC method in [8], the data availability for processing an individual operation gets smaller (falls asymptotically to 0), when the number of replicas of each data item is increased. This defeats the main objective of data replication, i.e., increasing data availability through replication. [9] provides a fully distributed QC method which guarantees that the data availability is increased (asymptotically to 1) as the number of replicas increases. But each operation process is forced to communicate with at least $\Omega(\sqrt{n} \log^{0.5} n)$ remote sites.

In this paper, we will present a new QC method. The proposed method is also fully distributed, and guarantees that the data availability goes asymptotically to 1. Further, we can show that using the proposed approach, each operation process may communicate to less than $\Omega(\sqrt{n} \log^{0.5} n)$ remote sites:

– If at the time when an operation is being processed, there are no (or a few) sites with failures, then the operation process will communicate with only $O(\sqrt{n} \log^{0.208} n)$ remote sites.

- If many sites with failures, an operation process will be forced to communicate with $O(\sqrt{n} \log^{0.792} n)$ remote sites.

The rest of the paper is organized as follows. In Section 2, we first give our environment assumptions, and then introduce a mathematical model to justify the degree of fault-tolerance of a QC method. In Section 3, we present our QC method. Section 4 provides a sketch of a rigorous performance analysis of our QC approach, and a comparison between our QC method and the related approaches. This is followed by a conclusion.

2 Preliminaries

In our distributed system environment, we assume that communication between different sites is through exchanging messages. Detection of a failure of a site by another site happens through sending a message, but receiving no reply. To simplify the analysis, we assume that communication links never fail, and that each site failure probability is independent. The networks are *fully connected*; that is, a message can be sent directly between any pair of sites. We follow the model where replicated data is represented by multiple copies and the transactions are either *simple* read or *simple* write (that is, each read or write manipulates only one object). So, in this paper we need only to consider a QC method with respect to one object. Without loss of generality, we assume *full data replication*; that is, a copy of each object exists at all sites.

Suppose that in a distributed system N, the probability of each site being alive is known. Given a site i and a QC method A, let $R_{r,A,i}$ ($R_{w,A,i}$) be the probability of assembling successfully a read (write) quorum group at i by the QC method A. The QC method A has a *high site resilience* [9] if

$$\min_{i \in N, q \in \{r,w\}} R_{q,A,i} \to 1 \text{ when } n \to \infty,$$

where n is the number of sites in the network N. Note that a QC method with a high site resilience can guarantee that in the asymptotical case, an alive site is always able to assemble a read (or write) quorum group to process the transactions issued at it; that is, the data availability is increased through increasing the number of replicas of data, unlike the one in [8].

Note that we are concerned only with the fault tolerance of transaction synchronization algorithms. So, as far as we are concerned, an operation process cannot successfully assemble a read (or write) quorum group *only* due to site failures. We do not consider the situation where an operation process fails because some other operation has already been granted permission to proceed.

3 A New QC method

Inspired by the QC method in [9], our QC method is built on the top of Kumar's approach [4] and Maekawa's approach [8]. We first briefly review the methods in [4,8], and then present our QC algorithm.

3.1 Kumar's Method

Suppose that in a distributed system, there are n sites. [4] proposed a hierarchical quorum consensus (HQC) method for update synchronization. Here, we introduce our special implementation of the HQC method, which will be adopted in our new QC method.

A logical organization T_n of n sites in a network is a *Kumar* tree if:

- T_n is a rooted 3-way tree, that is, each node has at most three children; and the leaf set of T_n consists of the n sites.
- The levels of the tree are ordered such that the root is located at level 0, and the level number of a parent is one level lower than that of its children. (L_i denotes the set of nodes at level i.)
- T_n has $m = \lceil \log_3 n \rceil + 1$ levels (i.e., the last level is L_{m-1}.)
- Each node in L_i, for $0 \le i \le m - 3$, has exactly three children; and each node in L_{m-2} either is a leaf or has at least two children.

Figure 1 illustrates the Kumar trees for $n = 6, 9$.

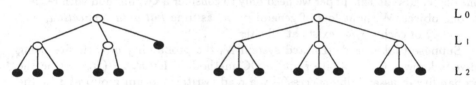

$$L_0$$
$$L_1$$
$$L_2$$

Fig. 1. Kumar trees

Our implementation of HQC performs as follows with respect to each site j:

Step 1: The n sites are logically organized as a Kumar tree. Each site keeps the Kumar tree as a map for assembling its read/write quorum groups. Go to Step 2.

Step 2: The regulations for assembling a read quorum and a write quorum group are the same:

The root asks (randomly) a pair of its children to take part in assembling the quorum group, and marks these two children. Recursively, each marked node at L_i (randomly) marks a pair of its children at L_{i+1}, and asks them to participate in making the quorum group. (If the number of children of a marked node is smaller than 3, then the marked node should mark all its children.)

Go to Step 3.

Step 3: A quorum group is formed by all the marked leaves. In case there are some marked leaves with a failure, we need to perform a quorum group again. In order to save communication cost among remote sites, we should keep as many marked non-failure leaves as possible in a new forming quorum group. So, instead of performing a new quorum group by Step 2 from the scratch, we can modify the failed formed quorum group to a new quorum group:

Suppose that a marked site l has a failure. The site j looks up the Kumar tree to find the parent $pa1$ of site l. Then re-do the Step 2 with respect only to $pa1$, where keeping another originally marked child of $pa1$ being marked in the new forming quorum group. After this, if we find that it is impossible to assemble the quorum group with respect to $pa1$, then j looks up the Kumar tree again to find the parent $pa2$ of $pa1$. Then re-do the Step 2 with respect to $pa2$, where keeping another originally marked child of $pa2$ being marked in the new forming quorum group, and so on. Finally, we can determine whether or not it is possible to assemble a quorum group with alive sites.

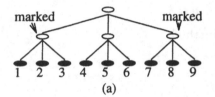
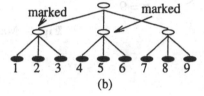

(a)   (b)

Fig. 2. HQC method

For instance, consider Figurer 2. It shows a collection of 9 sites organized into a three level Kumar tree. The various possible quorum groups are: $\{1, 2, 8, 9\}$ (see Figure 2(a)), $\{2, 3, 4, 6\}$ (see Figure 2(b)), etc. Suppose that $\{1, 2, 8, 9\}$ is initially formed, where the site 9 is not available. So by HQC, $\{1, 2, 7, 8\}$ is then formed. Again, we find site 7 is not available. By HQC, then one of $\{1, 2, 4, 5\}$, $\{1, 2, 4, 6\}$, and $\{1, 2, 5, 6\}$ will be formed.

It has been shown that the size of a quorum group created by using HQC is bounded by (not greater than) $2^{\lceil \log_3 n \rceil}$ ($\approx O(n^{0.63})$). This implies that a transaction process needs to communicate with $O(n^{0.63})$ remote sites to get a read (write) quorum group, if there are no sites with a failure.

Suppose that at the time when a transaction is issued, some sites are not available in the network due to their failures. As mentioned earlier, in a distributed environment a site failure is usually detected through sending a message. A transaction process, by HQC, may be forced to communicate with more than half of the sites in the network, through sending messages, to know whether or not it can successfully get a read (write) quorum group. Thus, in the worst case a read (write) quorum group is formed by communicating with $O(n)$ remote sites.

3.2 Maekawa's Method

[8] proposed another fully distributed quorum consensus method. A simple version of the method in [8] is to organize n sites into an approximate grid square, where the possible empty sites are placed at the grid positions as up and right as possible (see Figure 3). A (read or write) quorum group S_i of a site s_i ($1 \leq i \leq n$) consists of the sites which are, with respect to the grid, in the row and the column occupied by s_i. (Figure 3 illustrates the quorum groups of site 4 and site 1.) Clearly, each S_i has at most $2\lceil \sqrt{n} \rceil - 1$ sites, any two quorum groups have non-empty intersection, and each site belongs to $O(\sqrt{n})$ groups. If a failure of a site in S_i happens, then the process of a transaction from site s_i has to wait until the site recovers.

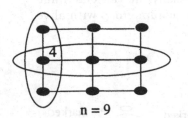
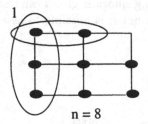

$$n = 9 \qquad\qquad n = 8$$

Fig. 3. Maekawa's method

3.3 Our QC Method: KMQC

In this sub-section, we present a new quorum consensus method - KMQC method, which is built on the top of Kumar and Maekawa's methods. In the KMQC method, the regulations for forming read and write quorum groups are the same. We first group n sites $\{s_i : 1 \leq i \leq n\}$ into k disjoint groups $\{G_i : 1 \leq i \leq k\}$ such that we try to make the sizes of these k groups as equal as possible. A quorum group is formed, in KMQC, through two layer constructions. KMQC first performs Maekawa's method on these k groups to obtain a "quorum group" A - a subset of $\{G_i : 1 \leq i \leq k\}$. By the application of HQC to each element G_i in A, we get a quorum group Q_i corresponding to each G_i in A; the union of all these Q_i forms a quorum group in KMQC. It can be precisely described as follows.

Step 1: Group the n sites into k disjoint subgroups $\{G_i : 1 \leq i \leq k\}$ such that

$$\lfloor \frac{n}{k} \rfloor \leq |G_i| \leq \lceil \frac{n}{k} \rceil, \tag{1}$$

and $\cup_{i=1}^{k} G_i = \{s_i : 1 \leq i \leq n\}$. A mapping g_1, from $\{s_i : 1 \leq i \leq n\}$ to $\{G_i : 1 \leq i \leq k\}$, is constructed, such that $g_1(s_i) = G_j$ if $s_i \in G_j$. Go to Step 2.

Step 2: By viewing each element G_i in $\{G_i : 1 \leq i \leq k\}$ as a "site", and by applying Maekawa's method [8] to these k "sites", we obtain the k "quorum groups" A_1, A_2, ..., A_k (i.e, each A_i is a subset of $\{G_i : 1 \leq i \leq k\}$) corresponding to theses k "sites" G_i for $1 \leq i \leq k$. Note $G_i \in A_i$. A mapping g_2 is constructed to identify the "quorum group" of "site" G_i: $g_2(G_i) = A_i$. (Note that $1 \leq i \leq k$, $|A_i| \leq 2\lceil\sqrt{k}\rceil - 1$.) Go to Step 3.

Step 3: For each site s_i, its (read or write) quorum group is formed, which consists of certain sites in group $g_2(g_1(s_i))$ (say A_j), such that for each element G_x in A_j, we use HQC to form a quorum (read or write) group Q_x^i in G_x. Then the quorum group of s_i is:

$$\cup_{G_x \in g_2(g_1(s_i))} Q_x^i.$$

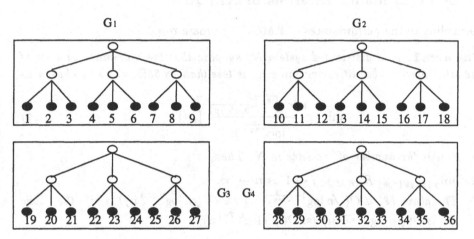

Fig. 4. KMQC approach

For example, consider Figure 4. There are 36 sites. Let $k = 4$, $G_1 = \{s_i : 1 \leq i \leq 9\}$, $G_2 = \{s_i : 10 \leq i \leq 18\}$, $G_3 = \{s_i : 19 \leq i \leq 27\}$, $G_4 = \{s_i : 28 \leq i \leq 36\}$. (Note that $g_1(s_{9i+k}) = G_{i+1}$ for $0 \leq i \leq 3$ and $1 \leq k \leq 9$.) By the application of Maekawa's algorithm, we have that $A_1 = \{G_1, G_2, G_3\}$, $A_2 = \{G_1, G_2, G_4\}$, $A_3 = \{G_1, G_3, G_4\}$, $A_4 = \{G_2, G_3, G_4\}$. (Note that $g_2(G_i) = A_i$ for $1 \leq i \leq 4$.) A transaction from site s_1 requires to form a read (write) quorum group consisting of certain sites in $g_2(g_1(s_1)) = A_1$, that is, certain sites in G_1, G_2, and G_3. Particularly, the quorum groups will be

$$\{s_2, s_3, s_5, s_6, s_{13}, s_{14}, s_{16}, s_{18}, s_{19}, s_{20}, s_{22}, s_{23}\},$$
$$\{s_4, s_5, s_7, s_8, s_{11}, s_{12}, s_{16}, s_{17}, s_{19}, s_{20}, s_{25}, s_{26}\}, \text{ etc.}$$

Our KMQC is a fully distributed method, since it is a combination of the method in [4] and the method in [8] which are fully distributed. It can be immediately verified that KMQC is also correct (i.e, each two quorum groups have at least one common site), based on the facts that both of the methods in [4,8] are correct.

Now, we estimate how many remote sites will be accessed by a transaction process, using KMQC. Clearly, if there are no sites with failures, a transaction process needs to access at most $(2\lceil\sqrt{k}\rceil - 1)2^{\lceil(\log_3\lceil\frac{n}{k}\rceil)\rceil}$ remote sites.

$$O(\lceil\sqrt{k}\rceil2^{\lceil(\log_3\lceil\frac{n}{k}\rceil)\rceil}) = O(\sqrt{k}\,2^{\log_3(\frac{n}{k})}) = O(\sqrt{k}\,(\frac{n}{k})^{\log_3 2}) \qquad (2)$$

If many sites are not available in the network, then to perform a quorum group Q_x^i with respect to each relevant G_x, $O(|G_x|)$ remote sites may be communicated with. So in the worst case, a transaction process will communicate with the following number of remote sites:

$$O(\lceil\sqrt{k}\rceil\lceil\frac{n}{k}\rceil) = O(\sqrt{k}\,\frac{n}{k}). \qquad (3)$$

4 A Performance Analysis of KMQC

Regarding to the performances of KMQC, the main result is:

Theorem 1. *In a distributed system N, suppose that the maximum value p of the site failure probabilities over all sites is less than 18.35%, and k is chosen as*

$$\min\{\lfloor\frac{n\log_2^{\frac{1}{\log_3 2}}\frac{1}{3p(2-p)}}{\log_2^{\frac{1}{\log_3 2}}n}\rfloor, n\}, \qquad (4)$$

where n is the number of the sites in N. Then,

1. *$\min_{i\in N, q\in\{r,w\}} R_{q,KMQC,i} \to 1$ as $n \to \infty$.*
2. *The above (2) is $O(\sqrt{n}\log^{1-\frac{1}{2\log_3 2}}n)$ ($\approx O(\sqrt{n}\log^{0.208}n)$), while the above (3) is $O(\sqrt{n}\log^{\frac{1}{2\log_3 2}}n)$ ($\approx O(\sqrt{n}\log^{0.792}n)$).*

Theorem 1 says:

In a class π of distributed systems where the maximum value of site failure probabilities through the whole class is p and $p \le 0.1835$, the KMQC method has a *high site resilience* if with respect to any n sites distributed system in π, k is always chosen as (4). Given any n sites distributed system in π, each transaction process needs only to communicate with $O(\sqrt{n}\log^{0.208}n)$ remote sites in the best case, and needs to communicate with $O(\sqrt{n}\log^{0.792}n)$ remote sites in the worst case.

The proof of 2 in Theorem 1 is very straight forward according to the choice of k, (2), and (3). The proof of 1 requires a detailed mathematic estimation. Due to the space limitation, we do not include the detailed proof in this paper. The interested readers may refer to [7] for details. A sketch of the proof 1 is as follows:

We first prove that HQC in [4] has a high site resilience through a mathematical estimation about the probability of failing to form any quorum groups at a site. Then we use the obtained mathematical estimation to prove 1. The main technique is to use some standard mathematic estimation approaches.

Next, we make a comparison between our KMQC method and the other three fully distributed algorithms in [4,8,9]. To simplify the description, we use HQC to denote the method in [4], GQC to denote the method in [8], HMV to denote the method in [9].

First, KMQC, HQC and HMV all have a high site resilience if the failure probability of each site is smaller than 18.35%, but GQC dose not have. So, there is only one loser, GQC, according to this index.

HQC requires to communicate with at least $\Omega(n^{0.63})$ remote sites, GQC requires $\Omega(\sqrt{n})$ remote sites, HMV requires at least $\Omega(\sqrt{n}\log^{0.5} n)$ remote sites. KMQC requires to communicate with at least $\Omega(\sqrt{n}\log^{0.208} n)$ remote sites. According to this index, GQC is the best, KMQC is the second best, HMV is the third.

Based on the above comparison, KMQC should be a winner, regarding to an integrated whole of remote sites number and fault-tolerance, in the reliable distributed systems where site failure probabilities are small.

5 Conclusion

In this paper, we present a fully distributed quorum consensus method, KMQC. KMQC method has a high site resilience like that in [9], while we can expect that in reliable distributed systems, KMQC should have a lower message overhead than that in [9].

Possible future studies could be on either further improving the message overhead or addressing the communication link failures.

Acknowledgement: The work of the first named author is partially supported by the grant IRG at UWA.

References

1. P. Bernstein, V. Hadzilocs and N. Goodman, *Concurrency Control and Recovery in Database Systems*, Addison-Wesley, Reading, Mass., 1987.
2. S. B. Davidson, H. Garcia-Molina and D. Skeen, Consistency in Partioned Networks, *ACM Computing Surveys*, 17(3), 341-370, 1985.
3. H. Garcia-Molina and D. Barbara, How to Assign Votes in a Distributed Systems, *J. ACM*, 32(4), 841-860, 1985.
4. A. Kumar, Hierarchical Quorum Consensus: A New Algorithm for Managing Replicated Data, *IEEE Transactions on Computers*, 40(9), 996-1004, 1991.
5. A. Kumar and A. Segev, Cost and Availability Tradeoffs in Replicated Data Concurrency Control, *ACM Transactions on Database Systems*, 18(1), 102-131, 1993.
6. X. Lin and M. E. Orlowska, An Optimal Voting Schema for Minimizing the Overall Communication Cost in Replicated Data Management, to appear in *Journal of Parallel and Distributed Computing*.
7. X. Lin and M. E. Orlowska, On High Resilience and Low Message Overhead in Replicated Data Management, *Technical Report*, Computer Science Department, The University of Western Australia, Australia.

8. M. Maekawa, A $\sqrt{N}$ Algorithm for Mutual Exclusion in Decentralized Systems, *ACM Transactions on Computer Systems*, 3(2), 145-159, 1985.
9. S. Rangarajan, S. Setia and S. K. Tripathi, A Fault-tolerant Algorithm for Replicated Data Management, *IEEE Proceedings of the 8th International Conference on Data Engineering*, 230-237, 1992.

Constructing Craig Interpolation Formulas

Guoxiang Huang

Mathematics Department, University of Hawaii at Manoa
Honolulu, HI 96822 (E-mail: huang@math.hawaii.edu)

Abstract. A Craig interpolant of two inconsistent theories is a formula which is true in one and false in the other. This paper gives an efficient method for constructing a Craig interpolant from a refutation proof which involves binary resolution, paramodulation, and factoring. This method can solve the machine learning problem of discovering a first order concept from given examples. It can also be used to find sentences which distinguish pairs of nonisomorphic finite structures.

1 Background and Introduction

Let Σ and Π be two inconsistent first order theories. Then by Craig's Interpolation Theorem, there is a sentence θ, called a *Craig interpolant*, such that θ is true in Σ and false in Π and every nonlogical symbol occurring in θ occurs in both Σ and Π. Craig interpolants can be used to solve the problem of learning a first order concept by letting Σ and Π be the lists of positive and negative examples of the concept to be learned.

The standard nonconstructive model-theoretic proof of Craig's Theorem is in [3]. Lyndon showed how to construct an interpolant from a special form of natural deduction (see [1]). We show how to construct an interpolant from a refutation proof which uses binary resolution, factoring and paramodulation. In our examples, we use OTTER (the standard text on OTTER is [4]) to generate such proofs.

Craig interpolants can be used to find a sentence which distinguishes two nonisomorphic finite structures. Let Σ and Π be the atomic diagrams of the two structures. Then they are inconsistent and any Craig interpolant for them is a sentence which is true in one structure and false in the other.

2 Constructing Interpolation Formulas from Refutations

Let L_Σ and L_Π be two languages, Σ a theory in L_Σ, and Π a theory in L_Π such that $\Sigma \cup \Pi$ is not consistent. In this paper we use $\Diamond$ to represent contradiction, use $\Box$ to indicate the end of a proof, and suppose P is a refutation of $\Sigma \cup \Pi \models \Diamond$ involving only binary resolutions, paramodulations, and factorings. The input clauses (clauses at the top of the refutation) are required to be instances of clauses from Σ and Π. For convenience, we will assume that different input clauses have disjoint sets of variables.

For any occurrence L in the proof P of a relational symbol in $L_\Sigma \cup L_\Pi$, we define L *is from* Σ recursively by:

(i). If the occurrence L is in an input clause from Σ, we say it is from Σ; otherwise, it is not.

(ii). If the occurrence L is in a non-input clause C, then it is from Σ if the corresponding occurrence in some parent clause is from Σ.

Similarly, we can define L *is from* Π. Since factoring is allowed in the proof, several occurrences of some literal may be factored into a single one. So it is possible that a literal in some clause may be from both Σ and Π.

Let T and F be the truth values of "truth" and "falsehood". For a binary resolution proof P we use the following recursive procedure to assign formulas to the clauses in P:

Interpolation Algorithm

(i). *If C is an input clause from Σ, its formula is F; if C is an input clause from Π, its formula is T.*

(ii). *If ϕ is assigned to $L \vee C$ and ψ is assigned to $\neg L' \vee D$, and if $(C \vee D)\pi$ is the resolvent of $L \vee C$ and $D \vee \neg L'$ resolving against $L\pi(= L'\pi)$, then the formula assigned to $(C \vee D)\pi$ is:*

(a). $(\phi \vee \psi)\pi$ *if the occurrences of both L and $\neg L'$ are from Σ alone;*

(b). $(\phi \wedge \psi)\pi$ *if the occurrences of both L and $\neg L'$ are from Π alone;*

(c). $((\neg L' \wedge \phi) \vee (L \wedge \psi))\pi$ *if neither (a) nor (b).*

Definition 1. A formula θ is a *relational interpolant* of Σ and Π relative to a clause C iff

(1). all relational symbols of θ are in $L_\Sigma \cap L_\Pi$,

(2). $\Sigma \models \theta \vee C$, and

(3). $\Pi \models \neg\theta \vee C$.

Theorem 2. *For each clause C of a binary resolution proof P of $\Sigma \cup \Pi \models \Diamond$, the formula assigned by the above algorithm is a relational interpolant of Σ and Π relative to C. In particular, the formula θ assigned to the final empty clause of the proof P is a relational interpolant between Σ and $\neg\Pi$.*

Proof. It is obvious that any assigned formula contains only relation symbols from $L_\Sigma \cap L_\Pi$. So condition (1) of the definition holds.

For any occurrence of a clause or subclause C in the proof P, let C_Σ (let C_Π) be C with all occurrences of literals not from Σ (not from Π) deleted. Then $C_\Sigma \models C$, $C_\Pi \models C$, and $C_\Sigma \vee C_\Pi \iff C$ are valid and $(C \vee D)_\Sigma = (C_\Sigma \vee D_\Sigma)$ and $(C_\Sigma)\pi = (C\pi)_\Sigma$ for any unifier π.

We prove by induction on the depth of C in P the following strengthenings of (2) and (3):

(2)'. $\Sigma \models \theta \vee C_\Sigma$,

(3)'. $\Pi \models \neg\theta \vee C_\Sigma$.

Suppose C is an input clause from Σ. Then θ is F and $C_\Sigma = C$. Thus (2)' and (3)' hold since $\Sigma \models F \vee C$ and $\Pi \models T \vee C$. The argument for an input clause from Π is similar.

Suppose $(2)'$ and $(3)'$ are true for clauses $L \vee C$ and $\neg L' \vee D$ of P whose resolvent in P is $(C \vee D)\pi$ where π is a unifier such that $L\pi = L'\pi$. Assume $L \vee C$ is assigned the formula ϕ and $\neg L' \vee D$ is assigned ψ. Thus we have

$$\Sigma \models \phi \vee (L \vee C)_\Sigma, \qquad \Sigma \models \psi \vee (\neg L' \vee D)_\Sigma,$$
$$\Pi \models \neg\phi \vee (L \vee C)_\Pi, \quad \Pi \models \neg\psi \vee (\neg L' \vee D)_\Pi.$$

Case (a). Suppose the occurrences of L and $\neg L'$ are both from Σ alone. Then $(L \vee C)_\Sigma = L \vee C_\Sigma$ and $(\neg L' \vee D)_\Sigma = \neg L' \vee D_\Sigma$. By resolution we get $(2)'$: $\Sigma \models ((\phi \vee C_\Sigma) \vee (\psi \vee D_\Sigma))\pi = (\phi \vee \psi)\pi \vee (C \vee D)\pi_\Sigma$. For $(3)'$ we have $(L \vee C)_\Pi = C_\Pi$ and $(\neg L' \vee D)_\Pi = D_\Pi$ and so $\Pi \models (\neg\phi \vee C_\Pi) \wedge (\neg\psi \vee D_\Pi)$ and $\Pi \models \neg(\phi \vee \psi)\pi \vee (C \vee D)\pi_\Pi$.

Case (b) for L and $\neg L'$ from Π alone is similar.

Case (c). In any model of Σ with any assignment of variables, if both $C_\Sigma\pi$ and $D_\Sigma\pi$ are false, then $(\phi \vee L)\pi$ and $(\psi \vee \neg L')\pi$ are true. So if $L\pi = L'\pi$ is true, then so is $\psi\pi$; if $L\pi$ is false, then $\phi\pi$ is true. Either way,

$((\neg L' \wedge \phi) \vee (L \wedge \psi) \vee (C \vee D)_\Sigma)$ is always true.

Similarly, $\Pi \models (((L \vee \neg\phi) \wedge (\neg L' \vee \psi)) \vee (C \vee D)_\Pi)\pi$.

Hence, by induction, the theorem holds. $\square$

Resolution provers often use paramodulation to handle equality. Given clauses $C(r)$ and $s = t \vee D$ with no variables in common and a unifier π such that $r\pi = s\pi$ or $r\pi = t\pi$, paramodulation infers the paramodulant $(C(t) \vee D)\pi$ or $(C(s) \vee D)\pi$ respectively.

Definition 3. For a deduction P in $L_\Sigma \cup L_\Pi$, a *noncommon term* is a term which begins with a symbol not in $L_\Sigma \cap L_\Pi$. Such a term is called a Σ-*term* if its initial symbol is from Σ, a Π-*term* if its initial symbol is from Π. An occurrence of a Σ (Π)-term is *maximal* if this occurrence is not a subterm of a larger Σ (Π)-term.

Now we extend the Interpolation Algorithm to proofs with paramodulation as follows:

(iii). If ϕ is assigned to $C(r)$ and ψ is assigned to $s = t \vee D$ and if π is a unifier such that $r\pi = s\pi$, then the formula assigned to the paramodulant is:

(d). $[(\phi \wedge s = t) \vee (\psi \wedge s \neq t)]\pi \vee (s = t \wedge h(s) \neq h(t))\pi$ provided r occurs in $C(r)$ as a subterm of a maximal Π-term $h(r)$ and there is more than one occurrence of $h(r)$ in $C(r) \vee \phi$.

(e). $[(\phi \wedge s = t) \vee (\psi \wedge s \neq t)]\pi \wedge (s \neq t \vee h(s) = h(t))\pi$ provided r occurs in $C(r)$ as a subterm of a maximal Σ-term $h(r)$ and there is more than one occurrence of $h(r)$ in $C(r) \vee \phi$.

(f). $((\phi \wedge s = t) \vee (\psi \wedge s \neq t))\pi$ if neither (d) nor (e).

Lemma 4. *If ϕ, ψ are the interpolants relative to $C(r)$ and $s = t \vee D$, respectively, then the above formula is an interpolant relative to the paramodulant $(C(t) \vee D)\pi$.*

Proof. We prove case (f). Since $\Sigma \models (C(r) \vee \phi)\pi$ and $\Sigma \models (s = t \vee D \vee \psi)\pi$, for any model A of Σ, if $s\pi = t\pi$ in A, then $A \models (C(t) \vee \phi)\pi$; otherwise if $s\pi \neq t\pi$ in A, then $A \models D\pi \vee \psi\pi$. Either way, we have $A \models (C(t) \vee D \vee \theta)\pi$.

Similarly, $\Pi \models (C(t) \vee D \vee \neg\theta)\pi$ in the two cases: For $s\pi = t\pi, \Pi \models (\neg\phi \vee C(t))\pi$; for $s\pi \neq t\pi, \Pi \models (\neg\psi \vee D)\pi$. Thus the assigned formula satisfies the requirement. □

The final rule of inference we need is factoring. Given a clause $L \vee L' \vee C$ and a unifier π such that $L\pi = L'\pi$, factoring infers the clause $(L \vee C)\pi$. We extend the Interpolation Algorithm to proofs with factoring as follows:

(iv). *If ϕ is assigned to $L \vee L' \vee C$ and π is a unifier as above, then we assign $\phi\pi$ to the factor clause $(L \vee C)\pi$.*

Clearly $\Sigma \models L \vee L' \vee C \vee \phi$ and $\Pi \models L \vee L' \vee C \vee \neg\phi$ imply $\Sigma \models (L \vee C)\pi \vee \phi\pi$ and $\Pi \models (L \vee C)\pi \vee \neg\phi\pi$.

Thus for a refutation proof P by a series of binary resolutions, factorings, and paramodulations, applying the above extended algorithm gives a formula, say θ, for the empty clause. Since $\Sigma \models \theta$ and $\Pi \models \neg\theta$, θ is a relational interpolant between Σ and $\neg\Pi$. Though θ does not contain any non-common relational symbol, it may contain noncommon terms with constants or function symbols which are not in $L_\Sigma \cap L_\Pi$. We now show how to get a Craig interpolant by replacing all noncommon terms in θ with appropriately quantified variables.

First we define a *binary tree deduction* to be a deduction in which any clause is used at most once. Such a deduction involving only binary resolutions, factorings, and paramodulations forms a binary tree.

Lemma 5. *Any refutation P using only binary resolutions, paramodulations, and factorings, lifts to a binary tree deduction P_b with the same conclusion.*

Proof. We prove this lemma by induction on the number $k(P)$ of clauses which are used more than once in the deduction P. If $k(P) = 0$, P is a binary tree deduction. Assume the lemma holds for all deductions with $k(P) \leq n$ and suppose $k(P) = n + 1$. Let C be a clause such that C is used $m \geq 2$ times in P but all the ancestors of C are used only once. We construct a new deduction P' from P such that P' has m copies of C and its ancestors, and each copy of C and its ancestors is used exactly once in P'. Finally, variables may be renamed if necessary, so that different input clauses have disjoint sets of variables. Otherwise P is the same as P' and has the same conclusion. Since P' is a deduction with $k(P') \leq n$, by induction, P' lifts to a binary tree deduction P_b with the same conclusion.

Suppose P is a binary tree deduction whose input clauses have disjoint sets of variables and whose substitutions are generated by the usual unification algorithm, then the following properties hold in P:

1. Every variable of any noninput clause in P occurs in exactly one parent clause and thus traces back to a unique ancestral input clause.
2. Any two incomparable (neither is the ancestor of the other) clauses have disjoint sets of variables.
3. For any substitution π of P and any variable x, either π is trivial on x, i.e., $\pi(x) = x$, or x does not occur in the term $\pi(x)$.
4. If π is nontrivial on x, x never appears in any clause below π.

Definition 6. Given a binary tree deduction P as above, for any variable x occurring in P, let π_p, the *composite substitution for* P, be the substitution such that $\pi_p(x)$ is the term resulting from applying to x the composition of all the substitutions along the path from the unique input clause which contains x to the bottom of P.

Lemma 7. *For any clause C in such a binary tree deduction P, $C\pi_p$ is the clause obtained by applying to C the composition of all the substitutions along the path from C to the bottom of P.*

Proof. Suppose x is a variable of C. Then x traces back to a unique ancestral input clause D. All of the substitutions along the path from D to C are trivial on x since otherwise, x would not occur in C. Hence $\pi(x) = $ the composition of all substitutions from D to the bottom. $\square$

We say a deduction is *propositional* if there are no nontrivial unifying substitutions involved in the deduction.

Lemma 8. *The Boolean operations $\vee, \wedge, \neg$ and propositional binary resolution, factoring, and paramodulation commute with substitution. That is, if π is a substitution, then $(A \vee B)\pi = A\pi \vee B\pi$, $(A \wedge B)\pi = A\pi \wedge B\pi$, $(\neg A)\pi = \neg(A\pi)$; and for any propositional binary resolution $\{A \vee L, B \vee \neg L\} \models A \vee B$, we have $\{A\pi \vee L\pi, B\pi \vee \neg L\pi\} \models A\pi \vee B\pi$; and for any propositional paramodulation $\{C(s), \ s = t \vee D\} \models C(t) \vee D$, we have $\{C(s)\pi, \ (s = t)\pi \vee D\pi\} \models C(t)\pi \vee D\pi$.*

Lemma 9. *Every binary tree proof P_b projects to a propositional proof P_p.*

Proof. Given a binary tree proof P_b, rename the variables if necessary so that the above four properties hold and let π_p be the composite substitution for P. Let P_p be the result of replacing each clause C of P_b with $C\pi_p$ and replacing each substitution with the trivial identity substitution. Then by Lemma 8 P_p is a projection of P_b and P_p is a propositional binary tree deduction. $\square$

Lemma 10. *Assume P_b projects to P_p as in Lemma 9. If we apply the Interpolation Algorithm to the propositional deduction P_p, and if a clause C' in P_p is assigned formula ϕ', and if its corresponding clause C in P_b is assigned formula ϕ, then $\phi' = \phi\pi_p$. In particular, the assignments to $\Diamond$ from both deductions are the same.*

Proof. Any occurrence of a literal L in a clause C' of P_p is from Σ or Π or both iff its corresponding occurrence in P_b is from Σ, Π, or both. So the corresponding clauses of P_b and P_p are assigned interpolants by the same case of the Interpolation Algorithm. Lemma 8 gives the result. $\square$

Let P_p be a propositional deduction in $L_\Sigma \cup L_\Pi$, and $t_1, ..., t_n$ be all the Π-terms with maximal occurrences in P_p. Let $x_1, ..., x_n$ be a set of new variables which do not occur in P_p. For any term or formula θ in P_p, define $\overline{\theta}(x_1, ..., x_n)$ to be the term or formula obtained by simultaneously replacing all maximal occurrences of the Π-terms t_j's by the new variables x_j's. We call $\overline{\theta}$ the *lifted* formula of θ from Π-terms.

Lemma 11.

$$\overline{(A \vee B)}(x_1, ..., x_n) \iff \overline{A}(x_1, ..., x_n) \vee \overline{B}(x_1, ..., x_n)$$
$$\overline{(A \wedge B)}(x_n, ..., x_n) \iff \overline{A}(x_1, ..., x_n) \wedge \overline{B}(x_1, ..., x_n)$$
$$\overline{(s = t)}(x_1, ..., x_n) \iff \overline{s}(x_1, ..., x_n) = \overline{t}(x_1, ..., x_n)$$
$$\overline{(\neg A)}(x_1, ..., x_n) \iff \neg \overline{A}(x_1, ..., x_n)$$
$$\theta = \overline{\theta}(t_1, ..., t_n)$$

Lemma 12. *If θ is the relational interpolant of Σ and Π relative to C by the Interpolation Algorithm for the propositional deduction P_p, then we have*
$$\Sigma \models \overline{(C \vee \theta)}(x_1, ..., x_n).$$

Proof. We prove this lemma by induction on P_p. If $\overline{C}(t_1, ..., t_n)$ is an instance of an input clause from Σ, then all the Π-terms in C come from free variables by the unifying substitutions in the original deduction. So by the construction of P_p we know that $\overline{C}(x_1, ..., x_n)$ is an instance of some input clause in Σ, and F is assigned to this clause. Thus $\Sigma \models \overline{C}(x_1, .., x_n) \vee F$. If $\overline{C}(t_1, .., t_n)$ is an instance of input clause from Π, then it has assigned formula T and $\Sigma \models \overline{C}(x_1, .., x_n) \vee T$ holds.

Now assume that $\Sigma \models \overline{(C \vee L \vee \phi)}(x_1, .., x_n)$ and $\Sigma \models \overline{(D \vee \neg L \vee \psi)}(x_1, .., x_n)$ and that $C \vee L$ and $D \vee \neg L$ resolving against L gives $C \vee D$ with interpolant θ. We show that $\Sigma \models \overline{(C \vee D \vee \theta)}(x_1, .., x_n)$.

Notice that by propositional deduction and Lemma 11 we have
$$\{\overline{C} \vee \overline{L} \vee \overline{\phi}, \quad \overline{D} \vee \neg\overline{L} \vee \overline{\psi}\} \models \overline{C} \vee \overline{D} \vee (\overline{\phi} \vee \overline{\psi}), \text{ and}$$
$$\{\overline{C} \vee \overline{L} \vee \overline{\phi}, \quad \overline{D} \vee \neg\overline{L} \vee \overline{\psi}\} \models \overline{C} \vee \overline{D} \vee (\neg\overline{L} \wedge \overline{\phi}) \vee (\overline{L} \wedge \overline{\psi}).$$

Using Lemma 11 again proves the Lemma for case (a) and case (c) of the Interpolation Algorithm definition of θ. For case (b), $\theta = \phi \wedge \psi$, and the occurrences of L and $\neg L$ are not from Σ. By the proof of Theorem 2 we know that $\Sigma \models \overline{C} \vee \overline{\phi}$ and $\Sigma \models \overline{D} \vee \overline{\psi}$ respectively. Thus we have $\Sigma \models \overline{C} \vee \overline{D} \vee (\overline{\phi} \wedge \overline{\psi})$.

Next assume $C(s)$ and $s = t \vee D$ gives $C(t) \vee D$ by paramodulation. Assume $\Sigma \models \overline{(C(s) \vee \phi)}(x_1, ..., x_n)$ and $\Sigma \models \overline{(s = t \vee D \vee \psi)}(x_1, ..., x_n)$. Here we consider case (d) of the assignment for paramodulation in which s occurs in $C(s)$ as a subterm of a maximal Π-term $h(s)$ which occurs more than once in $C(s) \vee \phi$. Then since $h(s), h(t)$ are distinct Π-terms, they will be replaced by distinct new variables $\overline{h(s)}, \overline{h(t)}$ in $\overline{C(t)} \vee \phi$. For any model of Σ and any assignment of all free variables in the lifted paramodulant and its assigned formula, if $\overline{C}(s)$ and $\overline{s} = \overline{t}$ are true but $\overline{C}(t)$ is false, then we must have $\overline{h(s)} \neq \overline{h(t)}$. So in this case we have: $\Sigma \models \overline{(s = t \wedge h(s) \neq h(t))}(x_1, ..., x_n)$. And hence, $\Sigma \models \overline{(C(t) \vee D \vee \theta)}(x_1, ..., x_n)$.

The arguments for the other cases are straightforward. □

We now assign a *dual* formula to each clause in the proof P as follows:

(i). If C is an input clause from Σ, its formula is T; if C is an input clause from Π, its formula is F.

(ii). If ϕ is assigned to $L \vee C$ and ψ is assigned to $\neg L' \vee D$, and if $(C \vee D)\pi$ is the resolvent of $L \vee C$ and $D \vee \neg L'$ against $L\pi = L'\pi$, then the formula assigned to $(C \vee D)\pi$ is:

(a). $(\phi \wedge \psi)\pi$ if the occurrences of both L and $\neg L'$ are from Σ alone;

(b). $(\phi \vee \psi)\pi$ if the occurrences of both L and $\neg L'$ are from Π alone;

(c). $((L \wedge \phi) \vee (\neg L' \wedge \psi))\pi$ if neither (a) nor (b).

(iii). If $(C(t) \vee D)\pi$ is the paramodulant from above described paramodulation, its formula is

(d). $[(\phi \vee s \neq t) \wedge (\psi \vee s = t)]\pi \wedge (s \neq t \vee h(s) = h(t))\pi$ provided the r is a subterm of a maximal Π-term $h(r)$ and there is more than one occurrence of $h(r)$ in $C(r) \vee \phi$.

(e). $[(\phi \vee s \neq t) \wedge (\psi \vee s = t)]\pi \vee (s = t \vee h(s) \neq h(t))\pi$ provided the r is a subterm of a maximal Σ-term $h(r)$ and there is more than one occurrence of $h(r)$ in $C(r) \vee \phi$.

(f). $((\phi \wedge s = t) \vee (\psi \wedge s \neq t))\pi$ if neither (d) nor (e).

(iv). If ϕ is assigned to $L \vee L' \vee C$ and π is a unifier such that $L\pi = L'\pi$, then we assign $\phi\pi$ to the factor clause $(L \vee C)\pi$.

By induction on the depth of a clause in the deduction we can show

Lemma 13. *The formula assigned by the dual method is the logical negation of that assigned by the original Interpolation Algorithm.*

Corollary 14. *Assume that $\bar{\theta}(s_1, ..., s_k)$ is the dual formula assigned to C in P_p by the dual assignment algorithm, where $s_1, ..., s_k$ are all the Σ-terms with maximal occurrences in P_p. If we let $\hat{\theta}(y_1, ..., y_k)$ be the formula obtained by simultaneously replacing all maximal occurrences of the Σ-terms $s_1, ..., s_k$ by the new variables $y_1, ..., y_k$, then we have $\Pi \models \hat{\theta}(y_1, ..., y_k)$.*

Now we are ready to quantify all the variables for noncommon terms in the relational interpolant θ of Σ and Π relative to the empty clause. Assume that all the maximal Σ-terms and Π-terms are $\{t_1, ..., t_n\}$, ordered by their lengths. Assume $\{t_1, ..., t_n\} = \{r_1, ..., r_k\} \cup \{s_{k+1}, ..., s_n\}$ where the r_i's are the maximal Π-terms and the s_j's are the maximal Σ-terms. If lifting θ from Π-terms gives $\bar{\theta}(x_1, ..., x_k)$, and lifting $\bar{\theta}(x_1, ..., x_k)$ from the Σ-terms gives $\theta^*(z_1, ..., z_n)$ where the z_i's are new variables for the t_i's, then we have

Theorem 15. $Q_1 z_1 ... Q_n z_n \theta^*(z_1, ..., z_n)$ *is a Craig interpolant separating Σ and Π, where Q_i is $\forall$ if t_i is a Π-term, otherwise Q_i is $\exists$.*

Proof. Clearly $Q_1 z_1 ... Q_n z_n \theta^*(z_1, ..., z_n)$ is a formula in $L_\Sigma \cap L_\Pi$. By Lemma 12 we have $\Sigma \models \forall x_1 ... x_k \bar{\theta}(x_1 ... x_k)$.

Each maximal Σ-term of $\bar{\theta}(x_1, ..., x_k)$ is a lifting $\bar{s}_j(x_1, ..., x_k)$ of one of the maximal Σ-terms $s_{k+1}, ..., s_n$ of θ. If x_i occurs in $\bar{s}_j(x_1, ..., x_k)$ then the term r_i which x_i replaces is a subterm of s_j and thus r_i occurs before s_j in the list $\{t_1, ..., t_n\}$ and the variable for r_i occurs before the variable for s_j in the prefix $Q_1 z_1 ... Q_n z_n$. Hence $\bar{s}_j$ is a witness for the quantifier $\exists y_j$ in $Q_1 z_1 ... Q_n z_n \theta^*(z_1, ..., z_n)$. Hence, $\Sigma \models Q_1 z_1 ... Q_n z_n \theta^*(z_1, ..., z_n)$.

On the other side, for the dual formula $\neg \theta$, using the same order among the noncommon terms and by the corollary above and the fact that θ^* is also the lifting from Π-terms of the lifting $\hat{\theta}$ of θ from Σ-terms, we also have
$$\Pi \models \overline{Q}_1 z_1 ... \overline{Q}_n z_n \neg \theta^*(z_1, ..., z_n) \text{ where } \overline{Q}_j = \forall \, (\exists) \text{ iff } Q_j = \exists \, (\forall).$$ Moving the negation symbol out we finally have $\Pi \models \neg Q_1 z_1 ... Q_n z_n \theta^*(z_1, ..., z_n)$. $\square$

The formula θ^* may contain free variables other than $z_1, ..., z_n$. We get a Craig interpolant sentence by quantifying these extra variables with the quantifier Q_1 or any other sequence of quantifiers.

Example 1. Let $\Sigma = \{R(x, a) \lor R(x, b)\}$, $\Pi = \{\neg R(c, y)\}$, where a, b and c are distinct constants. An OTTER resolution refutation for $\Sigma \cup \Pi \models \Diamond$ is:

1 $R(x, a) \lor R(x, b)$.
2 $-R(c, y)$.
3 [binary,1,2] $R(c, b)$.
4 [binary,3,2] .

The Interpolation Algorithm gives the formula $\theta = R(c, a) \lor R(c, b)$. Since a and b are Σ-terms and c is a Π-term, we replace a and b by the existentially quantified variables x and y, and replace c by the universally quantified variable z. Since the lengths of a, b, c are all 1, the order among the quantifiers does not matter. Thus the following three formulas are all Craig interpolants between Σ and Π:
$$\forall z \exists xy (R(z, x) \lor R(z, y)), \quad \exists x \forall z \exists y (R(z, x) \lor R(z, y)), \quad \exists xy \forall z (R(z, x) \lor R(z, y))$$

Example 2. Let $\Sigma = \{x \neq f(x), x \neq f(f(x))\}$, $\Pi = \{y = x \lor y = g(x)\}$, where both f and g are functions. Any model of Σ has a universe of size at least 3, while any model of Π has a universe of size at most 2. So Σ and Π are inconsistent. An OTTER resolution refutation for $\Sigma \cup \Pi \models \Diamond$ is:

1 $x \neq f(x)$.
2 $x \neq f(f(x))$.
3 $y = x \lor y = g(x)$.
4 [binary,3.1,2.1] $x = g(f(f(x)))$.
5 [binary,3.1,1.1] $x = g(f(x))$.
10 [para-into,4.1.2,5.1.2] $x = f(x)$.
11 [binary,10.1,1.1] .

The formula θ the Interpolation Algorithm gives is

$[(x \neq f(f(x)) \wedge x \neq g(f(f(x)))) \vee (x = g(f(f(x))) \wedge f(x) \neq f(f(x)))] \wedge x \neq f(x)$.

The noncommon terms, when sorted according to lengths, are $f(x), f(f(x)), g(f(f(x)))$, where $g(f(f(x)))$ is a Π-term and the others are Σ-terms. Replacing these terms with the variables u, v, w and quantifying them gives the formula

$\theta^* = \exists u v \forall w [(x \neq v \wedge x \neq w) \vee (x = w \wedge u \neq v)] \wedge (x \neq u)$.

Note that any model for θ^* contains at least three elements: It can not contain only one or two elements, for x, u, v must be distinct. Thus θ^* is a Craig interpolant separating Σ and Π.

3 Applications of Craig Interpolants

Given two finite structures, to show they are isomorphic, one finds an isomorphism. To show they are not, one gives a statement that separates them, i.e., a sentence which is true in one structure but false in the other. The Interpolation Algorithm can be used to find such a sentence.

For structures S_1 with elements $\{a_1, .., a_n\}$ and S_2 with elements $\{b_1, ..., b_n\}$, assume that the universes for S_1 and S_2 are disjoint, and all the elements a_i, b_j are named by new distinct constant symbols. Furthermore assume the diagrams (the collection of all atomic sentences and negations of atomic sentences which hold in the structure) for the structures are Δ_1 and Δ_2 respectively. Then each of the diagrams is a theory in some language. If the two structures are not isomorphic, then $\Sigma = \Delta_1 \cup \forall x (x = a_1 \vee ... \vee x = a_n)$ and $\Pi = \Delta_2 \cup \forall y (y = b_1 \vee ... \vee y = b_n)$ are inconsistent, and by completeness there exists a refutation proof for $\Sigma \cup \Pi \models \Diamond$. Applying the Interpolation Algorithm to this proof gives a first order sentence which separates the structures.

For example, let S_1 and S_2 be directed graphs. S_1 has vertices $\{a, b, c\}$ with edges $\{(a, b), (a, c)\}$. S_2 has vertices $\{a', b', c'\}$ with edges $\{(a', b'), (c', a')\}$. We use binary relation p to represent the edges of the graphs. Then the diagram for S_1 is

$\Delta_1 = \{p(a, b), p(a, c), \neg p(b, a), \neg p(b, c), \neg p(c, a), \neg p(c, b), a \neq b, a \neq c, b \neq c\}$.

And the diagram for S_2 is

$\Delta_2 = \{p(a', b'), p(c', a'), \neg p(a', c'), \neg p(b', a'), \neg p(b', c'), \neg p(c', b'), a' \neq b', a' \neq c', b' \neq c'\}$.

So $\Sigma = \Delta_1 \cup \forall x (x = a \vee x = b \vee x = c)$ and $\Pi = \Delta_2 \cup \forall x (x = a' \vee x = b' \vee x = c')$.

A refutation proof by OTTER is the following:

1 p(a,b).
2 p(a,c).
7 (a ≠ b).
8 (a ≠ c).
10 (x = a) ∨ (x = b) ∨ (x = c).
14 -p(b1,c1).

15 -p(c1,b1).

16 -p(a1,c1).

19 $(x = a1) \lor (x = b1) \lor (x = c1)$.

22 $(b1 \neq c1)$.

30 [para-from,10,7] $(a \neq x) \lor (x = a) \lor (x = c)$.

34 [para-from,10,1] $p(a,x) \lor (x = a) \lor (x = c)$.

44 [para-from,19,16] $-p(x,c1) \lor (x = b1) \lor (x = c1)$.

274 [para-from,30,8] $(a \neq x) \lor (x = a)$.

507 [para-from,34,2] $p(a,x) \lor (x = a)$.

699 [para-from,44,14] $-p(x,c1) \lor (x = c1)$.

711 [binary,699,507] $(a = c1) \lor (c1 = a)$.

783 [binary,711,274] $(c1 = a)$.

794 [para-from,783,22] $(b1 \neq a)$.

797 [para-from,783,15] $-p(a,b1)$.

816 [binary,794,507] $p(a,b1)$.

817 [binary,816,797] .

Applying the Interpolation Algorithm, and using some trivial logic rules such as $A \lor \neg A \iff T$, $(A \land \neg B) \lor B \iff A \lor B$, $A \land A \iff A$, $A \land \neg A \iff F$, $A \lor \neg A \iff T$, to simplify the formula, we get the following relational interpolant

θ : $(c' = a \lor p(a,c')) \land (b' = a \lor p(a,b'))$.

Note that a is a Σ-term, while b' and c' are Π-terms. If we replace a, b', c' by x, y, z, respectively, by Theorem 15, we get the following formulas. They all separate the two graphs:

$\exists x \forall y z [(z = x \lor p(x,z)) \land (y = x \lor p(x,y))]$,

$\forall y \exists x \forall z [(z = x \lor p(x,z)) \land (y = x \lor p(x,y))]$,

$\forall y z \exists x [(z = x \lor p(x,z)) \land (y = x \lor p(x,y))]$.

Note that there is a shorter formula $\exists x \forall y (x = y \lor p(x,y))$ which separates the two given structures. The generation of minimal length separating sentences is an open problem. We also need more efficient proof strategies for such problems since current resolution provers can not find refutation proof for pairs of structures with more than 6 elements.

Acknowledgements: The author thanks Dale Myers for his numerous suggestions and corrections to earlier versions of this paper.

References

1. Roger Lyndon, *Notes on Logic*, D. Van Nostrand Company, Princeton, 1966.
2. Chin-Liang Chang, Richard Char-Tung Lee, *Symbolic Logic and Mechanical Theorem Proving*, Academic Press, 1973.
3. C C. Chang, H. Jerome Keisler, *Model Theory*, North Holland, 1990.
4. Larry Wos, Overbeek, Lusk, Boyle, *Automated Reasoning: Introduction and Applications*, McGraw-Hill, 1992.

Currying of Order-Sorted Term Rewriting Systems

Yoshinobu Kawabe and Naohiro Ishii

Department of Intelligence and Computer Science
Nagoya Institute of Technology
Gokiso-cho, Showa-ku, Nagoya 466, Japan
E-Mail: {kawayosi,ishii}@egg.ics.nitech.ac.jp

Abstract. Term rewriting system is a helpful tool for implementing functional programming languages. We focus upon a transformation of term rewriting systems called currying. Currying transforms a term rewriting system with symbols of arbitrary arity into another one, which contains only nulary symbols with a single binary symbol called application. Currying in single-sorted case is explored in [1] but currying in typed case remains as a problem. This paper first proposes currying of order-sorted term rewriting systems. Then, we prove that compatibility and confluence of order-sorted term rewriting systems are preserved by currying.

1 Introduction

Currying of term rewriting systems(TRSs, for short) is a transformation from a TRS of 'functional' form to another one of 'applicative' form. Terms of functional form are constructed from function symbols of various arities and variables, like $f(x, 2, 3)$. Terms of applicative form consist of variables, constants and a binary symbol called application, like $Ap(Ap(Ap(f, x), 2), 3)$.

A famous example of applicative form of TRSs is Combinatory Logic(CL), which was developed by Shönfinkel and rediscovered by Curry. Klop [3] says that the TRS CL has 'universal computational power' and that CL is used to implement functional programming languages. However, applicative forms of term rewriting are not usually treated in the theory of term rewriting. Kennaway et.al [1] insisted that one need to study the relationship between the two style of presentation and they proved that properties of TRSs, SN,WN,CR,WCR,UN and completeness etc., are preserved by currying; the proof of preservation of CR is described in [2].

They provided a framework for explaing functional programming languages by single-sorted TRS. However, many functional programming languages have a type system. Moreover, relations between types are considered in recent ones such as Haskell. Thus, we shall explore typed cases, especially ones with subtypes. We focus upon order-sorted TRSs and define currying in the order-sroted case. Reasons why we selected order-sorted TRSs are: (i) They build a nice framework to handle partially defined functions and subtypes, (ii) They are simple systems so that they are relatively easy to treat.

This paper is organized as follows. We first summarize some basic results in order-sorted TRSs in the next section. Then, we define currying of order-sorted TRSs in section 3. In section 4, we analyze the curried order-sorted TRSs; the results obtained shall be used in section 5. Finally, in section 5, we prove that compatibility and confluence of order-sorted TRSs are preserved by currying.

2 Foundations

We write $\mathcal{N}$ for the set of natural numbers. We describe $\twoheadrightarrow_\alpha$ for the reflexive and transitive closure of a relation $\rightarrow_\alpha$ (α is a meta variable), and $\twoheadleftarrow_\alpha$ for the inverse relation, in [3]. Given a set A, A^* is the set of finite words over A; ε is an empty word. Given an element $a \in A$ and a natural number $n \in \mathcal{N}$, a^n is a word consisting of a of n pieces; a^0 is an empty word. $v \cdot w$ is concatenation of two words v and w. A^* is partially ordered by the prefix ordering $\preceq$ satisfying $p \preceq q \Leftrightarrow \exists w \in A^*.[p \cdot w = q]$, and additionally, we write $p \prec q$ if $p \neq q$. $\succ$ is the inverse relation of $\prec$. Two words p and q are independent, written $p|q$, iff $p \npreceq q \wedge q \npreceq p$.

Following [4][5][6][7], we summarize some basic results in order-sorted TRSs.

An order-sorted signature is a triple $(S, \leq, \Sigma)$, where S is a set of sorts, $\leq$ is a partial ordering over S, and Σ is a family $\{\Sigma_{w,s} | w \in S^*, s \in S\}$ of sets of function symbols. We shall often write $f : w \to s$ for $f \in \Sigma_{w,s}$, and $f :\to s$ for $f \in \Sigma_{\varepsilon,s}$. A function symbol $f :\to s$ is called a constant. We use Σ as an abbreviation for both $(S, \leq, \Sigma)$ and $\bigcup_{w,s} \Sigma_{w,s}$.

An S-sorted variable set is a family $V = \{V_s | s \in S\}$ of disjoint sets. We write $x : s$ for a variable $x \in V_s$. We abbreviate V for $\bigcup_{s \in S} V_s$. The variable set V is disjoint from Σ.

The set $PT_\Sigma(V)$ of pseudo-terms over Σ and V is the least set with the following properties: (i) $x \in PT_\Sigma(V)$ if $x : s \in V$, (ii) $f \in PT_\Sigma(V)$ if $f :\to s$, (iii) $f(t_1, \ldots, t_n) \in PT_\Sigma(V)$ if $f : s_1 \cdots s_n \to s$ and $t_i \in PT_\Sigma(V)$ for $1 \leq i \leq n$ where n is finite; the condition "n is finite" forces trees in $T_\Sigma(V)_s$ to be finitely branching. In [4][5][6][7], pseudo-terms are called extended terms.

The set $T_\Sigma(V)_s$ of terms over Σ and V with sort s is the least set with the following properties: (i) $x \in T_\Sigma(V)_s$ if $x : s_0 \in V$ and $s_0 \leq s$, (ii) $f \in T_\Sigma(V)_s$ if $f :\to s_0$ and $s_0 \leq s$, (iii) $f(t_1, \ldots, t_n) \in T_\Sigma(V)_s$ if $f : s_1 \cdots s_n \to s_0$ such that $s_0 \leq s$ and $t_i \in T_\Sigma(V)_{s_i}$ for every $1 \leq i \leq n$ where n is finite. We can see terms in $T_\Sigma(V)_s$ as a tree. $T_\Sigma(V) := \bigcup_{s \in S} T_\Sigma(V)_{s_i}$ denotes the set of all terms over Σ and V. The set $T_\Sigma(V)$ is a subset of $PT_\Sigma(V)$. If $t \in PT_\Sigma(V)$ is an element of $T_\Sigma(V)$, we say that t is a well-formed term, otherwise t is called ill-formed.

The set of all variables in a term $t \in T_\Sigma(V)$ is abbreviated by $Var(t)$.

$O(t)$ is the set of all occurrences of the pseudo-term t. Let t and t' be pseudo-terms in $PT_\Sigma(V)$ and $v \in O(t)$ an occurrence; the result of the replacement of the subterm at v in t by t' is denoted by $t[v \leftarrow t']$. Note that $t[v \leftarrow t']$ may be ill-formed, even if t and t' are well-formed terms. t/v is the subterm of t at v.

Following [Kah94], we define the function $last(O(t))$, which is the number of parameters of the outer-most function symbol in t. Given the set of occurrences

$O(t)$ of a pseudo-term t, $last(O(t))$ is a natural number $n \in \mathcal{N}$ such that $n = 0 \iff O(t) = \{\varepsilon\}$ and otherwise $n \in O(t) \wedge n + 1 \notin O(t)$.

The function $label_t : O(t) \to \Sigma \cup V$ for pseudo-terms $t \in PT_\Sigma(V)$ is a function defined as: (i) $label_t(\varepsilon) = head(t)$ where (a) $head(t) = x : s$ if t is a variable $x : s$, (b) $head(t) = f : s_1 \cdots s_n \to s$ if $t = f(t_1, \ldots, t_n) \wedge f : s_1 \cdots s_n \to s$, (ii) $label_{f(t_1, \ldots, t_i, \ldots, t_n)}(i \cdot v) = label_{t_i}(v)$ where $f : s_1 \cdots s_n \to s$. $label_t(v)$ for $t \in PT_\Sigma(V)$ and $v \in O(t)$ is a function symbol or a variable in t at v. The idea of viewing a term t as function from $O(t)$ to $\Sigma \cup V$ is fairly standard. The particular advantage is the simple addressing of a subterm. Another advangage is that infinite terms are allowd; we only consider finite terms for the purposes of this paper.

Let $(S, \leq, \Sigma)$ be an order-sorted signature. A $(S, \leq, \Sigma)$-algebra A consists of a family $\{A_s | s \in S\}$ of sets and a function $A_f : D_f^A \to C_A$ for every $f \in \Sigma$ such that the following conditions are fulfilled: (i) $A_s \subseteq A_{s'}$, if $s \leq s'$, (ii) D_f^A is a subset of $(C_A)^*$, where $C_A := \bigcup_{s \in S} A_s$, (iii) If $f \in \Sigma_{w,s}$, then $A_w \subseteq D_f^A$ and $A_f(A_w) \subseteq A_s$. An assignment ν from a variable set V into Σ-algebra A is a set of functions $\{\nu_s : V_s \to A_s | s \in S\}$. A substitution σ is an assignment from a variable set Y into the term algebra $T_\Sigma(X)$. We also write σ for the extension $\sigma^* : T_\Sigma(Y) \to T_\Sigma(X)$ defined by $\sigma^*(f(t_1, \ldots, t_n)) = f(\sigma^*(t_1), \ldots, \sigma^*(t_n))$. A substitution $\{x_1, \ldots, x_n\} \mapsto T_\Sigma(X)$ that maps the variables $x_1, \ldots, x_n$ to the terms $t_1, \ldots, t_n$, respectively, is written $\sigma = (x_1 \leftarrow t_1, \ldots, x_n \leftarrow t_n)$. Note that if y is any variable such that $y \notin \{x_1, \ldots, x_n\}$, then we define $\sigma(y) = y$.

A rewrite rule is a pair (l, r) $(l, r \in T_\Sigma(Y))$ where (i) l is not a variable, (ii) All variables appearing in r appear in l. We shall usually write $l \to_R r$ instead of (l, r). The label of the binary relations R is sometimes omitted.

An order-sorted TRS is a triple $((S, \leq, \Sigma), V, R)$, where $(S, \leq, \Sigma)$ is a signature, V is a variable set and R is a set of rewrite rules. $((S, \leq, \Sigma), V, R)$ is often abbreviated by R.

A term $t \in T_\Sigma(X)$ rewrites to $t' \in T_\Sigma(X)$ with a rewrite rule $l \to r$ at the occurrence v if the following conditions are satisfied: (i) There exists a substitution $\sigma : Var(l) \to T_\Sigma(X)$ such that $\sigma(l) = t/v$, (ii) $t' = t[v \leftarrow \sigma(r)]$, (iii) t' is well-formed. We abbreviate this by $t \xrightarrow{[v, l \to r]} t'$, possibly ommiting the indices.

Order-sorted rewriting differs from many-sorted rewriting in that the well-formedness of the resulting term must be explicitly checked. We can see an order-sorted TRS as a kind of nondeterministic computer program. Generally, well-typedness of programs is not guaranteed. We consider the following conditions. A rewrite rule $l \to r$ is called compatible, if for every $t \in T_\Sigma(X)$ and every $v \in O(t)$ we have: If $\sigma(l) = t/v$ for some substitution $\sigma : Var(l) \to T_\Sigma(X)$, then $t[v \leftarrow \sigma(r)]$ is a well-formed term. Compatibility corresponds to well-typedness of programs. If the signature $(S, \leq, \Sigma)$ is finite, then the compatibility of a rewrite rule is decidable [Wal89]. We say R is compatible if all rewrite rules in R are compatible. A rewrite rule $l \to r$ is called sort decreasing, if we have for every substitution $\sigma : Var(l) \to T_\Sigma(X)$ and every sort $s \in S$: If $\sigma(l) \in T_\Sigma(X)_s$, then $\sigma(r) \in T_\Sigma(X)_s$. Sort decreasingness implies compatibility. We say R is sort decreasing if all rewrite rules in R are sort decreasing.

3 Definition of Currying in Order-Sorted Case

In this section, we first introduce the signature and the terms of applicative TRSs. Then we define applicative TRSs and currying in order-sorted case.

Definition 1. Let $(S, \leq, \Sigma)$ be an order-sorted signature of functional form. The signature of applicative TRSs, $(Cu(S), \leq, Cu(\Sigma))$, is defined as:

1. $Cu(S) = Bas(Cu(S)) \cup Der(Cu(S))$ where $Bas(Cu(S))$ and $Der(Cu(S))$ are: (i) $Bas(Cu(S)) = S$. We call every sort in the set $Bas(Cu(S))$ a basic sort of $Cu(S)$, (ii) For every function symbols $f : s_1 \cdots s_n \to s (n \geq 1)$ and every $i \in \{0, \ldots, n-1\}$, a symbol $\left\langle \frac{f : s_{i+1} \to s_{i+2} \to \cdots \to s_n \to s}{f : s_1 \to s_2 \to \cdots \to s_n \to s} \right\rangle$ exists and $\left\langle \frac{f : s_{i+1} \to s_{i+2} \to \cdots \to s_n \to s}{f : s_1 \to s_2 \to \cdots \to s_n \to s} \right\rangle \in Der(Cu(S))$, where $\left\langle \frac{f : s_{i+1} \to s_{i+2} \to \cdots \to s_n \to s}{f : s_1 \to s_2 \to \cdots \to s_n \to s} \right\rangle \notin S$ $(0 \leq i \leq n-1)$. We call every sort in the set $Der(Cu(S))$ a derivation sort of $Cu(S)$.

2. $Cu(\Sigma)$, which is a family $\{Cu(\Sigma)_{w,s} | w \in Cu(S)^*, s \in Cu(S)\}$ of the set of function symbols, is the least set with the following properties: (i) $f \in Cu(\Sigma)_{\varepsilon, s}$ for every constant $f :\to s$, (ii) $f_0 \in Cu(\Sigma)_{\varepsilon, \left\langle \frac{f : s_1 \to s_2 \to \cdots \to s_n \to s}{f : s_1 \to s_2 \to \cdots \to s_n \to s} \right\rangle}$ for every function symbol $f : s_1 \cdots s_n \to s (n \geq 1)$, (iii) For every function symbol $f : s_1 \cdots s_n \to s$ $(n \geq 1)$, $Ap \in Cu(\Sigma)_{S_i s_{i+1}, S_{i+1}}$ $(\forall i \in \{0, \ldots, n-1\})$, where $S_i = \left\langle \frac{f : s_{i+1} \to s_{i+2} \to \cdots \to s_n \to s}{f : s_1 \to s_2 \to \cdots \to s_n \to s} \right\rangle$ is a derivation sort; we regard $S_{i+1} = s$ if $i + 1 = n$. We call the operators whose name is Ap Ap-symbols.

$f \in Cu(\Sigma)_{w,s}$ and $f \in Cu(\Sigma)_{\varepsilon, s}$ are abbreviated by $f : w \rightsquigarrow s$ and $f :\rightsquigarrow s$, respectively.

Definition 2. An applicative term over the signature $(S, \leq, \Sigma)$ and the set of variables V is a term over $(Cu(S), \leq, Cu(\Sigma))$ and V.

There exist no variables having a derivation sort. The set of all applicative terms over Σ and V is denoted by $AT_\Sigma(V)$.

Definition 3. The function $cur : PT_\Sigma(V) \to PT_{Cu(\Sigma)}(V)$ is defined by the induction on the structure of the pseudo-terms in $PT_\Sigma(V)$. For every pseudo-term $t \in PT_\Sigma(V)$, (i) $cur(f :\to s) = f :\rightsquigarrow s$ if $t = f :\to s$, (ii) $cur(x : s) = x : s$ if $t = x : s$, (iii) If $t = f(t_1, \ldots, t_n)$ for $f : s_1 \cdots s_n \to s$ and $t_i \in PT_\Sigma(V)$ where $i \in \{1, \ldots, n\}$ then $cur(f(t_1, \ldots, t_n))$ is an applicative term defined as: (a) $label_{cur(f(t_1,\ldots,t_n))}(1^i) = Ap : S_{n-i-1} s_{n-i} \rightsquigarrow S_{n-i}$ $(1 \leq i \leq n-1)$, (b) $label_{cur(f(t_1,\ldots,t_n))}(\varepsilon) = Ap : S_{n-1} s_n \rightsquigarrow s$, (c) $label_{cur(f(t_1,\ldots,t_n))}(1^n) = f_0 :\rightsquigarrow S_0$, and (d) $cur(f(t_1, \ldots, t_n))/1^i \cdot 2 = cur(t_{n-i}) \in PT_\Sigma(V)$ $(0 \leq i \leq n-1)$, where $S_i = \left\langle \frac{f : s_{i+1} \to s_{i+2} \to \cdots \to s_n \to s}{f : s_1 \to s_2 \to \cdots \to s_n \to s} \right\rangle$.

Intuitively, $cur(f(t_1, \ldots, t_n)) = Ap(\ldots Ap(f_0, cur(t_1)) \ldots, cur(t_n))$. Note that the function $cur : PT_\Sigma(V) \to PT_{Cu(\Sigma)}(V)$ is not surjective: e.g., there exist no pseudo-terms t such that $cur(t) = f_0 \rightsquigarrow \left\langle \frac{f : s_1 \to s_2 \to \cdots \to s_n \to s}{f : s_1 \to s_2 \to \cdots \to s_n \to s} \right\rangle$.

Definition 4. An applicative TRS (or ATRS) is a TRS. Terms of ATRS are applicative terms over some signature and some variable set. Every rewrite rules has LHS with the form of $cur(t)$ for some (non-applicative) term t.

Definition 5. Let $((S, \le, \Sigma), V, R)$ be an order-sorted TRS. We define currying of $((S, \le, \Sigma), V, R)$ as the ATRS $((Cu(S), \le, Cu(\Sigma)), V, Cu(R))$ where $Cu(R) = \{cur(l) \to_{Cu(R)} cur(r) | l \to r \in R\}$. Currying of R is abbreviated by $Cu(R)$.

4 Analysis of Curried Order-Sorted TRSs

We analyze the curried order-sorted TRSs in this section. The main results here, theorem 19, 22 and 23, will be used in section 5.

4.1 Restricted Occurrences and Curried Position

We first define the set $RO(cur(t))$ for every pseudo-term t, which is a subset of $O(cur(t))$. Then, we define the bijective function cp_t from $O(t)$ to $RO(cur(t))$. The function cp_t binds an occurrence of t with the corresponding one of $cur(t)$.

Definition 6. Let $t \in PT_\Sigma(V)$ be a pseudo-term over the signature $(S, \le, \Sigma)$ and the variable set V. The set of restricted occurrences $RO(cur(t))$ over $cur(t)$ is defined as: (i) $RO(cur(t)) = \{\varepsilon\}$ if t is a variable, (ii) $RO(cur(t)) = \{\varepsilon\} \cup \{1^{n-i} \cdot 2 \cdot o | 1 \le i \le n \wedge o \in RO(cur(t_i))\}$ if $t = f(t_1, \ldots, t_n)$ where $n \ge 1$.

We can prove that $RO(cur(t))$ is a subset of $O(cur(t))$ by the induction on the structure of pseudo-terms $t \in PT_\Sigma(V)$. The following lemma gives us the way to test if an occurrence in $O(cur(t))$ is in $RO(cur(t))$ or not(the proof is omitted).

Lemma 7. *For any pseudo-term $t \in PT_\Sigma(V)$ and any occurrence $v \in O(cur(t))$:*
(i) $v \in RO(cur(t)) \iff v = \varepsilon$ or $v = w \cdot 2$ for some w,
(ii) $v \in O(cur(t)) \backslash RO(cur(t)) \iff v = w \cdot 1$ for some w.

Seeing the tail of an occurrence in $O(cur(t))$ enables us to check if the occurrence is in $RO(cur(t))$ or not.

Definition 8. Let $t \in PT_\Sigma(V)$ be a pseudo-term. The function cp_t whose domain is $O(t)$ is defined as: (i) $cp_t(\varepsilon) = \varepsilon$, (ii) $cp_{f(t_1, \ldots, t_i, \ldots, t_n)}(i \cdot o) = 1^{n-i} \cdot 2 \cdot cp_{t_i}(o)$.

We can prove that the range of cp_t is $RO(cur(t))$ by the induction on the structure of pseudo-terms t, and cp_t is bijective; the inverse is $pc_t : RO(cur(t)) \to O(t)$, which is: (i) $pc_t(\varepsilon) = \varepsilon$, (ii) $pc_{f(t_1, \ldots, t_i, \ldots, t_n)}(1^{n-i} \cdot 2 \cdot o) = i \cdot pc_{t_i}(o)$.

The function cp_t binds an occurrence of a pseudo-term t with the corresponding occurrence of $cur(t)$, that is, $cur(t/v) = cur(t)/cp_t(v)$ holds for any pseudo-term t and any occurrence $v \in O(t)$ (by the induction on the structure of t). The results here are mainly used in section 4.3 or later.

4.2 The Uncurrying Function

We define the left-inverse function of *cur* here. In definition 9, we first define a TRS called semi partial parametrized TRS, which is complete.

Definition 9. Let $((S, \leq, \Sigma), V, R)$ be an order-sorted TRS. The order-sorted TRS $((Cu(S), \leq, SPP(\Sigma)), V, SPP(R))$ is defined as:

1. $SPP(\Sigma)$, which is a family $\{SPP(\Sigma)_{w,s} | w \in Cu(S)^*, s \in Cu(S)\}$ of the set of function symbols, is the least set with the following properties: (i) $f \in SPP(\Sigma)_{s_1 \cdots s_n, s}$ for every function symbol $f : s_1 \cdots s_n \to s (n \geq 0)$, (ii) $Ap \in SPP(\Sigma)_{s_\alpha s_\beta, s}$ for every Ap-symbol $Ap : s_\alpha s_\beta \leadsto s$, (iii) For each function symbol $f : s_1 \cdots s_n \to s$ $(n \geq 1)$, $f_i \in SPP(\Sigma)_{s_1 \cdots s_i, \left\langle \frac{f : s_{i+1} \to s_{i+2} \to \cdots \to s_n \to s}{f : s_1 \to s_2 \to \cdots \to s_n \to s} \right\rangle}$ ($\forall i \in \{0, \ldots, n-1\}$). We call function symbols $f_i (0 \leq i \leq n-1)$ for $f : s_1 \cdots s_n \to s$ imcomplete function symbols. $f \in SPP(\Sigma)_{w,s}$ and $f \in SPP(\Sigma)_{\varepsilon, s}$ are abbreviated by $f : w \leadsto' s$ and $f :\leadsto' s$, respectively.

2. Let $S_i = \left\langle \frac{f : s_{i+1} \to s_{i+2} \to \cdots \to s_n \to s}{f : s_1 \to s_2 \to \cdots \to s_n \to s} \right\rangle$ be a derivation sort in $Cu(S)$. For any function symbol $f : s_1 \cdots s_n \to s (n \geq 1)$ and every incomplete function symbol $f_i : s_1 \cdots s_i \leadsto' S_i (0 \leq i \leq n-1)$ of $f : s_1 \cdots s_n \to s$, there are rewrite rules $Ap(f_i(x_1, \ldots, x_i), y) \to_{SPP(R)} f_{i+1}(x_1, \ldots, x_i, y)$ in $SPP(R)$, where $Ap : S_i s_{i+1} \to S_{i+1}$; we regard $f_{i+1} = f$ and $S_{i+1} = s$ if $i+1 = n$. We call these rules uncurrying rules. There are only uncurrying rules in $SPP(R)$.

$((Cu(S), \leq, SPP(\Sigma)), V, SPP(R))$ is often abbreviated by $SPP(R)$. We describe $SPP_\Sigma(V)$ for the set of terms over the signature $(Cu(S), \leq, SPP(\Sigma))$ and the variable set V. $SPP(R)$ is compatible since $SPP(R)$ is sort decreasing.

The following lemma is needed for defining a function in definition 11. The proof is omitted due to space limitation.

Lemma 10. $SPP(R)$ *is complete.*

From lemma 10, we can define the function *uncur* given in the following.

Definition 11. Let $((S, \leq, \Sigma), V, R)$ be an order-sorted TRS. We define the uncurrying function $uncur : AT_\Sigma(V) \to SPP_\Sigma(V)$, which is defined as $uncur(t) = s$ where $t \twoheadrightarrow_{SPP(R)} s$ and s is a normal form in $SPP(R)$.

The following lemma says the function *uncur* is the left-inverse function of *cur*.

Lemma 12. $uncur(cur(t)) = t$ *for any term* $t \in T_\Sigma(V)$.

4.3 The Kernel of Curried Term Rewriting Systems

The following four lemmata, although non-trivial, are given here without proofs due to space limitation.

Lemma 13. *Let* $((S, \leq, \Sigma), V, R)$ *be an order-sorted TRS. For any applicative term* $t \in AT_\Sigma(V)$, *the pseudo-term* $uncur(t)$ *is well-formed. And the sort of* t *and the sort of* $uncur(t)$ *are the same.*

Lemma 14. *Let $((S, \leq, \Sigma), V, R)$ be an order-sorted TRS. The pseudo-term t is well-formed, iff the pseudo-term $cur(t)$ is well-formed. And the sort of t and the sort of $cur(t)$ are the same for any term $t \in T_\Sigma(V)$.*

Lemma 15. *Let $((Cu(S), \leq, Cu(\Sigma)), V, Cu(R))$ be a curried order-sorted TRS. For any applicative term $cur(t) \in AT_\Sigma(V)$ and any occurrence $v \in O(cur(t))$, $v \in O(cur(t)) \backslash RO(cur(t))$ implies that every rewrite rule in $Cu(R)$ cannot be applied for $cur(t)$ at v.*

Lemma 16. *Let $((Cu(S), \leq, Cu(\Sigma)), V, Cu(R))$ be a curried order-sorted TRS. For any applicative term $cur(t) \in AT_\Sigma(V)$ and any occurrence $v \in O(cur(t))$, if a rewrite rule in $Cu(R)$ can be applied for $cur(t)$ at v, then $v \in RO(cur(t))$.*

If $\sigma = (x_1 \leftarrow t_1, \ldots, x_n \leftarrow t_n)$, we have $cur(\sigma(x_1)) = cur(t_1), \cdots, cur(\sigma(x_n)) = cur(t_n)$. We write $(cur \circ \sigma) = (x_1 \leftarrow cur(t_1), \ldots, x_n \leftarrow cur(t_n))$ for this. We also write $(cur \circ \sigma)$ for the extension $(cur \circ \sigma)^* : AT_\Sigma(Y) \to AT_\Sigma(X)$ defined by $(cur \circ \sigma)^*(Ap(t_1, t_2)) = Ap((cur \circ \sigma)^*(t_1), (cur \circ \sigma)^*(t_2))$.

We will need the following easy technical lemmata(the proofs are omitted):

Lemma 17. $(cur \circ \sigma)(cur(t)) = cur(\sigma(t))$ *for any term $t \in T_\Sigma(V)$ and a substitution σ.*

Lemma 18. $cur(t[v \leftarrow s]) = cur(t)[cp_t(v) \leftarrow cur(s)]$ *for any term $t, s \in T_\Sigma(V)$ and any occurrence $v \in O(t)$.*

We get the following theorem, our main result in this section:

Theorem 19. *Let $((S, \leq, \Sigma), V, R)$ an order-sorted TRS.*

1. *The function cur is injective,*
2. *Let $t \in T_\Sigma(V)$ be a term, $v \in O(t)$ an occurrence, $l \to r \in R$ a rewrite rule and σ a substitution. $t/v = \sigma(l)$, iff $cur(t)/cp_t(v) = (cur \circ \sigma)(cur(l))$,*
3. *$t \to_R s$ is a rewrite in R, iff $cur(t) \to_{Cu(R)} cur(s)$ is a rewrite in $Cu(R)$,*
4. *Let $cur(t) \to_{Cu(R)} s$ be a rewrite in $Cu(R)$. There exists a term $s' \in T_\Sigma(V)$ such that $cur(s') = s$ and that $t \to_R s'$ is a rewrite in R,*
5. *t is a normal form in R, iff $cur(t)$ is a normal form in $Cu(R)$.*

Proof. 1. Follows immediately from lemma 12. □

2. "$\Rightarrow$" If $t/v = \sigma(l)$, $cur(t/v) = cur(\sigma(l))$. And, $cur(t/v) = cur(t)/cp_t(v)$ can be proved for any $t \in PT_\Sigma(V)$ and any $v \in O(t)$ by the induction on the length of v. Hence, $cur(\sigma(l)) = cur(t)/cp_t(v)$. From lemma 17, $cur(\sigma(l)) = (cur \circ \sigma)(cur(l))$. Therefore, $cur(t)/cp_t(v) = (cur \circ \sigma)(cur(l))$.

"$\Leftarrow$" Suppose $cur(t)/cp_t(v) = (cur \circ \sigma)(cur(l))$. $cur(t/v) = (cur \circ \sigma)(cur(l))$ because $cur(t/v) = cur(t)/cp_t(v)$. From lemma 17, $(cur \circ \sigma)(cur(l)) = cur(\sigma(l))$. Consequently, $cur(t/v) = cur(\sigma(l))$. $t/v = \sigma(l)$ beacuse the function cur is injective from *1* of this theorem. □

3. "$\Rightarrow$" Suppose $t \xrightarrow{[v,l\rightarrow r]}_R s$ is a rewrite in R. There exists a rewrite rule $cur(l) \rightarrow_{Cu(R)} cur(r)$ from definition 5. $\sigma(l) = t/v$ and $\sigma(r) = s/v$ for some substitution σ since the rewrite rule $l \rightarrow r$ is applied for the term t at the occurrence v. $(cur \circ \sigma)(cur(l))$ is a redex of $cur(t)$ at $cp_t(v)$ from 2 of this theorem. Thus, $cur(t)/cp_t(v) = (cur \circ \sigma)(cur(l)) \rightarrow_{Cu(R)} (cur \circ \sigma)(cur(r)) = cur(\sigma(r))$. The pseudo-term $cur(t)[cp_t(v) \leftarrow (cur \circ \sigma)(cur(t))]$ is well-formed because $cur(s)$ is well-formed from lemma 14. From lemma 17 and lemma 18, we have $cur(s) = cur(t[v \leftarrow \sigma(r)]) = cur(t)[cp_t(v) \leftarrow (cur \circ \sigma)(cur(r))]$. Consequently, $cur(t) \rightarrow_{Cu(R)} cur(s)$ is a rewrite in $Cu(R)$.

"$\Leftarrow$" Suppose $cur(t) \xrightarrow{[v,cur(l)\rightarrow_{Cu(R)}cur(r)]}_{Cu(R)} cur(s)$ is a rewrite in $Cu(R)$. $v \in RO(cur(t))$ from lemma 16 and there exists some $v' \in O(t)$ such that $v = cp_t(v')$; there exists only one v' such that $v = cp_t(v')$ because cp_t is bijective. Thus, $(cur \circ \sigma)(cur(l)) = cur(t)/v = cur(t)/cp_t(v')$. There exists the rewrite rule $l \rightarrow_R r$ in R from definition 5 and 1 of this theorem. Hence, $\sigma(l) = t/v'$ from 2 of this theorem, and $\sigma(l) = t/v' \rightarrow_R \sigma(r)$. We consider the pseudo-term $t[v' \leftarrow \sigma(r)]$; $cur(s) = cur(t)[cp_t(v') \leftarrow (cur \circ \sigma)(cur(r))] = cur(t[v' \leftarrow \sigma(r)])$ from lemma 17 and lemma 18. $s = t[v' \leftarrow \sigma(r)]$ from 1 of this theorem, and $s = uncur(cur(s))$ is well-formed from lemma 12 and lemma 13. Therefore, s is well-formed, and $t \rightarrow_R s$ is a rewrite in R. $\quad\square$

4. Let $cur(t) \xrightarrow{[v,cur(l)\rightarrow_{Cu(R)}cur(r)]}_{Cu(R)} s \in AT_\Sigma(V)$ be a rewrite in $Cu(R)$. We have $v \in RO(cur(t))$ from lemma 16, and there exists only one $v' \in O(t)$ such that $v = cp_t(v')$ since cp_t is bijective. Thus, from lemma 17 and lemma 18, we have $s = cur(t)[cp_t(v') \leftarrow (cur \circ \sigma)(cur(r))] = cur(t[v' \leftarrow \sigma(r)])$. There exists a term $uncur(cur(t[v' \leftarrow \sigma(r)])) = t[v' \leftarrow \sigma(r)] \in T_\Sigma(V)$, which is well-formed, from lemma 12 and lemma 13. There exists the rule $l \rightarrow_R r$ in R from definition 5 and 1 of this theorem. Therefore, there exists the rewrite $t \xrightarrow{[v',l\rightarrow r]}_R t[v' \leftarrow \sigma(r)] = s'$ in R. $\quad\square$

5. A term is a normal form, iff (a) The term has no redexes, or (b) All resulting terms from the term by rewrite are ill-formed. We shall consider both cases. "$\Rightarrow$" (a) If a redex of the term $cur(t)$ in $Cu(R)$ exists, a redex of the term t in R exists, from 2 of this theorem: if there exist no redexes of t in R, there exist no redexes of $cur(t)$ in $Cu(R)$. Therefore, if t is a normal form in R then $cur(t)$ is a normal form in $Cu(R)$. (b) Let $v \in O(t)$ be any occurrence, $l \rightarrow_R r \in R$ any rewrite rule and σ a substitution such that $t/v = \sigma(l)$. Suppose the pseudo-term $t[v \leftarrow \sigma(r)]$ is ill-formed. $cur(t)/cp_t(v) = (cur \circ \sigma)(cur(l))$ from 2 of this theorem. From lemma 17 and 18, we have $cur(t[v \leftarrow \sigma(r)]) = cur(t)[cp_t(v) \leftarrow (cur \circ \sigma)(cur(r))]$. The pseudo-term $t[v \leftarrow \sigma(r)]$ is ill-formed and hence $cur(t[v \leftarrow \sigma(r)])$ is ill-formed from lemma 14. Therefore, $cur(t)[cp_t(v) \leftarrow (cur \circ \sigma)(cur(r))]$ is ill-formed.

"$\Leftarrow$" (a) The proof is similar to that of (a) of "$\Rightarrow$". (b) Let $v \in O(cur(t))$ be any occurrence, $cur(l) \rightarrow_{Cu(R)} cur(r) \in Cu(R)$ any rewrite rule and $(cur \circ \sigma)$ a substitution such that $cur(t)/v = (cur \circ \sigma)(cur(l))$. We have $v \in RO(cur(t))$ from lemma 16, and there exists only one $v' \in O(t)$ such that $v = cp_t(v')$ since cp_t is bijective. Suppose $cur(t)[cp_t(v') \leftarrow (cur \circ \sigma)(cur(r))]$ is ill-formed. We have $cur(t)/cp_t(v') = (cur \circ \sigma)(cur(l))$ from 2 of this theorem. From lemma 17 and lemma 18, we have $cur(t)[cp_t(v') \leftarrow (cur \circ \sigma)(cur(r))]$ $= cur(t[v' \leftarrow \sigma(r)])$, which is ill-formed. Consequently, the pseudo-term $t[v' \leftarrow \sigma(r)]$ is ill-formed from lemma 14. □

Now, we define the abstract rewriting system $Ker(R)$ called the kernel of $Cu(R)$.

Definition 20. Let $((S, \leq, \Sigma), V, R)$ be an order-sorted TRS. We define the abstract rewriting system $Ker(((S, \leq, \Sigma), V, R))$ as $Ker(((S, \leq, \Sigma), V, R)) = (\{cur(t) | t \in T_\Sigma(V)\}, \to_{Cu(R)})$. We write the set of terms $\{cur(t) | t \in T_\Sigma(V)\}$ for $KT_\Sigma(V)$. $Ker(((S, \leq, \Sigma), V, R))$ is abbreviated by $Ker(R)$.

From theorem 19, we can say that $Ker(R)$ is an isomorphic copy of R which is closed under reduction.

4.4 Basic Sorts and Derivation Sorts

We can prove the following lemma by the induction on $i \in \{0, \ldots, n-1\}$.

Lemma 21. Let $t_{i+1} \in AT_\Sigma(V)$ be an applicative term. If $t_{i+1} = Ap(t_i, p_{i+1})$ where $label_{t_{i+1}}(\varepsilon) = Ap : \left\langle \frac{f:s_{i+1} \to s_{i+2} \to \cdots \to s_n \to s}{f:s_1 \to s_2 \to \cdots \to s_n \to s} \right\rangle s_{i+1} \rightsquigarrow s' (0 \leq i \leq n-1)$, then the term t_{i+1} has the form of $Ap(\ldots Ap(f_0, p_1), \ldots, p_{i+1})$.

Theorem 22. Let $((S, \leq, \Sigma), V, R)$ be an order-sorted TRS. $t \in KT_\Sigma(V)$, iff $t \in \{u | u \in AT_\Sigma(V) \wedge u \text{ has a basic sort.}\}$.

Proof. "⇒" Let $t \in KT_\Sigma(V)$ be an applicative term. We have $KT_\Sigma(V) \subset AT_\Sigma(V)$ from definition 20. For any applicative term $cur(t') \in AT_\Sigma(V)$, $cur(t')$ has a basic sort from lemma 14. Therefore, t has a basic sort.
   "⇐" We will use the induction on the structure of the terms in $AT_\Sigma(V)$. Let t be any applicative term such that $t \in \{u | u \in AT_\Sigma(V) \wedge u \text{ has a basic sort.}\}$. If t is a constant $f :\rightsquigarrow s$, $f :\rightsquigarrow s = cur(f :\to s) \in KT_\Sigma(V)$ since s is a basic sort. If t is a variable $x : s$, $x : s = cur(x : s) \in KT_\Sigma(V)$ since s is a basic sort. Suppose $t = Ap(p, q_n)$ where $label_t(\varepsilon) = Ap : \left\langle \frac{f:s_n \to s}{f:s_1 \to s_2 \to \cdots \to s_n \to s} \right\rangle s_n \rightsquigarrow s$ and p, q_n are terms. We have $t = Ap(\ldots Ap(f_0, q_1), \ldots, q_n)$ from lemma 21, where each $q_i (1 \leq i \leq n)$ is a term which has a basic sort. Each q_i is in $KT_\Sigma(V)$ by the induction hypothesis, and hence there is $q_i' \in T_\Sigma(V)$ such that $q_i = cur(q_i')$. Thus, $Ap(\ldots Ap(f_0, cur(q_1')), \ldots, cur(q_n')) = cur(f(q_1', \ldots, q_n')) \in KT_\Sigma(V)$. □

Theorem 23. Let $((S, \leq, \Sigma), V, R)$ be an order-sorted TRS. $t \in \{u | u \in AT_\Sigma(V) \wedge u \text{ has a derivation sort.}\}$, iff $t \in \{u/1^i | u \in KT_\Sigma(V) \wedge i \in \{1, \ldots, last(O(u))\}\}$.

Proof. "⇒" Let $t \in AT_\Sigma(V)$ be an applicative term which has a derivation sort. Two cases for t from definition 1: (i) Suppose $t = f_0 :\rightsquigarrow \left\langle \frac{f:s_1 \to s_2 \to \cdots \to s_n \to s}{f:s_1 \to s_2 \to \cdots \to s_n \to s} \right\rangle$. We have $cur(f(t_1, \ldots, t_n))/1^n = f_0 :\rightsquigarrow \left\langle \frac{f:s_1 \to s_2 \to \cdots \to s_n \to s}{f:s_1 \to s_2 \to \cdots \to s_n \to s} \right\rangle$ where $label_{t'}(\varepsilon) = f : s_1 \cdots s_n \to s$ and $t_i \in T_\Sigma(V)_{s_i} (1 \leq i \leq n)$. Therefore, $t = f_0 \in \{u/1^i | u \in$

$KT_\Sigma(V) \wedge i \in \{1, \ldots, last(O(u))\}\}$. (ii) Suppose $label_t(\varepsilon) = Ap : s_1 s_2 \rightsquigarrow s_3$ where s_3 is a derivation sort. From lemma 21, $t = Ap(\ldots Ap(f_0, p_1), \ldots, p_m)$ for a function symbol $f_0 \rightsquigarrow \left\langle \frac{f:s_1 \rightarrow s_2 \rightarrow \cdots \rightarrow s_n \rightarrow s}{f:s_1 \rightarrow s_2 \rightarrow \cdots \rightarrow s_n \rightarrow s} \right\rangle$ and terms $p_i (i \in \{1, \ldots, m\})$ where $1 \leq m \leq n$. Each q_i has a basic sort. Thus, each q_i is in $KT_\Sigma(V)$ from theorem 22 so that there exists some $p_i' \in T_\Sigma(V)$ such that $p_i = cur(p_i')$ for each $i \in \{1, \ldots, m\}$. Hence, $t = Ap(\ldots Ap(f_0, cur(p_1')), \ldots, cur(p_m'))$ and $t = cur(f(p_1', \ldots, p_m', \ldots, p_n'))/1^{n-m} \in \{u/1^i | u \in KT_\Sigma(V) \wedge i \in \{1, \ldots, last(O(u))\}\}$ from definition 3, where each p_i' $(m + 1 \leq i \leq n)$ is a term in $T_\Sigma(V)$.

"$\Leftarrow$" Let $t \in KT_\Sigma(V)$ be an applicative term. There exists a term $t' \in T_\Sigma(V)$ such that $t = cur(t')$. Terms $t/1^i$ for any $i \in \{1, \ldots, O(last(t))\}$ are in $AT_\Sigma(V)$ from definition 2. Each $t/1^i$ has a derivation sort from definition 3. $\square$

Every term in $Cu(R)$ has the form of $cur(t)$ or the form of subterm of $cur(t)$.

5 Preservation of Properties

We call a property P is preserved by currying if $R \models P$ implies $Cu(R) \models P$.

5.1 Compatibility

Theorem 24. *Compatibility is preserved by currying of order-sorted TRSs.*

Proof. Suppose the order-sorted TRS $((S, \leq, \Sigma), V, R)$ is compatible. The set of terms in $Cu(R)$ is $KT_\Sigma(V) \cup \{u/1^i | u \in KT_\Sigma(V) \wedge 1^i \in O(u)\}$ from theorem 22 and theorem 23. Consider $t \in AT_\Sigma(V)$. Two cases: (i) If $t \in KT_\Sigma(V)$, let $cur(l) \rightarrow_{Cu(R)} cur(r)$ be a rewrite rule in $Cu(R)$. There exists some term $t' \in T_\Sigma(V)$ such that $cur(t') = t$ since t is in $KT_\Sigma(V)$. Suppose $cur(t')/v = (cur \circ \sigma)(cur(l))$ for an occurrence $v \in O(cur(t'))$ and a substitution $(cur \circ \sigma)$. We have $v \in RO(cur(t'))$ from lemma 16, and there exists only one $v' \in O(t')$ such that $v = cp_{t'}(v')$ since cp_t is bijective. From 2 of theorem 19, $\sigma(l) = t'/v'$ for a rewrite rule $l \rightarrow_R r \in R$ and a substitution σ. And the term $t'[v' \leftarrow \sigma(r)]$ is well-formed because the order-sorted TRS R is compatible. $cur(t'[v' \leftarrow \sigma(r)])$ is well-formed from lemma 14, and we have $cur(t'[v' \leftarrow \sigma(r)]) = cur(t')[cp_{t'}(v') \leftarrow (cur \circ \sigma)(cur(r))]$ from lemma 17 and lemma 18. Therefore, the term $cur(t')[cp_{t'}(v') \leftarrow (cur \circ \sigma)(cur(r))]$ is well-formed. (ii) There are more two cases for t if $t \in \{u/1^i | u \in KT_\Sigma(V) \wedge 1^i \in O(u)\}$: (ii-a) If $t = f_0 :\rightsquigarrow s$ where s is a derivation sort, $t/v \neq (cur \circ \sigma)(cur(l))$ for any rewrite rule $cur(l) \rightarrow_{Cu(R)} cur(r)$ in $Cu(R)$, any occurrence $v \in O(t)$, and any substitution $(cur \circ \sigma)$. Thus, $t = f_0 :\rightsquigarrow s$ is a normal form. (ii-b) Suppose $t = Ap(\ldots Ap(f_0, cur(t_1)), \ldots, cur(t_m))$ where $f_0 \rightsquigarrow \left\langle \frac{f:s_1 \rightarrow s_2 \rightarrow \cdots \rightarrow s_n \rightarrow s}{f:s_1 \rightarrow s_2 \rightarrow \cdots \rightarrow s_n \rightarrow s} \right\rangle$ and $1 \leq m < n$. There are no substitution $(cur \circ \sigma)$ such that $(cur \circ \sigma)(cur(l)) = t/1^i$ for any rewrite rule $cur(l) \rightarrow_{Cu(R)} cur(r)$ in $Cu(R)$ and any $i \in \{1, \ldots, m\}$, because the sort of $t/1^i$ is a derivation sort while the LHS of every rewrite rule has a basic sort. Thus, whenever a rewrite rule in $Cu(R)$ is applied for t, a subterm of t, $cur(t_i)$ for some $i \in \{1, \ldots, m\}$, is rewritten. There exists a well-formed term u_i such that $cur(t_i) \rightarrow_{Cu(R)} u_i$ from case (i). And there exists a term $t' \in KT_\Sigma(V)$ such that $t' = Ap(\ldots Ap(t, cur(t_{m+1})), \ldots, cur(t_n))$ where each t_i $(m+1 \leq i \leq n)$ is a term in $T_\Sigma(V)$. Therefore, the pseudo-term $t'[1^{n-i} \cdot 2 \leftarrow u_i]$ is well-formed from case (i). Consequently, the pseudo-term $t'[1^{n-i} \cdot 2 \leftarrow u_i]/1^{n-m}$ is well-formed and hence $t[1^{m-i} \cdot 2 \leftarrow u_i]$ is well-formed. □

5.2 Confluence

Generally, confluence is not preserved by currying.

Example 1. Let $((S, \leq, \Sigma), V, R)$ be an order-sorted TRS, where $S = \{s_1, s_2, s_3\}$, $\leq = \phi$, $\Sigma = \{a :\rightarrow s_1, b :\rightarrow s_1, c :\rightarrow s_1, d :\rightarrow s_2, e :\rightarrow s_2, g :\rightarrow s_3, f : s_1 s_1 \rightarrow s_3\}$ and $R = \{a \rightarrow_R b, a \rightarrow_R c, b \rightarrow_R d, c \rightarrow_R e, d \rightarrow_R g, e \rightarrow_R g, f(x,y) \rightarrow_R g\}$. The currying $((Cu(S), \leq, Cu(\Sigma)), V, Cu(R))$ is an order-sorted TRS where:

1. $Cu(S) = \{s_1, s_2, s_3, \left\langle \frac{f : s_1 \rightarrow s_1 \rightarrow s_3}{f : s_1 \rightarrow s_1 \rightarrow s_3} \right\rangle, \left\langle \frac{f : s_1 \rightarrow s_3}{f : s_1 \rightarrow s_1 \rightarrow s_3} \right\rangle\}$,
2. $Cu(\Sigma) = \{a :\rightsquigarrow s_1, b :\rightsquigarrow s_1, c :\rightsquigarrow s_1, d :\rightsquigarrow s_2, e :\rightsquigarrow s_2, g :\rightsquigarrow s_3,$
$f_0 :\rightsquigarrow \left\langle \frac{f : s_1 \rightarrow s_1 \rightarrow s_3}{f : s_1 \rightarrow s_1 \rightarrow s_3} \right\rangle$, $Ap : \left\langle \frac{f : s_1 \rightarrow s_1 \rightarrow s_3}{f : s_1 \rightarrow s_1 \rightarrow s_3} \right\rangle s_1 \rightsquigarrow \left\langle \frac{f : s_1 \rightarrow s_3}{f : s_1 \rightarrow s_1 \rightarrow s_3} \right\rangle$,
$Ap : \left\langle \frac{f : s_1 \rightarrow s_3}{f : s_1 \rightarrow s_1 \rightarrow s_3} \right\rangle s_1 \rightsquigarrow s_3\}$,
3. $Cu(R) = \{a \rightarrow_{Cu(R)} b, a \rightarrow_{Cu(R)} c, b \rightarrow_{Cu(R)} d, c \rightarrow_{Cu(R)} e,$
$d \rightarrow_{Cu(R)} g, e \rightarrow_{Cu(R)} g, Ap(Ap(f_0, x), y) \rightarrow_{Cu(R)} g\}$.

In example 1, $((S, \leq, \Sigma), V, R)$ is confluent. However, the currying is not confluent. $Ap(f_0, a) \rightarrow_{Cu(R)} Ap(f_0, b)$ and $Ap(f_0, a) \rightarrow_{Cu(R)} Ap(f_0, c)$ are rewrites in $Cu(R)$; the terms $Ap(f_0, b)$ and $Ap(f_0, c)$ are normal forms of $Ap(f_0, a)$.

From theorem 19, we get the following corollary.

Corollary 25. $((S, \leq, \Sigma), V, R)$ *is confluent, iff* $Ker(((S, \leq, \Sigma), V, R))$ *is confluent for any order-sorted TRS* $((S, \leq, \Sigma), V, R)$.

We are capable of proving the following theorem with corollary 25.

Theorem 26. *Let* $((S, \leq, \Sigma), V, R)$ *be a compatible order-sorted TRS. If* R *is confluent, then* $Cu(R)$ *is confluent.*

Proof. Let $((S, \leq, \Sigma), V, R)$ be a compatible order-sorted TRS. Suppose R is confluent. Consider any applicative term $t \in AT_\Sigma(V)$. Two cases: (i) If t is in $KT_\Sigma(V)$, there exists some term $t' \in T_\Sigma(V)$ such that $t = cur(t')$. Let s and u be applicative terms in $AT_\Sigma(V)$ such that $s \leftarrow_{Cu(R)} cur(t') \rightarrow_{Cu(R)} u$. From 4 of theorem 19, there exist terms s' and u' in $T_\Sigma(V)$ such that $s = cur(s')$ and $u = cur(u')$, respectively. Thus, there exists some $cur(w') \in KT_\Sigma(V)$ such that $cur(s') \rightarrow_{Cu(R)} cur(w') \leftarrow_{Cu(R)} cur(u')$ from corollary 25. (ii) If t is not in $KT_\Sigma(V)$, t is in $\{u/1^i | u \in KT_\Sigma(V) \wedge 1^i \in O(u)\}$. There are more two cases. (ii-a) If $t = f_0 :\rightsquigarrow \left\langle \frac{f : s_1 \rightarrow s_2 \rightarrow \cdots \rightarrow s_n \rightarrow s}{f : s_1 \rightarrow s_2 \rightarrow \cdots \rightarrow s_n \rightarrow s} \right\rangle$, t is a normal form in $Cu(R)$. (ii-b) Suppose $t = Ap(\ldots Ap(f_0, cur(t_1)), \ldots, cur(t_m))$ where $f_0 :\rightsquigarrow \left\langle \frac{f : s_1 \rightarrow s_2 \rightarrow \cdots \rightarrow s_n \rightarrow s}{f : s_1 \rightarrow s_2 \rightarrow \cdots \rightarrow s_n \rightarrow s} \right\rangle$,

202

$1 \leq m < n$ and $t_i \in T_\Sigma(V)_{s_i}$ for $i \in \{1, \ldots, m\}$. Any rewrite rule in $Cu(R)$ cannot be applied for terms in $KT_\Sigma(V)$ at any occurrence 1^i ($i \geq 1$) from lemma 7 and lemma 15. Thus, any rewrite rule in $Cu(R)$ are not applied for t at occurrences 1^i ($0 \leq i \leq m$). Therefore, whenever t is rewritten, some subterm $cur(t_i)$ for $i \in \{1, \ldots, m\}$ of t is rewritten. From theorem 24, the term $Ap(\ldots Ap(f_0, u_1), \ldots, u_m)$ is well-formed where each u_i ($1 \leq i \leq m$) is an applicative term such that $cur(t_1) \twoheadrightarrow_{Cu(R)} u_1, \ldots, cur(t_m) \twoheadrightarrow_{Cu(R)} u_m$, respectively. From corollary 25, we have $\forall p, q \in KT_\Sigma(V)[\, p \leftarrow_{Cu(R)} cur(t_i) \twoheadrightarrow_{Cu(R)} q \Rightarrow \exists r \in KT_\Sigma(V).p \twoheadrightarrow_{Cu(R)} r \leftarrow_{Cu(R)} q]$. And, we have $v_1 | v_2$ for occurrences v_1 and v_2 such that $t/v_1 = cur(t_i)$ and $t/v_2 = cur(t_j)$, where $i \neq j$ for $i, j \in \{1, \ldots, m\}$. Therefore, if $p \leftarrow_{Cu(R)} t \twoheadrightarrow_{Cu(R)} q$ for any $p, q \in AT_\Sigma(V)$ then there exists some applicative term $w \in AT_\Sigma(V)$ such that $p \twoheadrightarrow_{Cu(R)} w \leftarrow_{Cu(R)} q$. □

6 Conclusion

In this study, we introduced the currying of order-sorted TRSs and we have analyzed it. We have shown (theorem 24) that currying preserves compatibility for arbitrary order-sorted TRSs. And we have shown (theorem 26) that confluence is preserved by currying of compatible order-sorted TRSs.

The analysis of the currying $Cu(R)$ is essentially the analysis of $Ker(R)$, because R and $Ker(R)$ are isomprphic(theorem 19), and every term in $Cu(R)$ has the form of $cur(t)$ or the form of its subterm from theorem 22 and 23.

Acknowledgements We are grateful to Professor Tohru Naoi, Dept. of Electronic and Computer Engineering, Gifu University, for several enlightening discussions.

References

1. J.R.Kennaway, J.W.Klop, M.R.Sleep, and F.J. de Vries. Comparing curried and uncurried rewriting. Technical Report, ftp://ftp.sys.uea.ac.uk/pub/kennaway/, 1994.
2. S.Kahrs. Confluence of Curried Term-Rewriting Systems. Technical Report, Laboratory for Foundations of Computer Science, University of Edinburgh, January 1994.
3. Jan Willem Klop. Term rewriting systems. In S.Abramsky, D.M. Gabbai,and T.S.E. Maibaum, editors, *Handbook of Logic in Computer Science, Volume 2*, pages 1-116. Oxford University Press, 1992.
4. G.Smolka, W.Nutt, J.A.Goguen, and J.Meseguer. Order-Sorted Equational Computation. In H.Aït-Kaci and M.Nivat, editors, *Resolution of Equations in Algebraic Structures*, volume 2, pages 297-367. Academic Press, 1989.
5. M.Schmidt-Schauß. Computational Aspects of an Order-Sorted Logic with Term Declarations. J.Siekmann, editor, LNAI 395. Springer-Verlag, 1989.
6. Uwe Waldmann. Unification in Order-Sorted Signatures. Technical Report 298, Universität Dortmund(Germany), 1989.
7. Uwe Waldmann. Compatibility of Order-Sorted Rewrite Rules. In S.Kaplan and M.Okada, editors, *2nd International CTRS Workshop, Conditional and Typed Rewriting Systems. Proceedings. Montreal, Canada*, LNCS 516, pages 407-416. Springer-Verlag, 1990.

Stack and Queue Number of 2-Trees

S. Rengarajan   C.E. Veni Madhavan

Dept. of Computer Science & Automation
Indian Institute of Science
Bangalore 560 012, India
email : {rengs,cevm}@csa.iisc.ernet.in

Abstract. We consider the two problems of embedding graphs in a minimum number of pages and ordering the vertices of graphs in the form of queue layouts. We show that the class of 2-trees requires 2-pages for a book embedding and 3-queues for a queue layout. The first result is new and the latter result extends known results on subclasses of planar graphs.

1 Introduction

A *book embedding* of a graph consists of an embedding of its vertices along the spine of a book(i.e., a linear ordering of the nodes) and an embedding of its edges on the pages so that edges embedded on the same page do not intersect. Each page of the book is a half-plane that has the spine as its boundary. The minimum number of pages required to book embed a graph is called the *page number* or *book thickness* of the graph. The book embedding problem for a graph is the determination of the page number and an embedding using the minimum number of pages.

The book embedding problem is of interest because it models many problems in computer science and VLSI theory. For more details on book thickness see the work of Rosenberg [R 83]. The book embedding problem also has other origins, such as the problem of realizing fixed permutations of $\{1, 2, \cdots, n\}$ with noninteracting stacks. The notion of book thickness and a characterization of graphs with page ≤ 2 were introduced by Bernhart and Kainen [BK 79]. Malitz [M 94], Moran and Wolfsthal [MW 93], Obrenic [B 91] have studied the book embedding problem for general graphs with respect to edge size, neighborhood constraints and specific families of graphs such as deBruijn and shuffle-exchange graphs.

It is NP-complete to decide if a planar graph can be embedded in two pages [CLR 87]. Every n-vertex d-ary tree can be embedded in a book having one page of width $(\lceil d/2 \rceil)logn$. Every outer planar graph can be embedded in one page. Therefore trees and maximal outer planar graphs can be embedded in one page. Maximal outer planar graphs (*mops*) form a sub-class of 2-trees. In this paper, we show that 2-trees can be book embedded using two pages.

2 Definitions and Notations

A *k-queue layout* of an undirected graph $G = (V, E)$ has two aspects. The first aspect is a linear order of V (which we think of as being on a horizontal line). The second aspect is an assignment of each edge in E to one of k-queues in such a way that the set of edges assigned to each queue obeys a first-in/first-out discipline. Consider scanning the vertices, in order, from left to right. When the left endpoint of an edge is encountered, the edge enters the assigned queue (at the back of the queue). When the right endpoint of an edge is encountered, the edge exits its assigned queue (and must, therefore, be at the front of the queue).

More formally, a *k-queue layout* QL of an n-vertex undirected graph $G = (V, E)$ consists of a linear order of V denoted $\sigma = 1, \ldots, n$, and an assignment of each edge in E to exactly one of k queues, $q_1, \ldots, q_k$. Each q_j operates as follows. The vertices of V are scanned in the imposed order σ. When vertex i is encountered, edges assigned to q_j that have vertex i as their right endpoint must be at the front of the queue; they are removed(dequeued). Edges assigned to q_j that have vertex i as left end point are placed on the back of that queue(enqueued), in ascending order of their right endpoints. k is the queue number of the layout. The queue number of G, $QN(G)$ is the smallest k such that, G has a k-queue layout, and G is said to be a k queue graph.

A *k-stack layout* SL of an n-vertex undirected graph $G = (V, E)$ consists of a linear order of V denoted $\sigma = 1, \ldots, n$, and an assignment of each edge in E to exactly one of k stacks, $s_1, \ldots, s_k$. Each stack s_j operates as follows. The vertices of V are scanned in the imposed order σ. When vertex i is encountered, edges assigned to s_j that have vertex i as their right endpoint must be on the top of the stack; they are removed(popped). Edges assigned to s_j that have vertex i as left end point are placed on the top of that stack(pushed), in descending order of their right endpoints. k is the stack number of the layout. The stack number of G, $SN(G)$ is the smallest k such that, G has a k stack layout, and G is said to be a k-stack graph.

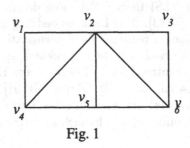

Fig. 1

The ordering corresponding to the 1-queue layout of the graph in Fig. 1 is $v_1 v_4 v_2 v_5 v_3 v_6$.

The ordering corresponding to the 1-stack layout of the same graph is $v_1 v_4 v_5 v_6 v_3 v_2$.

In this paper we consider finite, undirected loop less graphs without multiple edges. Let $G = (V, E)$ be such a *simple* graph. Denote the cardinality of

V by n. We use without definition the standard notions pertaining to chordal graphs namely, simplicial vertices, perfect elimination ordering, monotone adjacency set(i.e., the set of neighbors of a simplicial u that appear later than u in a peo), and minimal vertex separator.

3 Stack or Page number of 2-trees

Definition 1. A *k-tree* on k vertices consists of a clique on k vertices; given any k-tree T_n on n vertices, a k-tree on $n+1$ vertices is obtained by adjoining a new vertex by making it adjacent to some k-clique of T_n.

Thus trees are termed 1-tree and mops form a sub-class of 2-trees. Note that 2-trees are planar graphs; but are not outer planar.

We make use of the following theorem from [BK 79] in proving that 2-trees are 2-page embeddable.

Theorem 2. *A graph has a page number ≤ 2 if and only if it is a subgraph of a hamiltonian planar graph.*

We now show that

Theorem 3. *2-trees can be embedded in two pages.*

Proof: A graph has book thickness one if and only if the graph is outer planar [BK 79]. Since 2-trees are not outer planar they require at least two pages.

We show that every 2-tree, G, is two page embeddable by exhibiting a planar hamiltonian cycle in G by introducing some additional edges in G. While adding extra edges we maintain the planarity of G. The additional edges do not appear in the book embedding but help in understanding the construction of the embedding. The added edges can be placed along the spine without occupying any pages.

The minimal vertex separator of two nonadjacent vertices a and b in a 2-tree G is always an edge, say $e = (v_1, v_2)$. Let the set S of vertices adjacent to both the vertices v_1 and v_2 be $\{u_1, u_2, \ldots, u_k\}$ in some order. Now consider the edges $(u_i, v_1), (u_i, v_2)$ for some i $1 \leq i \leq k$. In G it is possible to have the simplicial vertices adjacent to (u_i, v_1) and/or (u_i, v_2). Based on these adjacency relationships three different cases are possible. For $u_i \in S, i = 1, \cdots, k$

1. The vertex $u_i \in S$ is adjacent only to v_1 and v_2 in the peo, i.e., the *monotone adjacency set* of $u_i \in S$ is precisely $\{v_1, v_2\}$.
2. The vertex u_i has a set $N_v(u_i)$ of adjacent vertices all of which are adjacent to v_1 or v_2, but not both. Let $N_v(u_i) = \{w_1, w_2, \ldots, w_l\}$.
3. The vertex u_i has two neighbor sets $N_{v_1}(u_i)$ adjacent to v_1 and $N_{v_2}(u_i)$ adjacent to v_2. The vertices in $N_{v_1}(u_i)$ are not adjacent to v_2 and the vertices in $N_{v_2}(u_i)$ are not adjacent to v_1. Let $N_{v_1}(u_i) = \{x_1, x_2, \ldots, x_{m_1}\}$ and $N_{v_2}(u_i) = \{y_1, y_2, \ldots, y_{m_2}\}$.

Each vertex adjacent to the vertices v_1 or v_2 of the edge (v_1, v_2), should belong to one of the above cases. The vertices $u_i \in S$ are partitioned into three groups U_1, U_2, U_3 depending on the adjacency relationship of u_i described above such that $|U_1| = k_1, |U_2| = k_2, |U_3| = k_3$ and $k_1 + k_2 + k_3 = k$. We assume that the vertices of each U_i appear consecutively. That is, the k_1 vertices of U_1 appear first and the k_2 vertices of U_2 appear next and so on. Also, we assume that the vertices of the neighbor set $N_v(u_i)$, $k_1 + 1 \leq i \leq k_2$, are all embedded in one face. The vertices of $N_{v_1}(u_i)$ and $N_{v_2}(u_i)$, $k_2 + 1 \leq i \leq k_3$, are embedded in *different* faces of the triangle v_1, v_2, u_i. This *ordering* is important to get a hamiltonian circuit. We will introduce additional edges between the vertices in U_1, U_2, U_3 and their neighbor sets and construct a planar hamiltonian cycle. The main idea is to traverse from v_1 to v_2 going through the vertices $u_i, N_v(u_i)(v = v_1 \text{ or } v_2)$, $N_{v_1}(u_i), N_{v_2}(u_i)$ by a procedure explained below.

- Add edges (u_i, u_{i+1}) for $1 \leq i < k_1$. This addition of edges generates a path $P_1 = \langle u_1, u_2, \ldots, u_{k_1} \rangle$.
- Add edges $(w_i, w_{i+1}), 1 \leq i < l$. This addition of edges creates a path $\langle w_1, w_2, \ldots, w_l \rangle$. For each $u_i, k_1 + 1 \leq i \leq k_2$, the vertices w_i in $N_v(u_i)$ are now connected. Also add edges between u_{k_1+i} and w_1 which is adjacent to u_{k_1+i+1}, $1 \leq i \leq k_2 - k_1$. The complete path from u_{k_1+1} to u_{k_2} is given by $P_2 = \langle u_{k_1+1}, w_1, w_2, \ldots, w_l, u_{k_1+2}, w_1, w_2, \ldots, w_l, \ldots, u_{k_2-1}, $ $w_1, w_2, \ldots, w_l, u_{k_2} \rangle$. The planarity of the given graph is preserved in while constructing such a path.
  Note: The number of vertices w_i incident with different edges may vary. We use a variable l, without a subscript for simplicity, to denote this number. The vertices w_i that appear between u_{k_1+1} and u_{k_1+2} and the similar u_i are adjacent respectively to corresponding u_i and v_1 or v_2. This is to avoid cumbersome notation.
- Add edges $(x_i, x_{i+1}), (y_j, y_{j+1})$ $1 \leq i < m_1, 1 \leq j < m_2$. This creates a path $\langle x_1, x_2, \ldots, x_{m_1}, y_1, y_2, \ldots, y_{m_2} \rangle$ without violating planarity as described earlier. Further edges are added between y_{m_1} and x_1. Here y_{m_1} is adjacent to u_{k_2+i} and x_1 is adjacent to u_{k_2+i+1}. This gives a path $P_3 = \langle u_{k_2-1}, x_1, x_2, \ldots, x_{m_1}, u_{k_2}, y_1, y_2, \ldots, y_{m_2} x_1, x_2, \ldots, x_{m_1} u_{k_2+1}, y_1, y_2, \ldots, $ $y_{m_2}, \ldots, x_1, \ldots u_{k_2+i}, y_1, y_2, \ldots, y_{m_2}, x_1, \ldots, x_{m-1}, u_{k_3}, y_1, y_2, y_{m_2} \rangle$.

Thus, in all the three cases the new edges do not cross the existing edges of E. Certainly we have a hamiltonian path from v_1 to v_2 by concatenating P_1, P_2, and P_3. Including the edge (v_1, v_2) we get a hamiltonian cycle starting and ending at v_1.

Actually while traversing from v_1 to v_2 passing through the other vertices we may encounter an edge (a, b) which is similar to (v_1, v_2), i.e., a new set S' of vertices incident with (a, b) is encountered. This set S' may be one of the above mentioned three types, U_1, U_2, U_3. Now for the edge (a, b), we can recursively follow the procedure outlined above and get a path from a to b visiting all the vertices in S'.

Hence it follows from Theorem 2 that, 2-trees have two page book embeddings.

Remark: The ordering of the vertices in the hamiltonian cycle gives an ordering of the vertices on the spine of the book.

4 Queue Number of mops and 2-trees

1-queue graphs are characterized by Heath and Rosenberg [HR 92]. In the following, we give a brief account of their characterization of 1-queue graphs.

A graph $G = (V, E)$ is a *leveled planar* graph if V can be partitioned into sets $V_1, V_2, \ldots, V_m$ such that

1. all the vertices of V_i belong to the level i; (represented on a line parallel to the abscissa)
2. each edge in E is embedded as a straight line segment between levels i and $i + 1$.

A leveled planar embedding induces an order on V as follows: As i takes the values $1, 2, \ldots, m$, scan the level from or left to right. Label the vertices $1, 2, \ldots, n$ as they are encountered. For $1 \leq i \leq m$ let b_i be the first vertex in level i and let t_i be the last vertex. let s_i be the first vertex in level i that is adjacent to some vertex in level $i+1$, or if there are no edges between levels i and $i+1$, let $s_i = t_i$.

The characterization captures graphs that are leveled planar and as well as leveled planar graphs augmented by certain additional edges. These edges are of the following nature. A *level-i-arch* for G is an edge connecting vertex t_i with vertex j where $b_i \leq j \leq min(t_i - 1, s_j)$. Because of the leveling, arches do not cross. A leveled planar graph augmented by (zero or more) arches is called an *arched-leveled* planar graph. The edges that are not arches are called *cross* edges. Note that these edges appear between successive levels.

Theorem 4. *[HR 92] A graph G is 1-queue graph if and only if G is an arched leveled planar graph.*

Definition 5. [HR 92] A k-rainbow is a set of edges $\{e_i = (a_i, b_i); 1 \leq i \leq k\}$ such that $a_1 < a_2 < \ldots < a_{k-1} < a_k < b_k < b_{k-1} < \ldots < b_2 < b_1$ in an ordering σ.

Note that a rainbow is a nested set of matching edges in G.

Theorem 6. *[HR 92] If σ has no rainbow of more than k edges, then there is a k-queue layout for σ. Such a σ can be found in time $O(|E|loglogn)$.*

4.1 2-queue layout of mops

In this section we show that a mop can be represented in a manner similar to, but not as, a leveled planar graph. Thus we show that mops require *at least* 2-queues. Also, we give an assignment of edges such that 2-queues are sufficient to layout mops.

Theorem 7. *2-queues are necessary and sufficient for a maximal outer planar graph.*

Proof: First we represent mops in a structure similar to leveled planar graphs. Consider a vertex of degree 2, say x, in a mop. Place it on level 1. Then keep the neighbors of x, say y and z, on level 2. The edges appearing between vertices in 1 and 2, i.e., (x, y) and (x, z) are called *cross* edges and the edges that appear between the vertices of the same level, such as (y, z) are termed as *level* edges. As a next step the neighbors of y and z form the level 3. Again the edges that appear between the vertices of 2 and 3 can be partitioned into cross edges and level edges. The rest of the vertices can be arranged into different levels by the above method. This method is described by the following two rules applied in succession.

1. if a vertex v adjacent to vertices both of which are at the same level i, then v is placed at the level $i + 1$.
2. if a vertex v is adjacent to a vertex on level i and to a vertex on level $i + 1$, then v is placed at level $i + 1$.

For any given edge (v_1, v_2) in a mop there can be at most two vertices a, b adjacent to both v_1 and v_2. Rule 1 is applied first to generate the next level and rule 2 is applied to arrange the vertices in the generated level. Therefore, for any level edge at level i, the two vertices a, b will appear at level $i - 1$ and $i + 1$. Similarly, for the cross edges the two vertices a, b adjacent to v_1, v_2 will appear at the left and right sides of the cross edge. Thus a mop can be represented in a manner similar to a leveled planar graph without ambiguity in levels.

Clearly, the above method stratifies the vertices of a mop into different levels. Let this stratified graph be denoted by G'. Let the total number of levels G' be m and let the vertices that appear at level i be labeled as $v_{i1}, v_{i2}, \ldots, v_{ik_i}$ where i denotes the level and k_i denotes the number of vertices in level i.

of any degree 2 vertex, the arguments above hold for any ordering σ.

For the necessity part, we present a graph which can not be laid out using 1-queue.

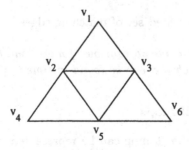

Fig. 2

We show that the graph in Fig.2 does not admit a 1-queue layout for any of the 720 possible permutations. We make use of the symmetry of the graph extensively. Let σ be a permutation of the six vertices. Let v_1 be the first vertex in σ. Then

v_2, v_3 should occur in any order immediately after v_1 in σ. Let v_2, v_3 be the order of the vertices. Then the order of the edges is: $(v_1, v_2), (v_1, v_3), (v_2, v_4), (v_2, v_5),$ $(v_3, v_5), (v_3, v_6)$. This order on the edges imposes an order on the vertices of V. That is, v_4 should be the successor of v_3 and v_5 can not be the successor of the v_3. If so, edge (v_2, v_5) would be placed before the edge (v_2, v_4) and (v_3, v_5) would be placed after (v_2, v_4). That means when v_5 is encountered while scanning the vertices in σ, (v_2, v_5) and (v_3, v_5) can not be removed from the queue as (v_2, v_4) would block (v_3, v_5). Thus, v_5, v_4, v_6 and v_5, v_6, v_4 ordering should be avoided in σ. For the same reason the ordering with v_6, v_4, v_5 and v_6, v_5, v_4 should also be avoided. This leaves with the two other orderings v_4, v_5, v_6 or v_4, v_6, v_5 to be considered. v_4, v_6, v_5 is ruled out because the edge (v_3, v_5) would appear after (v_3, v_6) and (v_6, v_5) would come next to (v_3, v_6), i.e., the edge ordering is $(v_1, v_2), (v_1, v_3), (v_2, v_4), (v_2, v_5), (v_3, v_6), (v_3, v_5)$ which is not admissible. Finally, $\sigma = (v_1, v_2, v_3, v_4, v_5, v_6)$ gives an edge ordering $(v_1, v_2), (v_1, v_3), (v_2, v_3), (v_2, v_4),$ $(v_2, v_5), (v_3, v_5), (v_3, v_6), (v_4, v_5), (v_5, v_6)$ which is also not possible because the last three edges can not be sorted using 1-queue while scanning σ. If the order of v_2 and v_3 is changed in σ then a similar situation will arise due to symmetric nature of the graph. That is, as in the above argument the only possible permutation to be considered is $\sigma = (v_1, v_3, v_2, v_6, v_5, v_4)$ which can not be edge sorted by 1-queue. Thus, by symmetry, any σ starting with v_1 or v_4 or v_6 can not produce a 1-queue layout. Next we consider permutations starting with v_2 or v_3 or v_5. Again by symmetry, if we prove that 1-queue layout is not possible for one of them it follows for other cases also. Let v_2 be the first vertex in σ. The vertex v_2 has four neighbors v_1, v_3, v_4, v_5. Next to v_2 either v_1 or v_4 should appear. This is because if v_3 or v_5 appear then the edge with the end vertex v_6 will appear before edges with end vertices v_4 or v_1, which is not admissible for a 1-queue layout. Let v_1 be the second vertex in σ and therefore v_3 should be the third vertex in σ. But this will impose the following order on the edges: $(v_2, v_1), (v_2, v_3)$ and edges starting with v_2 and ending with v_4 or v_5 in any order. When v_1 is scanned in σ the edge (v_1, v_3) will appear after (v_2, v_4) and (v_2, v_5). But (v_2, v_3) appears before (v_2, v_4) and (v_2, v_5). Thus, when v_3 is scanned these two edges can not be removed from the queue as they are not together. That means the second position can not be occupied by the vertices v_1, v_3, v_5 and v_6. By symmetry, and using the above arguments it is also clear that v_4 can not also appear in the second position in σ. That is, any σ starting with v_2 can not be a valid permutation. By symmetry, it is easy to see that the σ having the other two vertices v_3 and v_5 can not yield a valid permutation to sort the edges using 1-queue. Thus, 1-queue layout is not possible for the graph G shown in Fig.2. Since G is a mop, it is clear that mops require 2-queues.

We now present the sufficiency part of the theorem. The ordering of the vertices σ can be obtained by scanning the vertices in each level, starting from 1 to m. That is, $\sigma = \langle v_{11}, \ldots, v_{1k_1}, v_{21}, \ldots, v_{2k_2}, \ldots, v_{m1}, v_{mk_m} \rangle$. The assignment of edges to the two different queues is straightforward. Assign the *cross* edges to the first queue, q_1. Assign the *level* edges to the second queue, q_2. It is easy to verify that this assignment of edges with the ordering σ gives a 2-queue layout for a mop.

4.2   3-queue layout of a 2-tree

While generalizing the queue layout for 2-trees from mops, we relax the condition that for any edge (v_1, v_2) there can be at most two vertices a, b such that both a and b are adjacent to v_1 and v_2. This condition is required for a mop to maintain outer planarity. Since 2-trees are not outer planar, an edge (v_1, v_2) in a 2-tree can have any finite number of vertices adjacent to both v_1 and v_2. That is, the set $\{v_1, v_2\}$ can be the monotone adjacency set for many vertices and not just for two vertices as in the case of mops. This presents the difficulty of arranging vertices in different levels. We obtain a 3-queue layout procedure for 2-trees by imposing a certain ordering on the vertices of 2-trees.

We find a set C_1 of vertices called, *2-chain*, formally defined below, which forms the first level of vertices. The set C_1 is used as a basis for constructing other levels of vertices. For every edge e formed by the vertices of C_1, we find the set of vertices C_2 incident on e. Then we impose an ordering on these vertices of C_2 to get a layout. We proceed to find the neighbors of vertices of C_2 and so on.

The following analogy may be helpful in viewing the process of laying out the vertices of 2-trees using 3-queues.

We can view the vertices of a (rooted) tree T as if they are placed at different levels. Since a tree is a subgraph of a leveled planar graph the ordering of the vertices of T can be obtained by a breadth-first- search of the vertices starting from the root. This ordering yields a 1-queue layout for a tree.

Similarly, the vertices of a 2-tree can also be viewed as being at different levels. The vertices of a 2-chain can be placed at level 1, and their neighbors can be kept at level 2. This notion can be extended to all levels, because the vertices at level i are adjacent only to the vertices of $i - 1$ and $i + 1$.

Definition 8. Let x, y be two vertices of degree two and P_1, P_2 be the two vertex disjoint paths joining x and y. Then the subgraph induced by the vertices of P_1, P_2 and x, y is called a $2 - chain$.

Since every 2-chain C is an arched leveled planar graph, C has a 1-queue layout. Let the ordering of the vertices of a 2-chain C be $\langle v_1, v_2, \ldots, v_k \rangle = \langle V_C \rangle$. Edges are ordered in the following way. Let $(v_i, v_j), (v_k, v_l) \in E_C$. Then (v_i, v_j) precedes (v_k, v_l) if $v_i < v_k$ or $(v_i = v_k$ and $v_j < v_l)$. With this ordering σ_E on the edges, the edges are totally ordered based on the ordering σ of the vertices. All the remaining vertices of the graph G are connected to this 2-chain.

Each edge in E_C is scanned from left to right in the ordering of σ_E. For each edge, given its neighbors, we get a 2-queue layout. Combining this layout with the layout of C we get a three queue layout for the given graph. The procedure for getting a 2-queue layout for the neighbors for an edge of a 2-chain is described below.

Consider an edge $(x, y) \in E_C$ and its possible neighbors. We first label these neighbors. Let $w_1, \ldots, w_k$ be the vertices adjacent to both x and y. Let

$v_{11}^1, \ldots, v_{1k_1}^1$ be the vertices adjacent to x and w_1. Let $v_{21}^1, \ldots, v_{2k_2}^1$ be the vertices adjacent to x and v_{11}. In this way we can add many vertices $v_{i1}^1, \ldots, v_{ik_i}^1$ adjacent to x and to the vertex $v_{(i-1)1}$. Likewise, let $u_{11}^1, \ldots, u_{1k_1}^1$ be the vertices adjacent to y and w_1, and let $u_{21}^1, \ldots, u_{2k_2}^1$ be the vertices adjacent to y and u_{11}^1 and so on. In this way we can add many vertices $u_{i1}^1, \ldots, u_{ik_i}^1$ adjacent to y and to the vertex $u_{(i-1)1}$. The above vertices are thus added to the vertices x or y and to w_1 or v_{11} or u_{11} and so on. We can also add vertices adjacent to x or y and to other w_i, $2 \leq i \leq k$.

Note: All the vertices w_i's or v_{ii}'s or u_{ii}'s are adjacent to either x or y. We do not consider the edges between w_i's and v_{ii}'s or u_{ii}'s.

We give an ordering σ_x of the vertices such that it contains only one 2-rainbow, i.e., at most two nested matchings. Hence we can get a 2-queue layout.

Place the vertices x, y in the beginning of σ_1. The ordering of the vertices x and y is $x < y$ i.e., x precedes y. Next place all the vertices adjacent to both x and y, i.e., $w_1, \ldots, w_k$. Since w_i's form an independent set, the ordering among them is not important in the sense that placement of neighbors of w_i should be after the placement of neighbors of w_{i-1}. Let the ordering of the neighbors of x, y and w_1 be

$$\sigma_x = \langle v_{11}^1, \cdots, v_{1k_1}^1, v_{21}^1, \cdots, v_{2k_2}^1, \cdots, v_{m1}^1, \cdots, v_{mk_m}^1,$$
$$u_{11}^1, \cdots, u_{1k_1}^1, u_{21}^1, \cdots, u_{2k_2}^1, u_{m1}^1, \cdots, u_{mk_m}^1 \rangle.$$

All the edges incident with the vertex x can be partitioned into two groups (x, w_i), (x, v_{i1}^1), $1 \leq i \leq m$. All the edges (x, w_i) properly nest some of the edges which belong to the 2-chain C whose end vertices appear after x in σ and the edges (y, w_j), $j < i$. The other edges (x, v_{i1}) have one vertex x in common with the edge (x, w_i). So they are not properly nested. So they induce a 1-rainbow in the layout. Also, in the ordering σ_x, the vertices v_{i1} appear after w_i's. Therefore, (x, w_i) edges properly nest one more edge. Next we consider the edges (x, v_{i1}). Similar to the edges (x, w_i) these edges also properly nest some of the edges of the 2-chain C, and the edges (y, w_j) $j < i$, and the edges (w_i, v_{j1}), $j < i$. These two sets of edges, (y, w_j) and (w_i, v_{j1}), overlap one another and there are no proper nesting of edges amongst themselves. Thus, with this ordering σ_x the layout has a 2-rainbow.

Similarly, the edges incident on the vertex y can be partitioned into two groups (y, w_i), (y, u_{i1}^1), $1 \leq i \leq m$. The edges (y, w_i) need not be considered as they are nested above by the edges with x with an end vertex. The other set of edges (y, u_{i1}^1) follow a similar pattern of the edges (x, v_{i1}). The above layout is for an edge and its neighbors. The vertices of the other edges of the 2-chain C and their neighbors can be placed in a manner similar to a 2-rainbow layout. Let (y, z) be an edge in the 2-chain, C. The neighbors of z will appear after the neighbors of y. This may cause another nesting of the edge (y, z) by some of the edges (x, w_i). This *may* lead to a 3-rainbow. The above process gives a layout for the 2-chain, C, and its one neighbors a 3-queue layout.

We note that, there are no 3-rainbows in the layout after the vertices of the 2-chain,C, i.e., every edge properly nests at most one edge. There are no more edges from the vertices of C to vertices at a distance two from the vertices of C. Now all the one distance neighbors either form a 2-chain or are single vertices. Thus the same procedure can be carried out from the neighbors of C to the vertices which are at distance two. Thus we get a 2-queue layout. This process can be inductively carried out to get 3-queue layout for the graph. Hence we have,

Theorem 9. *2-trees can be embedded using 3-queues.*

5 Conclusions

We have shown that 2-trees need two pages to embed because they are subgraphs of planar hamiltonian graphs. In the case of planar-3-trees since not all of them are hamiltonian, the nonhamiltonian planar-3-trees require at least 3 pages to embed. What is not known about the stack number of planar-3-trees is whether or not they can be embedded using three pages. As far as the queue number is concerned, the lower bound on the queue number of 2-trees is open. If the queue number is two one should get a different way of embedding 2-trees to get a 2-queue layout. But we believe that 2-trees require three queues. The queue number of planar-3-trees is not determined yet. We believe it should be four. Inferring from the previous results for mops, 2-trees and planar-3-trees we conjecture that The queue number QN of planar graphs is $5 \le QN < \infty$. This above conjecture would strengthen the conjecture of Heath and Rosenberg by improving the lower bound. This should be compared with the stack number (page number) of planar graphs, which is proved to be four.

References

[R 83]   A.L.Rosenberg,"The DIOGENES approach to testable fault-tolerant arrays", *IEEE Transaction on Computer*, **C-32**, 1983, pp.902-910.

[CLR 87] F.Chung, T.Leighton, A.L.Rosenberg,"Embedding graphs in books: A Layout problem with applications to VLSI Design", *SIAM Jl. on Algebraic and Discrete Methods*, **8**, 1987, pp.33-58.

[BK 79] F.Bernhart, P.C.Kainen, "The book thickness of a graph", *J. Comb. Theory, Ser.B*, **27**, 1979, pp.320-331.

[HR 92] L.S. Heath, A. Rosenberg, "Laying out graphs using queues", *SIAM Jl. on Computing*, **21**, 1992, pp.927-958.

[M 94]   S.M.Malitz, "Graphs with E edges have pagenumber $O(\sqrt{E})$", *J. Algorithms*, **17**,1994, pp. 71-84, 85-109.

[MW 93] S. Moran and Y. Wolfsthal, "Two-page book embedding of trees under vertex-neighborhood constraints", *Discrete Appl. Math.*, **43**, 1993, pp. 233-41, .

[B 91]   B.Obrenic, "Embedding deBruijn and shuffle-exchange graphs in five pages", *Proceedings of the 3rd Annual ACM Symposium on Parallel Algorithms and Architectures*, pp.137-146, 1991.

Shortest Paths in Random Weighted Graphs

Scott K. Walley and Harry H. Tan

University of California, Irvine, Irvine CA 92717, USA

Abstract. We consider the probability distribution of the cost of short-est paths and the diameter in a complete, weighted digraph with non-negative random edge costs. Asymptotic results as the number of nodes goes to infinity are developed and applied to extend several probabilis-tic shortest path algorithms to edge cost distributions having a general Taylor's series at zero edge cost.

1 Introduction

There has been significant interest in shortest path algorithms and methods to reduce the computation time of these algorithms (see, for example, Gallo and Pallottino [5] for a good discussion of the classical algorithms). One approach which has been applied to reduce computation time is to make use of the proper-ties of the edge costs. With the assumption of integer edge costs, for example, the algorithms of Goldberg [6] and Ahuja *et al.* [1] can provide reduced computation time bounds compared to the general case. In some problems it is reasonable to model edge costs as following probability distributions. In this case, the algo-rithms of Spira [8], Bloniarz [2], Frieze and Grimmett [4], and Hassin and Zemel [7] have been proposed to reduce the expected computation time. In particular, the algorithms of Frieze and Grimmett [4] and Hassin and Zemel [7] assume that edge costs are independently distributed with Cumulative Distribution Function (CDF) F, where $F(0) = 0$ and $F'(0) > 0$, and obtain algorithms for determining shortest paths between all node pairs with expected time bound $O(n^2 \ln n)$ for n nodes. However, in some cases it may be appropriate to have an edge cost model with $F'(0) = 0$. The focus of this paper is the determination of asymptotic re-sults as the number of nodes goes to infinity for the cost of shortest paths and the diameter for the following random graph model which we apply to extend these algorithms to edge cost distributions having a general Taylor's series.

The random graph model considered here consists of a complete, labeled, and weighted digraph G on n nodes with fixed node set $V = \{v_1, v_2, \ldots, v_n\}$ and fixed edge set $E \subseteq V \times V$. Edges are then assigned random weights or costs which are determined in the following general manner. For $x, y \in V$, $x \neq y$, and for each unordered node pair (x, y), let W_1 and W_2 be non-negative random variables such that the variables (W_1, W_2) for each node pair are independent and identically distributed over the node pairs and have $\text{Prob}(W_1 = 0) = 0$ and $\text{Prob}(W_2 = 0) = 0$. We use the notation $(y, x) \in E$ to denote the directed edge from y to x. Then edges (x, y) and (y, x) in E are assigned random costs, where with probability .5, the cost of edge (x, y) is W_1 and the cost of edge (y, x) is W_2, and with probability .5, the cost of edge (x, y) is W_2 and the cost of edge (y, x) is W_1. This formulation randomizes the edge costs in different directions which allows us to make use of the probabilistic symmetry which results for the graph. If $W_1 = W_2$, then the resulting random graph model reduces to the important special case of an undirected graph with random edge costs.

For nodes $x, y \in V$, we define for the weighted graph G the directed distance $\text{dist}(x, y)$ from x to y as the cost of the shortest directed path from x to y, and say that y is of distance d from x if $\text{dist}(x, y) = d$. Moreover, we define $\text{dist}(x, x) = 0$. We also define the diameter of G as: $\text{diam}(G) = \max\{\text{dist}(x, y) : x, y \in V\}$, and equals the largest distance between any two nodes in G. We also use $\lfloor x \rfloor$ to denote the greatest integer of x and $\lceil x \rceil$ to denote the least integer which is $\geq x$. Also, for a given random variable Z, we use $\text{E}(Z)$ or $\overline{Z}$ to denote the expectation of Z. Finally, we use the notation $f^{(k)}(x)$ to denote the k-th derivative of $f(x)$ at x.

2 Shortest Path Cost and Diameter

Let $F(t)$ denote the CDF for an arbitrarily selected edge which may be a function of n, and let $W_1(t)$ and $W_2(t)$ be the CDF for W_1 and W_2, respectively. Then, $F(t) = .5W_1(t) + .5W_2(t)$. Upper and lower bounds on the CDF of the shortest path cost from a source node to an arbitrary destination node are determined in Theorem 1 below for a particular family of edge cost distributions $F(t)$. These bounds are then used in Theorem 5 to establish the main limit result for the shortest path cost and diameter distributions as $n \to \infty$.

Theorem 1. *Suppose for fixed integer $n > 1$, $c \in (0, \infty)$, $\mu \in (0, \infty)$, and integer m, $0 \leq m < \infty$, which all may depend on n, are given. Let*

$$\lambda = \frac{\ln(n - 1) + \ln \mu}{[(n - 1)c]^{\frac{1}{m+1}}}. \tag{1}$$

Furthermore, suppose $\lambda \geq 0$ and for $0 \leq t \leq \lambda$, that $F(t)$ is given by

$$F(t) = 1 - \exp\left\{-c\frac{t^{m+1}}{(m+1)!}\right\}. \tag{2}$$

Then as $n \to \infty$

$$\max\{0, L_1(\mu)\} \leq Prob(dist(x, y) \leq \lambda \text{ in } G) \leq \min\{1, U_1(\mu)\} \tag{3}$$

for $x, y \in V$, $x \neq y$, where

$$U_1(\mu) = \frac{\mu}{m + 1} + \frac{1}{n - 1}\left[\frac{m((n - 1)\mu)^{\cos(\frac{2\pi}{m+1})}}{m + 1} - 1\right] \tag{4}$$

$$L_1(\mu) = \begin{cases} 1 - \exp\left\{o(1) - (1 - o(1))\frac{(\ln \mu)^{m+1}}{5^{m+1}(m+1)!}\min\left\{\frac{1}{5}, \frac{\ln\ln n + \ln \mu}{5\ln n}\right\}\right\} & \mu \geq 1 \\ 0 & 0 < \mu < 1. \end{cases} \tag{5}$$

Also, for the special case of $m = 0$, let arbitrary t^* be given such that $0 \leq 2t^* < \lambda$. Then we have an improved bound $L_1(\mu)$ given by

$$L_1(\mu) = 1 - e^{ct^*n} \left[(n-1)\mu\right]^{-n/(n-1)} -$$
$$\frac{(\ln(n-1) + \ln\mu - (n-1)ct^*)(n-1)}{(\ln(n-1) + \ln\mu - 2(n-1)ct^*)^2} e^{-nct^*+1-2ct^*} \left[(n-1)\mu\right]^{1/(n-1)}. \quad (6)$$

Proof. We provide here an outline of the general proof of this theorem. Since the event $\text{dist}(x, y) \leq \lambda$ depends only on edge costs $\leq \lambda$, the values for $F(t)$, $t > \lambda$, have no effect on $\text{Prob}(\text{dist}(x, y) \leq \lambda$ in $G)$ in (3). So we assume in the proof without loss of generality that $F(t)$ is still given by (2) for $t > \lambda$. That is, we have

$$F(t) = 1 - \exp\left\{-c\frac{t^{m+1}}{(m+1)!}\right\} \quad (7)$$

for $t \in [0, \infty)$. Consider a modified weighted graph G' which differs from G in that if w is the edge cost from node i to node j in G, then the edge cost w' from node i to node j in G' has its value determined from w according to: $w' = \lfloor Nw \rfloor + 1$, where N is an arbitrary positive integer. The edges in G' are thus integer valued and provide a discrete approximation to the edges in G where $w'/N \to w$ as $N \to \infty$. For $x, y \in V$, $x \neq y$, it can be seen that if the number of edges in a shortest path from x to y in G' is h and $\text{dist}(x, y) = k$ in G', then in G we have $(k-h)/N \leq \text{dist}(x, y) < k/N$. Therefore, it follows that $\text{Prob}(\text{dist}(x, y) \leq k$ in $G') \leq \text{Prob}(\text{dist}(x, y) \leq k/N$ in $G) \leq \text{Prob}(\text{dist}(x, y) \leq k+h$ in $G')$. Since the number of edges in a shortest path is less than n, we have

$$\text{Prob}(\text{dist}(x, y) \leq k \text{ in } G') \leq \text{Prob}(\text{dist}(x, y) \leq k/N \text{ in } G) \leq$$
$$\text{Prob}(\text{dist}(x, y) \leq k+n \text{ in } G'). \quad (8)$$

In order to establish the bounds in (3) we consider the distribution of $\text{dist}(x, y)$ in G' and apply (8) in the limit as $N \to \infty$.

Let a_i be the probability distribution of an arbitrarily selected edge in G'. Then

$$a_i = \begin{cases} F(i/N) - F((i-1)/N) & i \geq 1 \\ 0 & i = 0. \end{cases} \quad (9)$$

Also define b_i to be the conditional probability that an arbitrarily selected edge has cost i given that it has cost $\geq i$:

$$b_i = \begin{cases} \dfrac{a_i}{1 - \sum_{j=0}^{i-1} a_j} & \text{if } i \geq 1, \sum_{j=0}^{i-1} a_j < 1 \\ 0 & \text{else.} \end{cases} \quad (10)$$

In particular, we have the property that $b_0 = 0$ which is used frequently throughout this paper. For a given source node $x \in V$ and non-negative integer k, define

$$\Gamma_k(x) = \{y \in V : \text{dist}(x, y) = k \text{ in } G'\}$$

as the set of nodes from V which have distance equal to k from x in G'. It can be seen in particular that $\Gamma_0(x) = \{x\}$. Let $n_k = |\Gamma_k(x)|$ be the number of these nodes, and let $X_k = \sum_{i=0}^{k} n_i$.

Now for a particular $1 \leq i < \infty$, let $n_i = j_i$ be given. Then from the symmetry of the random graph model each set of j_i distinct nodes chosen from $V - x$ is equally probable to be $\Gamma_i(x)$. Thus if $j_i \geq 1$, $\text{Prob}(y \in \Gamma_i(x)|y \neq x, n_i = j_i) = \binom{n-2}{j_i-1}/\binom{n-1}{j_i} = j_i/(n-1)$, and if $j_i = 0$ then $\text{Prob}(y \in \Gamma_i(x)|y \neq x, n_i = j_i) = 0$. The probability distribution of the distance from a source node x to any destination node $y \in V - x$ in G' can then be obtained from the distribution of n_k as follows (since $n_0 = 1$):

$$\text{Prob}(\text{dist}(x, y) = i \text{ in } G') = \text{Prob}(y \in \Gamma_i(x)|y \neq x) = \frac{E(n_i) - \delta_0(i)}{n - 1}. \quad (11)$$

Then (11) gives

$$\text{Prob}(\text{dist}(x, y) \leq i \text{ in } G') = \frac{1}{n - 1}\left(\sum_{j=0}^{i} E(n_j) - 1\right) = \frac{E(X_i) - 1}{n - 1}. \quad (12)$$

The proof of the theorem proceeds by determining upper and lower bounds for $E(X_k)$ as $k \to \infty$. In particular, we consider for fixed $t \in (0, \infty)$ that $k = \lfloor Nt \rfloor$ where $N \to \infty$. Denote $\delta_j(i) = 1$ if $j = i$, and $\delta_j(i) = 0$, otherwise. In order to derive the upper bound $U_1(\mu)$ we start with the following lemma from [9] for the joint distribution for the number of nodes with distances $k, k - 1, \ldots, 0$ from x in G':

Lemma 2. For $k \geq 1$,

$$\text{Prob}(n_k = j_k, \ldots, n_0 = j_0 \text{ in } G') = \delta_{j_0}(1) \prod_{m=1}^{k}\left[\prod_{i=0}^{m-1}(1 - b_{m-i})^{j_i}\right]^{n - \sum_{i=0}^{m} j_i}$$

$$\cdot \left[1 - \prod_{i=0}^{m-1}(1 - b_{m-i})^{j_i}\right]^{j_m}\binom{n - \sum_{i=0}^{m-1} j_i}{j_m}. \quad (13)$$

From (13) and the properties of $F(t)$, we can write

$$E(n_k|n_{k-1}, \ldots, n_0) = \sum_{i=0}^{k-1} b_{k-i} n_i (n - X_{k-1}) - O(N^{-2}). \quad (14)$$

An upper bound to $\overline{n}_k$ in terms of $\overline{X}_i$ and b_i for $1 \leq i < k$ follows from (14) by setting $(n - X_{k-1})$ to $(n - 1)$ and taking expectations on both sides. Substituting $F(t)$ into b_i and considering the limit as $N \to \infty$ with $k = \lfloor Nt \rfloor$ leads to a differential equation for an upper bounding function $U(t)$, namely

$$U^{(1)}(t) = (n - 1)c\frac{t^m}{m!} + (n - 1)\int_0^t c\frac{(t - \tau)^m}{m!}U^{(1)}(\tau)d\tau \quad (15)$$

with $U(0) = 1$ where $\lim_{N\to\infty} \overline{X}_{\lfloor Nt \rfloor} \leq U(t)$ for $t \in [0, \infty)$. An upper bound for $(U(\lambda)-1)/(n-1)$ is obtained from (15), and $U_1(\mu)$ is set to this bound. Then from (8) and (12) it follows that $\mathrm{Prob}(\mathrm{dist}(x,y) \leq \lambda \text{ in } G) \leq (\overline{X}_{\lfloor N\lambda \rfloor + 1 + n} - 1)/(n-1)$. So, in the limit as $N \to \infty$, $\mathrm{Prob}(\mathrm{dist}(x,y) \leq \lambda \text{ in } G) \leq (U(\lambda) - 1)/(n-1)$ for $\lambda \in [0, \infty)$ as desired, which outlines the proof for the upper bound in (3).

For the lower bound $L_1(\mu)$ in (3), we similarly make use of Lemma 2 except we derive a lower bound to $\overline{n}_k$ by picking an arbitrary integer $0 \leq i^* \leq k-1$ and considering $\mathrm{E}(n_k | X_{i^*})$. We then bound $\mathrm{E}(n_k | X_{i^*})$ further by considering that $X_i = 0$ for $0 \leq i < i^*$ and $X_i = X_{i^*}$ for $i^* \leq i < k$. Considering the limit as $N \to \infty$ of this bound leads to a differential equation in a manner similar to the above case for $U(t)$. The solution of this equation after unconditioning on X_{i^*} gives for arbitrary t^* where $0 \leq t^* < t$,

$$\frac{\overline{X}_{\lfloor Nt \rfloor} - 1}{n-1} \geq 1 - \mathrm{E}\left(\exp\left\{-c\frac{(t-t^*)^{m+1}}{(m+1)!} X_{\lfloor Nt^* \rfloor}\right\}\right) - O(N^{-1}). \quad (16)$$

In order to bound the right-hand side of (16) we make use of the following lemma which is stated without proof due to space limitations:

Lemma 3. *Let $1 \leq I \leq n$ be an arbitrary integer, and let $y(\tau)$ be the solution of*

$$y^{(m+1)}(\tau) = \left(1 - e^{-y(\tau)}\right)(n-I)c \quad (17)$$

with initial conditions $y(0) = 1/I$ and for $m \geq 1$, $y^{(i)}(0) = 0$ for $1 \leq i \leq m$. Then for $F(t)$ given by (7), we have for $t \geq 0$ that

$$\lim_{N\to\infty} \mathrm{Prob}\left(X_{\lfloor Nt \rfloor} \leq I\right) \leq e^{1-y(t)}. \quad (18)$$

The expression for $L_1(\mu)$ given by (5) is obtained by bounding the solution to $y(\tau)$ in (17) since it is difficult to obtain for all m and then using (18) and the bound $\mathrm{E}\left(\exp\left\{-rX_{\lfloor Nt^* \rfloor}\right\}\right) \leq \mathrm{Prob}(X_{\lfloor Nt^* \rfloor} \leq I) + \exp\{-rI\}$ for $r = c(t-t^*)^{m+1}/(m+1)!$ to bound the right-hand side of (16). It remains to select t^* and I, which we set to

$$t^* = \frac{\ln(n-1) + .5\ln\mu}{[(n-1)c]^{\frac{1}{m+1}}}$$

$$I = \left\lceil (n-1)\min\left\{\frac{1}{5}, \frac{\ln\ln(n-1) + \ln\mu}{5\ln(n-1)}\right\}\right\rceil. \quad (19)$$

This gives after simplifying for $\mu \geq 1$ and as $n \to \infty$,

$$\frac{\overline{X}_{\lfloor N\lambda \rfloor} - 1}{n-1} \geq 1 - O(N^{-1}) -$$

$$\exp\left\{o(1) - (1 - o(1))\frac{(\ln\mu)^{m+1}}{5^{m+1}(m+1)!}\min\left\{\frac{1}{5}, \frac{\ln\ln(n-1) + \ln\mu}{5\ln(n-1)}\right\}\right\}. \quad (20)$$

From (8) and (12), $\text{Prob}(\text{dist}(x,y) \leq \lambda \text{ in } G) \geq (\overline{X}_{\lfloor N\lambda \rfloor} - 1)/(n-1)$. So as $N \to \infty$, (20) gives the desired lower bound $L_1(\mu)$ in (5).

Finally, for the special case $m = 0$, we can solve for the exact value of $y(\tau)$ in (17) and apply this to the right-hand side of (16) by rewriting the expectation of the exponent in terms of $\text{Prob}\left(X_{\lfloor Nt^* \rfloor} \leq k\right)$ for each $1 \leq k \leq n$. Let arbitrary t^* be given such that $0 \leq 2t^* < t$. Then we have after simplifying,

$$\lim_{N \to \infty} \frac{\overline{X}_{\lfloor Nt \rfloor} - 1}{n - 1} \geq 1 - e^{-c(t-t^*)n} - \frac{t - t^*}{c(t - 2t^*)^2} e^{-nct^* + 1 + c(t - 2t^*)}. \qquad (21)$$

Setting $t = \lambda$ in (21) and using (8) and (12) as before gives the desired lower bound $L_1(\mu)$ in (6). $\qquad\qquad\qquad\qquad\qquad\qquad\qquad\qquad\qquad\qquad\qquad\qquad$ $\square$

The following theorem establishes bounds for the CDF of the distance between two nodes in the graph for more general distribution functions $F(t)$ and is presented without proof. This theorem extends Theorem 1 so that arbitrary edge cost distributions which include those having a Taylor's series expansion near zero edge cost can be considered.

Theorem 4. *For fixed integer $n > 1$, let $c_1, c_2 \in (0, \infty)$ and integers m_1, m_2, $0 \leq m_1, m_2 < \infty$ which may depend on n be given. Let*

$$\mu_1 = \frac{\exp\left\{[(n-1)c_1]^{\frac{1}{m_1+1}} t\right\}}{n - 1}$$

$$\mu_2 = \frac{\exp\left\{[(n-1)c_2]^{\frac{1}{m_2+1}} t\right\}}{n - 1}.$$

Suppose for $0 \leq \tau \leq t$, $F(\tau)$ is bounded by

$$1 - \exp\left\{-c_1 \frac{\tau^{m_1+1}}{(m_1 + 1)!}\right\} \leq F(\tau) \leq 1 - \exp\left\{-c_2 \frac{\tau^{m_2+1}}{(m_2 + 1)!}\right\}. \qquad (22)$$

Then for $x, y \in V$, $x \neq y$, as $n \to \infty$,

$$\max\{0, L_1(\mu_1)\} \leq \text{Prob}(\text{dist}(x,y) \leq t \text{ in } G) \leq \min\{1, U_1(\mu_2)\} \qquad (23)$$

where $U_1(\mu)$ and $L_1(\mu)$ are given by (4) and (5), (6) in Theorem 1, respectively.

The next theorem establishes the main limit result for the CDF of the distance between arbitrary two nodes and the diameter in the limit as $n \to \infty$.

Theorem 5. *Let $\epsilon \in (0, 1)$ be given. Suppose for fixed n, integer m, where $0 \leq m < \infty$ may depend on n, is given so that $F^{(m+1)}(0^+)$ exists and is finite. Let*

$$\lambda^* = \frac{\ln(n-1)}{[(n-1)F^{(m+1)}(0^+)]^{\frac{1}{m+1}}},$$

and let

$$g_m = \begin{cases} 6+\epsilon & m=0 \\ 1+\epsilon & m \geq 1 \end{cases} \qquad d_m = \begin{cases} 8+\epsilon & m=0 \\ 1+\epsilon & m \geq 1. \end{cases} \qquad (24)$$

Suppose for fixed n, $F(0^+) = 0, \ldots, F^{(m)}(0^+) = 0, F^{(m+1)}(0^+) \in (0, \infty)$, *and* $F^{(m+2)}(t) \in (-\infty, \infty)$, *for* $0 < t \leq d_m \lambda^*$, *where* $F(t)$ *may depend on* n, *and suppose* $\lambda^* \to 0$ *as* $n \to \infty$. *Then for* $x, y \in V$, $x \neq y$, *as* $n \to \infty$,

$$Prob\left((1-\epsilon)\lambda^* \leq dist(x,y) \leq (1+\epsilon)\lambda^* \text{ in } G\right)$$

$$\geq 1 - O\left(n^{-\epsilon} + n^{\cos\left(\frac{2\pi}{m+1}\right)-1} + \exp\left\{-(1-o(1))\frac{\epsilon^{m+2}(\ln n)^{m+1}}{5^{m+2}(m+1)!}\right\}\right) \quad (25)$$

and

$$Prob\left((1-\epsilon)\lambda^* \leq diam(G) \leq g_m \lambda^*\right) \geq 1 - o(1) \qquad (26)$$

$$Prob\left(diam(G) \leq d_m \lambda^*\right) \geq 1 - o(n^{-1}). \qquad (27)$$

Proof. We consider first the distance distribution. Since $\lambda^* \to 0$ as $n \to \infty$ and $F^{(m+2)}(t) \in (-\infty, \infty)$ for $0 < t \leq d_m \lambda^*$, then given an arbitrary $\delta \in (0,1)$, we can write for sufficiently large n and $0 < t \leq d_m \lambda^*$,

$$(1-\delta)F^{(m+1)}(0^+) < F^{(m+1)}(t) < (1+\delta)F^{(m+1)}(0^+). \qquad (28)$$

Let

$$F_1(t) = 1 - \exp\left\{-(1-\delta)F^{(m+1)}(0^+)\frac{t^{m+1}}{(m+1)!}\right\}$$

$$F_2(t) = 1 - \exp\left\{-(1+\delta)F^{(m+1)}(0^+)\frac{t^{m+1}}{(m+1)!}\right\}.$$

Then it can be seen from (28) that for sufficiently large n, $F_1(t) < F(t) < F_2(t)$ for $0 \leq t \leq d_m \lambda^*$. Theorem 4 then gives $Prob(dist(x,y) \leq (1-\epsilon)\lambda^*) \leq U(\mu_2)$ where $\mu_2 = (n-1)^{(1-\epsilon)(1+\delta)^{1/(m+1)}-1}$. Using (4) and since δ can be selected arbitrarily small compared to ϵ, this gives $Prob(dist(x,y) \leq (1-\epsilon)\lambda^*) = O\left(n^{-\epsilon} + n^{\cos\left(\frac{2\pi}{m+1}\right)-1}\right)$. Also, from Theorem 4, $Prob(dist(x,y) \leq (1+\epsilon)\lambda^*) \geq L_1(\mu_1)$ where $\mu_1 = (n-1)^{(1+\epsilon)(1-\delta)^{1/(m+1)}-1}$. Then from (5), it follows that $L_1(\mu_1) = 1 - O\left(\exp\left\{-(1-o(1))\epsilon^{m+2}(\ln n)^{m+1}/(5^{m+2}(m+1)!)\right\}\right)$. Combining the two bounds completes the proof for the distance distribution. The diameter distribution bounds (26), (27) for $m \geq 1$ as well as the lower limit in (26) for $m = 0$ follow immediately from (25) since $Prob(diam(G) > (1+\epsilon)\lambda^*) \leq n^2 Prob(dist(x,y) > (1+\epsilon)\lambda^*)$ and $diam(G) \geq dist(x,y)$. In order to show the upper limit of (26) when $m = 0$, we set $t^* = (3+\epsilon/4)\ln(n-1)/[(n-1)c]$ in (6). This gives

$$L_1(\mu) \geq 1 - \frac{(n-1)^{2+\epsilon/4}}{\mu} - \frac{[\ln \mu - (2+\epsilon/4)\ln(n-1)]\, e\mu^{1/(n-1)}}{[\ln \mu - (5+\epsilon/2)\ln(n-1)]^2 (n-1)^{2+\epsilon/4}} \qquad (29)$$

when $\ln \mu > (5 + \epsilon/2) \ln(n-1)$. Theorem 4 then gives $\text{Prob}(\text{dist}(x,y) \leq (6 + \epsilon)\lambda^*) \geq L_1(\mu_1)$ where $\mu_1 = (n-1)^{(6+\epsilon)(1-\delta)-1}$. Since δ can be made arbitrarily small, (29) gives $L_1(\mu_1) = 1 - o(n^{-2})$. So $\text{Prob}(\text{diam}(G) \leq (6+\epsilon)\lambda^*) = 1 - o(1)$ which shows (26). The upper limit in (27) for $m = 0$ follows similarly where we set $t^* = (4 + \epsilon/4) \ln(n-1)/[(n-1)c]$ in (6) and $\mu_1 = (n-1)^{(8+\epsilon)(1-\delta)-1}$, which completes the proof of the theorem. $\qquad \square$

We note that the bound (27) is a slight improvement over the bounds obtained for $m = 0$ by Frieze and Grimmett [4] and Hassin and Zemel [7] with values for d_0 of 12 and 27, respectively.

3   Shortest Path Algorithms

We consider a complete digraph G with edge cost distributions which have a Taylor's series near zero edge cost as in Theorem 5. Thus, in particular, the edge cost CDF $F(t)$ for an arbitrary edge is in the range $(1 \pm \delta) F^{(m+1)}(0^+) t^{m+1}/(m+1)!$ for $0 < t \leq d_m \lambda^*$ for some integer $m \geq 0$, any $\delta > 0$, and sufficiently large n. In the shortest path algorithm of Hassin and Zemel [7], the authors consider the problem of finding all the shortest paths from a given source node to the other nodes of G for the case $m = 0$. The approach used to reduce computations is to delete edges with cost larger than the graph diameter, $\text{diam}(G)$, since these are not part of any shortest path, and then apply a standard shortest path algorithm. Therefore, in $O(n^2)$ preprocessing time, form the graph $\hat{G} = (V, \hat{E})$ which differs from G in that for each $x, y \in V$, edge $(x,y) \in \hat{E}$ when $(x,y) \in E$ and the cost of edge $(x,y) \leq d_m \ln n/[n F^{(m+1)}(0^+)]^{1/(m+1)}$. Thus, $\text{Prob}((x,y) \in \hat{E}) = O((\ln n)^{m+1}/n)$ and Lemma 1 of Hassin and Zemel [7], or a standard binomial bound, gives that $|\hat{E}| = O(n(\ln n)^{m+1})$ with probability at least $1 - o(n^{-1})$. The shortest paths from each node to all other nodes are then determined for $\hat{G}$ using a standard algorithm such as Fredman and Tarjan [3] which runs in $O(|\hat{E}| + n \ln n)$ time for each node. Then $\text{diam}(\hat{G})$ is determined directly from these shortest paths, and if $\text{diam}(\hat{G}) \leq d_m \ln n/[n F^{(m+1)}(0^+)]^{1/(m+1)}$, the edge deletions where justified; otherwise, the procedure failed and the standard method applied to G can be used. This procedure applied to determine the shortest paths between all nodes pairs thus has time bound $O(n^2(\ln n)^{m+1} + n^2 \ln n) + o(n^{-1})O(n^3) = O(n^2(\ln n)^{m+1})$. It can be seen that the bound increases as m increases since fewer edges are initially deleted by the procedure.

The algorithm RANDOMSHORTPATH by Frieze and Grimmett [4] which also considered the case $m = 0$ is similar, but edges are eliminated according to the ordering of their costs. That is, for some integer p and each node $x \in V$, the edges rooted at x with costs among the p smallest for these edges are retained; the remainder are deleted. The resulting modified graph has np edges, and it remains to determine p. The derivation of p requires that Lemma 4.2 in Frieze and Grimmett [4] which determined the distribution of costs for deleted edges be modified for the new edge cost model using $F(t) = F^{(m+1)}(0^+) t^{m+1}/(m+1)!$ instead of $F(t) = t$ and requires that the new bound for the CDF of $\text{diam}(G)$ in

(27) be used. Appendix A gives the derivation, and we find that the value of p in algorithm RANDOMSHORTPATH can be modified to:

$$p = \begin{cases} 14.3 \ln n & m = 0 \\ 1.1(\ln n)^{m+1}/(m+1)! & m \geq 1. \end{cases} \tag{30}$$

For this modified graph, we again use a standard shortest path algorithm for all the node pairs. This then gives the same expected time bound for this algorithm as the previous one, namely, $O(n^2(\ln n)^{m+1})$.

Appendix A: Value of p for RANDOMSHORTPATH

In this appendix we determine integer p so that Prob(cost of any deleted edge $\leq$ diam$(G)) = o(n^{-1})$ where edges are deleted according to the ordering of their costs. For each node $x \in V$, the edges rooted at x with costs among the p smallest for these edges are retained; the remainder are deleted. Let e_{p+1} be the cost of the $(p+1)$-th smallest edge rooted at an arbitrary node, and let D_e be the cost of the smallest deleted edge. Then we can write for arbitrary $0 \leq T < \infty$,

$$\text{Prob}(D_e \leq \text{diam}(G)) = \text{E}\left[\text{Prob}(D_e \leq \text{diam}(G)|\text{diam}(G))\right]$$
$$\leq \text{E}\left[\text{Prob}(D_e \leq \text{diam}(G)|\text{diam}(G))|\text{diam}(G) \leq T\right]\text{Prob}(\text{diam}(G) \leq T) +$$
$$\text{Prob}(\text{diam}(G) > T)$$
$$\leq \text{E}\left[\text{Prob}(D_e \leq T|\text{diam}(G))|\text{diam}(G) \leq T\right]\text{Prob}(\text{diam}(G) \leq T) +$$
$$\text{Prob}(\text{diam}(G) > T)$$
$$\leq \text{E}\left[\text{Prob}(D_e \leq T|\text{diam}(G))\right] + \text{Prob}(\text{diam}(G) > T)$$
$$= \text{Prob}(D_e \leq T) + \text{Prob}(\text{diam}(G) > T). \tag{31}$$

The union bound gives from (31),

$$\text{Prob}(D_e \leq \text{diam}(G)) \leq n\text{Prob}(e_{p+1} \leq T) + \text{Prob}(\text{diam}(G) > T). \tag{32}$$

We can also write

$$\text{Prob}(e_{p+1} \leq T) = (n-1)\binom{n-2}{p}\int_0^T F(t)^p[1 - F(t)]^{n-p-2}F^{(1)}(t)dt \tag{33}$$

where the edge cost CDF $F(t)$ for an arbitrary edge is in the range $(1 \pm \delta)F^{(m+1)}(0^+)t^{m+1}/(m+1)!$ for $0 < t \leq d_m\lambda^*$ for some integer $m \geq 0$, any $\delta > 0$, and sufficiently large n. We set $T = d_m\lambda^*$ which gives from Theorem 5 that the last term in (32) is $o(n^{-1})$. For arbitrary $\xi \in (0, \infty)$, we also set $p = \xi(n-1)F(T)$ where ξ can depend on n so that p is an integer. Therefore, for some $\gamma \in [1-\delta, 1+\delta]$, we can write $F(T) = \gamma d_m^{m+1}[\ln(n-1)]^{m+1}/[(n-1)(m+1)!]$. Thus, we set

$$p = \frac{\xi\gamma d_m^{m+1}[\ln(n-1)]^{m+1}}{(m+1)!}. \tag{34}$$

Now, $x^p(1-x)^{n-p-2}$ is monotonically increasing for $0 \leq x \leq p/(n-p-1)$, so provided $p/[\xi(n-1)] = F(T) \leq p/(n-p-1)$ which we show to be true later, $F(t)^p[1-F(t)]^{n-p-2} \leq F(T)^p[1-F(T)]^{n-p-2}$ for $0 \leq t \leq T$. This gives from (33) and the bound $\binom{n-2}{p} \leq (n-1)^p/p! \leq ((n-1)e/p)^p$,

$$
\begin{aligned}
&\text{Prob}(e_{p+1} \leq T) \\
&\leq (n-1)\binom{n-2}{p} F(T)^{p+1}[1-F(T)]^{n-p-2} \\
&\leq \exp\left\{(p+1)\ln(F(T)) - (n-p-2)F(T) + p[\ln(n-1) + 1 - \ln(p)] + \right. \\
&\quad \left. \ln(n-1)\right\} \\
&\leq \exp\left\{(p+1)\left[\frac{\ln(p)}{p+1} - \frac{(n-p-2)p}{(p+1)(n-1)\xi} - \ln(\xi) + \frac{p}{p+1}\right]\right\} \\
&\leq \exp\left\{(p+1)\left[1 - \frac{1}{\xi} - \ln(\xi) + o(1)\right]\right\}. \tag{35}
\end{aligned}
$$

The constants ϵ from the definition of d_m in (24) and δ can be chosen arbitrarily small. Thus, for $m = 0$, numerical evaluation gives from (34) and (35) that $\text{Prob}(e_{p+1} \leq T) \leq o(n^{-2})$ for $\xi > 1.787$ or $p \geq 14.3\ln n$. This also implies that $p/[\xi(n-1)] = F(T) \leq p/(n-p-1)$ as assumed. Similarly, for $m \geq 1$, (34) and (35) give $\text{Prob}(e_{p+1} \leq T) \leq o(n^{-2})$ for $\xi > 1$ or $p \geq 1.1(\ln n)^{m+1}/(m+1)!$. This again gives $p/[\xi(n-1)] = F(T) \leq p/(n-p-1)$ as assumed. Therefore, from (32), $\text{Prob}(D_e \leq \text{diam}(G)) = o(n^{-1})$ as desired when p is chosen equal to these lower bounds. This establishes (30).

References

1. R. K. Ahuja, K. Mehlhorn, J. B. Orlin, and R. E. Tarjan: Faster Algorithms for the Shortest Path Problem. Technical Report CS-TR-154-88, Department of Computer Science, Princeton University. (1988)
2. P.A. Bloniarz: A shortest-path algorithm with expected time $O(n^2 \log n \log^* n)$. Technical Report 80-3, Dept. of Computer Science, State Univ. of New York at Albany. (1980)
3. M. L. Fredman and R. E. Tarjan: Fibonacci heaps and their uses in improved network optimization algorithms. J. Assoc. Comput. Mach. 34 (1987) 596-615
4. A. M. Frieze and G. R. Grimmett: The shortest-path problem for graphs with random arc-lengths. Discrete Appl. Math. 10 (1985) 57-77
5. G. Gallo and S. Pallottino: Shortest path algorithms. Annals of Oper. Res. 13 (1988) 3-79
6. A. V. Goldberg: Scaling algorithms for the shortest paths problem. Proc. 4th ACM-SIAM symposium of Discrete Algorithms. (1993) 222-231
7. Refael Hassin and Eitan Zemel: On shortest paths in graphs with random weights. Math. of Oper. Res. 10 (1985) 557-564
8. P. Spira: A new algorithm for finding all shortest paths in a graph of positive edges in average time $O(n^2 \log^2 n)$. SIAM J. Comput. 2 (1973) 28-32
9. Scott K. Walley, Harry H. Tan, and Audrey M. Viterbi. Shortest path cost distribution in random graphs with positive integer edge costs. Proc. of IEEE INFO-COM'93. (1993) 1023-1032

Simple Reduction of f-Colorings to Edge-Colorings

Xiao Zhou[1] and Takao Nishizeki[2]

[1] Education Center for Information Porcessing
[2] Graduate School of Information Sciences
Tohoku University, Sendai 980-77, Japan

Abstract. In an edge-coloring of a graph $G = (V, E)$ each color appears around each vertex at most once. An f-coloring is a generalization of an edge-coloring in which each color appears around each vertex v at most $f(v)$ times where f is a function assigning a natural number $f(v) \in \mathbf{N}$ to each vertex $v \in V$. In this paper we first give a simple reduction of the f-coloring problem to the ordinary edge-coloring problem, that is, we show that, given a graph $G = (V, E)$ and a function $f : V \to \mathbf{N}$, one can directly construct in polynomial-time a new simple graph whose edge-coloring using a minimum number of colors immediately induces an f-coloring of G using a minimum number of colors. As by-products, we give a necessary and sufficient condition for a graph to have an f-factorization, and show that the edge-coloring problem for multigraphs can be easily reduced to edge-coloring problems for simple graphs.

1 Introduction

An *edge-coloring* of a graph $G = (V, E)$ is to color all the edges of G so that no two adjacent edges are colored with the same color. The minimum number of colors needed for an edge-coloring is called the *chromatic index* of G and denoted by $\chi'(G)$. Throughout the paper the *maximum degree* of a graph G is denoted by $\Delta(G)$ or simply by Δ. König showed that $\chi'(G) = \Delta$ if G is bipartite [3, 6]. Vizing showed that $\chi'(G) = \Delta$ or $\Delta + 1$ if G is a simple graph, that is, G has no multiple edges or self-loops [3, 8]. The *edge-coloring problem* is to find an edge-coloring of G using $\chi'(G)$ colors. Let $f : V \to \mathbf{N}$ be a function which assigns a natural number $f(v) \in \mathbf{N}$ to each vertex $v \in V$. Then an f-*coloring* of G is to color all the edges of G so that, for each vertex $v \in V$, at most $f(v)$ edges incident to v are colored with the same color. Thus an f-coloring of G is a partition of the edge set of G into subsets, each inducing a spanning subgraph whose vertex-degrees are bounded by f. Figure 1(a) illustrates an f-coloring of a graph with three colors, which are indicated by solid, bold and dashed lines. An ordinary edge-coloring is a special case of an f-coloring such that $f(v) = 1$ for every vertex $v \in V$. The minimum number of colors needed for an f-coloring is called the f-*chromatic index* of G and denoted by $\chi'_f(G)$. The f-*coloring problem* is to find an f-coloring of G using $\chi'_f(G)$ colors for a given graph G. Let $\Delta_f(G) = \max_{v \in V} \lceil d(v)/f(v) \rceil$ where $d(v)$ is the *degree* of vertex v. It is known that $\chi'_f(G) = \Delta_f$ or $\Delta_f + 1$ for any simple graph G [4].

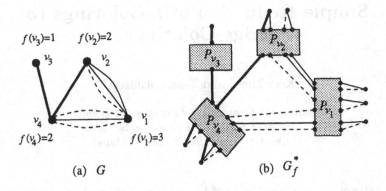

$f(v_3)=1$ $f(v_2)=2$

v_3 v_2

v_4 v_1

$f(v_4)=2$ $f(v_1)=3$

(a) G (b) G_f^*

Fig. 1. Transformation from G to G_f^*.

Since the ordinary edge-coloring problem is NP-complete [5], the f-coloring problem is also NP-complete. Therefore the theory of NP-completeness immediately implies that there exists a polynomial-time reduction of the f-coloring problem to the ordinary edge-coloring problem plausibly through another NP-complete problem, say 3-SAT. However, no simple direct reduction has been known so far.

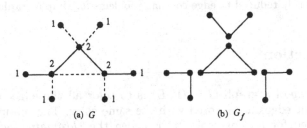

(a) G (b) G_f

Fig. 2. Trivial reduction.

We have given the following trivial reduction [9]. For each vertex $v \in V$ of a graph G, replace v with $f(v)$ copies $v_1, v_2, \cdots, v_{f(v)}$, and attach to the copies the $d(v)$ edges which were incident to v; attach $\lceil d(v)/f(v) \rceil$ or $\lfloor d(v)/f(v) \rfloor$ edges to each copy v_i, $1 \le i \le f(v)$. Let G_f be the resulting graph. Figure 2 illustrates G and G_f. Clearly $\Delta(G_f) = \Delta_f(G) = \max_{v \in V} \lceil d(v)/f(v) \rceil$. Since an edge-coloring of G_f immediately induces an f-coloring of G using the same number of colors, we have

$$\chi'_f(G) \le \chi'(G_f). \tag{1}$$

If G is simple then G_f is also simple, and if G is bipartite then G_f is also bipartite. Therefore, the results of Vizing and König together with the reduction above immediately imply that $\chi'_f(G) = \Delta_f(G)$ or $\Delta_f(G) + 1$ if G is simple and that $\chi'_f(G) = \Delta_f(G)$ if G is bipartite. Thus the reduction is trivial but very useful.

However, Eq.(1) does not always hold in equality. For example, $\chi'_f(G) = 2$ for a graph G in Figure 2(a) as indicated by solid and dashed lines, but $\chi'(G_f) = 3$ for a graph G_f in Figure 2(b).

In this paper we first give a very simple reduction of the f-coloring problem to the ordinary edge-coloring problem. That is, we show that, given a simple graph G together with a function $f : V \to \mathbf{N}$, one can directly construct in polynomial-time a new simple graph G_f^* such that $\chi'_f(G) = \chi'(G_f^*)$. We construct G_f^* from G by inserting an appropriate bipartite graph P_v for each vertex $v \in V$ as illustrated in Figure 1. It should be noted that the theory of NP-completeness does not imply the existence of such a single graph G_f^*. We then show that the f-coloring problem for a multigraph can be directly reduced in polynomial time to the edge-coloring problem for several simple graphs. Thus we show that the f-coloring problem is not more intractable than the ordinary edge-coloring problem although the former looks to be more difficult than the latter. Furthermore the simple reduction above immediately yields a necessary and sufficient condition for a graph G to have an f-factorization, that is, a partition of G to spanning subgraphs in each of which $d(v) = f(v)$ for every vertex v. Finding such a condition has been an open problem in graph theory [1]. The reduction above also implies that the edge-coloring problem for *multigraphs* can be easily reduced to the edge-coloring problem for *simple graphs*.

2 Preliminaries

In this section we give some definitions. Let $G = (V, E)$ denote a graph with vertex set V and edge set E. We often denote by $V(G)$ and $E(G)$ the vertex set and the edge set of G, respectively. We assume that G has no selfloops but may have multiple edges, that is, G is a so-called *multigraph*. If G has no multiple edges, then G is called a *simple graph*. An edge joining vertices u and v is denoted by (u, v). The *degree* of vertex $v \in V(G)$ is denoted by $d(v, G)$ or simply by $d(v)$. The *maximum degree* of G is denoted by $\Delta(G)$, or simply by Δ. Clearly $\chi'(G) \geq \Delta(G)$. A graph $G = (V, E)$ is *bipartite* if V is bipartitioned into two subsets U and W so that $v \in U$ and $w \in W$ for every edge $(v, w) \in E$. By König's theorem $\chi'(G) = \Delta(G)$ if G is bipartite [3, 6].

Let f be a function which assigns a natural number $f(v)$ to each vertex $v \in V$. One may assume without loss of generality that $f(v) \leq d(v)$ for each vertex $v \in V(G)$. Let $d_f(v, G) = \lceil d(G, v)/f(v) \rceil$ for $v \in V$, and let $\Delta_f(G) = \max\{d_f(v, G)|v \in V(G)\}$. We often denote $d_f(v, G)$ simply by $d_f(v)$. Clearly $\chi'_f(G) \geq \Delta_f(G)$. It is known that $\chi'_f(G) = \Delta_f(G)$ if G is bipartite [4].

3 Simple Reduction

In this section we give a simple and direct reduction of an f-coloring to an ordinary edge-coloring. Clearly a multigraph G satisfies $\chi'_f(G) = 1$ if $\Delta_f(G) = 1$. Furthermore one can easily observe the following lemma.

Lemma 1. *Let $G = (V, E)$ be a connected multigraph with $\Delta_f(G) = 2$. Then the following* (a) *and* (b) *hold.*

(a) $\chi'_f(G)$ *is either 2 or 3; and*

(b) $\chi'_f(G) = 3$ *if and only if*

  • $d(v, G) = 2f(v)$ *for every vertex* $v \in V$, *and*

  • $|E|$ *is odd.*

Thus the f-coloring problem can be easily solved in linear time if $\Delta_f(G) \leq 2$. Therefore, in the remaining of this paper, we may assume that $\Delta_f(G) \geq 3$.

In this section we give a sophisticated reduction for which Eq.(1) always holds in equality. That is, as the main result of this paper, we give the following theorem.

Theorem 2. *Given any simple graph* $G = (V, E)$ *and function* f *such that* $\Delta_f(G) \geq 3$, *one can directly construct a simple graph* $G^*_f = (V^*_f, E^*_f)$ *such that* $\chi'_f(G) = \chi'(G^*_f)$ *and* $|E^*_f|$ *is polynomial in* $|E|$.

It should be noted that, given an edge-coloring of G^*_f with $\chi'(G^*_f)$ colors, one can find in polynomial time an f-coloring of G using $\chi'_f(G)$ colors. Thus the f-coloring problem for a simple graph can be reduced to the ordinary edge-coloring problem for a simple graph in polynomial-time.

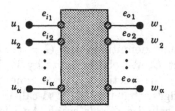

Fig. 3. (α, β)-permutation graph $P_{\alpha\beta}$.

We use the following graph $P_{\alpha\beta}$ called an (α, β)-permutation graph as a building-block to construct G^*_f from G. See Figure 3. For positive integers α and β, let $P_{\alpha\beta}$ be a bipartite simple graph such that

  • there are α input vertices $U = \{u_1, u_2, \cdots, u_\alpha\}$ and α input edges $E_i = \{e_{i1}, e_{i2}, \cdots, e_{i\alpha}\}$ incident to input vertices;
  • there are α output vertices $W = \{w_1, w_2, \cdots, w_\alpha\}$ and α output edges $E_o = \{e_{o1}, e_{o2}, \cdots, e_{o\alpha}\}$ incident to output vertices;
  •
$$d(v, P_{\alpha\beta}) = \begin{cases} 1 & \text{if } v \in U \cup W; \\ \beta & \text{otherwise.} \end{cases}$$

Thus $\Delta(P_{\alpha\beta}) = \beta$. Let $C = \{c_1, c_2, \cdots, c_\beta\}$ be any set of β colors. We call $P_{\alpha\beta}$ an (α, β)-*permutation graph* if

  (i) for any edge-coloring of $P_{\alpha\beta}$ with β colors, the sequence of colors of output edges is a permutation of that of input edges; and

(ii) for any sequence of colors $C_i = \{c_{i1}, c_{i2}, \cdots, c_{i\alpha}\}$, $c_{ij} \in C$, and for any permutation $C_o = \{c_{o1}, c_{o2}, \cdots, c_{o\alpha}\}$ of C_i, there is an edge-coloring of $P_{\alpha\beta}$ with the β colors such that input edge e_{ij} is colored c_{ij} and output edge e_{oj} is colored c_{oj} for each j, $1 \leq j \leq \alpha$.

The following lemma holds on $P_{\alpha\beta}$. The construction of $P_{\alpha\beta}$ is similar as that of a well-known Clos permutation network.

Lemma 3. *For any* $\alpha \geq 1$ *and* $\beta \geq 3$ *there is an* (α, β)*-permutation graph* $P_{\alpha\beta}$ *such that* $|E(P_{\alpha\beta})| = O(\alpha\beta^2 \lceil \log_\beta(\alpha + 1) \rceil)$.

For any integer $\beta \geq 3$, we construct $G^*_{f\beta}$ from G and copies of $P_{\alpha\beta}$ as follows (see Figure 1):

(a) replace each vertex $v \in V$ by a copy P_v of a $(d(v), \beta)$-permutation graph $P_{d(v)\beta}$;
(b) merge the $d(v)$ output vertices $w_1, w_2, \cdots, w_{d(v)}$ of P_v, $v \in V$, into $f(v)$ vertices $v_1, v_2, \cdots, v_{f(v)}$ so that $d(v_j, G^*_{f\beta}) = \lfloor d(v, G)/f(v) \rfloor$ or $\lceil d(v, G)/f(v) \rceil$ for every j, $1 \leq j \leq f(v)$;
(c) for each edge $e = (v, v') \in E$, identify, as a single edge, an input edge of P_v and an input edge of $P_{v'}$ which are surrogates of e.

Clearly

$$\Delta(G^*_{f\beta}) = \begin{cases} \beta & \beta \geq \Delta_f(G); \\ \Delta_f(G) & \text{otherwise.} \end{cases}$$

Furthermore $G^*_{f\beta}$ is a simple graph even if G has multiple edges. Figure 1(b) illustrates $G^*_{f\beta}$ for the graph G in Figure 1(a).

We have the following theorem on $G^*_{f\beta}$.

Theorem 4. *Let* $G = (V, E)$ *be any multigraph and let* β *be any integer with* $\beta \geq 3$. *Then* $\chi'_f(G) \leq \beta$ *if and only if* $\chi'(G^*_{f\beta}) = \beta$.

We denote $G^*_{f\beta}$ with $\beta = \Delta_f(G)$ simply G^*_f. Then $\Delta(G^*_f) = \beta = \Delta_f(G)$, and Theorem 2 immediately follows from Theorem 4.

The number of edges joining vertices u and v in a multigraph G is called the *edge-multiplicity* of (u, v), and denoted by $\mu(u, v)$. Let $\mu(u) = \max\{\mu(u, v) \mid (u, v) \in E\}$, and let $\mu(G) = \max\{\mu(u) \mid u \in V(G)\}$. Let $\mu_f(G) = \max\{\lceil \mu(v)/f(v) \rceil \mid v \in V\}$. Then, since $\Delta_f(G) \leq \chi'_f(G) \leq \Delta_f(G) + \mu_f(G)$ [4], by Theorem 4 we have the following corollary.

Corollary 5. *Any multigraph* G *satisfies*

$$\chi'_f(G) = \min\{\beta \mid \chi'(G^*_{f\beta}) = \beta \text{ and } \Delta_f(G) \leq \beta \leq \Delta_f(G) + \mu_f(G)\}.$$

Given an edge-coloring of $G^*_{f\beta}$ with $\chi'(G^*_{f\beta}) = \beta$ colors, one can find in polynomial time an f-coloring of G with at most β colors. Therefore the f-coloring problem for a multigraph is polynomial-time reducible to the ordinary edge-coloring problem for simple graphs. Indeed, solving the edge-coloring problem at most $\log_2 \mu_f(G)$ times, one can solve the f-coloring problem for a multigraph

G. However, an approximate edge-coloring of $G_{f\beta}^*$ using more than $\chi'(G_{f\beta}^*)$ colors does not yield an approximate f-coloring of G.

The edge-coloring problem looks to be more intractable for multigraphs than for simple graphs. However, since an edge-coloring is an f-coloring in which $f(v) = 1$ for each vertex, we have the following corollary.

Corollary 6. *An edge-coloring problem for multigraphs can be easily reduced to edge-coloring problems for simple graphs in polynomial-time.*

An f-factor of a graph $G = (V, E)$ is a spanning subgraph G' of G such that $d(v, G') = f(v)$ for every $v \in V$. For each vertex v of G, replace v with a complete bipartite graph $K_{d(v), f(v)}$ and attach to each of the $d(v)$ left vertices of $K_{d(v), f(v)}$ one of the $d(v)$ edges which were incident to v in G. Let G_T be the resulting graph. Tutte showed that G has an f-factor if and only if G_T has a 1-factor, that is, a complete matching [7]. An f-*factorization* of graph $G = (V, E)$ is a partition of set E into subsets each of which induces an f-factor. Thus an f-factorization of G is indeed an f-coloring with $\Delta_f(G)$ colors in which each of the color classes induces an f-factor. Therefore Theorem 4 immediately yields the following corollary.

Corollary 7. *A multigraph $G = (V, E)$ has an f-factorization if and only if*

(a) $d(v)/f(v) = \Delta_f(G)$ *for every vertex $v \in V$, and*
(b) $\chi'(G_f^*) = \Delta(G_f^*)$.

Corollary 7 has a resemblance to Tutte's classical result above, and is interesting in its own right since finding such a necessary and sufficient condition for a graph to have an f-factorization has been an open problem in graph theory [1].

A $[g, f]$-*factor* of a graph $G = (V, E)$ is a spanning subgraph G' of G such that $g(v) \leq d(v, G') \leq f(v)$ for each $v \in V$. A $[g, f]$-*factorization* of G is a partition of set E into subsets each of which induces a $[g, f]$-factor. Since the $[g, f]$-factorization problem is polynomial-time reducible to the f-coloring problem [2], the $[g, f]$-factorization problem is also polynomial-time reducible to the ordinary edge-coloring problem for simple graphs.

References

1. J. Akiyama and M. Kano. Factors and factorizations of graphs – a survey. *Journal of Graph Theory*, 9:1–42, 1985.
2. M-C. Cai. Private communication, 1995.
3. S. Fiorini and R. J. Wilson. *Edge-Colourings of Graphs*. Pitman, London, 1977.
4. S. L. Hakimi and O. Kariv. On a generalization of edge-coloring in graphs. *Journal of Graph Theory*, 10:139–154, 1986.
5. I. Holyer. The NP-completeness of edge-colouring. *SIAM J. Comput.*, 10:718–720, 1981.
6. D. König. Über graphen und iher anwendung auf determinantentheorie und menglehre. *Math. Ann.*, 77:453–465, 1916.
7. W. T. Tutte. A short proof of the factor theorem for finite graphs. *Canad. J. Math.*, 6:347–352, 1954.
8. V. G. Vizing. On an estimate of the chromatic class of a p-graph. *Discret Analiz*, 3:25–30, 1964.
9. X. Zhou and T. Nishizeki. Edge-coloring and f-coloring for various classes of graphs. In *Proc. of the Fifth International Symposium on Algorithms and Computation, Lect. Notes in Computer Science, Springer-Verlag*, 834:199–207, 1994.

Output-size Sensitiveness of OBDD Construction Through Maximal Independent Set Problem

Kazuyoshi HAYASE* , Kunihiko SADAKANE and Seiichiro TANI**

Department of Information Science, University of Tokyo
Hongo, Tokyo 113, Japan

Abstract. This paper investigates output-size sensitiveness of construction of OBDD by analyzing the maximal independent set problem of a graph, which would give several insights to efficient manipulation of Boolean functions by OBDD and graph theory.

1 Introduction

Ordered Binary Decision Diagrams (OBDD, in short) are useful representations for Boolean functions [1, 2]. OBDDs have been investigated and applied to various fields such as design and formal verification of digital systems, combinatorics, and so on.

In previous methods, an OBDD is obtained by applying Boolean operations between OBDDs iteratively [1]. One of difficulties in such a construction is that the target OBDD may not be obtainable when intermediate OBDDs become very large, even if its size (= the number of its nodes; we use the word "node" in OBDDs to distinguish from "vertex" in other graphs) is acceptable. In [5], a breadth first algorithm is proposed to resolve this difficulty by constructing the target OBDD directly. This algorithm can construct OBDDs in an output-size sensitive manner but requires good strategy for equivalence test between subproblems. This paper investigates the output-size sensitiveness of OBDD construction by adopting the maximal independent set problem of a graph as a good instance to reveal the intractable parts and tractable parts clearly, and to derive results useful to analyze prime implicants of Boolean functions. We now describe the meanings and importance of our research in more detail.

The maximal independent set problem has attracted much attention in graph theory. Several efficient algorithms for their generation have been proposed [6, 4, 3]. Independent sets and dominating sets have some interesting relations to maximal independent sets and they also form meaningful classes of Boolean functions. The class $\{f_{\mathrm{IS}(G)}\}$ of characteristic functions of independent sets is equal to the class of negative 2CNFs and the class $\{f_{\mathrm{DS}(G)}\}$ of characteristic functions of dominating sets is a subclass of positive CNFs of polynomial size.

---

* E-mail : hayase@is.s.u-tokyo.ac.jp
** Currently with Nippon Telegraph and Telephone Corporation

We can construct an OBDD of the characteristic function $f_{\mathrm{MIS}(G)}$ of maximal independent sets by applying Boolean AND between two OBDDs of $f_{\mathrm{IS}(G)}$ and $f_{\mathrm{DS}(G)}$. Furthermore, we can find two one-to-one correspondences of the family of maximal independent sets to both sets of the prime implicants of $f_{\mathrm{IS}(G)}$ and $f_{\mathrm{DS}(G)}$. Computing prime implicants and representing them in a compact way are required especially in the field of VLSI CAD, and OBDDs play an important role [2]. We expect that nice results for OBDD of $f_{\mathrm{MIS}(G)}$ might give suggestions to questions on relationship between a function and its prime implicants.

As mentioned above, the breadth first algorithm [5] requires efficient strategy for equivalence test. Unfortunately, it will be shown that the equivalence test between two subproblems of $f_{\mathrm{MIS}(G)}$ is still *co-NP* complete if we encode subproblems obediently. On the other hand, we can apply the breadth first algorithm to $f_{\mathrm{IS}(G)}$ and $f_{\mathrm{DS}(G)}$ efficiently. We discuss relationship among the size of these three OBDDs and show that the OBDD of $f_{\mathrm{MIS}(M_n)}$ can be obtained in output-size sensitive manner where M_n denotes a graph of $n \times n$ mesh.

2 OBDDs

We assume that a Boolean function is defined on variable set $\{x_1, x_2, \ldots, x_n\}$. An OBDD [1] is a labelled directed acyclic graph with a root. Each non-terminal node v is labelled by a Boolean variable $index[v] \in \{x_1, x_2, \ldots, x_n\}$ and called a variable node. Each terminal node v is labelled by a constant $value[v] \in \{0, 1\}$ and called a constant node. There is a total order on the variable set and we assume it to be $x_1 < x_2 < \ldots < x_n$. Each path from the root to a constant node follows this total order. Consequently we can consider variable nodes are levelled by the subscript of its label. From each variable node v, there are two outgoing edges labelled by 0 and 1. The directed two nodes are denoted by $edge(v, 0)$ and $edge(v, 1)$ respectively. We define a Boolean function $F[v]$ for each node v as follows: (i) $F[v] = value[v]$ if v is a constant node, (ii) $F[v] = \overline{x_i} \cdot F[edge(v, 0)] + x_i \cdot F[edge(v, 1)]$ if $index[v] = x_i$.

It has been known that ROBDD (Reduced OBDD) and QOBDD (Quasi-reduced OBDD) are unique up to isomorphism for a function. We can remove the following two kinds of nodes from an OBDD without changing its function: (i) Nodes u and v are equivalent nodes if the two subgraphs rooted by them are isomorphic, (ii) A node v is a redundant node if $edge(v, 0) = edge(v, 1)$. An OBDD is ROBDD if it has no redundant nodes and equivalent nodes. An OBDD is QOBDD if it has no equivalent nodes and every path from the root to a constant node consists of $n+1$ nodes. We will treat mainly QOBDDs rather than ROBDDs because QOBDDs are appropriate in discussing the time complexity and the size of a QOBDD is at most the size of its ROBDD times the number of variables n.

2.1 The Breadth First Algorithm

A breadth first algorithm for constructing QOBDDs is proposed in [5]. Figure 1 shows an outline. This algorithm uses an idea which resembles the branching operation of the branch-and-bound method. In fact, we can identify a subproblem in branch-and-bound with a node of the QOBDD. The algorithm proceeds from level 1 to level n in breadth first manner. To create nodes of level $i+1$, we make two subproblems of a node of level i by fixing the variable x_i to be 0 and 1 using RESTRICT. We keep the constructing OBDD *quasi-reduced* by checking *table* whether there exist equivalent nodes. Using this algorithm with a good strategy to test equivalence, we can construct QOBDDs in output-size sensitive manner in the following sense. An algorithm which constructs QOBDDs (for a specified problem) is called *output-size sensitive* if it has a time complexity of the order of a *polynomial* in the size of them and the size of the input.

procedure CONST(P):
    create a root node v in level 1
    for $i := 1$ **to** n **do** *table* $:= \phi$;
        for_all u of level i **do**
            for $b := 0$ **to** 1 **do begin**
                $P[u]|_{x_i := b} :=$ RESTRICT($P[u], x_i, b$);
                if $\exists P[u'] \in$ *table* s.t. ($P[u]|_{x_i := b} = P[u']$) **then** $edge(u, b) := u'$;
                else create a variable node u', set $edge(u, b)$ to u' and add $P[u]|_{x_i := b}$ to *table*;
            end;

Fig. 1. The Breadth First Algorithm for Constructing the QOBDD for P

3 Examinations of the Breadth First Algorithm

3.1 Basic Notations

Let $G = (V, E)$ be a simple undirected graph where $V = \{1, 2, \ldots, n\}$ is the vertex set and E is the edge set. $\Gamma(v)$ denotes the adjacent vertex set of v.

A vertex set $S \subseteq V$ is independent if any pair of vertices in S are not adjacent to each other. S is maximal independent if S is independent and there is no independent set that includes S properly. S is dominating if any vertex is either in S or has an adjacent vertex in S. Let IS(G), MIS(G), and DS(G) denote the families of independent sets, maximal independent sets and dominating sets of G respectively. The next fact is well known.

Fact 1. $S \subseteq V$ is maximal independent iff S is independent and dominating.

Here we introduce how we can treat a family of subsets by means of a Boolean function. Let U be a finite set $\{1, 2, \ldots, |U| = n\}$. The characteristic vector $\chi^U(S)$ of S on U is an n-dimensional Boolean vector in $\{0, 1\}^n$ such that $\chi_i^U(S) = 1$ iff i is in S ($S \subseteq U$). The characteristic function $f_{\mathcal{F}}$ of the family $\mathcal{F}$ of subsets of U is a Boolean function such that $f_{\mathcal{F}}(\chi^U(S)) = 1$ iff $S \in \mathcal{F}$.

The following proposition claims that we can identify MIS(G), PI($f_{IS(G)}$) and PI($f_{DS(G)}$) where PI(f) denotes the set of prime implicants of f. (A product of literal set m is a prime implicant of f iff it implies f and no product of proper subset of m implies f.)

Proposition 2. *There are two one-to-one correspondences of MIS(G) to both PI($f_{IS(G)}$) and PI($f_{DS(G)}$).*

3.2 QOBDDs of Maximal Independent Sets

The essential problem of the breadth first algorithm is that it requires equivalence test between subproblems. In the case of $f_{MIS(G)}$, its complexity is considered to be intractable because the following problem EQ is *co-NP* complete (we can reduce 3SAT to EQ using an idea in [3]). EQ formalizes equivalence test between two subproblems of level $j + 1$ of the QOBDD of $f_{MIS(G)}$ where subproblems are represented in a straightforward way. Note that we can not conclude that there is no possibility to construct the QOBDD of $f_{MIS(G)}$ in an output-size sensitive manner even if we assume both this theorem and the $NP \neq P$ proposition because we might be able to find a good representation of subproblems to solve this hard problem or devise another approach to construct the QOBDD.

Definition (Problem EQ). *Given a graph G, an integer j and two vertex sets V_1 and V_2. Let $U = \{j + 1, j + 2, \ldots, n\} \subseteq V$. We define two Boolean functions F_i ($i = 1, 2$) on the characteristic vector of a subset X of U such that $F_i(\chi^U(X)) = 1$ iff $(V_i \cup X)$ is in $MIS(G)$. Decide whether $F_1 = F_2$ or not.*

Theorem 3. *The problem EQ is co-NP complete.*

3.3 QOBDDs of Independent Sets and Dominating Sets

Here we show strategies for $f_{IS(G)}$ and $f_{DS(G)}$ to apply the breadth first algorithm efficiently. We use the following notations in this section. If a subproblem is found to be 0 or 1, it is represented by $\bot$ or $\top$. Though details will not be mentioned, $\bot$ and $\top$ can be easily processed as exceptions. $VAR[i] = \{x_j | j \geq i\}$ denotes the set of variables which are not fixed in subproblems of level i.

The QOBDD of Independent Sets. We explain a strategy for $f_{IS(G)}$. If we fix a variable x_i to be 1 in a subproblem P where some vertex x_j in $\Gamma(x_i)$ has been fixed to be 1 in P, the subproblem $P|_{x_i:=1}$ can be decided to be $\bot$. Thus we represent a subproblem P by the subset of vertices which are in $VAR[i]$ and have an adjacent vertex x_j which has been fixed to be 1 in P. We call these vertices *forbidden*. It is easy to see that we can construct OBDD represents $f_{IS(G)}$ by using the set of forbidden vertices as representation of a subproblem. Further we can confirm the OBDD to be *quasi-reduced* by considering the difference between the sets of forbidden vertices. Consequently two subproblems are equivalent iff their forbidden vertex sets are identical.

Theorem 4. *The QOBDD of $f_{IS(G)}$ can be constructed by the breadth first algorithm in output-size sensitive manner.*

The QOBDD of Dominating Sets. We show a strategy for $f_{DS(G)}$. $f_{IS(G)}$ can be expressed by a negative 2CNF and the strategy in the previous section can be seen to compute minimal CNFs of subproblems. So we expect that we could get a similar strategy for $f_{DS(G)}$ by computing minimal CNFs.

We consider a subset of vertices $\Pi_i(x_p) := \mathrm{VAR}[i] \cap (\Gamma(x_p) \cup \{x_p\})$ in place of a clause of the CNF. An equivalence relation $\equiv_i$ on V, which is defined such that $x_p \equiv_i x_q$ iff $\Pi_i(x_p) = \Pi_i(x_q)$, is introduced to identify equivalent clauses. In the strategy of Fig. 2, we represent a subproblem of level i with a characteristic vector $P[u]$ on $V/\!\!\equiv_i$. The function CANON, whose description is omitted, makes a CNF minimal.

function RESTRICT$(P[u], x_i, b)$;
   $Q := 0^{|V/\equiv_{i+1}|}$;
   if $b = 0$ **then begin**
     if $\{x_i\} \in V/\!\!\equiv_i$ and $P[u]_{\{x_i\}} = 1$ **then return** $\bot$;
     for_all $S \in V/\!\!\equiv_{i+1}$ **do**
       if $\exists T \in V/\!\!\equiv_i$ s.t. $(\Pi_{i+1}(S) = \Pi_i(T) - \{x_i\}$ and $P[u]_T = 1)$ **then** $Q_S := 1$;
   end;
   else begin {the case of b=1}
     for_all $S \in V/\!\!\equiv_{i+1}$ **do**
       if $\exists T \in V/\!\!\equiv_i$ s.t. $(\Pi_{i+1}(S) = \Pi_i(T)$ and $P[u]_T = 1)$ **then** $Q_S := 1$;
   end;
   return CANON$(Q, i + 1)$;

Fig. 2. The Strategy for Dominating Sets

Theorem 5. *The strategy indicated in Fig. 2 constructs the QOBDD representing $f_{\mathrm{DS}(G)}$ in output-size sensitive manner.*

Here we describe a strategy for $f_{\mathrm{DS}(G)}$ only, but it can be applied to a wider class, that is, the class of positive CNFs.

Theorem 6. *There exists an output-size sensitive algorithm for constructing the QOBDD of a positive CNF.*

Furthermore, we can obtain an OBDD of $f_{\mathrm{MIS}(G)}$ in top-down fashion using the strategies for $f_{\mathrm{IS}(G)}$ and $f_{\mathrm{DS}(G)}$, although it may not be in a purely output-size sensitive manner because this resulting OBDD may not be quasi-reduced due to its equivalent nodes.

3.4 Size of QOBDDs of Degree-Bounded Planar Graphs

It is known that size of QOBDD is seriously influenced by the total order of variables. Therefore method to compute good total order is required. The next theorem shows that a good order can be obtained using structure of input.

Theorem 7. *If G is planar and degree-bounded, there exist QOBDDs of $f_{\mathrm{MIS}(G)}, f_{\mathrm{IS}(G)}$ and $f_{\mathrm{DS}(G)}$ of size of $O(n2^{O(\sqrt{n})})$.*

It is considered that there might be some relationship between the size of QOBDD of PI(f) and that of f [2]. We show results concern this matter. Let M_n denote $n \times n$ mesh and its vertices be numbered in row-major manner (this order is fixed in the following). Let $\{F_n\}$ be a sequence of Fibonacci numbers such that $F_1 = 2, F_2 = 3$ and $F_n = F_{n-1} + F_{n-2}$. We can prove the next theorem which implies that the QOBDD of $f_{\text{MIS}(M_n)}$ can be obtained in output-size sensitive manner. It should be also noted that $|\text{MIS}(M_n)|$ is $2^{\Omega(n^2)}$ and this also shows a respect of efficiency in using OBDDs.

Theorem 8. *The size of the QOBDD of* $f_{\text{IS}(M_n)}$ *is equal to* $F_{n+1} + (n^2 - 2n + 3)F_n + n^2 - 7 \approx 2n^2 \left(\frac{1+\sqrt{5}}{2}\right)^n$. *The size of the QOBDD of* $f_{\text{DS}(M_n)}$ *is* $O(n3^n)$. *The size of the QOBDD of* $f_{\text{MIS}(M_n)}$ *is* $\Omega(nF_n)$ *and* $O(n3^n)$.

4 Conclusion

We have investigated effectiveness of the breadth first construction of QOBDD and discussed relationship between the size of QOBDD of a function and that of its prime implicants through maximal independent set problem.

As future works, we will study the possibility to construct the QOBDD of $f_{\text{MIS}(G)}$ in output-size sensitive manner by another algorithm. From the viewpoint of using conventional algorithm, we should find measures to provide good order of applying Boolean operations to make intermediate OBDDs smaller.

Acknowledgement

The authors would like to thank Associate Professor Imai and members of his laboratory for their helpful discussions and comments to this work.

References

1. R. E. Bryant. Graph-based algorithms for Boolean function manipulation. *IEEE Trans. on Computers*, C-35(8):677–691, 1986.
2. O. Coudert. Doing two-level logic minimization 100 times faster. In *Proc. ACM-SIAM Symposium on Discrete Algorithms*, pages 112–121, 1995.
3. D. S. Johnson, M. Yannakakis, and H. Papadimitriou. On generating all maximal independent sets. *Information Processing Letters*, 27:119–123, 1988.
4. E. L. Lawler, J. K. Lenstra, and A. H. G. Rinnooy Kan. Generating all maximal independent sets: NP-hardness and polynomial-time algorithms. *SIAM J. Comput.*, 9(3):558–565, 1980.
5. S. Tani and H. Imai. A reordering operation for an ordered binary decision diagram and an extended framework for combinatorics of graphs. In *ISAAC'94, Lecture Notes in Computer Science*, volume 834, pages 575–583, 1994.
6. S. Tsukiyama, M. Ide, H. Ariyoshi, and I. Shirakawa. A new algorithm for generating all the maximal independent sets. *SIAM J. Comput.*, 6(3):505–517, 1977.

Small Weight Bases for Hamming Codes *

John Tromp[1] and Louxin Zhang[2] and Ying Zhao[3]

[1] Dept. of Comp. Sci., University of Waterloo, Waterloo, Ont. N2L 3G1, Canada,
tromp@daisy.uwaterloo.ca
[2] Dept. of Comp. Sci., University of Waterloo, Waterloo, Ont. N2L 3G1, Canada,
lzhang@neumann.uwaterloo.ca
[3] Dept. of Math., Shanxi Teacher's University, China.

Abstract. We present constructions of bases for a Hamming code having small *width* and *height*, i.e. number of 1s in each row and column in the corresponding matrix. Apart from being combinatorially interesting in their own right, these bases also lead to improved embeddings of a hypercube of cliques into a same-sized hypercube.

1 Introduction

Let $n = 2^k - 1, k \geq 2$, and let A_k be the k by n matrix over $GF(2)$ whose i-th column, for $1 \leq i \leq n$, is the k-bit binary representation of i. For example,

$$A_3 = \begin{pmatrix} 1\,0\,1\,0\,1\,0\,1 \\ 0\,1\,1\,0\,0\,1\,1 \\ 0\,0\,0\,1\,1\,1\,1 \end{pmatrix}.$$

We denote by C_k the nullspace of A_k, i.e. the set of n-vectors x with $A_k \cdot x = 0^k$. We are interested in finding a basis of the nullspace, C_k, of A_k, that has small *height* and *width*. The height of a set of vectors is defined as the maximum number of ones in any vector, while width is defined as the maximum over all n positions, of the number of vectors in the set having a 1 in that position. A basis of height h and width w is called a (h, w)-basis. The pair (h, w) is called the *weight*.

Low weight bases for the nullspace C_k have applications in coding theory[7], combinatorial designs[2], network embeddings[1, 5], and distributing resources in hypercube computers[9]. In fact, C_k is a one-error-correcting code which was first discovered by Hamming[4] for words of length $2^k - k - 1$. More precisely, Hamming proved that the words of length $2^k - k - 1$ can be encoded as words of length $2^k - 1$ so that each word has Hamming distance at most 1 to exactly one codeword.

Recently, bases for C_k were shown useful for hypercube embeddings. An embedding of a network G into a network H consists of an assignment of nodes of G to nodes of H and a mapping from edges of G onto paths in H graph. Desirable properties of an embedding are small load (maximum number of nodes

---

* This work was supported in part by an NSERC International Fellowship and ITRC.

of G assigned to the same node in H), low dilation (maximum length of path that an edge is mapped to) and low congestion (maximum number of paths using an edge). In [1], Aiello and Leighton discovered that for any $k > 0$, a (h, w)-basis for C_k induces a one-to-one embedding of a hypercube of cliques $H_{2^k-k} \otimes K_k$ in a same-sized hypercube H_{2^k} with dilation h and congestion $2w+2$. Moreover, this embedding is useful in finding efficient embeddings of (dynamic) binary trees in the hypercube and reconfigurations of the hypercube around faults.

Although the existence of a height 3 basis for C_k is well known, the existence question for a $(3, 3)$-basis is open ([5], page 430). Towards this problem, only weak results were obtained in [1, 5, 8, 11]. In this paper, we present two classes of bases with small weight, which improve the existing bounds on weight. In Section 2, we present a $(3, 5)$-basis for C_k that has a very simple structure.

There are many constructions of codes from the incidence matrices of graphs, designs, etc. (for example, see [3, 8]). Using the approach observed in [8], we construct a class of $(3, 4)$-bases in Section 3. As a consequence, we obtain a better one-to-one embedding of a hypercube of cliques into a same-sized hypercube, with dilation 3 and congestion 10.

Finally, we propose a construction of $(3, 3)$-bases. In [1], Aiello and Leighton observed that a primitive trinomial of degree k induces a $(3, 3)$-basis for C_k. But, primitive trinomials do not always exist. This observation is generalized in Section 4. We show that the existence of a trinomial $f(x)$ such that $gcd(f(x), x^{2^k-1} + 1)$ is primitive of degree k implies a $(3, 3)$-basis for C_k. We present results of computations supporting our conjecture that such trinomials always exist.

2 A Simple Construction of a $(3, 5)$-Basis

Note that the rank of A_k equals k. It follows that C_k has rank $n - k$, and that a basis for it consists of $n - k$ linearly independent vectors. We identify a boolean n-vector with its *support*, i.e. the set of positions (as non-zero boolean k-vectors) where it has a 1. For example, the support of (0100101) is $\{010, 101, 111\}$. The product $A_k \cdot \{u, v, w\}$ is easily seen to equal the sum over GF(2) (bitwise exclusive-or) of u, v, and w. E.g. $A_3 \cdot \{010, 101, 111\} = 010 \oplus 101 \oplus 111 = 0^3$. To better visualize the exclusive-or operation, we sometimes write the vectors in the support below each other with the bits aligned:

$$\left\{ \begin{array}{l} 0\,1\,0, \\ 1\,0\,1, \\ 1\,1\,1 \end{array} \right\}.$$

For a bit b, we denote by $\bar{b}$ its complement $b \oplus 1$. For a binary string/vector x, $|x|$ denotes the length of x.

A basis of C_k is constructed as follows. For $x \in \{0, 1\}^i$, and $i + p + 2 \le k$, let $b_{x,p}$ be the vector

$$\left\{ \begin{array}{l} 0^{k-i-p-2}\,1\,x_1\,x_2\,\ldots\,x_{i-1}\,x_i\,1\,0^p, \\ 0^{k-i-p-2}\,0\,1\,x_1\,\ldots\,x_{i-2}\,\overline{x_i}\,1\,0^p, \\ 0^{k-i-p-2}\,1\,\overline{x_1}\,x_{1,2}\ldots x_{i-2,i-1}\,1\,0\,0^p \end{array} \right\},$$

where we write $x_{i,j}$ for $x_i \oplus x_j$. For definiteness, we have for the cases $i = 0, 1$:

$$b_{\epsilon,p} = \left\{ \begin{array}{l} 0^{k-p-2}\,1\,1\,0^p, \\ 0^{k-p-2}\,0\,1\,0^p, \\ 0^{k-p-2}\,1\,0\,0^p \end{array} \right\}, \qquad b_{x_0,p} = \left\{ \begin{array}{l} 0^{k-p-3}\,1\,x_1\,1\,0^p, \\ 0^{k-p-3}\,0\,\overline{x_1}\,1\,0^p, \\ 0^{k-p-3}\,1\,1\,0\,0^p \end{array} \right\}.$$

Note that $A_k \cdot b_{x,p} = 0^k$, so that any $b_{x,p}$ is in C_k.

Our proposed basis simply consists of the set B of all $b_{x,p}$. We must check that these vectors are indeed independent and that we have the right number of them.

To see the latter, partition B into $k - 1$ sets B_p, and each B_p into $k - p - 1$ sets $B_{p,i}$, containing all $b_{x,p}$ with $|x| = i$. Clearly, different pairs (x, p) define different vectors. Thus the size of B is

$$\sum_{p=0}^{k-2} \sum_{i=0}^{i=k-2-p} 2^i = \sum_{p=0}^{k-2} 2^{k-1-p} - 1 = 2^k - 2 - (k - 1) = n - k.$$

Thus, to prove that B is a basis, it remains to show that its elements are linearly independent.

2.1 Independence

Consider any nonempty subset C of B. We prove independence by showing that the sum of all vectors in C is not 0^k.

Let p be minimal such that $C \cap B_p \neq \emptyset$ and for this p, let i be maximal such that $C \cap B_{p,i} \neq \emptyset$, say $b_{x,p} \in C \cap B_{p,i}$. By definition, $b_{x,p}$ has $0^{k-i-p-2}1x10^p$ in its support. For any other $b_{x',p'}$ to have $0^{k-i-p-2}1x10^p$ in its support, would require either $p' = p + 1$ or $|x'| = |x| - 1$, so by minimality of p and maximality of i, such a $b_{x',p'}$ cannot be in C. Since $b_{x,p}$ is thus the only vector in C with $0^{k-i-p-2}1x10^p$ in its support, the sum of all vectors in C also has $0^{k-i-p-2}1x10^p$ in its support and hence is not 0^k.

2.2 Height and Width

The height of B is obviously 3, since each vector $b_{x,p}$ has exactly 3 one bits. We claim that the width of B is at most 5. To see this, consider any position z. If z is of the form $0^{k-q-1}10^q$ then it appears only in the support of $b_{\epsilon,q}$, $b_{\epsilon,q-1}$ (if $q > 0$), and $b_{1,q}$. Hence the width at such positions is no more than 3.

Otherwise, z is of the form $0^{k-j-q-2}1y_1y_2\ldots y_j 10^q$. Consider the $b_{x,p}$ that have this z in their support. We necessarily have one of the following three cases.

1. $z = 1x10^p$. This implies $p = q$ and $x = y$, and so accounts for one $b_{x,p}$.
2. $z = 1x_1\ldots x_{i-2}\overline{x_i}10^p$. This implies $p = q$, $x_{1:i-2} = y_{1:j-1}$ and $x_i = \overline{y_j}$, and so accounts for two $b_{x,p}$ (x_{i-1} can be 0 or 1).
3. $z = 1\overline{x_1}x_{1,2}\ldots x_{i-2,i-1}100^p$. This implies $p = q - 1$ and $x_1 = \overline{y_1}, x_2 = y_2 \oplus x_1 = \overline{y_1} \oplus y_2, x_3 = y_3 \oplus x_2 = \overline{y_1} \oplus y_2 \oplus y_3, \ldots, x_{i-1} = \overline{y_1} \oplus y_2 \oplus \cdots \oplus y_j$, and so accounts for two $b_{x,p}$ (x_i can be 0 or 1).

In total we find that at most five $b_{x,p}$ can have a one in position z, as claimed.

3 A (3,4) Basis

While the $(3,5)$ basis may be preferred in some applications for its simplicity, we can get a better $(3,4)$ basis by combining results from finite fields with an inductive construction based on finding Hamiltonian paths in complete bipartite graphs.

We start with the empty base B_1 for the null space $C_1 = \{0\}$ of A_1, which is the 1 by 1 matrix (1). To extend B_1 to a basis for the null space $C_2 = \{000, 111\}$ of

$$A_2 = \begin{pmatrix} 1 & 0 & 1 \\ 0 & 1 & 1 \end{pmatrix},$$

Next we explain how to extend B_k to a basis B_{k+1} for the null space C_{k+1}. A subset B_k' of $2^k - 1 - k$ vectors in B_{k+1} will be derived from the $2^k - 1 - k$ vectors in B_k. Namely, for each vector $\{u, v, w\}$ in B_k, where $u, v, w \in \{0,1\}^k$, we put $\{0u, 0v, 0w\}$ into B_k'.

We form B_{k+1} as the union of B_k' and a set B of $2^k - 1$ more vectors, to get the required number of $2^k - 1 - k + 2^k - 1 = 2^{k+1} - 1 - (k+1)$ vectors. These vectors will have a support consisting of one position in $X = 01\{0,1\}^{k-1}$ and one in each $Y_i = 1i\{0,1\}^{k-1}, i = 0, 1$. Note that, for such a vector $\{01x, 10y_0, 11y_1\}$ to be in the nullspace, it must satisfy $x = y_0 \oplus y_1$, so that it is determined by just the pair $(10y_0, 11y_1) \in Y_0 \times Y_1$. Our problem can thus be seen as the selection of $2^k - 1$ edges in the complete bipartite graph on $Y_0 \cup Y_1$. We'll consider X to be a set of colors and say that an edge between $10y_0$ and $11y_1$ has *color* $10(y_0 \oplus y_1) \in X$. Getting a low width basis corresponds to minimizing the maximum degree of any vertex and simultaneously minimizing the maximum number of edges of any color. Our construction is based on finding a Hamiltonian path in the graph (see [8]). Such a path contains exactly the required number $|Y_0 \cup Y_1| - 1 = 2^k - 1$ of edges $\{10y_0, 11y_1\}$, each corresponding to a basis vector $\{y_0 \oplus y_1, y_0, y_1\}$.

Suppose we have found a set B of $2^k - 1$ vectors corresponding to the edges in a Hamiltonian path. Since a path is acyclic, any non-empty subset of vectors in B induces a subgraph with at least one vertex of degree 1. Such a vertex is a position which is in the support of the subset vector sum, and furthermore, will remain so under the addition of any vectors in B_k', which have no support in $Y_0 \cup Y_1$. This proves that if B_k is basis of C_k, then B_{k+1} is a basis of C_{k+1}, as desired.

For $k = 1, 2, 3$, we use the following Hamiltonian paths:

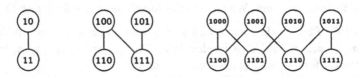

The following table gives the number of ones in each position of B_4, as the total number of basis vectors in which it appears as either a color $y_0 \oplus y_1$ or as a vertex y_i (its degree in the Hamiltonian path).

position	0001	0010	0011	0100	0101	0110	0111	1000	1001	1010	1011	1100	1101	1110	1111
color	1	2	1	2	3	0	2	0	0	0	0	0	0	0	0
degree	0	1	1	2	1	1	2	2	2	1	2	2	2	2	1
total	1	3	2	4	4	1	4	2	2	1	2	2	2	2	1

Since the maximum degree in a Hamiltonian path is 2, the width in positions $1\{0,1\}^{k-1}$ of any B_k will be at most 2. For $k \leq 4$, the table shows that the width in positions $0\{0,1\}^{k-1} \setminus \{0^k\}$ of B_k is at most 4. In order to continue our induction beyond $k = 4$, it suffices to find a Hamiltonian path in which each color $x \in X$ appears at most twice. Equivalently, we need to find a Hamiltonian *cycle* in which each $x \in X$ colors *exactly* two edges. The reason we make the first 3 induction steps explicit is that such a Hamiltonian cycle doesn't exist in the complete bipartite graph on $\{1000, 1001, 1010, 1011\} \cup \{1100, 1101, 1110, 1111\}$. Instead we compensated for the triple use of the color 0101 in the third path by limiting the degree of node 5 to 1 in the second path.

3.1 Hamiltonian Cycles

We turn to algebra to find the paths with the required color restrictions.

Let $GF(2)[x]$ denote the class of binary polynomials, that is, with coefficients 0 or 1, and addition and multiplication mod 2. We borrow a result from finite field theory which says that for any k, there exists a *primitive* binary polynomial $f(x)$ of degree k. This means that $GF(2)[x]/(f)$, the class of residues modulo f, is a finite field whose multiplicative group is generated by x. In other words, the set $\{x^0, x^1, \ldots x^{2^k-2}\}$ contains all $n = 2^k - 1$ non-zero elements.

We can bring $GF(2)[x]/(f)$ in one-one correspondence to each of the three position sets X, Y_0 and Y_1 in the inductive step from $k + 1$ to $k + 2$, where they each have size 2^k. A position $p = p_1 \ldots p_{k+2}$ will correspond to the the binary polynomial $\sum_{i=0}^{k-1} p_{k+2-i} x^i$, i.e. we ignore the two first bits that distinguish between $X, Y_0,$ and Y_1. For example, with $k = 4$, $101101 \in Y_0$ corresponds to $x^3 + x^2 + 1$.

Let $(x+1)^{-1}$, the inverse of $x+1$, be equal to x^r for some $r, 0 \leq r < 2^k - 1$. Note that $x^i \mapsto \sum_{j<i} x^j = (x^i + 1)(x+1)^{-1} = x^r(x^i + 1)$ is a bijection from all the non-zero elements of $GF(2)[x]/(f)$ to all elements except $x^r(0+1) = x^r$. Also, $\sum_{j<n} x^j = x^r(x^n + 1) = x^r(1+1) = 0$.

These facts are the basis of the following cycle decomposition (using $\overset{i}{\sum}$ as a shorthand for $\sum_{j=0}^{i} x^j$):

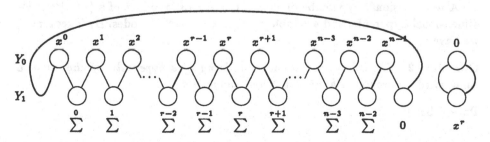

The left cycle uses every color $\sum\limits^{i}$, $0 \leq i < n$ exactly twice, once on the edge between $\sum\limits^{i-1}$ and x^i, and once on the edge between x^{i+1} and $\sum\limits^{i+1}$. The right cycle uses the single color not expressible as $\sum\limits^{i}$, namely x^r, exactly twice. A series of 5 edge swaps transform the two cycles into the following Hamiltonian cycle:

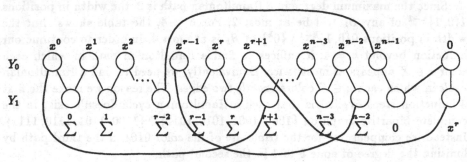

We'll refer to the 2-cycle decomposition as 2-cycle and to the Hamiltonian cycle as 1-cycle. The edge between $x^0 = 1$ and 0 in the 2-cycle has color 1, as does the edge between x^{r+1} and x^r in the 1-cycle, since $x^{r+1} + x^r = x^r(x+1) = 1$. The edge between $\sum\limits^{r-2}$ and x^{r-1} in the 2-cycle has color $\sum\limits^{r-1}$, as does the edge between $x^0 = 1$ and $\sum\limits^{r+1}$ in the 1-cycle, since $\sum\limits^{r+1} + 1 = \sum\limits^{r-1} + x^{r+1} + x^r + 1 = \sum\limits^{r-1}$. The edge between x^{r+1} and $\sum\limits^{r+1}$ in the 2-cycle has color $\sum\limits^{r}$, as does the edge between $\sum\limits^{r-2}$ and x^{n-1} in the 1-cycle, since $\sum\limits^{r-2} + x^{n-1} = \sum\limits^{r} + x^r + x^{r-1} + x^{-1} = \sum\limits^{r} + x^{-1}(x^{r+1} + x^r + 1) = \sum\limits^{r}$. The edge between $\sum\limits^{n-2}$ and x^{n-1} in the 2-cycle has color $\sum\limits^{n-1} = 0$, as does the edge between 0 and 0 in the 1-cycle. The edge between 0 and x^r in the 2-cycle has color x^r, as does the edge between x^{r-1} and $\sum\limits^{n-2}$ in the 1-cycle, since $x^{r-1} + \sum\limits^{n-2} = x^{r-1} + x^{-1} = x^r + x^{-1}(x^{r+1} + x^r + 1) = x^r$.

It remains to show that this transformation does not suffer from r being too close to 0 or $n-1$. Indeed, $x^{r+1} + x^r + 1 = 0$ implies that $r+1 \geq k \geq 3$, hence $r - 1 > 1$ and we're safe on the left. Similarly, $x^{n-r} + x + 1 = (x^r)^{-1} + x + 1 = 0$ implies that $n - r \geq k \geq 3$, hence $r + 1 \leq n - 2$, so we're safe on the right too.

Altogether, this shows

Theorem 1. *For any k, C_k has a $(3,4)$-basis.*

A n-dimensional hypercube of cliques is the cross product of a $(n - \lfloor \log r \rfloor)$-dimensional hypercube and a complete graph with $2^{\lfloor \log r \rfloor}$ nodes. By Theorem 1, we have

Corollary 2. *There is a one-to-one embedding of a hypercube of cliques in a same-sized hypercube with dilation 3 and congestion 10.*

Proof. See [5]. □

4 On (3, 3) Bases

In this section we give a sufficient condition for the existence of a $(3,3)$-basis for C_k. Suppose some degree k primitive polynomial $h(x)$ is the g.c.d. of a trinomial $f(x) = 1 + x^i + x^j$ and $x^n + 1$. Then C_k has a $(3,3)$-basis, constructed as follows. Consider the $n \times n$ circulant matrix F generated by f; the i'th column F_i of this matrix $(i = 0, \ldots, n-1)$, is formed by the coefficients of $x^i f(x) \bmod x^n + 1$. For example, with $n = 7$, $h(x) = f(x) = 1 + x + x^3$ generates the matrix

$$F = \begin{pmatrix} 1 & 0 & 0 & 0 & 1 & 0 & 1 \\ 1 & 1 & 0 & 0 & 0 & 1 & 0 \\ 0 & 1 & 1 & 0 & 0 & 0 & 1 \\ 1 & 0 & 1 & 1 & 0 & 0 & 0 \\ 0 & 1 & 0 & 1 & 1 & 0 & 0 \\ 0 & 0 & 1 & 0 & 1 & 1 & 0 \\ 0 & 0 & 0 & 1 & 0 & 1 & 1 \end{pmatrix}$$

We use the fact that $h(x)$ is primitive to define a column re-ordering of A_k, called A'_k, whose i'th column corresponds to $x^i \bmod h(x)$. Now $A'_k F_i$ corresponds to $x^i f(x) \bmod x^n + 1 \bmod h(x) = x^i f(x) \bmod h(x) = 0$, since $h(x)$ divides both $f(x)$ and $x^n + 1$. Thus, all columns of F are in the nullspace C'_k of A'_k.

From a theorem of König and Rados [6], it follows that the rank of F is $n - \deg(\gcd(f(x), x^n + 1)) = n - \deg(h(x)) = n - k$. Now if some column i is linearly dependent on columns $0, \ldots, i-1$, then, since F is circulant, column $i + 1$ is linearly dependent on columns $1, \ldots, i$ and therefore also on columns $0, \ldots, i-1$. Similarly, columns $i + 2, \ldots, n-1$ would be linearly dependent on the first i columns. Thus, the first $n - k$ columns of F must actually be linearly independent, else the rank of F would be less than $n - k$. This shows that $F_0, \ldots, F_{n-k-1}$ forms a basis of C'_k, and, by an appropriate permutation of dimensions, a basis of C_k.

The existence of degree k primitive polynomials $h(x)$ that are the g.c.d. of a trinomial $f(x) = 1 + x^i + x^j$ and $x^n + 1$, is demonstrated in Table 1 for $k \le 171$. Only those k for which there is no primitive trinomial of degree k are listed; see Stahnke [10] for a table of primitive binary polynomials up to degree 171. Therefore, we pose the following

Conjecture 1 *There always exists a trinomial $f(x)$ such that $\gcd(f(x), x^{2^k - 1} + 1)$ is a primitive polynomial of degree k on the finite field $GF(2^k)$ for any k.*

A subsequent effort by [12] shows the conjecture to hold through all $k \le 500$.

5 Acknowledgements

The authors would like to thank Dan Pritikin for careful reading of the manuscript and helpful suggestions.

k	i	j	k	i	j	k	i	j
8	13	1	67	76	29	120	174	31
12	19	6	69	75	23	122	128	15
13	16	3	70	82	15	125	128	3
14	17	2	72	93	7	126	141	70
16	29	6	74	80	39	128	131	50
19	22	3	75	77	4	131	138	61
24	55	6	76	88	43	133	136	43
26	29	12	77	80	9	136	139	30
27	29	1	78	89	2	138	183	23
30	41	12	80	83	23	139	142	3
32	59	29	82	85	19	141	148	71
34	37	6	83	85	14	143	147	1
37	43	4	85	93	28	144	159	14
38	42	1	86	91	22	146	149	6
40	43	3	88	154	37	147	149	19
42	51	7	90	111	28	149	151	2
43	53	2	91	99	13	152	155	38
44	52	15	92	103	39	154	157	22
45	59	12	96	123	1	155	158	75
46	58	9	99	101	13	156	188	59
48	70	27	101	103	2	157	164	25
50	54	7	102	115	3	158	167	54
51	53	4	104	109	9	160	177	19
53	61	28	107	109	8	162	166	27
54	93	23	109	118	21	163	171	70
56	67	31	110	117	19	164	189	68
59	61	26	112	133	1	165	173	42
61	66	17	114	118	7	166	186	53
62	77	30	115	125	6	168	179	38
64	74	21	116	136	1	171	173	10
66	83	20	117	123	31			

Table 1. Trinomials $f(x) = x^i + x^j + 1$ that imply the existence of $(3,3)$-bases.

References

1. W. Aiello and T. Leighton, Coding theory, hypercube embeddings, and fault tolerance, *Proc. of the 3rd Annual ACM Symposium on Parallel Algorithms and Architectures*, pp. 125–136, 1991. To appear in IEEE Trans. Computer.
2. E. Assmus, Jr and H. Mattson, On tactical configurations and error-correcting-codes, *J. of Combin. Theory*, vol. 2, pp. 243–257, 1967.
3. L. Babai, H. Oral and K. Phelps, Eulerian self-dual codes, *SIAM J. Discrete Math.*, vol. 7, pp. 325–330, 1994.

4. R. H. Hamming, Error detecting and error correcting codes, *Bell System Technical Journal*, vol. 29, pp. 147–160, 1950.
5. T. Leighton, *Introduction to Parallel Algorithms and Architectures: Arrays · Trees · Hypercubes*, Morgan Kaufmann, San Mateo, California, 1992, Chapter 3, p. 430.
6. R. Lidl and H. Niederreiter, *Finite Fields*, Addison-Wesley, California, 1983.
7. V. Pless, *Introduction to the Theory of Error-Correcting-Codes*, John Wiley and Sons, Inc., New York, 1989.
8. D. Pritikin, Graph embeddings from Hamming bases, *DIMACS Workshops on Interconnection Networks and Mapping and Scheduling Parallel Computations*, February 1994.
9. A. Reddy, Parallel input/output architectures for multiprocessors, *Ph. D. dissertation*. Dept. of Electronical and Computer Engineering, Univ. of Illinois, Urbana, 1990.
10. W. Stahnke, Primitive binary polynomials, *Mathematics of Computation*. vol. 124, pp. 977–980, 1973.
11. L. Zhang, A new bound for the width of Hamming codes, to appear in *Proceedings of the 6th Inter. Conference on Computing and Information*, 1994.
12. I. F. Blake, S.Gao and R. J. Lambert, Construction and distribution problem for irreducible trinomials over finite fields, To appear in *Proceedings of the Holloway Conference on Finite Fields*, Oxford University Press, 1995.

Toeplitz Words, Generalized Periodicity and Periodically Iterated Morphisms

(Extended Abstract)

Julien Cassaigne[1] and Juhani Karhumäki[2]

[1] LITP, Université Pierre et Marie Curie, F-75252 Paris Cedex 05, France
(Julien.Cassaigne@litp.ibp.fr)
[2] Department of Mathematics, University of Turku, SF-20500 Turku, Finland
(karhumak@cs.utu.fi)

Abstract. We consider so-called Toeplitz words which can be viewed as generalizations of one-way infinite periodic words. We compute their subword complexity, and show that they can always be generated by iterating periodically a finite number of morphisms. Moreover, we define a structural classification of Toeplitz words which is reflected in the way how they can be generated by iterated morphisms.

1 Introduction

Toeplitz introduced in [14] an iterative construction to define almost periodic functions on the real line. In [8] Jacobs and Keane modified this construction to define infinite words. Their motivation, however, was on topological aspects of words, in particular, on ergodic theory.

Starting from [13] these words, now referred to as *Toeplitz words*, have been considered from the combinatorial point of view. Surprising connections of these words to different combinatorial problems are discussed in [2]. The combinatorial research of Toeplitz words seems to have concentrated on a special case of our later classification, corresponding to so-called *paperfolding* words, cf. [2, 4, 12].

The Toeplitz words we are considering here are defined as follows: take an infinite periodic word w^ω on the alphabet $\Sigma \cup \{?\}$, where ? corresponds to a hole. Fill the holes iteratively by substituting the word itself into the remaining holes. As the limit all holes are filled and an infinite word is defined.

The goals of this paper are as follows. We intend to demonstrate that the above Toeplitz words are natural, and not too large, generalizations of periodic words. As a part of that we classify different types of Toeplitz words in terms of their subword complexity. We also point out connections between Toeplitz words and different types of iteration of morphisms, cf. [3, 10].

More precisely, let us call the above Toeplitz word (p, q)-Toeplitz word, if the length of the pattern, that is $|w|$, is p, and if it contains q holes. We show that

(i) if $q = 1$, then the word can be generated by iterating a morphism;
(ii) if q divides p, then it can be generated by iterating a morphism, and then mapping the result by another morphism (or, in fact, by a coding); and

(iii) otherwise, i.e. if q does not divide p, it can be generated by iterating periodically q morphisms.

In each of the above cases the word can be periodic, as we shall see. However, we are able to characterize when this takes place, and moreover, we can show that if this is not the case, then the subword complexity of a word in case (i) or (ii) is $\Theta(n)$ and in case (iii) is $\Theta(n^r)$ where $r = \frac{\log(p/d)}{\log(p/q)}$ with d equal to the greatest common divisor of p and q. Consequently the growth of the subword complexity does not depend on the pattern itself, that is on w, but on p and q only, assuming that the infinite word is not periodic. This is an evidence that Toeplitz words reflect the periodicity properties of words.

We want to finish this introduction with the following comment. As we said Toeplitz words were introduced when studying properties of real functions. Their natural environment, however, is that of words, and we believe they provide most elegant results on this area. This is exactly the same as what happened to the well known periodicity lemma of Fine and Wilf, cf. [6]: it was introduced for real functions, but became really fundamental in connection with words, cf. [11]!

2 Basic Definitions

For a finite alphabet Σ, let Σ^*, Σ^+ and Σ^ω denote the sets of all finite, finite non-empty, and one-way infinite words over Σ, respectively. We call an infinite word $\mathbf{w} \in \Sigma^\omega$ *periodic* or *ultimately periodic*, if it can be written in the form $\mathbf{w} = v^\omega$ or $\mathbf{w} = uv^\omega$ for some finite words u and v.

Let ? be a letter not in Σ. For a word $w \in \Sigma(\Sigma \cup \{?\})^*$ let

$$
\begin{aligned}
T_0(w) &= ?^\omega \\
T_{i+1}(w) &= F_w(T_i(w))
\end{aligned}
\tag{1}
$$

where $F_w(\mathbf{u})$, defined for any $\mathbf{u} \in (\Sigma \cup \{?\})^\omega$, is the word obtained from w^ω by replacing the sequence of all occurrences of ? by $\mathbf{u}$; in particular $F_w(\mathbf{u}) = w^\omega$ if w contains no ?.

Clearly,

$$
T(w) = \lim_{i \to \infty} T_i(w) \in \Sigma^\omega
\tag{2}
$$

is well-defined, and it is referred to as the *Toeplitz word determined by the pattern* w. Let $p = |w|$ and $q = |w|_?$ be the length of w and the number of ?'s in w, respectively. Then we call $T(w)$ a (p, q)-*Toeplitz word*.

The above definition emphasizes the *iterative nature* of Toeplitz words. There exists another equivalent definition, which points out the *selfreading nature* of these words: let $w \in \Sigma(\Sigma \cup \{?\})^*$ as above; we define a sequence of infinite words $I_i(w)$ as follows: $I_0(w) = w^\omega$ and $I_i(w)$ is the word obtained from $I_{i-1}(w)$ by replacing the first occurrence of ? in $I_{i-1}(w)$ by the i-th letter of $I_{i-1}(w)$.

Then, clearly,

$$
I(w) = \lim_{i \to \infty} I_i(w) \in \Sigma^\omega \;,
\tag{3}
$$

and it is a straightforward consequence of the definitions that

$$I(w) = T(w) \ .$$

Note that in the process of defining $I(w)$ holes are filled one by one by reading the prefix of $I(w)$. Consequently, $I(w)$ is *selfreading* in the same sense as the Kolakoski word, cf. [9, 3].

We proceed with a few examples.

Example 1. The periodic word $(112)^\omega$ is obtained as a Toeplitz word in a number of different ways :

$$(112)^\omega = T(112??????) = T(112?12?1?) = T(1?21?211?) \ .$$

Example 2. As is easy to see the only ultimately periodic Toeplitz words are the periodic ones. This, however, is true only since we required that w start with a symbol in Σ. Of course, if this is not required, and the definition (2) is used, then patterns of the form

$$w = ?^n u?^{|u|}$$

determine all ultimately periodic words, where the "initial mess" is arbitrary.

Example 3. The famous paperfolding sequence is the Toeplitz word determined by the pattern 1?0?, cf. [2].

Example 4. The pattern 12? determines a Toeplitz word which is the fixed point of the morphism $h: 1 \mapsto 121, 2 \mapsto 122$, i.e. $T(12?) = \lim\limits_{i\to\infty} h^i(1)$. From this it follows easily that this Toeplitz word is cube-free, although it contains arbitrarily long repetitions of the form $uauau$ with $a \in \{1, 2\}$.

We conclude this section by pointing out that Toeplitz words indeed possess certain kinds of periodicity properties. Consider a (p, q)-Toeplitz word $T(w)$. Hence the length of w is p and it contains q holes. Further let us call the word $T_i(w)$ in (1) the *i-th iterate* of w. Of course each $T_i(w)$ is a word over $\Sigma \cup \{?\}$, and moreover it is periodic in this alphabet. We can easily compute a period of $T_i(w)$ as well as the number of holes in it: if $\mathbf{u}$ has a period of length a containing b holes, then $F_w(\mathbf{u})$ has a period of length pa containing qb holes. Repeating inductively we conclude, by the iterative definition of the Toeplitz word, that $T_i(w)$ has a period of length p^i having exactly q^i holes.

It follows that with the exception of q^i positions, $T(w)$ is periodic with a period of length p^i. Moreover, the ratio between the numbers of non-fixed and fixed positions in this period tends to zero very rapidly.

Consequently, Toeplitz words are really, in a sense, natural extensions of periodic words. Moreover, note that the above considerations do not depend on the word w itself, nor on its alphabet Σ, but only on the numbers p and q. Note also that, as is easy to see, if p and q are relative primes then the period we computed for $T_i(w)$ is its smallest period.

It follows from above that Toeplitz words are *uniformly recurrent* in the sense of ergodic theory, cf. [7], i.e. for each number k there exists a constant $n(k)$ such that whenever u with length k occurs as a factor in $T(w)$, then it occurs in any factor of $T(w)$ of length $n(k)$.

3 Toeplitz Words and Iteration of Morphisms

In this section we show that all Toeplitz words can be obtained by iterating morphisms in a suitable way. Moreover, this allows to classify Toeplitz words.

Let w be a (p, q)-Toeplitz word. We define three different cases as follows:

(i) $q = 1$, i.e. Toeplitz words of type $(p, 1)$;
(ii) q divides p, i.e. Toeplitz words of type (tq, q);
(iii) q does not divide p.

Next we define three different ways of iterating morphisms corresponding to the above classification. We recall that a *DOL word* $\alpha \in \Sigma^\omega$ is a word which is a fixed point of a morphism, i.e.

$$\alpha = \lim_{i \to \infty} h^i(a)$$

for some morphism $h: \Sigma^* \to \Sigma^*$ such that $h(a) \in a\Sigma^+$, with $a \in \Sigma$. A *CDOL word* $\beta \in \Sigma^\omega$ is a morphic image of a DOL word under a letter-to-letter morphism (or a coding). Finally, a *p-DOL word* $\gamma \in \Sigma^\omega$ is a fixed point of the periodic iteration of p morphisms $h_1, \ldots, h_p: \Sigma^* \to \Sigma^*$, i.e.

$$\gamma = \lim_{i \to \infty} h^i(a)$$

where $h_1(a) \in a\Sigma^+$ and the mapping $h: \Sigma^* \to \Sigma^*$ is defined as follows: if

$$u = a_1 \ldots a_p a_{p+1} \ldots a_{2p} \ldots a_{(t-1)p+1} \ldots a_{tp} a_{tp+1} \ldots a_{tp+j}$$

with $t \geq 0$ and $j < p$, then

$$h(u) = \left(\prod_{k=0}^{t-1} h_1(a_{kp+1}) \ldots h_p(a_{kp+p}) \right) h_1(a_{tp+1}) \ldots h_j(a_{tp+j}) . \qquad (4)$$

Clearly, the mapping h above is a simple case of a deterministic GSM.

It is well known that there are CDOL words which are neither DOL words nor p-DOL words. On the other hand, to find a p-DOL word which is not a CDOL is not that easy. The existence of such words was proved in [10], and we shall obtain a new proof for that as a corollary of Theorem 7.

Now we are ready for our results of this section.

Theorem 1. *Every Toeplitz word of type $(p, 1)$ is a DOL word.*

Proof. Let the pattern be $w = a_1 \ldots a_{i-1}?a_{i+1} \ldots a_p$, for $2 \leq i \leq p$. Now, define the morphism $h: \Sigma^* \to \Sigma^*$ by the condition

$$h(a) = a_1 \ldots a_{i-1} a a_{i+1} \ldots a_p .$$

It follows from the selfreading definition of Toeplitz words that

$$I(w) = \lim_{i \to \infty} h^i(a_1) .$$

Note that as $|h(a)| = p$ for all $a \in \Sigma$, h is a uniform morphism. $\qquad \square$

Theorem 2. *Every Toeplitz word of type* (tq, q) *is a CDOL word.*

Proof. Let the pattern be $w = w_1 \ldots w_t$ with $|w_1| = |w_2| = \cdots = |w_t| = q$. Further for each $i = 1, \ldots, t$, let $\bar{w}_i$ be the word obtained from w_i by filling its holes in the generation of $T(w)$. Finally let

$$T(w) = \bar{w}_1 \bar{w}_2 \bar{w}_3 \ldots$$

with $|\bar{\bar{w}}_i| = tq$ for $i \geq 1$. Then it follows that, for $i = 1, \ldots, t$, $\bar{\bar{w}}_i$ is obtained from w by filling its holes with the letters of $\bar{w}_i$. Or more generally, consecutive blocks of $T(w)$ of length q determine consecutive blocks of $T(w)$ of length tq.

Therefore we are guided to define a morphism $h \colon (\Sigma^q)^* \to (\Sigma^q)^*$ by the condition: $h(a_1 \ldots a_q)$ is the word obtained from w by filling its holes on letters $a_1, \ldots, a_q$ in this order, viewed as a word of length t on the alphabet Σ^q.

It follows that

$$T(w) = f \left(\lim_{i \to \infty} h^i(\bar{w}_1) \right)$$

where $f \colon (\Sigma^q)^* \to \Sigma^*$ is the morphism satisfying

$$f(a_1 \ldots a_q) = a_1 \ldots a_q \ .$$

Consequently, we have proved that $T(w)$ can be generated by iterating a *uniformly growing* morphism, and by mapping the result by another morphism. But by a general result, this other morphism can be replaced by a letter-to-letter one, cf. also Example 3 below. □

Example 3 (continued). The construction of Theorem 2 yields the morphism

$$h \colon \begin{array}{l} 11 \longmapsto 11.01 \\ 01 \longmapsto 10.01 \\ 10 \longmapsto 11.00 \\ 00 \longmapsto 10.00 \end{array}$$

and the morphism $f \colon \{00, 01, 10, 11\}^* \to \{0, 1\}^*$ mapping each block of length two into itself.

Now a CDOL presentation for the word $T(w)$ is obtained by defining the two morphisms $\bar{h} \colon \{0, 1, \bar{0}, \bar{1}\}^* \to \{0, 1, \bar{0}, \bar{1}\}^*$ and $\bar{f} \colon \{0, 1, \bar{0}, \bar{1}\}^* \to \{0, 1\}^*$ such that

$$\bar{h} \colon \begin{array}{l} 1 \longmapsto 1\bar{1} \\ \bar{1} \longmapsto 0\bar{1} \\ 0 \longmapsto 1\bar{0} \\ \bar{0} \longmapsto 0\bar{0} \end{array}$$

and $\bar{f}$ just erases the bars. Then clearly

$$f \left(\lim_{i \to \infty} h^i(11) \right) = \bar{f} \left(\lim_{i \to \infty} \bar{h}^i(1) \right) \ .$$

Example 5. Toeplitz words of Theorem 2 need not be DOL words. Such an example is the word $\mathbf{w} = T(1??2) = 1112112211121222\ldots$.

Assume on the contrary that $\mathbf{w}$ is a fixed point of a morphism h. Clearly, $h(1)$ is a prefix of $\mathbf{w}$, and it is different from 1, 11 and 111. Therefore $h(1) = 111u$ for some non-empty word u; then $111u111u$ is a prefix of $\mathbf{w}$, and the form of $\mathbf{w}$ implies that the length of $111u$ is divisible by 4. Consequently, by the selfreading definition of $\mathbf{w}$ both the first and second half of $111u$ define the same word, namely $111u$. Hence we can write $111u = 111u'111u'$. Now, we can proceed by induction and conclude that $\mathbf{w}$ contains a prefix 11121112, a contradiction.

Actually, we believe that typically Toeplitz words of Theorem 2 are not DOL words.

Our last result shows that all Toeplitz words we have been considering are p-DOL words for some p. This extension is, however, necessary since as an application of the subword complexity results of Sect. 5 we see that there are Toeplitz words which are not CDOL words.

Theorem 3. *Every (p,q)-Toeplitz word is a q-DOL word.*

Proof. Consider a (p,q)-Toeplitz word $T(w)$ defined by the pattern

$$w = w_0?w_1?\ldots?w_q$$

with $w_0 \in \Sigma^+$ and $w_i \in \Sigma^*$ for $i = 1, \ldots, q$. Define the morphisms $h_i : \Sigma^* \to \Sigma^*$, for $i = 1, \ldots, q$, as follows. For $i < q$ set

$$h_i(a) = w_{i-1}a \text{ for } a \in \Sigma$$

and for $i = q$ set

$$h_q(a) = w_{q-1}aw_q \text{ for } a \in \Sigma .$$

It follows directly from the selfreading definition of Toeplitz words that

$$T(w) = \lim_{i \to \infty} h^i(a_1)$$

where h is defined as in (4) and a_1 denotes the first letter of w_0. □

4 Periodic Toeplitz Words

As we said, in each case of our classification of Toeplitz words we can obtain periodic words, cf. Example 1. In the case of one hole this is possible only if the alphabet Σ is unary. The goal of this section is to characterize when a Toeplitz word is periodic. Theorem 4 states that it is sufficient to examine a short prefix of a Toeplitz word to check if it is periodic:

Theorem 4. *Let $T(w)$ be a (p,q)-Toeplitz word, and $d = \gcd(p,q)$. Then $T(w)$ is periodic if and only if its prefix of length p is d-periodic. In this case, the minimal period of $T(w)$ is a divisor of d.*

Note that this means that the size of the alphabet (considering only letters actually appearing in the pattern w) cannot be more than d if $T(w)$ is periodic. In particular, when p and q are coprime, $d = 1$ and $T(w)$ is not periodic as soon as the pattern contains two different letters.

One implication is easy: if the prefix of length p is $v^{p'}$ with $|v| = d$ and $p' = p/d$, then by induction on n, $T(w)$ has same prefix of length n as v^ω for every n, and thus $T(w) = v^\omega$.

In this extended abstract, we shall only sketch the proof of the other implication. We first need some notations:

Let us denote by u_n the n-th letter of the Toeplitz word $T(w)$, and by p' and q' the quotients $p' = \frac{p}{d}$ and $q' = \frac{q}{d}$. Let $\tilde{w}$ be the word in $(\Sigma \cup \{h_1, \ldots, h_q\})^*$ obtained from w by replacing the j-th hole by the symbol h_j. Let w_i be the i-th letter in $\tilde{w}$.

We define a directed graph G with $d + \#\Sigma$ vertices labelled by the numbers $1, \ldots, d$ and the letters of Σ, and at most p edges defined as follows: there is an edge from the number i to the letter x whenever there is an integer k such that $w_{kd+i} = x$; and there is an edge from the number i to the number j whenever there are integers k and l such that $w_{kd+i} = h_{ld+j}$. Let Σ_i, for $1 \le i \le d$, be the set of letters x in Σ such that there is a directed path from i to x in G. Note that Σ_i is never empty.

Lemma 5. *The subsequence $(u_{nd+i})_{n \ge 0}$ of $T(w)$ has exactly alphabet Σ_i, for $1 \le i \le d$. Moreover, for every letter $x \in \Sigma_i$, there exists an integer k such that*

$$\forall n, u_{ndp'^l + kd + i} = x$$

where $l \le d$ is the length of the shortest path from i to x in G.

The proof of Lemma 5 is by induction on l.

In particular, Lemma 5 provides a useful criterium for testing whether $T(w)$ is d-periodic or not:

Lemma 6. *The Toeplitz word $T(w)$ is d-periodic if and only if Σ_i has exactly one element for all i.*

To finish the proof of Theorem 4, we first prove that if $T(w)$ is periodic then it has a period $t'd$ with $\gcd(t', p') = 1$. Because of Lemma 5, this implies that Σ_i only has one element for all i, which in turn by Lemma 6 implies that $T(w)$ is d-periodic.

5 Complexity of Toeplitz Words

In this section we compute the subword complexity of Toeplitz words. We recall that the *subword complexity*, or briefly *complexity*, of an infinite word $\mathbf{u}$ is the function $p_\mathbf{u} : \mathbb{N} \to \mathbb{N}$ such that

$$p_\mathbf{u}(n) = \text{the number of factors of } \mathbf{u} \text{ of length } n.$$

Recently quite a lot of research has been done on this field, cf. [1].

It follows from our Theorems 1 and 2, by a result in [5], that for our first two cases of Toeplitz words, i.e. when the number of holes divides the length of the pattern, the subword complexity is linear. In this section we show that arbitrary Toeplitz words have polynomial complexity, or more precisely:

Theorem 7. *Let $T(w)$ be a non-periodic (p,q)-Toeplitz word, and define $d = \gcd(p,q)$, $p' = \frac{p}{d}$, $q' = \frac{q}{d}$. Then the complexity $p(n)$ of $T(w)$ satisfies $p(n) = \Theta(n^r)$ with $r = \frac{\log p'}{\log p' - \log q'}$.*

First, we canonicalize w in two ways that decrease d but leave p', q', and the growth order of $T(w)$ unchanged.

1. We can suppose that w is a primitive word: if $w = v^k$, then $T(w) = T(v)$.
2. We can suppose that $\forall i, \#\Sigma_i \geq 2$. If this is not the case, it can be proven that there exist a pattern v of length $p'd'$ with $q'd'$ holes, where $d' < d$, and another pattern z, such that $T(w) = F_z(T(v))$. The complexities of $T(w)$ and $T(v)$ are linked, and in particular $p_{T(w)}(n) = \Theta(n^r)$ if and only if $p_{T(v)}(n) = \Theta(n^r)$.

We now suppose that w satisfies these two restrictions, and establish the following lemma:

Lemma 8. *Define $n_0 = dp'^{d+1}$, and let v be any factor of $T(w)$ such that $|v| \geq n_0$. Then the positions at which v occurs in $T(w)$ are all equal modulo p.*

Using Lemma 8, we can decompose $p(n)$, when $n \geq n_0$, as

$$p(n) = \sum_{i=1}^{d} p_i(n)$$

where $p_i(n)$ is the number of factors of length n occurring at a position equal to i modulo d in $T(w)$.

We then obtain the recurrence relation:

$$p_i(n) = \sum_{k=0}^{p'-1} p_{j(kd+i)}(l(kd+i, n)) \tag{5}$$

where j and l are two functions depending on w and satisfying $1 \leq j(i) \leq d$ and $q \left\lfloor \frac{n}{p} \right\rfloor \leq l(i,n) \leq q \left\lceil \frac{n}{p} \right\rceil$. This relation is enough to compute $p(n)$: in particular, we can deduce from it that $p(n) = \Theta(n^r)$.

6 Concluding Remarks

We have considered one-way infinite Toeplitz words over a finite alphabet. We recalled the fact that they provide a natural generalization of periodicity, and also pointed out a new evidence on that. Namely, by Theorem 7, the complexity of a Toeplitz word defined by a pattern of a certain length p and with a certain number q of holes is independent of this pattern, but depends only on p and q, provided that the Toeplitz word is not periodic.

Our main concern was, in one hand, to classify different types of Toeplitz words in terms of possibilities of defining them by iterated morphisms, and on the other hand, to compute the complexity of Toeplitz words.

We noted that the relative sizes of the above p and q play an important role here. In general, Toeplitz words can be generated by iterating periodically a finite number of morphisms. If there is only one hole, that is $q = 1$, then one morphism is enough. An interesting in-between case occurs when p divides q. Then one morphism is enough if after the iteration another one is allowed to translate the result, but as pointed out in Example 5 this other morphism is necessary.

Also in the case of complexity of Toeplitz words p and q are important. When computing the complexity in Theorem 7 we at the same time introduced a new possibility for the complexity of infinite words, cf. [1]. Moreover, Theorem 7 allows to reprove a result of Lepistö in [10]. Namely, taking $p = 5$ and $q = 3$ we obtain a Toeplitz word having the complexity more than quadratic, and hence by a result in [5] it cannot be generated by iterating a single morphism and then applying another one to the result.

References

1. J.-P. ALLOUCHE, Sur la complexité des suites infinies, *Bull. Belg. Math. Soc.* 1 (1994), 133–143.
2. J.-P. ALLOUCHE AND R. BACHER, Toeplitz sequences, paperfolding, towers of Hanoi and progression free sequences of integers, *Ens. Math.* 38 (1992), 315–327.
3. K. CULIK II, J. KARHUMÄKI, AND A. LEPISTÖ, Alternating iteration of morphisms and the Kolakoski sequence, in *Lindenmayer systems*, G. Rozenberg and A. Salomaa, eds., pp. 93–106, Springer-Verlag, 1992.
4. C. DAVIS AND D. E. KNUTH, Number representations and dragon curves I, II, *J. Recr. Math.* 3 (1970), 61–81 and 133–149.
5. A. EHRENFEUCHT, K. P. LEE, AND G. ROZENBERG, Subword complexity of various classes of deterministic languages whithout interaction, *Theoret. Comput. Sci.* 1 (1975), 59–75.
6. N. J. FINE AND H. S. WILF, Uniqueness theorem for periodic functions, *Proc. Am. Math. Soc.* 16 (1965), 109–114.
7. H. FURSTENBERG, *Recurrences in ergodic theory and combinatorial number theory*, Princeton Univ. Press, 1981.
8. K. JACOBS AND M. KEANE, 0-1 sequences of Toeplitz type, *Z. Wahr. verw. Geb.* 13 (1969), 123–131.

9. W. KOLAKOSKI, Self-generating runs, prob. 5304, *Amer. Math. Monthly* **73** (1966), 681–682. Solution by N. Ucoluk.

10. A. LEPISTÖ, On the power of periodic iteration of morphisms, in *ICALP '93*, pp. 496–506, *Lect. Notes Comput. Sci.* **700**, Springer-Verlag, 1993.

11. M. LOTHAIRE, *Combinatorics on Words*, vol. 17 of *Encyclopedia of Mathematics and its Applications*, Addison-Wesley, 1983.

12. M. MENDÈS FRANCE AND A. J. VAN DER POORTEN, Arithmetic and analytic properties of paperfolding sequences, *Bull. Austr. Math. Soc.* **24** (1981).

13. H. PRODINGER AND F. J. URBANEK, Infinite 0-1 sequences without long adjacent identical blocks, *Discr. Math.* **28** (1979), 277–289.

14. O. TOEPLITZ, Beispiele zur Theorie der fastperiodischen Funktionen, *Math. Ann.* **98** (1928), 281–295.

A Construction for Enumerating k-coloured Motzkin Paths

Elena Barcucci, Alberto Del Lungo, Elisa Pergola, Renzo Pinzani

Dipartimento di Sistemi e Informatica, Via Lombroso 6/17, 50134 Firenze, Italy, e-mail:pire@ingfi1.ing.unifi.it

Abstract. We illustrate a method for enumerating k-coloured Motzkin paths according to various parameters and, we give a recursive description of these paths from which we deduce the k-coloured Motzkin paths' generating function according to their length, area and last fall length.

1 Introduction

Various studies (see, for instance, [3, 4]) have been made on Motzkin paths. The k-coloured Motzkin paths are Motzkin paths whose horizontal steps are coloured by means of k colours, and these paths are in bijection with many other combinatorial objects such as: trees, words, polyominoes, permutations ... (see, [4]). As a consequence, the results obtained for k-coloured Motzkin paths can be easily transferred on the above-mentioned structures. For instance, these paths' area corresponds to the internal path length of some classes of planar trees (see, [1]). In this paper, we propose a method for enumerating these paths according to various parameters and we present some results for q-series and continued fractions. We determine an operator that allows us to construct each k-coloured Motzkin path P from another k-coloured Motzkin path Q by performing a "local expansion" on Q. Furthermore, every path P is obtained by only one path Q. Therefore, we obtain a recursive description of the k-coloured Motzkin paths from which we can deduce a functional equation verified by their generating function. By solving this functional equation, we obtain the k-coloured Motzkin paths' generating function according to various parameters. A similar method was used in [1, 2] for enumerating some subclasses of plane trees and column-convex polyominoes. In section 2, we describe the basic idea of our method, that is, we give the definition of an operator on a class S of combinatorial objects and two conditions on the operator and this allows us to deduce a recursive description of S's objects. Section 3 contains some definitions of the k-coloured Motzkin words and paths. In section 4, we determine a recursive construction of k-coloured Motzkin words, and, in section 5, we transpose this construction into the paths. We examine two definitions of a path's area: 1) an area related to the non-decreasing step of the path, and 2) a standard area. We use the construction to establish the k-coloured Motzkin paths' generating function according to their length and first area. This generating function is a continued fraction that can be expressed as the ratio between two formal power series. Moreover, we determine the k-coloured Motzkin paths' generating function according to their length and standard area. We find the continued fraction obtained by Flajolet [3] and we show that it can be expressed as a ratio between two formal power series whose coefficients can be calculated in a recursive way.

2 The method

Let S be a class of combinatorial objects. An operator θ on S is a function from S to 2^S, where 2^S is the power set of S.

Proposition 1. *Let θ be an operator on S. If θ satisfies the following conditions:*

1. $\forall Y \in S \; \exists X \in S$ such that $Y \in \theta(X)$,
2. *if $X_1, X_2 \in S$ and $X_1 \neq X_2$ then $\theta(X_1) \cap \theta(X_2) = \emptyset$,*

then the following family of sets: $\mathcal{F} = \{\theta(X) : \forall X \in S\}$ is a partition of S.

Given a class S of combinatorial objects, if we are able to define an operator θ which satisfies conditions 1 and 2, then Proposition 1 allows us to construct each object $Y \in S$ from another object $X \in S$ and every $Y \in S$ is obtained by only one $X \in S$. Therefore, if we have an operator on S which satisfies conditions 1 and 2, then we have a recursive description of S's elements from which we can deduce a functional equation verified by S's generating function. In the following sections, we examine the class of the k-coloured Motzkin paths and determine an operator which satisfies conditions 1 and 2. Consequently, we obtain a construction of these paths from which we deduce the k-coloured Motzkin paths' generating function according to various parameters.

3 Notations and Definitions

The paths we wish to consider here are the positive ones in the $x - y$ plane, and we can codify them by means of the $k-coloured$ Motzkin words. Let $X = \{a, b, c_1, c_2, \ldots, c_k\}$ be an alphabet with $k + 2$ letters. The letters $c_1, c_2, \ldots, c_k$ are called the colours of X. We use X^* to denote the free monoid generated by X, that is, the set of the finite words written with letters from X. The product of $u = u_1 \ldots u_p \in X^*$ and $v = v_1 \ldots v_q \in X^*$ is defined as the *concatenation* of these words: $uv = u_1 \ldots u_p v_1 \ldots v_q$. The word u is called a *left factor* of the word $w = uv$. The *empty word* is denoted by ε. The number of occurences of the letter $x \in X$ in the word w is denoted by $|w|_x$, and the length of w, by $|w| = \sum_{x \in X} |w|_x$.

The set of k-coloured Motzkin words is the set of words w in X^* characterized by the following two conditions : 1) for any left factor u of w, $|u|_a \leq |u|_b$, 2) $|w|_a = |w|_b$.

We denote the set of k-coloured Motzkin words by $\mathcal{W}_k$. If $k = 0$, then we obtain the Dyck words. Let us now introduce the definition of *positive path* (in the sense used by Goulden and Jackson [5, pag. 293]). Let $P = (i, j) \in Z^2$. Then j is called P's *altitude*, and we write $j = alt(P)$. Let $P_0 P_1 \ldots P_n$ be a sequence of

points in Z^2, such that 1) $P_0 = (0,0)$, 2) $P_{i+1} = P_i + (1,\alpha_i)$, $\alpha_i \in \{-1,0,1\}$
for $0 \leq i < n$, 3) $alt(P_i) \geq 0$ for $0 < i < n$ and $alt(P_n) = 0$. Then $P_0 P_1 \ldots P_n$
is called a *positive path* with *origin* P_0 and *terminus* P_n and is denoted by
$\alpha = \alpha_0 \alpha_1 \ldots \alpha_{n-1}$. Moreover, α_i is called a *step*, and is called, respectively, a
fall, a *level*, or a *rise* step when $\alpha_i = -1, 0$ or 1. If $\alpha_i = 0$ or 1, then P_{i+1} is
said to be a *non-decreasing point*. Thus a positive path is a sequence of steps
having no vertex with a negative altitude (see fig. 1). The length of a path is
the number of its steps. Each k-coloured Motzkin word $w = w_1 \ldots w_n$ codifies a

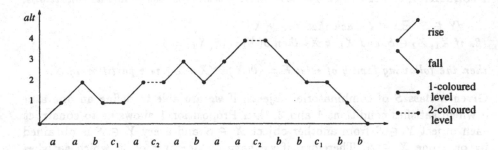

Fig. 1. A positive path

positive path $\alpha = \alpha_0 \alpha_1 \ldots \alpha_{n-1}$ in the following way: 1) if $w_i = a$ then $\alpha_{i-1} = 1$,
2) if $w_i = b$ then $\alpha_{i-1} = -1$, 3) if $w_i = c_j$ with $j \in [1..k]$ then, $\alpha_{i-1} = 0$ and
this step is labelled as c_j. The labelled level steps are called *level coloured* steps.
Therefore, α is a positive path in which each level step is coloured by a colour
of X (see fig. 1). The positive paths codified by the k-coloured Motzkin words
are called $k-coloured$ *Motzkin paths* and are denoted as $\mathcal{P}_k$.

4 An operator on $\mathcal{W}_k$

In this section, we define an operator on the set of the $k-$coloured Motzkin
words which satisfies conditions 1 and 2 of Proposition 1.

Definition 2. Let w be a word of $\mathcal{W}_k$. We define the operator $\theta : \mathcal{W}_k \to 2^{\mathcal{W}_k}$ as
follows:

$$\theta(w) = \{u \in \mathcal{W}_k : u = w' x w'', \text{ with } x \in \{ab, c_1, c_2, \ldots, c_k\}, w = w' w'', w'' \in \{b\}^*\}.$$

Example 1. Let $w \in \mathcal{W}_2$. If $w = ac_1 abbc_2$, then c_2 is the last letter of w, and so:
$\theta(w) = \{wab, wc_1, wc_2\}$. If $w = ac_2 abb$, then b is the last letter of w and, if we denote the left factor ac_2a of w by u, we obtain: $\theta(w) = \{uabbb, ubabb, ubbab, uc_1 bb,$
$ubc_1 b, ubbc_1, uc_2 bb, ubc_2 b, ubbc_2\}$.

Theorem 3. *The operator θ on $\mathcal{W}_k$ satisfies conditions 1 and 2 of Proposition 1.*

Proof. Let $w \in \mathcal{W}_k$. Let x be the last letter of w such that $x \neq b$. If $x = c_j$ with $j \in [1..k]$, then $w = w'c_jw''$, where $w'' \in \{b\}^*$. The word $\bar{w} = w'w'' \in \mathcal{W}_k$ and, from the definition of θ, it follows that $w \in \theta(\bar{w})$. If $x = a$, then $w = w'abw''$, where $w'' \in \{b\}^*$. The word $\bar{w} = w'w'' \in \mathcal{W}_k$ and from the definition of θ, we obtain $w \in \theta(\bar{w})$. Consequently, θ verifies condition 1 of Proposition 1. Let us now show that θ satisfies condition 2 of Proposition 1. We assume that $u, v \in \mathcal{W}_k$ and $\theta(u) \cap \theta(v) \neq \emptyset$. Let $w \in \theta(u) \cap \theta(v)$. From the definition of θ, we obtain $w = u'xu'' = v'yv''$, where $x, y \in \{ab, c_1, c_2, \dots, c_k\}$ and $u = u'u''$, $v = v'v''$ with $u'', v'' \in \{b\}^*$. Therefore, x and y include the last letter of w which is different from b and so $x = y$. Since $u'xu'' = v'yv''$, $x = y$ and $u'', v'' \in \{b\}^*$, we obtain $u'' = v''$. Consequently, $xu'' = yv''$ and so $u' = v'$. It follows that $u = v$.

5 An operator on $\mathcal{P}_k$

In the previous sections, we showed that each k–coloured Motzkin path is codified by a k–coloured Motzkin word. Therefore, we can use the operator θ defined on $\mathcal{W}_k$ to obtain a construction of $\mathcal{P}_k$'s paths. Let $P = P_0P_1\dots P_n \in \mathcal{P}_k$ and let $\alpha = \alpha_0\alpha_1\dots\alpha_{n-1}$ be its coding. An $h \leq n$ exists such that $\alpha_{h+1}, \alpha_{h+2}, \dots, \alpha_{n-1}$ are fall steps. This sequence is called *P's last fall*. If $h = n$, then the P's last fall is empty (i.e., P's last step is a level coloured step, see fig. 2a). A *peak* of P is a sequence of one rise step and one fall step. A peak is codified by the word ab. We transpose the operator θ into $\mathcal{P}_k$ and get the following:

- $\theta(P)$ is the set of paths obtained from P by inserting a peak or a level coloured step into P's the last fall (see fig. 2).

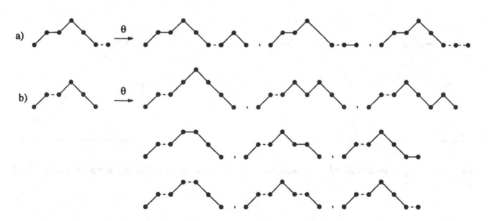

Fig. 2. The set of paths obtained by means of operator θ

From Theorem 3, we deduce that the operator θ allows us to construct each path $P \in \mathcal{P}_k$ from another path $P' \in \mathcal{P}_k$ and every $P \in \mathcal{P}_k$ is obtained by only one

$P^{'} \in \mathcal{P}_k$. Hence, we have a recursive description of these paths and use it to deduce a functional equation verified by the k−coloured Motzkin paths' generating function. We give two definitions of a path's area and use this construction for enumerating the class of the k−coloured Motzkin paths according to the area's definitions.

5.1 Area related to non-decreasing steps

Let $P = P_0 P_1 \ldots P_n \in \mathcal{P}_k$. P's *area* is defined as the sum of the heights of P's non-decreasing points. The area of the path in fig. 1 is 24. We denote P's length by $n(P)$, P's last fall length by $f(P)$ and P's area by $a_1(P)$. The k−coloured Motzkin paths' generating function according to the above-listed three parameters is the following:

$$S_1(x, y, q) = \sum_{P \in \mathcal{P}_k} x^{n(P)} y^{f(P)} q^{a_1(P)}.$$

Let $P \in \mathcal{P}_k$. From the results in section 4, it follows that θ inserts a peak or a level coloured step into P's last fall and there are $f(P)+1$ points where the step can be inserted. Furthermore, for each $i \in [0..f(P)]$, there is a point F in the last fall such that $alt(F) = i$ (see fig. 3). Let $\mathcal{F}(P) = \{F_0, F_1, \ldots, F_{f(P)}\}$, where F_i is the point of the last step having height i.

- If we insert a peak at point F_i, with $i \in [0..f(P)]$, then we obtain $P^{'} \in \mathcal{P}_k$ such that: $n(P^{'}) = n(P) + 2$, $f(P^{'}) = i+1$ and $a_1(P^{'}) = a_1(P) + i + 1$,
- if we insert a level j−coloured step with $j \in [1..k]$ at point F_i with $i \in [0..f(P)]$, then we obtain $P^{'} \in \mathcal{P}_k$ such that: $n(P^{'}) = n(P) + 1$, $f(P^{'}) = i$ and $a_1(P^{'}) = a_1(P) + i$.

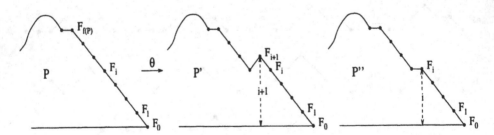

Fig. 3. The paths obtained by inserting a peak or a level coloured step at point F_i of P.

The translation of this construction into the generating function $S_1(x, y, q)$ gives us:

$$\sum_{P \in \mathcal{P}_k} \sum_{i=0}^{f(P)} x^{n(P)+2} y^{i+1} q^{a_1(P)+i+1} + k \sum_{P \in \mathcal{P}_k} \sum_{i=0}^{f(P)} x^{n(P)+1} y^i q^{a_1(P)+i} =$$

259

$$
= \frac{x^2 yq}{1 - yq}(S_1(x, 1, q) - yqS_1(x, yq, q)) + \frac{kx}{1 - yq}(S_1(x, 1, q) - yqS_1(x, yq, q)).
$$

and so:

Proposition 4. *The generating function $S(x, y, q)$ satisfies the following functional equation:*

$$
S_1(x, y, q) = 1 + x\,\frac{k + xyq}{1 - yq}S_1(x, 1, q) - x\,\frac{xy^2q^2 + kyq}{1 - yq}S_1(x, yq, q). \qquad (1)
$$

Solving equations Let us now solve the functional equation (1) by using the following Lemma [2]. We often denote the function $f(x, y, q)$ by $f(y)$ for brevity's sake.

Lemma 5. *Let $\mathcal{R} = \mathbb{R}[[x, y, q]]$ be the algebra of the formal power series in variables x, y and q with real coefficients, and let $\mathcal{A}$ be a sub-algebra of $\mathcal{R}$ such that the series converge for $y = 1$. Let $A(x, y, q)$ be a formal power series in $\mathcal{A}$. Let assume that:*

$$
A(y) = e(y) + xf(y)A(1) + xg(y)A(yq),
$$

where $e(y), f(y)$ and $g(y)$ are some given power series in $\mathcal{A}$. Then:

$$
A(y) = \frac{E(y) + E(1)F(y) - E(y)F(1)}{1 - F(1)},
$$

where

$$
E(y) = \sum_{n \geq 0} x^n e(yq^n) \prod_{k=0}^{n-1} g(yq^k), \qquad F(y) = \sum_{n \geq 0} x^{n+1} f(yq^n) \prod_{k=0}^{n-1} g(yq^k).
$$

By means of Lemma 5 and Proposition 4, we get the following:

Theorem 6. *The generating function $S_1(x, y, q)$ is given by:*

$$
S_1(y) = \frac{J_1(y)J_0(1) - J_1(1)J_0(y) + J_1(1)}{J_0(1)},
$$

with

$$
J_1(y) = \sum_{n \geq 0}(-1)^n \frac{x^n y^n q^{\frac{n(n+1)}{2}}}{(yq; q)_n} \prod_{i=0}^{n-1}(k + xyq^{i+1}),
$$

and

$$
J_0(y) = 1 - \sum_{n \geq 0}(-1)^n \frac{x^{n+1} y^n q^{\frac{n(n+1)}{2}}}{(yq; q)_{n+1}} \prod_{i=0}^{n}(k + xyq^{i+1}).
$$

where we denote $(a; q)_n = \prod_{i=0}^{n-1}(1 - aq^i).$

Let us now take generating function $S_1(x, 1, q)$ into consideration. For brevity's sake, we denote the functions $S_1(x, y, q)$, $J_1(x, y, q)$ and $J_0(x, y, q)$ as $S_1(x, q)$, $J_1(x, q)$ and $J_0(x, q)$. By means of some computations, we obtain:

Lemma 7. *The functions $J_0(x, q)$ and $J_1(x, q)$ satisfy the following equations:*

$$J_0(xq, q) - J_1(x, q) = x^2 q^2\, J_1(xq, q),$$
$$J_0(x, q) - J_1(x, q) = -x(k + xq)\, J_0(xq, q).$$

Let us now use this lemma to prove that:

Proposition 8. *The generating function of the $k-$coloured Motzkin paths according to their length and their area satisfies:*

$$S_1(x, q) = \cfrac{x(k + xq)}{1 - x(k + xq + xq^2) - \cfrac{x^3 q^3(k + xq^2)}{1 - xq(k + xq^2 + xq^3) - \cfrac{x^3 q^6(k + xq^3)}{\cdots}}} + 1.$$

Proof. From Lemma 7, we deduce that:

$$\frac{x^2 q^2\, J_1(xq, q)}{J_0(xq, q)} = \frac{J_0(x, q) - J_1(x, q) + x(k + xq)\, J_1(x, q)}{J_0(x, q) - J_1(x, q)},$$

and so we obtain:

$$x^2 q^2\, S_1(xq, q) = \frac{1 - S_1(x, q)(1 - kx - x^2 q)}{1 - S_1(x, q)}.$$

If $\bar{S}_1(x, q) = S_1(x, q) - 1$, then

$$\bar{S}_1(x, q) = \frac{x(k + xq)}{1 - x(k + xq + xq^2) - x^2 q^2\, \bar{S}_1(xq, q)},$$

and our Proposition follows from this.

This Proposition and Theorem 6 give the following identity:

Corollary 9.

$$\frac{\displaystyle\sum_{n \geq 0} (-1)^n \frac{x^n q^{\frac{n(n+1)}{2}}}{(q; q)_n} \prod_{i=0}^{n-1} (k + xq^{i+1})}{\displaystyle\sum_{n \geq 0} (-1)^n \frac{x^n q^{\frac{n(n-1)}{2}}}{(q; q)_n} \prod_{i=0}^{n-1} (k + xyq^{i+1})} =$$

$$= \cfrac{x(k + xq)}{1 - x(k + xq + xq^2) - \cfrac{x^3 q^3(k + xq^2)}{1 - xq(k + xq^2 + xq^3) - \cfrac{x^3 q^6(k + xq^3)}{\cdots}}} + 1.$$

5.2 Standard area

Let us now give the standard definition of a path's area. Let $P = P_0 P_1 \ldots P_n \in$ $\mathcal{P}_k$. P's *area* is defined as $\sum_{i=1}^{n-1} alt(P_i)$. We denote P's length by $n(P)$, P's last fall length by $f(P)$ and P's area by $a_2(P)$. The k-coloured Motzkin paths' generating function according to the above-listed three parameters is the following:

$$S_2(x, y, q) = \sum_{P \in \mathcal{P}_k} x^{n(P)} y^{f(P)} q^{a_2(P)}.$$

Let $P \in \mathcal{P}_k$. Let $\mathcal{F}(P) = \{F_0, F_1, \ldots, F_{f(P)}\}$, where F_i is the point of P's last step having height i.

- If we insert a peak at point F_i, with $i \in [0..f(P)]$, then we obtain $P' \in \mathcal{P}_k$ such that: $n(P') = n(P) + 2$, $f(P') = i + 1$ and $a_2(P') = a_2(P) + 2i + 1$,
- if we insert a level j-coloured step with $j \in [1..k]$ at point F_i with $i \in [0..f(P)]$, then we obtain $P' \in \mathcal{P}_k$ such that: $n(P') = n(P) + 1$, $f(P') = i$ and $a(P') = a(P) + i$.

We proceed as in the previous section, and obtain:

Proposition 10. *The generating function $S_2(x, y, q)$ satisfies the following functional equation:*

$$S_2(x, y, q) = 1 + x \left(\frac{k}{1 - yq} + \frac{xyq}{1 - yq^2} \right) S_2(x, 1, q) - \frac{kxyq}{1 - yq} S_2(x, yq, q) +$$

$$- \frac{x^2 y^2 q^3}{1 - yq^2} S_2(x, yq^2, q)).$$

Solving equations Let us now solve the above functional equation by using a generalization of Lemma 5:

Lemma 11. *Let $\mathcal{R} = \mathbb{R}[[x, y, q]]$ be the algebra of the formal power series in variables x, y and q with real coefficients, and let $\mathcal{A}$ be a sub-algebra of $\mathcal{R}$ such that the series converge for $y = 1$. Let $A(x, y, q)$ be a formal power series in $\mathcal{A}$. Let assume that:*

$$A(y) = e(y) + x f(y) A(1) + \sum_{i=1}^{m} x^i g_i(y) A(yq^i),$$

where $e(y), f(y)$ and $g_i(y)$ for $i \in [1..m]$ are some given power series in $\mathcal{A}$. Then:

$$A(y) = \frac{E(y) + E(1)F(y) - E(y)F(1)}{1 - F(1)},$$

where, $E(y) = \sum_{n \geq 0} x^n g_1^{(n)}(y)e(yq^n),$ $F(y) = \sum_{n \geq 0} x^{n+1} g_1^{(n)}(y)f(yq^n),$ *and*

$$g_i^{(n)}(y) = g_{i+1}^{(n-1)}(y) + g_1^{(n-1)}(y)g_i(yq^{n-1}) \qquad \text{for } i \in [1..m-1],$$
$$g_m^{(n)}(y) = g_1^{(n-1)}(y)g_m(yq^{n-1}),$$
$$g_i^{(1)}(y) = g_i(y), \qquad g_i^{(0)}(y) = 1 \qquad \text{for } i \in [1..m-1].$$

By means of this Lemma and Proposition 10, we get the following:

Theorem 12. *The generating function* $S_2(x, y, q)$ *is given by:*

$$S_2(y) = \frac{J_1(y)J_0(1) - J_1(1)J_0(y) + J_1(1)}{J_0(1)},$$

where

$$J_1(y) = \sum_{n \geq 0} x^n g_1^{(n)}(y),$$

and

$$J_0(y) = 1 - \sum_{n \geq 0} x^{n+1} \left(\frac{k}{1 - yq^{n+1}} + \frac{xyq^{n+1}}{1 - yq^{n+2}} \right) g_1^{(n)}(y)$$

with

$$g_1^{(n)}(y) = g_1^{(n-1)}(y)g_1(yq^{n-1}) + g_1^{(n-2)}(y)g_2(yq^{n-2}),$$
$$g_1^{(0)}(y) = 1$$

and

$$g_1(y) = \frac{kyq}{yq - 1} \qquad g_2(y) = \frac{y^2 q^3}{yq^2 - 1}.$$

Let us now take generating function $S_2(x, 1, q)$ into consideration. For brevity's sake, we denote the functions $S_1(x, y, q)$, $J_1(x, y, q)$ and $J_0(x, y, q)$ as $S_2(x, q)$, $J_1(x, q)$ and $J_0(x, q)$. By means of some computations, we obtain:

Lemma 13. *The functions* $J_0(x, q)$ *and* $J_1(x, q)$ *satisfy the following equations:*

$$J_1(x, q) = J_0(xq, q), \qquad x^2 q\, J_1(xq, q) = (1 - xk)\, J_1(x, q) - J_0(x, q).$$

By means of this lemma we find, for $k = 0$ or 1, Flajolet's result [3]:

Proposition 14. *The generating function of the* k*−coloured Motzkin paths according to their length and area satisfies:*

$$S_2(x, q) = \cfrac{1}{1 - kx - \cfrac{x^2 q}{1 - kxq - \cfrac{x^2 q^3}{1 - kxq^2 - \cfrac{x^2 q^5}{\cdots}}}}.$$

Proof. From Lemma 13, we deduce that:

$$\frac{x^2 q \, J_1(xq, q)}{J_0(xq, q)} = \frac{(1 - kx) \, J_1(x, q) - J_0(x, q)}{J_1(x, q)},$$

and we have:

$$S_2(x, q) = \frac{1}{1 - kx - x^2 q S_2(xq, q)}.$$

and our Proposition follows from this.

This Proposition and Theorem 12 give the following identity:

Corollary 15.

$$\frac{\sum_{n \geq 0} x^n g_1^{(n)}(1),}{\sum_{n \geq 0} \left(\frac{x}{q}\right)^n g_1^{(n)}(1)} = \cfrac{1}{1 - kx - \cfrac{x^2 q}{1 - kxq - \cfrac{x^2 q^3}{1 - kxq^2 - \cfrac{x^2 q^5}{\cdots}}}}.$$

where

$$g_1^{(n)}(1) = g_1^{(n-1)}(1) g_1(q^{n-1}) + g_1^{(n-2)}(1) g_2(q^{n-2}),$$
$$g_1^{(0)}(1) = 1$$

and

$$g_1(z) = \frac{kzq}{zq - 1} \qquad g_2(z) = \frac{z^2 q^3}{zq^2 - 1}.$$

References

1. Barcucci, E., Del Lungo, A., Pergola, E., Pinzani, R.: Towards a methodology for tree enumeration. Proc. of the 7th FPSAC, Marne-la-Vallée, 1995 (to appear)
2. Bousquet-Mélou, M.: A method for the enumeration of various classes of column-convex polygons. Report LaBRI **378-93**, Université Bordeaux I (1993).
3. Flajolet P.: Combinatorial aspects of continued fractions. Discrete Mathematics **32** (1980) 125-161.
4. Donaghey, R., Shapiro, L.: Motzkin numbers. J. Combin. Theory, A **23** (1977) 291-301.
5. Goulden, I. P., Jackson, D. M.: Combinatorial enumeration. John Wiley and Sons, New York (1983).

On Public-Key Cryptosystem Based on Church-Rosser String-Rewriting Systems (Extended Abstract)

Vladimir A. Oleshchuk

Department of Electrical Engineering and Computer Science
Agder College, N-4890 Grimstad, Norway

Abstract. We propose an approach toward public-key cryptosystems based on finite string-rewriting systems with Church-Rosser property. The approach utilizes an existence of unique normal form for any congruence class modulo such a system and possibility to find it in linear time. Such cryptosystems can be used in the case we are dealing with a large network of communicating parties when it is impractical to use a distinct secret method signing for every pair users and we would like to have a unified secret method for all senders sending to a receiver.

1 Introduction

By now there exist numerous public-key cryptosystems based on quite diverse concepts [12]. Many of them depend heavily on number-theoretic concepts and complexity some specific number-theoretic problems such as, for instance, factoring of numbers or primality testing. Some other cryptographic techniques, not dangerously depending on the difficulty of specific number-theoretic problems, have been proposed recently [14, 15, 18]. They are based on the theories of automata and formal languages and depend on complexity language-theoretic problems some of which are unsolvable in general. The last fact usually is negative more to the cryptoanalysis than to the construction of cryptosystems of that type.

In this work we continue development of cryptographic techniques base on language and automata theories and propose an approach towards public-key cryptosystems based on string-rewriting systems (SRS's) with Church-Rosser property.

Informally, a SRS is a set of ordered pair (l, r) of string over a finite alphabet. The rewriting of a given string w is performed by (non-deterministically) replacing some occurrence of the string l in w by the string r or by replacing some occurrence of string r in w by string l. For a given system T a string x reduces to a string y if y can be obtained from x by applying a sequence of length-decreasing rules of T. The system T has the Church-Rosser property if for every x and y, x and y are congruent if and only if there exists a z such that both x and y reduce to z. The Church-Rosser property of finite SRS's means that there is a unique normal form for every congruence class in such systems and such a normal form can be found in linear time. We will use that property as a basis to construct a trapdoor function in the case when decryption a message is considered as finding normal form for the message.

2 Problem description

Consider a network of communicating entities presented by graph $G = (V, E)$ where nodes V represent communicating entities and edges E represent channels that elements from V can communicate through by sending to and receiving from messages. Let V' be a set of senders, $V' \subseteq V$, that can send messages to receiver $r \in V, r \notin V'$ by broadcasting them into the network. When sender $s \in V'$ sends a message to r it passes through many nodes that can read the message and, hence "snoop" it. Since there is no way to prevent snooping attacks, a sender s has to encrypt messages. In case we are dealing with a large network of communicating parties then it is impractical to use a distinct secret method of signing for every pair users. As example of such a case can be elections that are held over a computer network [10]. Because of V' can be very large we would like to use the same cryptic key to decrypt messages from all members of V'. At the same time we want to distinguish between messages from distinct senders and prevent decryption of messages by other members of V'.

3 Notations and basic definitions

Here we provide formal definitions of SRS's and related notions. For additional information and comments regarding the various notions introduced, the reader is asked to consult [3, 4].

If Σ is a finite alphabet, then Σ^* is a set of all finite words in alphabet Σ with empty word λ. If $w \in \Sigma^*$, then the *length* of a word w is denoted by $|w|$ where $|\lambda| = 0, |a| = 1$ for $a \in \Sigma$, and $|wa| = |w| + 1$ for $w \in \Sigma^*, a \in \Sigma$. The concatenation of words u and v is written as uv. If $u, w \in \Sigma^*$ such that $w = vut$ for some $v, t \in \Sigma^*$, then u is a *subword* of w. If $A, B \subseteq \Sigma^*$, then the concatenation of A and B, denoted AB, is defined to be $\{xy | x \in A, y \in B\}$.

Let Σ be a finite alphabet. A *SRS T* on an alphabet Σ is a subset of $\Sigma^* \times \Sigma^*$ and each element (u, v) of T is a *rewriting rule*. A SRS T induces a congruence on Σ^*. The *congruence generated by* T is the reflexive transitive closure $\leftrightarrow_T^*$ of the relation defined as follows: if $(u, v) \in T$ or $(v, u) \in T$, then for every $x, y \in \Sigma^*$, $xuy \leftrightarrow_T xvy$. The *congruence class* of $z \in \Sigma^*$ with respect to T is a set $[z]_T = \{w \in \Sigma^* \mid w \leftrightarrow_T^* z\}$. This notation is extended to sets $A \subseteq \Sigma^*$ as $[A]_T = \{y \in \Sigma^* \mid \text{there exits an } x \in A \text{ such that } x \leftrightarrow_T^* y\}$, so that $[A]_T = \cup \{[x] | x \in A\}$. If T_1 and T_2 are SRS's on Σ such that for all $x, y \in \Sigma^*$, $x \leftrightarrow_{T_1}^* y$ implies $x \leftrightarrow_{T_2}^* y$, then T_1 *refines* T_2, denoted $T_1 \subseteq T_2$.

For a SRS T, write $x \rightarrow_T y$ if $x \leftrightarrow_T y$ and $|x| > |y|$; write $\rightarrow_T^*$ for the reflexive transitive closure of $\rightarrow_T$; write $\rightarrow_T^+$ for the irreflexive transitive closure of $\rightarrow_T$. A string x is *irreducible* if there is no y such that $x \rightarrow_T y$ and $IRR(T)$ denotes the set of all strings that are irreducible with respect to T. For finite T the set

$IRR(T)$ is a regular set and a finite state acceptor recognizing $IRR(T)$ can be effectively constructed from T. The SRS T is *Church-Rosser* if $x \leftrightarrow_T^* y$ implies that, for some z, $x \to_T^* z$ and $y \to_T^* z$. The problem that finite SRS T has a Church-Rosser property can be solved in polynomial time [7]. If the SRS T is Church-Rosser, then each congruence class has a unique irreducible element and its word problem can be solved in linear time [2].

We call nonempty set C such that $C \subseteq \Sigma^*$ a *code* if for all words $x_{i_1}, x_{i_2}, ..., x_{i_n}$, $x_{j_1}, x_{j_2} ..., x_{j_m}$ from C equality $x_{i_1} x_{i_2} ... x_{i_n} = x_{j_1} x_{j_2} ... x_{j_m}$ implies $x_{i_1} = x_{j_1}$. It means that $m = n$ and $x_{i_k} = x_{j_k}$, $k = 1, ..., n$. Thus if C is a code then any sequence from C^* can be uniquely presented as concatenation of words from C, i.e. uniquely decoded. Directly from definition follows that empty word $\lambda \notin C$ and any nonempty subset of C is also a code [13]. Note that for finite set C a property to be a code can be effectively tested [17, 13].

4 Method description

Suppose that Σ is a plaintext alphabet, without loss of generality, we can assume that $\Sigma = \{x_0, x_1\}$. Let Δ be a cryptotext alphabet that supposed to be bigger than Σ, and T be a Church-Rosser SRS over Δ.

System T can be used as a classical cryptosystem in the following fashion.

For letters from Σ we choose two words $u_0, u_1 \in IRR(T)$ such that $u_i u_j \in IRR(T), i, j = 0, 1$ and the set $\{u_0, u_1\}$ is a code. A letter x_i is encrypted as a word w_i such that $w_i \to_T^* u_i$, i.e. $w_i \in [u_i]_T$. Church-Rosser and code properties guarantee that decryption will be unique. However in the case it is used as a classical cryptosystem, T has to be kept secret.

In the public-key case the system T is used as a basis for a public-key cryptosystem in the way described further this section. For the sake of simplicity we first present a one-sender version of our public-key system, i.e. the case with one sender and one receiver. Then we describe a many-senders case.

Let $u_1, u_2, ..., u_t$ be a set of irreducible in T strings such that $u_i u_j \in IRR(T), i, j = 1, ..., t$ and set $\{u_1, u_2, ..., u_t\}$ is a code. Let R_0 and R_1 be two nonempty finite sets such that $R_0, R_1 \subset \{u_1, u_2, ..., u_t\}$ and $R_0 \cap R_1 = \emptyset$. Denote L_0 and L_1 two regular sets such that $L_i \subseteq [R_i]_T$, $i = 0, 1$ and grammars for L_i can be effectively constructed. Note that by construction $L_0 \cap L_1 = \emptyset$. Let S be a SRS over Δ such that for each rule (u, v) from S holds that $u \leftrightarrow_T^+ v$, i.e. S refines T. In case T is Church-Rosser and S is finite then the property "S refines T" can be easily tested.

The SRS S and sets L_0 and L_1 are publicized as the encryption key. The encryption of a letter x_i is an arbitrary word w_i from the set $[L_i]_S$. The encryption of a string $x_{i_1} x_{i_2} ... x_{i_n}$, where $x_{i_k} \in \Sigma$, $k = 1, ..., n$ is an arbitrary word from the set $[L_{i_1} L_{i_2} ... L_{i_n}]_S$.

The decryption y leads to finding a word from $\{L_0 \cup L_1\}^*$ that is equivalent to y with respect to rewriting system S, i.e. to finding w such that $y \leftrightarrow_S^* w$ and $w = u_{i_1} u_{i_2} ... u_{i_k}$, where $u_{i_j} \in L_0 \cup L_1$. The problem is that S may have undecidable word problem. Further, according to [1] the decidability of word problem for S may not guarantee that the problem is computationally feasible. In fact one should not expect that decidable word problem can be solved in polynomial time [1].

The secret trapdoor consists of the Church-Rosser SRS T and finite sets R_0 and R_1 that gives an easy way to find such a word w. Because of Church-Rosser property of T there is a uniquely defined word $w \in IRR(T)$ such that $y \to_T^* w$ and such an irreducible form w for y can be found in linear time. Code property of selected sets R_0 and R_1 guarantees a unique presentation w as $u_{i_1} u_{i_2} ... u_{i_k}$, where $u_{i_j} \in R_0 \cup R_1$, that can be also found effectively. Then every subword $u_{i_j} \in R_{i_j}$ is decrypted as symbol x_{i_j} from Σ.

The following observations can be considered to argue that cryptoanalysis has to be difficult.

Let S be a non-monadic system. We may define a cryptosystem that encodes x_i as elements from the set $\langle L_i \rangle_S = \{w | \exists u \in L_i$ such that $w = w_t \leftrightarrow_S w_{t-1} \leftrightarrow_S$ ... $\leftrightarrow_S w_0 = u$ and $|w_i| \leq |w_{i+1}|, i = 0, ..., t\}$. By construction, $\langle L_i \rangle_S \subseteq [L_i]_S$. In this simplified case $\langle L_i \rangle_S$ is a context-sensitive language and a grammar G_i such that $L(G_i) = \langle L_i \rangle_S$ can be constructed from the grammar of L_i and from S. Therefore to decrypt w is equivalent to test $w \in L(G_i)$. The last problem is known to be P-SPACE-complete even for deterministic context-sensitive grammars [6].

Let w be a string that cryptoanalyst has to decrypt. It means that he has to find a sequence $i_1, i_2, ..., i_k$ such that $w \in [L_{i_1} L_{i_2} ... L_{i_k}]_S$ where $i_1, i_2, ..., i_k \in \{0, 1\}$. In our system we don't use boundary markers between symbols. This is an advantage of our system since according to [15] the use of boundary markers certainly weakens such kind of systems. Therefore to decrypt w a cryptoanalyst has to find boundaries between words, i.e. present w in form $w_1 w_2 ... w_k$ such that $w_t \in L_{i_t}, t \in \{1, ..., k\}$. There is $2^{|w|-1}$ possible way to do that.

Consider another possible attack on a cryptosystem of this type.

The eavesdropper has to solve the word problem in S which may often be untractable even if it is decidable [1]. The legal receiver decrypts by solving the easy problem: finding an irreducible representation in T instead of S and decoding with R_0 and R_1 instead of L_0 and L_1. In fact, it is not necessary for a cryptoanalyst to find the system T that actually used by the cryptosystem designer. Any Church-Rosser system T' such that

(i) S refines T'
(ii) $[L_0]_{T'} \cap [L_1]_{T'} = \emptyset$
(iii) sets $([L_0]_{T'} \cup [L_1]_{T'}) \cap IRR(T')$ are codes, $i = 0, 1$
(iv) for any $u'_0, u'_1 \in ([L_0]_{T'} \cup [L_1]_{T'}) \cap IRR(T')$ holds $u'_i u'_j \in IRR(T'), i, j = 0, 1$

can be used by a cryptoanalyst to decrypt a message. To show that cryptoanalysis by preprocessing is difficult we have to show that for given $\langle S, L_0, L_1 \rangle$ it is difficult to find such a system T'.

First, we note that the secret system T cannot been found from S since S doesn't contain all information about T in general. There is no algorithm to decide whether a finite system S is equivalent to any finite Church-Rosser system T' [11]. From the other side the question whether exists a finite system T' such that $S \subseteq T'$ is solvable because for the special Church-Rosser system $T' = \{(a, \lambda) \,|\, a \in \Delta\}$ refinement $S \subseteq T'$ holds for any S over Δ. Of course that fact does not help to find an appropriate T'. Therefore a cryptoanalyst has to analyze all finite Church-Rosser SRS's T' such that $S \subseteq T'$. For a given T' the property "S refines T'" for any finite system S (property (i)) can be tested in linear time. Therefore a question about how many Church-Rosser systems might be analyzed by a cryptoanalyst is important. For the case of systems containing only one rule, that also may be used by a cryptosystems designer, Book and Squier in [5] shows that almost all one-rule SRS's are Church-Rosser. The phrase "almost all" is used in the following sense. Let Δ be a finite alphabet of cardinality k, $|\Delta| = k$. For all such alphabets let $p_{k,n}$ denote the number of one-rule SRS's of the form $T = \{(u, v)\}$ where $|u| > |v|$ and $|u| = n$, and let $q_{k,n}$ denote the number of those SRS's that are Church-Rosser. It is shown in [5] that as k and n go to infinity, the ration $q_{k,n}/p_{k,n}$ goes to one. Because of the number of one-rule SRS's is $\Theta(k^n)$ and T' is not necessary one-rule system it is supposed that the number of the Church-Rosser systems a cryptoanalyst has to analyze grows exponentially with growing k and n.

Suppose that T' such that $S \subseteq T'$ is found. Then a cryptoanalyst has to test the properties (ii)-(iv). The difficulty to test property (ii) comes from the fact that in the case of arbitrary finite Church-Rosser system T' a set $[L]_{T'}$ is not a context-free language in general, and there is no algorithm to test whether $[L]_{T'}$ is context-free (or some other restricted type of context-free language, e.g. deterministic context-free, linear context-free, regular, etc.) even for one-element set L (see [9]).

Let T and $\{u_1, u_2, ..., u_t\}$ be as above. Consider a receiver r that receives messages from senders $s_1, s_2, ..., s_n$. For each sender s_i we select nonempty finite sets $R_0^{(i)}, R_1^{(i)} \subset \{u_1, u_2, ..., u_t\}$ such that $R_0^{(i)} \cap R_1^{(j)} = \emptyset$, $i, j = 1, ..., n$. We denote $L_0^{(i)}$ and $L_1^{(i)}$ two regular languages such that $L_k^{(i)} \subseteq \left[R_k^{(i)}\right]_T$, $k = 0, 1$, $i = 1, ..., n$ and grammars for $L_0^{(i)}$ and $L_1^{(i)}$ can be constructed effectively. Let $S^{(i)}$ be a finite SRS such that $S^{(i)}$ refines T. Then a triple $\left\langle S^{(i)}, L_0^{(i)}, L_1^{(i)} \right\rangle$ is publicized as the encryption key for sender s_i. The encryption of a letter x_k by a sender s_i is an arbitrary word w_k from the set $\left[L_k^{(i)}\right]_{S^{(i)}}$, $k = 0, 1$, $i = 1, ..., n$. The secret trapdoor consists of the Church-Rosser system T and two finite sets $\bigcup_{i=1}^{n} R_0^{(i)}$ and $\bigcup_{i=1}^{n} R_1^{(i)}$ and gives an easy and uniform way to decrypt messages from the senders s_i, $i = 1, ..., n$. In the case the receiver r wants to distinguish between messages from distinct senders the sets $R_0^{(i)}$ and $R_1^{(i)}$ should be chosen such that $R_k^{(i)} \cap R_k^{(j)} = \emptyset$, $i, j = 1, ..., t$, $i \neq j$, $k = 0, 1$. Then a decrypted message w is sent by s_i iff $w \in \left(R_0^{(i)} \cup R_1^{(i)}\right)^*$. Thus information about sender is incorporated into messages structure. Making secret the sets $L_0^{(i)}$ and $L_1^{(i)}$ from the public key of sender s_i we can add authentication property to our system.

5 Conclusion

The proposed public-key cryptosystems apply an idea that called in literature encryption by coloring [14, 16, 18] that associates a color to each plaintext letter. The public encryption key provides a method of generating arbitrarily many elements colored each plaintext letter. In such cryptosystems elements are strings of symbols. For different occurrence of a letter different strings colored the letter may be chosen from infinite sets. The number of strings possible for encryption of each sequence of letters is potentially infinite. It means that a cryptoanalyst cannot generate all encryptions in advance and use a table search in order to encrypt. Of course more research is still needed as regards issues concerning the complexity of cryptoanalysis in general.

References

1. Bauer, G., Otto, F.: Finite complete rewriting systems and the complexity of the word problem. Acta Informatica **21** (1984) 521-540
2. Book, R.: Confluent and other types of Thue systems. J. ACM **29** (1982) 171-183
3. Book, R.: Thue systems as rewriting systems. J. Symb. Comput. **3** (1987) 39-68
4. Book, R., Otto, F.: *String-Rewriting Systems.* Springer: New-York, 1993
5. Book, R., Squier, C.: Almost all one-rule Thue systems have decidable word problem. Discrete Mathematics **49** (1984) 237-240.
6. Garey, M., Johnson, D.: *Computer and Intractability: A Guide to the Theory of NP-Completeness.* Freeman, San Francisco, CA, 1979.
7. Kapur, D., Krishnamoorthy, M., McNaughton, R., Narendran, P.: An $O(|T|^3)$ algorithm for testing the Church-Rosser property of Thue systems. Theor. Comp. Sci. **35** (1985) 109-114
8. Kari, J.: Observations concerning a public-key cryptosystem based on iterated morphisms. Theoretical Computer Science **66** (1989) 45-53
9. Narendran, P., O'Dunlaing, C., Rolletschek, H.: Complexity of certain decision problems about congruential languages. J. Comp. Syst. Sci. **30** (1985) 343-358
10. Nurmi, H., Salomaa, A.: Conducting secret ballot elections in computer networks: problems and solutions, Annals of Operations Research **5** (1994) 185-190
11. O'Dunlaing, C.: Undecidable questions related to Church-Rosser Thue systems. Theoretical Computer Science **23** (1983) 339-345
12. Rivest, R.: Cryptography. Handbook of Theoretical Computer Science, Vol. **A**, J. van Leeuwen, ed., (1990) 717-755
13. Salomaa, A.: Jewels of formal language theory. Comp. Sci. Press, Rockville, 1981
14. Salomaa, A.: A public-key cryptosystem based on language theory. Computer and Security **7** (1988) 83-87
15. Salomaa, A., Yu, S.: On a public-key cryptosystem based on iterated morphisms and substitutions. Theoretical Computer Science **48** (1989) 283-246
16. Salomaa, A.: *Public-Key Cryptography.* EATCS Monographs on Theoretical Computer Science **23**, Springer-Verlag, 1993
17. Sardinas, A., Patterson, G.: A necessary and sufficient condition for the unique decomposition of coded messages, I.R.E. Int. Conv. Rec. **8** (1953) 104-108
18. Wagner, N. R., Magyarik, M. R.: A public-key cryptosystem based on the word problem. Lecture Notes in Computer Science **196** (1985) 19-37

Extending the Hong-Kung Model to Memory Hierarchies*

John E. Savage

Brown University,Providence, Rhode Island 02912

Abstract. The speed of CPUs is accelerating rapidly, outstripping that of peripheral storage devices and making it increasingly difficult to keep CPUs busy. Consequently multi-level memory hierarchies, scaled to simulate single-level memories, are increasing in importance. In this paper we introduce the Memory Hierarchy Game, a multi-level pebble game that simulates data movement in memory hierarchies in terms of which we study space-time tradeoffs.

We provide a) a common generalization of the Hong-Kung and Paterson-Hewitt pebble models to the Memory Hierarchy Game, b) a greatly simplified proof of the Hong-Kung lower bound on I/O complexity that makes their result readily accessible, c) straight-line algorithms for a representative set of problems that are simultaneously optimal at each level in the memory hierarchy in their use of space and I/O and computation time, and d) an extension the game to block transfers of data between memories.

1 Introduction

In this paper we study tradeoffs between the number of storage locations (*space*) at each level of a memory hierarchy and the number of data movements (*I/O time*) between levels in the hierarchy. We develop upper and lower bounds on I/O time in terms of space. We model computations as pebblings of straight-line programs according to the rules of the **Memory Hierarchy Game** (MHG), a pebble game in which different kinds of pebbles represent storage locations at different levels in a memory hierarchy. The MHG generalizes to L levels the two-level game introduced by Hong and Kung [8]. Straight-line computations are modeled by directed acyclic graphs (dags). Not only are many important computational science algorithms described by dags, efficient prefetching in large memory hierarchies may require that all large computations be straight-line.

The rules of the MHG assume that data migrate up and down the hierarchy. Input data resides initially in the highest level memory (the Lth) and the values of all output vertices reside there at the end of a computation. The location of a datum in the jth memory is denoted by placing a level-j pebble on that vertex. Movement up and down the hierarchy is modeled by either replacing a

---

* This work was supported in part by the Office of Naval Research under contract N00014-91-J-4052, ARPA Order 8225 and by NSF under Grant MIP-902570.

pebble at one level with one at an adjacent level or by possibly adding such a pebble. Level-1 pebbles model data storage in a register. A level-1 pebble can be placed on a vertex (a **computation step**) that has no pebbles only if all of its predecessors carry level-1 pebbles, thereby modeling the requirement that a value of an operation can be computed by a CPU only if all of its operands are present in registers. A **level-l I/O operation** is the placement of a level-$(l-1)$ pebble on a vertex carrying a level-l pebble (an input from level-l) or the placement of a level-l pebble on a vertex carrying a level$(l-1)$ pebble (an output to level-l). We allow an unlimited number of level-L pebbles.

We consider two variants of the MHG, the **standard game** in which highest level pebbles be used on intermediate vertices of a dag $G = (V, E)$ and the **I/O-limited game** in which highest level pebbles cannot be used this way. The latter is appropriate when there is a large gap between the access times of the highest and next-highest level memory units because in this case data-independent prefetching is essential to avoid the large delays required by data-dependent fetching. The two-level I/O-limited game is actually the Paterson-Hewitt [11] **red-pebble game** while the two-level standard game is the Hong-Kung **red-blue pebble game**. Thus, the MHG combines elements of both games and generalizes them to multiple levels.

A pebble strategy is **minimal** if the number of highest level I/O operations is minimized after which the number of I/O operations is minimized at successively lower levels and, finally, the number of computation steps is minimized. This definition of minimality reflects the fact that the time to access data on memories increases very rapidly with their level in the hierarchy.

We develop methods for deriving upper and lower bounds on performance that are applied to a representative set of problems consisting of matrix multiplication, the Fast Fourier Transform and permutation and merging networks. We also generalize the model to block transfers. If the storage unit at level l holds p_l words, and the I/O time at level l, T_l, is the number of times blocks of b_l words move between the storage units at levels $l-1$ and l, then we establish a framework in which to derive lower bounds on T_l in terms of b_l and s_{l-1}, the storage capacity of all units up to and including the $(l-1)$st. Under weak conditions on b_l and for the problems under consideration, we show that our lower bounds on T_l can be achieved up to multiplicative constants for all levels simultaneously.

This approach is illustrated by the problem of multiplying matrices. We show that using any variant of the classical algorithm to multiply two $n \times n$ matrices, T_l is proportional to $\Theta(n^3/(b_l\sqrt{s_{l-1}}))$ when $s_{l-1} = O(n^2)$. If t_l is the time (relative to that of the fastest memory) for a level-l block I/O operation, a memory hierarchy will behave as single flat memory if $t_l \ll b_l\sqrt{s_{l-1}}$ for $2 \leq l \leq L$.

Related Research Aggarwal and Vitter [3] examined a two-level memory in which P B-item blocks can be transferred in each step, obtaining tight bounds for sorting-related problems. They did not handle the I/O-limited case or multiple levels. Aggarwal, Alpern, Chandra and Snir [1] introduced the hierarchical memory model (HMM), which treats memory as a linear array with cost $f(x)$ to access location x in the array, and obtained tight bounds for a number of problems and a number of cost functions. They don't handle blocks, nor handle the I/O-limited case or large discontinuities in the storage access time between levels.

Aggarwal, Chandra and Snir [2] introduced the BT model, an extension of the HMM model supporting block transfers in which the time to move a block of size b ending at location x is $f(x) + b$. They establish tight bounds on computation time for problems including matrix transpose, FFT, and sorting using a number of cost functions. They allow blocks to be arbitrarily large and problem dependent but do not handle the I/O-limited case nor large discontinuities in access time.

Alpern, Carter and Feig [4] introduced the uniform memory hierarchy (UMH) which has uniform exponential values for memory capacity, block size, and the time to move a block between levels. They allow I/O overlap between levels and determine conditions under which matrix transposition, matrix multiplication and Fourier transforms can and cannot be done efficiently.

Savage and Vitter [12] extend the one-level Paterson-Hewitt model [11] to support parallel pebbling and the Hong-Kung model to support contiguous block I/O. Vitter and Shriver [16] examine block transfers in three parallel disk memory systems and present a randomized version of distribution sort that meets the lower bounds for these models of computation. Nodine and Vitter [10] give an optimal deterministic sorting algorithm for these memory models. The models have the limitations described above.

2 The Memory Hierarchy Game

The Memory Hierarchy Game (MHG) formally defined below captures the essential features of serial computers that use storage units organized into levels and in which data moves between levels from the highest to the lowest level and back. The highest level storage unit models an archival store with large access time whereas the lowest level unit models a fast memory used by the CPU for all its computations.

The L-level Memory Hierarchy Game (MHG) is played on dags with p_l pebbles at level l, $1 \leq l \leq L - 1$, and an unlimited number of pebbles at level L. It has **resource vector** $\underline{p} = (p_1, p_2, \ldots, p_{L-1})$, where $p_j \geq 1$ for $1 \leq j \leq L - 1$, and uses $s_l = \sum_{j=1}^{l} p_j$ pebbles at level l or less. Its rules are given below.

R1. (Computation Step) A first-level pebble can be placed on or moved from a predecessor to any vertex all of whose immediate predecessors carry first-level pebbles.

R2. (Pebble Deletion) Except for level-L pebbles on output vertices, a pebble at any level can be deleted from any vertex.

R3. (Initialization) A level-L pebble can be placed on an input vertex at any time.

R4. (Input from Level-l) For $2 \leq l \leq L - 1$, a level-$(l - 1)$ pebble can be placed on any vertex carrying a level-l pebble.

Standard Game

R5. (Output to Level-l) For $2 \leq l \leq L$, a level-l pebble can be placed on any vertex carrying a level-$(l-1)$ pebble.

I/O-limited Game

R5. (Output to Level-l) For $2 \leq l \leq L-1$, a level-l pebble can be placed on any vertex carrying a level-$(l-1)$ pebble.

R6. (I/O-limitation) Level-L pebbles can only be placed on input vertices and output vertices carrying level-$(L-1)$ pebbles.

3 Computational Inequalities

In this section we derive generic lower bounds on the number of I/O and computation steps required by a minimal pebbling of a dag $G = (V, E)$ in the MHG. We assume throughout that all vertices of a dag $G = (V, E)$ are reachable from input and output vertices.

Consider a pebbling $\mathcal{P}(\underline{p}, G)$ of the dag G in the L-level MHG with resource vector p. A minimal pebbling is one that successively minimizes the number of level-l I/O operations at decreasing levels starting at level L. The **computation time** of a pebbling, $T_1^{(L)}(\underline{p}, G)$, is the minimal number of lowest level pebblings in a minimal pebbling and the **level-l I/O time**, $T_l^{(L)}(\underline{p}, G)$, is the minimal number of level-l I/O operations, $2 \leq l \leq L$.

The following result shows that if the number of pebbles available at or below a given level, is large enough, no I/O operations at the next level are necessary except on input and output vertices.

Lemma 1. *Let S_{min} be the minimum number of pebbles to pebble $G = (V, E)$ in the red-pebble game. If the number of pebbles at level $k < L$ or less, s_k, exceeds $S_{min} + (k-1)$, a minimal pebbling $\mathcal{P}(\underline{p}, G)$ with resource vector $\underline{p}$ in the L-level MHG does not perform I/O operations at level $k+1$ or higher except on inputs and outputs.*

Because every input must be read from level L and every output written to level L, $T_l^{(L)}(\underline{p}, G)$ is at least equal to the number of input and output vertices of G for $2 \leq l \leq L$. Also $T_1^{(L)}$ is at least the number of non-input vertices $|V^*|$ of G.

The following definition of the S-span of a graph abstracts ideas used by Hong and Kung [8] and is used to derive lower bounds on the I/O time of minimal pebblings of dags.

Definition 2. *Given a dag $G = (V, E)$ the S-span of G, $\rho(S, G)$, is the number of computation steps on G with the red-pebble game in a minimal pebbling maximized over all initial placements of S red pebbles. The pebblings are done assuming that pebbles are left on output vertices.*

The following theorem generalizes the Hong-Kung [8] lower bound on I/O time for the two-level MHG and provides a new and much simpler proof. Aggarwal and Vitter [3] have given a simple proof of the Hong-Kung lower bound for the FFT dag.

Theorem 3. *Consider a minimal pebbling of the dag $G = (V, E)$ in the standard MHG with resource vector $\underline{p}$ using $s_l = \sum_{j \leq l} p_l$ pebbles at level l or less. Let $T_l^{(L)}(\underline{p}, G)$ be the number of I/O operations at level l, $2 \leq l \leq L$, and let $T_1^{(L)}(\underline{p}, G)$ be the number of computations steps used in this pebbling. Then, the following lower bound on $T_l^{(L)}(\underline{p}, G)$, $2 \leq l \leq L$, must be satisfied whether the MHG is I/O-limited or not:*

$$\lceil T_l^{(L)}(\underline{p}, G)/s_{l-1} \rceil \rho(2s_{l-1}, G) \geq T_1^{(L)}(\underline{p}, G) \geq |V^*|$$

Proof Sketch. Since fewer pebblings are done at each level for the standard game, lower bounds are derived for this case. s_{l-1} is the number of pebbles at all levels up to and including level $l-1$. Let C be a minimal pebbling of G. Divide C into consecutive sequential sub-pebblings $\{C_1, C_2, \ldots, C_h\}$ where each sub-pebbling has s_{l-1} level-l I/O operations except possibly the last which has no more such operations. Thus $h = \lceil T_l^{(L)}(\underline{p}, G)/s_{l-1} \rceil$.

We develop an upper bound Q to the number of computation steps in each sub-pebbling. This number multiplied by the number h of sub-pebblings is an upper bound to the total number of computation steps, $T_1^{(L)}(\underline{p}, G)$, performed by the pebbling C. It follows that $Qh \geq T_1^{(L)}(\underline{p}, G)$.

The upper bound on Q is developed by adding s_{l-1} new level-$(l-1)$ pebbles and showing that we may use these new pebbles to move all I/O operations at level l or higher in a sub-pebbling C_t to either the beginning or end of the sub-pebbling without changing the number of computation steps or I/O operations at level $l-1$ or less. Thus, we move without changing them all computation steps and I/O operations at level $l-1$ or lower to a *middle interval* of C_t in between the higher-level I/O operations.

An upper bound to Q is obtained by observing that at the start of the pebbling of the middle interval of C_t there will be at most $2s_{l-1}$ pebbles at level $l-1$ or less on G, at most s_{l-1} original pebbles plus a like number of new pebbles. Any output vertex pebbled in this interval must have a level-$(l-1)$ or lower pebble on it at the end of the interval since the pebbling is minimal. Clearly, the number of computation steps on vertices pebbled in the middle interval is largest when all $2s_{l-1}$ pebbles of levels $l-1$ or less on G are treated as level-1 pebbles. It follows that at most $\rho(2s_{l-1}, G)$ computation steps can be done on G in C_t since C_t is part of a minimal pebbling. This completes the proof of this theorem.

In the standard game each vertex of G is pebbled once in a computation step. However, when the I/O limitation applies, some vertices may have to be repebbled. The following theorem relates the number of moves in an L-level game to the number in a two-level game and allows us to use prior results. The proof is by simulation of an L-level MHG with a two-level MHG.

Theorem 4. *Let $S = s_{L-1} = \sum_{j=1}^{L-1} p_j$. The following inequalities hold for $2 \leq l \leq L - 1$ when the graph G is pebbled in the L-level MHG, whether I/O limited or not, with resource vector $\boldsymbol{p}$:*

$$T_l^{(L)}(\boldsymbol{p}, G) \geq T_L^{(L)}(\boldsymbol{p}, G) \geq T_2^{(2)}(S, G)$$
$$T_1^{(L)}(\boldsymbol{p}, G) \geq T_1^{(2)}(S, G) \geq |V^*|$$

Here $T_1^{(2)}(S, G)$ and $T_2^{(2)}(S, G)$ are the number of computation and I/O operations, respectively, in the red-blue pebble game played on G with S red pebbles. If MHG is played with the I/O-limitation, these two measures are computed under the I/O-limitation. $|V^*|$ is the number of non-input vertices of G.

4 Matrix Multiplication

Consider a matrix multiplication algorithm conforming to the classical algorithm. All products of pairs of entries in the two matrices A and B are formed and combined in independent binary addition trees. We develop tight upper and lower bounds on the I/O and computation time for the standard MHG as well as the I/O-limited MHG. Developing good upper bounds under the I/O limitation is related to the challenging problem of deriving fast algorithms for matrix multiplication.

Lemma 5 [8]. *The S-span of any graph G associated with the classical algorithm to multiply two $n \times n$ matrices with the binary operations of addition and multiplication of component values satisfies $\rho(S, G) \leq 2S^{3/2}$ for $S \leq n^2$.*

Theorem 6. *Let G be any graph consistent with the classical algorithm to multiply two $n \times n$ matrices. Let it be pebbled in the standard MHG with the resource vector $\underline{p}$. Let $s_l = \sum_{j=1}^{l} p_j$ and let k be the largest integer such that $s_k \leq 3n^2$. When $p_1 \geq 3$ there is a pebbling of G such that the following bounds hold simultaneously:*

$$T_l^{(L)}(\underline{p}, G) = \begin{cases} \Theta(n^3) & l = 1 \\ \Theta(n^3/\sqrt{s_{l-1}}) & 2 \leq l \leq k+1 \\ \Theta(n^2) & k+2 \leq l \leq L \end{cases}$$

Proof Sketch. We note that any graph G consistent with the classical matrix multiplication algorithm has $|V^*| = \Theta(n^3)$ non-input vertices. The first lower bound follows from the fact that we have to pebble $\Theta(n^3)$ vertices with level-1 pebbles to pebble the graph. The second follows from Theorem 3 and Lemma 5. The third lower bound follows from Theorem 4.

The pebbling strategy that simultaneously achieves these lower bounds up to constant multiplicative factors uses the standard representation of a $n \times n$

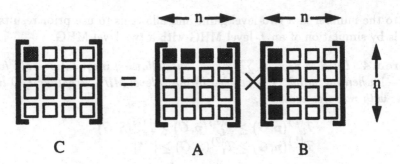

Fig. 1. Pebbling schema for matrix multiplication.

as an $n/m \times n/m$ matrix of $m \times m$ matrices when m divides n, as suggested by Figure 1. In turn the submatrices are themselves recursively decomposed as submatrices where the matrix sizes are chosen to minimize the amount of memory fragmentation. The innermost matrices are $r_1 \times r_1$ and are contained in $r_2 \times r_2$ matrices which are finally contained in $r_k \times r_k$ where the r_i are set as shown below and k is chosen to be the largest integer such that $r_k^2 \leq n^2$. (Without loss of generality we assume that $r_k = n$, that is, we assume that A, B and C are $r_k \times r_k$ matrices.)

$$r_i = \begin{cases} \lfloor \sqrt{s_1/3} \rfloor & i = 1 \\ r_{i-1}\lfloor \sqrt{(s_i - i + 1)}/(\sqrt{3}r_{i-1})\rfloor & i \geq 2 \end{cases}$$

Our pebbling strategy is recursive. It is done using two sets of pebbles, a reserve set containing one pebble per level except the first, and the remainder. With the remaining pebbles, recursively pebble the product of two $r_j \times r_j$ submatrices A_1 and B_1 of A and B under the assumption that A_1 and B_1 have pebbles at level j or lower on their entries initially. When pebbles at levels j or lower are placed on inputs of A_1 and B_1, we exhaust pebbles at each level before using pebbles at a lower level. Since $s_j \geq 3r_j^2 + j - 1$, this guarantees that if a submatrix of A_1 or B_1 does not have all its input vertices covered with pebbles at level $j - 1$ or less, there will be low-level pebbles that can be moved to them using the reserve pebbles without having to remove pebbles from other vertices.

We close this section with a lower bound for the I/O-limited version of the game using a new lower bound [13] of the Grigoryev style [7] on the I/O time in the red-pebble game for matrix multiplication of the form $T_2^{(2)}(S, G) \geq n^3/\sqrt{32(S+1)}$.

Theorem 7. *Let G be any dag associated with matrix multiplication of $n \times n$ matrices and let it be pebbled under the **I/O limitation**. If $S = s_{L-1} \leq n$, then the time to pebble G at the lth level, $T_l^{(L)}(p, G)$, must satisfy the following relation for $1 \leq l \leq L$:*

$$T_l^{(L)}(\underline{p}, G) = \Omega(\frac{n^3}{\sqrt{S}})$$

5 The Fourier Transform

The discrete Fourier transform (DFT) on n inputs is defined by the matrix-vector multiplication $A\underline{x}$ with a Vandermonde matrix. The fast Fourier transform (FFT) graph is the well known fast implementation of the DFT. As suggested in Figure 2, for $1 \leq j \leq d-1$, the FFT graph $F^{(d)}$ on 2^d inputs can be represented as the composition of 2^{d-j} disjoint "top" FFT graphs on 2^j inputs $\{F_{t,p}^{(j)} \mid 0 \leq p \leq 2^{d-j}-1\}$ with 2^j disjoint "bottom" FFT graphs on 2^{d-j} inputs $\{F_{b,m}^{(d-j)} \mid 0 \leq m \leq 2^j - 1\}$, where the mth input vertex of $F_{t,p}^{(j)}$ is the pth output vertex of $F_{b,m}^{(d-j)}$ for $0 \leq m \leq 2^j - 1$ and $0 \leq p \leq 2^{d-j} - 1$.

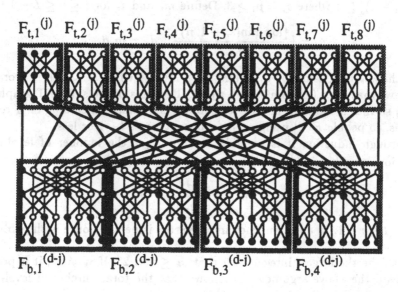

$F_{t,1}^{(j)}$ $F_{t,2}^{(j)}$ $F_{t,3}^{(j)}$ $F_{t,4}^{(j)}$ $F_{t,5}^{(j)}$ $F_{t,6}^{(j)}$ $F_{t,7}^{(j)}$ $F_{t,8}^{(j)}$

$F_{b,1}^{(d-j)}$ $F_{b,2}^{(d-j)}$ $F_{b,3}^{(d-j)}$ $F_{b,4}^{(d-j)}$

Fig. 2. Decomposition of the FFT graph $F^{(5)}$. Edges between shaded boxes identify vertices common to two FFT subgraphs. The ordering of their endpoints defines a matrix transposition.

The following bound on the S-span of the FFT dag is implicit in the work of Hong and Kung [8] and explicit in the work of Aggarwal and Vitter [3].

Lemma 8. $\rho(S, F^{(d)})$ on the 2^d-input FFT graph $F^{(d)}$ satisfies $\rho(S, F^{(d)}) \leq 2S \log S$ when $S \leq n$.

Theorem 9. Let the FFT graph on $n = 2^d$ inputs, $F^{(d)}$, be pebbled in the standard MHG with resource vector $\underline{p}$. Let $s_l = \sum_{j=1}^{l} p_j$ and let k be the largest *integer such that $s_k \leq n$. When $p_1 \geq 3$ there is a pebbling of $F^{(d)}$ such that the following relations are simultaneously satisfied:*[2]

$$T_l^{(L)}(\underline{p}, F^{(d)}) = \begin{cases} \Theta(n \log n) & l = 1 \\ \Theta(n \log n / \log s_{l-1}) & 2 \leq l \leq k+1 \\ \Theta(n) & k+2 \leq l \leq L \end{cases}$$

Proof. The lower bounds follow directly from elementary considerations and Theorem 3 and Lemma 8. We exhibit a pebbling strategy giving upper bounds matching these lower bounds for all $1 \leq l \leq L$. Our pebbling strategy is based on the decomposition of Figure 2.

If e divides d, decompose $F^{(d)}$ into a collection of subgraphs $F^{(e)}$ each of which can be pebbled level by level with $2^e + 1$ level-1 pebbles without repebbling any vertex. This observation is used in our hierarchical pebbling strategy.

The following non-decreasing sequence $\underline{d} = (d_1, d_2, \ldots, d_{L-1})$ of integers is used to describe an efficient pebbling strategy for $F^{(d)}$. Let $d_0 = 1$ and $d_1 = \lfloor \log(s_1 - 1) \rfloor \geq 1$ where $s_1 = p_1 \geq 3$. Define m_r and d_r for $1 \leq r \leq L - 1$ by

$$m_r = \lfloor \frac{\lfloor \log \min(s_r - 1, n) \rfloor}{d_{r-1}} \rfloor, \quad d_r = m_r d_{r-1}$$

Note that d_r is divisible by d_{r-1} and $s_r \geq 2^{d_r} + 1$ when $s_r \leq n + 1$. From this it follows that $d_r \geq (\log \min(s_r - 1, n))/4$. The sizes of the sub-FFT graphs are chosen so that a sufficient number of pebbles is available, including $l - 1$ reserve pebbles, to pebble $F^{(d_r)}$ using s_{l-1} pebbles at levels $l - 1$ or less.

Through induction it is possible to show that the number of level-l I/O operations on $F^{(d_r)}$, $n_l^{(r)}$, satisfies the following inequality

$$n_l^{(r)} \leq 2 \frac{d_r}{d_{l-1}} 2^{d_r}$$

for $2 \leq l \leq r + 1$. In addition, each input vertex is pebbled once with pebbles at each level.

Let k be the largest integer such that $s_k \leq n = 2^d$. If $s_k \neq 2^d$, it is possible to extend the above argument. It follows that the total number of level-l I/O operations, $2 \leq l \leq k + 1$, is at most

$$(\lceil d/d_k \rceil) 2^{d-d_k} \frac{2d_k}{d_{l-1}} 2^{d_k} = \lceil d/d_k \rceil \frac{2d_k}{d_{l-1}} 2^d \leq \frac{4d 2^d}{d_{l-1}}$$

The desired conclusions follows from the observation above that $d_l \geq (\log(s_l - 1))/4$ when $s_l - 1 \leq n$.

We now state lower bounds on the number of I/O and computation steps for DFTs realized by straight-line algorithms for the DFT. Corresponding upper bounds are stated for FFT graphs. The lower bounds are larger than the lower bounds of Theorem 9 when the total number of pebbles at all levels but the

---

[2] All logarithms are to base 2.

highest, S, satisfies $S \leq n/\log n$ where n is the number of inputs to the DFT. The lower bounds apply to any straight-line program for the DFT, not just linear straight-line programs, as shown by Tompa [14], due to a new unpublished result [13].

Theorem 10. *Let $FFT(n)$ be any dag associated with the DFT on n inputs when realized by a linear straight-line program. Let $FFT(n)$ be pebbled under the I/O limitation with resource vector $\underline{p}$. Let $s_l = \sum_{j=1}^{l} p_j$. If $S = s_{L-1} \leq n$, then the time to pebble $FFT(n)$ at the lth level, $T_l^{(L)}(\underline{p}, FFT(n))$, must satisfy the following relation for $1 \leq l \leq L$:*

$$T_l^{(L)}(\underline{p}, FFT(n)) = \Omega(\frac{n^2}{S})$$

Also, when $n = 2^d$, there is a pebbling of the FFT graph $F^{(d)}$ such that the following relations hold simultaneously:

$$T_l^{(L)}(\underline{p}, F^{(d)}) = \begin{cases} O(\frac{n^2}{S} + n\log S) & l = 1 \\ O(\frac{n^2}{S} + n\frac{\log S}{\log s_{l-1}}) & 2 \leq l \leq L \end{cases}$$

when $S \geq 2\log n$.

6 Permutation and Merging Networks

Consider merging networks $BS^{(d)}$ based on bitonic sorting networks [9, pp. 632] and permutation networks $P^{(d)}$ based on three back-to-back FFT graphs [17]. Replacing comparators in $BS^{(d)}$ with two-input butterfly graphs produces an FFT with edge directions reversed. It is immediate from the layered cross-product representation of FFT graphs [6] that $BS^{(d)}$ is isomorphic to the FFT graph. Thus, the bounds for the FFT apply directly to $BS^{(d)}$.

It can be shown that lower bounds for the FFT apply to $P^{(d)}$ because $P^{(d)}$ reduces to $F^{(d)}$ be eliminating edges and coalescing chains to single edges. Since our pebbling strategy for the FFT graph in the standard MHG advances pebbles level-by-level, using higher-level pebbles sparingly to achieve the lower bounds, it follows that the pebbling strategy for the FFT graph is directly applicable to this graph.

Theorem 11. *Let $P^{(d)}$ be pebbled in the standard L-level MHG with resource vector $\underline{p}$. Let $s_l = \sum_{j=1}^{l} p_j$ and let k be the largest integer such that $s_k \leq n$. When $p_1 \geq 3$ there is a pebbling of $P^{(d)}$ such that the number of pebblings at each of the L levels, $\{T_l^{(L)} \mid 1 \leq l \leq L\}$, simultaneously satisfy the following relations:*

$$T_l^{(L)}(\underline{p}, P^{(d)}) = \begin{cases} \Theta(n\log n) & l = 1 \\ \Theta(n\log n/\log s_{l-1}) & 2 \leq l \leq k+1 \\ \Theta(n) & k+2 \leq l \leq L \end{cases}$$

When the I/O limitation applies and s_{L-1} is too small, the I/O and computation time to pebble the permutation and sorting networks can be much larger than that to pebble the FFT graph. For example, Carlson and Tompa together show the following result, which implies that $P^{(d)}$, which has three FFT graphs back-to-back, requires at least as much I/O time:

Lemma 12 [5,14,15]. *Let G be the graph consisting of two back-to-back copies of the FFT graph $F^{(d)}$ on $n = 2^d$ inputs. Then the number of second-level I/O operations when the **I/O limitation** applies in the red-pebble game when played with S level-1 pebbles satisfies the following inequality:*

$$T_2^{(2)}(S, G) \geq \Omega(n^3/S^2 + (n^2 \log n)/S)$$

7 Generalization to Block-I/O

Data is typically moved in blocks between memories in a hierarchy; data must fetched from the same block in which it was stored. Our lower bounds can be generalized to the block-I/O case by dividing the number of I/O operations by the size b_l of blocks moving between levels $l - 1$ and l. This lower bound can be achieved for matrix multiplication because data is always read from the higher-level memory in the same way every time. For the FFT graph in the standard MHG instead of pebbling FFT subgraphs on 2^{d_r} inputs, we pebble b_l FFT subgraphs on $2^{d_r}/b_l$ inputs (assuming that b_l is a power of 2). This allows all the data moving back and forth between memories in blocks to be used and accommodates the transposition mentioned in the caption to Figure 2. This provides an upper bound of $O(n \log n/(b_{l-1} \log(s_{l-1}/b_{l-1})))$ on the I/O time at level l. Clearly, when b_{l-1} is much smaller than s_{l-1}, say $b_{l-1} = O(\sqrt{s_{l-1}})$, the upper and lower bounds match.

8 Conclusions

The Memory Hierarchy Game has been introduced and new lower bounds developed on the computation and I/O time needed at each level of a memory hierarchy to compute functions from straight-line programs. We have studied two variants of the game, one in which the highest level memory can be used for intermediate results and another in which it cannot. We have demonstrated the utility of this new game by showing that our lower bounds can be met by pebbling strategies for matrix multiplication, the Fourier transform, as well as merging and permutation networks.

Since it is very expensive to increase the speed of CPUs, parallel machines are becoming increasingly more attractive. To fully exploit parallelism it will be necessary to provide high-speed parallel memory hierarchy systems. We need a much better understanding of parallel I/O systems if we are going to meet this challenge.

References

[1] A. Aggarwal, B. Alpern, A. Chandra, and M. Snir, "A Model for Hierarchical Memory," *Procs. 21st Annual ACM Symposium on Theory of Computing* (May 15-17, 1989), 305–314.

[2] A. Aggarwal, A. Chandra, and M. Snir, "Hierarchical Memory with Block Transfer," *Proc. 28th Annl. Symp. on Foundations of Computer Science* (October 1987), 204–216.

[3] A. Aggarwal and J. S. Vitter, "The Input/Output Complexity of Sorting and Related Problems," *Communications of the ACM* 31 (September 1988), 1116–1127.

[4] B. Alpern, L. Carter, and E. Feig, "Uniform Memory Hierarchies," *Proc. 31st Annual Symposium on Foundations of Computer Science* (October 22-24, 1990), 600–608.

[5] D. A. Carlson, "Time-Space Tradeoffs for Back-to-Back FFT Algorithms," *IEEE Trans. Computing* C-32 (1983), 585–589.

[6] S. Even and A. Litman, "Layered Cross Product - A Technique to Construct Interconnection Networks," *Proc. 4th Ann. ACM Symp. on Parallel Algorithms and Architectures* (June 29 - July 1, 1992), 60–69.

[7] D. Y. Grigoryev, "An Application of Separability and Independence Notions for Proving Lower Bounds of Circuit Complexity," *Notes of Scientific Seminars, Steklov Math. Inst.* 60 (1976), 35–48.

[8] J. -W. Hong and H. T. Kung, "I/O Complexity: The Red-Blue Pebble Game," *Proc. 13th Ann. ACM Symp. on Theory of Computing* (May 11-13, 1981), 326–333.

[9] F. T. Leighton, in *Introduction to Parallel Algorithms and Architectures*, Morgan Kaufmann Publishers, Inc., San Mateo, CA, 1992.

[10] M. H. Nodine and J. S. Vitter, "Large-Scale Sorting in Parallel Memories (Extended Abstract)," *Procs. 3rd Annual ACM Symposium on Parallel Algorithms and Architectures* (July 21-24, 1991), 29–39.

[11] M. S. Paterson and C. E. Hewitt, "Comparative Schematology," *Proc. Proj. MAC Conf. on Concurrent Systems and Parallel Computation* (June 1970), 119–127.

[12] J. E. Savage and J. S. Vitter, "Parallelism in Space-Time Tradeoffs," in *Advances in Computing Research*, F. P. Preparata, ed., 1987, 117–146.

[13] J. E. Savage, *A Generalization of Grigoryev's Space-Time Tradeoff Method*, Unpublished manuscript, March 1995.

[14] M. Tompa, "Time-Space Tradeoffs for Computing Functions, Using Connectivity Properties of Their Circuits," *JCSS* 20 (1980), 118–132.

[15] M. Tompa, "Corrigendum: Time-Space Tradeoffs for Computing Functions, Using Connectivity Properties of Their Circuits," *JCSS* 23 (1981), 106.

[16] J. S. Vitter and E. A. M. Shriver, "Optimal Disk I/O with Parallel Block Transfer," *Procs. 22nd Annual ACM Symposium on Theory of Computing* (May 1990), 159–169.

[17] C. L. Wu and T. Y. Feng, "The Universality of the Shuffle-Exchange Network," *IEEE Trans. Computing* C-30 (May 1981), 324–332.

On log-Time Alternating Turing Machines of Alternation Depth k
(Extended Abstract)

Liming Cai[1] * and Jianer Chen[2] **

[1] East Carolina University, Greenville, NC 27858, USA
[2] Texas A&M University, College Station, TX 77843, USA

Abstract. Several input read-modes for alternating Turing machines have been proposed in the literature. For each input read-mode and for each fixed integer $k \geq 1$, a precise circuit characterization is established for log-time alternating Turing machines of k alternations, which is a nontrivial refinement of Ruzzo's circuit characterization of alternating Turing machines. Complete languages in strong sense for each level of the log-time hierarchy are presented, refining a result by Buss. The class $GC(s(n), \Pi_k^B)$ is investigated, which is the class of languages accepted by log-time alternating Turing machines of k alternations enhanced by an extra ability of guessing a string of length $s(n)$. A systematic technique is developed to show that for many functions $s(n)$ and for every integer $k > 1$, the class $GC(s(n), \Pi_k^B)$ has natural complete languages. Connections of these results to computational optimization problems are exhibited.

1 Introduction

Sublinear-time Turing machines have proved to be a very useful computational model. In particular, the parallel complexity classes NC and AC can be characterized by sublinear-time alternating Turing machines [18, 12]. To make sublinear-time Turing machines meaningful, we allow a Turing machine to have a *random access input tape* plus an *input address tape*, such that the Turing machine has access to the bit of the input tape denoted by the contents of the input address tape.

A number of input read-modes for sublinear-time alternating Turing machines have appeared in the literature. The original input read-mode [10] has no restrictions on input reading so each computation path of an $O(\log n)$-time alternating Turing machine can read up to $\Theta(\log n)$ input bits. Ruzzo [18] proposed an input read-mode in which each computation path can read at most one input bit and the reading must be performed at the end of the path. An input read-mode studied by Sipser [19] insists that the input address tape be always reset to blank after each input reading so that each input reading takes time $\Omega(\log n)$. Another interesting input read-mode was proposed by Aggarwal, Chandra, and Snir [1] in which access to location i takes time $\log i$ and consecutive locations can be read one unit of time per bit after the initial access time.

* Supported in part by Engineering Excellence Award from Texas A&M University.
** Supported in part by the United States National Science Foundation grant CCR-9110824 and by a P.R.China HTP-863 grant.

We first study the relationship between uniform circuit families and alternating Turing machines under different input read-mode. For each of the input read-modes and for each integer $k \geq 1$, we establish a precise circuit characterization for $O(\log n)$-time alternating Turing machines of k alternations. Our result is a refinement of Ruzzo's circuit characterization of alternating Turing machines [18] as well as a refinment of a theorem by Buss et al. [5] who showed the equivalence of $O(\log n)$-time alternating Turing machines of constant alternations and uniform families of circuits of constant depth. Our circuit characterizations indicate clearly the differences among these input read-modes. Moreover, these characterizations give directly natural complete languages in strong sense for each level of the log-time hierarchy, refining a result by Buss [4].

We then study the model $GC(s(n), \Pi_k^B)$ for each $k \geq 1$, which is an $O(\log n)$-time alternating Turing machine of k alternations that is allowed to make extra $O(s(n))$ amount of nondeterminism. For functions $s(n)$ larger than $\Theta(\log n)$, the model $GC(s(n), \Pi_k^B)$ has guessing ability presumely stronger than and verifying ability provably weaker than deterministic polynomial-time Turing machines. A systematic method is developed to show that for many functions $s(n)$ and for all integers $k > 1$, the class $GC(s(n), \Pi_k^B)$ has natural complete languages.

The importance of the class $GC(s(n), \Pi_k^B)$ is its close connections to computational optimization problems. We show that the optimization classes $LOGSNP$ and $LOGNP$ introduced by Papadimitriou and Yannakakis [17] can be precisely characterized by $GC(\log^2 n, \Pi_2^B)$ and $GC(\log^2 n, \Pi_3^B)$, respectively. A proof is given that the class $GC(\log^2 n, \Pi_2^B)$ is a proper subclass of the class $GC(\log^2 n, \Pi_3^B)$, which gives a strong evidence that the optimization classes $LOGSNP$ and $LOGNP$ are distinct. We explain based on our characterization of the class $LOGSNP$ why the problems Log^{2}SAT, Log Clique, and Log Chordless Path do not seem to be complete for the class $LOGSNP$. This partially answers a question posed by Papadimitriou and Yannakakis [17]. Our characterizations also give a restricted version of the satisfiability problem that is polynomial-time equivalent to the problem Tournament Dominating Set, improving a result of Meggido and Vishkin [15]. An application of the class $GC(s(n), \Pi_2^B)$ in theory of fixed-parameter tractability is also described: the class $GC(s(n), \Pi_2^B)$ is a subclass of the class P for some function $s(n) = \omega(\log n)$ if and only if a large class of NP-hard optimization problems, including Dominating Set and 0-1 Integer Programming, are fixed parameter tractable.

We assume reader's familiarity with the basic definitions in circuit complexity theory, such as circuit depth and size [2], U_D-uniformity of circuit families [18], and alternating Turing machines (ATMs) [10]. The reader may read the original papers for interesting discussions on these topics.

To simplify the expressions, we will denote $2^{\lceil \log \log n \rceil}$ by $\ell(n)$. A string x of length n can be partitioned into $\lceil n/\ell(n) \rceil$ segments of length $\ell(n)$. These segments will be called the ℓ-blocks of x.

We will concentrate on log-time ATMs. The read-modes of ATMs proposed in the literature [1, 10, 18, 19] motivate the following definitions.

Definition 1. A log-time ATM M is of *read-mode* $\mathcal{R}$ if each computation path p of M reads at most one input bit and the reading is performed at the end of p; M is of *read-mode* $\mathcal{S}_c$ if each computation path of M reads at most c input bits; M is of *read-mode* B_c if every computation path of M reads input from at most c ℓ-blocks of the input; and M is of *read-mode* U if there is no restriction on input reading.

A $\Pi_k^{S_c}$-*LOGTIME ATM* (resp. Π_k^R-, Π_k^U-, $\Pi_k^{B_c}$-*LOGTIME ATM*) is a log-time ATM of read-mode S_c (resp. R, U, B_c) that makes at most k alternations and must begin with $\wedge$ states.

Definition 2. [2] A circuit is a Π_k^s-*circuit* if it contains at most s gates organized into at most k levels with an AND-gate at the output. A circuit is a $\Pi_k^{s,c}$-*circuit* if it contains at most s gates organized into at most $k+1$ levels with an AND-gate at the output and level-1 gates of fan-in at most c. A family $\{\alpha_n \mid n \geq 1\}$ of circuits is a Π_k^{poly}-*family* (resp. $\Pi_k^{poly,c}$-*family*) if there is a polynomial p such that for all $n \geq 1$, α_n is a $\Pi_k^{p(n)}$-circuit (resp. $\Pi_k^{p(n),c}$-circuit).

2 Circuit characterizations of log-time ATMs of k alternations

Definition 3. A log-time ATM M is *simple* if it satisfies the following conditions:
1. no phases except the last phase of any computation path of M read the input;
2. For each input length n, the structure of the $\wedge$-$\vee$ tree of M and the set of input variables read by each computation path in the $\wedge$-$\vee$ tree are independent of the contents of the input; and
3. Each last phase of M either is an "accept/reject" decision independent of the contents of the input, or computes a function of form either $\wedge_{i=1}^l z_i$ or $\vee_{i=1}^l z_i$, where z_i is an input literal.

We say that a log-time ATM M is of *read-mode* B_c *in the last phase* if for every computation path of M on any input, the last phase reads input from at most c ℓ-blocks of the input. Similarly, we define read-mode R, S_c, and U in the last phase of a log-time ATM.

Theorem 4. *For any log-time ATM M of k alternations, there is a simple log-time ATM M' of k alternations accepting the same language such that M and M' have the same read-mode in the last phase, and that M and M' begin with the same type of branching on each input.*

We point out that Theorem 4 improves previous results. Ruzzo [18] described a method to convert an alternating Turing machine M of read-mode U into one M' of read-mode R, in which the number of alternations can be as many as twice of the number of input bits read along a computation path of M (a more careful implementation of Ruzzo's method may make M' at most double the number of alternations of M). Boppana and Sipser [2] proposed a different method so that the number of alternations of the simulating machine M' is the same as that of the simulated machine M. However, the read-mode of their simulating machine M' *in the last phase* is equal to the read-mode of the simulated machine M *in the whole computation*. (Note that M may have read-mode U in the whole computation but have read-mode R in the last phase.)

Now we are ready to derive the circuit characterizations of log-time ATMs under variety of input read-modes. We first consider the log-time ATMs of read-mode $\mathcal{B}_c$. Let $x = x_1 x_2 \cdots x_n$ be a string of n Boolean variables. A *specimen* of an ℓ-block $x_{b+1} \cdots x_{b+\ell(n)}$ of x is a string $y_{b+1} \cdots y_{b+\ell(n)}$, where $b = i\ell(n)$ for some integer $i \geq 0$, and y_j is either x_j or $\overline{x}_j$.

Definition 5. A family $F = \{\alpha_n \mid n \geq 1\}$ of circuits is a $\Pi_k^{poly, \mathcal{B}_c}$-*family* if there is a polynomial p such that for all $n \geq 1$, α_n is a $\Pi_k^{p(n), c\ell(n)}$-circuit in which the input of each level-1 gate can be partitioned into exactly c groups such that each group is a specimen of an ℓ-block of $x_1 x_2 \cdots x_n$.

Theorem 6. *A language $L \in \{0, 1\}^*$ is accepted by a $\Pi_k^{\mathcal{B}_c}$-LOGTIME ATM if and only if it is accepted by a U_D-uniform $\Pi_k^{poly, \mathcal{B}_c}$-family of circuits.*

A circuit family $F = \{\alpha_n \mid n \geq 1\}$ is U_D^*-*uniform* if F is U_D-uniform and the set of depth-1 subcircuits of the circuits in F is computed by a linear-time deterministic Turing machine.

Theorem 7. *A language $L \in \{0, 1\}^*$ is accepted by a $\Pi_k^{\mathcal{U}}$-LOGTIME ATM if and only if L is accepted by a U_D^*-uniform Π_{k+1}^{poly}-family of circuits.*

Theorem 8. *Let $c, k \geq 1$ be arbitrary integers. A language $L \in \{0, 1\}^*$ is accepted by a $\Pi_k^{\mathcal{S}_c}$-LOGTIME ATM if and only if L is accepted by a U_D-uniform $\Pi_k^{poly, c}$-family of circuits.*

Corollary 9. *For any integer $k \geq 1$, a language $L \in \{0, 1\}^*$ is accepted by a $\Pi_k^{\mathcal{R}}$-LOGTIME ATM if and only if L is accepted by a U_D-uniform Π_k^{poly}-family of circuits.*

We give two direct applications of the above results.

Sipser [19] has shown that the log-time hierarchy *based on Sipser's model* does not collapse. With our circuit characterization, this result can be extended to the log-time hierarchy based on the standard model.

Theorem 10. *Let $\Pi_k^{\mathcal{U}}$ be the class of languages accepted by $\Pi_k^{\mathcal{U}}$-LOGTIME ATMs. Then for all $k \geq 1$, the class $\Pi_k^{\mathcal{U}}$ is a proper subclass of the class $\Pi_{k+1}^{\mathcal{U}}$.*

Let $h \geq 1$ be an integer. A Boolean sentence s is a Π_0^h-sentence or a Σ_0^h-sentence if it is either 0 or 1. Inductively, for $k \geq 1$, a Boolean sentence s is a Π_k^h-sentence (resp. Σ_k^h-sentence) if it is a product of 2^h Σ_{k-1}^h-sentences (resp. a sum of 2^h Π_{k-1}^h-sentences). A Boolean sentence s is a *depth-k Boolean sentence* if it is a Π_k^h-sentence for some integer h. The depth-k Boolean sentences defined here are essentially the same as the functions $F_k^{2^h}$ introduced by Sipser [19].

The *Depth-k Boolean Sentence Value Problem* (k-BSVP) is to determine the truth value of a depth-k Boolean sentence.

Theorem 11. *For all integer $k > 1$, the k-BSVP problem is complete under log-time reduction for the class $\Pi_k^{\mathcal{U}}$.*

Buss [4] has shown that the *balanced Boolean sentence value problem* is complete for ALOGTIME under log-time reduction. Since each balanced Boolean sentence has depth $O(\log n)$ (where n is the length of the sentence), Theorem 11 is actually a refinement of Buss' result.

Each level of the log-time hierarchy does not seem to be closed under the log-time reduction. Buss [4] has observed that if L is in the class $\Pi_k^{\mathcal{U}}$ and L' is log-time reduced to L, then L' is in the class $\Pi_{k+1}^{\mathcal{U}}$. We say that a language L is complete for the class $\Pi_k^{\mathcal{U}}$ under log-time reduction *in strong sense* if L is complete for $\Pi_k^{\mathcal{U}}$ under log-time reduction and every language log-time reducible to L is in the class $\Pi_k^{\mathcal{U}}$.

Theorem 12. *The k-BSVP problem is complete for the class $\Pi_k^{\mathcal{U}}$ under log-time reduction in strong sense.*

3 Log-time ATMs that can guess $s(n)$ extra bits

In this section, we study log-time ATMs that are enhanced by an extra guessing ability. A log-time ATM is a Π_k^B-*LOGTIME ATM* if it is a $\Pi_k^{B_d}$-LOGTIME ATM for some integer $d \geq 1$. Define Π_k^B to be the class of languages accepted by Π_k^B-LOGTIME ATMs.

Definition 13. Let $s(n)$ be a function and let $k \geq 1$ be an integer. A language L is in the class $GC(s(n), \Pi_k^B)$ if there are a Π_k^B-LOGTIME ATM M and an integer $c > 0$ such that for all $x \in \{0,1\}^*$, $x \in L$ if and only if $\exists y \in \{0,1\}^*$, $|y| = c \cdot s(|x|)$, and M accepts $\langle x, y \rangle$.

The GC class can be generalized by replacing the class Π_k^B by any complexity class. A number of classes with limited nondeterminism [3, 13, 20] can be studied based on the generalized GC models [7].

Definition 14. $BWCS(s(n), k)$ is the set of pairs $x = \langle \alpha, w \rangle$, where $w \leq s(|x|)$, and α is a circuit of depth at most k with an AND gate at the output such that α accepts an input of weight w.

Theorem 15. *Let $s(n) \leq n$ be a non-decreasing function computable in deterministic $O(\log n)$ space. Then for all $k > 1$, the language $BWCS(s(n), k)$ is complete for the class $GC(s(n)\ell(n), \Pi_k^B)$ under log-space reduction.*

Proof. We omit the proof that $BWCS(s(n), k)$ is in the class $GC(s(n)\ell(n), \Pi_k^B)$.

To prove the hardness, let L be a language in $GC(s(n)\ell(n), \Pi_k^B)$. By the definition, there are a Π_k^B-ATM M and an integer $c > 0$ such that $x \in L$ if and only if there is a $y \in \{0,1\}^*$, $|y| = c\,s(|x|)\ell(|x|)$, and M accepts $\langle x, y \rangle$. By Theorem 4, we can assume that in the $\wedge$-$\vee$ tree of M, no phases except the last phases have access to input. By Theorem 6, there is a U_D-uniform Π_k^{poly, B_h}-family $\{\tau_m \mid m \geq 1\}$ of circuits that accepts $L(M)$, where h is a constant.

Given an instance x of the language L, we show how to reduce x to an instance $z = \langle \alpha(x), w(x) \rangle$ for the language $BWCS(s(n), k)$ such that $x \in L$ if and only if z is in $BWCS(s(n), k)$. Let $|x| = n$, and $|\langle x, y \rangle| = m$, where y is a binary string of length

$c\,s(n)\ell(n)$. Let $\tau_m(x)$ be the circuit τ_m with the first part of the input assigned by the value of x. $\tau_m(x)$ is a circuit with $c\,s(n)\ell(n)$ inputs, and $x \in L$ if and only if the circuit $\tau_m(x)$ is satisfiable.

Since $|y| = c\,s(n)\ell(n) = O(n\log n)$, so $|x| = n \le m = |\langle x,y\rangle| \le n^2$ for n sufficiently large. Therefore, we have either $\ell(m) = \ell(n)$, or $\ell(m) = 2\ell(n)$. Thus, the input length of the circuit $\tau_m(x)$ can be written as $as(n)\ell(m)$, where a is a constant.

Recall that the circuit τ_m is a Π-circuit of depth $k+1$ in which the input of each level-1 gate is exactly h specimens of ℓ-blocks of $\langle x,y\rangle$. Moreover, each computation path of M reads input bits from at most one ℓ-block from the string y. Therefore, at most one specimen in the input of each level-1 gate of τ_m is from an ℓ-block from the string y. Thus, the circuit $\tau_m(x)$ is a Π-circuit of depth $k+1$ and input length $as(n)\ell(m)$ in which the input of each level-1 gate is a specimen of an $\ell(m)$-block of its input. As shown in [6], we can construct a Π-circuit $\gamma_m(x)$ of depth $k+1$ computing the same function as $\tau_m(x)$ in which the input of each level-1 gate is a specimen of an $(a\ell(m))$-block of its input. Moreover, the size of the circuit $\gamma_m(x)$ is bounded by a polynomial of m, for all $m \ge 1$.

Now we are ready to describe the circuit $\alpha(x)$.

The input $I(\alpha(x)) = v_1 \cdots v_{s(n)2^{a\ell(m)}}$ of the circuit $\alpha(x)$ is of length $s(n)2^{a\ell(m)}$ and partitioned into $s(n)$ $2^{a\ell(m)}$-blocks. Similarly, the input $I(\gamma_m(x))$ of the circuit $\gamma_m(x)$ is partitioned into $s(n)$ $(a\ell(m))$-blocks. The circuit $\alpha(x)$ will be constructed from the circuit $\gamma_m(x)$ by replacing each level-1 gate in $\gamma_m(x)$ by an input node of $\alpha(x)$. For this, we will use a position in the qth $2^{a\ell(m)}$-block of $I(\alpha(x))$ to represent an assignment to the qth $(a\ell(m))$-block of $I(\gamma_m(x))$. To simplify the discussion, we describe the construction of the first $2^{a\ell(m)}$-block of $I(\alpha(x))$ based on the first $(a\ell(m))$-block of $I(\gamma_m(x))$. The construction of the qth $2^{a\ell(m)}$-block of $I(\alpha(x))$ for general q can be done similarly.

Let g be a level-1 gate of the circuit $\gamma_m(x)$ whose input is a specimen of the first $(a\ell(m))$-block of $I(\gamma_m(x))$. If k is even, then the gate g is an AND gate. Thus, there is a unique Boolean assignment to the first $(a\ell(m))$-block of $I(\gamma_m(x))$ that makes the gate g have value 1. Let b be the unique Boolean string of length $a\ell(m)$ that makes the gate g have value 1. Regarding b as a binary number between 0 and $2^{a\ell(m)} - 1$, we replace the gate g in $\gamma_m(x)$ by the positive input node v_{b+1} in $I(\alpha(x))$. If k is odd, then the gate g is an OR gate. There is a unique assignment to the first $(a\ell(m))$-block of $I(\gamma_m(x))$ that makes the gate g to have value 0. Let b be the binary number corresponding to this assignment. We replace the gate g by the negative input node $\bar{v}_{b+1}$ of $I(\alpha(x))$. In either case, we will perform the corresponding replacement on each level-1 gate in the circuit $\gamma_m(x)$. The resulting circuit $\alpha(x)$ has depth k.

By the construction, each assignment of the qth $(a\ell(m))$-block of $I(\gamma_m(x))$ in $\gamma_m(x)$ can be implemented by a weight 1 assignment of the qth $2^{a\ell(m)}$-block of the input $I(\alpha(x))$ in the circuit $\alpha(x)$. Therefore, each assignment of $I(\gamma_m(x))$ in $\gamma_m(x)$ can be implemented by a weight $s(n)$ assignment of $I(\alpha(x))$ in $\alpha(x)$, in which the assignment to each $2^{a\ell(m)}$-block has weight 1. We conclude that if circuit $\gamma_m(x)$ is satisfiable, then circuit $\alpha(x)$ accepts an input of weight $s(n)$.

The construction has not been completed yet. Note that if in an assignment of $I(\alpha(x))$, some $2^{a\ell(m)}$-block has weight different from 1, then the assignment does not implement any assignment of $I(\gamma_m(x))$ in $\gamma_m(x)$. To ensure that each $2^{a\ell(m)}$-block of $I(\alpha(x))$ is assigned exact one 1, we use the following function:

$$\phi(v_1, \cdots, v_t) = (v_1 \vee \cdots \vee v_t) \bigwedge_{i,j} (\bigwedge (\bar{v}_i \vee \bar{v}_j))$$

It is easy to see that $\phi(v_1, \cdots, v_t) = 1$ if and only if exactly one v_i is 1. The function $\phi(v_1, \cdots, v_t)$ can be implemented by a Π-circuit of depth 2. Now for each $2^{a\ell(m)}$-block of $I(\alpha(x))$, we add a Π-subcircuit of depth 2 that implements the function ϕ with $2^{a\ell(m)}$ variables, and connect the output of this subcircuit to the output gate of the circuit $\alpha(x)$ (which is the output gate of the circuit $\gamma_m(x)$). Note that this would not increase the depth of the circuit $\alpha(x)$ since both the output gate of such a subcircuit and the output gate of the circuit $\alpha(x)$ are AND gates, and the depth of the circuit $\alpha(x)$ is at least 2. This completes the construction of the circuit $\alpha(x)$.

Now if the circuit $\alpha(x)$ accepts an assignment of weight $s(n)$ then each $2^{a\ell(m)}$-block of the assignment must have weight 1. Thus, the assignment implements a satisfying assignment to the circuit $\gamma_m(x)$.

Since the circuit $\gamma_m(x)$ is satisfiable if and only if $x \in L$, we conclude that the circuit $\alpha(x)$ accepts a weight $s(n)$ input if and only if $x \in L$. Consequently, the mapping from x to $\langle \alpha(x), s(n) \rangle$ is a many-one reduction from the language L to the language $BWCS(s(n), k)$. Note that we can always make $|\langle \alpha(x), s(n) \rangle| \geq n$, so we always have $s(n) \leq s(|\langle \alpha(x), s(n) \rangle|)$.

By the assumption, the function $s(n)$ can be constructed in deterministic $O(\log n)$ space. Since the circuit family $\{\tau_m \mid m \geq 1\}$ is U_D-uniform, the circuit τ_m can be constructed in deterministic $O(\log m) = O(\log n)$ space. It is also easy to see that the circuit constructions from τ_m to $\tau_m(x)$, from $\tau_m(x)$ to $\gamma_m(x)$, and from $\gamma_m(x)$ to $\alpha(x)$ can all be done in deterministic $O(\log n)$ space. Therefore, the reduction can be implemented in deterministic $O(\log n)$ space. This completes the proof of the theorem. $\square$

In many cases, the function $\ell(n)$ in $GC(s(n)\ell(n), \Pi_k^B)$ can be replaced by $\log n$, as shown by the following theorem.

Theorem 16. *If the function $s(n)$ is computable in deterministic $O(\log n)$ time, and $|s(n)| = O(\log n / \log \log n)$ for all n, then for all $k \geq 1$*

$$GC(s(n)\ell(n), \Pi_k^B) = GC(s(n) \log n, \Pi_k^B)$$

4 GC classes and optimization problems

The GC classes have a number of interesting connections to computational optimization problems. We will present two of them here. For a more complete discussion, we refer our readers to [8, 9].

We first consider the relationship between the GC classes and the optimization classes $LOGSNP$ and $LOGNP$ introduced by Papadimitriou and Yannakakis [17].

Following [17], define $LOGNP_0$ to be the class of all problems described as follows:

$$\{I : \exists S \in [n]^{\log n} \forall x \in [n]^p \exists y \in [n]^q \forall j \in [\log n] \phi(I, s_j, x, y, j)\} \tag{1}$$

where ϕ is a quantifier-free first-order expression involving the relation symbol I, and the variables in x, y, as well as the variables j and s_j.

A weaker class $LOGSNP_0$ contains all problems definable by one less alternation of quantifiers:

$$\{I : \exists S \in [n]^{\log n} \forall x \in [n]^p \exists j \in [\log n]\phi(I, s_j, x, j)\} \tag{2}$$

The class $LOGNP$ is defined to be the class of all languages polynomial-time reduced to a problem in $LOGNP_0$, and the class $LOGSNP$ is defined to be the class of all languages polynomial-time reduced to a problem in $LOGSNP_0$. The reader should realize that the classes $LOGNP$ and $LOGSNP$ are defined very much in the spirit of the classes $MAXNP$ and $MAXSNP$ that have received considerable attention recently [16].

The complexity of a number of interesting optimization problems can be nicely characterized by the classes $LOGNP$ and $LOGSNP$. For instance, it has been shown [17] that Tournament Dominating Set and Rich Hypergraph Cover are complete under polynomial-time reduction for the class $LOGSNP$, and that the problem V-C Dimension is complete under polynomial-time reduction for the class $LOGNP$.

Theorem 17. *A language L is in the class $LOGSNP$ if and only if L is polynomial-time reduced to a language in $GC(\log^2 n, \Pi_2^B)$.*

Theorem 18. *A language L is in the class $LOGNP$ if and only if L is polynomial-time reduced to a language in $GC(\log^2 n, \Pi_3^B)$.*

One may not expect an easy proof that the classes $LOGSNP$ and $LOGNP$ are different because that would imply $P \neq NP$. However, based on the GC model characterizations of these classes, we can show a strong evidence that these two classes are distinct.

Theorem 19. *For all integer $k \geq 1$, the class $GC(\log^2 n, \Pi_k^B)$ is a proper subclass of the class $GC(\log^2 n, \Pi_{k+1}^B)$.*

Papadimitriou and Yannakakis [17] have shown that the following three problems are in $LOGSNP$.

Log²SAT: "Given a CNF Boolean formula with n clauses and $\log^2 n$ variables, does it have a satisfying truth assignment?"
Log Clique: "Given a graph with n vertices, does it have a clique of size $\log n$?"
Log Chordless Path: "Given a graph with n vertices, does it have a chordless path of length $\log n$?"

It was asked in [17] whether these three problems are complete for the class $LOGSNP$. By Theorems 17 and 19, we show that these problems are unlikely to be complete for the class $LOGSNP$.

The problem Log Clique is actually in the class $GC(\log^2 n, \Pi_1^B)$. In fact, let G be a graph and let y encode $\log n$ vertices of G, then on input $\langle G, y \rangle$, a Π_1^B-ATM can universally pick a pair (v, w) of vertices from the string y and check with G if there is an edge connecting them. Similarly, the problem Log Chordless Path can be shown to be in the class $GC(\log^2 n, \Pi_1^B)$. By Theorem 19, the class $GC(\log^2 n, \Pi_1^B)$ is a proper subclass of the class $GC(\log^2 n, \Pi_2^B)$. Therefore, the problems Log Clique and Log Chordless Path seem not hard enough to be complete for the class $LOGSNP$. Similarly, the problem Log^{2}SAT can be shown to belong to the class $GC(\log^2 n, \Pi_2^{B*})$, where Π_2^{B*} is the class of languages accepted by restricted Π_2^B-ATMs in which the last phase of each computation path runs at most $O(\log\log n)$ steps and reads at most constant number of input bits. It can also be shown that $GC(\log^2 n, \Pi_2^{B*})$ is a proper subclass of $GC(\log^2 n, \Pi_2^B)$. Therefore, the problem Log^{2}SAT also seems not hard enough to be complete for the class $GC(\log^2 n, \Pi_2^B)$.

Meggido and Vishkin [15] studied the problem Tournament Dominating Set and proved that the problem Log^{2}SAT is polynomial-time reduced to Tournament Dominating Set, and that Tournament Dominating Set is polynomial-time reduced to a generalization of the Log^{2}SAT problem. They asked whether there is a version of the satisfiability problem that precisely characterizes the problem Tournament Dominating Set. Our Theorem 17 concludes that the problem Tournament Dominating Set is polynomial-time equivalent to the problem BWCS($\log n$, 2), which is the standard satisfiability problem with a weight restriction on the truth assignment.

Finally, we describe an application of the GC model in the study of fixed-parameter tractability of optimization problems.

It has been observed [14] that many NP-hard optimization problems can be parameterized, while their time complexity behaves very differently with respect to the parameter. For example, with a fixed parameter q, the problem Vertex Cover ("Given a graph G, does G have a vertex cover of size q?") can be solved in time $O(n^c)$, where c is a constant independent of q, while the problem Dominating Set ("Given a graph G, does G have a dominating set of size q?") has the contrasting situation where essentially no better algorithm is known than the "trivial" one of time $\Theta(n^{q+1})$ that just exhaustively tries all possible solutions.

Downey and Fellows [14] have introduced a new framework to discuss the *fixed-parameter tractability* of NP-hard problems. A hierarchy, called the *W-hierarchy*, has been established to classify the fixed-parameter intractability of NP-hard optimization problems. At the bottom of the W-hierarchy (the zeroth level) is the class of NP-hard problems whose parameterized version can be solved in time $O(n^c)$, where c is a constant independent of the parameter q. Vertex Cover, for example, is in the bottom level of the W-hierarchy. Independent Set, Clique, and many others are in the first level of the W-hierarchy; and in the second level, there are Dominating Set, 0-1 Integer Programming, and many others. A fundamental question is whether the W-hierarchy collapses.

Theorem 20. *For any integer $k > 1$, the kth level of the W-hierarchy collapses to the zeroth level if and only if $GC(s(n), \Pi_k^B) \subseteq P$ for some function $s(n) = \omega(\log n)$.*

The above theorem in particular shows that the problems such as Dominating Set and 0-1 Integer Programming are unlikely to be fixed-parameter tractable.

Corollary 21. *Dominating Set and 0-1 Integer Programming are fixed parameter tractable if and only if $GC(s(n), \Pi_2^B)$ is a subset of P for some function $s(n) = \omega(\log n)$.*

References

1. Aggarwal, A., Chandra, A. K., and Snir, M.: Hierarchical memory with block transfer. *Proc. 28th Annual IEEE Symposium on Foundations of Computer Science* (1987) 204-216
2. Boppana, R. B. and Sipser, M.: The complexity of finite functions. in *Handbook of Theoretical Computer Science Vol. A*, J. van Leeuwen, ed. (1990) 757-804
3. Buss, J. F. and Goldsmith, J.: Nondeterminism within *P*. *SIAM J. Comput.* **22** (1993) 560-572
4. Buss, S. R.: The Boolean formula value problem is in ALOGTIME. *Proc. 19th Annual ACM Symposium on Theory of Computing* (1987) 123-131
5. Buss, S., Cook, S., Gupta, A., and Ramachandran, V.: An optimal parallel algorithm for formula evaluation. *SIAM J. Comput.* **21** (1992) 755-780
6. Cai, L. and Chen, J.: On input read-modes of alternating Turing machines. *Theoretical Computer Science* (1995) to appear
7. Cai, L. and Chen, J.: On the amount of nondeterminism and the power of verifying. *Lecture Notes in Computer Science* **711** *(MFCS'93)* (1993) 311-320
8. Cai, L. and Chen, J.: Fixed parameter tractability and approximability of *NP*-hard optimization problems. *Proc. 2rd Israel Symposium on Theory of Computing and Systems* (1993) 118-126
9. Cai, L., Chen, J., Downey, R., and Fellows, M: On the structure of parameterized problems in *NP*. *Lecture Notes in Computer Science* **775** *(STACS'94)* (1994) 509-520. Journal version to appear in *Information and Computation*
10. Chandra, A. K., Kozen, D. C., and Stockmeyer, L. J.: Alternation. *J. Assoc. Comput. Mach.* **28** (1981) 114-133
11. Chen, J.: Characterizing parallel hierarchies by reducibilities. *Information Processing Letters* **39** (1991) 303-307
12. Cook, S.: A taxonomy of problems with fast parallel algorithms. *Information and Control* **64** (1985) 2-22
13. Díaz, J. and Torán, J.: Classes of bounded nondeterminism. *Math. System Theory* **23** (1990) 21-32
14. Downey, R. G. and Fellows, M. R.: Fixed-parameter intractability. *Proc. 7th Structure in Complexity Theory Conference* (1992) 36-49
15. Meggido, N. and Vishkin, U.: On finding a minimum dominating set in a tournament. *Theoretical Computer Science* **61** (1988) 307-316
16. Papadimitriou, C. H. and Yannakakis, M.: Optimization, approximation, and complexity classes. *J. Comput. System Sci.* **43** (1991) 425-440
17. Papadimitriou, C. H. and Yannakakis, M.: On limited nondeterminism and the complexity of the V-C dimension. *Proc. 8th Structure in Complexity Theory Conference* (1993) 12-18
18. Ruzzo, W. L.: On uniform circuit complexity. *J. Comput. System Sci.* **22** (1981) 365-383
19. Sipser, M.: Borel sets and circuit complexity. *Proc. 15th Annual ACM Symposium on Theory of Computing* (1983) 61-69
20. Wolf, M. J.: Nondeterministic circuits, space complexity, and quasigroups. *Theoretical Computer Science* **125** (1994) 295-313

New Bound for Affine Resolvable Designs and Its Application to Authentication Codes

Kaoru KUROSAWA[1] and Sanpei KAGEYAMA[2]

[1] Department of Electrical and Electronic Engineering,
Tokyo Institute of Technology,
2-12-1 Ookayama, Meguro-ku, Tokyo 152, Japan
kkurosaw@ss.titech.ac.jp
[2] Department of Mathematics, Hiroshima University,
Higashi-Hiroshima 739, Japan
ksanpei@ue.ipc.hiroshima-u.ac.jp

Abstract. In this paper, we first prove that $b \leq v+t-1$ for an "affine" $(\alpha_1, \cdots, \alpha_t)$-resolvable design, where b denotes the number of blocks, v denotes the number of symbols and t denotes the number of classes. Our inequality is an opposite to the well known inequality $b \geq v+t-1$ for a "balanced" $(\alpha_1, \cdots, \alpha_t)$-resolvable design. Next, we present a more tight lower bound on the size of keys than before for authentication codes with arbitration by applying our inequality. Although this model of authentication codes is very important in practice, it has been too complicated to be analyzed. We show that the receiver's key has a structure of an affine α-resolvable design ($\alpha_1 = \cdots = \alpha_t = \alpha$) and v corresponds to the number of keys under a proper assumption. (Note that our inequality is a lower bound on v.)

1 Introduction

The importance of resolvability in the context of experimental plannings is well known (for example, see [1] ~ [8]). A block design is said to be $(\alpha_1, \cdots, \alpha_t)$-resolvable ([5] ~ [8]) if the blocks can be grouped into t classes (resolutions) such that, in the ith class, each symbol appears in α_i blocks ($i = 1, \cdots, t$). Then, it is known [5, 6] that, for a "balanced" $(\alpha_1, \cdots, \alpha_t)$-resolvable design, $b \geq v+t-1$ holds, where b denotes the number of blocks, v denotes the number of symbols and t denotes the number of classes. This inequality is important in experimental plannings because b corresponds to the number of experiments and the inequality shows a lower bound on b. Suppose that each block contains a constant number of symbols, say k. Then, the block design is said to be balanced if any pair of symbols appear in λ blocks, where λ is a constant.

On the other hand, an $(\alpha_1, \cdots, \alpha_t)$-resolvable block design is said to be affine $(\alpha_1, \cdots, \alpha_t)$-resolvable ([6] ~ [8], [3]) if any pair of blocks belonging to the same class contain q_1 symbols in common and any pair of blocks belonging to different classes contain q_2 symbols in common.

In this paper, we first prove that $b \leq v+t-1$ holds for an "affine" $(\alpha_1, \cdots, \alpha_t)$-resolvable design. Our inequality is an opposite to the above mentioned inequality for "balanced" $(\alpha_1, \cdots, \alpha_t)$-resolvable designs. Next, we provide a more tight lower bound on the size of keys than before for authentication codes with arbitration (A^2-codes) ([9] $\sim$ [15]) by applying our inequality. Although this model of authentication codes is very important in practice, it has been too complicated to be analyzed. We show that the receiver's key has a structure of an affine α-resolvable design ($\alpha_1 = \cdots = \alpha_t = \alpha$) and v corresponds to the number of keys under a proper assumption. (Note that our inequality is a lower bound on v.)

In the model of authentication codes (A-codes) [16], there are three participants, a transmitter T, a receiver R and an opponent O. From impersonation attack and substitution attack of O, an A-code protects messages, under the assumption that T and R are both honest and trust each other. This model has been studied extensively so far [16, 17].

However, it is not always the case that T and R want to trust each other. In a new model, the model of authentication codes with arbitration (A^2-codes) [9, 10], there is a fourth person, an arbiter. In this model, caution is taken against deception of T and R as well as that O. The arbiter has access to all key information of T and R and solves disputes between them. In this model, there are essentially five different kinds of attacks, impersonation attacks by O, T and R, and substitution attacks by O and R. Recently, some progress has been made for this model which is much more complicated than the model of A-codes. Johansson [11, 12] showed a lower bound on the size of keys in terms of the cheating probabilities which is tight for small $|S|$, where S denotes the set of data to be transmitted (source states). Kurosawa showed a more tight lower bound on the size of keys for large $|S|$ and for "the separable case" [14]. Quite recently, a combinatorial lower bound on the cheating probabilities was given by Kurosawa and Obana [15].

The proposed lower bound on the size of keys is more tight than Johansson [11, 12] for large $|S|$ and for "the general case". In our bound, cheating probalilities take the minimum values shown by [15]. The bound is also larger than that of Kurosawa [14].

2 Preliminaries

2.1 Notation and definitions

An incomplete block design is a pair (V, X), where $V = \{a_1, \cdots, a_v\}$ is a set of symbols and $X = \{B_1, \cdots, B_b\}$ with $B_j \subsetneq V$, $j = 1, \cdots, b$. B_j is called a block. Let $v \overset{\triangle}{=} |V|$ (the number of symbols) and $b \overset{\triangle}{=} |X|$ (the number of blocks). Furthermore, we assume that $|B_j| = k$ for any j.

Definition 1. An incomplete block design is said to be $(\alpha_1, \cdots, \alpha_t)$-resolvable if the blocks can be grouped into t classes $S_1, \cdots, S_t$ such that, in the ith class, every symbol appears in α_i blocks ($i = 1, \cdots, t$). That is, $|\{B_j | a \in B_j\} \cap S_i| = \alpha_i$ for $\forall a \in V$, where $i = 1, \cdots, t$.

Let $|S_i| = \beta_i$. We then have $b = \sum_{i=1}^{t} \beta_i$.

Definition 2. An $(\alpha_1, \cdots, \alpha_t)$-resolvable incomplete block design is called α-resolvable if $\alpha_1 = \cdots = \alpha_t = \alpha$.

Definition 3. An $(\alpha_1, \cdots, \alpha_t)$-resolvable incomplete block design is called affine $(\alpha_1, \cdots, \alpha_t)$-resolvable if any pair of blocks of the same class intersects in q_1 symbols and any pair of blocks from different classes intersects in q_2 symbols.

Definition 4. An incomplete block design is balanced if, for $\forall a_1 \in V$ and $\forall a_2 \in V$ $(a_1 \neq a_2)$, $|\{B_j | \{a_1, a_2\} \subseteq B_j\}| = \lambda (= \text{constant})$.

Definition 5. An incidence matrix of an incomplete block design is a $v \times b$ matrix $N = (n_{ij})$ such that

$$n_{ij} = \begin{cases} 1 \text{ if } a_i \in B_j \\ 0 \text{ otherwise} \end{cases}$$

where $a_i \in V$ and $B_j \in X$.

2.2 Known results

Proposition 6. *[5, 6] For a balanced $(\alpha_1, \cdots, \alpha_t)$-resolvable incomplete block design, $b \geq v + t - 1$.*

Proposition 7. *[2, 3] If an α-resolvable incomplete block design has any two of the following properties, then it has the third:*
   1. Affine α-resolvability.   2. Balance.   3. $b - v = t - 1$.

Proposition 8. *[8] In an affine $(\alpha_1, \cdots, \alpha_t)$-resolvable design with parameters $v, b = \sum_{i=1}^{t} \beta_i, k, q_1$ and q_2, it holds that $q_1 = k(\alpha_i - 1)/(\beta_i - 1)$ and $q_2 = k\alpha_i/\beta_i$ for $1 \leq \forall i \leq t$.*

Proposition 9. *[?] Let $u, s_1, \cdots, s_u$ be positive integers, and consider the $s \times s$ matrix*

$$A = \begin{bmatrix} a_1 I_{s_1} + b_{11} J_{s_1 s_1} & b_{12} J_{s_1 s_2} & \cdots & b_{1u} J_{s_1 s_u} \\ b_{21} J_{s_2 s_1} & a_2 I_{s_2} + b_{22} J_{s_2 s_2} & \cdots & b_{2u} J_{s_2 s_u} \\ \vdots & \vdots & \ddots & \vdots \\ b_{u1} J_{s_u s_1} & b_{u2} J_{s_u s_2} & \cdots & a_u I_{s_u} + b_{uu} J_{s_u s_u} \end{bmatrix}$$

where $s = s_1 + \cdots + s_u$ and the $u \times u$ matrix $B = (b_{ij})$ is symmetric. Then the eigenvalues of A are a_i with multiplicity $s_i - 1 (1 \leq i \leq u)$ and $\mu_1^*, \cdots, \mu_u^*$, where $\mu_1^*, \cdots, \mu_u^*$ are the eigenvalues of $\Delta = D_a + D_s^{1/2} B D_s^{1/2}$ with $D_a = \text{diag}(a_1, \cdots, a_u)$ and $D_s^{1/2} = \text{diag}(s_1^{1/2}, \cdots, s_u^{1/2})$.

Here I_s denotes the $s \times s$ identity matrix. J_{st} denotes the $s \times t$ matrix whose elements are all 1.

3 New inequality for affine resolvable design

3.1 New inequality

In this section, we show an opposite inequality to Proposition 6 for "affine" $(\alpha_1, \cdots, \alpha_t)$-resolvable designs.

Theorem 10. *For an affine $(\alpha_1, \cdots, \alpha_t)$-resolvable incomplete block design,*

$$b \leq v + t - 1.$$

The equality holds if the design is balanced.

The proof will be given in the next subsection.

Corollary 11. *For an affine α-resolvable incomplete block design, $b \leq v + t - 1$. The equality holds if and only if the design is balanced.*

Proof. The last line comes from Proposition 2.2. □

3.2 Proof of Theorem 10

Lemma 12. *Let N be the incidence matrix of an affine $(\alpha_1, \cdots, \alpha_t)$-resolvable design with parameters $v, b = \sum_{i=1}^{t} \beta_i, k, q_1, q_2$. Then $N^T N$ has at least the eigenvalues; $k - q_1$ with multiplicity $b - t$ and $q_2 b$ with multiplicity at least 1.*

Proof. It is easy to see that

$$N^T N = \begin{bmatrix} kI_{\beta_1} + q_1(J_{\beta_1\beta_1} - I_{\beta_1}) & q_2 J_{\beta_1\beta_2} & \cdots & q_2 J_{\beta_1\beta_t} \\ q_2 J_{\beta_2\beta_1} & kI_{\beta_2} + q_1(J_{\beta_2\beta_2} - I_{\beta_2}) & \cdots & q_2 J_{\beta_2\beta_t} \\ \vdots & \vdots & \ddots & \vdots \\ q_2 J_{\beta_t\beta_1} & q_2 J_{\beta_t\beta_2} & \cdots & kI_{\beta_t} + q_1(J_{\beta_t\beta_t} - I_{\beta_t}) \end{bmatrix}$$

By Proposition 9, the eigenvalues of this matrix are $k - q_1$ with multiplicity $\sum_{i=1}^{t}(\beta_i - 1) = b - t$ and $\mu_1^*, \cdots, \mu_t^*$, when $\mu_1^*, \cdots, \mu_t^*$ are the eigenvalues of

$$\Delta = (k - q_1)I_t + D(q_1 I_t + q_2(J_t - I_t))D \text{ with } D = \text{diag}(\sqrt{\beta_1}, \cdots, \sqrt{\beta_t}).$$

We will show that Δ has an eigenvalue $\mu_1^* = q_2 b$. Let

$$E \triangleq (\beta_1, \cdots, \beta_t)^T, \quad E^{\frac{1}{2}} \triangleq (\sqrt{\beta_1}, \cdots, \sqrt{\beta_t})^T.$$

Then
$$\Delta E^{\frac{1}{2}} = (k - q_1)E^{\frac{1}{2}} + D(q_1 I_t + q_2(J_t - I_t))DE^{\frac{1}{2}}$$
$$= (k - q_1)E^{\frac{1}{2}} + D(q_1 I_t + q_2(J_t - I_t))E.$$

Let $(q_1 I_t + q_2(J_t - I_t))E = (c_1, \cdots, c_t)^T$. Then

$$c_i = \beta_i q_1 + q_2(\sum_{j=1}^{t} \beta_j - \beta_i) = \beta_i q_1 + q_2(b - \beta_i).$$

From Proposition 8, $q_1(\beta_i - 1) = k(\alpha_i - 1)$. $q_1\beta_i = k(\alpha_i - 1) + q_1$. $q_2\beta_i = k\alpha_i$.

Therefore $c_i = q_2 b + q_1\beta_i - q_2\beta_i = q_2 b + q_1 - k$. Hence

$$\Delta E^{\frac{1}{2}} = (k - q_1)E^{\frac{1}{2}} + D(q_2 b + q_1 - k)J_{t1}$$
$$= (k - q_1)E^{\frac{1}{2}} + (q_2 b + q_1 - k)E^{\frac{1}{2}} = q_2 b E^{\frac{1}{2}}.$$

Thus, Δ has an eigenvalue $q_2 b$. $\qquad\square$

Proof of Theorem 10. Note that $k - q_1 > 0$ for the present design. Then, by Lemma 12, $N^T N$ has at least $b - t + 1$ nonzero eigenvalues. Therefore, rank$(N^T N) \geq b - t + 1$.
On the other hand, rank$(N^T N)$ = rank$(N^T) \leq v$. Hence, $b - t + 1 \leq v$. That is, $b \leq v + t - 1$. Further, from Proposition 6, for a balanced $(\alpha_1, \cdots, \alpha_t)$-resolvable design, $b \geq v + t - 1$. Therefore, if an affine $(\alpha_1, \cdots, \alpha_t)$-resolvable design is balanced, then $b = v + t - 1$. $\qquad\square$

4 Model of authentication codes with arbitration

4.1 Authentication code (A-code)

In the model of authentication codes (A-codes) [16], the transmitter T and the receiver R share a common encoding rule e. On input a source state s (a data), T computes a message $m = e(s)$ and sends m to R. R accepts or rejects m based on e. We assume independent probability distributions on source states and on encoding rules, respectively. The opponent O tries to cheat R by impersonation attack or substitution attack. In the impersonation attack, O sends a message m to R. O succeeds if m is accepted by the receiver as authentic. The impersonation attack probability P_I is defined by

$$P_I \triangleq \max_m \Pr[R \text{ accepts } m]. \qquad (4.1)$$

In the substitution attack, O observes a message m that is transmitted by T and substitutes m with another message $\hat{m}$. O succeeds if $\hat{m}$ is accepted by the receiver as authentic. The substitution attack probability P_S is defined by

$$P_S \triangleq \sum_m \Pr(M = m) \max_{\hat{m}} \Pr[R \text{ accepts } \hat{m} | R \text{ accepts } m], \qquad (4.2)$$

where the maximum is taken over $\hat{m}$ whose source state is different from that of m. Let $S = \{s\}$, $M = \{m\}$ and $E = \{e\}$. We denote an A-code by (S, M, E). An A-code is called an A-code without splitting if $\forall e$ generates just one message m for $\forall s$. Further, an A-code is called an A-code without secrecy if a souce state s is uniquely determined from a message m.

This model has been studied extensively so far [16, 17]. For example, the following proposition is known.

Proposition 13. *[17] In an A-code without splitting (S, M, E), $P_I \geq |S|/|M|$. The equality holds if and only if $\Pr[R \text{ accepts } m] = |S|/|M|$ for $\forall m \in M$.*

4.2 Authentication codes with arbitration (A^2-code)

In the model of A-codes, T and R are both honest and trust each other. However, it is not always the case that the two parties want to trust each other. Inspired by this problem, Simmons introduced an extended model, A^2-code model, in which there is a fourth person, an arbiter [9, 10]. In this model, caution is taken against deception of T and R as well as that O. The arbiter has access to all key information of T and R, and solves disputes between them.

We denote an A^2-code by (S, M, E_R, E_T), where $S = \{s\}$ is the set of source state, $M = \{m\}$ is the set of messages, $E_R = \{f\}$ is the set of the receiver's encoding rules and $E_T = \{e\}$ is the set of the transmitter's encoding rules. On input s, T sends m such that $m = e(s)$ to R. R accepts m iff $f(m)$ is valid. The arbiter accepts m as authentic iff e can generate m.

The selection of e and f may be done in several ways. One choice is to let the receiver R choose his f and then secretly passes this on to the arbiter. In this case, the arbiter constructs e and passes this on to the transmitter T. Another choice is to do the other way around and the third approach is to let the arbiter construct both rules.

In this model, there are essentially five different kinds of attacks.

Impersonation by the opponent. The cheating probability P_I is defined in the same way as eq.(4.1).

Substitution by the opponent. The cheating probability P_S is defined in the same way as eq.(4.2).

Impersonation by the transmitter. The transmitter sends a message to the receiver and denies having sent it. The transmitter succeeds if the message is accepted by the receiver as authentic and if the message is not one of the messages that the transmitter could have generated due to his encoding rule. This cheating probability P_T is defined as follows.

$$P_T \triangleq \max_e \max_m \Pr[\text{R accepts } m \text{ and } m \text{ is not generated by } e | \text{T has } e \in E_T].$$
(4.3)

Impersonation by the receiver. The receiver claims to have received a message from the transmitter. The receiver succeeds if the message could have been generated by the transmitter due to his encoding rule. This cheating probability P_{R_0} is defined by

$$P_{R_0} \triangleq \max_f \max_m \Pr[\text{Arbiter accepts } m | \text{R has } f \in E_R].$$
(4.4)

Substitution by the receiver. The receiver receives a message from the transmitter but claims to have received another message. The receiver succeeds if this other message could have been generated by the transmitter due to his encoding rule. This cheating probability P_{R_1} is defined by

$$P_{R_1} \triangleq \max_f \sum_m \Pr(m) \max_{\hat{m} \neq m} \Pr[\text{Arbiter accepts } \hat{m} | \text{R has } f \in E_R \text{ and T sends } m].$$

5 New bound for A^2-codes

Suppose that E_T is no splitting. That is, $\forall e \in E_T$ generates just one message $m \in M$ for $\forall s \in S$. For simplicity, we restrict ourselves to the bound on $|E_R|$ in this paper. A similar result for the bound on $|E_T|$ will be given in the final paper.

5.1 Previous bounds

Johansson [11, 12] showed the following bound.

Proposition 14. $|E_R| \geq (P_{\mathrm{I}} P_{\mathrm{S}} P_{\mathrm{T}})^{-1}$, $\qquad\qquad |M| \geq (P_{\mathrm{I}} P_{\mathrm{R}_0})^{-1} |S|$.

Kurosawa [14] showed the following lower bound on $|E_R|$ for the "separable case" of A^2-codes without secrecy. The "separable case" means that (A4) of Proposition 16 holds. (An A^2-code without secrecy is defined in the same way as in A-codes. That is, a source state s is uniquely determined from a message m.) Let

$$M_s \triangleq \{m \mid \text{the source state of } m \text{ is } s\}. \tag{5.1}$$

Definition 15. Let $E_T(f_i) \triangleq \{e_j \mid \Pr(e_j, f_i) > 0\}$, $E_R(e_i) \triangleq \{f_j \mid \Pr(e_i, f_j) > 0\}$. We say that each key is uniformly distributed if $E_T, E_R, E_T(f_i)$ and $E_R(e_j)$ are all uniformly distributed.

Proposition 16. *[14] In an A^2-code without secrecy, suppose that*

(A1) $P_{\mathrm{I}} = P_{\mathrm{S}} = P_{\mathrm{R}_0} = P_{\mathrm{R}_1} = P_{\mathrm{T}} = 1/\delta$,
(A2) *each key is uniformly distributed,*
(A3) $|M| = \delta^2 |S|$,
(A4) *(Separable case) for $\forall s$, there exist $A_1, A_2, \cdots$ such that*

$$M_s = A_1 \cup A_2 \cup \cdots, \quad A_i \cap A_j = \phi, \quad |A_i| = \text{constant},$$

and $\forall f_h \in E_R$ accepts just one message in $\forall A_i$.

Then $\qquad\qquad\qquad |E_R| \geq |S|(\delta^2 - \delta) + 1. \tag{5.2}$

(Remarks)
(1) A^2 codes shown by [13] and [14] satisfy that $P_{\mathrm{I}} = P_{\mathrm{S}} = P_{\mathrm{R}_0} = P_{\mathrm{R}_1} = P_{\mathrm{T}} = 1/\delta$. In this case, Proposition 14 becomes

$$|E_R| \geq \delta^3. \quad (5.3) \qquad\qquad\qquad |M| \geq \delta^2 |S|. \quad (5.4)$$

(2) Eq.(5.2) is more tight than eq.(5.3) if $|S| > \delta + 1 + (1/\delta)$.
(3) (A3) means that the equality of eq.(5.4) is satisfied.

On the other hand, a combinatoril lower bound on "the cheating probabilities" was recently shown by Kurosawa and Obana [15]. For $f \in E_R$ and $s \in S$, define $M(f, s) \triangleq \{m \mid f \text{ accepts } m \text{ and the source state of } m \text{ is } s\}$.

Proposition 17. *[15] In an A^2-code without secrecy, suppose that*

(B1) $|M(f,s)| = \delta$ *for* $\forall f \in E_R$ *and* $\forall s \in S$,
(B2) $|M| = \delta^2|S|$,

Then $P_I \geq 1/\delta$, $P_{R_0} \geq 1/\delta$, *and* $P_T \geq 1/(\delta+1)$. *When* $P_I = 1/\delta$, $P_S \geq 1/\delta$.
When $P_{R_0} = 1/\delta$, $P_{R_1} \geq 1/\delta$.

This proposition is tight because all the bounds are satisfied with equalities by an A^2-code given by Johansson [12].

5.2   New bound

Now, we show a new lower bound on $|E_R|$ for the "general case" of A^2-codes without secrecy by using Cororally 11. The "general case" means that (A4) of Proposition 16 is eliminated.

Theorem 18. *In an A^2-code without secrecy, suppose that*

(C1) $|M(f,s)| = \delta$ *for* $\forall f \in E_R$ *and* $\forall s \in S$,
(C2) $|M| = \delta^2|S|$,
(C3) $P_I = P_S = P_{R_0} = P_{R_1} = 1/\delta$ *and* $P_T = 1/(\delta+1)$.
(C4) *each key is uniformly distributed (see Def.15).*

Then
$$|E_R| \geq |S|(\delta^2 - 1) + 1. \tag{5.5}$$

The inequality holds if and only if, for $\forall f_1 \in E_R$ *and* $\forall f_2 \in E_R$,

$$|\{m \mid m \text{ is accepted by both } f_1 \text{ and } f_2\}| = constant.$$

The proof will be given in the next subsection.

(Remarks)

(1) (C3) means that $P_I, P_S, P_{R_0}, P_{R_1}, P_T$ take the minimum values from Proposition 17.
(2) Suppose that $P_I, \cdots, P_T$ take the minimum values. That is, $P_I = P_S = P_{R_0} = P_{R_1} = 1/\delta$ and $P_T = 1/(\delta+1)$. Then, Proposition 14 becomes $|E_R| \geq \delta^2(\delta+1)$. Eq.(5.5) is more tight than this bound if $|S| > \delta + 1 + 1/(\delta+1) + 1/(\delta^2-1)$.
(3) The right hand side of eq.(5.5) is larger than that of eq.(5.2).

5.3 Proof of Theorem 18

- For a matrix $X = \{x_{ij}\}$, x_j denotes the j-th column vector of X.
- For a column vector x, $w(x) \triangleq$ the number of nonzero elements of x.
- For $x_i = (x_{1i}, x_{2i}, \cdots)^T$ and $x_j = (x_{1j}, x_{2j}, \cdots)^T$, $x_i \odot x_j \triangleq (x_{1i}x_{1j}, x_{2i}x_{2j}, \cdots)^T$.

Definition 19. A skelton matrix for (E, M) is a $|E| \times |M|$ matrix $X = \{x_{ij}\}$ such that
$$x_{ij} = \begin{cases} 1 & \text{if } e_i \text{ accepts (or could generate) } m_j \\ 0 & \text{otherwise.} \end{cases}$$
where $e_i \in E$ and $m_j \in M$.

The following claim was shown by [14] to prove Proposition 16. For $f \in E_R$, further, let $\quad M_f \triangleq \{m \mid f \text{ accepts } m\}, \quad M(f, s) = \{m \mid m \in M_f, m \in M_s\}$. (For the definition of M_s, see eq.(5.1).) Let $X = \{x_{ij}\}$ be the skelton matrix for (E_R, M).

Claim 20. [14] In an A^2-code without secrecy, suppose that $P_I = P_S = P_{R_0} = P_{R_1} = 1/\delta$, each key is uniformly distributed and $|M| = \delta^2|S|$. Then,

(1) $w(x_i) = |E_R|/\delta$, $i = 1, \cdots, |M|$.
(2) $|M_s| = \delta^2$ for $\forall s \in S$.
(3) For $\forall s, \forall s'$ such that $s \neq s'$, and for $\forall m_i \in M_s, \forall m_j \in M_{s'}$, $w(x_i \odot x_j) = |E_R|/\delta^2$.

Note that this claim holds under (C1)~(C4).

We will prove that E_R has a structure of an affine α-resolvable design with parameters $v = |E_R|$, $\quad t = |S|$ and $b = \delta^2|S|$ (see Lemma 21, 22, and 23 below). Then Theorem 18 is obtained from Corollary 11.

For (E_R, M), there exists an incomplete block design $(\hat{V}, \hat{X})$ whose incidence matrix coincides with the skelton matrix of (E_R, M). That is, $\hat{X} = \{\hat{B}_i\}$ such that $\hat{B}_i = \{f \mid f \in E_R, f \text{ accepts } m_i \in M\}$.

Lemma 21. For the above $(\hat{V}, \hat{X})$, under (C1)~(C4),

$$v \triangleq |\hat{V}| = |E_R|, \quad b \triangleq |\hat{X}| = |M| = \delta^2|S|, \quad |\hat{B}_i| = |E_R|/\delta(= \text{constant}).$$

Proof. The second equation comes from (C2). For the third equation, from Claim 20 (1), $|\hat{B}_i| = w(x_i) = |E_R|/\delta$. $\quad\square$

Lemma 22. $(\hat{V}, \hat{X})$ is a δ-resolvable incomplete block design with $t = |S|$ classes.

Proof. We can group M into $|S|$ classes $M_{s_1}, M_{s_2}, \cdots$. We then group $\hat{X}$ correspondingly. From (C1), each $f \in E_R$ accepts δ messages belonging to M_s for $\forall s$. Therefore, in $(\hat{V}, \hat{X})$, $\forall f \in \hat{V}$ appears in δ blocks in each class. $\quad\square$

Lemma 23. $(\hat{V}, \hat{X})$ *is an affine δ-resolvable incomplete block design with parameters*

$$q_1 = |E_R|/\delta(\delta + 1). \quad (5.6) \qquad\qquad q_2 = |E_R|/\delta^2. \quad (5.7)$$

Proof. Let $X = \{x_{ij}\}$ be the skelton matrix for (E_R, M). From Claim 20 (3), eq.(5.7) holds. For eq.(5.6), it is enough to show that

$$w(x_i \odot x_j) = |E_R|/\delta(\delta + 1)$$

for $\forall s \in S, \forall m_i \in M_s$ and $\forall m_j \in M_s$. For $\forall s \in S$ and $\forall m \in M_s$, we can consider an A-code without splitting $(\hat{S}, \hat{M}, \hat{E})$ such that

$$\hat{M} = M_s \backslash \{m\}, \quad \hat{E} = \{f \mid f \in E_R, \; f \text{ accepts } m\}.$$
$$|\hat{S}| = |M(f, s)| - 1 \; = \; \delta - 1 \quad (\text{from (C1)}).$$

in a natural way, where the probability distribution on $\hat{E}$ is conditioned by the fact that R accepts m. From Claim 20 (2), $|\hat{M}| = |M_s| - 1 = \delta^2 - 1$. The impersonation attack probability for this A-code is given by

$$\max_{\hat{m} \in M_s \backslash \{m\}} \Pr[\text{R accepts } \hat{m} | \text{R accepts } m]$$

from eq.(4.1). Then from Proposition 13,

$$\max_{\hat{m} \in M_s \backslash \{m\}} \Pr[\text{R accepts } \hat{m} | \text{R accepts } m]$$
$$\geq |\hat{S}|/|\hat{M}| = (\delta - 1)/(\delta^2 - 1) = 1/(\delta + 1). \quad (5.8)$$

Now, from (C3) and eq.(5.8),

$$1/(\delta + 1) \; = \; P_{\mathbf{T}}$$
$$\geq \max_{\hat{m} \in M_s \backslash \{m\}} \Pr[\text{R accepts } \hat{m} | \text{R accepts } m] \geq 1/(\delta + 1).$$

This means that $\max_{\hat{m} \in M_s \backslash \{m\}} \Pr[\text{R accepts } \hat{m} | \text{R accepts } m] = 1/(\delta + 1)$ for $\forall s$ and $\forall m \in M_s$. Then, again, from Proposition 13,

$$\Pr[\text{R accepts } \hat{m} | \text{R accepts } m] = 1/(\delta + 1)$$

for $\forall s, \forall m \in M_s$ and $\forall \hat{m} \in M_s \backslash \{m\}$. That is, $\forall s \in S, \forall m_i \in M_s$ and $\forall m_j \in M_s \backslash \{m_i\}$,

$$\max_{m_j \in M_s \backslash \{m_i\}} \Pr[\text{R accepts } m_j | \text{R accepts } m_i] = 1/(\delta + 1).$$

From (C4), $\Pr[\text{R accepts } m_j | \text{R accepts } m_i] = w(x_i \odot w_j)/w(x_i)$. Therefore

$$w(x_i \odot w_j)/w(x_i) = 1/(\delta + 1)$$

Hence $\qquad\qquad w(x_i \odot w_j) = w(x_i)/(\delta + 1) = |E_R|/\delta(\delta + 1)$

from Claim 20 (1). $\qquad\qquad\qquad\qquad\qquad\qquad\qquad\qquad\qquad\qquad\qquad\qquad\qquad\qquad\quad$ □

References

1. R.C.Bose, "A note on the resolvability of balanced incomplete block designs", *Sankhyā*, 6, 105–110. (1942)
2. S.S.Shrikhande and D.Raghavarao, "Affine α-resolvable incomplete block designs", Contributions to Statistics. Presented to Prof. P.C.Mahalanobis on the occasion of his 70th birthday. Pergamon Press, pp.471–480. (1963)
3. D.Raghavarao, "Constructions and combinatorial problems in design of experiments", John Wiley & Sons, Inc. (1971).
4. S.Kageyama, "On μ-resolvable balanced incomplete block designs", *Ann. Statist.* 1, 195–203. (1973)
5. S.Kageyama, "Resolvability of block designs", *Ann. Statist.* 4, 655–61. (1976)
6. D.R.Hughes & F.C.Piper, "On resolutions and Bose's theorem", *Geom. Dedicata* 5, 129–33. (1976)
7. S.Kageyama, "Some properties on resolvability of variance-balanced designs", *Geom. Dedicata* 15, 289–92. (1984)
8. R.Mukerjee & S.Kageyama, "On resolvable and affine resolvable variance balanced designs", *Biometrika*, 72, 1, pp.165–172. (1985)
9. G.J.Simmons, "Message Authentication with Arbitration of Transmitter/Receiver Disputes", Proceedings of Eurocrypt'87, Lecture Notes in Computer Science, LNCS 304, Springer Verlag, pp.150–16 (1987)
10. G.J.Simmons, "A Cartesian Product Construction for Unconditionally Secure Authentication Codes that Permit Arbitration", Journal of Cryptology, Vol.2, no.2, 1990, pp.77–104 (1990).
11. T.Johansson, "Lower Bounds on the Probability of Deception in Authentication with Arbitration", In *Proceedings of 1993 IEEE International Symposium on Information Theory*, San Antonio, USA, January 17–22, pp.231. (1993)
12. T.Johansson, "Lower Bounds on the Probability of Deception in Authentication with Arbitration", IEEE Trans. on IT, vol.40, No.5, pp.1573-1585 (1994)
13. T.Johansson, "On the construction of perfect authentication codes that permit arbitration", Proceedings of Crypto'93, Lecture Notes in Computer Science, LNCS 773, Springer Verlag, pp.341-354 (1993).
14. K.Kurosawa, "New bound on authentication code with arbitration", Proceedings of Crypto'94, Lecture Notes in Computer Science, LNCS 899, Springer Verlag, pp.140–149. (1994)
15. K.Kurosawa and S.Obana, "Combinatorial Bounds for Authentication Codes with Arbitration", Eurocrypt'95 (to be presented)
16. G.J.Simmons, "A survey of Information Authentication", in *Contemporary Cryptology, The science of information integrity*, ed. G.J.Simmons, IEEE Press, New York, (1992).
17. G.J.Simmons, "Authentication theory/coding theory", Proceedings of Crypto'84, Lecture Notes in Computer Science, LNCS 196, Springer Verlag, pp.411–431 (1985).

Dense Packings of 3k(k+1)+1 Equal Disks in a Circle
for k = 1, 2, 3, 4, and 5

B. D. Lubachevsky   bdl@research.att.com
R. L. Graham   rlg@research.att.com

AT&T Bell Laboratories,
Murray Hill, New Jersey 07974, USA

ABSTRACT

For each $k \geq 1$ and corresponding hexagonal number $h(k) = 3k(k + 1) + 1$, we introduce $m(k) = (k - 1)!$ packings of $h(k)$ equal disks inside a circle which we call the *curved hexagonal* packings. The curved hexagonal packing of 7 disks ($k = 1$, $m(1) = 1$) is well known and the one of 19 disks ($k = 2$, $m(2) = 1$) has been previously conjectured to be optimal. New curved hexagonal packings of 37, 61, and 91 disks ($k = 3$, 4, and 5, $m(3) = 2$, $m(4) = 6$, and $m(5) = 24$) were the densest we obtained on a computer using a so-called "billiards" simulation algorithm. A curved hexagonal packing pattern is invariant under a 60° rotation. When $k \geq 3$, the curved hexagonal packings are not mirror symmetric but they occur in $m(k)/2$ image-reflection pairs. For $k \rightarrow \infty$, the density (covering fraction) of curved hexagonal packings tends to $\frac{\pi^2}{12}$. The limit is smaller than the density of the known optimum disk packing in the infinite plane. We present packings that are better than curved hexagonal packings for 127, 169, and 217 disks ($k = 6$, 7, and 8).

In addition to new packings for $h(k)$ disks, we present new packings we found for $h(k) + 1$ and $h(k) - 1$ disks for k up to 5, i.e., for 36, 38, 60, 62, 90, and 92 disks. The additional packings show the "tightness" of the curved hexagonal pattern for $k \leq 5$: deleting a disk does not change the optimum packing and its quality significantly, but adding a disk causes a substantial rearrangement in the optimum packing and substantially decreases the quality.

1. Introduction

Patterns of dense geometrical packings are sensitive to the geometry of the enclosing region of space. In particular, dense packings of equal nonoverlapping disks in a circle are different from those in a regular hexagon, as one might expect. In this paper, for any $k \geq 1$ and corresponding hexagonal number $h(k) = 3k(k + 1) + 1$, we present a pattern of packings of $h(k)$ equal disks in a circle which can be viewed a "curved" analogue of the densest packing of $h(k)$ disks in a regular hexagon. For a particular k, there exists a set of $m(k) = (k - 1)!$ equivalent different *curved hexagonal packings*. A curved hexagonal packing pattern is invariant under a 60° rotation. When $k \geq 3$, the curved hexagonal packings are not mirror symmetric but they occur in $m(k)/2$ "image-reflection" pairs. The density (covering fraction) of a curved hexagonal packing tends to $\frac{\pi^2}{12}$ as $k \rightarrow \infty$. Because the limit density is smaller than the density of the best (hexagonal) packing of equal disks on an infinite plane, it is natural to expect the curved hexagonal packings not to be optimal for all sufficiently large k. It might come as a surprise, though, that for several initial values of k there seems to be no better packing than the curved hexagonal ones. Indeed, for 7 disks ($k = 1$, $m(1) = 1$) the curved hexagonal packing is well known to be optimal and the one for 19 disks ($k = 2$, $m(2) = 1$) has been previously conjectured as such [K]. For 37, 61, and 91 disks ($k = 3$, 4, and 5, $m(3) = 2$, $m(4) = 6$, $m(5) = 24$), the curved hexagonal packings were the densest we obtained by computer experiments using the so-called "billiards" simulation algorithm.

The "billiards" simulation algorithm [L] [LS] has so far proved to be a reliable method for generating optimal packings of disks in an equilateral triangle [GL1]. Our experiments with this algorithm for packings in a circle (a detailed account will be reported elsewhere [GL2]) either confirmed or improved the best previously reported packings for $n \leq 25$. We are unaware of any published conjectures for packing $n > 25$ disks in a circle, but the "billiards" algorithm kept producing packings for many $n > 25$, specifically, for $n = h(3) = 37$, $n = h(4) = 61$, and $n = h(5) = 91$. The latter three sets of packings happened to have the curved hexagonal pattern and they were the best found for their value of n. As for $n = h(6) = 127$, $n = h(7) = 169$, and $n = h(8) = 217$, the algorithm found the packings which are *better* than the corresponding curved hexagonal ones.

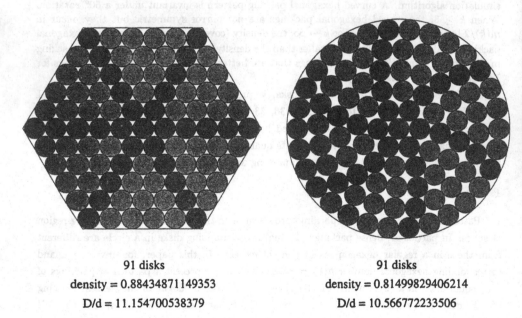

91 disks	91 disks
density = 0.88434871149353	density = 0.81499829406214
D/d = 11.154700538379	D/d = 10.566772233506

Figure 1.1: *Left.* The well known best packing of $h(5) = 91$ disks in a regular hexagon. *Right.* One of the 24 best (that we found) packings of $h(5) = 91$ disks in a circle. (The different shading of the disks is auxiliary; it is not a part of the pattern).

For the values $k \leq 5$, for which the densest packings of $h(k)$ disks in a circle apparently have the curved hexagonal patterns, these packings look "tight." To test our intuition of their "tightness" we compared these packings with packings obtained for $h(k) - 1$ and $h(k) + 1$ disks. Thus, we generated dense packings for 36, 38, 60, 62, 90, and 92 disks and verified that deleting a disk does not change the optimum packing and its quality significantly, but adding a disk causes a substantial rearrangement in the optimum packing and substantially decreases the quality. This tightness may be considered an analogue of the similar tightness property for the infinite classes of packings in an equilateral triangle as noted in [GL1]. In particular, the variations in the packing pattern and quality when one disk is added or subtracted are similar to those observed for packings of $\frac{k(k+1)}{2}$ disks in the triangles.

2. Packings of 7, 19, 37, and 61 disks in a circle

The four best packings are presented in Fig. 2.1. The packing of 7 disks is well known

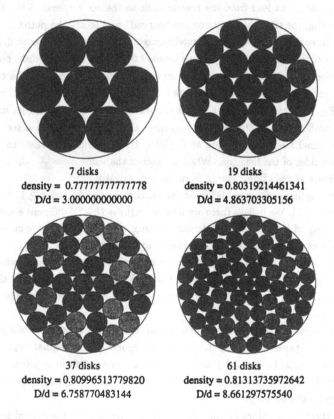

7 disks
density = 0.77777777777778
D/d = 3.000000000000

19 disks
density = 0.80319214461341
D/d = 4.863703305156

37 disks
density = 0.80996513779820
D/d = 6.758770483144

61 disks
density = 0.81313735972642
D/d = 8.661297575540

Figure 2.1: *Top*. The well known best packing of $h(1) = 7$ disks and the best previously conjectured packing of $h(2) = 19$ disks. *Bottom*. The best packing (one of the two best that we found) of $h(3) = 37$ disks and the best packing (one of the six best that we found) of $h(4) = 61$ disks.

to be optimal, and that of 19 disks is conjectured as such in [K]. The packings shown of 37 and 61 disks have not been reported before; they are the best we found for these numbers of disks. The density 7/9 of the 7-disk packing is presented in decimal form in conformance with the other three densities; an alternative finite form of the parameters for all the packings in Fig. 2.1 also exists.

3. The curved hexagonal pattern

The pattern can be explained by comparing it with the corresponding hexagonal pattern. Fig. 1.1 depicts the two patterns side-by-side for $h(5) = 91$ disks. Each is composed of six sections which we tried to emphasize by shading more heavily the disks on the boundaries between them. In the true hexagonal packing, the sections are triangular, the boundaries are six straight "paths" that lead from the central disk to the six extreme disks. In the curved hexagonal packing, the triangular sections are "curved" and so are the paths.

To define the entire structure of the curved hexagonal packing, it suffices to define positions of disks on one path. Let the central disk be labeled 0 and the following disks on the path be successively labeled 1, 2, ... k. Consider k straight segments connecting the centers of the adjacent labeled disks: 0 to 1, 1 to 2, ... , $(k-1)$ to k. Given a direction of rotation (it is clockwise in Fig. 1.1), each following segment is rotated at the same angle, let us call it α, in this direction with respect to the previous segment. Consider α as a parameter. When $\alpha = 0$ we have the original hexagonal packing at the left. If we gradually increase the α, "humps" grow on the six sides of the hexagon. When α reaches the value $\alpha_k = \frac{\pi}{3k}$, all the disks of the last, kth, layer are at the same distance from the central disk.

A curved hexagonal packing of $h(k)$ disks can be constructed for any $k \geq 1$. Fig. 3.1 depicts an instance for $k = 13$. We believe there are total of $m(k) = (k-1)!$ different equivalent curved hexagonal patterns of $h(k)$ disks. In particular, we suggest that a distinctive curved hexagonal packing of k layers can be identified by a permutation in the sequence $1, 2, \ldots, k-1$. For each permuted sequence, a path of $k+1$ disks can be constructed and this path, when completed with layers, happens to form a curved hexagonal packing which is as tight as the basic one. Permutation $i_1, \ldots, i_{k-1}$ produces the mirror reflection to the pattern produced by permutation $k - i_1, \ldots, k - i_{k-1}$.

We conjecture $m(5) = 24$ equivalent different curved hexagonal packings of $h(5) = 91$ disks. Two packings are represented in Fig. 1.1; they correspond to the original sequence 1, 2, 3, 4 or its reflection 4, 3, 2, 1. Fig. 3.2 presents four more image-reflection pairs. Each packing is accompanied by its two generating sequences and its bond diagram. The bond diagrams clearly distinguish the packings.

For $k > 5$ there are packings of $h(k)$ disks that are better than the curved hexagonal, but we believe there is no $(m(k) + 1)$st packing equal in quality to a curved hexagonal and distinct from any one of $m(k)$ of them.

Because each internal ith layer of disks has triangles (as seen on bond diagrams) that con-

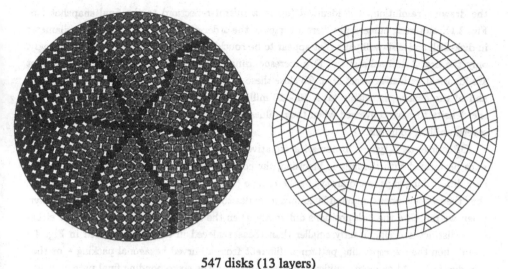

547 disks (13 layers)

density = 0.81954773488724   D/d = 25.834885197829

Figure 3.1: *Left.* A curved hexagonal packing of $h(13) = 547$ disks in a circle. In this packing the sense of rotation is flipped for layers 6,7,8, and 9. *Right.* The bond diagram for the packing at the left. The diagram contains a straight segment between centers of any pair of touching disks.

nect it to the corresponding outer layer $i + 1$, and because the outermost kth layer is obviously rigid, a curved hexagonal packing is rigid. The rigidity means that the only continuous motion of a subset of disks which is possible without violating the no-overlap condition is rotation of the entire assembly as a whole.

4. How the "billiards" algorithm produces packings

A detailed description of the philosophy, implementation and applications of this event-driven algorithm can be found in [L], [LS]. Essentially, the algorithm simulates a system of n perfectly elastic disks. In the absence of gravitation and friction, the disks move along straight lines, colliding with each other and the region walls according to the standard laws of mechanics, all the time maintaining a condition of no overlap. To form a packing, the disks are uniformly allowed to gradually increase in size, until no significant growth can occur. Fig. 4.1 displays four successive snapshots in an experiment with 61 disks. We took the snapshots beginning the time when a local order begins to form and till the time when disk positions are fixed.

The latest snapshot shown in Fig. 4.1, the one at 441704 collisions, looks dense and, within the drawing resolution, it is identical (up to a mirror-reflection) to the final snapshot (see Fig. 2.1). However, numerically there are gaps of the order of 10^{-5} to 10^{-3} of the disk diameter in disk-disk and disk-wall pairs that appear to be touching each other in Fig. 4.1. Accordingly, only the first 3 decimal digits of D/d corresponding to the latest shown in Fig. 4.1 snapshot are identical with the correct D/d. To close these gaps and to achieve full convergence of D/d and of the density, it usually takes 10 - 20 million further collisions. We consider D/d and the density to have converged when their values do not change with full double precision for several million collisions.

Of course, as is typical in numerical iterative convergent procedures, if the computations were performed with the infinite precision, the convergence would be never achieved and the ever diminishing gaps would always be there. The "experimental" converged values agree with the "theoretical" ones computed by known formulas to 14 or more significant digits. Moreover, when we initialize the disk positions differently, then the final parameters achieved are either quite distinct and significantly smaller than those achieved in the run presented in Fig. 4.1 – and then the corresponding pattern is different from a curved hexagonal packing – or they are identical to 14 or more significant digits – and then the corresponding final pattern is one of the six known curved hexagonal ones. This makes us suspect that we have found the best possible packing and that its parameters D/d and density are correct to 14 significant digits.

5. Packings of 127, 169, and 217 disks

We ran the "billiards" algorithm for $n = h(6) = 127$, $n = h(7) = 169$, and $n = h(8) = 217$ disks. For these n the algorithm produced better packings than the curved hexagonal ones, although these packing are probably not the best because of the presence of multiple local minima. An example is the packing of 127 disk in Fig. 5.1.

6. Tightness of curved hexagonal packings

Depictions and parameter values of the best found packings of $n = h(k) - 1$ and $h(k) + 1$ disks for $k = 2, 3, 4$, and 5, that is, for $n = 18, 20, 36, 38, 60, 62, 90$, and 92 demonstrate the tendency also exhibited by known packings of $n = 6$ and 8 disks, namely:

(a) the pattern of dense packing of $n = h(k) - 1$ disks is obtained by removing one disk from the pattern of dense packing of $h(k)$ disks (which is a curved hexagonal packing for the considered $k \leq 5$) and, for $k \geq 3$, by a small rearrangement of the disks; its parameter D/d is either not changed (for $k = 1$ and 2) or is decreased only slightly (for $k = 3, 4$, and 5);

(b) the pattern of dense packing of $n = h(k) + 1$ disks differs significantly from the pattern of dense packing of $h(k)$ disks and its parameter D/d is increased substantially for all $k = 1$, 2, 3, 4, and 5).

The changes in D/d are "slight" and "substantial" only in comparison to each other.

7. Discussion

It is natural to expect for a sufficiently large n the dense packing of n equal disks in a circle to be of an "uninteresting" pattern like that in Fig. 5.1, with a large, perhaps disturbed, core of hexagonally packed disks and irregularly placed disks along the periphery. The fraction of the peripheral irregularity disks and the perturbation in the hexagonally packed core would diminish with n. On the other hand, for $n \leq 25$ symmetric patterns of dense packing have been previously observed that do not obey the general description for "uninteresting" packings given above for large n. Our computer experiments show that at least for a particular class of $n = h(k) = 3k(k+1) + 1$, the transition from the structured "interesting" pattern to the "uninteresting" core-hexagonal one occurs between $n = 91$ and $n = 127$.

References

[CFG]   H. T. Croft, K. J. Falconer and R. K. Guy, *Unsolved Problems in Geometry*, Springer Verlag, Berlin, 1991, 107–111.

[FG]   J. H. Folkman and R. L. Graham, A packing inequality for compact convex subsets of the plane, *Canad. Math. Bull.* **12** (1969), 745–752.

[GL1]   R. L. Graham and B. D. Lubachevsky, Dense packings of equal disks in an equilateral triangle: from 22 to 34 and beyond, *The Electronic Journ. of Combinatorics* **2** (1995), #A1.

[GL2]   R. L. Graham and B. D. Lubachevsky, Dense packings of equal disks in a circle: from 25 to 61 and beyond (*In preparation.*)

[G]   M. Goldberg, Packing of 14, 16, 17 and 20 circles in a circle, *Math. Mag.* **44** (1971), 134-139.

[K]   S. Kravitz, Packing cylinders into cylindrical containers, *Math. Mag.* **40** (1967), 65-71.

[L]   B. D. Lubachevsky, How to simulate billiards and similar systems, *J. Computational Physics* **94** (1991), 255–283.

[LS]   B. D. Lubachevsky and F. H. Stillinger, Geometric properties of random disk packings, *J. Statistical Physics* **60** (1990), 561–583.

[O]   N. Oler, A finite packing problem, *Canad. Math. Bull.* **4** (1961), 153–155.

[R]   G. E. Reis, Dense packing of equal circles within a circle, *Math. Mag.* **48** (1975), 33–37.

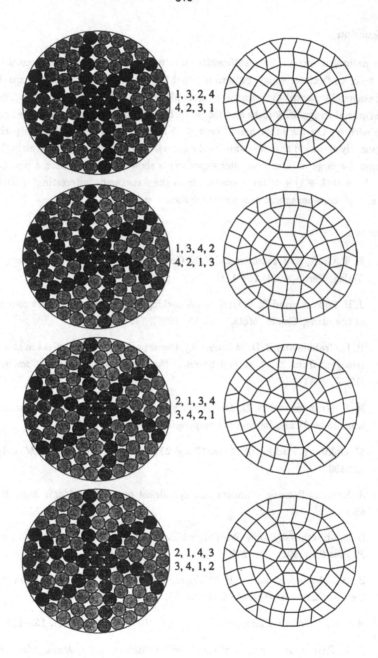

1, 3, 2, 4
4, 2, 3, 1

1, 3, 4, 2
4, 2, 1, 3

2, 1, 3, 4
3, 4, 2, 1

2, 1, 4, 3
3, 4, 1, 2

Figure 3.2: Four (out of 12 existing) densest, that we found, image-reflection pairs of packings of 91 disks. Each pattern at the left represents a pair. It is is accompanied by its two generating sequences in the middle and the bond diagram at the right.

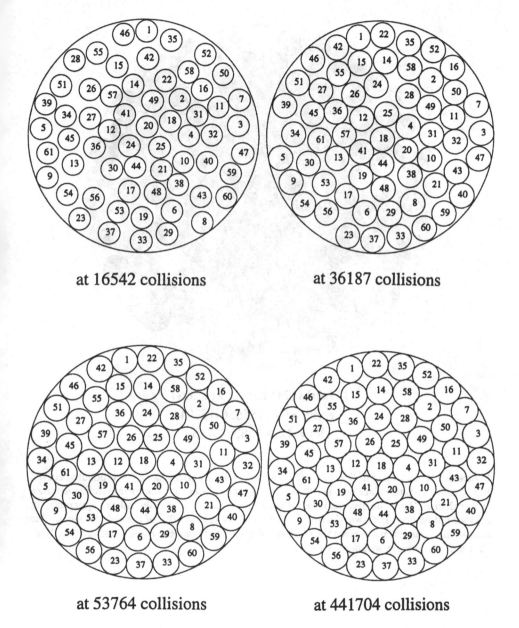

at 16542 collisions

at 36187 collisions

at 53764 collisions

at 441704 collisions

Figure 4.1: Successive snapshots of simulating expansion of 61 disks inside a circle. The progress is monitored by counting collisions. Disks are labeled 1 to 61 arbitrarily but the same disk carries the same label in all four snapshots. The last pattern is a mirror-reflection of the packing of 61 disks in Fig. 2.1

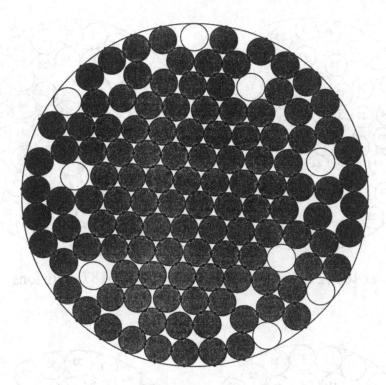

127 disks (244 bonds; 9 rattlers)
density = 0.81755666415904
D/d = 12.463583540213

Figure 5.1: A packing of $h(6) = 127$ disks in a circle that is better than a corresponding curved hexagonal one. Little black dots are "bonds"; a bond indicates that the corresponding distance is less than 10^{-13} of the disk diameter. Where a pair disk-disk or disk-wall are apparently in contact but no bond is shown, the computed distance is at least 10^{-5} of the disk diameter. The shaded disks can not move given the positions of their neighbors, the non-shaded are "rattlers" that are free to move within their confines.

Efficient Parallel Algorithms for some Tree Layout Problems *

J. Díaz[1] A. Gibbons[2] G. Pantziou[3,4] M. Serna[1] P. Spirakis[4,5] J. Toran[1]

[1] Departament de Llenguatges i Sistemes, Universitat Politècnica Catalunya
[2] Department of Computer Science, University of Warwick
[3] University of Central Florida
[4] Computer Technology Institute
[5] University of Patras

Abstract. The minimum cut and minimum length linear arrangement problems usually occur in solving wiring problems and have a lot in common with job sequencing questions. Both problems are NP-complete for general graphs and in P for trees. We present here two parallel algorithms for the CREW PRAM. The first solves the minimum length linear arrangement problem for trees and the second solves the minimum cut arrangement for trees. We prove that the first problem belongs to NC for trees, and the second problem also is in NC for bounded degree trees.

1 Introduction

Given a graph $G = (V, E)$ with $|V| = n$, a *layout* of G is a one-to-one mapping φ from V to the first n integers $\{1, 2, \cdots, n\}$. The term layout is also known as *linear arrangement* [Yan83], [Shi79]. Notice that a layout φ on V determines a linear ordering of the vertices. Given a natural i, the *cut* of the layout at i is the number of edges that cross over i; i.e. the number of edges $\{u, v\} \in E$ with $\varphi(u) < i \leq \varphi(v)$. The *cutwidth* of φ, denoted $\gamma(\varphi)$, is the maximum cut of φ over all integers from 1 to n. The *length* of φ, denoted $\lambda(\varphi)$, is the sum over all edges (u, v) of $|\varphi(u) - \varphi(v)|$.

Graph layout problems are motivated as simplified mathematical models of VLSI layout. We can model a VLSI circuit by means of a graph, where the edges of the graph represent the wires, and the nodes represent the modules. Of course, this graph is an over-simplified model of the circuit, but understanding and solving problems in this simple model can help us to obtain better solutions for the real-world model (see the survey by Shing and Hu [SH86] and [Di92]).

In this paper we consider two layout problems. The first problem is called *the minimum linear arrangement*, MINLA. Given a graph $G = (V, E)$, find the layout φ which minimizes $\lambda(\varphi)$. The MINLA problem is NP-complete for general graphs [GJ79]. Moreover, due to the importance of the problem, there has been some work trying to obtain polynomial time algorithms for particular types of graphs. For instance, Adolph and Hu gave a $O(n \log n)$ algorithm for the case the graph is a rooted tree [AH73]. Even and Shiloach proved the problem is also NP-complete for bipartite graphs [ES78]. Finally, Shiloach proved that the MINLA can be solved for the case of an unrooted tree with n nodes by a deterministic algorithm running in time $O(n^{2.2})$ [Shi79].

The second problem is called *the minimum cut problem*, MINCUT. Given a graph $G = (V, E)$, find the layout φ that minimizes the cutwidth $\gamma(\varphi)$. The MINCUT problem is NP-complete for

* This research was supported by the ESPRIT BRA Program of the EC under contract no. 7141, project ALCOM II.

general graphs [Gav77], weighted trees and planar graphs [MS86]. As in the case on the MINLA, the MINCUT has a history of results for particular types of graphs. Harper gave a polynomial time algorithm for the n-dimensional hypercube [Har66]. Chung, Makedon, Sudborough and Turner presented a $O(n(\log n)^{d-2})$ time algorithm to solve the MINCUT problem on trees, where d is the maximum degree of any node in the tree [CMST82]. The MINCUT can be solved for the case of an unrooted tree with n nodes in time $O(n \log n)$ [Yan83].

We present here two parallel algorithms. The first one solves the MINLA for unrooted trees in $O(\log^2 n)$ time and $O(n^2 3^{\log n})$ processors on a CREW PRAM. The second algorithm solves the MINCUT for unrooted trees of maximum degree d in time $O(d \log^2 n)$ and $O(n^2/\log n)$ processors in the CREW PRAM model.

Thus we classify the two problems as members of the class NC, but for the MINCUT we are forced to consider only bounded degree trees. As it happens for the sequential case, when first was obtained a polynomial time algorithm for bounded trees [CMST82] and later the boundeness was removed [Yan83], it would be nice to put in NC, the problem for unbounded degree trees, which remains open.

2 A parallel algorithm for the MINLA problem on trees

Let v be a vertex of T. Deleting v and its incident edges from T, yields several subtrees of T. Each of them is called a *subtree of T mod v*. For each edge (v', v) there is a unique subtree of T mod v, say T', such that $v' \in T'$. The vertex v' is *the root of T' mod v*.

Given $v \in V$, let $T_1, \ldots, T_k$ be subtrees of T mod v, $T - (T_1, \ldots, T_k)$ denotes the tree obtained by removing the vertices of $T_1, \ldots, T_k$ and their incident edges. When $T_0, T_1, \ldots, T_k$ are all the subtrees of T mod v we will assume that they are numbered in such a way that $n_0 \geq n_1 \geq \ldots \geq n_k$ where n_i denotes the size of T_i, $i = 0, 1, \ldots, k$.

A *central vertex* is a vertex v_* such that if $T_0, T_1, \ldots, T_k$ are all the subtrees of T mod v_*, then $n_i \leq \lfloor n/2 \rfloor$ for $i = 0, 1, \ldots, k$. It is straightforward to compute in $O(\log n)$ parallel steps a central vertex for a tree T, using $O(n^2)$ processors. Such a vertex plays a central role in the sequential algorithm as well as in the parallel one.

Let T_0 be a subtree of T mod v, and let v_0 be its root mod v. When computing layouts for T_0 and $T - T_0$ separately, we have no control on the length of the edge (v_0, v). In order to take into account this edge we consider *right (left) anchored trees*. A tree right (left) anchored at a node v has an extra edge that covers the distance between v and the right (left) most vertex of T_0. Actually, we will not consider the extra edge, we incorporate the existence of such edge in the length definition.

Let T be an n-vertex tree, right (left) anchored at v, and let φ be a layout of T. Its length is defined by $\lambda(\varphi, v) = \lambda(\varphi) + n - \varphi(v)$ $(\lambda(\varphi, v) = \lambda(\varphi) + \varphi(v) - 1)$. Notice that finding a minimum length layout for right and left anchored trees is equivalent, since by reversing the order of the vertices a right anchored tree becomes a left anchored tree, while the total length remains unchanged. We will denote a right anchored tree as $\overrightarrow{T}(v)$ and a left anchored tree as $\overleftarrow{T}(v)$. When considering all the subtrees mod v, all the anchored subtrees will be anchored at their root mod v. In such a case we will not state explicitly the corresponding root.

In the following we will use $T(\alpha)$ to denote a tree, whith $\alpha = 0$ for free trees and $\alpha = 1$ for anchored trees. Further, $\lambda(T, v, \alpha)$ will denote the minimum length of a layout for $T(\alpha)$ where v is either the vertex at which the anchor is connected, or a central vertex for free trees. In both cases we refer to vertex v as the root of the tree.

In order to simplify notation we will use $(T_1(\alpha_1), \ldots, T_k(\alpha_k), \ldots, T_r(\alpha_r))$ to represent the layout obtained from optimal layouts for each one of the tree components composed together in a sequential way, shifting when necessary. The length of this type of layout is easy to compute from the length of the optimal layouts for the trees and the tree sizes.

Given a tree $T(\alpha)$ let v be its root, we define $p(T, v, \alpha)$ as the value of the greatest integer p satisfying $n_i > \lfloor \frac{n_*+2}{2} \rfloor + \lfloor \frac{n_*+2}{2} \rfloor$ for $i = 1, 2, \ldots, 2p - \alpha$ where $n_* = n - \sum_{i=0}^{2p-\alpha} n_i$ and n is the size of T, if any, otherwise we set $p(T, v, \alpha) = 0$. We will denote by T_* the tree $T(\alpha) - (T_1, \ldots, T_{2p-\alpha})$, for $p = p(T, v, \alpha)$.

We state now the main result given in [Shi79]. The design of the sequential algorithm is based on the decomposition given in the following theorem.

Theorem 1 Shiloach. *Let $T(\alpha)$ be a tree with root v_*, and let $T_0, \ldots T_k$ be all its subtrees mod v_*. Let $p = p(T, v, \alpha)$ and $T_* = T - (T_1, \ldots, T_{2p-\alpha})$. Then,*
(a) If $p = 0$ then $T(\alpha)$ has an optimal layout of type $A = (\overrightarrow{T}_0, \overleftarrow{T - T_0})$ (see figure 1a), for $\alpha = 0$ or type $A = (\overrightarrow{T}_0, T - T_0)$ (see figure 1b) for $\alpha = 1$.
(b) If $p > 0$ then $T(\alpha)$ has an optimal layout of type A (defined as in case (a)) or of type $B = (\overrightarrow{T}_1, \overrightarrow{T}_3, \ldots, \overrightarrow{T}_{2p-1}, T_*, \overleftarrow{T}_{2p-2\alpha}, \ldots, \overleftarrow{T}_4, \overleftarrow{T}_2)$ (see figures 2a and 2b).

The sequential algorithm recursively computes an optimal layout from optimal layouts of subtrees. Notice that in case $p(T, v, \alpha) > 0$ the algorithm computes both types of layouts and takes the one with the smaller length. Our parallel algorithm will be divided into two stages. In the first stage, we decompose the tree into subtrees until all subtrees have size 1. At the same time we record the expressions that will allow us to compute the layout from the smallest trees. In the second stage, we reconstruct the layout, until we get a minimum layout for the whole tree.

In the sequel, we comment on the main points of the parallel algorithm which computes the length of an optimal layout for an undirected tree. It is easy to handle the modifications to get also the layout. We only present the decomposition stage and specifically, the decomposition of both free and anchored trees. The reconstruction stage can be easily derived in view of the decomposition one. **Tree representation.** Each tree will be represented by a linked list keeping an Euler tour representation, together with a mask that keeps which nodes of the original tree are present in it. This mask will also contain pointers to the linked list. To distinguish between free and anchored trees we keep the parameter α and the corresponding root for anchored trees. We record in a matrix, pointers to the subtree masks that form part of any of the layouts in a decomposition phase, together with the additional information required to trace back the cost of any layout.

Free tree decomposition.

The decomposition is based on theorem 1 and exploits the properties of type A and B layouts. Let us first consider another type of layout. A *balanced layout* of $T_1, \ldots, T_k, T(\alpha)$ where $\alpha = 0$ for k even and $\alpha = 1$ for k odd, is the layout $(\overrightarrow{T}_1, \overrightarrow{T}_3, \overrightarrow{T}_5, \ldots, T(\alpha), \ldots, \overleftarrow{T}_6, \overleftarrow{T}_4, \overleftarrow{T}_2)$. Notice that when the balanced layout corresponds to a tree in which $T_1, \ldots, T_k$ are subtrees of a node (the root in case $\alpha = 1$) in $T(\alpha)$ the extra length created by this connexions can be easily computed from the tree sizes.

Balanced layouts play a crucial role in the decomposition of free trees.

Lemma 2. *Let T be a free tree, and v_* be a central vertex of T. Let $T_0, \ldots, T_k$ be all the subtrees of T mod v_*. Let β be the first index for which $| T - (T_0, \ldots, T_\beta) | \leq n/2$, let $p_i = p(T - (T_0, \ldots, T_i), v_*, \alpha_i)$ where $\alpha_i = 1$ for i even, and $\alpha_i = 0$ for i odd. Let T_*^i be $T - (T_0, \ldots, T_{i-1}, T_{i+1}, \ldots T_{2p_i - \alpha_i})$, for each $i = 1, \ldots, \beta$. Let A_i be the balanced layout for $(T_0, \ldots, T_{i-1}, T_{i+1}, \ldots T_{2p_i - \alpha_i}, T_*^i(\alpha_i))$. The layout A_i, $i = 0, \ldots, \beta + 1$ which attains minimum length is a minimum length layout for T. Furthermore, the size of all subtrees that appear in any such layouts is at most $3|T|/4$.*

The decomposition of a free tree T of size n, is obtained as follows.

1. Compute a central vertex v_* of T.
2. Compute a representation of all subtrees $T_0, \ldots, T_k$ of T mod v_*.
3. For each subtree in parallel compute its size.
4. Record in a matrix subtree sizes sorted in decreasing order together with pointers to the corresponding representation.
5. Compute β the first index for which $| T - (T_0, \ldots, T_\beta) | \leq n/2$.
6. For each $i = 1, \ldots, \beta$ in parallel

    6.1. Compute $p_i = p((T - (T_0, \ldots, T_i), v_*, \alpha_i)$

    6.2. Compute a representation of $T_*^i = T - (T_0, \ldots, T_{i-1}, T_{i+1}, \ldots T_{2p_i - \alpha_i})$, where $\alpha_i = 1$ for i even, and $\alpha_i = 0$ for i odd

    6.3. Record the balanced layout $(T_0, \ldots, T_{i-1}, T_{i+1}, \ldots T_{2p_i - \alpha_i}, T_*^i(\alpha_i))$

    6.4. Compute the extra length of the anchors in the balanced layout.

Implementation. The central vertex can be computed in time $O(\log n)$ using $O(n^2)$ processors Once we have the central vertex, we have to compute subtree sizes (now the tree is rooted) using the Euler tour technique, and sort subtrees by size. From the tree sizes using suffix sums we compute $\beta, p_0, p_1, \ldots, p_\beta$. Consider the decomposition given in lemma 2. There are two ways to create new subtrees at each decomposition phase. First, trees obtained just removing an edge, i.e., all subtrees mod v for a given root v. Second, the union of some of the subtrees mod v rooted at a "new" copy of v. In the first case, we compute the corresponding subtrees by removing the root and running a rooting algorithm (in parallel) for each root mod v, in order to separate subtrees. This part will be the basic step for the second case; now we just have to merge the corresponding trees adding a new vertex as root. So for a tree of size k, we can mantain the tree representation using $O(k^2)$ processors in time $O(\log k)$.

Anchored tree decomposition. Before we give the anchored tree decomposition we introduce some additional notation. Let $\overrightarrow{T}(v_*)$ be an anchored tree. We will denote by $T_0^0, T_0^1, \ldots$ its subtrees mod v_* sorted by size, and by $v_0^0, v_0^1, \ldots$ the corresponding roots mod v_*. Recursively, whenever we have an anchored tree $\overrightarrow{T_0^i}(v_0^i)$, by $T_0^{i+1}, T_1^{i+1} \ldots$ we will denote its subtrees mod v_0^i, and $v_0^{i+1}, v_1^{i+1} \ldots$ will denote their corresponding roots mod v_0^i. The decomposition of an anchored tree is based on the following lemma.

Lemma 3. *Let $\overrightarrow{T}(v_*)$ be an anchored tree and let γ be the first index for which $\mid T_0^\gamma \mid \leq n/2$. Let B_i be the layout $(\overrightarrow{T}_0(v_0^0), \ldots, \overrightarrow{T_0^{i-1}}(v_0^{i-1}), \overrightarrow{T_1^i}(v_1^i), T_0^{i-1} - (T_1^i))$ for $i = 0, \ldots, \gamma - 1$ and $B_\gamma = (\overrightarrow{T}_0(v_0), \ldots, T_0^{\gamma-1} - (T_0^\gamma))$. The layout B_i, $i = 0, \ldots, \gamma$ which attains minimum length is a minimum length layout for T.*

Notice that all subtrees appearing in the B_i layouts for $i = 0, \ldots, \gamma - 1$ have size smaller than $3|T|/4$. The only tree that can have big size in the B_γ layout is $T - (T_0^0, \ldots, T_0^\gamma - 1)$, in such a case we decompose it according to lemma 2, taking as a parameter the size of the original tree. Furthermore, the extra length created by the anchors depends only of the size of trees and can be computed easily.

Finally, note that the trees for which we have to compute an optimal layout in the decomposition for a given tree, have size less than $3|T|/4$. Furthermore, we obtain in any case less than $2n$ trees of size less than $3n/4$, but the total sum of the subtree sizes is at most $3n$.

The decomposition is obtained as follows.

1. Compute the size of all subtrees (the tree is rooted)

2. Compute γ the first index for which $\mid T_0^\gamma \mid \leq n/2$.

3. For each $i = 1, \ldots, \gamma - 1$ in paralell

3.1. Compute a representation of the subtrees $\overrightarrow{T_0^i}(v_1^i), \overrightarrow{T_1^i}(v_1^i)$

3.2. Compute a representation of $T_0^{i-1} - T_1^i$

3.3. Record $B_i = (\overrightarrow{T}_0(v_0), \ldots, \overrightarrow{T_0^{i-1}}(v_0^{i-1}), \overrightarrow{T_1^i}(v_1^i), T_0^{i-1} - T_1^i)$.

3.4. Compute the extra length of the anchors in layout B_i.

4. In case that $T_0^{\gamma-1} - T_0^\gamma$ is bigger than $3n/4$ do the free tree decomposition until all subtrees have size at most $3n/4$, taking into account that the layout is

$$(\overrightarrow{T}_0(v_0), \ldots, \overrightarrow{T_0^{\gamma-1}}(v_0^{\gamma-1}), T - (T_0^0, \ldots, T_0^\gamma - 1))$$

Implementation. Now we have a rooted tree, we first compute subtree sizes using the standard Euler Tour technique and then again with the same technique, we find a path of roots of trees of maximum cardinality. Finally, using suffix sums we compute the index γ. From the decomposition given in lemma 3, it is easy to compute the representation of each subtree. Again, the length of each layout can be computed as the sum of the lengths of the corresponding subtrees, adding an expression that depends only on the sizes.

As it is mentioned above, the computation of the layout is based on the above decomposition of the input tree. Thus, we have the following theorem.

Theorem 4. *There is an NC algorithm to compute a minimum sum layout for an undirected tree. The algorithm uses $O(n^{3.6})$ processors and $O(\log^2 n)$ time, where n is the number of vertices in the tree.*

Proof. (idea): The correctness follows from theorem 1 and the fact that all the subtrees in a decomposition phase have sizes at most $3/4$ of the original size.

Note that as all subtrees in a given phase have size at most $3/4$ of the size of the subtrees in the previous phase, the number of decomposition phases in the first stage is $O(\log n)$. Taking into account that the sum of the sizes of the trees obtained in the decomposition of a tree T is at most $3|T|$, the number of processors needed in any phase is at most 3 times the number of processors in the previous one. Thus, the maximum number of processors needed by the algorithm is $O(3^{\log n} n^2) = O(n^{3.6})$. Furthermore, the time used in each phase in the first and second stage is $O(\log n)$. Thus, the total requirements of the algorithm are $O(n^{3.6})$ processors and $O(\log^2 n)$ time.

3 A parallel algorithm for the MINCUT problem on trees

In this section we give an $O(n^2/\log n)$-processor, $O(d \log^2 n)$-time parallel algorithm which finds a minimum cutwidth linear arrangement of a tree T_o of the maximum degree d). This is achieved by using the parallel tree contraction technique on the binary tree T corresponding to T_o. The parallel tree-contraction algorithm (see [ADKP89]) evaluates the root of a tree T by processing a local operation, the *shunt*, to a subset of the leaves of T_{i-1}. The *shunt* operation in our algorithm involves the construction of an *optimal* layout for a subtree T_v, rooted at a node v, using optimal layouts of the subtrees $T_{v_1}, \cdots, T_{v_d}$ rooted at v's children, $v_1, \cdots, v_d$.

Let us first introduce the notion of optimality that we will use through the section, it is based on the definition of *cost-sequence* for a given layout, together with a criterium to compare cost-sequences.

Given a layout φ for the subtree rooted at u, $\langle leftcost(\varphi) \rangle$ is a sequence $\langle \gamma_1, \eta_1, \gamma_2, \eta_2, ... \rangle$ where parameters γ_i and η_i are defined as follows: γ_1 is the largest cut (in φ) occuring on the left side of u. Let w_1 be the point where the cut of γ_1 occurs. If w_1 is immediately to the left of u then $\langle leftcost(\varphi) \rangle = \langle \gamma_1 \rangle$. Otherwise, let η_1 be the smallest cut between w_1 and u and let w_2 be the point closest to u where η_1 occurs. Suppose that γ_2 is the maximum cut between w_2 and u and w_3 is the point closest to u where γ_2 occurs. If $\gamma_2 = \eta_1$ or w_3 is immediately to the left of u then $\langle leftcost(\varphi) \rangle = \langle \gamma_1, \eta_1, \gamma_2 \rangle$. Otherwise, we continue similarly by taking the smallest cut between w_3 and u. $\langle rightcost(\varphi) \rangle$ is a sequence $\langle \gamma_1', \eta_1', \gamma_2', ... \rangle$ where γ_1' is the largest cut in φ occuring on the right side of u. The rest of the sequence is defined in a way similar to that of $\langle leftcost(\varphi) \rangle$ but we now work on the right side of u. Clearly, $\gamma_1 \geq \gamma_2 \geq ..., \eta_1 \leq \eta_2 \leq ..., \gamma_1' \geq \gamma_2' \geq ..., \eta_1' \leq \eta_2' \leq ....$ Also, $\gamma_1 \geq \eta_1, \gamma_2 \geq \eta_2$, etc., and $\gamma_1' \geq \eta_1', \gamma_2' \geq \eta_2'$, etc. Notice that the cutwidth of φ, $\gamma(\varphi) = max\{\gamma_1, \gamma_1'\}$

When $\gamma_1 = \gamma_1'$ we say that the layout φ is *balanced*; otherwise, it is *unbalanced*.

Definition 5. Given a layout φ for a tree rooted at u The *cost-sequence* of the layout φ is $cost(\varphi) = (leftcost(\varphi), *, rightcost(\varphi))$, where the "$*$" denotes the position of u.

In order to compare two cost-sequences, we first define how to compare the left and right parts of the cost-sequence, once we know which is bigger, we generate a new sequence, called *compare*, and finally we determine which of the two compare sequences is bigger.

Let a and b be the two subsequences of a cost-sequence $cost$. If $a \neq b$, and neither is a prefix of the other, then $a > b$ iff a is lexicographically larger than b. If a is a prefix of b and a ends with a γ_i entry, then $a > b$, while if a ends with a η_i entry, then $a < b$.

By *heavy side* (*light side*) we denote the subsequence $leftcost(\varphi_u)$, $rightcost(\varphi_u)$ that is bigger (smaller). Let $heavyside = \gamma_1, \eta_1, \gamma_2, \eta_2, \cdots$ and $lightside = \gamma_1', \eta_1', \gamma_2', \eta_2', \cdots$. We construct the sequence $compare = \langle \gamma_1^c, \eta_1^c, \gamma_2^c, \eta_2^c, \cdots \rangle$, as follows: If $\gamma_1 \neq \gamma_1'$ then $\gamma_1^c = \gamma_1$ and $compare = \langle \gamma_1^c \rangle$. If $\gamma_1 = \gamma_1'$ and there are no next entries η_1, η_1' in $heavyside$, $lightside$ respectively, then let $\gamma_1^c = \gamma_1$ and $compare = \langle \gamma_1^c, \gamma_1^c \rangle$. If only one of the $heavyside$ or $lightside$ has an entry following γ_1 or γ_1^c then call that entry η_1^c and let $compare = \langle \gamma_1^c, \eta_1^c \rangle$. If $\eta_1 \neq \eta_1'$ then let $\eta_1^c = \eta_1'$ and $compare = \langle \gamma_1^c, \eta_1^c \rangle$. If $\eta_1 = \eta_1'$ then in the case that $\gamma_2 = \eta_1$ or $\gamma_2 \neq \gamma_2'$, let $\eta_1^c = \eta_1$, $\gamma_2^c = \gamma_2$ and $compare = \langle \gamma_1^c, \eta_1^c, \gamma_2^c \rangle$. In the case that $\gamma_2 = \gamma_2'$, we continue in the same way as in the case (above) where $\gamma_1 = \gamma_1'$.

Let $cost_1 = (leftcost_1, *, rightcost_1)$, $cost_2 = (leftcost_2, *, rightcost_2)$, be two cost-sequences corresponding to two layouts for the same tree. Consider the respective sequences $compare_1$ and $compare_2$. We say that $cost_1 = cost_2$ iff $compare_1 = compare_2$. If the sequences $compare_1 \neq compare_2$ and neither is a prefix of the other, then $cost_1 \prec cost_2$ iff $compare_1$ is lexicographically smaller than $compare_2$. If $compare_1$ is a prefix of $compare_2$ and $compare_1$ is of odd length then $cost_1 \prec cost_2$ while if $compare_1$ is of even length then $cost_2 \prec cost_1$. Note that $\prec$ is a transitive relation.

Definition 6. Let T_u be a tree rooted at a node u and φ_u a layout of it. φ_u is *optimal* iff there is no other layout φ_u' of T_u such that $cost(\varphi_u') \prec cost(\varphi_u)$.

Notice that $cost(\varphi_u') \prec cost(\varphi_u)$ implies that $\gamma(\varphi_u') \leq \gamma(\varphi_u)$.

For each node $v \in T_o$, the parallel algorithm proceeds as follows: Let T_{vo} be the tree that contains all the nodes of T_o and is rooted at v. The algorithm converts T_{vo} into a binary tree T_v and then applies the parallel tree contraction technique to T_v to compute an optimal layout of T_{vo}. Let φ_v be the optimal layout of T_v. The algorithm outputs the layout φ_v, $v \in T_o$, with the minimum cutwidth.

A layout φ_i is the restriction of the layout φ of T on T_1 if and only if the nodes of T_i appear in φ_i at the same order as in φ. The correctness of our approach is based on the following theorem proved by Yannakakis ([Yan83]):

Theorem 7 Yannakakis. *Let T be a tree that consists of two rooted trees T_1, T_2 rooted respectively at nodes v_1, v_2, and the edge $\{v_1, v_2\}$. There is an minimum cutwidth layout of T such that its restrictions on T_1, T_2 are optimal layouts of T_1, T_2 respectively.*

To convert a tree T_o into a binary one T we proceed as follows. Let v be a vertex of T_o with degree d and let $w_1, ..., w_d$ be its children. Then, the vertex set of T includes vertices $v^1, ..., v^{d+1}$. For $1 \leq i \leq d$, v^{i+1} is the right child of v^i in T (see figure 3). We will say that the vertices $v^i, 1 \leq i \leq d$, are of the *same label* since they are coming from the same vertex of T_o (e.g., in figure 3 the vertices v^2 and v^{d+1} are of the same label while w_d^1 and v^2 are not).

With each node $u \in T$, we associate two pieces of information: i) A layout φ_u, realizing the layout of the subtree rooted at u and u's position in this layout and ii) the cost-sequence, $cost(\varphi_u)$ of the layout φ_u. The *shunt* operation involves two merge-operations on the layout sequences:

Merge-operation A: Let T_{uv} be a tree rooted at a node u. T_{uv} consists of two trees T_u, T_v (rooted at u, v respectively) and the edge $\{u, v\}$. Suppose that φ_u and φ_v are optimal layouts of T_u, T_v, respectively. The merge-operation A computes an optimal layout φ_{uv} of T_{uv}.

Merge-operation B: Let T_u be a tree rooted at u with children $u_1, ..., u_d$ and T_v a tree rooted at v with children $v_1, ..., v_{d'}$. Suppose that we are given the optimal layouts of T_u, T_v. The merge-operation B computes an optimal layout φ_{uv} realizing a layout of the tree T_{uv} which is rooted at u and has as children the children of both T_u and T_v.

We give now the shunt operation of the tree contraction technique. Suppose that l_i is the leaf which is ready to perform the *shunt* and that f_i is the father of l_i, $p(f_i)$ is the father of f_i and f_j is the other child of f_i. Suppose also that optimal layouts of $l_i, f_i, f_j, p(f_i)$ are given. For the sake of simplicity in the notation, assume that the layouts are denoted by $c(l_i), c(f_i), c(f_j), c(p(f_i))$ respectively. In the sequel, f_{ij} will be the node which is the result of the *shunt* operation. A and B will denote the merge operations.

Case 1. l_i is the left leaf of f_i. (l_i, f_j cannot be of the *same label*.)

$c(f_i l_i) = A(c(l_i), c(f_i))$

Case 1.a. f_i, f_j are of the *same label*. $c(f_{ij}) = B(c(f_i l_i)), c(f_j))$.

Case 1.b. f_i, f_j are not of the *same label*. $c(f_{ij}) = A(c(f_i l_i)), c(f_j))$.

Case 2. l_i is the right leaf of f_i. (f_i, f_j cannot be of the *same label*.)

Case 2.a. l_i, f_i are of the *same label*. $c(f_i l_i) = B(c(l_i), c(f_i))$

Subcase 2.a.a. f_i, $p(f_i)$ are also of the *same label*.

$c(p(f_i)) = B(c(f_i l_i), c(p(f_i)))$

$c(f_{ij}) = c(f_j)$

Subcase 2.a.b. f_i, $p(f_i)$ are not of the *same label*.

$c(f_{ij}) = A(c(f_i l_i), c(f_j))$

Suppose that the resulting sequence $c(f_{ij})$ is $(A, *, B)$ as in the figure 4. From $c(f_{ij})$ we easily take the layout sequence $c'(f_{ij}) = (C, *, D)$, where the "$*$" denotes the position of f_j (see figure 4). Let T_{fij} be the subtree - of the current T_{v0} - rooted at f_{ij}. In the sequel, every merge operation of f_{ij} with a vertex w of T_{fij} is done using the layout-sequence $c'(f_{ij})$ while every merge operation of f_{ij} with a vertex of $T_{v0} - T_{fij}$ is done using the layout-sequence $c(f_{ij})$.

Case 2.b. l_i, f_j are not of the *same label*. This case is similar to the above ones.

The efficient parallel implementation of the merge operation A considers a series of cases depending on whether the leaf that is going to perform the shunt is a right or a left leaf and whether or not it has the same label with its father.

Lemma 8. *Let T_{uv} be a tree rooted at a node u that consists of the trees T_u, T_v rooted at u, v respectively, and the edge $\{u, v\}$. Let also φ_u, φ_v be optimal layouts of T_u and T_v respectively. Then, the merge-operation A correctly computes an optimal layout φ_{uv} of T_{uv} in $O(\log n)$ time using $O(n)$ CREW PRAM processors (n is the maximum of the lengths of φ_u, φ_v).*

Let $v_1, \cdots, v_k$ be the children of a node u and let $T_1, \cdots, T_k$ be the subtrees rooted at $v_1, \cdots, v_k$ respectively. Let also $\varphi_{v1}, \cdots, \varphi_{vk}$ be optimal layouts of the above subtrees. Suppose that φ_u is the layout obtained by merging with the merge-operation A, the layouts $\varphi_{v1}, \cdots, \varphi_{vk}$ with the layout φ_u. From the optimality of the merge-operation A we have that the cost-sequence of $c(u)$ is optimal whichever is the order with which we merge the trees T_{v_i}, $i = 1, \cdots, k$, with the layout sequence of u.

The proof of the following lemma is based on lemma 8 and the definition of merge-operation B.

Lemma 9. *Let T_u be a tree rooted at u, T_v a tree rooted at v and φ_u, φ_v be optimal layout of T_u, T_v respectively. Then, the merge-operation B correctly computes an optimal layout φ_{uv} realizing a layout of the tree T_{uv} which is rooted at u and has as children the children of both T_u and T_v, in $O(d \log n)$ time using $O(n)$ CREW PRAM processors, where d is the minimum of the degrees of u, v and n is the maximum of the sizes of T_u, T_v.*

Theorem 10. *Given a tree T with n nodes, the above algorithm constructs an optimal layout of T in time $O(d \log^2 n)$ using $O(n^2 / \log n)$ CREW PRAM processors, where d is the maximum degree of T.*

Proof. The proof is based on the theorem 3, lemma 8, lemma 9, and the description of the shunt operation.

References

[ADKP89] K. Abrahamson, N. Dadoun, D. Kirkpatrick, and K. Przytycka. A simple parallel tree contraction algorithm. *Journal of Algorithms*, 10:287–302, 1989.

[AH73] D. Adolphson and T.C. Hu. Optimal linear ordering. *SIAM J. on Applied Mathematics*, 25(3):403–423, Nov 1973.

[CMST82] M. Chung, F. Makedon, I.H. Sudborough, and J. Turner. Polynomial time algorithms for the min cut problem on degree restricted trees. In *FOCS*, volume 23, pages 262–271, Chicago, Nov 1982.

[Di92] J. Díaz. Graph layout problems. In I.M. Havel and V. Koubek, editors, *Mathematical Foundations of Computer Science*, volume 629, pages 14–24. Springer-Verlag, Lecture Notes in Computer Science, 1992.

[ES78]   S. Even and Y. Shiloach. NP-completeness of several arrangements problems. Technical report, TR-43 The Technion, Haifa, 1978.

[Gav77] F. Gavril. Some NP-complete problems on graphs. In *Proc. 11th. Conf. on Information Sciences and Systems*, pages 91–95, John Hopkins Univ., Baltimore, 1977.

[GJ79]   M.R. Garey and D.S. Johnson. *Computers and Intractability: A Guide to the Theory of NP-Completeness.* Freeman, San Francisco, 1979.

[Har66] L.H. Harper. Optimal numberings and isoperimetric problems on graphs. *Journal of Combinatorial Theory*, 1(3):385–393, 1966.

[MS86]   B. Monien and I.H. Sudborough. Min cut is NP-complete for edge weighted trees. In L. Kott, editor, *Proc. 13th. Coll. on Automata, Languages and Programming*, pages 265–274. Springer-Verlag, Lectures Notes in Computer Science, 1986.

[SH86]   M.T. Shing and T. C. Hu. Computational complexity of layout problems. In T. Ohtsuki, editor, *Layout design and verification*, pages 267–294, Amsterdam, 1986. North-Holland.

[Shi79] Yossi Shiloach. A minimum linear arrangement algorithm for undirected trees. *SIAM J. on Computing*, 8(1):15–31, February 1979.

[Yan83] Mihalis Yannakakis. A polynomial algorithm for the min cut linear arrangement of trees. In *IEEE Symp. on Found. of Comp. Sci.*, volume 24, pages 274–281, Providence RI, Nov. 1983.

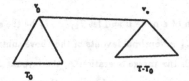

Figure 1a:

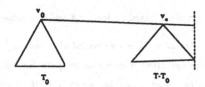

Figure 1b:

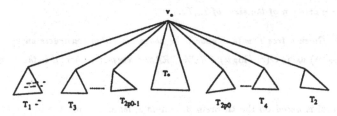

Figure 2a:

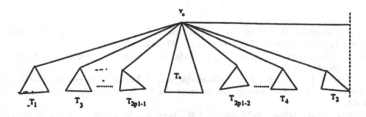

Figure 2b:

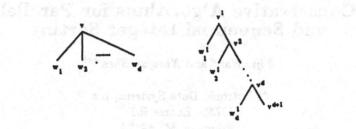

Figure 3:

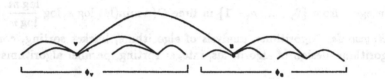

Figure 4:

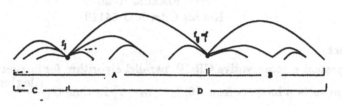

Figure 5:

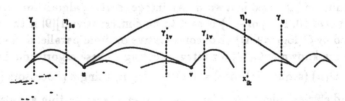

Figure 6:

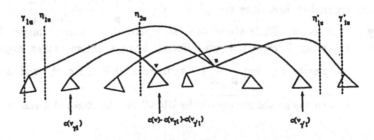

Conservative Algorithms for Parallel and Sequential Integer Sorting

Yijie Han* and Xiaojun Shen**[1]

*Electronic Data Systems, Inc.
37350 Ecorse Rd.
Romulus, MI 48174

**Computer Science Telecommunications Program
University of Missouri — Kansas City
5100 Rockhill Road
Kansas City, MO 64110

Abstract

We present a conservative CRCW parallel algorithm for integer sorting. This algorithm sorts n integers from $\{0, 1, ..., m-1\}$ in time $O(\frac{n \log \log \min(m, n)}{p} + \log n)$ using p processors. The simulation of our parallel algorithm on the sequential machine yields a sequential algorithm for integer sorting which sorts n integers from $\{0, 1, ..., m-1\}$ in time $O(n \min(\log \log n, \log \frac{\log m}{\log n}))$.

Keywords: Algorithms, analysis of algorithms, bucket sorting, conservative algorithms, design of algorithms, integer sorting, parallel algorithms.

1 Introduction

Parallel integer sorting has been studied by many researchers [1][3][5][8][13][14] [15][17]. An important parameter in integer sorting is the word length which is the number of bits used in a word. An integer sorting algorithm sorting n integers in the range $\{0, 1, ..., m-1\}$ is said to be conservative[1][9] if the word length is bounded by $O(\log(m+n))$. Nonconservative optimal parallel and sequential algorithms are known[8][9][14] for integer sorting. These algorithms have time complexity $O(n)$ (sequential case) and $O(\frac{n}{p} + \log n)$ using p processors (parallel case). Fast and efficient algorithms for conservative integer sorting are also sought. The previous best deterministic parallel algorithm for integer sorting is due to Bhatt et al.[3]. For the problem of sorting n integers in the range $\{0, 1, ..., m-1\}$ their algorithm has time complexity $O(\frac{n \log \log m}{p} + \frac{\log n}{\log \log n} + \log \log m)$ using p processors. Their algorithm *was* optimal when m is large. For sequential integer sorting Kirkpatrick and Reisch have a conservative algorithm[9] with time complexity $O(n \log \frac{\log m}{\log n})$. Fredman and Willard showed time complexity

---

[1] This author is partially supported by UMKC Faculty Research Grant K-2-11191.

$O(n\dfrac{\log n}{\log\log n})$ for conservative sequential integer sorting[4]. In this paper we improve on these results. We present a conservative parallel algorithm with time complexity $O(\dfrac{n\log\log\min(m,n)}{p}+\log n)$. Our algorithm is more efficient than the algorithm of Bhatt et al.. When $\log\log m > \log n$ or $\dfrac{n\log\log m}{p} > \log n$ our algorithm is faster than the algorithm of Bhatt et al.. The simulation of our parallel algorithm on the sequential machine yields a conservative sequential algorithm with time complexity $O(n\min(\log\log n, \log\dfrac{\log m}{\log n}))$. This improves the result by Kirkpatrick and Reisch[9] and the result by Fredman and Willard[4]. The most interesting feature of our algorithm is that its time complexity does not grow as m grows. No matter how large m is, the processor time product of our algorithm remains at $O(n\log\log n)$.

The computation model used in the design of our parallel algorithms are the CREW PRAM and ARBITRARY CRCW PRAM. In these models any processor can access any memory cell. On the CREW PRAM the concurrent read of a memory cell by several processors is allowed but concurrent write to a memory cell is prohibited. On the ARBITRARY CRCW PRAM the concurrent read and concurrent write of a memory cell are allowed. In the case of concurrent write an arbitrary processor succeeds in writing. The ARBITRARY CRCW PRAM is also the model used in the algorithm of Bhatt et al.. The computation model used in the design of our sequential algorithm is the Random Access Machine.

The integer sorting problem we consider is to stably sort n records each has an integer key in the range $\{0, 1, ..., m-1\}$ using word length $O(\log(m+n))$. The n input integers are stored in array $N[0..n-1]$. Integer $N[i]$ precedes integer $N[j]$ if $N[i] < N[j]$ or if $N[i] = N[j]$ and $i < j$. The input n records are sorted if they are arranged in an array of size n with their keys in nondecreasing order. The sorting is stable if nonidentical records with the same integer key occur in the same order in the sorted array as in the input array.

2 Parallel Bucketing

In this section we explain the parallel bucketing technique. Our presentation of the parallel bucketing technique can be viewed as an reinterpretation of the recursive chaining technique used in [3][5].

Let $\log m$ be a power of 2. Let T be a complete binary tree with 2^m leaves. The nodes of the tree will be labeled as follows. The nodes at level i of T will be labeled with an i-bit binary string. The root is labeled with ϵ. The left son of the root is labeled with 0 and the right son of the root is labeled with 1. Let u be a node in the tree labeled with s. The left son of u is labeled with $s0$ and the right son of u is labeled with $s1$. For a label l we use $num(l)$ to denote the number obtained by viewing l as a binary number.

We associate one bucket and one memory cell with each node of T. The labeling of nodes of T applies to the buckets and the memory cells.

Initially there are $k \leq m$ distinct integers, each having $\log m$ bits. There is a processor associated with each integer. We associate a pair (b, r) called sorting pair with each integer i, where b is the label of a bucket which indicates that integer i is in bucket b, and r is called the remaining number which indicates the remaining integer to be sorted in bucket b. Initially all k integers are in bucket ϵ and therefore an integer i is represented by (ϵ, i), where i has $\log m$ bits. A parallel bucketing step can be executed if the number of bits in the remaining number is > 1. A parallel bucketing step is as follows:

First empty all buckets. Let i be an integer with sorting pair (b, r). This indicates that i was in bucket b and has remaining number r. Let the binary representation of r be $r_{t-1} r_{t-2} ... r_1 r_0$ (r is represented with t bits, these t bits may contain some leading zeros). Let b' be the label formed by concatenating b and $r_{t-1} r_{t-2} ... r_{t/2}$ (viewed as a string). In a parallel step the processor associated with integer i will write its processor id into node b' in tree T. If the processor succeeds in writing its id into node b' it will throw integer i into bucket b and change the remaining number r to $r' = r_{t-1} r_{t-2} ... r_{t/2}$ (thus the remaining number has only $t/2$ bits). Thus if the processor succeeds, it merely change the sorting pair (b, r) of integer i to (b, r'). If the processor fails in writing its id into node b' in tree T (because of contention), it will throw integer i into bucket b' and change the remaining number to $r'' = r_{t/2-1} r_{t/2-2} ... r_1 r_0$. Thus if the processor fails, it merely change the sorting pair (b, r) to (b', r'') indicating that integer i is now in bucket b'. Note that r'' has only $t/2$ bits.

Please note the following facts:

(1) A parallel bucketing step can be done in constant time with k processors, one for each integer.

(2) A parallel bucketing step reduces the number of bits for the remaining number by half.

(3) If two integers i_1, i_2 are both thrown into bucket b and their remaining numbers after the parallel bucketing step are r_1 and r_2 then $r_1 \neq r_2$. Note that by assumption $i_1 \neq i_2$.

(4) Let r_1, r_2 be the remaining numbers for integers i_1, i_2, respectively. If both i_1 and i_2 are in bucket b and $i_1 > i_2$ then $r_1 > r_2$.

3   Parallel Chaining

The parallel chaining problem was formulated in [3][5]. We give a review of parallel chaining[3] in terms of parallel bucketing.

Let $i_1, i_2, ..., i_k$ be k integers in bucket b with sorting pairs (b, r_1), (b, r_2), ..., (b, r_k). By the observation we have made in the previous section we assume the all r_t's, $1 \leq t \leq k$, are distinct. The parallel chaining for bucket b is in parallel for each r_t to compute $r = \min\{r_j | r_j > r_t, 1 \leq j \leq k\}$ if $r_t \neq \max\{r_j | 1 \leq k\}$ (r is called the successor of r_t). In the case $r_t = \max\{r_j | 1 \leq j \leq k\}$ the successor to r_t is null.

If the remaining number for each integer in bucket b has only 1 bit, bucket b has at most 2 integers in it because all integers in bucket b have distinct remaining numbers. Therefore parallel chaining for bucket b is simply to chain the at most 2 integers and it can be done in constant time.

If the remaining number for each integer in bucket b has t bits, where t is a power of 2, the parallel chaining for bucket b can be done by a recursive procedure. This technique is given in [3]. First execute a parallel bucketing step. The parallel bucketing step redistributes the integers originally in bucket b into several buckets. It also reduces the number of bits for the remaining numbers by half. After the parallel bucketing we execute a recursion which accomplishes parallel chaining for all buckets. The original parallel chaining for bucket b is now accomplished by a chain joining step. The chain joining step[3] inserts the integer which succeeded in the parallel write in the parallel bucketing step back into the chain computed in the recursion and then joins all chains computed in the recursion. The parallel bucketing step and the chain joining step takes constant time if a processor is associated with each integer[3].

If the integers in a bucket b has t bits in the remaining number then the above parallel chaining process has $\log t$ levels of recursion. That is, the parallel chaining can be done in $\log t$ steps[3].

4 The Main Algorithm

We first present a parallel algorithm for parallel chaining.

Parallel Chaining Problem: n distinct integers stored in an array $N[0..n-1]$. Each integer i has the sorting pair (ϵ, i), where i has $\log m$ bits. Also assume that $\log m$ and $\log n$ are powers of 2. The problem is to parallel chaining all integers in bucket ϵ.

Algorithm Chaining: (Using n processors)

Stage 1: Execute $\min\{3\log\log n, \log\log m\}$ parallel bucketing steps. This stage reduces the bits for the remaining numbers to $\lceil \log m/(\log n)^3 \rceil$. If $3\log\log n > \log\log m$ then do parallel chaining for each bucket (in this case there are at most two elements in each bucket because the remaining number has only 1 bit) and go to stage 4 else go to stage 2.

Stage 2: Sort all integers by their remaining numbers. Note that we mix all integers in all buckets together and sort them all together by their remaining numbers. The sorting is performed on an array of size n with all the integers and sorting pairs stored in it (this array is essentially the input array). Note that the sorting is to sort n integers each having $\log m/(\log n)^3$ bits with word length $\log m$. The sorting is accomplished by the nonconservative sorting algorithm presented in the next section.

Stage 3: Consider all integers in a bucket after stage 1. After stage 2 these integers are stored in the sorted array in the sorted order. That is, if $i_1 <$

i_2 then i_1 precedes i_2 in the sorted array. Note that these integers are not necessarily adjacent. If integer i is in position j in the sorted array we change its remaining number to j while keeping its bucket label. Note that this change does not alter the order of the integers in a bucket while reducing the bits for the remaining number to $\log n$. We then execute, for each bucket, the parallel chaining algorithm with time complexity $O(\log \log n)$ (because the remaining number has only $\log n$ bits now) mentioned in the previous section. After this step the parallel chaining for every bucket is done.

Stage 4: Execute $\min(3 \log \log n, \log \log m)$ chain joining steps to join the chains in all buckets together.

We have explained stages 1 and 4 in the previous section. The only novelty of the algorithm is stages 2 and 3. Instead of continuing the recursion we choose to use a different sorting algorithm to sort on the remaining numbers and the sorting reduces the number of bits for the remaining numbers to $\log n$. Therefore a parallel chaining for each bucket can be done in an additional $O(\log \log n)$ steps.

In the next section we will prove the following theorem.

Theorem 1: n integers in the range $\{0, 1, ..., m-1\}$ can be sorted in $O(n/p + \log n)$ time using p processors and word length $O((\log n)^3 \log m)$. $\square$

The correctness of Algorithm Chaining is obvious if the parallel chaining, parallel bucketing and chain joining techniques are well understood. Now let us analyze the time complexity of Algorithm Chaining. We use $p \leq n$ processors. Stages 1 and 4 take $O(\dfrac{n \log \log \min(m, n)}{p} + \log \log n)$ time. By Theorem 1 stage 2 takes $O(n/p + \log n)$ time. Stage 3 executes the parallel chaining algorithm in section 3 for remaining numbers of $\log n$ bits. Therefore stage 3 takes $O(\dfrac{n \log \log n}{p} + \log \log n)$ time. Adding the time complexity for all 4 stages and note that stages 2 and 3 will not be executed if $\log \log m < 3 \log \log n$ we conclude that the time complexity for Algorithm Chaining is $O(\dfrac{n \log \log \min(m, n)}{p} + \log n)$.

Theorem 2: The Parallel Chaining Problem can be solved in $O(\dfrac{n \log \log \min(m, n)}{p} + \log n)$ time. $\square$

We now solve the integer sorting problem.

Let the n input integers be stored in array $N[0..n-1]$. Each integer has $\log m$ bits. Assume that both $\log m$ and $\log n$ are powers of 2.

Algorithm Sort:

Stage 1: Compute sorting pair for each integer. The bucket label for $N[i]$ is the $\log n$ bit binary string obtained from $b = \lfloor (N[i] * n + i)/m \rfloor$ by adding some leading 0's to the binary representation of b. The remaining number for $N[i]$ is $r = (N[i] * n + i) \bmod m$. The remaining number has $\log m$ bits. What this stage accomplishes is to put the input integers into n buckets. Let b_1, b_2 be the labels of two buckets. If $num(b_1) < num(b_2)$ then all integers in bucket b_1

precede all integers in bucket b_2. For any two integers i_1, i_2 in the same bucket, if the remaining number of i_1 is less than the remaining number of i_2 then i_1 precedes i_2.

Stage 2: For all buckets (each bucket is labeled with a $\log n$ bit binary number b, $0 \leq b < n$) do in parallel: solve the parallel chaining problem for all integers in bucket b. This is accomplished by Algorithm Chaining. Note that the sorting in stage 2 of Algorithm Chaining is done by mixing all remaining numbers in all buckets and by sorting them all together.

Stage 3: After stage 2 a chain is computed for each nonempty bucket formed in stage 1. In this stage we join all these chains together. We use an array $B[0..n-1]$. Let b be a bucket, if b is not empty we write 1 in $B[num(b)]$ else we write 0 in $B[num(b)]$. Array B then allows us to join the chains in all n buckets in time $O(n/p + \log n)$ by using an array prefix algorithm[10][11].

Stage 4: Rank the chain obtained in stage 3 by using a parallel linked list prefix algorithm [2][6][7]. Then route each integer to its final sorted position.

Theorem 3: Algorithm Sort sorts n integers in the range $\{0, 1, ..., m-1\}$ using word length $\log m$ in time $O(\frac{n \log \log \min(m, n)}{p} + \log n)$ time.

Proof: Algorithm Sort uses the scheme of chaining the integers in a linked list and then rank the linked list. This scheme was used before in [3][5]. Thus the correctness of the algorithm can be easily seen. We now analyze the complexity of the algorithm.

Stage 1 involves $O(n)$ operations and can be done in constant time. Therefore stage 1 takes $O(n/p)$ time using p processors. Stage 2 calls Algorithm Chaining and therefore can be done in time $O(\frac{n \log \log \min(m, n)}{p} + \log n)$ time by Theorem 2. Stage 3 uses an array prefix algorithm which has time complexity $O(n/p + \log n)$ [10][11]. Stage 4 uses a linked list prefix algorithm which has time complexity $O(n/p + \log n)$ [2][6][7]. Thus the time complexity of stage 2 dominates. The overall time complexity for Algorithm Sort is $O(\frac{n \log \log \min(m, n)}{p} + \log n)$. □

5 A Nonconservative Sorting Subroutine

In this section we will prove Theorem 1, namely to sort n integers in the range $\{0, 1, ..., m-1\}$ with word length $O((\log n)^3 \log m)$ in time $O(n/p + \log n)$ using p processors. The main technique used in our algorithm is the encoding technique. The encodeing technique has been used before [1][8][9][14] in the design of integer sorting algorithms. Because the word length is $O((\log n)^3 \log m)$ we can encode $(\log n)^3$ integers in one word with one integer taking $\log m$ bits. However, we have also to encode the ranks of integers in a word. A rank of an integer i is the number of integers which precede i. Thus each rank can take $O(\log n)$ bits. For our purpose we encode $(\log n)^2$ integers into one word. Each integer takes

$O(\log(m + n))$ bits (need $\log m$ bits for the integer, $\log n$ bits for the rank and then some auxiliary bits). We will design our algorithm on the CREW PRAM.

In stage 1 we sort every $(\log n)^2$ integers into one word. Thus after stage 1 the n integers are encoded into $n/(\log n)^2$ words with each word containing the sorted sequence of $(\log n)^2$ integers. This stage has $2 \log \log n$ steps. In step i, $0 \le i < 2 \log \log n$, we merge $n/2^{i+1}$ pairs of words with each word containing a sorted sequence of 2^i integers. Step i can be done using $n/2^i$ processors in time $O(i)$[1]. Thus step i takes $O(n \cdot i/2^i)$ operations. Therefore stage 1 can be done in time $O(n/p + (\log \log n)^2)$.

In stage 2 we work with only $n/(\log n)^2$ words. We assign $\log n$ processors to each word. Thus the total number of processors used is $n/\log n$. There are $O(\log n/ \log \log n)$ steps in stage 2. Each step takes $O(\log \log n)$ time. Thus stage 2 takes $O(n/p + \log n)$ time. Each step is to merge every $\log n$ sorted sequences into one sorted sequence. Thus after $O(\log n/ \log \log n)$ steps all sequences are merged into one sorted sequence. The algorithm described here is an adaptation of Preparata's parallel sorting algorithm[12].

To merge $\log n$ sequences in a step, we do a pairwise merge for these sequences. Because we assign $\log n$ processors to each word, each processor will participate in one pairwise merge. Let s_1, s_2 be two sequences. Let x be an integer in s_1. The merge of s_1 and s_2 will tell how many integers in s_2 which precede x. The number of integers in s_2 which precede x is called the rank of x in s_2. After pairwise merge of all $\log n$ sequences we can tell the ranks of x in all $\log n$ sequences. The sum of these ranks tells the sorted position of x. Thus what we need to do is to do a pairwise merge to get the ranks for each integer, to sum the ranks for each integer, and then to route each integer to its sorted position.

We use k processors to merge two sequences s_1 and s_2 each containing k words (therefore $k(\log n)^2$ integers) in the pairwise merge mentioned in the above paragraph. We first sample the two sequences. The first integer in each word will be sampled. In addition the last integer in the last word of a sequence will be sampled. These $2k + 2$ sampled integers form 2 sorted sequences. We put one sampled integer in one word. Using $2k$ processors we first merge these samples. Valiant's parallel merge algorithm[16] can be used here. The merge takes $O(\log \log n)$ time. Let x be a sampled integer from s_1. The merge of sampled integers will tell which word in s_2 should x be inserted in the merge of s_1 and s_2. Let w be the word in s_2 into which x should be inserted. By a binary search we can find out how many integers in w which precede x. This binary search takes one processor and $O(\log \log n)$ time because there are only $(\log n)^2$ integers in a word. We do such binary search for all sampled integers. After the binary search all sampled integers from s_1 know their rank in s_2 and all sampled integers from s_2 know their rank in s_1. These ranks partition s_1 and s_2 into $2k + 1$ pairs of subsequences, as shown in Figure 1 We then do a pairwise merge of these subsequences. Each pairwise merge is to merge integers in two words. And therefore each pairwise merge can be done in $O(\log \log n)$ time using one processor[1]. We have outlined the scheme of merging two sequences s_1 and s_2.

After merging, each integer in s_1 is ranked in s_2 and vice versa. These ranks need to be routed back to the original positions of the integers. This can be done by reversing the computation and route back each integer from its merged position to its original position. The overall time complexity is $O(\log\log n)$.

After the pairwise merge we need to add the ranks for each integer. Since each word has length $(\log n)^3 \log m$, we can store $(\log n)^2$ ranks in each word, one rank for each integer. By adding $\log n$ words of ranks we have computed rank for each integer. $\log n$ words can be added in $O(\log\log n)$ time because there are $\log n$ processors for the addition.

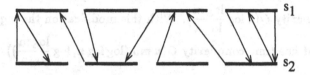

Figure 1: Partition of s1 and s2 into 2k+1 pairs.

Now the rank for the $\log n$-way merge is computed for each integer i and stored in the corresponding position for i as shown in Figure 2 For each sequence of k words, we expand it to $k \log n$ words to move each integer in the sequence to its ranked position. Vacant positions in the $k \log n$ words are filled with 0's (see Figure 3). This expansion can be done in $O(\log\log n)$ time. After expansion we have $\log n$ expanded sequences. We then OR these expanded sequence together to get the sorted sequence. This OR operation is easily seen to take $O(\log\log n)$ time.

word 0				word 1			
1	3	4	6	7	8	9	11

Figure 2: Ranks of four integers in a word.

0	1	0	3	4	0	6	7	8	9	0	11

Figure 3: Expanding the ranks.

Summarizing the above paragraphs, we have shown how to do $\log n$-way merge using $n/\log n$ processors in time $O(\log\log n)$.

Because stage 2 has $O(\log n/\log\log n)$ steps. The time complexity for stage 2 is $O(\log n)$ using $n/\log n$ processors or, in other words, $O(n/p + \log n)$ time.

We have described a nonconservative sorting subroutine which sorts n integers in the range $\{0, 1, ..., m-1\}$ using word length $(\log n)^3 \log m$ in time $O(n/p + \log n)$. This proves Theorem 1.

6 Adaptation to Sequential Machine

Straightforward simulation of this parallel algorithm on a sequential machine yields time complexity $O(n \log \log \min(m, n))$. We modify this straightforward simulation by taking advantage of the conservative sequential integer sorting algorithm of Kirkpatrick and Reisch[9]. We compare $\log \log n$ with $\log \frac{\log m}{\log n}$. If $\log \log n < \log \frac{\log m}{\log n}$ we do a sequential simulation of our parallel algorithm, otherwise we simply use Kirkpatrick and Reisch's algorithm[9]. Their algorithm has time complexity $O(n \log \frac{\log m}{\log n})$. With this modification the sequential algorithm obtained has time complexity $O(n \min(\log \log n, \log \frac{\log m}{\log n}))$.

References

[1] S. Albers and T. Hagerup. Improved parallel integer sorting without concurrent writing. *Proc. The Third Annual ACM-SIAM Symp. on Discrete Algorithms*, pp. 463-472.

[2] R. Anderson and G. Miller. Deterministic parallel list ranking. *Algorithmica* 6: 859-868(1991).

[3] P.C.P. Bhatt, K. Diks, T. Hagerup, V.C. Prasad, T. Radzik, S. Saxena. Improved deterministic parallel integer sorting. *Information and Computation* 94, 29-47(1991).

[4] M. L. Fredman and D. E. Willard. Blasting through the information theoretic barrier with fusion trees. *Proc. 1990 ACM Symp. on Theory of Computing*, pp. 1-7(1990).

[5] T. Hagerup. Towards optimal parallel bucket sorting. *Inform. and Comput.* 75, pp. 39-51(1987).

[6] Y. Han. Designing fast and efficient parallel algorithms. Ph.D. Thesis. Department of Computer Science, Duke University, Durham, North Carolina, 1987.

[7] Y. Han. Parallel algorithms for computing linked list prefix. *J. of Parallel and Distributed Computing* 6 537-557(1989).

[8] T. Hagerup and H. Shen. Improved nonconservative sequential and parallel integer sorting. *Infom. Process. Lett.* 36, pp. 57-63(1990).

[9] D. Kirkpatrick and S. Reisch. Upper bounds for sorting integers on random access machines. *Theoretical Computer Science* 28, pp. 263-276(1984).

[10] P. M. Kogge and H. S. Stone. A parallel algorithm for the efficient solution of a general class of recurrence equations. *IEEE Trans. Comput.*, Vol C-22, pp. 786-792(Aug. 1973).

[11] R. E. Ladner and M. J. Fischer. Parallel prefix computation. *J. ACM*, pp. 831-838(Oct. 1980).

[12] F. Preparata. New parallel-sorting schemes. *IEEE Transactions on Computers*, Vol. c-27, No. 7, pp. 669-673(July 1978).

[13] S. Rajasekaran and J. Reif. Optimal and sublogarithmic time randomized parallel sorting algorithms. *SIAM J. Comput.* 18, pp. 594-607.

[14] S. Rajasekaran and S. Sen. On parallel integer sorting. *Acta Informatica* 29, 1-15(1992).

[15] R. Raman. The power of collision: randomized parallel algorithms for chaining and integer sorting. *Proc. 10th Conf. on Foundations of Software Technology and Theoretical Computer Science*, Springer Lecture Notes in Computer Science, Vol. 472, pp. 161-175.

[16] L. Valiant. Parallelism in comparison problems. *SIAM J. Comput.*, 4, pp. 348-355(1975).

[17] R.A. Wagner and Y. Han. Parallel algorithms for bucket sorting and the data dependent prefix problem. *Proc. 1986 International Conf. on Parallel Processing*, pp. 924-930.

An Optimal Algorithm for Proper Learning of Unions of Two Rectangles with Queries

Zhixiang Chen *

Computer Science Department
Boston University
Boston, MA 02215, USA
Email: zchen@cs.bu.edu

Abstract. We study the problem of proper learning of unions of two discretized axis-parallel rectangles over the domain $\{0, n-1\}^d$ in the on-line model with equivalence and membership queries. An obvious approach to this problem would use two equivalence queries to find one example in each of the two rectangles contained in the target concept and then use membership queries to find end points of the rectangles. However, there is one substantial difficulty: For any two end points, how to decide whether they belong to the same rectangle? In this paper, we develop some strategies to overcome the above difficulties and construct an algorithm that properly learns unions of two rectangles over the domain $\{0, n-1\}^d$ with at most two equivalence queries and at most $(11d + 2) \log n + d + 3$ membership queries. We also show that this algorithm is optimal in terms of query complexity.

1 Introduction

Since Valiant's seminal paper [V] great efforts have been devoted to the study of the efficient proper (or *proper* for short) learnability of DNF formulas in both pac-learning model and exact learning model with queries. Pitt and Valiant [PV] have shown that, for any constant $k \geq 2$, k-term DNF formulas are not properly learnable in the PAC model when $RP \neq NP$. Their result implies that k-term DNF formulas, for constant $k \geq 2$, is not proper learnable in the exact learning model using equivalence queries under the assumption that $P \neq NP$. Bshouty *et. al.* [BGHM] have shown that $\sqrt{\log n}$-term DNF formulas are properly online learnable using equivalence and membership queries. It was shown in [PRa] that this positive result cannot be significantly improved in the exact model or the PAC model allowing membership queries, given certain standard theoretical complexity assumptions.

Although unions of rectangles are generalizations of DNF formulas, no significant progress has been made on the proper learnability of unions of rectangles. In this paper, we will study the efficient proper learnability of unions of two rectangles in the on-line model with queries. We assume that N is the set of all natural numbers. $\forall i, j \in N$, we use $[i, j]$ to denote the set $\{i, \ldots, j\}$ if $i \leq j$

* The author was supported by NSF grants CCR-9103055 and CCR-9400229.

or ϕ otherwise. We define the class of all discretized *axis-parallel rectangles* (or *rectangles* for short) over the domain $[0, n-1]^d$ as follows,

$$BOX_n^d = \{\prod_{i=1}^{d}[a_i, b_i] \mid 0 \le a_i \le b_i \le n-1, \forall i \in [1, d]\} \cup \{\phi\}.$$

We consider the following concept class of unions of two rectangles over the domain $[0, n-1]^d$

$$TWO_n^d = \{C_1 \cup C_2 | C_1, C_1 \in BOX_n^d\}.$$

Recently, the problem of learning unions of rectangles has been studied by many researchers. A good survey of the relevant results can be found in [BGGM, CHb].

2 The Learning Model

Our model is the on-line learning model with equivalence and membership queries [A]. The goal of a learning algorithm (or learner) for a concept class **C** over a domain **X** is to learn any unknown target concept $C_t \in$ **C** that has been fixed by the teacher (or environment). In order to obtain information about C_t the learner can ask equivalence queries by proposing hypotheses H from a fixed hypothesis space **H** with **C** $\subseteq$ **H** $\subseteq 2^{\mathbf{X}}$ to the equivalence oracle $EQ()$. $EQ(H) = $ "*yes*" if $H = C_t$, so the learner learns it. If $H \ne C_t$, then $EQ(H) = x$ for some $x \in (H - C_t) \cup (C_t - H)$, called a counterexample. x is called a *positive counterexample* (PCE) if $x \in C_t - H$, or a *negative counterexample* (NCE) otherwise. The learner can also ask membership queries by presenting examples in the domain to the membership oracle $MQ()$. For each membership query for an example x, $MQ(x) = $ "*yes*" if x is in the target concept, otherwise $MQ(x) = $ "*no*". Each new hypothesis issued by the leaner may depend on the earlier hypotheses and the received CE's as well as the previous membership queries. We say that a learning algorithm for the concept class **C** is polynomial if its time complexity is polynomial in the logarithm of the size of the domain and the size of the target concept. **C** is *efficiently properly* (or *properly* for short) learnable if there is a polynomial time algorithm for **C** which uses hypotheses from **C**.

3 The Upper Bound and The Lower Bound

For any example $x = (x_1, \ldots, x_d)$, $y = (y_1, \ldots, y_d)$, and $a \in [0, n-1]$, we define

$$x|_a^i = (x_1, \ldots, x_{i-1}, a, x_{i+1}, \ldots, x_d),$$

$$rec(x, y) = \prod_{i=1}^{d}[min\{x_i, y_i\}, max\{x_i, y_i\}].$$

Consider a target concept $A \cup B = \prod_{i=1}^{d}[a_i, b_i] \cup \prod_{i=1}^{d}[e_i, f_i] \in TWO_n^d$. Given $x \in A \cup B$, we describe the following two procedures.

```
Rsearch(x):
  begin
  for (j = 1; j ≤ d; j = j + 1)  {
    s_j = x_j;
    t_j = n - 1;
    while (s_j < t_j)  {
      z_j = s_j + ⌈(t_j−s_j)/2⌉;
      if (MQ(x|^j_{z_j}) = "yes")
      then s_j = z_j else t_j = z_j - 1;  } }
  output (s_1, ..., s_d);
  end

Rscan(x):
  begin
  for (j = 1; j ≤ d; j = j + 1)  {
    s_j = x_j;
    t_j = n - 1;
    while (s_j < t_j)  {
      z_j = s_j + ⌈(t_j−s_j)/2⌉;
      if (MQ((s_1, ..., s_{j−1}, z_j, x_{j+1}, ..., x_d)) = "yes")
      then s_j = z_j else t_j = z_j - 1;  } }
  output (s_1, ..., s_d);
  end
```

The above two procedures employ binary search at each dimension to find respectively some right end point of one of A and B. Those two procedures are illustrated in figure 1. Note that the example produced by $Rscan(x)$ is in $A \cup B$. However, the example produced by $Rsearch(x)$ may not be in $A \cup B$.

Remark 4.1. *We can define with the similar manner analogous versions of Rsearch(x) and Rscan(x), Lsearch(x) and Lscan(x), which find respectively some left end point of one of A and B at each dimension.*

We now establish several lemmas for the above procedures. Proofs for Lemma 4.2 and 4.3 are omitted due to space limitation.

Lemma 4.2. *Assume $y \in A \cup B$. Let $\theta = (\theta_1, \ldots, \theta_d) = Rsearch(y)$, and $\theta' = (\theta'_1, \ldots, \theta'_d) = Lsearch(y)$. Then, the following properties hold.*

(1) *If $y_j \in [e_j, f_j] - [a_j, b_j]$, then $\theta_i = f_i$ and $\theta'_i = e_i$ for any $i \neq j$. In particular, if $y_j < a_j$ then $\theta'_i = e_i$, while if $y_j > b_j$ then $\theta_i = f_i$, $\forall i \in [1, d]$.*
(2) *If there are distinct j and k such that $y_j \in [e_j, f_j] - [a_j, b_j]$ and $y_k \in [e_k, f_k] - [a_k, b_k]$, then $\theta_i = f_i$ and $\theta'_i = e_i$, $\forall i \in [1, d]$.*

In the following we assume that $x \in A \cup B$. Define $\alpha = Rscan(x)$ and use α_i to denote the i-th component of α. Similarly, define $\beta = Rscan(\alpha)$, $\gamma = Rscan(\beta)$, $\delta' = Lsearch(\gamma)$, $\alpha' = Lscan(x)$, $\beta' = Lscan(\alpha')$, $\gamma' = Lscan(\beta')$, and $\delta = Rsearch(\gamma')$. It is easy to see from the constructions of $Rscan$ and $Lscan$ that, $\forall i \in [1, d]$,

$$\gamma'_i \leq \beta'_i \leq \alpha'_i \leq x_i \leq \alpha_i \leq \beta_i \leq \gamma_i.$$

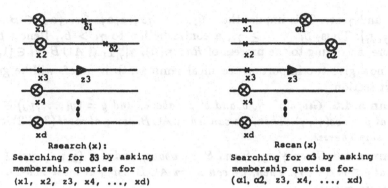

Rsearch(x):
Searching for δ3 by asking
membership queries for
(x1, x2, z3, x4, ..., xd)

Rscan(x)
Searching for α3 by asking
membership queries for
(α1, α2, z3, x4, ..., xd)

Fig. 1. The Searching Mechanism of Rsearch and Rscan

From now on, the above relation will be used implicitly.

Lemma 4.3. *Assume that $A \cup B$ is a single rectangle, say, $A \cup B = B$. Then,*

(1) $\alpha = \beta = \gamma = \delta = (f_1, \ldots, f_d)$, *and*
(2) $\alpha' = \beta' = \gamma' = \delta' = (e_1, \ldots, e_d)$.

Lemma 4.4. *For γ and γ', the following properties hold.*

(1) $\forall i \in [1, d], \gamma|_{\gamma_i + 1}^{i} \notin A \cup B$. *(Hence, either $\gamma = (b_1, \ldots, b_d)$ or $\gamma = (f_1, \ldots, f_d)$.)*
(2) $\forall i \in [1, d], \gamma'|_{\gamma_i' - 1}^{i} \notin A \cup B$. *(Hence, either $\gamma' = (a_1, \ldots, a_d)$ or $\gamma' = (e_1, \ldots, e_d)$.)*

Proof. We only prove (1), since (2) is symmetric to (1). We analyze the following cases.

Case 4.4.1: $\alpha = \gamma$. This implies that $\beta = \gamma$. Thus, $\gamma|_{\gamma_i + 1}^{i} \notin A \cup B$, $\forall i \in [1, d]$, according to the process of $Rscan(\beta)$.

Case 4.4.2: $\alpha \neq \gamma$ and, α and γ are not in the same rectangle. Say, $\alpha \in A - B$, and $\gamma \in B - A$. $\alpha \neq \gamma$ implies $\alpha \neq \beta$. So, we can fix j such that $\alpha_i = \beta_i$ for $i = 1, \ldots, j-1$, but $\alpha_j < \beta_j$. Suppose that $\exists k \in [j+1, d]$ such that $x_k \notin [a_k, b_k]$. Consider the process of finding α_j in the execution of $Rscan(x)$. We know that $(\alpha_1, \ldots, \alpha_{j-1}, \alpha_j, x_{j+1}, \ldots, x_d) \in A \cup B$, i.e., $(\alpha_1, \ldots, \alpha_{j-1}, \alpha_j, x_{j+1}, \ldots, x_d) \in B - A$ since $x_k \notin [a_k, b_k]$. Hence, $\forall z \geq \alpha_j$, $(\alpha_1, \ldots, \alpha_{j-1}, z, x_{j+1}, \ldots, x_d) \in A \cup B$ if and only if $z \leq f_j$. Thus, $\alpha_j = f_j$ according to the process of $Rscan(x)$. Note that $\alpha_j < \beta_j \leq \gamma_j$. So, $\gamma \notin B$, a contradiction to the fact that $\gamma \in B - A$. This follows that $\forall k \in [j+1, d]$, $x_k \in [a_k, b_k]$. In this case, $(\alpha_1, \ldots, \alpha_{j-1}, z, x_{j+1}, \ldots, x_d) \in A$, $\forall z \in [\alpha_j, b_j]$. This implies that $\alpha_j = b_j$, since $\alpha \in A - B$. Thus, $b_j < \beta_j$. Hence, $\beta \in B - A$. Now, $\forall i \in [1, d], \forall z \geq \beta_i$, $(\gamma_1, \ldots, \gamma_{i-1}, z, \beta_{i+1}, \ldots, \beta_d) \in A \cup B$ if and only if $(\gamma_1, \ldots, \gamma_{i-1}, z, \beta_{i+1}, \ldots, \beta_d) \in B - A$. Therefore, $\gamma = (f_1, \ldots, f_d)$ and, $\forall i \in [1, d], \gamma|_{\gamma_i + 1}^{i} \notin A \cup B$, since $f_j \geq \beta_j > b_j$.

Case 4.4.3: $\alpha \neq \gamma$ and they are in the same rectangle, say, B. If $\gamma \neq \beta$, then we can choose j such that $\gamma_i = \beta_i$ for $i = 1, \ldots, j-1$, but $\gamma_j > \beta_j$. β is in B, since α and γ are. This implies that $(\beta_1, \ldots, \beta_{j-1}, z, \alpha_{j+1}, \ldots, \alpha_d) \in B$ for any $z \in [\alpha_j, f_j]$. Thus, $\beta_j \geq f_j \geq \gamma_j$, a contradiction to $\gamma_j > \beta_j$. Hence, $\beta = \gamma$. Therefore, according to the process of $Rscan(\beta)$, $\gamma|_{\gamma_i+1}^i \notin A \cup B$, $\forall i \in [1, d]$. $\square$

We now give two important technical lemmas, their proofs will be given in the next section.

Lemma 4.5. *Given γ', γ, δ and δ' as above, and $y = (y_1, \ldots, y_d) \in A \cup B$ such that $y \notin rec(\gamma', \gamma)$. Then, we can learn $A \cup B$ using at most $(3d+2)\log n + 3$ membership queries.*

Lemma 4.6 *Given γ', γ, δ and δ' as above, and $y = (y_1, \ldots, y_d) \notin A \cup B$ such that $y \in rec(\gamma', \gamma)$. Then, we can learn $A \cup B$ using at most $2d\log n + d + 2$ membership queries.*

Theorem 4.7. *There is a poly-time algorithm that properly learns TWO_n^d using at most 2 equivalence queries and at most $(11d+2)\log n + d + 3$ membership queries.*

Proof. Given $A \cup B = \prod_{i=1}^d [a_i, b_i] \cup \prod_{i=1}^d [e_i, f_i] \in TWO_n^d$, we construct the learning algorithm as follows.

At step 1, set $x = EQ(\phi)$. If $x =$ "*yes*" then stop, otherwise find
$$\alpha = Rscan(x), \quad \beta = Rscan(\alpha), \gamma = Rscan(\beta),$$
$$\alpha' = Lscan(x), \quad \beta' = Lscan(\alpha'), \quad \gamma' = Lscan(\beta'),$$
$$\delta = Rsearch(\gamma'), \quad \delta' = Lsearch(\gamma).$$
At this step, we use one equivalence query and at most $8d\log n$ membership queries.

Recall that $\forall i \in [1, d], \gamma_i' \leq \gamma_i$. Set $H = rec(\gamma', \gamma)$. Ask an equivalence query for H and let $y = EQ(H)$. If $y =$ "*yes*" then stop. When y is a PCE, then by Lemma 4.5, we can learn $A \cup B$ using two equivalence queries and at most $(11d + 2)\log n + 3$ membership queries. When y is a NCE, by Lemma 4.6, we can learn $A \cup B$ using two equivalence queries and at most $10d\log n + d + 2$ membership queries. Hence, in either case, we can learn $A \cup B$ using at most two equivalence queries and at most $(11d + 2)\log n + d + 3$ membership queries. $\square$

Theorem 4.8. *Assume that L is any poly-time algorithm for learning TWO_n^d using equivalence and membership queries. Then, (1) L requires at least two equivalence queries; and (2) the number of equivalence and membership queries required by L is at least $2d(\log n - 1)$.*

Proof Sketch. (1) Let p be the polynomial such that the time complexity of L is bounded by $p(d, \log n)$. Then, the number of membership queries required by L is bounded by $p(d, \log n)$. Choose d and n large enough such that $n^d - p(d, \log n) \geq 2$. Thus, there are at least two examples which have not used by any membership queries, so the adversary can use these two examples to defeat the leaner.

(2) The adversary constructs rectangles $A \subset X_1 = [0, \lfloor \frac{n-1}{2} \rfloor]^d$ $B \subset X_2 = [\lfloor \frac{n-1}{2} \rfloor + 1, n-1]^d$ and lets all the d left points of A be 0 and all the d right end points of B be $n - 1$. Since A and B have $2d$ end points which are not fixed, so the learner needs at least $2d(\log n - 1)$ equivalence and membership queries to find those end points.

4 Proofs of Two Technical Lemmas

Proof of Lemma 4.5. By Lemma 4.4, $y \notin rec(\gamma', \gamma)$ implies that $A - B \neq \phi$ and $B - A \neq \phi$. $y \notin rec(\gamma', \gamma)$ also implies that $\exists j$, $y_j < \gamma'_j$ or $y_j > \gamma_j$. We only consider $y_j < \gamma'_j$. The proof for $y_j > \gamma_j$ can be done in the same manner. By Lemma 4.4, we assume without loss of generality that $\gamma' = (a_1, \ldots, a_d)$. Find $\sigma' = Lsearch(y)$ and $\sigma = Rsearch(y)$. This takes at most $2d \log n$ membership queries. By Lemma 4.2, $\sigma' = (e_1, \ldots, e_d)$ and $\sigma_i = f_i$ for any $i \neq j$. Now, in order to learn the target we only need to find f_j and $(b_1, \ldots, b_d)$. We analyze the following cases.

Case 4.5.1: $\exists k \neq j$, $\gamma_k \neq \sigma_k$. In this case, $\gamma = (b_1, \ldots, b_d)$ by Lemma 4.4. So, to learn the target we only need to find f_j. If $\sigma_j \neq \gamma_j$, then $\sigma_j = f_j$, since $\gamma = (b_1, \ldots, b_d)$. Now, we consider that $\sigma_j = \gamma_j$. If $\exists m \neq j$ such that $\gamma_m < \sigma_m$, then f_j is the largest $z \in [y_j, n]$ such that $\sigma|_z^j \in A \cup B$. Similarly, if $\exists m \neq j$ such that $\sigma'_m < \gamma'_m$, then f_j is the largest $z \in [y_j, n]$ such that $\sigma'|_z^j \in A \cup B$. In either of the above two subcases, we can find f_j using binary search with at most $\log n$ membership queries. Now, the remaining subcase is that $\forall i \neq j$, $\gamma'_i \leq \sigma'_i \leq \sigma_i \leq \gamma_i$ and, $\sigma_j = \gamma_j$. When $\sigma'|_{\gamma'_j - 1}^j \in A \cup B$, then $f_j = \gamma_j$. Otherwise, it is the largest $z \in [\sigma'_j, \gamma'_j - 1]$ such that $\sigma'|_{\gamma'_j - 1}^j \in A \cup B$, so we can find it using binary search. Combining all the subcases, with at most $(2d + 1)\log n + 1$ membership queries we learn the target.

Case 4.5.2: $\forall i \neq j$, $\sigma_i = \gamma_i$ and, $\sigma_j < \gamma_j$. In this case, $f_j = \sigma_j$, since $\sigma_j \geq f_j$ according to the process of $Rsearch(y)$. Thus, by Lemma 4.4, $\gamma = (b_1, \ldots, b_d)$. Hence, in this case, we need at most $2d \log n$ membership queries to find the target.

Case 4.5.3: $\forall i \neq j$, $\sigma_i = \gamma_i$ and, $\gamma_j < \sigma_j$. We consider the following subcases.

Subcase 4.5.3a: $\exists m \neq j$, $\delta_m \neq \gamma_m$. Note that $\gamma_m = \sigma_m = f_m$. By Lemma 4.4, $\gamma = (f_1, \ldots, f_d)$. Thus, $\sigma_j > \gamma_j = f_j$. Hence, by Lemma 4.2, $(b_1, \ldots, b_d) = Rsearch(\gamma'|_{\sigma_j}^j)$. So, we find the target with at most $3d \log n$ membership queries.

Subcase 4.5.3b: $\forall i \neq j$, $\delta_i = \gamma_i$. In this subcase, there must be some $m \neq j$ such that $\gamma'_m < \sigma'_m = e_m$. Suppose that this is not true. Then, $\forall i \neq j$, $e_i \leq a_i \leq b_i \leq f_i$. Recall also that $e_j < a_j$. Thus, $A \cup B = rec(\sigma', \sigma)$, a contradiction to the fact that $A \cup B$ is not a rectangle. Now, since $\gamma'_m < \sigma_m$, b_j is the largest $z \in [\gamma'_j, n]$ such that $\gamma'|_z^j \in A \cup B$. So, we can find b_j using binary search with at most $\log n$ membership queries. Note that γ_j and σ_j are two right end points of the target $A \cup B$ according to the processes of Rscan and Rsearch. We have $b_j = \gamma_j$ or $b_j = \sigma_j$. If $b_j = \gamma_j$, then by Lemma 4.4, $\gamma = (b_1, \ldots, b_d)$, thus $f_j = \sigma_j$. If $b_j = \sigma_j$, then by Lemma 4.4, $\gamma = (f_1, \ldots, f_d)$ and, by Lemma 4.2, $(b_1, \ldots, b_d) = Rsearch(\gamma'|_{b_j}^j)$. So, with at most $3d \log n$ membership queries, we can find the target.

Case 4.5.4: $\forall i$, $\sigma_i = \gamma_i$. We analyze the following subcases.

Subcase 4.5.4a: $\exists m, \delta_m > \gamma_m$. By Lemma 4.4, $(f_1, \ldots, f_d) = \gamma$. By Lemma 4.2, $(b_1, \ldots, b_d) = Rsearch(\gamma'|_{\delta_m}^m)$. Thus, we can find the target with at most $3d \log n$ membership queries.

Subcase 4.5.4b: $\forall i, \delta_i \leq \gamma_i$ and, $\forall k \neq j, \sigma'_k \leq \gamma'_k$. In this subcase, $\forall i \neq j$, $e_i \leq a_i \leq b_i \leq f_i$. If $\sigma'|^j_{\gamma'_j-1} \in A \cup B$, then $A \cup B = rec(\sigma', \sigma)$, a contradiction to the fact that $A \cup B$ is not a rectangle. Hence, $\sigma'|^j_{\gamma'_j-1} \notin A \cup B$. This implies that f_j is the largest $z \in [\sigma'_j, \gamma'_j - 1]$ such that $\sigma'|^j_z \in A \cup B$. By Lemma 4.2, $(b_1, \ldots, b_d) = Rsearch(\gamma') = \delta$. So, we can find the target with at most $(2d + 1) \log n$ membership queries.

Subcase 4.5.4c: $\forall i, \delta_i \leq \gamma_i$ and, $\exists m \neq j, \gamma'_m < \sigma'_m$. By Lemma 4.2, $\forall i \neq m, \delta_i = b_i$. So, we only need to find b_m and f_j.

Assume $\exists k, \delta_k < \gamma_k$. Then, by Lemma 4.2, $f_j = \gamma_j$. We consider how to find b_m. When $\exists m' \neq m, j, \gamma'_{m'} < \sigma'_{m'}$, then by Lemma 4.2, $b_m = \delta_m$. When $\forall i \neq m, j, \sigma'_i \leq \gamma'_i$, then $b_m = f_m$ if $\gamma'_m|^m_{\sigma'_m-1} \in A \cup B$. Otherwise, b_m is the largest $z \in [\gamma'_m, \sigma_m - 1]$ such that $\gamma'|^m_z \in A \cup B$, so we can find it using binary search. Hence, with at most $(2d + 1) \log n + 1$ membership queries we can find the target.

Now, assume $\forall i, \delta_i = \gamma_i$. First, we consider that $\exists m' \neq m, j, \gamma'_{m'} < \sigma'_{m'}$. By Lemma 4.2, $b_m = \delta_m$. So, we only need to find f_j. When $\exists k \neq m, j, \sigma'_k < \gamma'_k$, then f_j is the largest $z \in [\sigma'_j, n]$ such that $\sigma'|^j_z \in A \cup B$, so we can find it using binary search. When $\forall i \neq m, j, \gamma'_i \leq \sigma'_i$, then $f_j = b_j$ if $\sigma'|^j_{\gamma'_j-1} \in A \cup B$. Otherwise, f_j is the largest $z \in [\sigma'_j, \gamma'_j - 1]$ such that $\sigma'|^j_z \in A \cup B$. Hence, with at most $(2d + 1) \log n + 1$ membership queries we can find the target.

Second, we consider that $\forall i \neq m, j, \sigma'_i \leq \gamma'_i$ and, $\exists m' \neq m, j, \sigma'_{m'} < \gamma'_{m'}$. Then, f_j is the largest $z \in [\sigma'_j, n]$ such that $\sigma'|^j_z \in A \cup B$, so we can find it using binary search. When $f_j < b_j = \gamma_j$, then by Lemma 4.4, $b_m = \gamma_m$. When $f_j = b_j = \gamma_j$, then $b_m = f_m$ if $\gamma'|^m_{\sigma'_m-1} \in A \cup B$. Otherwise, b_m is the largest $u \in [\gamma'_m, \sigma'_m - 1]$ such that $\gamma'|^m_u \in A \cup B$, so we can also find b_m using binary search. Therefore, with at most $(2d + 2) \log n + 1$ membership queries we can find the target.

Third, we consider that $\forall i \neq m, j, \sigma'_i = \gamma'_i$. When $\sigma'|^j_{\gamma'_j-1} \notin A \cup B$, then f_j is the largest $z \in [\sigma'_j, \gamma'_j - 1]$ such that $\sigma'|^j_z \in A \cup B$. Moreover, by Lemma 4.2, $b_m = \delta_m$. Thus, we can find f_j and b_m with at most $\log n + 1$ membership queries. Similarly, when $\gamma'|^m_{\sigma'_m-1} \notin A \cup B$, we can find f_j and b_m using at most $\log n + 1$ membership queries.

Now, suppose that $\sigma'|^j_{\gamma'_j-1} \in A \cup B$ and $\gamma'|^m_{\sigma'_m-1} \in A \cup B$. In this case, either $b_m = f_m = \gamma_m$ and $b_j = f_j = \gamma_j$, or $b_m < f_m = \gamma_m$ and $f_j < b_j = \gamma_j$. Note that $b_m < f_m = \gamma_m$ and $f_j < b_j = \gamma_j$ if and only if $(\gamma'|^m_{\gamma_m})|^j_{\gamma_j} \notin A \cup B$. When $(\gamma'|^m_{\gamma_m})|^j_{\gamma_j} \notin A \cup B$, then f_j is the largest $z \in [\sigma'_j, \gamma_j]$ such that $(\gamma'|^m_{\gamma_m})|^j_z \in A \cup B$, and b_m is the largest $u \in [\gamma'_m, \gamma_m]$ such that $(\gamma'|^m_u)|^j_{\gamma_j} \in A \cup B$. Hence, with at most $(2d + 2) \log n + 3$ membership queries we can find b_m and f_m.

Put all the cases together, we can find the target with at most $(3d+2) \log n+3$ membership queries. $\square$

Proof of Lemma 5.6. Since $y \notin A \cup B$ and $y \in rec(\gamma', \gamma)$, $A - B \neq \phi$ and $B - A \neq \phi$ by Lemma 4.3. By Lemma 4.4, we may assume without loss of generality that $\gamma' = (a_1, \ldots, a_d)$ and $\gamma = (f_1, \ldots, f_d)$. In order to learn $A \cup B$, we only need to find $(b_1, \ldots, b_d)$ and $(e_1, \ldots, e_d)$. We analyze the following cases.

Case 4.6.1: $\exists m$, such that $\delta'_m < \gamma'_m$. In this case, $\delta'_m = e_m < \gamma'_m = a_m$. By Lemma 4.3, with at most $d \log n$ membership queries we can find $Lsearch(\delta') = (e_1, \ldots, e_d)$. So, we only need to find $(b_1, \ldots, b_d)$. If $\exists j$ such that $\delta_j > \gamma_j$, then $\delta_j = b_j > \gamma_j = f_j$. By Lemma 4.2, $Rsearch(\delta) = (b_1, \ldots, b_d)$, i.e., with at most $d \log n$ membership queries we can find $(b_1, \ldots, b_d)$.

Now, we consider that $\delta_i \leq \gamma_i, \forall i \in [1, d]$. Since $A - B \neq \phi$, there are at least one j such that $\gamma'_j = a_j < e_j$. When $\exists j_1, j_2, j_1 \neq j_2$, such that $\gamma'_{j_1} < e_{j_1}$ and $\gamma'_{j_2} < e_{j_2}$, then by Lemma 4.2, $\delta = (b_1, \ldots, b_d)$. Suppose that there is exactly one j such that $\gamma'_j < e_j$. Then, by Lemma 4.2, $\forall i \neq j, \delta_i = b_i$. Moreover, if $\gamma'|^j_{e_j-1} \in A \cup B$ then $b_j = f_j$, otherwise b_j is the largest $z \in [\gamma'_j, e_j - 1]$ such that $\gamma'|^j_z \in A \cup B$, so we can find it using binary search with at most $\log n$ membership queries. Combining the above analysis, in this case we can find using at most $2d \log n + 1$ membership queries.

Case 4.6.2: $\forall i, \gamma'_i \leq \delta'_i$. If $\exists m$ such that $\gamma_m = f_m < \delta_m$, then we can find $(b_1, \ldots, b_d)$ and $(f_1, \ldots, f_d)$ in the same manner as we did in case 4.6.1, using at most $2d \log n + 1$ membership queries. Now, we assume that $\forall i \in [1, d], \delta_i \leq \gamma_i$. We analyze the following subcases.

Subcase 4.6.2a: $\exists j_1, j_2, j_1 \neq j_2, \gamma'_{j_1} < \delta'_{j_1}$, and $\gamma'_{j_2} < \delta'_{j_2}$. Then, by Lemma 4.2, $\delta = (b_1, \ldots, b_d)$. So, we only need to find $(e_1, \ldots, e_d)$. Because $B - A \neq \phi$, either there is exactly one m such that $\delta_m < \gamma_m$, or there are $m_1, m_2, m_1 \neq m_2$, such that $\delta_{m_1} < \gamma_{m_1}$ and $\delta_{m_2} < \gamma_{m_2}$. In the second case, by Lemma 4.2, $\delta' = (e_1, \ldots, e_d)$. In the first case, by Lemma 4.2, $\forall i \neq m, \delta'_i = e_i$. Moreover, $e_m = \gamma'_m$ if $\gamma|^m_{\delta_m+1} \in A \cup B$. Otherwise e_m is the smallest $z \in [\delta_m + 1, \gamma_m]$ such that $\gamma|^m_z \in A \cup B$. Thus, we can find it using binary search. Hence, with at most $\log n + 1$ membership queries we can find the target.

Subcase 4.6.2b: There is exactly one j such that $\gamma'_j < \delta'_j$. In this subcase, $\forall i \neq j, \delta'_i = \delta'_j$. By Lemma 4.2, $\forall i \neq j, \delta_i = b_i$. So, we only need to find $(e_1, \ldots, e_d)$ and b_j. First, we consider that there are distinct m_1 and m_2 such that $\delta_{m_1} < \gamma_{m_1}$ and $\delta_{m_2} < \gamma_{m_2}$. By Lemma 4.2, $\forall i, \delta'_i = e_i \leq a_i$. Recall that $\forall i, \delta_i \leq \gamma_i$. So, $b_j = \gamma_j$ if $\gamma'|^j_{\delta'_j-1} \in A \cup B$, otherwise b_j is the largest $z \in [\gamma'_j, \delta'_j - 1]$ such that $\gamma'|^j_z \in A \cup B$. Hence, we can find the target with at most $\log n + 1$ membership queries.

Second, we consider that there is exactly one m such that $\delta_m < \gamma_m$. By Lemma 4.2, $e_i = \delta'_i, \forall i \neq m$. According to the previous assumption, $\forall i \neq m$, $\delta_i = \gamma_i$ and, $\forall i \neq j, \delta'_i = \gamma'_i$. When $m = j$, then $\delta'_j = e_j$ and $\delta_j = b_j$, since $\delta'_j > \gamma'_j = a_j$ and $\delta_j < \gamma_j = f_j$. Now, suppose that $m \neq j$. If $\gamma'|^j_{\delta'_j-1} \notin A \cup B$, then b_j is the largest $z \in [\gamma'_j, \delta'_j - 1]$ such that $\gamma'|^j_z \in A \cup B$, and by Lemma 4.2, $e_m = \delta'_m$. When $\gamma'|^j_{\delta'_j-1} \in A \cup B$, we have $b_j = \gamma_j = f_j$ and we can find e_m as follows. $e_m = \gamma'_m$ if $\gamma|^m_{\delta_m+1} \in A \cup B$, otherwise it is the largest $z \in [\delta_m + 1, \gamma_m]$ such that $\gamma|^m_z \in A \cup B$. Combining the above analysis, we can find the target with at most $\log n + 2$ membership queries.

Third, we consider that $\forall i, \delta_i = \gamma_i$. Recall that $\forall i \neq j, \delta'_i = \gamma'_i$ and, $\gamma'_j < \delta'_j$. We must have $b_j < \gamma_j = f_j$, since $A - B \neq \phi$. This implies by Lemma 4.2 that

$\forall i \neq j$, $\delta'_i = e_i = a_i$. Hence, if $\gamma'|^j_{\delta'_j-1} \in A \cup B$, then $b_j = \gamma_j = f_j$, a contradiction to the fact that $A \cup B$ is not a rectangle. Thus, $\gamma'|^j_{\delta'_j-1} \notin A \cup B$. That is, b_j is the largest $z \in [\gamma'_j, \delta'_j - 1]$ such that $\gamma'|^j_z \in A \cup B$. So, we can find it using binary search. Hence, with at most $\log n + 1$ membership queries we can find the target.

Subcase 4.6.2c: $\forall i \in [1, d]$, $\gamma'_i = \delta'_i$. If there is at least one j such that $\delta_j < \gamma_j$, then we can find $(e_1, \ldots, e_d)$ and $(b_1, \ldots, b_d)$ with at most $\log n + 2$ membership queries in the similar manner as we did in subcase 4.6.2a and 4.6.2b. Now, consider that $\forall i \in [1, d]$, $\delta_i = \gamma_i$. Since $A - B \neq \phi$, there is one m such that $\gamma'_m = a_m < e_m$. By Lemma 4.2, $\forall i \neq m$, $\delta_i = b_i = \gamma_i = f_i$. Since $B - A \neq \phi$, $b_m < f_m = \gamma_m$. This implies by Lemma 4.2 that $\forall i \neq m$, $\delta'_i = e_i = a_i = \gamma'_i$. Because $y \notin A \cup B$ and $y \in rec(\gamma', \gamma)$, we must have that $b_m < y_m < e_m$, i.e., $\gamma'|^m_{y_m} \notin A \cup B$. Hence, We can find y_m by asking at most d membership queries for $\gamma'|^i_{y_i}$, $\forall i \in [1, d]$. After finding y_m, we can find b_m using binary search, which is the largest $z \in [\gamma'_m, y_m]$ such that $\gamma'|^m_z \in A \cup B$. Similarly, we can also find e_m using binary search, which is the smallest $z \in [y_m, \delta'_m]$ such that $\gamma|^m_z \in A \cup B$. Hence, with at most $2 \log n + d$ membership queries, we can find the target.

Combining all the cases together, we can find the target with at most $2d \log n + d + 2$ membership queries. □

5 Open Problems

Can one design an efficient algorithm which properly learns unions of k axis-parallel rectangles over the domain $[0, n-1]^d$ with equivalence and membership queries for any nonconstant k? It seems that this problem is not easy even if d is fixed.

References

[A]   D. Angluin, "Queries and concept learning", *Machine Learning*, 2, 1988, pages 319-342.

[Am]   F. Ameur, "A space-bounded learning algorithm for axis-parallel rectangles", EuroCOLT'95.

[Au]   P. Auer, "On-line learning of rectangles in noisy environment", *Proc of the 6th Annual Workshop on Computational Learning Theory*, 1993, pages 253-261.

[BEHW]   A. Blumer, A. Ehrenfeucht, D. David, and M. Warmuth, "Learnability and the Vapnik-Chervonenkis dimension", *J. ACM*, pages 929-965, 1989.

[BCH]   N. Bshouty, Z. Chen, S. Homer, "On learning discretized geometric concepts", *Proc of the 35th Annual Symposium on Foundations of Computer Science*, pages 54-63, 1994.

[BGGM]   N. Bshouty, P. Goldberg, S. Goldman, and D. Mathias, "Exact learning of discretized geometric concepts", Technical Report WUCS-94-19, Dept of Computer Science, Washington University at St. Louis, July, 1994.

[BM]   W. Bultman, W. Maass, "Fast identification of geometric objects with membership queries", *Proc of the 4th Annual ACM Workshop on Computational Learning Theory*, pages 337-353, 1991.

[C]   Z. Chen, "Learning unions of two rectangles in the plane with equivalence queries", *Proc of the 6th Annual ACM Conference on Computational Learning Theory*, pages 243-253, 1993.

[CHa]   Z. Chen, S. Homer, "Learning unions of rectangles with queries", Technical Report BUCS-93-10, Dept of Computer Science, Boston University, July, 93.

[CHb]   Z. Chen, S. Homer, "The bounded injury priority method and the learnability of unions of rectangles", accepted to publish in *Annals of Pure and Applied Logic*.

[CMa]   Z. Chen, W. Maass, "On-line learning of rectangles", *Proc of the 5th Annual Workshop on Computational Learning Theory*, pages 16-28, 1992.

[CMb]   Z. Chen, W. Maass, "On-line learning of rectangles and unions of rectangles", *Machine Learning* vol. 17, pages 201-223, 1994.

[GGM]   P. Goldberg, S. Goldman, and D. Mathias, "Learning unions of rectangles with membership and equivalence queries", *Proc of the 7th annual ACM Conference on Computational Learning Theory*, pages 198-207, 1994.

[J]   J. Jackson, "An efficient membership-query algorithm for learning DNF with respect to the uniform distribution", *Proc of the 35th Annual Symposium on Foundations of Computer Science*, pages 42-53, 1994.

[L]   N. Littlestone, "Learning quickly when irrelevant attributes abound: a new linear threshold algorithm", *Machine Learning*, 2, 1987, pages 285-318.

[LW]   P. Long, M. Warmuth, "Composite geometric concepts and polynomial predictability", *Proc of the 3th Annual Workshop on Computational Learning Theory*, pages 273-287, 1991.

[MTa]   W. Maass, G. Turán, "On the complexity of learning from counterexamples", *Proc of the 30th Annual Symposium on Foundations of Computer Science*, 1989, pages 262-267.

[MTb]   W. Maass, G. Turán, "On the complexity of learning from counterexamples and membership queries", *Proc of the 31th Annual Symposium on Foundations of Computer Science*, 1990, pages 203-210.

[MTd]   W. Maass, G. Turán, "Algorithms and lower bounds for on-line learning of geometric concepts", *Machine Learning*, 1994, pages 251-269.

[MW]   W. Maass, M. Warmuth, "Efficient learning with virtual threshold gates", Technical Report 395 of the Institutes for Information Processing Graz, August, 1994.

[PRa]   K. Pillaipakkamnatt, V. Raghavan, "On the limits of proper learnability of subclasses of DNF formulas", *Proc of the 7th annual ACM Conference on Computational Learning Theory*, pages 118-129, 1994.

[PRb]   K. Pillaipakkamnatt, V. Raghavan, "Read-twice DNF formulas are properly learnable", Technical Report TR-93-58, Department of Computer Science, Vanderbilt University, 1993.

[PV]   L. Pitt, L. G. Valiant, "Computational limitations on learning from examples", *J. of the ACM*, 35, 1988, 965-984.

[V]   L. Valiant, "A theory of the learnable", *Comm. of the ACM*, 27, 1984, pages 1134-1142.

Disjunctions of Negated Counting Functions Are Efficiently Learnable with Equivalence Queries

Zhixiang Chen *

Computer Science Department
Boston University
Boston, MA 02215, USA
Email: zchen@cs.bu.edu

Abstract. One open problem regarding learning counting functions is whether disjunctions of negated counting functions with a constant prime modulus p are efficiently learnable with equivalence queries. We give a positive solution to this problem by showing that for any constant prime p, conjunctions of counting functions with modulus p over the domain Z_p^n is efficiently learnable with at most $(n+1)^{p-1}+1$ equivalence queries. We further prove that any disjunctions of counting functions and negated counting functions with a constant prime modulus p over the domain Z_p^n are also efficiently learnable with at most $(n + 1)^{p-1} + 1$ equivalence queries.

1 Introduction

Counting functions, especially parity functions and modulo functions, have received certain attention in computational learning theory. It was shown in Helmbold, Sloan and Warmuth [HSW] and, Fisher and Simon [FS], that parity functions over the domain Z_2^n are efficiently learnable with at most n equivalence queries. Blum, Chalasani and Jackson [BCJ] proved that for any prime p, modulo p functions over the domain Z_p^n is also efficiently learnable with at most n equivalence queries. Chen and homer [CH] studied the learnability of counting functions, they proved that for any prime p, disjunctions of counting functions with modulus p over the domain Z_p^n is efficiently learnable with at most $n + 1$ equivalence queries. It was further proved in Bshouty *et al.* [BCDH] that disjunctions of counting functions with modulus q (q may be composite) over the domain Z_q^n is learnable with at most $n \log q + 1$ equivalence queries.

The problem whether disjunctions of negated counting functions with a prime modulus p over the domain Z_p^n are efficiently learnable was first brought to our attention by Blum and Rivest in early 1994. It was proved in [BCDH] that for arbitrary p this problem is harder than learning DNF formula, but disjunctions of no more than $O(\frac{\log n}{\log(p-1)})$ negated counting functions are efficiently learnable with both equivalence and membership queries. However, when p is constant the problem is still open.

* The author was supported by NSF grants CCR-9103055 and CCR-9400229.

Recently, Bertoni, Cesa-Bianchi and Fiorino [BCF] have shown that for any constant prime p, conjunctions of modulo p functions over the domain Z_p^n are efficiently learnable with equivalence queries. Enlightened by their technique, we show in this paper that conjunctions of counting functions with a constant prime modulus p over the domain Z_p^n are also efficiently learnable with equivalence queries, thus we give a positive solution to the above open problem in the case of constant prime modulus p.

2 Notations and Definitions

We assume that Z is the set of all integers. Let $Z_p = \{0, \ldots, p-1\}$ for any prime p. A counting function with modulus $p \geq 2$ is defined as follows:

$$C_{\mathbf{a},b}^p(x_1, \ldots, x_n) = \begin{cases} 0 \text{ if } \sum_{j=1}^n a_i x_j \equiv b \pmod{p}, \\ 1 \text{ otherwise,} \end{cases}$$

where $\mathbf{a} = (a_1, \ldots, a_n) \in Z_p^n$ and $b \in Z_p$. For convenience, we may use $C_{\mathbf{a},b}^q$ to stand for $C_{\mathbf{a},b}^p(x_1, \ldots, x_n)$. When $b = 0$, we say that $C_{\mathbf{a},0}^p$ is a modulo p function. A disjunction of negated counting functions is

$$F = \neg C_{\mathbf{a}_1,b_1}^p \vee \cdots \vee \neg C_{\mathbf{a}_m,b_m}^p.$$

Our learning model is the on-line learning with equivalence queries [A]. The goal of a learning algorithm (or learner) for a class of boolean-valued function $\mathbf{C}$ over a domain X^n is to learn any unknown target function $f \in \mathbf{C}$ that has been fixed by a teacher. In order to obtain information about f, the learner can ask equivalence queries by proposing hypotheses h from a fixed hypothesis space $\mathbf{H}$ of functions over X^n with $\mathbf{C} \subseteq \mathbf{H}$ to an equivalence oracle $EQ()$. If $h = f$, then $EQ(h) = $ "yes", so the learner succeeds. If $h \neq f$, then $EQ(h) = \mathbf{x}$ for some $\mathbf{x} \in X^n$ such that $h(\mathbf{x}) \neq f(\mathbf{x})$, called a counterexample. A learning algorithm learns $\mathbf{C}$, if for any target function $f \in \mathbf{C}$, it can find a $h \in \mathbf{H}$ that is logically equivalent to f. $\mathbf{C}$ is efficiently learnable, if there is a learning algorithm that learns any target function in $\mathbf{C}$ and runs in time polynomially in the size of the domain and the size of the target function.

3 The Learning Algorithm

Given a conjunction of counting functions with modulus p

$$F = C_{\mathbf{a}_1,b_1}^p \wedge \cdots \wedge C_{\mathbf{a}_m,b_m}^p,$$

for any $\mathbf{x} \in Z_p^n$, $F(\mathbf{x}) = 1$ if and only if

$$(1) \qquad \mathbf{a}_i \cdot \mathbf{x} \not\equiv b_i \pmod{p}, \quad i = 1, \ldots, m.$$

According to Fermat Little Theorem (which says that if p is prime then for any integer a not divisible by p, $a^{p-1} \equiv 1 \pmod{p}$), (1) is equivalent to

(2) $$(a_i \cdot x - b_i)^{p-1} \equiv 1 \pmod{p}, \quad i = 1, \dots, m,$$

When $b_i = 0$ for all $i = 1, \dots, m$, Bertoni, Cesa-Bianchi and Fiorino [BCF] developed an elegant transformation $\lambda_{n,m} : Z_m^n \to Z_m^{n^{m-1}}$ such that (2) is equivalent to

(3) $$\lambda_{n,p}(a_i) \cdot \lambda_{n,p}(x) \equiv 1 \pmod{p}, \quad i = 1, \dots, m.$$

Thus, by employing the following Theorem 1 obtained in [CH], they proved that for constant p, any conjunction of modulo p functions over the domain Z_p^n is efficiently learnable with at most $n^{p-1} + 1$ equivalence queries.

Theorem 1 [CH]. *There is an algorithm IHS that efficiently learns linear systems over the domain Z_p^n with at most $n + 1$ equivalence queries.*

We now extend the transformation $\lambda_{n,m}$ to a new one such that we can cope with the case of $b_i \neq 0$. As in [BCF], for any $n, m \geq 1$ and $p > 1$, let $\mu : \{1, \dots, n\}^m \to \{1, \dots, n^m\}$ be a bijection. We define $\Psi_{n,p}^m : Z_p^n \to Z_p^{n^m}$ as follows: For any $x \in Z_p^n$,

$$\Psi_{n,p}^m(x) = (y_1, \dots, y_{n^m}), \text{ where } y_{\mu(k_1, \dots, y_m)} = x_{k_1} \cdots x_{k_m}.$$

It is easy to see that for any $x, y \in Z_p^n$, $\Psi_{n,p}^m(x) \cdot \Psi_{n,p}^m(y) = (x \cdot y)^m$. Furthermore, one can choose a suitable μ such that $\Psi_{n,p}^m$ is computable in time polynomial in n when m and p are constant.

We expand the right hand side of (2) and obtain

(4) $$(a_i \cdot x - b_i)^{p-1} = \sum_{j=0}^{p-1} \binom{p-1}{j}(-b_i)^{p-1-j}(a_i \cdot x)^i$$

$$= (-b_i)^{p-1} + \sum_{j=1}^{p-1} \binom{p-1}{j}(-b_i)^{p-1-j}[\Psi_{n,p}^i(a_i) \cdot \Psi_{n,p}^i(x)]$$

Now, we define a transformation

$$\Upsilon_{n,p}^w : Z_p^n \to Z_p^{(n+1)^{p-1}}$$

such that for any $x \in Z_p^n$ and $w \in Z_p$,

$$\Upsilon_{n,p}^w(x) = ((-w)^{p-1}, \binom{p-1}{1}(-w)^{p-2}\Psi_{n,p}^1(x), \dots, \binom{p-1}{p-1}(-w)^0 \Psi_{n,p}^{p-1}(x)).$$

We define another transformation

$$\Gamma_{n,p} : Z_p^n \to Z_p^{(n+1)^{p-1}}$$

such that for any $x \in Z_p^n$,

$$\Gamma_{n,p}(x) = (1, \Psi_{n,p}^1(x), \dots, \Psi_{n,p}^{p-1}(x)).$$

$\Upsilon_{n,p}^w$ and $\Gamma_{n,p}$ are both polynomial time computable when p is a constant because of the polynomial computability of Ψ. According to (1), (2) and (4), $F(x) = 1$ if and only if $\Gamma_{n,p}(x)$ is a solution of the linear system

(5) $$\Upsilon_{n,p}^{b_i}(a_i) \cdot z \equiv 1 \pmod{p}, \quad i = 1, \dots, m,$$

over the domain $Z_p^{(n+1)^{p-1}}$, where z is a $(n+1)^{p-1} \times 1$ variable vector.

We are now ready to prove the following theorem.

Theorem 2. *For any constant prime p, conjunctions of counting functions with modulus p over the domain Z_p^n are efficiently learnable with at most $(n+1)^{p-1}+1$ equivalence queries.*

Proof. For any conjunction of counting functions $F = C_{a_1,b_1}^p \wedge \cdots \wedge C_{a_m,b_m}^p$, with the above analysis we can learn the set of all $x \in Z_p^n$ such that $F(x) = 1$ via learning the set of all $x \in Z_p^n$ such that $\Gamma_{n,p}(x)$ is a solution of the linear system (5) over the domain $Z_p^{(n+1)^{p-1}}$. In other words, we can learn F via learning (5).

We employ the algorithm IHS of Theorem 1 to learn (5). For the i-th hypothesis H_i issued by the algorithm IHS for the linear system (5), $i \geq 1$, we construct

$$S_i = \{x \in Z_p^n | \Gamma_{n,p}(x) \in H_i\}$$

as the i-th hypothesis for F. We ask an equivalence query for S_i. If we receive a "yes" then stop, otherwise we obtain a counterexample y_i. By (5), $\Gamma_{n,p}(y_i)$ is also a counterexample to H_i. Then, we use the algorithm IHS to produce the $(i+1)$-th hypothesis for the linear system (5). Similarly, we construct from H_{i+1} the $(i+1)$-th hypothesis S_{i+1} for F. By Theorem 1, the algorithm IHS learns (5) efficiently with at most $(n+1)^{p-1}+1$ equivalence queries, this implies that we can learn F efficiently with $(n+1)^{p-1}+1$ equivalence queries. $\square$

Given any disjunction of negated counting function $F = \neg C_{a_1,b_1}^p \vee \cdots \vee \neg C_{a_m,b_m}^p$, $\neg F = C_{a_1,b_1}^p \wedge \cdots \wedge C_{a_m,b_m}^p$. In order to learn F, we only need to learn $\neg F$. Hence, by Theorem 2, we can efficiently learn any disjunctions of negated counting functions with a constant prime modulus p over the domain Z_p^n with at most $(n+1)^{p-1}+1$ equivalence queries.

4 When Both Counting Functions and Negated Counting Functions Occur

In this section, we extend Theorem 2 to cope with any disjunctions of counting functions and negated counting functions. As before we assume that p is a constant prime.

Theorem 3. *Any disjunction F of counting functions and negated counting functions with modulus p over the domain Z_p^n is efficiently learnable with at most $(n+1)^{p-1}+1$ equivalence queries.*

Proof. Let F be as follows:

$$F = \neg C_{a_1,b_1}^p \vee \cdots \vee \neg C_{a_m,b_m}^p \vee C_{a_{m+1},b_{m+1}}^p \vee \cdots \vee C_{a_{m+k},b_{m+k}}^p.$$

In order to learn F we only need to learn the set of all examples $x \in Z_p^n$ such that $F(x) = 0$, i.e., we only] need to learn the set of examples $x \in Z_p^n$ which are solutions of the following system

(6) $$a_i \cdot x \not\equiv b_i \pmod{p}, \quad i = 1, \ldots, m, \text{ and}$$

$$\mathbf{a}_i \cdot \mathbf{x} \equiv b_i \quad (\bmod\ p),\ i = m+1, \ldots, m+k.$$

Similar to (2), (6) is equivalent to (7) according to Fermat Little Theorem.

(7)
$$(\mathbf{a}_i \cdot \mathbf{x} - b_i)^{p-1} \equiv 1 \quad (\bmod\ p),\ i = 1, \ldots, m,\ \text{and}$$
$$\mathbf{a}_i \cdot \mathbf{x} \equiv b_i \quad (\bmod\ p),\ i = m+1, \ldots, m+k.$$

Besides the two transformations $\Upsilon_{n,p}^w$, $\Gamma_{n,p}: Z_p^n \rightarrow Z_p^{(n+1)^{p-1}}$, we need to define another transformation $\Pi_{n,p}: Z_p^n \rightarrow Z_p^{(n+1)^{p-1}}$ such that for any $\mathbf{x} \in Z_p^n$,

$$\Pi_{n,p}(\mathbf{x}) = (0, \mathbf{x}, 0, \ldots, 0).$$

Note that

$$\Pi_{n,p}(\mathbf{a}_i) \cdot \Gamma_{n,p}(\mathbf{x}) = \mathbf{a}_i \cdot \Psi_{n,p}^1(\mathbf{x}) = \mathbf{a}_i \cdot \mathbf{x}.$$

In the same manner as (1), (2) and (4), we know that for any $\mathbf{x} \in Z_p^n$, $F(\mathbf{x}) = 0$ if and only if $\Gamma_{n,p}(\mathbf{x})$ is a solution to the following linear system

(8)
$$\Upsilon_{n,p}^{b_i}(\mathbf{a}_i) \cdot \mathbf{z} \equiv 1 \quad (\bmod\ p),\ i = 1, \ldots, m,\ \text{and}$$
$$\Pi_{n,p}(\mathbf{a}_i) \cdot \mathbf{z} \equiv b_i \quad (\bmod\ p),\ i = m+1, \ldots, m+k,$$

over the domain $Z_p^{(n+1)^{p-1}}$, where $\mathbf{z}$ is a $(n+1)^{p-1} \times 1$ variable vector.

As we did in the proof of Theorem 2, we can learn F via learning the linear system (8) by employing a copy of the algorithm IHS of Theorem 1. Thus, by Theorem 1, we can learn F efficiently with at most $(n+1)^{p-1} + 1$ equivalence queries. $\qquad \square$

Acknowledgment. We think Blum and Rivest for drawing our attention to the problem of efficiently learning disjunctions of negated counting functions. We thank Cesa-Bianchi for sending us the paper [BCF].

References

[A]   D. Angluin, "Queries and concept learning", *Machine Learning*, 2, 1988, pages 319-342.

[BCF]   A. Bertoni, N. Cesa-Bianchi, G. Fiorino, "Efficient learning with equivalence queries of conjunctions of modulo functions", submitted to *Information Processing Letters*, 1995.

[B]   A. Blum, Personal Communication, 1994.

[BCJ]   A. Blum, P. Chalasani, J. Jackson, "On learning embedded symmetric concepts", *Proc. of the 6th ACM Annual Conference on Computational Learning Theory*, pages 337-346, 1993.

[BCDH] N. Bshouty, Z, Chen, S. Decatur, S. Homer, "On the learnability of Z_N-DNF formulas", *Proc. of the 8th ACM Annual Conference on Computational Learning Theory*, 1995.

349

[CH]   Z. Chen, S. Homer, "On learning counting functions with queries" *Proc. of the 7th ACM Annual Conference on Computational Learning Theory*, pages 218-227, 1994.

[HSW]   D. Helmbold, R. Sloan, M. Warmuth, "Learning integer lattices", *SIAM J. Comput.*, 1992, pages 240-266.

[R]   R. Rivest, Personal Communication, 1994.

Non-empty Cross−3−Intersection Theorems of Subsets *

Shiquan Wu

System Engineering and Mathematics Department
National University of Defense Technology
Changsha,Hunan 410073,China

Abstract. Maximum families of subsets satisfying some specified conditions are widely studied in combinatorics on set systems. In this paper, an extremal problem on subsets is considered. Let C_n^k denote the set of all k−subsets of an n−set. Assume $\mathcal{A} \subseteq C_n^a$ and $\mathcal{B} \subseteq C_n^b$. $(\mathcal{A}, \mathcal{B})$ is called a cross−t−intersecting family if $| A \cap B | \geq t$ for any $A \in \mathcal{A}$, $B \in \mathcal{B}$. For $t = 3$, maximum non-empty cross−3−intersecting families of a− and b−subsets are obtained.

1 Introduction

Combinatorics in set systems widely studies the following type of extremal problems on subsets: What are the maximum families of subsets of a finite set with the property that the intersection of any two sets in the family satisfies some specified conditions ? In this paper, we study cross−3−intersecting families and obtain maximum non-empty cross−3−intersecting families of subsets.

Throughout , $X = \{1, 2, \cdots, n\}$ denotes a finite set. For an integer k with $0 \leq k \leq n$, C_n^k denotes the collection of all k−subsets of X. For subset A, $| A |$ denotes the cardinality of A. For subsets $A, B(A \cap B = \emptyset)$, and $x \in X$, we denote $A + B = A \cup B$, $A + x = A \cup \{x\}$, and $A - x = A - \{x\}$. Most of our other notation and terminologies are the same as those in [1] .

Definition 1 . Let $\mathcal{A} \subseteq C_n^a$ and $\mathcal{B} \subseteq C_n^b$. Assume $a + b \leq n + t - 1$. $(\mathcal{A}, \mathcal{B})$ is called a cross−t−intersecting family if $| A \cap B | \geq t$ for any $A \in \mathcal{A}, B \in \mathcal{B}$.

This problem is studied by several authors. For $t = 1$, maximum cross−intersecting families of subsets are obtained by Hilton and Milner for the case $\mathcal{A}, \mathcal{B} \subseteq C_n^k$ in [3], and also by Frankl and Tokushige for the further case $\mathcal{A} \subseteq C_n^a$ and $\mathcal{B} \subseteq C_n^b$ in [2]. In [4], a similar kind of maximum cross−intersecting families of subsets is studied. Recently, Simpson reproved Frankl and Tokushige's theorems in a new way (see [5]). In [6], Wu studies non-empty cross−2−intersecting families of subsets for some a and b. In [7], Wu obtains maximum non-empty cross−2−intersecting families of subsets for

*Partially supported by National Natural Science Foundation of China (No.19401008) and by Postdoctoral Science Foundation of China.

$a + b \le n - 1$, $a < b$, and $(n, a, b) \ne (2i, i - 1, i)$, and proves that if $\mathcal{A}$ is a single-ton , i.e.,unique a−subset $\{A\}$, and $\mathcal{B}$ consists of the collection of all b−subsets excluding those b−subsets B with $|A \cap B| \le 1$, then $(\mathcal{A}, \mathcal{B})$ is a maximum non-empty cross−2−intersecting families of a− and b−subsets.

In this paper, we obtain the following main result.

Main Theorem. *Let* $a + b \le n - 1$ *and* $a < b$. *Assume* $(n, a, b) \ne$ $(2i, i - 1, i)$ *for any integer* i, $\mathcal{A} \subseteq C_n^a$ *and* $\mathcal{B} \subseteq C_n^b$. *If* $(\mathcal{A}, \mathcal{B})$ *is a non-empty cross−3−intersecting families of subsets, i.e.,both $\mathcal{A}$ and $\mathcal{B}$ are not empty, then*

$$| \mathcal{A} + \mathcal{B} | \le 1 + \binom{n}{b} - \binom{n - a}{b} - a\binom{n - a}{b - 1} - \binom{a}{2}\binom{n - a}{b - 2}. \tag{1}$$

and the bound is sharp, which is achieved by the cross-3-intersecting family $(\{A\}, C_n^b - \{B \mid |B| = b, |A \cap B| \le 2\})$ *with* $|A| = a$.

2 Proof of The Main Theorem

In this section, we prove the main theorem by replacing $\mathcal{A} - \{A\}$ by some family of b−subsets. First of all , we introduce the following concepts and prove some lemmas.

Definition 2 . Let $A \in \mathcal{A} \subseteq C_n^a$. Denote

$$\begin{aligned} C_t(A, b) &= \{B \mid |B| = b, |A \cap B| \le t - 1\} \\ \mathcal{C}_t(\mathcal{A}, b) &= \{B \mid B \in C_t(A, b) \text{ for some } A \in \mathcal{A}\} \end{aligned}$$

From the definition, we can easily have

Lemma 1 . *Let* $\mathcal{A} \subseteq C_n^a$ *and* $\mathcal{B} \subseteq C_n^b$ *with* $a < b$. *Then*

(i) If $(\mathcal{A}, \mathcal{B})$ *is a cross−t−intersecting family of subsets, then* $\mathcal{C}_t(\mathcal{A}, b) \cap \mathcal{B} = \emptyset$,

(ii) For $A \in \mathcal{A}$, $|C_t(A, b)| = \sum_{i=0}^{t-1} \binom{a}{i}\binom{n-a}{b-i}$,

(iii) For $B \in \mathcal{B}$, $|C_t(B, a)| = \sum_{i=0}^{t-1} \binom{b}{i}\binom{n-b}{a-i}$.

In the following lemma, we calculate $|C_3(A_0, b) \cap C_3(A, b)|$.

Lemma 2 . *Let* $A_0, A \in \mathcal{A}$ *with* $p = |A_0 \cap A|$. *Denote* $G(p) = |C_3(A_0, b) \cap C_3(A, b)|$, *then*

$$G(p) = \binom{n - 2a + p}{b} + (2a - p)\binom{n - 2a + p}{b - 1} + \binom{2a - p}{2}\binom{n - 2a + p}{b - 2}$$

$$+ [p(a - p)^2 + 2(a - p)\binom{a - p}{2}]\binom{n - 2a + p}{b - 3}$$

$$+\binom{a-p}{2}^2\binom{n-2a+p}{b-4}.$$

Proof. For any $B \in A_0 \cap A$, denote $q = |(A_0 \cup A) \cap B|$, then $0 \leq q \leq 4$.

If $q \leq 2$, then for each q, there exist $\binom{2a-p}{q}\binom{n-2a+p}{b-q}$ such B's.

If $q = 3$, denote $i = |B \cap (A_0 - A)|$, $j = |B \cap (A - A_0)|$ and $r = |B \cap A_0 \cap A|$, then we have three cases: $(i, j, r) = (2, 1, 0)$, $(i, j, r) = (1, 2, 0)$, and $(i, j, r) = (1, 1, 1)$. The numbers of such B's are $\binom{a-p}{2}\binom{a-p}{1}\binom{n-2a+p}{b-3}$, $\binom{a-p}{1}\binom{a-p}{2}\binom{n-2a+p}{b-3}$, and $\binom{a-p}{1}\binom{p}{1}\binom{a-p}{1}\binom{n-2a+p}{b-3}$, respectively.

If $q = 4$, then $(i, j, r) = (2, 2, 0)$ and there exist $\binom{a-p}{2}\binom{a-p}{2}\binom{n-2a+p}{b-4}$ such B's.

Therefore, we can obtain $G(p)$ by taking the sum for all q. $\qquad\square$

We prove the following inequalities so as to compare $G(p+1)$ with $G(p)$.

Lemma 3 . *Let* $0 \leq p \leq a - 2$. *Denote*

$$\begin{aligned}
g_0(p) &= 1 \\
g_1(p) &= 2a - p \\
g_2(p) &= \binom{2a-p}{2} \\
g_3(p) &= p(a-p)^2 + 2(a-p)\binom{a-p}{2} \\
g_4(p) &= \binom{a-p}{2}^2.
\end{aligned}$$

Then

$$\begin{aligned}
g_0(p+1) &\geq g_0(p) \\
g_1(p+1) &\geq g_1(p) - g_0(p) \\
g_2(p+1) &\geq g_2(p) - g_1(p) + g_0(p)
\end{aligned}$$

$$\begin{aligned}
g_3(p+1) &\; -[g_3(p) - g_2(p) + g_1(p) - g_0(p)] \\
&= \binom{2a-p-1}{2} - (2a - 2p - 1)(a - 1),
\end{aligned}$$

$$\begin{aligned}
g_4(p+1) &\; -[g_4(p) - g_3(p) + g_2(p) - g_1(p) + g_0(p)] \\
&= p(a-p)^2 + 2(a-p-1)^2 + (a-p-1) - \binom{2a-p-1}{2}.
\end{aligned}$$

Proof . The first three inequalities are trivial. We prove the last two equalities as follows.

(i) It is obvious $g_2(p) - g_1(p) + g_0(p) = \binom{2a-p-1}{2}$. Since

$$\begin{aligned}
&g_3(p+1) - g_3(p) \\
&= (p+1)(a-p-1)^2 + 2(a-p-1)\binom{a-p-1}{2} \\
&\quad - [p(a-p)^2 + 2(a-p)\binom{a-p}{2}] \\
&= p(a-p-1)^2 - p(a-p)^2 + (a-p-1)^2 + 2(a-p-1)\binom{a-p-1}{2} \\
&\quad - 2(a-p-1)\binom{a-p}{2} - 2\binom{a-p}{2} \\
&= -p(2a-2p-1) + (a-p-1)^2 - 2(a-p-1)\binom{a-p-1}{1} - 2\binom{a-p}{2} \\
&= -p(2a-2p-1) - (a-p-1)^2 - (a-p)(a-p-1) \\
&= -(2a-2p-1)(a-1).
\end{aligned}$$

It follows that

$$g_3(p+1) - [g_3(p) - g_2(p) + g_1(p) - g_0(p)] = \binom{2a-p-1}{2} - (2a-2p-1)(a-1).$$

(ii) For each p, we have $g_4(p+1) - g_4(p) = \binom{a-p-1}{2}^2 - \binom{a-p}{2}^2 = -(a-p-1)^3$.
On the other hand, we have $g_2(p) - g_1(p) + g_0(p) = \binom{2a-p-1}{2}$ and

$$\begin{aligned}
g_3(p) &= p(a-p)^2 + 2(a-p)\binom{a-p}{2} \\
&= p(a-p)^2 + (a-p)(a-p)(a-p-1) \\
&= p(a-p)^2 + (a-p-1)^3 + 2(a-p-1)^2 + (a-p-1).
\end{aligned}$$

Therefore, we have

$$\begin{aligned}
&g_4(p+1) - [g_4(p) - g_3(p) + g_2(p) - g_1(p) + g_0(p)] \\
&= p(a-p)^2 + 2(a-p-1)^2 + (a-p-1) - \binom{2a-p-1}{2}. \qquad \square
\end{aligned}$$

With Lemma 3, we can compare $G(p+1)$ with $G(p)$ and obtain the maximum value of $G(p)$.

Lemma 4 . *Let $0 \leq p \leq a-2$. Assume $a+b \leq n-1$, then*

$$G(p) \leq G(p+1) \leq G(a-1). \qquad (2)$$

Proof . For each p, we have

$$G(p+1) - G(p)$$

$$\begin{aligned}
=\ & g_0(p+1)\binom{n-2a+p+1}{b} + g_1(p+1)\binom{n-2a+p+1}{b-1} + g_2(p+1)\binom{n-2a+p+1}{b-2} \\
& + g_3(p+1)\binom{n-2a+p+1}{b-3} + g_4(p+1)\binom{n-2a+p+1}{b-4} - [g_0(p)\binom{n-2a+p}{b} \\
& + g_1(p)\binom{n-2a+p}{b-1} + g_2(p)\binom{n-2a+p}{b-2} + g_3(p)\binom{n-2a+p}{b-3} + g_4(p)\binom{n-2a+p}{b-4}] \\[4pt]
=\ & \binom{n-2a+p+1}{b} - [\binom{n-2a+p}{b} + \binom{n-2a+p}{b-1}] + g_1(p+1)\binom{n-2a+p+1}{b-1} \\
& -[g_1(p) - g_0(p)][\binom{n-2a+p}{b-1} + \binom{n-2a+p}{b-2}] + [g_2(p+1)\binom{n-2a+p+1}{b-2} \\
& -(g_2(p+1) - g_1(p) + g_0(p))[\binom{n-2a+p}{b-2} + \binom{n-2a+p}{b-3}] \\[4pt]
& + g_3(p+1)\binom{n-2a+p+1}{b-3} - (g_3(p) - g_2(p) + g_1(p) - g_0(p)) \\
& \times [\binom{n-2a+p}{b-3} + \binom{n-2a+p}{b-4}] + [g_4(p+1)\binom{n-2a+p+1}{b-4} \\
& -(g_4(p) - g_3(p) + g_2(p) - g_1(p) + g_0(p))\binom{n-2a+p}{b-4}] \\[4pt]
\geq\ & [g_3(p+1) - (g_3(p) - g_2(p) + g_1(p) - g_0(p))]\binom{n-2a+p+1}{b-3} + [g_4(p+1) \\
& -(g_4(p) - g_3(p) + g_2(p) - g_1(p) + g_0(p)]\binom{n-2a+p}{b-4} \\[4pt]
\geq\ & [\binom{2a-p-1}{2} - (2a-2p-1)(a-1)]\binom{n-2a+p}{b-3}
\end{aligned}$$

$$+[p(a-p)^2 + 2(a-p-1)^2 + (a-p-1) - \binom{2a-p-1}{2}]\binom{n-2a+p}{b-4}.$$

Since $a+b \leq n-1$, we have $n-2a+p \geq 2(b-3)$. It follows that $\binom{n-2a+p}{b-3} \geq \binom{n-2a+p}{b-4}$.

Define a real function $h(p) = p(a-p)^2 + 2(a-p-1)^2 + (a-p-1) - (2a-2p-1)(a-1)$ for $0 \leq p \leq a-1$. We then have $\frac{d}{dp}h(p) = (p+2)(a-p)^2 + (2a+1)(a-p) + a = 3(p - \frac{a-1}{3})(p - (a-1))$. It follows that for $0 \leq p \leq \frac{a-1}{3}$, $\frac{d}{dp}h(p) \geq 0$, which implies $h(p) \geq h(0) = 0$.

Similarly, for $\frac{a-1}{3} \leq p \leq a-1$, we have $\frac{d}{dp}h(p) \leq 0$ and $h(p) \geq h(a-1) = 0$. Therefore, for $0 \leq p \leq a-1$, we can obtain $h(p) \geq 0$, i.e.,

$$p(a-p)^2 + 2(a-p-1)^2 + (a-p-1) \geq (2a-2p-1)(a-1).$$

It implies

$$
\begin{aligned}
G(p+1) \quad &-G(p) \\
&\geq [\binom{2a-p-1}{2} - (2a-2p-1)(a-1) + p(a-p)^2 \\
&\qquad +2(a-p-1)^2 + (a-p-1) - \binom{2a-p-1}{2}]\binom{n-2a+p}{b-4} \\
&\geq h(p)\binom{n-2a+p}{b-4} \\
&\geq 0.
\end{aligned}
$$

The inequality (2) holds.   □

In addition, we prove some combinatorial inequalities.

Lemma 5 . Let $a+b \leq n-1$ and $a < b$. Assume $(n,a,b) \neq (2i, i-1, i)$ for any integer i, then

(i) $\binom{n-a}{b} \geq \binom{n-b}{a} + \binom{n-a-1}{b}$,

(ii) $a\binom{n-a}{b-1} \geq b\binom{n-b}{a-1} + (a+1)\binom{n-a-1}{b-1} + \binom{n-a-1}{b-2}$,

(iii) $\binom{a}{2}\binom{n-a}{b-2} \geq \binom{b}{2}\binom{n-b}{a-2} + \binom{a+1}{2}\binom{n-a-1}{b-2} + (a-1)\binom{n-a-1}{b-3}$.

Proof. (i) Since $\binom{n-b}{a} = \binom{n-b}{n-a-b} \leq \binom{n-a-1}{n-a-b} = \binom{n-a-1}{b-1}$, we then have

$$\binom{n-b}{a} + \binom{n-a-1}{b} \leq \binom{n-a-1}{b-1} + \binom{n-a-1}{b} = \binom{n-a}{b}.$$

(ii) Since $a+b \leq n-1$, $a < b$, and $(n,a,b) \neq (2i, i-1, i)$ for any integer i, we have

$$(n-a)\cdots(n-b+1) \geq b(b-1)\cdots(a+1)+2$$
$$(n-a)\cdots(n-b+1) \geq b(b-1)\cdots(a+1)+\frac{a+1}{a}\frac{n-a-b+1}{n-a}+\frac{b-1}{a(n-a)}$$
$$a\frac{(n-a)!}{(b-1)!(n-a-b+1)!} \geq b\frac{(n-b)!}{(a-1)!(n-a-b+1)!}$$
$$+(a+1)\frac{n-a-b+1}{n-a}\frac{(n-a)!}{(b-1)!(n-a-b+1)!}$$
$$+\frac{b-1}{(n-a)}\frac{(n-a)!}{(b-1)!(n-a-b+1)!}$$

This means $\quad a\binom{n-a}{b-1} \geq b\binom{n-b}{a-1}+(a+1)\binom{n-a-1}{b-1}+\binom{n-a-1}{b-2}.$

(iii) Also by $a+b \leq n-1$, $a < b$, and $(n,a,b) \neq (2i, i-1, i)$ for any integer i, we have

$$(n-a)\cdots(n-b+1) \geq b(b-1)\cdots(a+1)+2$$
$$(n-a)\cdots(n-b+1) \geq b(b-1)\cdots(a+1)+\frac{a+1}{a-1}\frac{n-a-b+2}{n-a}+\frac{2(b-2)}{a(n-a)}$$
$$\binom{a}{2}\binom{n-a}{b-2} \geq \binom{b}{2}\binom{n-b}{a-2}+\binom{a+1}{2}\frac{n-a-b+2}{n-a}$$
$$\times\binom{n-a}{b-2}+(a-1)\frac{b-2}{n-a}\binom{n-a}{b-2}.$$

This implies the inequality. $\qquad\qquad\qquad\qquad\qquad\qquad\qquad\qquad\qquad\square$

With the inequalities in Lemma 4 and 5, we can prove the following important lemma.

Lemma 6 . *Let $a+b \leq n-1$ and $a < b$. Assume $(n,a,b) \neq (2i, i-1, i)$ for any integer i, $A \subseteq C_n^a$ and $B \subseteq C_n^b$. If (A,B) is a cross$-3-$intersecting family of subsets, then for any $A_0 \in A$,*

$$|A-\{A_0\}| \leq |C_3(A,b)-C_3(A_0,b)|. \tag{3}$$

Proof . Consider the pairs (A,B) with $A \in A - \{A_0\}$ and $B \in C_3(A,b) - C_3(A_0,b)$. We count these pairs in two ways.

For each $A \in A$, it follows, by Lemma 1 , that there exist $\binom{n-a}{b}+a\binom{n-a}{b-1}+\binom{a}{2}\binom{n-a}{b-2}$ ($= |C_3(A,b)|$) such subsets $B \in C_3(A,b)$.

By Lemma 4, we have $|C_3(A_0,b) \cap C_3(A,b)| \leq G(a-1)$. It follows that the number of such pairs (A,B) is at least $[\,|C_3(A,b)|-G(a-1)\,]\,|A-\{A_0\}|$.

On the other hand, for each $B \in C_n^b$, there exist $\binom{n-b}{a}+b\binom{n-b}{a-1}+\binom{b}{2}\binom{n-b}{a-2}$ $(=|C_3(B,a)|)$ subsets $A \in C_n^a$ such that $B \in C_3(A,b)$. We then obtain that the number of such pairs (A,B) is at most $|C_3(B,a)|\,[\,|C_3(A,b)-C_3(A_0,b)|\,]$.

Furthermore, we get

$$[\,|C_3(A,b)|-G(a-1)\,]\,|A-\{A_0\}| \leq |C_3(B,a)|\,[\,|C_3(A,b)-C_3(A_0,b)|\,].$$

By (i), (ii) and (iii) of Lemma 5, we have $|C_3(A,b)|-G(a-1) \geq |C_3(B,a)|.$ This means that

$$|A-\{A_0\}| \leq |C_3(A,b)-C_3(A_0,b)|.$$

This completes the proof. $\qquad\qquad\qquad\qquad\qquad\qquad\qquad\qquad\qquad$ $\square$

With lemma 6, we can prove our main theorem.

Proof of the main theorem. Let $A \in \mathcal{A}$, by Lemma 6, it follows

$$| \mathcal{A} - \{A\} | \leq | C_3(A, b) - C_3(A, b) | .$$

By Lemma 1, $\mathcal{B} \cap (C_3(A, b) - C_3(A, b)) = \emptyset$. Therefore, we have

$$
\begin{aligned}
| \mathcal{A} + \mathcal{B} | &= |\{A\} + \mathcal{B} + \mathcal{A} - \{A\}| \\
&\leq | \{A\} + \mathcal{B} + C_3(A, b) - C_3(A, b) | \\
&\leq | \{A\} + C_n^b - C_3(A, b) | \\
&\leq 1 + \binom{n}{b} - \binom{n-a}{b} - a\binom{n-a}{b-1} - \binom{a}{2}\binom{n-a}{b-2}.
\end{aligned}
$$

We can verify that $\{A\} + C_n^b - C_3(A, b)$ is a cross$-3-$intersecting family of $a-$ and $b-$subsets, which achieves the bound. $\qquad\qquad\qquad$ $\square$

Acknowlegdement. The author is much indebted to Prof.Dingzhu Du for his encouragement and guidance.

References

1. Anderson,I.,Combinatorics of finite sets, Oxford ,1987
2. Frankl, P. and Tokushige, N., Some best possible inequalities concerning cross$-$intersecting families, J. Combin. Th. Ser. **A61** (1992), 87-97.
3. Hilton, A. J. W. and Milner, E.C., Some intersection theorems for systems of finite sets, Quart. J. Math. Oxford **18 (2)** (1967),369-384.
4. Matsumoto,M., The exact bound in Erdös-Ko-Rado Theorem for cross$-$intersecting families, J. Comb. Theory **A52** (1989), 90-97.
5. Simpson,J.E., A bipartite Erdös-Ko-Rado Theorem, Discrete Mathematics 113(1993)277-280.
6. Wu Shiquan , Non-empty cross$-2-$intersecting families of subsets , Applied Mathematics: A Journal of Chinese Universities, **B8**(1993)2,176-182.
7. Wu Shiquan , Cross$-t-$intersection theorems of subsets, Technical Report of the Center of Mathematical Sciences,Zhejiang University,n1(1994).

Convexity of Minimal Total Dominating Functions in Graphs

Bo Yu

Department of Combinatorics and Optimisation
University of Waterloo, Waterloo, Ontario
Canada, N2L 3G1

Abstract. A *total dominating function* (TDF) of a graph $G = (V, E)$ is a function $f : V \to [0, 1]$ such that for each $v \in V$, the sum of f values over all neighbours of v (i.e., all vertices adjacent to v) is at least one. Integer-valued TDFs are precisely the characteristic functions of total dominating sets of G. A *minimal* TDF (MTDF) is one such that decreasing any value of it makes it non-TDF. An MTDF f is called *universal* if convex combinations of f and any other MTDF are minimal. We give a sufficient condition for an MTDF to be universal which generalises previous results. Also we define a *splitting operation* on a graph G as follows: take any vertex v in G and a vertex w not in G and join w with all the neighbours of v. A graph G has a universal MTDF if and only if the graph obtained by splitting G has a universal MTDF. A corollary is that graphs obtained by the operation from paths, cycles, complete graphs, wheels, and caterpillar graphs have a universal MTDF.

1 Introduction

A *total dominating function* (TDF) of a graph $G = (V, E)$ is a non-negative real function f from the vertex set V to the unit interval $[0, 1]$ such that for each $v \in V$, the sum of f values over all neighbours of v is at least one. That is $f(N(v)) \geq 1$, where $N(v)$ is the set of neighbours of v, $N(v) = \{w \in V | vw \in E\}$ and $f(N(v))$ denotes $\sum_{u \in N(v)} f(u)$. Zero-one valued TDFs are precisely the characteristic functions of total dominating sets of G, where a subset X of vertices is called a *total dominating set* of G if every $v \in V$ has a neighbour in X. The reader is referred to [9] for an excellent bibliography concerning domination in graphs. Total dominating functions and total dominating sets have been studied in [2-9]. In particular, the convexity of minimal total dominating functions has been examined [6, 7, 10, 5]. A *minimal* total dominating function (MTDF) f is a TDF such that f does not remain a TDF if any $f(v), v \in V$ is decreased. It is easily shown that a convex combination of TDFs is still a TDF. But, a convex combination of MTDFs is not necessarily an MTDF. Figure 1 shows two MTDFs f and g such that a convex combination of f and g, say $\frac{1}{2}f + \frac{1}{2}g$, is not an MTDF. This is easily verified using Theorem 2.

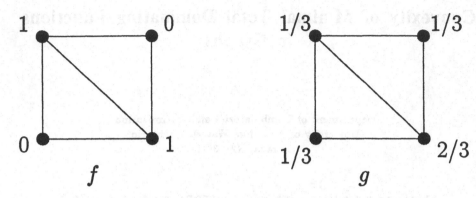

FIG. 1. *A convex combination of f and g is not an MTDF*

Motivated by this fact, [6] introduced the notion of a universal MTDF. An MTDF is called *universal* if any convex combination of g and any other MTDF is also an MTDF. Another reason for the study of convexity and universal MTDFs is the following interpolation problem due to Hedetniemi [8] (the interpolation problem in [8] concerns dominating functions [6]). The *aggregate* of a TDF f is the sum of f values over all vertices. Suppose that we are given MTDFs f and g with aggregates α and β, and a number $t \in (\alpha, \beta)$. Does G have an MTDF with aggregate t? It is easily seen that the answer is "yes" provided G has a universal MTDF.

In Section 2, we give a sufficient condition for an MTDF to be universal, which generalizes previous results due to Cockayne et al [6]. In Section 3, we present an operation on graphs such that a graph G has a universal MTDF if and only if the graph obtained by applying the operation has a universal MTDF. This operation gives several classes of graphs that have universal MTDFs. Finally, we give some open problems.

2   A sufficient condition for an MTDF to be universal

In this section, we introduce two types of vertices, dominating vertices and low vertices. In a tree, dominating vertices and low vertices are equivalent to so-called *short vertices* and *hot vertices* defined in [6], respectively. Our main theorem, Theorem 6, gives a sufficient condition for an MTDF to be universal. This generalizes one of the main theorems (Theorem 18) in [6]. If we restrict the graph in Theorem 6 to be a tree, then the condition in our theorem is necessary and sufficient, see [5].

For a TDF f, we define B_f, the *boundary* of f, to be the set $\{v \in V | f(N(v)) = 1\}$, and P_f, the *positive set* of f, to be the set $\{v \in V | f(v) > 0\}$. For subsets A, B of V, we write $A \to B$ if every vertex in B has a neighbour in A, i.e., B is a subset of $\bigcup_{v \in A} N(v)$.

We need the following two theorems, Theorem 1 and Theorem 2, from [6].

Theorem 1. *[6] A TDF f is minimal if and only if $B_f \rightarrow P_f$.*

For two MTDFs f and g, we define a *convex combination* of f and g to be $h_\lambda = \lambda f + (1-\lambda)g$ where $0 < \lambda < 1$. Note that λ is not allowed to be 0 or 1. The following theorem gives a necessary and sufficient condition for the convex combination to be minimal.

Theorem 2. *[6] Let f and g be MTDFs. A convex combination of f and g, $h_\lambda = \lambda f + (1-\lambda)g$ where $0 < \lambda < 1$, is an MTDF if and only if $B_f \bigcap B_g \rightarrow P_f \bigcup P_g$.*

This theorem shows that the minimality of h_λ is independent of λ; that is, either all convex combinations of f and g are minimal or none are minimal. Using this theorem, it is easily shown that in Figure 1 convex combinations of f and g are not minimal.

Recall that an MTDF g is called *universal* if any convex combination of g and any other MTDF is also an MTDF. The following theorem essentially states that a convex combination of two universal MTDF is also a universal MTDF.

Theorem 3. *A graph G has either no universal MTDF or a unique universal MTDF or infinitely many universal MTDFs.*

Proof. It suffices to show that G has infinitely many universal MTDFs if it has two distinct universal MTDFs g_1 and g_2. We claim that $h_\lambda = \lambda g_1 + (1-\lambda)g_2$ $(0 < \lambda < 1)$, a convex combination of g_1 and g_2, is universal. Let f be any MTDF. By definition, we only need to show that $g = \beta h_\lambda + (1-\beta)f$ $(0 < \beta < 1)$ is an MTDF. We have that

$$g = \beta h_\lambda + (1-\beta)f = \beta\lambda g_1 + \beta(1-\lambda)g_2 + (1-\beta)f$$

$$= \beta\lambda g_1 + (1-\beta\lambda)\left(\frac{\beta(1-\lambda)}{1-\beta\lambda}g_2 + \frac{1-\beta}{1-\beta\lambda}f\right).$$

Since g_2 is universal, by definition $\frac{\beta(1-\lambda)}{1-\beta\lambda}g_2 + \frac{1-\beta}{1-\beta\lambda}f$, a convex combination of g_2 and f, is an MTDF. This implies that g is an MTDF because g_1 is universal and g is a convex combination of g_1 and $\frac{\beta(1-\lambda)}{1-\beta\lambda}g_2 + \frac{1-\beta}{1-\beta\lambda}f$. Therefore h_λ is universal for any $\lambda, 0 < \lambda < 1$. □

A vertex v is called a *dominating vertex* if there exists a vertex u such that the open neighbourhood of u is properly contained in the open neighbourhood of v, i.e., $N(u) \subset N(v)$, where "$\subset$" means the relation of proper containment. We denote the set of dominating vertices of the graph G by $D(G)$.

Proposition 4. *For an MTDF f of G and a dominating vertex v in the boundary of f, if $u \in V(G)$ has $N(u) \subset N(v)$, then*
(1) u is in the boundary of f, and
(2) $f(w) = 0$ for all $w \in N(v) \backslash N(u)$.

Proof. Obvious. ☐

Now we introduce a type of vertex that is important in our study of universal MTDFs.

Let f be an MTDF of G. Vertex v is called *f-low* if $B_f \bigcap N(v) \subseteq D(G)$. An *f-low* vertex v may have $B_f \bigcap N(v)$ empty. Further, v is called *low* if it is *f-low* for some MTDF f.

Proposition 5. *If f is an MTDF and vertex v is f-low, then $f(v) = 0$.*

Proof. First, we claim that if vertex v is *f-low*, then for any $u \in B_f \bigcap N(v)$, there exists a vertex w such that $N(w) \subseteq N(u) \backslash \{v\}$.

Since v is *f-low*, by the definition, any vertex $u \in B_f \bigcap N(v)$ is a dominating vertex. Then there exists a vertex x such that $N(x) \subset N(u)$. If $x \notin N(v)$, then we are done. Otherwise let $x \in N(v)$. Since $u \in B_f$, Proposition 4 shows that $x \in B_f$. Since $x \in B_f \bigcap N(v)$, x is also a dominating vertex. We repeat the previous argument to get a squence of vertices $u, x, x_1, \ldots$, such that $N(u) \supset N(x) \supset N(x_1) \ldots$. See Figure 2. Since $N(u)$ is finite, the sequence must be finite. Hence, there exists a vertex w such that $N(w) \subseteq N(u) \backslash \{v\}$. This proves the claim.

Now we use the claim to prove the proposition. If $f(v) > 0$, then there exists $u \in B_f \bigcap N(v) \subseteq D(G)$. By the claim, there exists a vertex w such that $N(w) \subseteq N(u) \backslash \{v\}$. By Proposition 4, $w \in B_f$ and $f(v) = 0$. This is a contradiction. Therefore $f(v) = 0$. ☐

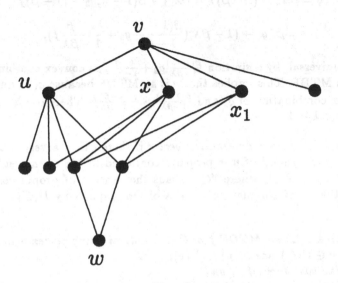

FIG. 2.

Before we state the main theorem of this section, we introduce another type of vertex.

Let W be the subset of vertices v such that for each $u \in N(v)$, there exists t in $N(u) \backslash \{v\}$ with $t \in B_f$ for every MTDF f. We now state the sufficient condition for an MTDF to be universal.

Theorem 6. *An MTDF g is a universal MTDF if*
(i) all vertices except dominating vertices and vertices in W are in the boundary of g, and
(ii) all low vertices v have $g(v) = 0$.

Proof. Let f be any MTDF. By Theorem 2, we only need to show $B_g \bigcap B_f \to P_f \bigcup P_g$. Let $v \in P_f \bigcup P_g$. It suffices to show $B_g \bigcap B_f \to \{v\}$. First, we claim that v is not f-low, for otherwise $f(v) = 0$ by Proposition 5 and $g(v) = 0$ by (ii), contradicting $v \in P_f \bigcup P_g$. Therefore, by the definition of low vertices, there exists a vertex $x \in B_f \bigcap N(v)$ but $x \notin D(G)$. If furthermore $x \notin W$, then $x \in B_g$ by (i). It follows that $x \in B_g \bigcap B_f \bigcap N(v)$. So $B_g \bigcap B_f \to \{v\}$. Now we may assume $x \in W$. Since $v \in N(x)$, by the definition of W there must exist a vertex $t \in N(v) \backslash \{x\}$ such that t is in the boundary of every MTDF. Therefore $t \in B_g \bigcap B_f \bigcap N(v)$; that is $B_g \bigcap B_f \to \{v\}$. Thus $B_g \bigcap B_f \to P_f \bigcup P_g$. □

The following theorem states that the second condition of Theorem 6 is also a necessary condition.

Theorem 7. *If an MTDF g is universal, then $g(v) = 0$ for all low vertices v.*

Proof. Let v be f-low for some MTDF f. Since g is universal, a convex combination h_λ of f and g is an MTDF, where $h_\lambda = \lambda f + (1 - \lambda)g$ and $0 < \lambda < 1$. Also $B_{h_\lambda} = B_f \bigcap B_g$ and $P_{h_\lambda} = P_f \bigcup P_g$. Hence,

$$B_{h_\lambda} \bigcap N(v) = (B_f \bigcap B_g) \bigcap N(v) \subseteq B_f \bigcap N(v) \subseteq D(G).$$

It follows that v is h_λ-low. Therefore $h_\lambda(v) = 0$ by Proposition 5. This implies that $g(v) = 0$. □

Notice that Theorem 6 together with Theorem 7 generalizes one of the main theorems (Theorem 18) in [6] by the fact that short vertices [6] are a subset of dominating vertices and low vertices [6] are a subset of low vertices.

Theorem 7 has an easy but useful corollary concerning graphs without universal MTDFs.

Corollary 8. *If a graph G has a vertex v such that all its neighbours are low vertices, then G has no universal MTDF.*

Proposition 9. *If a regular graph has no low vertex, then it has a universal MTDF.*

Proof. Suppose that the graph is r-regular. The function f having $f(v) = 1/r$ for every vertex v is an MTDF with $B_f = V$. Since the graph has no low vertex, f is universal by Theorem 6. □

We close this section by showing a 3-regular graph that has no universal MTDF.

Example 1: Using Theorem 1, it is easy to check that g in Figure 3 is an MTDF. Note that B_g consists of the solid square vertices. Notice that x is g-low since $B_g \bigcap N(x) = \emptyset$. Since x, y and z are vertex transitive, therefore y and z are also low vertices. Hence all neighbours of v are low vertices. By Corollary 8, this graph has no universal MTDF.

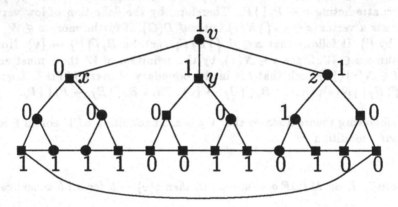

FIG. 3. *A regular graph without a universal MTDF*

3 The splitting operation

In this section, we define an operation on a graph G that leads to a class of graphs, denoted by $T(G)$, such that if any graph in $T(G)$ has a universal MTDF, then all graphs in $T(G)$ have universal MTDFs.

Let $v \in V(G)$ be any vertex and $w \notin V(G)$. Then v is *split* when we add w and the edges $\{wx | x \in N(v)\}$ to G. Note that $N(w) = N(v)$. Let the resulting graph be denoted by $G(v; w)$.

The proof of the following theorem applies to 0-1 universal MTDFs too. This is stated in Corollary 12 below.

Theorem 10. *G has a universal MTDF if and only if $G(v; w)$ has a universal MTDF.*

Proof. Let f be a universal MTDF of $G(v; w)$. Given f, define the function $g : V(G) \to [0, 1]$ to be:

$$g(t) = \begin{cases} f(w) + f(v) \text{ if } t = v \\ f(t) \quad\quad\quad \text{otherwise.} \end{cases}$$

To see that $g(v) \leq 1$, suppose that $f(v) > 0$ or $f(w) > 0$. Then by Theorem 1 there is a vertex $x \in N(v)$ $(N(w) = N(v))$ such that $x \in B_f$. Hence $g(v) = f(w) + f(v) \leq f(N(x)) = 1$.

By Theorem 1, it is easy to see that g is an MTDF of G by checking $B_g \to P_g$. Now we claim that g is a universal MTDF of G. Let g_1 be any MTDF of G. By Theorem 2, it suffices to show that $B_{g_1} \bigcap B_g \to P_{g_1} \bigcup P_g$. First, extend g_1 to be an MTDF f_1 of $G(v; w)$ by

$$f_1(t) = \begin{cases} 0 \quad\quad \text{if } t = w \\ g_1(t) \text{ otherwise.} \end{cases}$$

Since f is universal for $G(v; w)$, we have $B_f \bigcap B_{f_1} \to P_f \bigcup P_{f_1}$. Moreover, since $N(w) = N(v)$, therefore $(B_f \bigcap B_{f_1})\backslash\{w\} \to P_f \bigcup P_{f_1}$.

From the construction, it follows that $B_g = B_f\backslash\{w\}$, $P_g = P_f\backslash\{w\}$ and $B_{g_1} = B_{f_1}\backslash\{w\}$, $P_{g_1} = P_{f_1}$. So $B_{g_1} \bigcap B_g = (B_{f_1} \bigcap B_f)\backslash\{w\} \to P_{f_1} \bigcup P_f \supseteq P_{g_1} \bigcup P_g$. Therefore g is an MTDF of G.

Conversely, suppose g is a universal MTDF of G. Extend g to be an MTDF f of $G(v; w)$ by:

$$f(t) = \begin{cases} 0 \quad\quad \text{if } t = w \\ g(t) \text{ otherwise.} \end{cases}$$

We claim that f is universal for $G(v; w)$. Let f_1 be any MTDF of $G(v; w)$. We must show that $B_{f_1} \bigcap B_f \to P_{f_1} \bigcup P_f$. As before, we restrict f_1 to be an MTDF g_1 of G by:

$$g_1(t) = \begin{cases} f_1(v) + f_1(w) \text{ if } t = v \\ f_1(t) \quad\quad\quad \text{otherwise.} \end{cases}$$

Then $B_{g_1} \bigcap B_g \to P_{g_1} \bigcup P_g$. Using the same technique as above and the fact that $N(v) = N(w)$, it can be easily shown that $B_{f_1} \bigcap B_f \to P_{f_1} \bigcup P_f$. Hence f is a universal MTDF of $G(v; w)$. $\square$

The inverse of the splitting operation is to delete a vertex v from G if there exists another vertex w such that $N(v) = N(w)$. Let $T(G)$ denote the class of graphs obtained by starting from G and successively doing the splitting operation or inverse splitting operation. Then by Theorem 10, we have the following corollary.

Corollary 11. *G has a universal MTDF if and only if every graph in $T(G)$ has a universal MTDF.*

Further, from the proof of Theorem 10, we have the following corollary concerning 0-1 universal MTDFs.

Corollary 12. *G has a 0-1 universal MTDF if and only if every graph in $T(G)$ has a 0-1 universal MTDF.*

Let P_n, C_n, and K_n denote a path with n nodes, a cycle with n nodes, and a complete graph with n nodes, respectively. We also let W_n denote a wheel with n nodes (i.e. $W_n = K_1 + C_{n-1}$) and F denote caterpillar graphs [7].
In [6, 7], the following theorem was proved.

Theorem 13. P_n, C_n, K_n, W_n *and F all have universal MTDFs.*

Corollary 14. $T(P_n), T(C_n), T(K_n), T(W_n)$ *and $T(F)$ all have universal MTDFs.*

Proposition 15. *The complete n-partite graph $K_{m_1,m_2,\ldots,m_n}$ has a universal MTDF.*

Proof. The proposition follows from the above corollary and the fact that $K_{m_1,m_2,\ldots,m_n}$ is in $T(K_n)$. □

4 Open problems

We conclude with some open problems.

(1) If g is a universal MTDF, then does it necessarily satisfy the first condition in Theorem 6? The condition was shown to be necessary and sufficient for trees [5].
(2) Give a direct characterization of low vertices. A characterization of low vertices of trees is given in [7].
(3) Which graphs have universal MTDFs?
(4) Is there a good algorithm for deciding whether a graph has a universal MTDF?

Acknowledgement: The author would like to thank Joseph Cheriyan, Penny Haxell and Hugh Hind for helpful discussions.

References

1. R.B. Allan, R. Laskar and S.T. Hedetniemi, A note on total domination, *Discrete Math.* 49 (1984) 7-13.
2. A.A. Bertossi, Total domination in internal graphs, *Inform. Process. Lett.* 23(3) (1986) 131-134.
3. G.J. Chang, Total domination in block graphs, *Oper. Res. Lett.* 8(1) (1989) 53-57.
4. E.J. Cockayne, R.M. Dawes and S.T. Hedetniemi, Total domination in graphs, *Networks* 10 (1980) 211-219.

5. E.J. Cockayne and C.M. Mynhardt, A characterisation of universal minimal total dominating functions in trees, *Research Report*, Department of Mathematics, University of South Africa, 130/92(16).
6. E.J. Cockayne, C.M. Mynhardt and B. Yu, Universal minmal total dominating functions in graphs, *Networks* 24 (1994) 83-90.
7. E.J. Cockayne, C.M. Mynhardt and B. Yu, Total dominating functions in trees: minimality and convexity, *Journal of Graph Theory* 19(1) (1995) 83-92.
8. S.T. Hedetniemi, (*private communication*, 1990)
9. S. T. Hedetniemi and R. Laskar, Bibliography on domination in graphs and some basic definitions of domination parameters, *Discrete Math.* 86 (1990) 257-277.
10. B. Yu, *Convexity of minimal total dominating functions in graphs*, Master's Thesis, University of Victoria, 1992.

Transformations for Maximal planar graphs with minimum degree five

Short paper

Jean Hardouin Duparc
Philippe Rolland

IRIN
Université de Nantes
2, rue de la Houssinière
44072 Nantes cedex 03, France
{hardouin,rolland}@irin.univ-nantes.fr

Abstract. In this paper, we present some results on maximal planar graphs with minimum degree five, denoted by $MPG5$ graphs [4]. We describe a method to generate the $MPG5$ graphs with n fixed vertices. We prove that this proceed can produce all $MPG5$ graphs.

1 Introduction

We restrict ourselves to undirected, connected and planar graphs G which have no loops or multiple edges. Given a simple graph $G = (X, E)$ with node set X and edge set E. A graph G is said to be *embeddable* on a surface S if it can be drawn on S that its edges intersect only at their end vertices; and *planar* if it can be embedded on a plane. If we consider a planar graph with no loops or faces, bounded by two edges, it may be possible to add a new edge to the given representation of G, such that these properties are preserved. When no such adjunction can be made, we call G maximal planar. A planar graph is called triangulated when all faces have three corners. We have in [5], that a planar graph is maximal if, and only if, it is triangulated. In [4], a graph G is called an $MPG5$ graph if G is a Maximal Planar Graph with minimum degree five; we denote by $MPG5_n$ the set of the $MPG5$ graphs with n vertices.

The efficient generation of unrooted triangulations has received some attention in the literature. This appears to be harder than generating all rooted triangulations, and isomorphism testing is required by current algorithms. Recently, Avis [1] has proposed an efficient algorithm which uses the famous *Reverse Search* [2] procedure to generate all 2 and 3-connected r-rooted triangulations without repetitions. Thus, an idea to generate all maximal planar graphs with minimum degree five, will be to use the 3-rooted triangulations algorithm and with a filter to extract the $MPG5$ graphs. One drawback of this solution is its very expensive computational time. For example, for $n=20$, experimentally, we have found only 73 $MPG5$ non-isomorphic graphs and by Tutte's formula, we have approximately 10^{13} 3-rooted triangulations. Thus we have at least one good reason to introduce two algorithms to generate, directly, the $MPG5$ graphs: the expansion an! d reduction methods. The expansion method has been used by Marble and Matula [3], and despite the fact that it has undergone no formal analysis (it is not even known whether every $MPG5$ graph can be generated by this approch), it is useful because it tends to generate graphs that contain vertices of high degree. More recently, Morgenstern and Shapiro [4] have introduced a new planar graph generation method, the reduction method, which has the virtue of generating all $MPG5$ graphs with finite probability, though the graphs it generates have a more uniform degree distribution. The drawback of these solutions is that they do not generate the $MPG5$ graphs for an order n fixed. We are going to describe an algorithm to generate the $MPG5$ graphs with a number of vertices fixed. We prove that this algorithm can produce all $MPG5$ graphs.

2 Basic definitions

Let $N(x)$ be the set of neighbors of node x, $N^i(x)$ represent the i-neighborhood of x, for example,

$$N^2(x) = \bigcup_{y \in N(x)} N(y) - N(x) - \{x\}$$

The degree of node x, denoted $dg(x)$, is defined by the size $card(N(x))$. Δ (resp. δ) is the degree maximum (minimum) of the vertices in a graph.

Definition 1 *Consider a simple graph $G = (X, E)$. We denote*

$$X_{sup6} = \{x \in X / dg(x) \geq 6\}$$
$$X_{inf6} = \{x \in X / dg(x) \leq 5\}$$

Definition 2 *Let $G = (X, E)$ be an $MPG5$ graph. G is called, type A graph, if $X_{sup6} \neq \emptyset$ and $\forall x \in X_{sup6}, \{N(x) \cup N^2(x)\} \subseteq X_{inf6}$.*

By definition, the smallest type A graph has 14 vertices and $\Delta = 6$. We define one transformation called T and the inverse transformation called T^{-1}.

Definition 3 *Let G be an $MPG5$ graph and v be a vertex in X_{sup6}. We call $K = N(v)$. $T(v)$ is the graph G after the explosion of v in two new vertices adjacents z, t. We distribute K between $N(z)$ and $N(t)$ such that $T(v)$ is $MPG5$; i.e. the internal face of K (which contained v) is planar, maximum with $dg(z) \geq 5$, $dg(t) \geq 5$.*

We can remark $T(v)$ contains, at least two vertices in $K \cap X_{sup6}$ since by triangulation $\{N(z) \cap N(t)\} = \{x, y\} \subset X_{sup6}$.

Definition 4 *Let G be an $MPG5$ graph and $e = (z, t)$ be an edge. Since G is maximal, e bounds two triangles z, t, x and z, t, y. If $x, y \in X_{sup6}$, we denote $[z, t; x, y]$ this polygon. Let $[z, t; x, y]$ be a polygon, we call $T^{-1}(z, t)$ the graph G after contraction of z, t in a single vertex v.*

3 Generate all $MPG5$ graphs

We call a *wheel*, a cycle of vertices in X_{inf6} with at least five vertices, or exactly five. Two *wheels* C_1, C_2 are denoted *contiguous* is there are neighbors and concentrics.

Lemma 5 *Let G be an $MPG5$ graph. G contains at most two wheels contiguous.*

Proof Let G be an $MPG5$ graph with at least 3 wheels contiguous. We describe these wheels of G, internal to external wheels, by $\ldots, C_i, C_{i+1}, C_{i+2}, \ldots$. We consider x, y, z a sequence in clockwise order of vertex in an internal wheel like C_{i+1}. By definition of a *wheel* we have $\{N(x) \cup N(y) \cup N(z)\} \subseteq X_{inf6}$. This configuration is unique, it is exactly the smallest $MPG5$ graph, i.e. the $MPG5$ graph with 12 vertices of degree five. We have a contradiction, since by hypothesis, there exists at least 3 wheels, and so at least 3×5 vertices in G. Thus, an $MPG5$ graph has at most two wheels contigous. $\quad\Box$

Lemma 6 *Let G be an $MPG5$ graph with n vertices. If G contains two wheels contiguous then G is a type A graph.*

Proof Let C_1, C_2 two wheels contiguous. Suppose C_1 in the internal face of C_2. Since G is maximal, the subgraph between C_1 and C_2 is maximal and so triangulated. Let $H = (X_H, E_H)$ be the subgraph induced by the vertex set in C_1, C_2. We have by triangulation between C_1 and C_2, $\forall x \in X_H, dg(x) \geq 3$. By definition a *wheel* contains only vertices with degree five. So, in G, since the outer face of C_2 (resp. internal face of C_1) are triangulated, we have $\forall x \in X_H, dg(x) = 4$. Therefore, in G, the internal face of C_1 contains exactly one vertex a, and the external face of C_2 contains exactly one vertex b. This implicate that the order of G is even and $dg(a) = dg(b) = \frac{n-2}{2}$; i.e. $card(C_1) = card(C_2) = \frac{n-2}{2}$. And so we find that G is a type A graph. $\quad\Box$

By this symmetry between a, b, we can produce the type A graphs without problems. We consider the algorithm (i).

---

(0)	Let G be an $MPG5$ graph not type A with $n > 14$
(1)	$G^0 = G; i = 0$
(2)	**While** there exists $[z, t; x, y]$ in G^i **do**
(3)	$\quad G^{i+1} = T^{-1}(z, t); i = i + 1$
(4)	**End while**
(5)	$k = i; G' = G^k$

Algorithm (i): Generation of a type A graph.

Theorem 7 *G' is a type A graph.*

Proof We prove this theorem by using the lemma 8. We can observe that, after the **while** instruction, at the line (5) in the algorithm (i), we have

$$\forall x \in X_{sup6}, \{N^2(x) \cap X_{sup6}\} = \emptyset \tag{1}$$

Lemma 8 *In G', $\forall x \in X_{sup6}$, we have $N(x) \subset X_{inf6}$.*

Proof Consider $x \in X_{sup6}$ with $\{N(x) \cap X_{sup6}\} \neq \emptyset$. Let y be a neighbor of x with $dg(y) \geq 6$. Since G' is maximal, the edge (x, y) bounds two triangles (x, y, z) and (x, y, t). We have 3 cases:

- $z, t \in X_{inf6}$. We consider the subgraph, called H, induced by the following vertex set, $\{N(x) \cup N(y)\} \cup \{N^2(x) \cup N^2(y)\}$. Let F the outer face of H defined by $N^2(merge(x, y))$ i.e. some vertices in $N^2(x) \cup N^2(y)$. In H, by triangulation, all vertices in F are degree four excepted $\{N^2(x) \cap N^2(y)\} = \{a, b\}$ for which $dg(a) = dg(b) = 5$. But in G, by 1, all vertices of F are degree five; so it is impossible, by maximality near a, b, that an $MPG5$ graph contains H, like a subgraph.
- $z, t \in X_{sup6}$. We have a contradiction with 1, since $z \in N^2(t)$.
- $z \in X_{sup6}$ and $t \in X_{inf6}$. By 1, we have

$$\text{By } N^2(x): \{N(x) - \{y, z\}\} \subset \{N^2(y) \cup N^2(z)\} \subset X_{inf6} \tag{2}$$
$$\text{By } N^2(y): \{N(y) - \{x, z\}\} \subset \{N^2(x) \cup N^2(z)\} \subset X_{inf6} \tag{3}$$
$$\text{By } N^2(x) \cup N^2(y): \{N(z) - \{x, y\}\} \subset \{N^2(x) \cup N^2(y)\} \subset X_{inf6} \tag{4}$$

So, there exists two wheels called C_1 and C_2 defined respectively by

$$\{N(x) \cup N(y) \cup N(z)\} - \{x, y, z\}$$
$$\{N^2(x) \cup N^2(y) \cup N^2(z)\} - \{N(x) \cup N(y) \cup N(z)\}$$

Thus we have exactly two contiguous wheels and by the lemma 6 we have a type A graph. We have a contradiction: the internal face of C_1 contains three vertices, so the graph cannot be a type A graph.

Thus we have proved that in G', $\forall x \in X_{sup6}$, $N(x) \subset X_{inf6}$. $\qquad\square$

By 1, we have in G', $\forall x \in X_{sup6}$, $\{N(x) \cup N^2(x)\} \subseteq X_{inf6}$. So, by definition, G' is a type A graph, this prove the theorem 7. $\qquad\square$

We recall that the graph set $MPG5_{14}$ contains a single graph, defined by the smallest type A graph. Let G be a type A graph, with $X_{sup6} = \{x, y\}$. We consider the graph set called $\mathcal{G}_A^{odd}$ defined by the graphs which can be reach by $T(x)$ or $T(y)$ by starting on G. By definition the smallest order of the $\mathcal{G}_A^{odd}$ graphs could have 15 vertices, and for $n = 15$, there exits only one $MPG5$ graph; it is easy to verify that this unique graph is $\mathcal{G}_A^{odd}$. We have the following properties:

Corollary 9 *Let G be a $\mathcal{G}_A^{odd}$ graph with n vertices. Then $X_{sup6} = \{a, b, c, d, e\}$ with $[a, b; c, d]$ a polygon and $dg(c) = dg(d) = 6$, $dg(a) + dg(b) - 4 = dg(e) = \frac{n-3}{2}$. $T^{-1}(a, b)$ is the type A graph with $n - 1$ vertices.*

Theorem 10 *Let G be an $MPG5$ graph not type A, with $n > 14$ vertices. By starting on G, the algorithm (i) can reach G' such that G' will be the smallest type A graph.*

Proof By the theorem 7, we know that all $MPG5$ graphs can be reduced in a type A graph only by using the transformation T^{-1}. Now suppose G not type A with n vertices, then G' contains n' vertices with $n' < n$. We prove that we can have, always, $n' = 14$. We consider the graph G^{k-1}. This graph has a particular topology, since $G^{k-1} \in \mathcal{G}_A^{odd}$. We use the notation like in corollary 9. We describe all the polygons in a $\mathcal{G}_A^{odd}$ graph. It is easy to see that there exists, only and exactly, three differents polygons. For each polygon we can realize one transformation T^{-1}. We describe each of these three polygons.

1. We have the polygon $[a, b; c, d]$ and we can apply $T^{-1}(a, b)$. We call x the vertex after contraction of a, b. We obtain a type A graph, since $X_{sup6} = \{x, e\}$, and the first and second neighborhood of x, e are in X_{inf6}.
2. We describe two polygons which produce an $MPG5$ graph not type A.
   (a) We consider a_1, b_1 defined by $N(c) \cap N(e)$. We have $[a_1, b_1; c, e]$, and we can realize $T^{-1}(a_1, b_1)$.
   (b) We consider a_2, b_2 defined by $N(d) \cap N(e)$. We have $[a_2, b_2; d, e]$, and we can realize $T^{-1}(a_2, b_2)$.
   These two polygons are symmetricals in a $\mathcal{G}_A^{odd}$ graph, and we can verify without problem that $T^{-1}(a_1, b_1)$ and $T^{-1}(a_2, b_2)$ are isomorphics. For $n > 15$, at least a or b belong to X_{sup6}, since $dg(a) + dg(b) > 10$. And so in $T^{-1}(a_1, b_1)$ (resp. $T^{-1}(a_2, b_2)$), $d \in X_{sup6}$ with $\{N(d) \cap X_{sup6}\} \neq \emptyset$, (idem for c, i.e. $c \in X_{sup6}$ with $\{N(c) \cap X_{sup6}\} \neq \emptyset$).
   In these two cases, if $n > 15$, $\exists v \in X_{sup6}, \{N(v) \cap X_{sup6}\} \neq \emptyset$. Therefore the graph reduced by one of these polygons is not a type A graph.

So, if we select, always, the second (or third) polygon, in a $\mathcal{G}_A^{odd}$ graph, we delay the production of a type A graph. The smallest $\mathcal{G}_A^{odd}$ graph is the $MPG5$ graph with 15 vertices($\Delta = 6$); for this graph, the three polygons are equivalents and give the smallest type A graph. □

By using the inverse transformation of T^{-1}, called T, we have the following theorem,

Theorem 11 *Let $G = (X, E)$ be an $MPG5$ graph not a type A, with $n > 14$ vertices. By starting on the smallest type A graph and by some applications T we can reach G.*

4 Conclusion

We have developed some results about the generation of $MPG5$ graphs. These results have been found after to have defined a particular and interesting partitioning, called $\mathcal{Z}$, defined by the graphs where $\forall x \in X_{sup6}, N(x) \subset X_{inf6}$. We have described a new method to generate the $MPG5$ graphs with a number of vertices fixed. We have proved that this proceed can reach all $MPG5$ graphs.

An open problem is to realize this without repetitions like the Avis's algorithm for the r-rooted triangulations 2 and 3-connected, i.e. to generate all the $MPG5$ graphs with n vertices fixed without the need for isomorphism testing.

References

1. D. Avis. Generating rooted triangulations without repetitions. Technical Report SOCS-94.2, Mc Gill Montreal, 1994.
2. D. Avis and K. Fukuda. Reverse search for enumeration. Technical Report 92-5, University of Tsukuba, 1994.
3. G. Marble D.W.Matula and J.D. Isaacson. Graph coloring algorithms. In *R. C. Read, Graph Theory and Computing, Academic Press, New York*, pages 108–122, 1972.
4. C.A. Morgenstern and H.D. Shapiro. *Heuristics for Rapidly Four-Coloring Large Planar Graphs*, pages 869–891. Algorithmica. Springer-Verlag New York, 1991.
5. Oystein Ore. *The four-color problem*. Academic Press, 1967.

An Asynchronous Parallel Method for Linear Systems

You Zhaoyong*   Wang Chuanglong

The Research Center for Applied Mathematics

Xi'an Jiaotong University

Xi'an, 710049, Shaanxi, CHINA

Abstract   In this paper, we present a new asynchronous parallel method for solving linear system of equations $Ax = b$, where A is a nonsingular H-matrix. The method conbines usual asynchonous method [2] and multisplittings of matrices [5]. Convergence is established, rate of convergence is discussed, the region of relaxation parameter ω is given.

1. Introduction

Consider the solution of a large linear system of equations

$$Ax = b \tag{1.1}$$

on parallel computer, where A is a nonsingular H-matrix.

A great deal of research is currently being focused on the efficient implementation of iterative algorithms on parallel computers, especially for solving large linear equations. Many parallel methods were presented [1]-[9]. The typtical parallel methods are usual asynchronous algorithm [2] and O'Leary and White's multisplitting parallel algorithm, many parallel methods were respecively generalization and improvement such as [6]-[9]. We will establish a new asynchronous parallel method, which combines above two typtical methods so that it includes some asynchronous parallel methods [1]-[9].

2. The algorithms

We use $< A >$ denote the comparison matrix of A, and k denote the set of positive integers.

Definition 1 Let A be a nonsigular H-matrix. The splitting $A = M - N$ is called H-splitting if $< M > -|N|$ is a nonsingular M-matrix; and is called H-compatible splitting if $< A >=< M > -|N|$.

---

*Chairman of Shaanxi SIAM

Definition 2 For $k = 1, 2, \cdots$ and $i = 1, 2, \cdots, l$, let there be given sets $I_i^k \subset \{1, \cdots, n\}$ and n-tuples $(S_1^i(k), \cdots, S_n^i(k)) \in (k \bigcup \{0\})^n$. Suppuse that the following assumptions hold.

(i) For $i \in \{1, \cdots, l\}$ and $j \in \{1, \cdots, n\}$ $\quad S_j^i(k) \leq k - 1$,

(ii) For $i \in \{1, \cdots, l\}$ and $j \in \{1, \cdots, n\}$ $\quad \lim\limits_{k \to \infty} S_j^i(k) = \infty$,

(iii) For $i \in \{1, \cdots, l\}$ and $j \in \{1, \cdots, n\}$ the set $\{k : j \in I_i^k\}$ includes infinite elements.

Let A be a nonsingular H-matrix, $A = M_i - N_i$ $(i = 1, \cdots, l)$ be H-splittings. Then x^* is the solution of (1.1) iff x^* is a fixed point of following equations

$$x = \sum_{i=1}^{l} E_i M_i^{-1} N_i x + \sum_{i=1}^{l} E_i M_i^{-1} b \tag{2.1}$$

(a) Asynchronous parallel algorithm

Let $P^{(i)} = E_i M_i^{-1} N_i$ $(i = 1, \cdots, l)$ and suppose $j \in I_i^k \Longrightarrow (E_i)_{jj} \neq 0$. $x^{(0)}$ ia intial value.

$$x_i^{k+1} = \begin{cases} (E_i x^{(k)})_j & j \notin I_i^k, \\ p_{j1}^{(i)} x^{S_1^i(k)} + \cdots + p_{jl}^{(i)} x^{S_l^i(k)} + (E_i M_i^{-1})_j b & j \in I_i^k \end{cases}$$

$$x^{k+1} = \sum_{i=1}^{l} x_i^{(k+1)}$$

where $(E_i M_i^{-1})_j$ represents j-row of matrix $E_i M_i^{-1}$, $(E_i x_j^{(k)})$ represents j-element of vector $E_i x^{(k)}$.

(b) The relaxed asynchronous parallel algorithm

We will change $x^{(k+1)}$ generated by algorithm (a) as following

$$x^{(k+1)} = \omega \sum_{i=1}^{l} x_i^{(k+1)} + (1 - \omega) x^{s^i(k)} \quad \omega > 0$$

Remark: The above asynchronous parallel algorithms are usual algorithm [1]-[4], when $l = 1$. For $l > 1$, it is a combination of multisplitting parallel algorithm and usual algorithm. So it is an improvement to some asynchronous parallel methods such as [6]-[9]. In those methods, the power matrices $\{E_i\}$ were usual chosen satisfied $E_i E_j = \delta_i \delta_j I$, otherwise computing amount might be great difference from different from splittings, this results into that waiting time is increased. On the other hand, the condition $E_i E_j = \delta_i \delta_j I$ might be enable $\rho(\sum_{i=1}^{l} E_i < M_i >^{-1} |N_i|)$ not to be greatly decreased. According to above consideration, we establish the new method, power matrices $\{E_k\}$ mainly decrease spectral radius of iterative matrix, asynchronous computing mainly decreases waiting time.

In order to discuss convergent rate of the algorithm, we revise assumptions of $\{I_k\}$ and $\{S_j^i(k)\}$ as follow

(i) $I_i^k = I^k$ $i = 1, \cdots, l$,

(ii) For $i \in \{1, \cdots, l\}$ and $j \in \{1, \cdots, n\}$, there exists a positive integer T_l such that

$$k - T_l \le S_j^i(k) \le k - 1$$

(iii) For $j \in \{1, \cdots, n\}$, there exists a positive integer T_2 such that $j \in \{I^k, \cdots, I^{k-T_2}\}$.

Under revised conditions, the asynchronous parallel algorithm (a) is called algorithm (c), the relaxed asynchronous parallel algorithm (b) is called algorithm (d).

3. Main Results

Definition 3 Let $\{x^{(k)}\}_{k=0}^{\infty}$ be a sequence in R^n such that $\lim_{k \to \infty} x^{(k)} = x^*$. Then $\sigma(\{x^{(k)}\}_{k=0}^{\infty}) = \limsup_{k \to \infty} \|x^{(k)} - x^*\|^{\frac{1}{k}}$ is the R_1-factor of the sequence $\{x^{(k)}\}_{k=0}^{\infty}$. This factor is independent of the choice of the norm $\|\cdot\|$.

Theorem 4 Let A be a nonsigular H-matrix, and $A = M_i - N_i$ $(i = 1, \cdots, l)$ be H-splittings. If there exists a nonsingular M-matrix C such that $C \ll M_i > -|N_i|$ $(i = 1, \cdots, l)$. Then the sequence $\{x^{(k)}\}$ generated by algorithm (a) converges to the solution of (1.1).

Proof: Let x^* be the solution of (1.1). First, we assume that C is a strictly diagonally dominant M-matrix. Hence, $< M_i >$, $< M_i > -|N_i|$ $(i = 1, \cdots, l)$ are all strictly diagonally dominant M-matrices.

$$\| < M_i >^{-1} |N_i| \|_{\infty} \le \max_{1 \le k \le n} \frac{\sum_{j=1}^{n} (|N_i|)_{kj}}{(< M_i >)_{kk} - \sum_{j \neq k} (< M_i >)_{kj}}$$

$$\le \max_{1 \le k \le n} \frac{\sum_{j \neq k} |c_{kj}| + (|N_i|)_{kk}}{c_{kk} + (|N_i|)_{kk}}$$

$$= r_i \le \max_{1 \le i \le l} r_i = r < 1$$

Let $\mu = \min\{(E_i)_{jj} | (E_i)_{jj} \neq 0, i = 1, \cdots, l, j = 1, \cdots, n\}$ $e^{(k)} = x^{(k)} - x^*$.

Let $j_0 \in \{1, \cdots, n\}$ satisfy $\|e^{(k+1)}\|_{\infty} = |e_{j_0}^{(k+1)}|$

(1) When $j_0 \notin I_i^k$ $i = 1, \cdots, l$. By condition (iii) of definition 2, there exists a positive integer $k_1 < k$ such that $j_0 \in I_{i_0}^{k_1}$ $i_0 \in \{1, \cdots, l\}$.

(2) There exists at least $i_0 \in \{1, \cdots, n\}$ such that $j_0 \in I_{i_0}^k$

$$|e_{j_0}^{(k+1)}| \le (1 - \mu)\|e^{(k)}\|_{\infty} + r\mu \max_{1 \le j \le n} \{\|e^{S_j^{i_0}(k)}\|\}$$

Combine (1) and (2) as well as condition (iii), it is obtained

$$\lim_{k \to \infty} \sup \|e^{(k+1)}\|_{\infty} = 0$$

i.e. $\quad \lim_{n \to \infty} x^{(k)} = x^*$

By the property of H-matrix, there exists a optimal scaling matrix S such that CS is a strictly diagonally dominant matrix. Then $S^{-1} < M_i > S, S^{-1}(< M_i > -|N_i|)S$ $(i = 1, \cdots, l)$ are all strictly diagonally dominant matrices.

Similarily proving proceeding as (3.1), we can obtain $\dfrac{S_j}{S_k}$

$$\|S^{-1} < M_i >^{-1} |N_i|S\|)_\infty \leq \max_{1 \leq k \leq n} \frac{\sum\limits_{j \neq k} |c_{kj}| + (|N_i|)_{kk}}{c_{kk} + (|N_i|)_{kk}}$$

$$= r_i \leq \max_{1 \leq i \leq l} r_i = r < 1 \tag{3.2}$$

Let $e^{(k)} = S^{(-1)}(x^{(k)} - x^*)$, analysis $e^{(k)}$ as (1) and (2), we can obtain

$$\lim_{k \to \infty} \sup \|e^{(k+1)}\|_\infty = 0$$

That is $\quad \lim_{n \to \infty} x^{(k)} = x^*$

Colloally 5 *Let A be a nonsingular H-matrix, and $A = M_i - N_i$ $(i = 1, \cdots, l)$ be H-compatible splittings. Then the sequence $\{x^{(k)}\}$ generated by algorithm (a) coverges to the solution of (1.1).*

Theorem 6 *Let A be a nonsingular H-matrix, and $A = M_i - N_i$ $(i = 1, \cdots, l)$ be H- splittings, and $diag(M_i)=diag(A)$. If there exitsts a nonsingular M-matrix C such that $C << M_i > -|N_i|$ $(i = 1, \cdots, l)$, $0 < \omega \leq 1$. Then the sequence $\{x^{(k)}\}$ generated by algorithm (b) converges to the solution of (1.1).*

Proof, Similar the proof of theorem 4, there exists a scaling matrix S such that CS is a strictly diagonally dominant matrix, and $\|S^{-1} < M_i >^{-1} |N_i|S\|_\infty \leq r < 1$.

Let $j_0 \in \{1, \cdots, n\}$ satisfy $\|e^{(k+1)}\|_\infty = |e_{j_0}^{(k+1)}|$

(1) $j_0 \notin I_i^k$, $i = 1, \cdots, l$.

By condition (iii) of definition 2, there exists a positive integer $k_1 < k$ such that $j_0 \in I_{i_0}^{k_1}$ $\quad i_0 \in \{1, \cdots, l\}$.

(2) There exists at least $i_0 \in \{1, \cdots, n\}$ such that $j_0 \in I_{i_0}^k$

$$|e_{j_0}^{(k+1)}| \leq \omega[(1 - \mu)\|e^{(k)}\|_\infty + r\mu \max_{1 \leq j \leq n} \|e^{S_j^{i_0}(k)}\|_\infty] + (1 - \omega)\|e^{S^{i_0}(k)}\|_\infty$$

$$\leq [\omega((1 - \mu) + r\mu) + (1 - \omega)]\max\{\|e^{(k)}\|_\infty, \|e^{S_j^{i_0}(k)}\|_\infty, \|e^{S^i(k)}\|_\infty\}$$

when $0 < \omega \leq 1$,

$$\omega((1 - \mu) + r\mu) + (1 - \omega) = 1 - \omega(1 - r)\mu' < 1$$

Hence $\lim_{k \to \infty} x^{(k)} = x^*$

Theorem 7 *Let A be a nonsingular H-matrix, and $A = M_i - N_i$ $(i = 1, \cdots, l)$ be H-splittings, and $diag(M_i) = diag(A)$. If there exists a nonsingular M-matrix C such that $C \leq\; < M_i > -|N_i|$ $(i = 1, \cdots, l)$. Then the sequence $\{x^{(k)}\}$ generated by algorithm (c) converges to the solution of (1.1), and $\sigma\{x^{(k)}\}_{k=0}^{\infty} \leq \sqrt[T]{\rho(J_c)}$, where J_c is Jacobi matrix of C, $T = T_1 + T_2$.*

Proof, Let S be the optimal scaling matrix of C, then

$$\|S^{-1} J_c S\|_{\infty} = \rho(J_c) \tag{3.3}$$

$$\|S^{-1} < M_i >^{-1} |N_i| S\|_{\infty} \leq \max_{1 \leq k \leq n} \frac{\sum_{j \neq k} |c_{kj}| \frac{S_j}{S_k} + (|N_i|)_{kk}}{c_{kk} + (|N_i|)_{kk}}$$

By assumption, for every $i \in \{1, \cdots, l\}$ the diagonal entries of N_i are zeros. So

$$\|S^{-1} < M_i >^{-1} |N_i| S\|_{\infty} \leq \|S^{-1} J_c S\|_{\infty}$$

It implies that

$$\|S^{-1} \sum_{i=1}^{l} E_i < M_i >^{-1} |N_i| S\|_{\infty} \leq \|S^{-1} J_c S\|_{\infty}$$

Let $e^{(k)} = S^{-1}(x^{(k)} - x^*$ and $j_0 \in \{1, \cdots, n\}$

$$\|e^{(k+1)}\|_{\infty} = |e_{j_0}^{(k+1)}|$$

(1) $j_0 \notin I^k$. There exists a positive integer K_1 such that $j_0 \in I^{k_1}$, $k - T_2 \leq k_1 \leq k - 1$.

(2) $j_0 \in I^k$

$$\|e^{(k+1)}\|_{\infty} = |e_{j_0}^{k+1}| \leq \|S^{-1} J_c S\|_{\infty} \max_{k-T_1 \leq t \leq k} \|e^{(t)}\|_{\infty}$$

combine (1) and (2), we obtain

$$((3.4)) \qquad \|e^{(k+1)}\|_{\infty} \leq \rho(J_c) \max_{k-T \leq t \leq k} \|e^{(t)}\|_{\infty} \quad (T = T_1 + T_2)$$

Hence, $\sigma(\{x^{(k)}\}_{k=0}^{\infty} = \limsup_{k \to \infty} \|e^{(k)}\|^{\frac{1}{k}}$

$$\leq \lim_{k \to \infty} \sup (\rho(J_c)^{\frac{k}{T}} \max_{0 \leq t \leq T} e^{(t)})^{\frac{1}{k}} = \sqrt[T]{\rho(J_c)}$$

Theorem 8 *Let A be a nonsingular H-matrix, and $A = M_i - N_i$ $(i = 1, \cdots, l)$ be H-splittings, and $diag(M_i) = diag(A)$. If there exists a nonsingular M-matrix C such that $C \leq\; < M_i > -|N_i|$ $(i = 1, \cdots, l)$, $0 < \omega < \dfrac{2}{1 + \rho(J_c)}$. Then the sequence $\{x^{(k)}\}$ generated by algorithm (d) converges to the solution of (1.1).*

Proof, Similar proving proceeding as (3.4), we have

$$\|e^{(k+1)}\|_\infty \le (\omega\rho(J_c) + |1 - \omega|) \max_{k-T \le t \le k}\{\|e^{(t)}\|_\infty, \|e^{S(k)}\|_\infty\}$$

Obviously, if $\omega\rho(J_c) + |1 - \omega| < 1$, $\{x^{(k)}\}$ is convergent.

Theorem 6 gives the convergence under $0 < \omega \le 1$, when $\omega > 1$, solving inequality as following

$$\omega\rho(J_c) + \omega - 1 < 1$$

we obtain

$$1 < \omega < \frac{2}{1 + \rho(J_c)}$$

Hence , when $0 < \omega < \frac{2}{1+\rho(J_c)}$

$$\lim_{k\to\infty} x^{(k)} = x^*$$

Corollary 9 *Let A be a nonsingular H-matrix, and $A = M_i - N_i$ ($i = 1, \cdots, l$) be H- compatible splittings and diag(M_i)=diag(A). Then the sequence $\{x^{(k)}\}$ generated by algorithm (c) converges to the solution of (1.1), and $\sigma\{x^{(k)}\}_{k=0}^\infty \le \sqrt[T]{\rho(|J|)}$, where J is Jacobi matrix of A.*

Corollary 10 *Let A be a nonsingular H-matrix, and $A = M_i - N_i$ ($i = 1, \cdots, l$) be H- compatible splittings, and diag(M_i)=diag(A), $0 < \omega < \dfrac{2}{1 + \rho(|J|)}$. Then the sequence $\{x^{(k)}\}$ generated by algorithm (d) converges to the solution of (1.1).*

References

[1] D.Chazan and W.Miranker, Chaotic relaxation, Linear Algebla Appl. 2 (1969):199-222.

[2] C.Bauded, Asynchronous iterative methods for multiprocessors, J. Assoc. Comput. Math. 25(1978):226-244.

[3] R.Barlow and D.Evan, Synchronous and asynchronous iterative parallel algorithms for linear systems, Comput.J.25 (1982):56-60.

[4] M.El.Tarazi, Some convergence results for asychronous algorithms, Numer.Math. 39(1982):325-340.

[5] D.P.O'leary and R.E.White, Multi-splittings of matrices and parallel solution of linear systems, SIAM J.Algebraic Discrete Methods 6 (1985):630-640.

[6] R.Bru, L.Elsner and M.Neumann, Models of parallel chaotic reaxation methods, Linear Algebra Appl. 103(1988):175-192.

[7] Wang Deren, On the convergence of parallel multispitting AOR algorithm, Linear Algebra Appl.154-156(1991):473-486.

[8] A.Frommer and G.Mayer, Convergence of relaxed parallel multisplitting methods, Linear Algebra Appl. 119(1989):141-152.

[9] A.Bhaya, E.Kaszkurewicz and F.Mota, Asynchronous block-iterative methods for almost linear equations, Linear Algebra Appl. 154-156(1991):486-495.

[10] A.Frommer and D.Bszyls, H-splittings and two-stage iterative methods, Numer Math.63(1992):345-356.

On a Kind of Sequence of Polynomials

Zhang Xiangde*

Northeastern University, Shenyang 110006, China

Abstract. In this paper, motivated by a conjecture concerning the q-derangement polynomials $D_n(q)$ which are defined by Brenti, we study a kind of sequence of polynomials $\{S_n(q)\}$. The $S_n(q)$ is a polynomial of degree n with nonnegative real coefficients and satisfy some initial conditions and the following recurrence relation:

$$S_n(q) = a_n q S_{n-1}(q) + b_n q (1 + c_n q) S'_{n-1}(q) + d_n q S_{n-2}(q) \ ,$$

where $a_n > 0, b_n > 0, d_n > 0$ and $n \geq 0$. We show that $S_n(q)$ has n distinct real roots(≤ 0), separated by the roots of $S_{n-1}(q)$. As a consequence, the conjecture is proved.

1   Introduction

Let $P = \{1, 2, \cdots, n \cdots\}$ and S_n denote the symmetric group on n elements. For $n \in P$ and $\sigma \in S_n$, an element $i \in [n]$ (where $[n] = \{1, 2, \cdots, n\}$) is called an excedance of σ if $\sigma(i) > i$. We denote the number of excedances of σ by $e(\sigma)$. The permutation $\sigma \in S_n$ is called a derangement if $\sigma(i) \neq i$ for $i = 1, 2, \cdots, n$. Let the set of all derangements of S_n be denoted by D_n. It is known that Brenti [2] has defined polynomials $D_n(q)$ by

$$D_n(q) = \sum_{\sigma \in D_n} q^{e(\sigma)} \ ,$$

where $D_0(q) = 1$ and $D_1(q) = 0$. Since $D_n(1) = |D_n|$, Brenti considers $D_n(q)$ as q-derangement polynomials.

In [2], Brenti posed the following conjecture: for $n \in P$, the polynomials $D_n(q)$ have only real zeros. It is well-known that if a polynomial with nonnegative coefficients has only real roots(≤ 0), it is log-concave and unimodal. A polynomial $a_0 + a_1 q + \cdots + a_n q^n$ with real coefficients is said to be unimodal if for some $0 \leq j \leq n$ we have $a_0 \leq a_1 \leq \cdots \leq a_j \geq a_{j+1} \geq \cdots \geq a_n$, and is said to be log-concave if $a_i^2 \geq a_{i-1} a_{i+1}$ for all $1 \leq i \leq n-1$. Polynomials being unimodal, log-concave or with only real roots arise often from combinatorics, algebra, computer science and geometry and have been the subject of considerable research in recent years. It has now become apparent that the proof of these properties of polynomials can sometimes be a difficult task requiring the use of intricate combinatorial constructions or of refined mathematical tools.

* Supported by the Natural Science Foundation of China, Liaoning Province and Northeastern University.

The tools include, for example, classical analysis, the theory of total positivity, the theory of symmetric functions, the representation theory of Lie algebra and superalgebra, and algebraic geometry. [7] is an excellent survey of many of these techniques.

In this paper, motivated by the conjecture of Brenti, we study a kind of interesting sequence of polynomials $\{S_n(q)\}$, and find that the polynomial $S_n(q)$ has n distinct real roots. As corollaries, the conjecture is proved. The main results are as follows :

Theorem 1. *The polynomials $D_n(q)$ satisfy the following properties:*

1) $$D_n(q) = \sum_{k=0}^{n} (-1)^{n-k} \binom{n}{k} A_k(q)/q$$

where $A_0(q) = q$, for $n \in P$, $A_n(q)$ is the nth Eulerian polynomial.

2) $\quad D_n(q) = (n-1)qD_{n-1}(q) + q(1-q)D'_{n-1}(q) + (n-1)qD_{n-2}(q), n \geq 2.$

3) $\quad$ *The polynomial $D_n(q)$, i.e. $\sum_{k=0}^{n}(-1)^{n-k} \binom{n}{k} A_k(q)/q$ has $n-1$ distinct real roots (≤ 0), separated by the roots of $D_{n-1}(q)$.*

Theorem 2. *Let $S_n(q)$ be a polynomial of degree n with nonnegative real coefficients, and let $\{S_n(q)\}$ satisfy the following properties:*

4) $S_n(q) = a_n q S_{n-1}(q) + b_n q(1 + c_n q)S'_{n-1}(q) + d_n q S_{n-2}(q)$, *where $a_n > 0, b_n > 0, d_n \geq 0$ and $n \geq 2$.*

5) *For $n \geq 1$, zero is a simple root of $S_n(q)$.*

6) $S_0(q) = e, S_1(q) = e_1 q$ *and $S_2(q)$ has two real roots(≤ 0), where $e \geq 0, e_1 \geq 0$.*

Then the polynomial $S_n(q)$ has n distinct real roots (≤ 0), separated by the roots of $S_{n-1}(q)$.

2 The Proof of the Main Results

First, let us prove the theorem 2.

Proof. Let us use induction on n. For $n = 1, 2$ the theorem is easily checked. Assume that it holds for $n - 1(n > 3)$, and we proceed to show that it holds for n. Noting that $S_n(q)$ is a polynomial of degree n with nonnegative real coefficients, we obtain that $S_n(q)$ has no positive real roots. By assumption $S_{n-1}(q)$ and $S_{n-2}(q)$ have only distinct real roots. Denote the roots of $S_{n-1}(q)$ and $S_{n-2}(q)$ by $\alpha_1 = 0, \alpha_2, \cdots, \alpha_{n-1}$ and $\beta_1 = 0, \beta_2, \cdots, \beta_{n-2}$, respectively, where $0 = \alpha_1 > \alpha_2 > \cdots > \alpha_{n-1}$ and $0 = \beta_1 > \beta_2 > \cdots > \beta_{n-2}$, by assumption, we have

$$\alpha_{i+1} \leq \beta_i \leq \alpha_i, \quad i = 1, 2, \cdots, n - 2. \tag{1}$$

By Rolle's theorem, we see that $S'_{n-1}(q)$ has $n - 2$ distinct real roots and that the roots of $S'_{n-1}(q)$ separate those of $S_{n-1}(q)$. Denote the roots of $S'_{n-1}(q)$ by $\gamma_1, \gamma_2, \cdots, \gamma_{n-2}$, where $\gamma_1 > \gamma_2 > \cdots > \gamma_{n-2}$. Then we have

$$\alpha_{i+1} < \gamma_i < \alpha_i, \quad i = 1, 2, \cdots, n - 2. \tag{2}$$

Hence, we obtain

$$S_n(q) = a_n q^2 (q - \alpha_2) \cdots (q - \alpha_{n-1})$$
$$+ (n - 1) b_n q (1 + c_n q) (q - \gamma_1)(q - \gamma_2) \cdots (q - \gamma_{n-2})$$
$$+ d_n q^2 (q - \beta_2) \cdots (q - \beta_{n-2}) \tag{3}$$

Denote

$$a = -min \left\{ |\gamma_1/4|, 1/(1 + |c_n|), \left[(n-1) b_n |\gamma_1/2| \prod_{i=2}^{n-2} (\frac{\gamma_1}{2} - \gamma_i) \right] / N \right\},$$

where

$$N = (1 + a_n + d_n)(1 + |c_n|) \left(\prod_{i=2}^{n-1} |\alpha_i| + \prod_{i=2}^{n-2} |\beta_i| \right).$$

Then we see that $\alpha_2 < \gamma_1/4 \leq a < 0$. Using (3), we have

$$S_n(a) < 0 \tag{4}$$

By(1), (2) and (3), the following inequalities can be easily proved

$$(-1)^k S_n(\alpha_k) > 0, \quad k = 2, ..., n-1.$$

Since $\lim_{q \to -\infty} (-1)^n S_n(q) = +\infty$, there exists $M > 0$ ($M > 1 + |\alpha_{n-1}|$) such that if $q \leq -M$ then $(-1)^n S_n(q) > 0$ so in particular $(-1)^n S_n(-M-1) > 0$. Hence, by the intermediate value theorem of continuous functions, we get that $S_n(q)$ has only one root in each open interval (α_2, a), (α_{i+1}, α_i) and $(-M - 1, \alpha_{n-1})$, where $i = 2, 3, \cdots, n-2$. Recall that zero is a simple root of $S_n(q)$, thus we see that $S_n(q)$ has n distinct real roots and its roots are obviously separated by those of $S_{n-1}(q)$. Hence, we have proved the theorem by induction. $\square$

Second, we give the proof of theorem 1.

Proof. Recall the following relation between $D_n(q)$ and $A_n(q)$ [2]:

$$A_n(q)/q = \sum_{k=0}^{n} \binom{n}{k} D_k(q), \quad n \in P \tag{5}$$

Let $A_0(q) = q$, so that (5) also holds for $n = 0$. Using the following binomial inversion formulas

$$f_n = \sum_{k=0}^{n} \binom{n}{k} g_k \quad \Longleftrightarrow \quad g_n = \sum_{k=0}^{n} (-1)^{n-k} \binom{n}{k} f_k,$$

we obtain $D_n(q) = \sum_{k=0}^{n} (-1)^{n-k} \binom{n}{k} A_k(q)/q$.

Recall that the Eulerian polynomials $A_n(q)$ satisfy the following recurrence relation:

$$A_n(q) = q(1 - q) A'_{n-1}(q) + nq A_{n-1}(q), n \geq 2.$$

Clearly the recurrence relation also holds for $n = 1$, since $A_0(q) = q$, and therefore we have

$$
\begin{aligned}
D_n(q) &= \sum_{k=0}^{n}(-1)^{n-k}\binom{n}{k}A_k(q)/q \\
&= \sum_{k=0}^{n}(-1)^{n-k}\left(\binom{n-1}{k}+\binom{n-1}{k-1}\right)A_k(q)/q \\
&= -D_{n-1}(q) + \sum_{k=1}^{n}(-1)^{n-k}\binom{n-1}{k-1}A_k(q)/q \qquad \left(\binom{n-1}{-1}=0\right) \\
&= -D_{n-1}(q) + \sum_{k=1}^{n}(-1)^{n-k}\binom{n-1}{k-1}\left((1-q)A_{k-1}'(q)+k\,A_{k-1}(q)\right) \\
&= -D_{n-1}(q) + (1-q)\left(q\,D_{n-1}(q)\right)' + \sum_{k=1}^{n}(-1)^{n-k}\binom{n-1}{k-1}k\,A_{k-1}(q) \\
&= (n-1)\,q\,D_{n-1}(q) + q\,(1-q)\,D_{n-1}'(q) + (n-1)\,q\,D_{n-2}(q),
\end{aligned}
$$

where $n \geq 2$. Hence, 2) is proved.

Using the following property [2]:

$$
\sum_{n\geq 0} D_n(q)\frac{t^n}{n!} = \frac{1}{1 - \sum_{k\geq 2}(q+q^2+\cdots+q^{k-1})\frac{t^k}{k!}},
$$

we obtain that zero is a simple root of $D_n(q)$. So, by theorem 1, 3) is true. $\quad\square$

From theorem 1, we have the following two corollaries:

Corollary 3. *The conjecture is true.*

Corollary 4. *For $n \in P$, the polynomials $D_n(q)$ are log-concave.*

From theorem 1, we can also obtain the following well-known fact: The Eulerian polynomial $A_n(q)$ has n roots, separated by those of $A_{n-1}(q)$.

Notes. The polynomials $D_n(q)$ are different from the q-derangement numbers for the major index that have been previously considered in the literature (see, e.g., [3], [5], [6], [8] and [9]).

References

1. Brenti,F.: Unimodal,log-concave and Polya frequency sequences in combinatorics. Memoirs.Amer. Math. Soc. no. 413. 1989
2. Brenti,F.: Unimodal polynomials arising from symmetric functions. Proc. Amer. Math. Soc. **108** (1990)1133-1141

3. Chen,W., Rota,G.: q-Analogs of the inclusion - exclusion principle and permutation with restricted position. Discrete. Math. **104** (1992). 7-22
4. Comet,L.: Advanced combinatorics. Reidel Dordrecht/Boston. 1974
5. Garsia,A., Remmel,J.: A combinatorial interpretation of q-derangement and q-Laguerre numbers. European. J. Combin. **1** (1980) 47-59
6. Gessel,I., Reutenauer,C.: Counting permutations with given cycle structure and dessent set. J. Combina. Theory. Ser. A **64** (1993)189-215
7. Stanley,R.: Log-concave and unimodal sequences in algebra, combinatorics and geometry. Annals of the New York Academy of Sciences **576** (1989) 500-534
8. Wachs,M.: On q-derangement numbers. Proc. Amer. Math. Soc. **106** (1989) 273-278
9. Zhang,X.: On Some q-analysis problems in combinatorics. Ph. D. Thesis. Dalian University of Technology. January 1994

Hamiltonian Cycles in 2-generated Cayley Digraphs of Abelian Groups

Jixiang Meng

Department of Mathematics, Xinjiang University
Urumuchi 830046, China

Abstract. We give a necessary and sufficient condition for 2-generated Cayley digraphs of abelian groups to be hamiltonian or to be hamiltonian decomposable. As applications, we derive the counting formula for the number of the hamiltonian cycles in the conjunction of an undirected cycle and a directed cycle and that in the cartesian product of two directed cycles.

1 Introduction

This paper is motivated by the problem of characterizing hamiltonian double loops [1] or hamiltonian decomposable double loops. Let G be a finite group and S a subset of G not containing the identity 1 of G. The vertices of the Cayley digraph $D(G, S)$ are elements of G and there is an edge from x to y if and only if $x^{-1}y \in S$. If S is inverse-closed; that is, if $S^{-1} = S$, then $D(G, S)$ corresponds to an undirected graph, called a Cayley graph. It is known that any connected Cayley graph on an abelian group is hamiltonian [2]. Defferent from undirected Cayley graphs, Cayley digraphs of abelian groups are not necessarily hamiltonian. Here, we will charecterize 2-generated hamiltonian Cayley digraphs and 2-generated hamiltonian decomposable Cayley digraphs of abelian groups (a Cayley digraph $D(G, S)$ is called k-generated if $|S| = k$). Based on these results, we derive the counting formulae for the number of hamiltonian cycles in the conjunction $C_n \cdot \vec{C}_m$ and the cartesian product $\vec{C}_n \times \vec{C}_m$.

2 Hamiltonian Cycles in 2-generated Cayley Digraphs

In the following, we always suppose that G is a finite abelian group and $S = \{a, b\}$ is a generating subset of G. Thus the Cayley digraph $D(G, S)$ is strongly connected.

A sequence $T = [t_1, t_2, \cdots, t_r]$ on S is a sequence with $t_i \in S (1 \leq i \leq r)$. For $k(1 \leq k \leq r)$, define $S_k(T) = \sum_{i=1}^{k} t_i$. Let l be a positive integer and T a sequence on S. Then T^l is the sequence obtained by joining l copies of T.

A sequence T on S is called a H-sequence of G on S if it satisfies the following two conditions:

i) $r = n$ and $S(T) = 0$, and

ii) $S_i(T) \neq S_j(T)$ for any i and j with $1 \leq i \neq j \leq n$ where $n = |G|$.

Lemma 1. *There is one -one correspondence between the set of hamiltonian cycles and the set of H-sequences of G on S.*

In the following, we use $H(G, S)$ to denote the set of H-sequences of G on S.

Lemma 2. *Let* $S = \{a, b\}$ *be a generating subset of a finite abelian group* G, $K = < (a - b)$ *and* $[G : K] = m$. *If* $T = [s_1, s_2, \cdots, s_n] \in H(G, S)$ $(n = |G|)$, *then* $T = |K| * [s_1, s_2, \cdots, s_m]$.

Proof. Let $M = [s_1, s_2, \cdots, s_m]$ and $t = |K|$. Then we have

$$\{S_k(T), S_{m+k}(T), \cdots, S_{(t-1)m+k}(T)\} = S_k(T) + K \qquad (1)$$

Setting $k = m$ and $k = 1$ in (1) respectively, we have

$$\{S_m(T), S_{im}(T), \cdots, S_{tm}(T)\} = K \qquad (2)$$

and

$$\{s_1, s_{m+1} + S_m(T), \cdots, s_{(t-1)m+1} + S_{(t-1)m}(T)\} = s_1 + K \qquad (3)$$

By taking sum of the elements in both hands of (3) and combining (2), we have

$$s_1 + s_{m+1} + \cdots + s_{(t-1)m+1} = ts_1 \qquad (4)$$

Suppose that the number of a and b in $\{s_{jm+1} : 0 \leq j \leq t - 1\}$ is k_1 and k_2 respectively, then

$$k_1 + k_2 = t \qquad \text{and} \qquad k_1 a + k_2 b = ts_1$$

Without loss of generality, we assume that $s_1 = a$, then $(t - k_1)(a - b) = 0$. Since $\text{ord}(a - b) = t$, we have $t = k_1$. Thus

$$s_1 = s_{m+1}, \cdots, = s_{(t-a)m+1} = a$$

The result follows by noting the fact that $T' = [s_2, s_3, \cdots, s_n, s_1] \in H(G, S)$ whenever $T = [s_1, s_2, \cdots, s_n] \in H(G, S)$.

Theorem 3. *Under the notations of lemma 2, $D(G,S)$ contains hamiltonian cycles if and only if there exists a non-negative integer $x \leq m$ such that $\operatorname{ord}(xa + (m - x)b) = |K|$.*

Proof. Let $t = |K|$ and $M = [t_1, t_2, \cdots, t_m]$ be a sequence on S such that the number of $a's$ in $\{t_i : 1 \leq i \leq m\}$ is x and Let $T = t * M$. Then $T \in H(G, S)$. Conversely, if $T = [t_1, t_2, \cdots, t_n] \in H(G, S)$, then $T = t * [t_1, t_2, \cdots, t_m]$ by lemma 2 and by the definition of H-sequence we can deduce that the number x of $a's$ in $\{t_i : 1 \leq i \leq m\}$ satisfies the equation $\operatorname{ord}(xa + (m - x)b) = |K|$.

Under the notations of lemma 2, we set $N(G, S)$ to denote the set of solutions of the equation $\operatorname{ord}(xa + (m - s)b) = |K|$ with $0 \leq x \leq m$. Then by theorem 3, we have

Theorem 4. *Under the notations of lemma 2, we have*

$$|H(G,S)| = \sum_{x \in N(G,S)} \binom{m}{x} \tag{1}$$

Theorem 5. *Under the notations of lemma 2, $D(G, S)$ can be decomposed into hamiltonian cycles if and only if there exists a non-negative integer $x \leq m$ such that*

$$\operatorname{ord}(xa + (m - s)b) = \operatorname{ord}((m - x)a + xb) = |K| \tag{2}$$

Proof. Let x be a non-negative integer satisfying the specified conditions. Let $M_1 = [t_1, t_2, \cdots, t_m]$ and $M_2 = [t'_1, t'_2, \cdots, t'_m]$ be two sequences on S such that $t_i \neq t'_i (1 \leq i \leq m)$ and the number of $a's$ in M_1 and M_2 is x and m-x respectively. Then it is easy to check that the two hamiltonian cycles corresponding to $T_1 = |K| * M_1$ and $T_2 = |K| * M_2$ are arc-disjoint. The converse can be shown directly.

Let n be a positive integer. $C_n (n \geq 3)$ and $\overrightarrow{C}_n (n \geq 2$ denote the undirected cycle and directed cycle of length n respectively. By $C_n \cdot \overrightarrow{C}_m$, we mean the conjunction of the associated digraph C'_n of C_n and the digraph $\overrightarrow{C}_n$.

By applying Theorem 3 — Theorem 5 we can deduce the following results.

Theorem 6. *If $C_n \cdot \vec{C}_m$ is strongly connected, then the number $N_1(n,m)$ of hamiltonian cycles in $C_n \cdot \vec{C}_m$ is*
i) if n is odd, then

$$N_1(n,m) = \sum_k \binom{m}{k}$$

where $0 \leq k \leq m$ and $\gcd(m - 2k, n) = 1$
ii) if n is even, then

$$N_1(n,m) = \sum_k \binom{2m}{k}$$

where $0 \leq k \leq 2m$ and $\gcd(k - m, \frac{n}{2}) = 1$

Corollary 7. *If $m = n$ is an odd prime, then the number of hamiltonian cycles in $C_n \cdot \vec{C}_m$ is $2^m - 2$.*

Theorem 8. *The number $N_2(n,m)$ of hamiltonian cycles in $\vec{C}_n \times \vec{C}_m$ is*

$$N_2(n,m) = \sum_k \binom{(m,n)}{k}$$

where $0 < k < (m,n), (m,n) = \gcd(m,n), \gcd(k,n) = \gcd(k',m) = 1,$ and $k' = \gcd(m,n) - k$.

Theorem 9. *If $C_n \cdot \vec{C}_m$ is strongly connected, then it can be decomposed into hamiltonian cycles.*

References

1. Y.H.Xu, Double loop networks with minimum delay, jour Discrete Math. 66(1987), 109-118.
2. D.Witte and J.A.Gallian, A survey: Hamiltonian cycles in Cayley graphs, Discrete Math. 5(1984),283-304.

Pandiagonal Magic Squares

Cheng-Xu Xu and Zhun-Wei Lu[1]

Department of Mathematics and Mechanics
Taiyuan University of Technology
Taiyuan, Shanxi, China, 030024

Abstract. In 1972, Hudson [2] gave a method for the construction of pandiagonal magic squares of order $6t \pm 1$. In this paper we give the definition of pandiagonal Latin squares and the methods for the construction of self orthogonal pandiagonal Latin squares of order $4, 8, 9, 27$ and prime $p \geqslant 5$. One can use the technique of Kronecker products for the construction of self orthogonal pandiagonal Latin squares and pandiagonal magic squares of order n provided $n \neq 4t + 2$, $9t + 3$, $9t + 6$.

Let A be an $n \times n$ array. Index its rows and columns by $0, 1, \cdots, n - 1$. By the jth right diagonal of A we mean the following n cells of $A : (i, j + i); i = 0, 1, \cdots, n - 1;$ (mod n). Also, we define the jth left diagonal of A to be the following n cells of $A : (i, j - i); i = 0, 1, \cdots, n - 1;$ (mod n).

A Latin square A of order n is an $n \times n$ array such that every row and every column of A is a permutation of $0, 1, \cdots, n - 1$.

Two Latin squares of order n $A = (a_{ij})$ and $B = (b_{ij})$ are orthogonal if the n^2 pairs $(a_{ij}, b_{ij}) (i, j = 0, 1, \cdots, n - 1)$ are distinct.

If A and its transpose A' are orthogonal, A is said to be a self orthogonal Latin square.

If every right diagonal and every left diagonal of Latin square A is a permutation of $0, 1, \cdots, n - 1$, A is said to be a Knut Vik design. Hedayat[1] proved that Knut Vik designs of order n exist if and only if $n = 6t \pm 1$.

Theorem 1. If prime $p \geqslant 5$, $a_{ij} = 2i - j \pmod{p}$, $A_p = (a_{ij})$, then A_p is a self orthogonal Knut Vik design of order p.

Proof. If $2i - j_1 = 2i - j_2 \pmod{p}$, then $j_1 = j_2 \pmod{p}$. If $2i_1 - j = 2i_2 - j \pmod{p}$, then $i_1 = i_2 \pmod{p}$. A_p is a Latin square. $a_{ij+i} = i - j \pmod{p}$, $a_{ij-i} = 3i - j \pmod{p}$. If $i_1 - j = i_2 - j \pmod{p}$, then $i_1 = i_2 \pmod{p}$, If $3i_1 - j = 3i_2 - j \pmod{p}$, then $i_1 = i_2 \pmod{p}$. A_p is a Knut Vik design. If $(2i_1 - j_1, 2j_1 - i_1) = (2i_2 - j_2, 2j_2 - i_2) \pmod{p}$, then $2i_1 - j_1 + 2(2j_1 - i_1) = 2i_2 - j_2 + 2(2j_2 - i_2) \pmod{p}$, $j_1 = j_2 \pmod{p}$, $2(2i_1 - j_1) + 2j_1 - i_1 = 2(2i_2 - j_2) + 2j_2 - i_2 \pmod{}$

---

[1] The project supported by Natural Science Foundation of Shanxi Province, China

$p)$, $i_1 = i_2 (\mathrm{mod}\ p)$. A_p is self orthogonal.

If the sum of the numbers in each right diagonal and each left diagonal of Latin square A of order n is $\dfrac{n(n-1)}{2}$, A is said to be a pandiagonal Latin square.

Example 1.

$$A_4 = \begin{pmatrix} 0 & 3 & 1 & 2 \\ 1 & 2 & 0 & 3 \\ 2 & 1 & 3 & 0 \\ 3 & 0 & 2 & 1 \end{pmatrix}$$ is a self orthogonal pandiagonal Latin square of order 4.

Example 2.

$$A_8 = \begin{pmatrix} 0 & 4 & 6 & 2 & 3 & 7 & 5 & 1 \\ 1 & 5 & 7 & 3 & 2 & 6 & 4 & 0 \\ 2 & 6 & 4 & 0 & 1 & 5 & 7 & 3 \\ 3 & 7 & 5 & 1 & 0 & 4 & 6 & 2 \\ 4 & 0 & 2 & 6 & 7 & 3 & 1 & 5 \\ 5 & 1 & 3 & 7 & 6 & 2 & 0 & 4 \\ 6 & 2 & 0 & 4 & 5 & 1 & 3 & 7 \\ 7 & 3 & 1 & 5 & 4 & 0 & 2 & 6 \end{pmatrix}$$ is a self orthogonal pandiagonal Latin square of

order 8.

Example 3.

$$A_9 = \begin{pmatrix} 0 & 4 & 8 & 1 & 5 & 6 & 2 & 3 & 7 \\ 1 & 5 & 6 & 2 & 3 & 7 & 0 & 4 & 8 \\ 2 & 3 & 7 & 0 & 4 & 8 & 1 & 5 & 6 \\ 3 & 7 & 2 & 4 & 8 & 0 & 5 & 6 & 1 \\ 4 & 8 & 0 & 5 & 6 & 1 & 3 & 7 & 2 \\ 5 & 6 & 1 & 3 & 7 & 2 & 4 & 8 & 0 \\ 6 & 1 & 5 & 7 & 2 & 3 & 8 & 0 & 4 \\ 7 & 2 & 3 & 8 & 0 & 4 & 6 & 1 & 5 \\ 8 & 0 & 4 & 6 & 1 & 5 & 7 & 2 & 3 \end{pmatrix}$$ is a self orthogonal pandiagonal Latin

square of order 9.

Example 4. Let

$$B_1 = \begin{pmatrix} 0 & 9 & 18 & 12 & 21 & 3 & 24 & 6 & 15 \\ 1 & 10 & 19 & 13 & 22 & 4 & 25 & 7 & 16 \\ 2 & 11 & 20 & 14 & 23 & 5 & 26 & 8 & 17 \\ 3 & 12 & 21 & 15 & 24 & 6 & 18 & 0 & 9 \\ 4 & 13 & 22 & 16 & 25 & 7 & 19 & 1 & 10 \\ 5 & 14 & 23 & 17 & 26 & 8 & 20 & 2 & 11 \\ 6 & 15 & 24 & 9 & 18 & 0 & 21 & 3 & 12 \\ 7 & 16 & 25 & 10 & 19 & 1 & 22 & 4 & 13 \\ 8 & 17 & 26 & 11 & 20 & 2 & 23 & 5 & 14 \end{pmatrix} \quad B_2 = \begin{pmatrix} 4 & 13 & 22 & 16 & 25 & 7 & 19 & 1 & 10 \\ 5 & 14 & 23 & 17 & 26 & 8 & 20 & 2 & 11 \\ 3 & 12 & 21 & 15 & 24 & 6 & 18 & 0 & 9 \\ 7 & 16 & 25 & 10 & 19 & 1 & 22 & 4 & 13 \\ 8 & 17 & 26 & 11 & 20 & 2 & 23 & 5 & 14 \\ 6 & 15 & 24 & 9 & 18 & 0 & 21 & 3 & 12 \\ 1 & 10 & 19 & 13 & 22 & 4 & 25 & 7 & 16 \\ 2 & 11 & 20 & 14 & 23 & 5 & 26 & 8 & 17 \\ 0 & 9 & 18 & 12 & 21 & 3 & 24 & 6 & 15 \end{pmatrix}$$

$$
B_3 = \begin{pmatrix}
8 & 17 & 26 & 11 & 20 & 2 & 23 & 5 & 14 \\
6 & 15 & 24 & 9 & 18 & 0 & 21 & 3 & 12 \\
7 & 16 & 25 & 10 & 19 & 1 & 22 & 4 & 13 \\
2 & 11 & 20 & 14 & 23 & 5 & 26 & 8 & 17 \\
0 & 9 & 18 & 12 & 21 & 3 & 24 & 6 & 15 \\
1 & 10 & 19 & 13 & 22 & 4 & 25 & 7 & 16 \\
5 & 14 & 23 & 17 & 26 & 8 & 20 & 2 & 11 \\
3 & 12 & 21 & 15 & 24 & 6 & 18 & 0 & 9 \\
4 & 13 & 22 & 16 & 25 & 7 & 19 & 1 & 10
\end{pmatrix}
\quad
B_4 = \begin{pmatrix}
9 & 18 & 0 & 21 & 3 & 12 & 6 & 15 & 24 \\
10 & 19 & 1 & 22 & 4 & 13 & 7 & 16 & 25 \\
11 & 20 & 2 & 23 & 5 & 14 & 8 & 17 & 26 \\
12 & 21 & 3 & 24 & 6 & 15 & 0 & 9 & 18 \\
13 & 22 & 4 & 25 & 7 & 16 & 1 & 10 & 19 \\
14 & 23 & 5 & 26 & 8 & 17 & 2 & 11 & 20 \\
15 & 24 & 6 & 18 & 0 & 9 & 3 & 12 & 21 \\
16 & 25 & 7 & 19 & 1 & 10 & 4 & 13 & 22 \\
17 & 26 & 8 & 20 & 2 & 11 & 5 & 14 & 23
\end{pmatrix}
$$

$$
B_5 = \begin{pmatrix}
13 & 22 & 4 & 25 & 7 & 16 & 1 & 10 & 19 \\
14 & 23 & 5 & 26 & 8 & 17 & 2 & 11 & 20 \\
12 & 21 & 3 & 24 & 6 & 15 & 0 & 9 & 18 \\
16 & 25 & 7 & 19 & 1 & 10 & 4 & 13 & 22 \\
17 & 26 & 8 & 20 & 2 & 11 & 5 & 14 & 23 \\
15 & 24 & 6 & 18 & 0 & 9 & 3 & 12 & 21 \\
10 & 19 & 1 & 22 & 4 & 13 & 7 & 16 & 25 \\
11 & 20 & 2 & 23 & 5 & 14 & 8 & 17 & 26 \\
9 & 18 & 0 & 21 & 3 & 12 & 6 & 15 & 24
\end{pmatrix}
\quad
B_6 = \begin{pmatrix}
17 & 26 & 8 & 20 & 2 & 11 & 5 & 14 & 23 \\
15 & 24 & 6 & 18 & 0 & 9 & 3 & 12 & 21 \\
16 & 25 & 7 & 19 & 1 & 10 & 4 & 13 & 22 \\
11 & 20 & 2 & 23 & 5 & 14 & 8 & 17 & 26 \\
9 & 18 & 0 & 21 & 3 & 12 & 6 & 15 & 24 \\
10 & 19 & 1 & 22 & 4 & 13 & 7 & 16 & 25 \\
14 & 23 & 5 & 26 & 8 & 17 & 2 & 11 & 20 \\
12 & 21 & 3 & 24 & 6 & 15 & 0 & 9 & 18 \\
13 & 22 & 4 & 25 & 7 & 16 & 1 & 10 & 19
\end{pmatrix}
$$

$$
B_7 = \begin{pmatrix}
18 & 0 & 9 & 3 & 12 & 21 & 15 & 24 & 6 \\
19 & 1 & 10 & 4 & 13 & 22 & 16 & 25 & 7 \\
20 & 2 & 11 & 5 & 14 & 23 & 17 & 26 & 8 \\
21 & 3 & 12 & 6 & 15 & 24 & 9 & 18 & 0 \\
22 & 4 & 13 & 7 & 16 & 25 & 10 & 19 & 1 \\
23 & 5 & 14 & 8 & 17 & 26 & 11 & 20 & 2 \\
24 & 6 & 15 & 0 & 9 & 18 & 12 & 21 & 3 \\
25 & 7 & 16 & 1 & 10 & 19 & 13 & 22 & 4 \\
26 & 8 & 17 & 2 & 11 & 20 & 14 & 23 & 5
\end{pmatrix}
\quad
B_8 = \begin{pmatrix}
22 & 4 & 13 & 7 & 16 & 25 & 10 & 19 & 1 \\
23 & 5 & 14 & 8 & 17 & 26 & 11 & 20 & 2 \\
21 & 3 & 12 & 6 & 15 & 24 & 9 & 18 & 0 \\
25 & 7 & 16 & 1 & 10 & 19 & 13 & 22 & 4 \\
26 & 8 & 17 & 2 & 11 & 20 & 14 & 23 & 5 \\
24 & 6 & 15 & 0 & 9 & 18 & 12 & 21 & 3 \\
19 & 1 & 10 & 4 & 13 & 22 & 16 & 25 & 7 \\
20 & 2 & 11 & 5 & 14 & 23 & 17 & 26 & 8 \\
18 & 0 & 9 & 3 & 12 & 21 & 15 & 24 & 6
\end{pmatrix}
$$

$$
B_9 = \begin{pmatrix}
26 & 8 & 17 & 2 & 11 & 20 & 14 & 23 & 5 \\
24 & 6 & 15 & 0 & 9 & 18 & 12 & 21 & 3 \\
25 & 7 & 16 & 1 & 10 & 19 & 13 & 22 & 4 \\
20 & 2 & 11 & 5 & 14 & 23 & 17 & 26 & 8 \\
18 & 0 & 9 & 3 & 12 & 21 & 15 & 24 & 6 \\
19 & 1 & 10 & 4 & 13 & 22 & 16 & 25 & 7 \\
23 & 5 & 14 & 8 & 17 & 26 & 11 & 20 & 2 \\
21 & 3 & 12 & 6 & 15 & 24 & 9 & 18 & 0 \\
22 & 4 & 13 & 7 & 16 & 25 & 10 & 19 & 1
\end{pmatrix}
. \text{ Then } A_{27} = \begin{pmatrix} B_1 & B_2 & B_3 \\ B_4 & B_5 & B_6 \\ B_7 & B_8 & B_9 \end{pmatrix} \text{ is a self}
$$

orthogonal pandiagonal Latin square of order 27.

If $A = (a_{ir})$ and $B = (b_{ji})$ are Latin squares of order n and m respectively the Kronecker product is defined as the Latin square $A \times B = C(c_{kt})$ of order nm given by: $c_{kt} = a_{ir} + nb_{ji}$, if $k = i + nj$, $t = r + ns$.

Theorem 2. If A and B are two self orthogonal Latin squares, then C is also a self orthogonal Latin square.

proof. If $(c_{k,i_1}, c_{t,k_1}) = (c_{k,s_1}, c_{t,k_1})$, that is $a_{i,r_1} + nb_{j,t_1} = a_{i,r_1} + nb_{j,t_1}$, $a_{r,i_1} + nb_{s,j_1}$
$= a_{r,i_1} + nb_{s,j_1}$. Since $0 \leqslant a_{ir} \leqslant n - 1$, then $(a_{i,r_1}, a_{r,i_1}) = (a_{i,r_1}, a_{r,i_1})$, $(b_{j,t_1}, b_{s,j_1}) =$
(b_{j,s_1}, b_{s,j_1}). Then $i_1 = i_2$, $r_1 = r_2$, $j_1 = j_2$, $s_1 = s_2$, $k_1 = k_2$, $t_1 = t_2$. C is a self
orthogonal Latin square.

Theorem 3. If A and B are two pandiagonal Latin squares, then C is also a
pandiagonal Latin square.

proof. $\displaystyle\sum_{k=0}^{nm-1} C_{kt \pm k} = \sum_{i=0}^{n-1} \sum_{j=0}^{m-1} (a_{ir \pm i} + nb_{j t \pm j}) = m \sum_{i=0}^{n-1} a_{ir \pm i} + n^2 \sum_{j=0}^{m-1} b_{j t \pm j} =$
$\dfrac{mn(n-1)}{2} + \dfrac{n^2 m(m-1)}{2} = \dfrac{nm(nm-1)}{2}$. C is a pandiagonal Latin square.

Now we can use the Kronecker products of A_4, A_8, A_9, A_{27}, A_p for the
construction of self orthogonal pandiagonal Latin squares of order $n = 2^r \, 3^s \prod_{i=1}^{m} p_i^{t_i}$,
provided $r \neq 1$, $s \neq 1$, prime $p_i \geqslant 5$, $t_i \geqslant 1$. That is $n \neq 4t + 2$, $9t + 3$, $9t + 6$.

A pandiagonal magic square of order n is an $n \times n$ array of n^2 distinct integers
$0, 1, \cdots, n^2 - 1$ with the property that the sums of the numbers in every row, every
column, every right diagonal, every left diagonal are the same.

It is easy to verify that if A is a self orthogonal pandiagonal Latin square of
order n, then $nA + A'$ is a pandiagonal magic square of order n.

Thus we can construct the pandiagonal magic squares of order n, provided $n \neq$
$4t + 2$, $9t + 3$, $9t + 6$.

References

1. A. Hedayat, A Complete Solution to the Existence and Nonexistence of Knut Vik Designs
   and Orthogonal Knut Vik Designs, J. Combinatorial Theory(A) **22**(1977), 331 – 337.
2. C. B. Hudson, On pandiagonal magic squares of order $6t \pm 1$, Math. Mag. **45**(1972). 94
   – 96.

PFFM and Quasi-Morishima Matrices

Hua Wang and Zhao-yong You

Xi'an Jiao-tong University,Certre for Applied Mathematics, Xi'an, China

Abstract. In 1983, Greenberg [1] advanced an open problem "We do not have simple criterion that will enable us to characterize which elements of PFFM are quasi-Morishima and which are not." In this paper, two algorithms of time complexity $O(e)$ are provided.The algorithms can be used to decide which PFFM is quasi-Morishima and which is not. Here e denotes the number of edges of the researched graph. So we give an answer for the open problem.

1 Introduction

This paper continues the development of the theory and use of graphs and digraphs associated with rectangular matrices which Greenberg initiated in [1], Greenberg showed some properties of quasi-Morishima matrices and examined the application of the results to several classes of matrices occuring in energy economic models—PFM and PFFM.

Before introducing the PFFM we remind the reader some basic concepts.

Given the m by n matrix A we define two sets of points,$R = \{r_1, r_2, \cdots, r_m\}$ and $C = \{c_1, c_2, \cdots, c_n\}$ as represent the rows and columns of A, respectively.We then have the following definitions.

Bigraph(BG): is a bigraph on the points sets R and C, the line belongs to BG if $\alpha_{ij} \neq 0$.

Row graph(RG): is defined on R, the line $[r_i, r_j]$ belongs to RG if there exists $c_k \in C$ such that $[r_i, c_k]$ and $[r_j, c_k]$ are in BG.

Coumn graph(CG): is defined on C, the line $[c_i, c_j]$ belongs to CG if there exists $r_k \in R$ such that $[c_i, r_k]$ and $[c_j, r_k]$ are in BG.

It is clear that the sign information in the real matrix A can be immediately incorportated into the bigraph BG. We label the line $[r_i, c_j]$ *positive* if $\alpha_{ij} > 0$ and *negative* if $\alpha_{ij} < 0$. The resulting signed graph will be denoted by BG^+, similarly we can define RG^+ and CG^+.

A rectangular matrix A will be called *regular* if each row and column of A contains at least one nonzero element.

Physical flows with feedback matrix (PFFM).The matrix A is PFFM iff there exist permutation matrices P and Q, such that

$$PAQ = \begin{bmatrix} A_{11} & 0 & A_{13} \\ A_{21} & A_{22} & 0 \\ 0 & A_{32} & A_{33} \end{bmatrix} \tag{1}$$

Where $A_{11} \leq 0$, $A_{22} \leq 0$, $A_{33} \leq 0$, $A_{13} \geq 0$, $A_{21} \geq 0$, $A_{32} \geq 0$.

The matrix A will be called a *regular PFFM* if each of blocks A_{11}, A_{13}, A_{21}, A_{22}, A_{32}, A_{33} is nonempty and regular.

2 Signed Matrices

Let $X = (x_1, \cdots, x_N)$ and $Y = (y_1, \cdots, y_N)$ be vectors in the Euclidean space R^N, X and Y is called *conformal* if $x_i y_i \geq 0, 1 \leq i \leq N$, and *anticonformal* if $x_i y_i \leq 0$, $1 \leq i \leq N$. Let $A = [a_{ij}]$ be an m by n real matrix, we will call A *row signed* if the row vectors of A are pairwise either conformal or anticonformal, similarly we can define A to be *column signed*.

Lemma 1[1]   *A is column signed iff A is row signed.*

From lemma 1, we shall say A is *signed* without using the adjectives column or row, we recall that a signed graph is called *balanced* if every cycle is positive. It is easy to gain the following result

Corollary 2   *The signed graph G^+ is balanced iff the points of G^+ can be partitioned into disjoint subsets S_1 and S_2 such that every edge joining two points in the same set is positive and every edge joining two points in different sets is negative, and iff every cycle in G^+ is positive, and iff G^+ is a 2-colorable graph.*

Definition 3[1]   The m by n matrix A will be called a *quasi-Morishima matrix* if BG^+ is balanced.

From the theorem 3 in [1], we know that A is a quasi Morishima matrix iff BG^+ is balanced, and iff CG^+ is balanced, and iff RG^+ is balanced

PFM matrices are quasi-Morishima matrices, but this is not true for PFFM, However the PFFM

$$A = \begin{bmatrix} -1 & 0 & 0 & 1 & 0 \\ 0 & 1 & -1 & 0 & 0 \\ 0 & 0 & 1 & 0 & -1 \end{bmatrix}$$

is a quasi-Morishima matrix.

Theorem 4   *Suppose the points of graph G^+ can be partitioned into disjoint three subsets S_1, S_2 and S_3 such that every edge joining two points in the same set is positive and every edge joining two points in different set is negative, then* *the number of edges joining two different sets in a cycle is odd iff the cycle contains odd edges joining any two sets S_i and $S_j (i \neq j)$.*

Proof.Since we do not consider the number of edges in same set, if there is an odd cycle passing three point sets S_1, S_2, S_3, there is an odd edges joining S_i and S_j. Assume $i = 1, j = 2$, and because the number of edges entering S_1 equals the number of edges Coming out from S_1, then the number of edges joining S_1, S_3 and the number of edges joining S_2, S_3 are odd. Conversely. If a cycle contains odd edges joining two sets S_1 and S_2, the number of edges joining S_1 and S_3 is odd too, so the cycle must be odd. □

Corollary 5 *Suppose A satisfies the conditions of theorem 4. Then A is quasi-Morishima iff there is no cycle that contains odd edges joining two sets R_i and $R_j (i \neq j)$.*

Theorem 6 *There is an algorithm of $O(e)$ to find an odd cycle in an undirected graph $G = (V, E), |E| = e, V = \{c_1 \cdots, c_n\}$*

Proof. Suppose G is connected,we can give an algorithm.
Step 1: $k := 0, S_0 = \{c_1\}$,
Step k: $k := k + 1, S_k = \{c_j^k | [c_j^k, c_i^{k-1}]$ is an edge in G, and $c_i^{k-1} \in S_{k-1}\}$, then delete these edges.
......
Continue until all edges in G are deleted. Suppose $S_0, S_1, \cdots, S_l$ are found, here $l \leq e$. Let

$$M_0 = \bigcup_{\substack{i=2k \\ i \leq l}} S_i \qquad M_1 = \bigcup_{\substack{i=2k+1 \\ i \leq l}} S_i$$

if there is a point c_k in both M_0 and M_1, there is an odd cycle; Otherwise, there is no odd cycle.

It is clear that the time complexity is $O(e)$, e is the number of edges. □

From theorem 6, we get an algorithm of time $O(e)$ to decide whether a PFFM matrix is a quasi-Morishima matrix or not. Suppose the PFFM has the form (1)

Algorithm 1

Let us see BG^+, we have an algorithm for BG^+.

- Step 1 Find out the set of the connected sets of $BG^+(A_{ii})$, let it is $CS(A_{ii}), i = 1, 2, 3$.
- Step 2 Find out the set of pair of connected sets of $CS(A_{11})$, $CS(A_{22})$ with $A_{21} : \{\{A_{11}^1, A_{22}^1\}, \{A_{11}^2, A_{22}^2\}, \cdots\}$; and the set of pair of connected sets of $CS(A_{22})$, $CS(A_{33})$ with $A_{32} : \{\{\bar{A}_{22}^1, \bar{A}_{33}^1\}, \{\bar{A}_{22}^2, \bar{A}_{33}^2\}, \cdots\}$. Without loss of generality, assume that $\{\bar{A}_{22}^1, \bar{A}_{33}^2, \cdots\}$ is a subsequence of $\{A_{22}^1, A_{33}^2, \cdots\}$

Step 3   if $\{\bar{A}_{11}^1, \bar{A}_{11}^2, \cdots\}$ and $\{\bar{A}_{33}^1, \bar{A}_{33}^2, \cdots\}$ are not connected with A_{13}, the PFFM is quasi-Morishima; otherwise suppose $\{\bar{A}_{11}^{i1}, \bar{A}_{33}^{i2}\}$, $\{\bar{A}_{11}^{i2}, \bar{A}_{33}^{i3}\}, \cdots, \{\bar{A}_{11}^{ik}, \bar{A}_{33}^{i1}\}$ is a sequence of connected sets and $i1, i2, \cdots ik, i1$ is a cycle, If the cycle is odd ,the PFFM is not a quasi-Morishima;if there is no such an odd cycle,the PFFM is a quasi-Morishima.   □

Let R_1, R_2, R_3 denote the rows of A_{11}, A_{22}, A_{33} respectively, then the edge in $R_i (i = 1, 2, 3)$ is positive,the edge joining R_i with $R_j (i \neq j)$ is negative. We have an algorithm for RG^+

Algorithm 2

Step 1.   Find out the set of the connected sets of $RG^+(A_{ii})$, let it is. $CS(R_i), i = 1, 2, 3$

Step 2.   Find out the set of pair of connected sets of $CS(R_1), CS(R_2)$ not passing $R_3 : \{\{R_1^1, R_2^1\}, \{R_1^2, R_2^2\}, \cdots\}$, and the set of pair of connected sets of $CS(R_2), CS(R_3)$ not passing $R_1 : \{\{\bar{R}_2^1, \bar{R}_3^1\}, \{\bar{R}_2^2, \bar{R}_3^2\}, \cdots\}$. Without loss of generality, we can assume that $\{\bar{R}_2^1, \bar{R}_2^2, \cdots\}$ is a subsquence of $\{R_2^1, R_2^2, \cdots\}$.

Step 3.   if $\{\bar{R}_1^1, \bar{R}_1^2, \cdots\}, \{\bar{R}_3^1, \bar{R}_3^2, \cdots\}$ is not connected, the PFFM is a quasi-Morishima matrix; otherwise, suppose $\{\{\bar{R}_1^i, \bar{R}_3^j\}, \{\bar{R}_1^j, \bar{R}_3^k\}, \cdots, \{\bar{R}_1^h, \bar{R}_3^i\}\}$ is a sequence pair of connected sets not passing R_2 and $i, j, k, \cdots, i$ is a cycle, if it is an odd cycle, the PFFM is not a quasi-Morishima matrix; if there is no such an odd cycle, the PFFM is a quasi-Morishima matrix .   □

Theorem 7   *The algorithm 1 and 2 are the time complexity $O(e)$.*

References

1. Greenberg , H., Richard , J ., Maybee , J .:Rectangular matrices and signed graphs ,SIAM. J. ALG. DISC. Meth. 4(1983)50-61

2. Gondran, M .. Graphs and algorithms . John Wiley and Sons (1979) New York

3. Harary, F ., Normar , R ., Cartwright , D.:Structrual Models :An Introduction to the theory of Directed graphs . John Wiley and Sons (1965) .New York

4. Harary.F .:On the notion of balance of a signed graph. Michingan Math. J. 2(1953-54)143-146

Edge-Face Total Chromatic Number of Outerplanar Graphs with Δ (G) = 6

C. F. Chang[1], J. X. Chang[1], X. C. Lu[1], Peter C. B. Lam[2] and J. F. Wang[3]

[1] Lanzhou Railway Institute, 730070, China

[2] Dept. of. Math. Hong Kong Baptist University

[3] Ins. Appl. Math. Academic Sinica, Beijing 100080, China

Abstract. Let G be a planar graph without cut vertex, let $\chi_e(G)$ be the edge-face total chromatic number of G. This paper proves $\chi_e(G) = 6$ if G is an outerplanar graph with $\Delta(G) = 6$.

1 Introduction

In this paper, We consider only simple planar graph $G(V, E, F)$, where $V(G)$, $E(G)$ and $F(G)$ are the set of vertices, the set of edges and the set of faces in G. Two faces f_1, f_2 are said to be adjacent if and only if their boundaries contain at least one common edge. The edges at the boundary of a given face are called incident to the face.

Definition 1 The properly edge-face total colouring of a planar graph G is a colouring of its edges and faces in which any two adjacent or incident elements of $E(G) \bigcup F(G)$ are coloured by different colours, sometimes, which is, in short, also called the edge-face total colouring of G. If G can be coloured with k colours, such that each pair of adjacent or incident elements of $E(G) \bigcup F(G)$ receive different colours, then G is said to be a k-edge-face total colouring graph, and the corresponding colouring is called a k-edge-face total colouring; and

$$\chi_e(G) = \min\{k\}$$

is called the edge-face total chromatic number of G.

Definition 2[1]. A planar graph is said to be an outerplanar graph if its vertices are at the boundary of a given face f_0. The face f_0 is called the out face, other faces are called the inner face. The edges at the boundary of the face f_0 are called the out edge, other edges are called the inner edge.

In [2], the following conjecture:

$$\chi_e(G) \leqslant \Delta(G) + 3$$

is given and proved for any outerplanar graph G without cut vertex and with $\Delta(G) \geqslant 7$.

2 Main results

In this section, we consider only graph G without cut vertex.

Lemma 1. [3] Let G be an outerplanar graph with $\Delta(G) \geqslant 4$. Then at least one of the followring statements is true:

    (i) There are two adjacent vertices of degree 2;

    (ii) there are two adjacent vertices whose degrees are 2 and 3 respectively;

    (iii) there are two vertices of degree 2 adjacent to a vertex of degree 4.

Lemma 2. [3] Let G be an outerplanar graph with $\Delta(G) = 6$. If both the statements (i) and (ii) in lemma 1 do not hold, then at least one of the following statements is true:

    (iv) There exist two vertices, say u and v, of degree 2 adjacent to a vertex, say w, of degree 4, and at most one of u and v is adjacent to an other vertex of degree 6. If $y \in V(G)$, y is adjacent to u or v, $y \neq w$, $d(y) < 6$. Then y is adjacent to w.

    (v) there exist two vertices, say u and v, of degree 2 adjacent to a vertex, say w, of degree 4. If $w_1 \in V(G)$, $w_2 \in V(G)$, w_1 and w_2 are adjacent to u and v respectively, $d(w_1) = d(w_2) = 6$. Then w_1 is adjacent to w_2, and w_1 and w_2 are adjacent to w.

Theorem 1. Let G be an outerplanar graph with $\Delta(G) = 5$, then
$$5 \leqslant \chi_e(G) \leqslant 6$$
The proof of theorem 1 is similar to theorem 2 and it is omitted

Theorem 2. Let G be an outerplanar graph with $\Delta(G) = 6$, then
$$\chi_e(G) = 6$$

Proof. Apply induction on $p = |V(G)|$. First when $p = 7$, G is a fan graph F_7 and immediately $\chi_e(G) = 6$. We now assume that for any outerplanar graph H with $p(p \geqslant 8)$ vertices less than has $\chi_e(H) = 6$ and let G be an outerplanar graph with p vertices, $\Delta(G) = 6$. According to lemma 1 and lemma 2, we divide the discussion in to three cases:

Case 1. There are two adjacent vertices of degreey 2. In this case, we consider the graph G^* obtained by contracting two adjacent vertices, say u and v, of degree 2 to a new vertex and simultaneously removing the edge between u and v. Since $|V(G^*)| = p - 1$, $\Delta(G) = \Delta(G^*) = 6$, then it follows thus from the induction hypothesis that G^* has a 6-edge-face total colouring σ^*. On the basic of σ^* of G^*, a 6-edge-face total colouring σ of G can be easily constructed. Therefore, the conclusion is true.

Case 2. There is a vertex of degree 2 and adjacent to a vertex of degree 3. Let $d(u) = 2$, $d(v) = 3$, $ux, uv, vw_1, vw_2 \in E(G)$ and $x \neq v$, $w_1 \neq w_2$.

Subcase 2.1. If $vx \in E(G)$, i.e. $x = w_1 (or \ w_2)$, then consider the graph $G^* = G - u$.

Subcase 2.2. If $vx \notin E(G)$, then consider the graph $G^* = G - u + \{vx\}$.

Obviously, $|V(G^{\cdot})| = p - 1$ for the above two subcases, and $5 \leqslant \Delta(G^{\cdot})$ $\leqslant 6$ in subcase 2.1 and $\Delta(G^{\cdot}) = 6$ in subcase 2.2. By the induction hypothesis or theorem 1, $G^{\cdot}$ has a 6-edge-face total colouring $\sigma^{\cdot}$. We can form a 6-edge-face total colouring σ of G on the basic of $\sigma^{\cdot}$ of $G^{\cdot}$.

Consequently, the conclusion is true for case 2.

Case 3. There are two vertices of degree 2 adjacent to a vertex of degree 4. By lemma 2, consider the following two subcases:

Subcase 3.1. Assume $d(u) = 2, d(v) = 4, d(w) < 6, uv, uw, vw \in E(G)$. Consider the graph $G^{\cdot} = G - u$. Obviously, $|V(G^{\cdot})| = p - 1$, $\Delta(G^{\cdot}) = 6$. By the induction hypothesis, $G^{\cdot}$ has a 6-edge-face total colouring $\sigma^{\cdot}$ with the set B of colours. Then $|B| = 6$. Let f be the inner face of uvw, let $f^{\cdot}$ be the inner face in G adjacent to f and $f_0, f_0^{\cdot}$ the out face of $G, G^{\cdot}$ respectively. Moreover we denote by $\sigma(x)$ the set of colours with which the vertices and the edges are incident to x can be coloured under the given edge-face total colouring σ. We can form a 6-edge-face total colouring σ of G on the basic of $\sigma^{\cdot}$ of $G^{\cdot}$ as follows:

$\sigma(uw) \in B \backslash (\{\sigma^{\cdot}(w)\}, \sigma(f_0) = \sigma^{\cdot}(f_0^{\cdot}),$

$\sigma(uv) \in B \backslash (\{\sigma^{\cdot}(v)\} \cup \{\sigma(uw), \sigma(f_0)\}),$

$\sigma(f) \in B \backslash \{\sigma(f_0)\}, \sigma^{\cdot}(f^{\cdot}), \sigma(uv), \sigma(uw), \sigma^{\cdot}(vw)\}.$

Since $|B| = 6$, above colouring is feasible. The other uncoloured elements of G are coloured with the same colours as in $\sigma^{\cdot}$ of $G^{\cdot}$. It is easily seen that σ is a 6-edge-face total colouring of G and thus the conclusion is true for p

Subcase 3.2. If $d(u) = d(v) = 2, d(w) = 4, d(w_1) = d(w_2) = 6, uw_1, uw, vw,$ $vw_2, ww_1, ww_2, w_1w_2 \in E(G)$. In this case. the number of vertex is at least 12. Then consider the graph $G^{\cdot} = G - \{u, v, w\}$. Obviously, $|V(G^{\cdot})| < p$, $\Delta(G^{\cdot}) = 6$. By the induction hypothesis and subcase 3.1, $G^{\cdot}$ has a 6-edge-face total colouring $\sigma^{\cdot}$ with the set B of colours. Under $\sigma^{\cdot}$, we can make up a 6-edge-face total colouring σ of G as follows:

We use f_1, f_2, f_3 to denote the inner faces of G whose boundary contains edges ww_1 and ww_2, vw and vw_2, uw and uw_1 respectively.

Subcase 3.2.1. If $\sigma^{\cdot}(f_0^{\cdot}) \in \sigma^{\cdot}(w_1) \cap \sigma^{\cdot}(w_2)$, then

$\sigma(f_0) = \sigma^{\cdot}(f_0^{\cdot}), \sigma(ww_1) \in B \backslash \sigma^{\cdot}(w_1),$

$\sigma(ww_2) \in B \backslash (\sigma^{\cdot}(w_2) \cup \{\sigma(ww_1)\}),$

$\sigma(uw_1) \in B \backslash (\sigma^{\cdot}(w_1) \cup \{\sigma(ww_1)\}),$

$\sigma(vw_2) \in B \backslash (\sigma^{\cdot}(w_2) \cup \{\sigma(ww_2)\}),$

$\sigma(uw) \in B \backslash \{\sigma(f_0), \sigma(uw_1), \sigma(ww_1), \sigma(uw_2)\},$

$\sigma(vw) \in B \backslash \{\sigma(f_0), \sigma(uw), \sigma(ww_1), \sigma(ww_2), \sigma(vw_2)\},$

$\sigma(f_1) \in B \backslash \{\sigma^{\cdot}(f), \sigma^{\cdot}(w_1w_2), \sigma(ww_1), \sigma(ww_2)\},$

$\sigma(f_3) \in B \backslash \{\sigma(f_0), \sigma(f_1), \sigma(uw), \sigma(uw_1), \sigma(ww_1)\},$

$\sigma(f_2) \in B \backslash \{\sigma(f_0). \sigma(f_1), \sigma(vw), \sigma(vw_2), \sigma(ww_2)\}.$

Subcase 3. 2. 2. If $\sigma^*(f_0^*) \in \sigma^*(w_1)$, *and* $\sigma^*(f_0^*) \not\in \sigma^*(w_2)$ or $\sigma^*(f_0^*) \in \sigma(w_2)$, *and* $\sigma^*(f_0^*) \not\in \sigma^*(w_1)$. Assume $\sigma^*(f_0^*) \in \sigma^*(w_1)$, and $\sigma^*(f_0^*) \not\in \sigma^*(w_2)$, then

$$\sigma(f_0) = \sigma^*(f_0^*), \sigma(ww_1) \in B\backslash\sigma^*(w_1),$$
$$\sigma(uw_1) \in B\backslash(\sigma^*(w_1) \cup \{\sigma(ww_1)\}), \sigma(ww_2) = \sigma(f_0),$$
$$\sigma(vw_2) \in B\backslash(\sigma^*(w_2) \cup \{\sigma(f_0)\}),$$
$$\sigma(uw) \in B\backslash\{\sigma(uw_1), \{\sigma(ww_1), \sigma(f_0)\},$$
$$\sigma(vw) \in B\backslash\{\sigma(uw), \sigma(ww_1), \sigma(f_0), \sigma(vw_2)\},$$
$$\sigma(f_1) \in B\backslash\{(\sigma^*(f), \sigma^*(w_1w_2), \sigma(ww_1), \sigma(ww_2)\},$$
$$\sigma(f_3) \in B\backslash\{\sigma(f_0), \sigma(f_1), \sigma(uw), \sigma(uw_1), \sigma(ww_1)\},$$
$$\sigma(f_2) \in B\backslash\{\sigma(f_0), \sigma^*(f), \sigma(vw), \sigma(vw_2)\}.$$

If $\sigma^*(f_0^*) \in \sigma^*(w_2)$, *and* $\sigma^*(f_0^*) \not\in \sigma^*(w_1)$, the proof is similar.

Subcase 3. 2. 3. If $\sigma^*(f_0^*) \not\in \sigma^*(w_1) \cup \sigma^*(w_2)$, then

$$\sigma(f_0) = \sigma^*(f_0^*), \sigma(ww_1) = \sigma(f_0),$$
$$\sigma(uw_1) = \sigma(ww_2) = \sigma^*(w_1w_2),$$
$$\sigma(ww_1) \in B\backslash((\sigma^*(w_1) \cup \{\sigma(f_0)\}),$$
$$\sigma(vw_2) \in B\backslash(\sigma^*(w_2) \cup \{\sigma(f_0)\}),$$
$$\sigma(uw) \in B\backslash\{(\sigma(f_0), \sigma(uw_1), \sigma(ww_1), \sigma(ww_2)\},$$
$$\sigma(vw) \in B\backslash\{\sigma(f_0), \sigma(uw), \sigma(ww_1), \sigma(ww_2), \sigma(vw_2)\},$$
$$\sigma(f_1) \in B\backslash\{\sigma^*(f), \sigma(f_0), \sigma(ww_1), \sigma(ww_2)\},$$
$$\sigma(f_3) \in B\backslash\{\sigma(f_0), \sigma(f_1), \sigma(uw), \sigma(ww_1), \sigma(uw_1)\},$$
$$\sigma(f_2) \in B\backslash\{\sigma(f_0), \sigma(f_1), \sigma(vw), \sigma(ww_2), \sigma(vw_2)\}.$$

Since $|B| = 6$, the above colourings (subcase 3. 2. 1~subcase 3. 2. 3) are feasible and the other uncoloured elements of G are coloured with the same colours as in σ^* of G^*. Obviously, σ is a 6-edge-face total colouring of G and therefore the conclusion is true for p.

To sum up, we have proved the conclusion for p and consequently

$$\chi_e(G) = 6. \qquad\qquad \square$$

Acknowledgments: This paper project supported by the NNSF of China. GP-NSF and KMNSF.

References

1. G. Chartrand and L. lesniak, "Graphs and Digraph", Second edition, Wadswordth and Brooks/Cole, Monterey, Calif, 1986
2. Hu Guanzhang, Zhang zhongfu, The Edge-face total colouring of planar graphs, Journal of Tsinghua University, Vol 32, 3(1992),18~23
3. Zhang zhongfu e. t. c. The Complete chromatic number of some graphs, Science in China (Series A), Vol 36,10 (1993),1169~1177

Sets Computable in Polynomial Time on Average

Rainer Schuler[1] and Tomoyuki Yamakami[2]

[1] Abteilung Theoretische Informatik, Universität Ulm, D-89069 Ulm, Germany
[2] Department of Computer Science, University of Toronto, Toronto, Canada M5S 1A4

Abstract. In this paper, we discuss the complexity and properties of the sets which are computable in polynomial-time on average. This study is motivated by Levin's question of whether all sets in NP are solvable in polynomial-time on average for every reasonable (i.e., polynomial-time computable) distribution on the instances. Let $P_{\text{P-comp}}$ denote the class of all those sets which are computable in polynomial-time on average for every polynomial-time computable distribution on the instances. It is known that $P \subsetneq P_{\text{P-comp}} \subsetneq E$. In this paper, we show that $P_{\text{P-comp}}$ is not contained in $\text{DTIME}(2^{cn})$ for any constant c and that it lacks some basic structural properties: for example, it is not closed under many-one reducibility or for the existential operator. From these results, it follows that $P_{\text{P-comp}}$ contains P-immune sets but no P-bi-immune sets; it is not included in P/cn for any constant c; and it is different from most of the well-known complexity classes, such as UP, NP, BPP, and PP. Finally, we show that, relative to a random oracle, NP is not included in $P_{\text{P-comp}}$ and $P_{\text{P-comp}}$ is not in PSPACE with probability 1.

§1 Introduction. The theory of average-case complexity which was introduced by Levin [11] allows us to study the polynomial-time computability of NP problems in a more general setting. Several famous NP-complete problems have turned out to be easy on average with respect to some particular but natural distribution on the inputs [7].

Our main interest here is raised by the open question of whether some, or even all, NP-complete problems can be solved in polynomial-time on average with respect to every reasonably realistic probability distribution. Levin suggested that *polynomial-time computable* (P-computable, for short) distributions are reasonable (see [7]); he shows that a randomized version of the *bounded tiling problem* is complete for NP decision problems with P-computable distributions [11]. Later, a more general notion of *polynomially samplable* distributions has been proposed and studied in [4, 8]. In this paper, however, we restrict our attention to P-computable distributions.

A first partial answer to the above question – whether NP can be solved deterministically in polynomial-time on average – was given by Ben-David, et al. [4]. They showed that this situation would be unlikely since it would imply E = NE (its negation implies P ≠ NP). In this paper, we take a closer look at $P_{\text{P-comp}}$, the class of problems solvable in polynomial-time on average for every P-computable distribution. Our attempt is to embed $P_{\text{P-comp}}$ into the structure of well-known complexity classes with the hope that this helps answer questions like the one stated above, whether NP $\subseteq$ $P_{\text{P-comp}}$. To achieve our goal, we discuss the following two questions.

1. Does $P_{\text{P-comp}}$ contain complex sets ?
2. Does $P_{\text{P-comp}}$ have (or lack) structural properties ?

As the first step, to answer Question 1, we look at the worst-case time-complexity of sets in $P_{P\text{-comp}}$. We are able to show that $P_{P\text{-comp}}$ is not contained in $\text{Dtime}(2^{cn})$ for any fixed constant c. The idea is to define a set A such that A contains infinitely many strings which occur with very low probability, and $A \cap B$ is computable in average polynomial-time for any set B chosen from E. Since all elements of A occur with very low probability, we can diagonalize (on the elements of A), in average polynomial-time, against all sets in $\text{Dtime}(2^{cn})$. Note that this is an optimal bound since it is known that $P_{P\text{-comp}}$ is contained in E [16].

The second example of complex sets considered in this paper is that of immune and bi-immune sets. An infinite set is P-*immune* if it contains no infinite subset in P. We show that $P_{P\text{-comp}}$ contains P-immune sets but no P-bi-immune sets. This contrasts $P_{P\text{-comp}}$ with the class E that has both P-immune and P-bi-immune sets. A *(polynomial) complexity core* for a set A is a collection of strings that are hard to compute independent of the algorithm chosen to compute A. Thus the density of a complexity core gives a measure on the number of hard instances in a set. Since every P-immune set is a complexity core for itself, and the P-immune set constructed here is non-sparse, it follows that there are sets in $P_{P\text{-comp}}$ that have non-sparse complexity cores.

From those results, it follows, in particular, that $P_{P\text{-comp}}$ is neither P-close nor contained in APT, where APT is the class of "almost polynomial-time" sets defined by Meyer and Paterson [13]. Although the question of whether $P_{P\text{-comp}}$ is contained in P/poly is left open, the positive answer to this question is of special interest since it would imply that $PH \subseteq \Sigma_2^p$ under the assumption that $NP \subseteq P_{P\text{-comp}}$ and thus strengthen the result of [4] that $NP \subseteq P_{P\text{-comp}}$ implies that $E = NE$. The best result we show here is that $P_{P\text{-comp}}$ is not included in P/cn for every constant c.

It is not difficult to see that all P-*printable* sets in $P_{P\text{-comp}}$ are also in P. Since (as mentioned above) $P_{P\text{-comp}}$ is not contained in $\text{DTIME}(2^{cn})$, it seems, in general, hard to embed $P_{P\text{-comp}}$ into the known world of complexity classes; nevertheless, we can show that $P_{P\text{-comp}}$ is "different" from most of the well-known complexity classes. Along with this direction, we discuss Question 2. It turns out that $P_{P\text{-comp}}$ lacks some basic structural properties: for example, it is not closed under (polynomial-time) many-one reductions which immediately implies that $P_{P\text{-comp}}$ is different from UP, NP, BPP, and PP. Furthermore, $P_{P\text{-comp}}$ is not closed under the existential operator nor under similarly defined unique existential and probabilistic operators. Note that it is still open whether $P_{P\text{-comp}}$ has a complete problem under many-one reducibility.

Since $P_{P\text{-comp}}$ is not closed under polynomial-time many-one reductions, we propose honest many-one "increasing" reduction functions and observe that $P_{P\text{-comp}}$ is closed under those reductions. Note that all NP-complete problems which allow padding remain NP-complete under this reducibility. Thus, if such an NP-complete problem is in $P_{P\text{-comp}}$, then it follows that $NP \subseteq P_{P\text{-comp}}$.

Finally, using the results of Bennett and Gill [5], we show that, relative to a random oracle X, $NP^X \not\subseteq P_{P\text{-comp}}^X$ and $P_{P\text{-comp}}^X \not\subseteq PSPACE^X$ with probability 1.

§2 Preliminaries. In this paper, we follow the standard notations and definitions of computational complexity theory (see e.g. [1]). Let $\Sigma = \{0, 1\}$ be fixed and let $s\Sigma^n$ denote the set $\{sy \mid y \in \Sigma^n\}$. Let $x < y$ denote that a string x is smaller than a string y in the standard lexicographic order on Σ^*. For every set $A \subseteq \Sigma^*$, let $A^n = A \cap \Sigma^n$. A set of strings is identified with its *characteristic function*. We use a standard pairing function on strings, denoted by $\langle x, y \rangle$, which is polynomial-time computable and invertible with the property that $\langle x, y \rangle < \langle x', y' \rangle$ if either $x < x'$ and $|y| \le |y'|$ or $x = x'$ and $y < y'$. Denote by $\mathbb{N}$ and $\mathbb{R}^+$, respectively, the sets of nonnegative integers and nonnegative real numbers. We often identify a nonempty string $a_0 a_1 \cdots a_k$, with $a_i \in \{0, 1\}$, with a real number $\sum_{i \le k} a_i \cdot 2^{-i}$. In particular, the empty string, λ, represents 0. Hence, a Turing machine with one input tape and one output tape (called a transducer) is identified with a function from strings to the unit interval $[0, 1]$. We use $\log^k n$ to denote $(\log_2 n)^k$ for $k > 0$ and define $\log^* n = \min\{i \mid \log^i n \le 1\}$.

For a Turing machine M, let $\mathrm{Time}_M(x)$ denote the time used by M on input x. We use E to denote $\mathrm{DTIME}(2^{\text{linear}})$ and NE to denote $\mathrm{NTIME}(2^{\text{linear}})$.

A *tally* set is a subset of $\{0\}^*$, and a set S is *sparse* if, for some fixed polynomial q and for all n, $|S^n| \le q(n)$. A set S is *P-printable* if a P-computable function, on input 0^n, produces a list of all elements of A^n.

A function f is *increasing* if $x < y$ implies $f(x) \le f(x)$ for any x, y in the domain of f, and a *strictly increasing* function is defined by replacing the condition $f(x) \le f(y)$ by $f(x) < f(y)$. A function f is *unbounded* if, for all x, there exists a y such that $x < y$ and $f(x) < f(y)$.

Let $A \le_m^p B$ denote that A is *polynomial-time many-one reducible* (many-one reducible, for short) to a set B. If the reduction function is *(polynomially) honest*, i.e., for some fixed k and all x, $|f(x)|^k \ge |x|$, then A is said to be *honest (polynomial-time) many-one reducible* to B, denoted by $A \le_m^{hp} B$.

A *(probability) density function* μ is a function from Σ^* to $[0, 1]$ such that, for all x, $\mu(x) \ge 0$, and $\sum_x \mu(x) = 1$. The *(probability) distribution* μ^* of μ is defined by $\mu^*(x) = \sum_{y \le x} \mu(y)$. For a set A, let $\mu(A) = \sum_{x \in A} \mu(x)$. Note that $\mu(x) = \mu^*(x) - \mu^*(x^-)$, where x^- is the predecessor of x. Similarly, for every set $A = \{sy \mid y \in \Sigma^{n+1}\}$, it holds that $\mu(A) = \mu^*(s1^{n+1}) - \mu^*(s1^n)$.

A function g from Σ^* to $\mathbb{R}^+$ is *weakly polynomial-time computable* [9] if there exists a deterministic polynomial-time Turing machine M such that $|g(x) - M(x, 1^i)| \le 2^{-i}$ for all $x \in \Sigma^*$ and all $i \in \mathbb{N}$. Let P-comp be the class of density functions whose *distributions* are weakly polynomial-time computable. The class of exponential-time computable distributions, E-comp, is defined similarly.

Now we give Levin's definition of "polynomial-time on average." Note, however, that there are more obvious definitions of an average-case complexity measure, for example, $\sum_{x:|x|=n} \mathrm{Time}_M(x) \cdot \mu_n(x) \le n^k$ for all $n \in \mathbb{N}$, where $\mu_n(x) = \mu(x)/\sum_{z:|z|=n} \mu(z)$ and k is a positive constant. As pointed out by Levin [11] and discussed in [7, 4], those definitions do not have several important properties, such as robustness under functional composition and machine-independence.

Definition 1. [11] A Turing machine M is *polynomial-time on μ-average* if there exists a real number $\delta > 0$ such that $\sum_{x:x \ne \lambda} |x|^{-1} \cdot \mathrm{Time}_M(x)^\delta \cdot \mu(x) < \infty$.

In this paper, we consider a problem to be "efficiently solvable" if, for every P-computable distribution μ, then the problem is solved by a Turing machine which is polynomial-time on μ-average.

Definition 2. [16] Let $P_{P\text{-comp}}$ be the collection of all sets S such that, for every density function $\mu \in P\text{-comp}$, S is computed by a deterministic Turing machine in polynomial-time on μ-average. Similarly $P_{E\text{-comp}}$ is defined.

Note that there exist exponential-time computable distributions such that "average polynomial-time" collapses to "worst-case polynomial-time"; namely, $P_{E\text{-comp}} = P$ [16].

Ben-David, et al. [4] have first established a relationship between well-known worst-case complexity classes and average-case classes by proving (in our notation) that E = NE if and only if TALLY $\cap$ NP $\subseteq P_{P\text{-comp}}$. This follows from the fact that TALLY $\cap P_{P\text{-comp}} \subseteq P$. This property of TALLY sets can be extended to P-printable sets.

Proposition 3. *Let S be a P-printable set. Then, for every set A, if A is in $P_{P\text{-comp}}$, then $A \cap S$ is in P.*

Since the class P-comp is a broad class of real-valued density functions from Σ^* to $[0, 1]$, we may not enumerate all those functions effectively. Instead, here we consider a restricted concept of polynomial-time "approximation" of these density functions. Recall that we identify a machine T with one input tape and one output tape with a function from Σ^* to $[0, 1]$. For this T, let T^* denote the associated function of T defined as $T^*(x) = \sum_{z \leq x} T(z)$ for each $x \in \Sigma^*$.

Definition 4. A deterministic Turing machine T with one input tape and one output tape *approximates a density function* if T^* is polynomial-time bounded and $\sum_x T(x) \leq 1$. For a density function μ, T *approximates* μ if T approximates a density function and $|\mu(x) - T(x)| \leq 2^{-|x|}$ for all x.

There exists an effective enumeration of all deterministic polynomial-time Turing machines which approximate density functions such that, for all i, T_i^* is $n^i + i$ time bounded [15]: to get this enumeration, first enumerate all polynomial-time deterministic Turing machines $M_i's$ and modify them to satisfy the conditions that M_i is increasing and does not exceed 1, and then compute $M_i(x) - M_i(x^-)$.

Lemma 5. *A set S is in $P_{P\text{-comp}}$ if, for every deterministic Turing machine T which approximates a density function, there exists a deterministic Turing machine M such that M computes S in polynomial-time on T-average.*

§3 Pruning strings which occur with high probability.

In [15], a set is defined which is in $P_{P\text{-comp}}$ but not in P. A crucial idea used in the construction is to select one string of each length n which occurs with very low probability for all the "first $\log n$" P-computable distributions so that we can diagonalize on the string against any set in P in average polynomial time with respect to those distributions. This is done by partitioning Σ^n recursively into two parts and selecting one part which has the smaller probability. Thus, any potentially long computation is pruned on strings which (might) occur with high probability in early stages of the computation.

Here we refine this algorithm and construct a set in a slightly more general manner and prove that a set accepted by the algorithm belongs to $P_{P\text{-comp}}$. Furthermore, we show a tradeoff between the number of hard instances and the hardness of the instances.

Let $T_1, T_2, \ldots$ be an effective enumeration of all deterministic polynomial-time Turing machines which approximate density functions such that each T_i^* is $n^i + i$ time bounded. Let t and k be any time-constructible functions on $\mathbb{N}$ such that k is increasing and unbounded, $k(n) \leq \log n$, and $t(n) + k(n)^2 \log n \leq n$. Let $l(n) = n - k(n)^2 \log n - t(n) + (k(n) - 1) \log n - \log k(n)$. For every set $T \subseteq \Sigma^*$, define A_T to be the set which is accepted by the following algorithm:

> **input** x (let $n = |x|$)
> $y = \lambda$
> **for** $i = 1$ **to** $k(n)$ **do**
>   **repeat** $\lfloor k(n) \log n \rfloor$ **times**
>   let $L = y0\Sigma^{n-|y|-1}$ and $R = y1\Sigma^{n-|y|-1}$
>   **if** $T_i(L) < T_i(R)$ **then** $y = y0$ **else** $y = y1$
>   **if** y is not a prefix of x **then** reject
>   **repeat** $n - k(n)\lfloor k(n) \log n \rfloor - t(n)$ **times**
>   let $L = y0\Sigma^{n-|y|-1}$ and $R = y1\Sigma^{n-|y|-1}$
>   **if** $\sum_{i=1}^{k(n)} T_i(L) < \sum_{i=1}^{k(n)} T_i(R)$ **then** $y = y0$ **else** $y = y1$
>   **if** y is not a prefix of x **then** reject
>   **if** $x \in T$ **then** accept **else** reject

The strings left unrejected after the second repeat loop are called *candidates* (or *candidate strings*). The first for loop is referred to as Phase I, the second repeat loop as Phase II, and the last if line as Phase III.

Lemma 6. *Let A_T be the set defined above. If $T \in \mathrm{DTIME}(2^{c \cdot l(n)})$ for some constant $c > 0$, then $A_T \in P_{P\text{-comp}}$. In particular, if $T = \Sigma^*$, then $A_T \in \mathrm{DTIME}(n^{k(n)+2})$ and $|A_T \cap \Sigma^n| = 2^{t(n)}$ for all but finitely many n.*

Proof. For simplicity, let $s(n) = \lfloor k(n) \log n \rfloor$ and $q(n) = n - k(n)s(n) - t(n)$. Now let μ be any density function in P-comp. Since $\mu \in$ P-comp, there exists a $k \in \mathbb{N}$ such that T_k approximates μ. Take any sufficiently large integer n. It suffices to show that the average computation time, over all strings of length n, $\sum_{x:|x|=n} |x|^{-1} \cdot \mathrm{Time}_{A_T}(x)^{1/(c+k+2)} \cdot T_k(x)$ is at most $\leq c \cdot n^{-2}$ for some nonnegative constant c independent of n.

Phase I. Let $C_0, C_1, \ldots, C_{k(n)}$ be the subsets of Σ^n such that each C_i, $i \leq k(n)$, is the set of strings which remain unrejected after the ith transducer T_i is considered in the for loop. We have $T_k(C_i) \leq 2^{-s(n)}$ for all $i > k$. This is seen as follows: in the beginning, probability $T_k(C_{i-1})$ is smaller than or equal to 1, and in each iteration of the first repeat loop, the computation continues on either set L or set R, whichever has the smaller probability, and therefore the probability is reduced to at most half of its previous value. This gives the desired result after $s(n)$ iterations.

For the case that $i \leq k$, the average computation time required on all strings through Phase I is $O(T_k(\Sigma^n))$ since the time spent on input x is at most $O(n^k)$.

Next consider the case $i > k$. For each $x \in C_{i-1} - C_i$, the time spent on input x after i iterations of the for loop is $O(k(n)s(n)n^i)$. Hence, the average time over all strings through Phase I is $k(n) \cdot n^{-1} \cdot O(k(n)s(n)n^{k(n)})^{1/2} \cdot 2^{-s(n)} \subseteq O(n^{-2})$.

Phase II. Let $D_0, D_1, \ldots, D_{q(n)}$ be the subsets of $C_{k(n)}$ such that each D_i, $i \leq q(n)$, is the set of strings which remain unrejected after j iterations of the second repeat loop. As seen in Phase I, we have $\sum_{i=1}^{k(n)} T_i(D_j) \leq 2^{-j} \sum_{i=1}^{k(n)} T_i(C_{k(n)})$ for all $j \leq q(n)$; thus, $T_k(D_j) \leq k(n)2^{-s(n)-q(n)}$.

For all $x \in C_{k(n)}$, we need at most $O(k(n)q(n)n^{k(n)})$ steps to go through Phase II. Hence, the average computation time over all strings rejected in Phase II is $q(n) \cdot n^{-1} \cdot O(k(n)q(n)n^{k(n)})^{1/2} \cdot k(n)2^{-s(n)-q(n)} \subseteq O(n^{-2})$ since $q(n) \leq n$ and $k(n) \leq \log n$.

Phase III. Let C_T be the set of candidates. Note that $|C_T| = 2^{t(n)}$. The computation in Phase III is done in time $O(2^{c \cdot l(n)})$, but $T_k(C_T) \leq k(n)2^{-s(n)-q(n)} \leq n^{-1}2^{-l(n)}$. So, the average computation time over all candidates of length n is $n^{-1} \cdot O(2^{c \cdot l(n)})^{1/c} \cdot n^{-1}2^{-l(n)} \subseteq O(n^{-2})$.

If $T = \Sigma^*$, A_T is computed, through Phase I and II, in time $O(k(n)s(n)n^{k(n)}) + O(k(n)q(n)n^{k(n)}) \subseteq O(n^{k(n)+2})$ since $s(n) \leq n$, $q(n) \leq n$, and $k(n) \leq n$.

Choosing functions k and t carefully ($k(n) = \log^* n$ and $t(n) = \lfloor \log p(n) \rfloor$ below), we establish our main theorem which allows us to show that $P_{\text{P-comp}}$ contains sets which are hard to compute.

Theorem 7. *Let p be any function on $\mathbb{N}$ such that $\lambda x.p(|x|)$ is polynomial-time computable and $\log p(n) \in o(n)$. Then, there exists a set A in $\text{DTIME}(n^{2 \log^* n})$ such that $|A^n| = p(n)$ for all n, and $A \cap B \in P_{\text{P-comp}}$ for every set $B \in E$.*

§4 Hard sets in $P_{\text{P-comp}}$. We now apply Theorem 7 to discuss Question 1 stated in Section 1. The first application is to show that $P_{\text{P-comp}}$ contains sets with very high time-bounded complexity. The second application shows that $P_{\text{P-comp}}$ contains sets which are not even $\text{DTIME}(2^{cn})$-close, where a set S is C-close for a class C if there exists a set $B \in C$ such that $A \triangle B$ (i.e., the symmetric difference of A and B) is sparse.

Corollary 8. *For every $c > 0$, $\text{SPARSE} \cap P_{\text{P-comp}} \not\subseteq \text{DTIME}(2^{c \cdot n})$, and $P_{\text{P-comp}}$ is not $\text{DTIME}(2^{c \cdot n})$-close.*

From Proposition 3, it follows that P-printable $\cap P_{\text{P-comp}} \subseteq P$; however, Corollary 8 implies that $\text{SPARSE} \cap P_{\text{P-comp}} \not\subseteq P$.

We denote by NT the collection of *near-testable* sets, where a set is *near-testable* if there is an $f \in \text{FP}$ such that $f(x) \equiv S(x) + S(x^-) \pmod 2$ [6]. Take a set A, using Theorem 7, such that A is in $P_{\text{P-comp}} - \text{DTIME}(2^n)$ and $|A^n| \leq 1$ for all $n > 0$. It is not difficult to see that A is not near-testable. Hence we get:

Proposition 9. $P_{\text{P-comp}} \not\subseteq \text{NT}$.

Another corollary of Theorem 7 gives a relationship between $P_{\text{P-comp}}$ and non-uniform complexity classes defined by advice functions. For a function f from $\mathbb{N}$ to Σ^*, a set S is in $P/f(n)$ if there is a set $B \in P$ such that $S^n = \{x \in \Sigma^n | \langle x, f(n) \rangle \in B\}$ for each $n \in \mathbb{N}$.

Corollary 10. For every $c > 0$, $P_{P\text{-comp}} \not\subseteq P/cn$.

Proof. Take $p(n) = n^2$ and let A be the set defined in Theorem 7. Assume that $A^n = \{x_1^n, \ldots, x_{n^2}^n\}$ for each n. Let $\{M_i\}_{i \in \mathbb{N}}$ be an effective enumeration of all polynomial-time deterministic Turing machines (with repetition) such that M_i is $n^i + i$ time-bounded. Let $acc(D, x)$ (resp. $rej(D, x)$) be the collection of strings z in D such that $M_{\lfloor \log |x| \rfloor}$ accepts (resp. rejects) $\langle x, z \rangle$. Define B to be the set accepted by the following algorithm: on input x, $x = x_k^n$, in A, let $D_0 = \Sigma^{cn}$; for each $i < k$, recursively choose either set, $acc(D_i, x_i^n)$ or $rej(D_i, x_i^n)$, whichever has the smaller cardinality (if both cardinalities are the same, then choose $acc(D_i, x_i^n)$); let D_{i+1} be this set; and accept the input if and only if $|acc(D_k, x)| \leq |rej(D_k, x)|$. This set B ($\subseteq A$) is in $P_{P\text{-comp}}$ but not in P/cn.

However, we do not know whether $P_{P\text{-comp}} \subseteq P/poly$, where $P/poly = \bigcup_{k \in \mathbb{N}} P/n^k$.

Immune sets are another typical example of sets which are hard to compute. A set S is P-*immune* if S is infinite and S has no infinite subsets in P, and S is P-*bi-immune* if S and its complement $\overline{S}$ are both P-immune. Note that E has P-bi-immune sets; moreover, E has strongly P-bi-immune sets [2].

Theorem 11. There exists a non-sparse P-immune set in $P_{P\text{-comp}}$.

Proof. We use a technique by Ko and Moore [10]. Let $\{M_i\}_{i \in \mathbb{N}}$ be the enumeration of machines that is used in the proof of Corollary 10. Take a set $A \in \text{DTIME}(n^{2 \log^* n})$ in Theorem 7 such that $|A^n| = 2^{\lfloor n/\log n \rfloor}$ and $A \cap B \in P_{P\text{-comp}}$ for every set B in E. Now consider the following algorithm: on input x in A, if there exists a nonnegative integer $i \leq \log |x|$ such that $x \in L(M_i)$ and $z \notin L(M_i)$ for all z with $0^{2^i} \leq z < x$, then accept x, or else reject x. Let B denote the set accepted by this algorithm. Note that $B \in E$ and $|B^n| \leq \log n$. Let $C = A - B$. It is not hard to see that B and C are both in $P_{P\text{-comp}}$. Since $|A^n| = 2^{\lfloor n/\log n \rfloor}$, we have $|C^n| \geq 2^{\lfloor n/\log n \rfloor} - \log n$ for all n. Hence, C is infinite. In particular, C is non-sparse. The P-immunity of C is easily seen as in [10].

For a recursive set S, a set C is called a *polynomial complexity core for S* if, for any deterministic Turing machine M computing S and any polynomial p, the set $\{x \in C | \text{Time}_M(x) \leq p(|x|)\}$ is finite. Note that any P-immune set is a polynomial complexity core for itself, and a recursive set S is P-bi-immune if and only if Σ^* is a polynomial complexity core for S.

A set A is in APT [13] if there exists a polynomial p and a deterministic Turing machine M which accepts A such that the set $\{x \mid \text{Time}_M(x) \geq p(|x|)\}$ is sparse. Since a recursive set S is in APT if and only if any polynomial complexity core for S is sparse [14], we immediately obtain the following corollary of Theorem 11.

Corollary 12. $P_{P\text{-comp}} \not\subseteq \text{APT}$.

Since E contains P-bi-immune sets, E contains a set for which Σ^* is a complexity core. However, as stated in the next proposition, there are no P-bi-immune sets in $P_{P\text{-comp}}$. This follows from the fact, by Proposition 3, that every set in $P_{P\text{-comp}}$ is easy on every P-printable subset.

Proposition 13. *There are no P-bi-immune sets in* $P_{\text{P-comp}}$.

Therefore, if $\text{NP} \subseteq P_{\text{P-comp}}$, then NP has Lebesgue measure 0 in E since the class of non-P-bi-immune sets has Lebesgue measure 0 in E [12].

A similar argument used in Proposition 13 yields that no sets in $P_{\text{P-comp}}$ has an infinite polynomial complexity core in P unless $P \neq \text{NP}$.

§5 Structural properties of $P_{\text{P-comp}}$. This section considers the structural properties of $P_{\text{P-comp}}$. We show that $P_{\text{P-comp}}$ is not closed downward under many-one reducibility and not closed under the existential, the unique existential, or a probabilistic operators. The lack of these properties indicates a structural difference between $P_{\text{P-comp}}$ and many other complexity classes, such as P, UP, NP, BPP, and PP. Finally we show that, relative to a random oracle, neither $P_{\text{P-comp}}$ is contained in PSPACE nor PSPACE is contained in $P_{\text{P-comp}}$ with probability 1.

Theorem 14. $P_{\text{P-comp}}$ *is not closed under many-one reducibility.*

Proof. Choose a set A, using Theorem 7, and a set B such that $A \cap B$ is in $P_{\text{P-comp}}$ but not in $\text{DTIME}(2^n)$, and $|A^n| = 1$ if $n = k^{\log k}$ for some k, and $|A^n| = 0$ otherwise. Recall that the candidates of length n are computed in time $O(n^{2 \log^* n})$.

Let $T = \{0^k | f(0^k) \in A \cap B\}$, where the reduction function f is defined as follows: $f(0^k)$ is the candidate of length n (i.e., the element of A^n) if $k = n^{\log n}$ for some n, and 0 otherwise. Note that f is not honest.

Clearly $T \leq^p_m A \cap B$ via f. Note that f is computable in time $O(n)$. It remains to show that $T \notin P_{\text{P-comp}}$. Assume otherwise. Since T is tally, T belongs to P. Then, $A \cap B$ can be decided as follows: on input x, $|x| = n$, accept x if and only if $x \in A$ and $0^k \in T$, where $k = n^{\log n}$. This concludes that $A \cap B$ is in $\text{DTIME}(2^n)$, a contradiction. Hence $T \notin P_{\text{P-comp}}$. This completes the proof.

Corollary 15. $P_{\text{P-comp}}$ *is different from* UP, NP, BPP, *and* PP.

Unfortunately, it is not known whether $P_{\text{P-comp}}$ is closed under p-isomorphism, i.e., $A \cong^P B$ and $B \in P_{\text{P-comp}}$ implies $A \in P_{\text{P-comp}}$ (see [3]). It is possible, however, to show the existence of an *incomparable pair* in $P_{\text{P-comp}}$ with respect to the honest many-one reducibility.

Theorem 16. *There exist sets* $A, B \in P_{\text{P-comp}}$ *such that* $A \not\leq^{hp}_m B$ *and* $B \not\leq^{hp}_m A$.

Proof. Recall that there is a P-immune set in $P_{\text{P-comp}}$. Let A be such a set, and let $B = \{0\}^*$. Note that $A \notin P$. If $A \leq^{hp}_m B$, then $A \in P$ since $B \in P$. If $B \leq^{hp}_m A$, then B should be P-immune since the set of P-immune sets is closed downward under honest many-one reducibility. Both cases induce contradictions.

Assume that a many-one reduction f is honest and increasing. Now, if a density function μ is in P-comp, then the induced function ν, where $\nu(x) = \mu(f^{-1}(x))$, is also in P-comp since $\nu^*(x) = \mu^*(y)$ holds for y the largest string such that $f(y) \leq x$, and thus ν^* is P-computable by binary search. Consequently, we get the following proposition (see also [3]).

Proposition 17. P$_{\text{P-comp}}$ *is closed under honest many-one increasing reductions.*

The following result points out the difference between P$_{\text{P-comp}}$ and most of the well-known complexity classes by demonstrating that P$_{\text{P-comp}}$ is not closed under many-one reductions. Here we discuss the following two types of existential operators and a probabilistic operator, and show that P$_{\text{P-comp}}$ is not closed under such operators.

Definition 18. Let C be any complexity class. (1) A set S is in $\exists \cdot C$ if there exists a polynomial p and a set $B \in C$ such that $S = \{x \mid \exists y[|y| = p(|x|) \wedge \langle x, y \rangle \in B]\}$. (2) A set S is in U$\cdot C$ if S is in $\exists \cdot C$ via B and p, and, in addition, for all $x \in S$, there exists a unique $y \in \Sigma^{p(|x|)}$ such that $\langle x, y \rangle \in B$. (3) A set S is in P$\cdot C$ if there is a polynomial p and a set $B \in C$ such that $S = \{x \mid \#\{y \in \Sigma^{p(|x|)} \mid \langle x, y \rangle \in B\} > \frac{1}{2} \cdot 2^{p(|x|)}\}$.

Note that $\exists \cdot \text{NP} = \text{NP}$, $\text{U} \cdot \text{UP} = \text{UP}$, and $\text{P} \cdot \text{PP} = \text{PP}$.
For Theorem 20, we need the following lemma.

Lemma 19. *For every $A \in$ P$_{\text{P-comp}}$ and every polynomial p, the set $B = \{\langle 0^n, x \rangle \mid n = p(|x|) \wedge x \in A\}$ is also in P$_{\text{P-comp}}$.*

Theorem 20. $\text{U} \cdot$ P$_{\text{P-comp}} \nsubseteq$ P$_{\text{P-comp}}$ *and* $\text{P} \cdot$ P$_{\text{P-comp}} \nsubseteq$ P$_{\text{P-comp}}$.

Proof. We prove only the first claim. Assume that $\text{U} \cdot$ P$_{\text{P-comp}} \subseteq$ P$_{\text{P-comp}}$. Let A be a set in P$_{\text{P-comp}} - \text{DTIME}(2^n)$, using Theorem 7, which has the following property: for each length n, A contains at most one candidate which is computed in time $O(n^{2 \log^* n})$. Now define $B = \{\langle 0^n, x \rangle \mid n = p(|x|) \wedge x \in A\}$. By Lemma 19, we have $B \in$ P$_{\text{P-comp}}$. Let $C = \{0^n \mid \exists y[|y| = n \wedge \langle 0^n, y \rangle \in A]\}$. Clearly $C \in \text{U} \cdot$ P$_{\text{P-comp}}$.

By our assumption, the set C is also in P$_{\text{P-comp}}$. Since C is tally, there exists a deterministic polynomial-time Turing machine M which computes C. Then, A can be decided as follows: on input x, $|x| = n$, accept x if and only if $0^n \in L(M)$ and $x \in A$. This shows that A is in $\text{DTIME}(2^n)$; this is a contradiction.

As an immediate consequence of Theorem 20, we have $\exists \cdot$ P$_{\text{P-comp}} \nsubseteq$ P$_{\text{P-comp}}$, and thus P$_{\text{P-comp}} \neq \text{NP}$ and P$_{\text{P-comp}} \neq \text{UP}$. Similarly, we have P$_{\text{P-comp}} \neq \text{PP}$.

We note, however, that no *inclusion* relationship is known between the P$_{\text{P-comp}}$ and those complexity classes. At the end of this section, we give evidence of a negative answer to the question of whether $\text{NP} \subseteq$ P$_{\text{P-comp}}$ using random oracles.

Definition 21. Let P$^X_{\text{P-comp}}$ denote the collection of sets A such that, for every density function μ in P-comp, there exists a deterministic oracle Turing machine M such that M^X is polynomial-time on μ-average and $A = L(M, X)$.

Theorem 22. *For a random oracle X, $\text{NP}^X \nsubseteq$ P$^X_{\text{P-comp}}$ and P$^X_{\text{P-comp}} \nsubseteq \text{PSPACE}^X$ with probability 1.*

Proof. Note that Proposition 13 can be relativized for any oracle; however, it is known that there exists P-bi-immune sets in NP for a random oracle [5]. Hence, we reach the conclusion that $\text{NP}^X \nsubseteq$ P$^X_{\text{P-comp}}$ for a random oracle X.

For the second claim, recall that we identify a string with a nonnegative integer. Let $\eta_A(x) = A(x10)A(x10^2) \cdots A(x10^{|x|})$ for a set A and consider the set $\{w|w$ is a candidate, $\eta_X(0^w) \in X\}$. This set belongs to $P_{\text{P-comp}}$ for any oracle X, but, by the same argument used in [5], it does not belong to PSPACE^X for a random oracle X.

§6 Conclusions. Towards the NP versus $P_{\text{P-comp}}$ question, we have studied complex sets and structural properties of $P_{\text{P-comp}}$. Many hardness results known for E hold for $P_{\text{P-comp}}$; however, $P_{\text{P-comp}}$ has several structural properties different from E; e.g., P-bi-immunity. We have also shown that sets in $P_{\text{P-comp}}$ are polynomial-time computable on P-printable subsets, but there exist sparse sets in $P_{\text{P-comp}}$ that are not in $\text{DTIME}(2^{cn})$ for each constant c. However, there are many open questions left open: is $P_{\text{P-comp}}$ contained in P/poly ?; is $P_{\text{P-comp}}$ in $\oplus$P ?; are all sets in $P_{\text{P-comp}} - P$ almost P-immune ?; and does $P_{\text{P-comp}}$ have complete sets for E ? Elsewhere [16], we have defined the nondeterministic class $\text{NP}_{\text{P-comp}}$. The properties shown in Lemma 6, Theorem 7, and Corollary 8 hold for $\text{NP}_{\text{P-comp}}$ as well.

References

1. J.L. Balcázar, J. Díaz, and J. Gabarró, *Structural complexity I+II*, Springer, Berlin, 1989(I), 1990(II).
2. J.L Balcázar and U. Schöning, *Bi-immune sets for complexity classes*, Math. Systems Theory, 18 (1985), pp.1–10.
3. J. Belanger and J. Wang, *Isomorphisms of NP complete problems on random instances*, in Proc., 8th Conf. on Structure in Complexity Theory, 1993, pp.65–74.
4. S. Ben-David, B. Chor, O. Goldreich, and M. Luby, *On the theory of average case complexity*, J. Comput. System Sci., 44 (1992), pp.193–219.
5. C.H. Bennett and J. Gill, *Relative to a random oracle A, $P^A \neq NP^A \neq$ co-NP^A with probability 1*, SIAM J. Comput., 10 (1981), pp.96–113.
6. J. Goldsmith, L.A. Hemachandra, D. Joseph, and P. Young, *Near-testable sets*, SIAM. J. Comput., 20 (1991), pp.506–523.
7. Y. Gurevich, *Average case complexity*, J. Comput. System Sci., 42 (1991), pp.346–398.
8. R. Impagliazzo and L. Levin, *No better ways to generate hard NP instances than picking uniformly at random*, in Proc., 31st FOCS, 1990, pp.812–821.
9. K.I. Ko and H. Friedman, *Computational complexity of real functions*, Theoret. Comput. Sci., 20 (1982), pp.323–352.
10. K.I. Ko and D. Moore, *Completeness, approximation and density*, SIAM J. Comput., 10 (1981), pp.787–796.
11. L. Levin, *Average case complete problems*, SIAM J. Comput., 15 (1986), pp.285–286. A preliminary version appeared in Proc., 16th STOC, 1984, p.465.
12. E. Mayordomo, *Almost every set in exponential time is P-bi-immune*, in Proc., 17th MFCS, LNCS, Vol.629, 1992, pp.392–400.
13. A.R. Meyer and M. Paterson, *With what frequency are apparently intractable problems difficult*, Technical Report TM-126, MIT, 1979.
14. P. Orponen and U. Schöning, *On the density and complexity of polynomial cores for intractable sets*, Inform. Control, 70 (1986), pp.54–68.
15. R. Schuler, *On average polynomial time*, Technical Report, Universität Ulm, 1994.
16. R. Schuler and T. Yamakami, *Structural average case complexity*, in Proc., 12th FST & TCS, LNCS, Vol.652, 1992, pp 128–139.

Rankable Distributions Do Not Provide Harder Instances Than Uniform Distributions

Jay Belanger *
Div. of Math & Computer Science
Northeast Missouri State Univ.
Kirksville, MO 63501
belanger@cs-sun1.nemostate.edu

Jie Wang **
Dept. of Mathematical Sciences
Univ. of North Carolina at Greensboro
Greensboro, NC 27412
wang@uncg.edu

Abstract. We show that polynomially rankable distributions, proposed in [RS93], do not provide harder instances than uniform distributions for NP problems. In particular, we show that if Levin's randomized tiling problem is solvable in polynomial time on average, then every NP problem under any p-rankable distribution is solvable in average polynomial time with respect to rankability. One of the motivations for polynomially rankable distributions was to get average-case hierarchies, and we present a reasonably tight hierarchy result for average-case complexity classes under p-time computable distributions.

1 Introduction

When finding a solution to an NP-complete problem one would be satisfied if one could find an algorithm to solve the problem in expected polynomial time with respect to the underlying distribution on instances. Instance distributions are an important factor affecting average-case behaviors of computational problems. There is strong evidence to believe, as hypothesized by Levin (see [Joh84]), that any natural probability distribution used in practice either has a polynomial-time computable distribution function or else is dominated by a probability distribution that does.[3] Such distributions are referred to as p-time computable distributions. Several NP-complete problems have been proved to be solvable in average polynomial time with respect to their underlying natural distributions [Joh84, GS87]. On the other hand, there are NP-complete problems under p-time computable distributions which cannot have polynomial time on average algorithms unless every NP-complete problem under any p-time computable distribution has one [Lev86, Gur91, VL88, BG94, VR92, WB95, Wan95]. The theory of average-case completeness, initiated by Levin [Lev86], studies how likely hard instances may be in a problem.

P-time computable distributions are simple and natural. But they may seem somewhat restrictive or not precise enough in some situations as noted in [BCGL92, RS93]. Other types of instance distributions have thus been proposed recently in the theory of average-case completeness from different aspects. Among them are p-time samplable distributions [BCGL92] and polynomially-rankable distributions [RS93]. P-time samplable distributions are a natural generalization of p-time computable distributions. A distribution $\mu(x)$ is p-samplable if there is a p-time bounded probabilistic Turing machine that starts

---

* Supported in part by NSF under grant CCR-9503601.
** Supported in part by NSF under grants CCR-9396331 and CCR-9424164.
[3] This property holds for all commonly used distributions, e.g., in [JK69, Joh84, Bol85, GS87, Jai91].

with no input and outputs x with probability $\mu(x)$. All p-time computable distributions are p-samplable. But the inverse is not true if there exists a p-time computable function which is hard to invert on most instances [BCGL92]. P-samplable distributions define a new type of average-case NP-complete problems. But these problems are no harder than average-case NP-complete problems with p-time computable distributions. In particular, Impagliazzo and Levin [IL90] proved that every NP search problem complete for p-time computable distributions is also complete for all p-samplable distributions. So p-samplable distributions do not generate harder instances than p-time computable distributions. Similar investigation for randomized NP decision problems is currently undertaken by Blass and Gurevich [BG].

We investigate rankable distributions in this paper. Rankable distributions were introduced in [RS93] due to the consideration that all distributions with the same rankability should be treated in the same way. Two distributions μ and ν are said to have the same rank if for all x and y, $\mu(x) \leq \mu(y)$ iff $\nu(x) \leq \nu(y)$. The notion of average polynomial time is now with respect to all distributions with the same ranking. In so doing, a new type of average-case complexity class is defined and a tight hierarchy result is obtained for these classes. A new type of average-case NP-complete problems with respect to p-time rankable (p-rankable, in short) distribution is also constructed. A distribution μ is p-rankable if its ranking function $\mathrm{rank}_\mu(x) = |\{y : \mu(y) \geq \mu(x)\}|$ is one-to-one, and p-time computable.

There are two natural questions about any new type of distribution. First, one would like to know whether there are NP-complete problems that can be solved in average polynomial time under distributions in the new type which occur in practice. Second, one would like to know whether the new type of distributions can provide harder instances of a computationally difficult problem than p-time computable distributions on the average case. Regarding the first question, it is not known whether there are natural NP-complete problems which are solvable in average polynomial time with respect to the rank of a practical distribution. Regarding the second question, we provide a negative answer in this paper. In particular, we show that if Levin's randomized tiling problem is solvable in polynomial time on average, then every NP problem under any p-rankable distribution is solvable in average polynomial time with respect to rankability. This result holds for both decision and search problems in NP. So p-rankable distributions do not provide harder instances than p-time computable distributions. Randomizing reductions are employed to prove these results.

Finally, we present a reasonably tight hierarchy result for standard average-case complexity classes under p-time computable distributions. So the notion of p-time computable distributions is robust in that it provides the hardest instances of computationally difficult problems on the average case compared to p-samplable and p-rankable distributions, and it provides reasonably tight hierarchy results for average-case complexity classes. Moreover, the notion of p-time computable distributions is simple and all the commonly used distributions are p-time computable.

2   Rankable Distributions

We use $\Sigma = \{0, 1\}$ as the alphabet for languages and use $|x|$ to denote the length of x. Let A be a set and μ_A be a probability distribution on random instances. An instance x can be positive, meaning $x \in A$, or negative, meaning $x \notin A$. A randomized (or distributional) decision problem is a pair (A, μ_A). If $A \in$ NP, then (A, μ_A) is called a randomized NP decision problem. A function f is T-average on μ if $\sum_x \frac{T^{-1}(f(x))}{|x|} \mu(x) = O(1)$, where $T^{-1}(n) = \min\{m : T(m) \geq n\}$. This definition is due to Levin [Lev86] for polynomial T and extended by Ben-David et al [BCGL92] for arbitrary T, which overcomes inappropriate consequences of other more obvious definitions of the concept of polynomial time on average (see [Gur91] for more details).

Probability distributions on instances are important toward learning about average-case completeness. Let μ be a probability distribution (distribution, in short). The (cumulative) distribution function of μ is defined by $\mu^*(x) = \sum_{y \leq x} \mu(y)$, where $\leq$ is the standard lexicographical order on Σ^*. μ^* is p-time computable if there exists a deterministic algorithm $\mathcal{A}$ such that for every string x and every positive integer k, $\mathcal{A}$ outputs a finite binary fraction y with $|\mu^*(x) - y| \leq 2^{-k}$, and the running time of $\mathcal{A}$ is polynomially bounded on $|x|$ and k. A distribution μ is dominated by a distribution ν if $\mu(x) \leq f(x)\nu(x)$ and $f(x) \leq p(|x|)$ for some fixed polynomial p. A more liberal notion of domination is for f to be polynomial on μ-average. Denote by P-comp the class of all probability distributions which have p-time computable distribution function or else are dominated by a probability distribution that does. These probability distributions are called p-time computable distributions for simplicity. Most of the average-case complexity papers are built on p-time computable distributions. Denote by DNP the class of all randomized decision problems (A, μ_A), where $A \in$ NP and $\mu_A \in$ P-comp. A DNP problem is average-case NP-complete if every other DNP problem is reducible to it. So if an average-case NP-complete problem is solvable in average polynomial time, then so is every NP problem under any p-time computable distribution.

In their effort in studying hierarchies of average-case complexity classes, Reischuk and Schindelhauer [RS93] introduced a new type of distribution in which all distributions with the same rankability are treated in the same way based on the assumption that only the ranking of the inputs by decreasing weights matters. Recall that two distributions μ and ν are said to have the same rankability if for all x and y, $\mu(x) \leq \mu(y)$ iff $\nu(x) \leq \nu(y)$.

Define $\operatorname{rank}_\mu(x)$ to be $|\{y \in \Sigma^* : \mu(y) \geq \mu(x)\}|$. A function f is T-average with respect to ranking function rank_μ if for every real-valued monotone function m with $\sum_x m(\mu(x)) \leq 1$, $\sum_x \frac{T^{-1}(f(x))}{|x|} m(\mu(x)) = O(1)$. This condition depends only on rank_μ and not on μ itself. That is, it depends on all probability distributions that have the same ranking function as μ. f is polynomial on μ-average with respect to rankability if there is a polynomial p such that f is p-average with respect to rank_μ. A randomized NP problem (A, μ_A) is solvable in average polynomial time with respect to rankability if there exists a deterministic Turing machine that computes A in time polynomial on μ_A-average with respect to rankability.

Polynomially rankable distributions are used to define average-case NP-completeness with respect to rankability [RS93]. Let p-rankable denote the set of all probability distributions μ such that rank_μ is one-to-one, and p-time computable. The injectivity of ranking functions provides a unique rank for distributions. By a slight perturbation of the probability distributions, this can always be achieved.

Studying average polynomial time with respect to rankability directly from definition is difficult due to the fact that arbitrary real-valued function m is involved. This obstacle is overcome by the following lemma due to [RS93].

Lemma 1 [RS93]. *A function f is T-average with respect to ranking function* rank_μ *if and only if* $\forall l : \sum_{\text{rank}_\mu(x) \le l} \frac{T^{-1}(f(x))}{|x|} \le l.$

3 Randomizing Reductions

Let μ be a p-rankable distribution. Then rank_μ transforms the distribution of inputs into a monotone distribution on the outputs. However, while rank_μ may be p-time computable, rank_μ may not transform an NP problem into an NP problem if rank_μ is not p-honest.[4] To prove that p-rankable distributions do not provide harder instances than uniform distributions, we need to construct a hardest problem in (NP, p-rankable) with respect to rankability such that its ranking function is p-honest,[5] where (NP, p-rankable) is the class of all randomized NP decision problems with p-rankable distributions. This will be done using randomizing reductions.

Deterministic reductions are defined in [RS93] for randomized decision problems with respect to rankable distributions with a restriction that the reductions are required to be injective. It does not pose a real restriction for natural NP-complete problems as they are all complete under injective reductions, so we will follow this restriction in defining reductions with respect to rankability. Notice that a ranking function has small values for likely instances and has large values for unlikely instances. This is used in defining the notion of domination with respect to rankability. The following definition of reduction is due to [RS93].

(A, μ_A) is p-time reducible to (B, μ_B) with respect to rankability if $A \le^p_m B$ via a one-to-one reduction f and satisfies the domination property: $\text{rank}_{\mu_B}(f(x)) \le p(|x|)\text{rank}_{\mu_A}(x)$ for some fixed polynomial p.

This definition has the properties that one would want a reduction to have, namely that it is transitive, and if (B, μ_B) can be solved in average-polynomial time with respect to rankability, then so can (A, μ_A).

A randomized NP decision problem is *rankably complete* if its distribution is p-rankable and every other randomized NP decision problem with p-rankable distribution is p-time reducible to it. Rankably complete randomized NP decision problems were constructed in [RS93].

Notice that in the average case measure with respect to rankability, (X, μ) and (X, rank_μ) denote the same randomized problems, where X is a language.

Randomizing reductions for randomized NP problems with p-time computable distributions were first defined in [VL88] and were further studied in [BG93, BG94], which have been applied successfully to obtain a number of

---

[4] A function f is *p-honest* if there is a polynomial p such that for all x, $p(|f(x)|) \ge |x|$ whenever $f(x)$ is defined.

[5] Note that the ranking functions of the rankably complete problems constructed in [RS93] are not p-honest.

average-case NP complete problems under flat distributions.[6] They were also used by Impagliazzo and Levin [IL90] when they showed that p-samplable distributions did not generate harder instances than p-computable distributions. We will follow the same idea to define randomizing reductions for problems in (NP, p-rankable) with respect to rankability by allowing coin flips in algorithms, resulting in randomizing algorithms.

We assume that a randomizing algorithm does not flip a coin unless the computation requires another random bit. For simplicity, coins are assumed to be unbiased. So a randomizing algorithm on A with probability distribution μ_A can be viewed as a deterministic algorithm on inputs x and a sequence of coin flips s, which form a dilation Δ with probability distribution μ_Δ such that the following conditions are satisfied [Gur91].

1. Δ is a subset of $\Sigma^* \times \Sigma^*$ with the following property. For every x with $\mu_A(x) \neq 0$: $\Delta(x) \neq \emptyset$, and no string in $\Delta(x)$ is a prefix of a different string in $\Delta(x)$, where $\Delta(x) = \{s : (x, s) \in \Delta\}$.
2. For all $(x, s) \in \Delta$, the length of (x, s) is defined as the length of x.
3. For all x and s, $\mu_\Delta(x, s)$ is defined as $\mu_A(x)2^{-|s|}r_\Delta(x)$ if $s \in \Delta(x)$ and 0 otherwise, where $r_\Delta(x) = 1/\Sigma_{t \in \Delta(x)}2^{-|t|}$ is called the rarity function of Δ.

Yet for rankability an extra condition is required to make sure that the ranking of a dilation will solely depend on the ranking of the distribution. Otherwise, different distributions with the same rank may result in dilations with different ranks. This condition is formulated as below. This condition is not needed if one can live with dilations with different ranks generated by different distributions with the same rank.

4. If $\text{rank}_A(x') \leq \text{rank}_A(x)$ and $(x, s), (x', s') \in \Delta$, then $|s'| \leq |s|$.

We will only need the simplest randomizing algorithms and dilations in this paper, namely, the rarity function of the underlying dilation is always equal to 1. Such a dilation is called an "almost total" dilation.

A randomized decision problem (A, μ_A) is considered solvable efficiently with respect to rankability if there is a randomizing algorithm that decides A in average polynomial time with respect to ranking function rank_{μ_A}.

Definition 2. A randomized decision problem (A, μ_A) is solvable in average polynomial time with respect to rankability if there is an almost total dilation (Δ, μ_Δ) of (A, μ_A) and a deterministic Turing machine on Δ which decides A in average polynomial time with respect to ranking function rank_{μ_Δ}.

Notice that a deterministic algorithm can be thought of as a special case of randomizing algorithm with the dilation containing only the empty string as a coin toss for any input x. A similar notion can be defined for solvability in T-time on average with respect to rankability. More liberal notions of solvability on average with respect to rankability (namely, the rarity function may not always equal to 1) can be similarly defined following [BG93, BG94] and all our results presented in this section are still true.

---

[6] A distribution μ is flat if there exists an $\epsilon > 0$ such that for all x, $\mu(x) \leq 2^{-|x|^\epsilon}$. Randomized NP problems with flat distributions cannot be complete under deterministic many-one reductions unless nondeterministic exponential time collapses to deterministic exponential time. See [Gur91] for more details.

Definition 3. (A, μ_A) is p-time randomizing reducible to (B, μ_B) with respect to rankability if there is an almost total dilation (Δ, μ_Δ) of (A, μ_A) and a p-time computable, one-to-one function f such that

1. For each $(x, s) \in \Delta$: $x \in A$ iff $f(x, s) \in B$.
2. $\text{rank}_{\mu_B}(f(x, s)) \leq p(|x|) \cdot \text{rank}_{\mu_\Delta}(x, s)$ for some fixed polynomial p.

It can be shown that if (A, μ_A) is p-time randomizing reducible to (B, μ_B) and (B, μ_B) is solvable in average polynomial time with respect to rankability, then so is (A, μ_A). We can similarly define a completeness notion for randomized NP decision problems under randomizing reductions with respect to rankability.

We consider a bounded version of a randomized halting problem with respect to p-rankable distribution, where the ranking function is p-honest.

Let $\mathcal{N} = \{0, 1, 2, ...\}$ be the set of all natural numbers. Let $\langle \cdot, \cdot \rangle$ be a standard pairing function from $\Sigma^* \times \Sigma^*$ to Σ^* in lexicographical order which is both p-time computable and invertible. We can recursively define $\langle \cdot, \cdot, \cdot \rangle$. Let f be a function and we write $f(\cdot, \cdot)$ for $f(\langle \cdot, \cdot \rangle)$. Let β be a standard function that maps all binary strings to all binary numbers in $\mathcal{N}$ in lexicographical order, and β is both linear-time computable and invertible. Let $M_0, M_1, M_2, ...$ be a fixed enumeration of all (deterministic/nondeterministic) Turing machines.

When a ranking function rank is defined for a particular problem, we assume that $\text{rank}(x) = \infty$ for x not being an instance of the problem. Let $K = \{\langle i, x, 1^n \rangle : M_i \text{ accepts } x \text{ within } n \text{ steps}\}$, and $\text{rank}_K(i, x, 1^n) = \beta(i, x, n)$. Let $K' = \{\langle i, x, w \rangle : M_i \text{ accepts } x \text{ within } |w| \text{ steps}\}$, and $\text{rank}_{K'}(i, x, w) = \beta(i, x, w)$. It is easy to see that K' is NP-complete and $\text{rank}_{K'}$ is p-time computable, p-honest, and p-time invertible. It was shown in [RS93] that (K, rank_K) is rankably complete for (NP, p-rankable). But rank_K is not p-honest. We will show that (K, rank_K) is p-time randomizing reducible to $(K', \text{rank}_{K'})$ with respect to rankability. [7]

Theorem 4. $(K', \text{rank}_{K'})$ *is rankably complete for* (NP, p-rankable) *under p-time randomizing reductions.*

Proof. Let μ be a p-rankable distribution with ranking function rank_K such that $\mu(y) = 0$ if y is not in the form $\langle i, x, 1^n \rangle$, and $\mu(i', x', 1^{n'}) \geq \mu(i, x, 1^n)$ if $i' \leq i$, $x' \leq x$, and $n' \leq n$ while maintaining injectivity.

Clearly, K is nondeterministic linear time computable. Let M be a nondeterministic Turing machine that accepts K in linear time. Let M' be a nondeterministic Turing machine such that M' accepts input z iff there is a $y = \langle i, x, 1^n \rangle$ such that $\text{rank}_K(y) = \beta(z)$ and M accepts y. It is easy to see that there is a linear polynomial p such that M' accepts z iff there is a computation of M' that accepts z in $p(|y|)$ steps. So $y \in K$ iff M' accepts $\beta^{-1}(\text{rank}_K(y))$ in time $p(|y|)$. Let j be an index such that $M_j = M'$.

Define a dilation (Γ, μ_Γ) of (K, μ) by $\Gamma = \{(y, s) : \mu(y) > 0 \text{ and } |s| = p(|y|)\}$. Clearly, Γ is p-time computable and $\sum_{t \in \Gamma(y)} 2^{-|t|} = 1$ for all y with $\mu(y) \neq 0$. In particular, condition 4 is satisfied by noticing that $\text{rank}_K(y') \leq \text{rank}_K(y)$ iff $y' \leq y$, and $y' \leq y$ implies that $|y'| \leq |y|$, and so $p(|y'|) \leq p(|y|)$.

---

[7] It can also be shown that $(K', \mu_{K'})$ is average-case NP-complete in Levin's sense under randomizing reductions, where $\mu_{K'}(i, x, w)$ is flat, and is defined as $c \cdot \frac{2^{-(|i|+|x|+|w|)}}{(|i||x||w|)^2}$ for an appropriate constant c.

Define a reduction $f : \Gamma \to K'$ as follows. For all y and s with $\mu(y) > 0$:
$f(y, s) = \langle j, \beta^{-1}(\mathrm{rank}_K(y)), s \rangle$. It is easy to see that f is one-to-one and p-time computable since both β^{-1} and rank_K are one-to-one and p-time computable. Clearly, for all $(y, s) \in \Gamma$: $y \in K$ iff $f(y, s) \in K'$.

Now we check the domination property. Let $\mathrm{rank}_{\mu_\Gamma}$ denote the ranking function of μ_Γ, where $\mu_\Gamma(y, s) = \mu(y)2^{-|s|}$ for $(y, s) \in \Gamma$, and 0 otherwise. For $(y, s) \in \Gamma$, write $y = \langle i, x, 1^n \rangle$. Notice that p is a linear polynomial, we have $\mathrm{rank}_{\mu_\Gamma}(y, s) = |\{(y', s') \in \Gamma : \mu(y')2^{-|s'|} \geq \mu(y)2^{-|s|}\}| > |\{(y', s') : |i'| = |i| - 1, |x'| = |x| - 1, n' = n, \text{ and } |s'| = p(|\langle i', x', 1^{n'} \rangle|)\}|$ (where $y' = \langle i', x', 1^{n'} \rangle$) $\geq O(2^{|i| + |x| + |s|}) \geq O(\beta(i, x, n)2^{|s|}/|\langle i, x, 1^n \rangle|) = O(\mathrm{rank}_K(y)2^{|s|}/|y|)$. We know that $\mathrm{rank}_{K'}(f(y, s)) = \beta(f(y, s))$. By construction, $\beta(f(y, s)) = \beta(j, \beta^{-1}(\mathrm{rank}_K(y)), s) = O(\beta(\beta^{-1}(\mathrm{rank}_K(y))2^{|s|})) = O(\mathrm{rank}_K(y)2^{|s|}) \leq O(|y|\mathrm{rank}_{\mu_\Gamma}(y, s))$. This completes the proof. ∎

4 Rankable Instances Are Not Harder

We prove in this section that rankable distributions do not provide harder instances than uniform distributions for NP decision problems.

The standard uniform probability distribution μ is given by $\mu(x) = \frac{2^{-|x|}}{|x|(|x|+1)}$ or $\mu(x) = \frac{6}{\pi|x|^2}2^{-|x|}$, although this is often replaced by $\mu(x) = \frac{c}{|x|^k}2^{-|x|}$ for some $k > 1$ and appropriate c, or even $\mu(x) = \frac{c}{|x|\log^{1+\epsilon}|x|}2^{-|x|}$ for some $\epsilon > 0$, where $\log^\epsilon n$ denotes $(\log n)^\epsilon$. For notational convenience, we simply use $\frac{2^{-|x|}}{|x|^2}$ as the default uniform probability distribution of binary strings.

Lemma 5. *There is an NP-complete set S and a p-rankable distribution rank_S such that $(K', \mathrm{rank}_{K'})$ is p-time reducible to (S, rank_S) with respect to rankability, where $|\mathrm{rank}_S(x)|^3 \leq |x|$, and rank_S is p-honest. Moreover, rank_S is one-to-one.*

Proof. We use an easy fact that K' and SAT are p-isomorphic [BH77], meaning that there is a p-time computable and invertible bijection f such that K' is reducible to SAT via f. Pad the boolean formula generated by $f(i, x, w)$ such that the length of it is greater than the cube root of the length of $\mathrm{rank}_{K'}(i, x, w)$. Let g denote this new reduction, which is one-to-one, p-time computable, and p-time invertible. Let $S = g(K')$ and define rank_S as follows: For all instances F of $g(\Sigma^*)$ (positive or negative), let $\mathrm{rank}_S(F) = \mathrm{rank}_{K'}(i, x, w)$, where $g^{-1}(F) = \langle i, x, w \rangle$. This completes the proof. ∎

Lemma 6. *Let (S, rank_S) be from Lemma 5. Let $L = \mathrm{rank}_S(S)$ and $\nu(y)$ be the uniform distribution $2^{-|y|}/|y|^2$. If (L, ν) can be solved in T time on ν-average, then (S, rank_S) from Lemma 5 can be solved in average $O(T + p)$ time with respect to rankability for some polynomial p.*

Proof. Assume that (L, ν) can be solved in time T on ν-average. This means that L can be solved by a deterministic algorithm with running time t and the following is satisfied: $\sum_y T^{-1}(t(y))|y|^{-1}\nu(y) = O(1)$. Let $M = \sum_y T^{-1}(t(y))|y|^{-1}\nu(y)$. So S can be solved by a deterministic algorithm with running time $(t \circ \mathrm{rank}_S) + p$, where p is a polynomial time bound for computing rank_S. We will show that $t \circ \mathrm{rank}_S$ is T-average with respect to rank_S. For any natural number ℓ, let $R_\ell = \{y : y = \mathrm{rank}_S(x) \leq \ell\}$. We know that $2^{|\mathrm{rank}_S(x)|} \leq \mathrm{rank}_S(x) < 2 \cdot 2^{|\mathrm{rank}_S(x)|}$.

We get $M \geq \sum_{y \in R_\ell} T^{-1}(t(y))|y|^{-1}\nu(y) = \sum_{y \in R_\ell} T^{-1}(t(y))2^{-|y|}|y|^{-3} = \sum \text{rank}_S(x) \leq \ell \frac{(T')^{-1}(t(\text{rank}_S(x)))}{2^{|\text{rank}_S(x)|}|\text{rank}_S(x)|^3} \geq \sum \text{rank}_S(x) \leq \ell \frac{T^{-1}(t \circ \text{rank}_S(x))}{\text{rank}_S(x)|x|} \geq$

$\sum \text{rank}_S(x) \leq \ell \frac{T^{-1}(t \circ \text{rank}_S(x))}{\ell|x|}$. So $t \circ \text{rank}_S$ is $M \cdot T$-average with respect to rank_S from Lemma 1. This completes the proof. ∎

Corollary 7. *If an average-case NP-complete decision problem can be solved in T time on average, then (S, rank_S) from Lemma 5 can be solved in average $O(T \circ q)$ time with respect to rankability for some polynomial q.*

Theorem 8. *If the randomized Tiling problem can be solved in average polynomial time, then any NP decision problem under any p-rankable distribution is solvable in average polynomial time with respect to rankability.*

Similar results in this section also hold for search problems, which will be published in a more detailed version of this paper.

5 Average-case Hierarchies

The study of hierarchies among complexity classes is a fruitful area in complexity theory, yet surprisingly little has been done to investigate hierarchies among average-case complexity classes. Studying natural distributions that can provide hard instances of problems and finding suitable reductions to identify more naturally occurred NP problems to be average-case complete have been the major concerns in the theory of average-case complexity. Nevertheless, there is a strong interest to investigate hierarchy properties among interesting average-case complexity classes.

Let t be a time-constructible function. Denote by $\text{AvDTime}(t(n))$ the class of randomized decision problems which can be decided by a Turing machine whose running time is t on average (in Levin's sense). It can be seen that a problem which requires, for almost all inputs, n^2 time to solve, where n is the length of the input, will be, with the uniform distribution, in $\text{AvDTime}(n^{1+\epsilon})$ for every $\epsilon > 0$. This may seem not precise enough and to prevent this from happening, Reischuk and Schindelhauer [RS93] proposed the notions of rankable distributions and average time with respect to rankability (see Section 2), and they established a rather tight hierarchy for their average time complexity classes with respect to rankability. However, it would be more desirable to have hierarchy results using the standard notions.

It is known that under the universal distribution, the average-case complexity of a problem is the same as the worst-case complexity [LV92],[8] and so if no restrictions are put on the distributions, any hierarchy results for $\text{DTime}(t(n))$ apply to $\text{AvDTime}(t(n))$. Ben-David *et al* [BCGL92] have a similar result, using a non-standard definition of worst-case complexity. However, the distributions used in these results require super-polynomial time to compute, and we would like to restrict ourselves to distributions which can be computed in polynomial time. With this restriction, we obtain the following hierarchy result.

Theorem 9. *Let t and T be time-constructible functions. If for some $\epsilon > 0$ $t(n \log^\epsilon n) \log t(n \log^\epsilon n) = o(T(n))$, then there is a randomized decision problem $(L, \mu) \in \text{AvDTime}(T(n)) - \text{AvDTime}(t(n))$ for a uniform distribution μ.*

---

[8] Actually, this is shown for a different notion of average time, but it will imply it for our notion of average time for a large class of time-complexity functions.

418

Proof. Let t and T be as above. We immediately have AvDTime($t(n)$) is included in AvDTime($T(n)$). Define U by $U(n) = t(n \log^\epsilon n)$, so $U(n) \log U(n) = o(T(n))$. It was shown by Goldmann, Grape and Håstad [GGH94] that there exists a language L in DTIME($T(n)$) such that if T_M is the running time of a Turing machine M which decides L, then for sufficiently large n, say $n \geq N$, $T_M(x) \geq U(|x|) = t(|x| \log^\epsilon |x|)$ for a constant fraction c_M of instances x of length n. For these x, we have $t^{-1}(T_M(x)) \geq |x| \log^\epsilon |x|$. Letting $\mu(x) = \frac{c}{|x| \log^{1+\epsilon} |x|} 2^{-|x|}$ for the appropriate c, we get $\sum_x \frac{t^{-1}(T_M(x))}{|x|} \mu(x) = \sum_x \frac{t^{-1}(T_M(x))}{|x|} \frac{c}{|x| \log^{1+\epsilon} |x|} 2^{-|x|} \geq \sum_{|x| \geq N} \frac{t^{-1}(T_M(x))}{|x|} \frac{c}{|x| \log^{1+\epsilon} |x|} 2^{-|x|} \geq \sum_{n=N}^\infty \frac{c_M 2^n n \log^\epsilon n}{n} \frac{c}{n \log^{1+\epsilon} n} 2^{-n} = \sum_{n=N}^\infty \frac{c c_M}{n \log n}$, which diverges. So, (L, μ) cannot be in AvDTime($t(n)$). Since L is in DTIME($T(n)$), (L, μ) will be in AvDTime($T(n)$), and so AvDTime($t(n)$) is properly included in AvDTime($T(n)$). ∎

It is often useful to restrict our attention to a smaller class of distributions in order to obtain completeness results (e.g., see [WB93]). Let t be a time-constructible function and $\mathcal{F}$ a class of distributions. Then AvDTime($t(n), \mathcal{F}$) is the class of randomized decision problems with distributions in $\mathcal{F}$ which can be solved in average t time. Similar to the proof of Theorem 9, we can show the following hierarchy result. (A weaker result was shown in [SY92].)

Theorem 10. *Let $\mathcal{F}$ be a class of distributions containing $\mu(x) = c|x|^{-k} 2^{-|x|}$ for some k and suitable c, and let t and T be time-constructible functions such that $t(n^k) \log t(n^k) = o(T(n))$. Then AvDTime($t(n), \mathcal{F}$) is properly included in AvDTime($T(n), \mathcal{F}$).*

Corollary 11. *Let t and T be time-constructible functions such that, for some $\epsilon > 0$, $t(n \log^\epsilon n) \log t(n \log^\epsilon n) = o(T(n))$. Then AvDTime($t(n)$, P-comp) is properly included in AvDTime($T(n)$, P-comp).*

6 Final Remarks

It has been observed in [CS95] that average-case 2^n time complexity class cannot be separated from average-case c^n time class for any constant $c > 0$. So the log factor in our hierarchy results cannot be replaced by any constant. Under a somewhat restricted notion of time functions based on Hardy's logarithmico-exponential functions [Har11], Cai and Selman [CS95] obtain an average-case time hierarchy within polynomial bounds under Levin's definition with respect to p-time computable distributions, which is as tight as the Hartmanis-Sterns deterministic time hierarchy [HS65]. Further requiring that the time expectation converge in an appropriate fast rate, Cai and Selman [CS95] obtain an average-case time hierarchy for arbitrary average-case time classes as tight as the deterministic time hierarchy. Moreover, under their new definition, if a language L requires greater than $T(|x|)$ time to solve for almost all inputs x, then (L, μ) cannot be solved in average $T(n)$ time for any p-time computable distribution μ, thus fixing an earlier concern of the standard notion of average time.

Acknowledgement. The second author thanks Leonid Levin for his constructive comments.

References

[BCGL92] S. Ben-David, B. Chor, O. Goldreich, and M. Luby. On the theory of average case complexity. *J. Comp. Sys. Sci.*, 44:193–219, 1992.

[BG]   A. Blass and Y. Gurevich. Randomizing reductions of decision problems (tentative title). Personal communication.

[BG93]   A. Blass and Y. Gurevich. Randomizing reductions of search problems. *SIAM J. Comput.*, 22:949–975, 1993.

[BG94]   A. Blass and Y. Gurevich. Matrix decomposition is complete for the average case. *SIAM J. Comput.*, 1994. to appear.

[BH77]   L. Berman and J. Hartmanis. On isomorphisms and density of NP and other complete sets. *SIAM J. Comput.*, 6:305–321, 1977.

[Bol85] B. Bollobás. *Random Graphs*. Academic Press, 1985.

[CS95]   J.-Y. Cai and A. Selman. Average time complexity classes. Manuscript.

[GGH94] M. Goldmann, P. Grape, and J. Hastad. On average time hierarchies. *Inf. Proc. Lett.*, 49:15–20, 1994.

[GS87]   Y. Gurevich and S. Shelah. Expected computation time for hamiltonian path problem. *SIAM J. Comput.*, 16:486–502, 1987.

[Gur91] Y. Gurevich. Average case completeness. *J. Comp. Sys. Sci.*, 42:346–398, 1991.

[Har11] G. Hardy. Properties of logarithmico-exponential functions. *Proc. London Math. Soc.*, (2),10:54–90, 1911.

[HS65]   J. Hartmanis and R. Sterns. On the computational complexity of algorithms. *Trans. Amer. Math. Soc.*, 117:285–306, 1965.

[IL90]   R. Impagliazzo and L. Levin. No better ways to generate hard NP instances than picking uniformly at random. In *Proc. 31st FOCS*, pages 812–821, 1990.

[Jai91] R. Jain. *The Art of Computer Systems Performance Analysis*. John Wiley & Sons, 1991.

[JK69]   N. Johnson and S. Kotz. *Distributions in Statistics–Discrete Distributions*. John Wiley & Sons, 1969.

[Joh84] D. Johnson. The NP-completeness column: an ongoing guide. *Journal of Algorithms*, 5:284–299, 1984.

[Lev86] L. Levin. Average case complete problems. *SIAM J. Comput.*, 15:285–286, 1986.

[LV92]   M. Li and P. Vitányi. Average case complexity under the universal distribution equals worst-case complexity. *Inf. Proc. Lett.*, 42:145–149, 1992.

[RS93]   R. Reischuk and C. Schindelhauer. Precise average case complexity. *STACS'93*, vol 665 of *Lect. Notes in Comp. Sci.*, pages 650–661, 1993.

[SY92]   R. Schuler and T. Yamakami. Structural average case complexity. *FSTTCS'92*, vol 652 of *Lect. Notes in Comp. Sci.*, pages 128–139, 1992.

[VL88]   R. Venkatesan and L. Levin. Random instances of a graph coloring problem are hard. In *Proc. 20th STOC*, pages 217–222, 1988.

[VR92]   R. Venkatesan and S. Rajagopalan. Average case intractability of diophantine and matrix problems. In *Proc. 24th STOC*, pages 632–642, 1992.

[Wan95] J. Wang. Average-case completeness of a word problem for groups. In *Proc. 27th STOC*, 1995. To appear.

[WB93]   J. Wang and J. Belanger. On average-P vs. average-NP. In K. Ambos-Spies, S. Homer, and U. Schönings, editors, *Complexity Theory—Current Research*, pages 47–67. Cambridge University Press, 1993.

[WB95]   J. Wang and J. Belanger. On the NP-isomorphism problem with respect to random instances. *J. Comp. Sys. Sci.*. 50:151–164, 1995.

Transformations that Preserve Malignness of Universal Distributions

Kojiro Kobayashi

Tokyo Institute of Technology
Oh-okayama 2-12-1, Meguro–ku, Tokyo 152, JAPAN

Abstract. A function $\mu(x)$ that assigns a nonnegative real number $\mu(x)$ to each bit string x is said to be *malign* if, for any algorithm, the worst-case computation time and the average computation time of the algorithm are functions of the same order when each bit string x is given to the algorithm as an input with the probability that is proportional to the value $\mu(x)$. M. Li and P. M. B. Vitányi found that functions that are known as "universal distributions" are malign. We show that if $\mu(x)$ is a universal distribution and t is a positive real number, then the function $\mu(x)^t$ is malign or not according as $t \geq 1$ or $t < 1$. For $t > 1$, $\mu(x)^t$ is an example of malign functions that are not universal distributions.

1 Introduction

In [2], M. Li and P. M. B. Vitányi found that functions that are known as universal distributions have one pathological property concerning the relation between the worst-case computation time and the average computation time of algorithms, and P. B. Miltersen ([4]) used the word "malign" to denote this property. A more precise definition of malign functions is as follows.

Let Σ, Σ^n, Σ^* denote the set $\{0,1\}$ of bits 0, 1, the set of all bit strings of length n, and the set of all bit strings, respectively. Let the letter A denote algorithms that accept elements of Σ^* as inputs and halt for all inputs, and let $\mu(x)$ be a function from Σ^* to nonnegative real numbers. Let $t_A^{\text{wo}}(n)$ denote the worst-case computation time of A for inputs of length n, and let $t_A^{\text{av},\mu(x)}(n)$ denote the average computation time of A for inputs of length n under the assumption that an input x of length n is given to A with the probability $\mu(x)/\mu(\Sigma^n)$. We say that $\mu(x)$ is *malign* if for any algorithm A there exists a positive constant c such that $t_A^{\text{wo}}(n) \leq c\, t_A^{\text{av},\mu}(n)$ for any n.

Let $\mu(x)$ be a universal distribution. The result by M. Li and P. M. B. Vitányi shows that $\mu(x)$ is a malign function. The main result of this paper is that, for a positive constant t, the function $\mu(x)^t$ is malign or not according as $t \geq 1$ or $t < 1$. For $t > 1$ the function $\mu(x)^t$ is not a universal distribution. Hence, for such t, $\mu(x)^t$ is an example of malign functions that are not universal distributions. (Such an example was first shown in [1].)

2 Preliminaries

Let N, Q, R denote the set of all natural numbers, the set of all rational numbers, and the set of all real numbers, respectively. For functions $f(x), g(x)$ from a set to R, we write $f(x) \preceq g(x)$ if there exists a positive constant c such that $f(x) \leq cg(x)$ for any x, and write $f(x) \simeq g(x)$ if both of $f(x) \preceq g(x), g(x) \preceq f(x)$ hold.

By a *semi-distribution*, we mean a function $\mu(x)$ from Σ^* to R such that $\mu(x) \geq 0$ for any x. We will use the letter A to denote algorithms that accept elements of Σ^* as inputs and halt for all input. Let $\text{time}_A(x)$ denote the computation time of A for an input x. Let $t_A^{wo}(n)$ denote the worst-case computation time of A for inputs of length n, that is, the value $\max\{\text{time}_A(x) \mid x \in \Sigma^n\}$. For a semi-distribution $\mu(x)$ such that $\mu(\Sigma^n) > 0$ for any n, let $t_A^{av,\mu(x)}(n)$ denote the average computation time of A for inputs of length n under the assumption that an input x of length n is given to A with the probability $\mu(x)/\mu(\Sigma^n)$, that is, the value $\Sigma\{\text{time}_A(x)(\mu(x)/\mu(\Sigma^n)) \mid x \in \Sigma^n\}$. (For a subset X of Σ^*, $\mu(X)$ denotes $\Sigma\{\mu(x) \mid x \in X\}$.) We say that a pseudo-distribution $\mu(x)$ is *malign* if $\mu(\Sigma^n) > 0$ for any n and $t_A^{wo}(n) \preceq t_A^{av,\mu(x)}(n)$ for any algorithm A.

By a *distribution*, we mean a semi-distribution $\mu(x)$ such that $0 < \mu(\Sigma^*) < \infty$. We say that a distribution $\mu(x)$ is *enumerable* if there exists a recursive function $f(x,n)$ from $\Sigma^* \times N$ to Q such that $f(x,n) \leq f(x,n')$ for $n \leq n'$ and $\lim_{n \to \infty} f(x,n) = \mu(x)$. We say that a distribution $\mu(x)$ is *universal* if $\mu(x)$ is enumerable and $\mu'(x) \preceq \mu(x)$ for any enumerable distribution $\mu'(x)$. It is well-known that universal distributions exist (see, for example, [3]). If $\mu(x)$ and $\mu'(x)$ are universal distributions, then obviously $\mu(x) \simeq \mu'(x)$. The result by Li and Vitányi says that universal distributions are malign as semi-distributions.

In [1], we introduced a notion called "strongly malign semi-distributions" from the following motivation. The intuitive notion of malignness allows several variations for its formal definition other than the one mentioned above depending on the following factors:

(1) to what resource we pay attention (computation time or computation space),
(2) how we define the size of each input x (usually the size of inputs are determined by the algorithm that uses them as inputs).

Nevertheless, Li and Vitányi's result seems to hold true for any of these variations if the variation is reasonable. We introduced the notion of "strongly malign semi-distributions" as one candidate for notions that will capture all of the malignnesses represented by these variations.

We say that a pseudo-distribution $\mu(x)$ is *strongly malign* if it satisfies the following condition:

$(\forall \phi)(\forall \phi')$ $[\phi(n), \phi'(x)$ are partial recursive functions from N to Σ^* and from Σ^* to N, respectively
$$\Longrightarrow$$
$(\exists c > 0)(\forall n)$ $[\phi(n)$ is defined $\Longrightarrow \mu(\phi'^{-1}(n)) \leq c\mu(\phi(n))]]$.

Note that $\phi'^{-1}(n)$ is a subset of Σ^*. Intuitively, $\phi(n)$ is the bit string that satisfies some specification (such as "the worst-case input") corresponding to a parameter (such as "size") n and $\phi(x)$ is the value of the parameter of a bit string x. We do not require $\phi(n)$ to be in $\phi'^{-1}(n)$. The above condition may be replaced with a more simple one without changing the defined notion:

$$(\forall\phi)\ [\phi(x)\text{ is a partial recursive function from }\Sigma^*\text{ to }\Sigma^*$$
$$\Longrightarrow$$
$$(\exists c > 0)(\forall x)\ [\mu(\phi^{-1}(x)) \leq c\mu(x)]].$$

In [1] we proved two implications:

$\mu(x)$ is a universal distribution
$\Longrightarrow \mu(x)$ is a strongly malign semi-distribution
$\Longrightarrow \mu(x)$ is a malign semi-distribution.

Corollary 2 of the present paper implies the first implication as a special case. The second implication can be easily proved using the function

$$\phi(x) = (\text{the worst-case input of } A \text{ of length } |x|)$$

in the simpler definition of strong malignness for each algorithm A.

Let $f(x)$ be a function from R to R such that $f(0) = 0$, $f(x) > 0$ for $x > 0$ and $\lim_{x\to 0} f(x) = 0$. We say that $f(x)$ *preserves strong malignness* (or *malignness*) *of universal distributions*, or *preserves strong malignness* (or *malignness*, respectively) for short if $f(\mu(x))$ is strongly malign (or malign, respectively) for any universal distribution $\mu(x)$.

Functions $f(x)$ can be represented as $f(x) = x^{g(x)}$ for $x > 0$ using a function $g(x)$ from positive real numbers to real numbers. In this paper, we will consider only $f(x) = x^{g(x)}$ such that $\lim_{x\to 0} g(x)$ exists (possibly infinite). For such $f(x) = x^{g(x)}$ to satisfy $\lim_{x\to 0} f(x) = 0$, $g(x)$ must satisfy one of the following conditions:

- $\lim_{x\to 0} g(x) = \infty$,
- $0 < \lim_{x\to 0} g(x) < \infty$,
- $\lim_{x\to 0} g(x) = 0$ and $\lim_{x\to 0} g(x)\log(1/x) = \infty$.

The third case includes, for example,

$$g(x) = 1/(\log(1/x))^r, 0 < r < 1,$$
$$g(x) = 1/(\log^m(1/x))^r, 0 < r, 2 \leq m$$

but excludes, for example,

$$g(x) = 1/(1/x)^r = 1 + x^r, 0 < r,$$
$$g(x) = 1/(\log(1/x))^r, 1 \leq r,$$
$$g(x) = -1/(\log(1/x))^r, 0 < r < 1,$$
$$g(x) = -1/(\log^m(1/x))^r, 0 < r, 2 \leq m.$$

Here, $\log^m x$ means $\log\log\ldots\log x$ (m log's).

The main results of this paper are summarized as follows.

Case 1: $\lim_{x\to 0} g(x) = \infty$. If $g(x)$ is non-increasing then $f(x)$ does not preserve malignness.

Case 2: $1 < \lim_{x\to 0} g(x) < \infty$. We know several examples of $f(x)$ that preserve strong malignness including $f(x) = x^t$ ($1 < t$). However, we do not know examples that do not preserve malignness.

Case 3: $\lim_{x\to 0} g(x) = 1$. We know both examples of functions $f(x)$ that preserve strong malignness and functions that do not preserve malignness. However, for some simple functions such as $f(x) = x^{1+x}$ we do not know wether they preserve malignness or not.

Case 4: $0 < \lim_{x\to 0} g(x) < 1$. The function $f(x)$ does not preserve malignness.

Case 5: $\lim_{x\to 0} g(x) = 0$. If $g(x) \geq 0$ then the function $f(x)$ does not preserve malignness.

In the proofs of some results we need the notion of Kolmogorov complexity and another characterization of universal distributions.

A subset X of Σ^* is said to be *prefix-free* if there exist no x, y in Σ^* such that $x \in X$, $xy \in X$, and $y \neq \lambda$ (λ denotes the empty string). A partial function $\phi(x)$ from Σ^* to Σ^* is said to be *prefix-free* if its domain is prefix-free. Let $\phi_0(x)$, $\phi_1(x)$, $\ldots$ be an enumeration of all prefix-free partial recursive functions from Σ^* to Σ^*. We may assume that there exists one prefix-free partial recursive function $\phi_U(x)$ from Σ^* to Σ^* such that $\phi_U(0^i 1x) = \phi_i(x)$ for any i, x. We call such $\phi_U(x)$ a *universal* prefix-free partial recursive function. We fix one such partial function and use the expression $\phi_U(x)$ to denode it.

Let $K(x)$ be the function from Σ^* to N defined by $K(x) = \min\{|s| \mid s \in \Sigma^*, \phi_U(s) = x\}$ and let $\tilde{\mu}(x)$ be the function from Σ^* to R defined by $\tilde{\mu}(x) = \Sigma\{2^{-|s|} \mid s \in \Sigma^*, \phi_U(s) = x\}$. For these functions, we know that both of $2^{-K(x)}$, $\tilde{\mu}(x)$ are universal distributions, and hence $\tilde{\mu}(x) \simeq 2^{-K(x)}$. The value $K(x)$ is usually called the *Kolmogorov complexity* of a bit string x.

3 Functions that Preserve Strong Malignness

In this section we show several functions that preserve strong malignness.

Theorem 1. *Suppose that $f(x)$ satisfies the two conditions:*

(1) *there exists $a_0 > 0$ such that $f(x)/x$ is non-decreasing in the interval $(0, a_0]$,*
(2) *for any $c > 0$ there exist $a' > 0$ and $c' > 0$ such that $f(cx) \geq c' f(x)$ in the interval $(0, a']$.*

Then $f(x)$ preserves strong malignness.

Proof. First we note that, if $x_0, x_1, \ldots$ are nonnegative real numbers such that $\Sigma_i x_i \leq a_0$ then $f(\Sigma_i x_i) \geq \Sigma_i f(x_i)$ by the condition (1). This can be shown as follows. We may assume that each x_i is positive because $f(0) = 0$. Then we have

$$
\begin{aligned}
f(\Sigma_i x_i) &= (\Sigma_i x_i)(f(\Sigma_j x_j)/\Sigma_j x_j) \\
&= \Sigma_i(x_i(f(\Sigma_j x_j)/\Sigma_j x_j)) \\
&\geq \Sigma_i(x_i(f(x_i)/x_i)) \\
&= \Sigma_i f(x_i).
\end{aligned}
$$

Let $\mu(x)$ be an arbitrary universal distribution. We show that $f(\mu(x))$ is strongly malign. We will use the simpler definition of strong malignness.

Let $\phi(x)$ be an arbitrary partial recursive function from Σ^* to Σ^*. We will show the existence of a positive constant c such that $f(\mu(\phi^{-1}(x))) \leq cf(\mu(x))$ for any x. (By $f(\mu(\phi^{-1}(x)))$ we mean $\Sigma_y \{f(\mu(y)) \mid y \in \phi^{-1}(x)\}$, not $f(z)$ with $z = \mu(\phi^{-1}(x))$.)

Let $g(u)$ be the partial function defined by $g(u) = \phi(\phi_U(u))$. This $g(u)$ is a prefix-free partial recursive function, and hence there exists i such that $g(u) = \phi_i(u)$. Note that $f(x)$ is nondecreasing in the interval $(0, a_0]$ by the condition (1).

We have the following formula for all bit strings x except a finite number of bit strings. Here, $c_1, \ldots, c_5$ denote some appropriate positive constants.

$$
\begin{aligned}
f(\mu(x)) &\geq f(c_1 \tilde{\mu}(x)) \geq c_2 f(\tilde{\mu}(x)) \\
&= c_2 f(\Sigma\{2^{-|u|} \mid u \in \Sigma^*, \phi_U(u) = x\}) \\
&\geq c_2 f(\Sigma\{2^{-|0^i 1v|} \mid v \in \Sigma^*, \phi_U(0^i 1v) = x\}) \\
&= c_2 f(2^{-i-1} \Sigma\{2^{-|v|} \mid v \in \Sigma^*, \phi_i(v) = x\}) \\
&= c_2 f(2^{-i-1} \Sigma\{2^{-|v|} \mid v \in \Sigma^*, \phi(\phi_U(v)) = x\}) \\
&= c_2 f(2^{-i-1} \Sigma\{\Sigma\{2^{-|v|} \mid v \in \Sigma^*, \phi_U(v) = w\} \mid w \in \phi^{-1}(x)\}) \\
&= c_2 f(2^{-i-1} \Sigma\{\tilde{\mu}(w) \mid w \in \phi^{-1}(x)\}) \\
&\geq c_2 c_3 f(\Sigma\{\tilde{\mu}(w) \mid w \in \phi^{-1}(x)\}) \\
&\geq c_2 c_3 f(c_4 \Sigma\{\mu(w) \mid w \in \phi^{-1}(x)\}) \\
&\geq c_2 c_3 c_5 f(\Sigma\{\mu(w) \mid w \in \phi^{-1}(x)\}) \\
&\geq c_2 c_3 c_5 \Sigma\{f(\mu(w)) \mid w \in \phi^{-1}(x)\} \\
&= c_2 c_3 c_5 f(\mu(\phi^{-1}(x))).
\end{aligned}
$$

To justify the first inequality $f(\mu(x)) \geq f(c_1 \tilde{\mu}(x))$ we used the property $\tilde{\mu}(x) \preceq \mu(x)$, the monotonicity of $f(x)$ in $(0, a_0]$ and the fact that the number of x such that $\mu(x) > a_0$ is finite. We use similar arguments in deriving other inequalities.

The above inequality might not be true for some finite number of x. However, for any x we have $f(\mu(x)) > 0$. Hence there exists a positive constant c such that $(1/c)f(\mu(\phi^{-1}(x))) \leq f(\mu(x))$ for any x. $\qquad\square$

As an example of application of this theorem, consider the following functions $g(x)$ satisfying $1 \leq \lim_{x \to 0} g(x) < \infty$:

Type 1: $g(x) = t + 1/(1/x)^r (= t + x^r)$, $1 \le t$, $0 < r$,
Type 2: $g(x) = t + 1/(\log(1/x))^r$, $1 \le t$, $1 < r$,
Type 3: $g(x) = t + 1/(\log(1/x))^r$, $1 \le t$, $0 < r \le 1$,
Type 4: $g(x) = t + 1/(\log^m(1/x))^r$, $1 \le t$, $2 \le m$, $0 < r$,
Type 5: $g(x) = t$, $1 \le t$,
Type 6: $g(x) = t - 1/(\log^m(1/x))^r$, $1 \le t$, $2 \le m$, $0 < r$,
Type 7: $g(x) = t - 1/(\log(1/x))^r$, $1 \le t$, $0 < r < 1$,
Type 8: $g(x) = t - 1/(\log(1/x))^r$, $1 \le t$, $1 \le r$,
Type 9: $g(x) = t - 1/(1/x)^r (= t - x^r)$, $1 \le t$, $0 < r$.

For each of these functions $g(x)$ we can easily show the followings for the correspoinding $f(x) = x^{g(x)}$:

(1) $f(x) > 0$ for $x > 0$,
(2) $\lim_{x \to 0} f(x) = 0$ if we define $f(0)$ to be 0,
(3) $f(x)$ satisfies the condition (1) of Theorem 1 except the cases where $t = 1$ and $g(x)$ is of Types 1, 2, 6, 7.
(4) $f(x)$ satisfies the condition (2) of Theorem 1.

Hence, by Theorem 1 we know that these functions $f(x)$ except the four cases mentioned in (3) preserve strong malignness. Especially, from the result for Type 5 $(g(x) = t, 1 \le t)$ we have the following corollary.

Corollary 2. *If $t \ge 1$ then $f(x) = x^t$ preserves strong malignness.*

Note that the conditions (1), (2) of Theorem 1 concern the behavior of $f(x)$ in a small interval of the form $(0, a]$, and a may be arbitrarily small. Hence, if $f(x)$ has a Maclaurin expansion $b_0 + b_1 x + b_2 x^2 + \ldots$ with a non-zero radius of convergence, then $f(x)$ satisfies these conditions for the following cases:

- $b_0 = 0, b_1 > 0, b_2 = b_3 = \ldots = 0$,
- $b_0 = 0, b_1 > 0, b_2 = \ldots = b_{s-1} = 0, b_s > 0$ $(s \ge 2)$,
- $b_0 = b_1 = \ldots = b_{s-1} = 0, b_s > 0$ $(s \ge 2)$.

The following functions are examples of such functions.

$f(x) = ax$ $(a > 0)$,
$f(x) = x/(1-x) = x + x^2 + x^3 + \ldots$,
$f(x) = e^x - 1 = x + x^2/2 + x^3/6 + \ldots$,
$f(x) = \tan x = x + x^3/3 + \ldots$,
$f(x) = 1 - \cos x = x^2/2 - x^4/24 + \ldots$.

For each of these functions $f(x)$, if $\mu(x)$ is a universal distribution and $f(\mu(x)) > 0$ for each x then $f(\mu(x))$ is strongly malign.

Lemma 3. *If $\mu(x)$ is a universal distribution and $t > 1$ then $\mu(x)^t$ is a distribution that is not universal.*

Proof. First we show that $f(\mu(x))$ is a distribution, that is $0 < f(\mu(\Sigma^*)) < \infty$, if $f(x)$ satisfies the conditions (1), (2) of Theorem 1. The property $0 < f(\mu(\Sigma^*))$ is obvious because $f(x) > 0$ for any $x > 0$ and $\mu(x) > 0$ for any x. The property $f(\mu(\Sigma^*)) < \infty$ follows from the following inequation:

$$f(\mu(\Sigma^*)) = \Sigma\{f(\mu(x)) \mid x \in \Sigma^*, \mu(x) > a_0\} + \Sigma\{f(\mu(x)) \mid x \in \Sigma^*, \mu(x) \le a_0\}$$
$$\le \Sigma\{f(\mu(x)) \mid x \in \Sigma^*, \mu(x) > a_0\} + \Sigma\{(f(a_0)/a_0)\mu(x) \mid \mu(x) \le a_0\}.$$

The first term of the last formula is finite because the number of x such that $\mu(x) > a_0$ is finite and the second term is finite because it is at most $(f(a_0)/a_0)$ $\mu(\Sigma^*)$.

Next we show that $\mu(x)^t$ is not universal. Suppose that $\mu(x)^t$ is universal. Then there exists a positive constant c such that $\mu(x) \le c\mu(x)^t$ and hence $1/c^{1/(t-1)} \le \mu(x)$ for any x. This constadicts $\mu(\Sigma^*) < \infty$. $\square$

From Corollary 2 and Lemma 3, we know that, for each universal distribution $\mu(x)$ and each $t > 1$, $\mu(x)^t$ is an example of strongly malign distributions that are not universal.

For the case $1 < \lim_{x\to 0} g(x) < \infty$, $f(x) = x^{g(x)}$ preserves strong malignness for each $g(x)$ in the above mentioned list of nine functions. At present we do not know whether there exist functions $f(x) = x^{g(x)}$ with $1 < \lim_{x\to 0} g(x) < \infty$ that do not preserve strong malignness or not.

For the case $\lim_{x\to 0} g(x) = 1$, as we mentioned above, Theorem 1 fails to show that $f(x) = x^{g(x)}$ preserves strong malignness for the following four $g(x)$:

$$g(x) = 1 + 1/(1/x)^r (=1 + x^r),\ 0 < r,$$
$$g(x) = 1 + 1/(\log(1/x))^r,\ 1 < r,$$
$$g(x) = 1 - 1/(\log^m(1/x))^r,\ 2 \le m, 0 < r,$$
$$g(x) = 1 - 1/(\log(1/x))^r,\ 0 < r < 1.$$

In the following section we show that $f(x) = x^{g(x)}$ does not preserve malignness for the latter two $g(x)$ of these four. Hence, we know that in the class of functions $f(x) = x^{g(x)}$ with $\lim_{x\to 0} g(x) = 1$ some preserve strong malignness and some do not preserve malignness. At present we do not know whether $f(x) = x^{g(x)}$ for the former two $g(x)$ of the four (especially $f(x) = x^{1+x}$) preserve malignness or not.

4   Functions that do not Preserve Malignness

In this section we show several functions that do not preserve malignness. First we consider functions $f(x) = x^{g(x)}$ such that $0 \le \lim_{x\to 0} g(x) < 1$.

Theorem 4. *If either* $\lim_{x\to 0} g(x) = 0$ *and* $g(x) \ge 0$ *in an interval of the form* $(0, a]$ *or* $0 < \lim_{x\to 0} < 1$ *then* $f(x) = x^{g(x)}$ *does not preserve malignness.*

Proof. Let $\mu(x)$ be an arbitrary universal distribution. We will show that $f(\mu(x))$ is not malign.

There are two constants t_1, t_2 such that $0 \leq t_1 \leq g(\mu(x)) \leq t_2 < 1$ for all sufficiently long x.

For a natural number n, let $\text{bin}(n)$ denote the binary representation of n. It is easy to show that there is a constant c_1 such that $K(0^n) \geq K(\text{bin}(n)) + c_1$. Moreover, it is well-known that there are infinitely many bit strings y such that $K(y) \geq |y|$ (see [3]). Hence, there are infinitely many n such that $K(0^n) \geq \log n + c_1$ (the base of logarithm is 2). For such n, we have $2^{-K(0^n)} \leq 2^{-c_1}/n$. We know that $2^{-K(x)}$ is a universal distribution. Hence, there is a positive constant c_2 such that $\mu(x) \leq c_2 2^{-K(x)}$, and hence $\mu(0^n) \leq c_2 2^{-c_1}/n$, and $f(\mu(0^n)) = \mu(0^n)^{g(\mu(0^n))} \leq (c_2 2^{-c_1}/n)^{t_1} = c_2^{t_1} 2^{-c_1 t_1}/n^{t_1}$ if n is sufficiently large.

On the other hand, for all sufficiently large n, for any $x \in \Sigma^n$ we have $K(x) \leq n + 1.01 \log n$ and consequently $2^{-K(x)} \geq 1/(2^n n^{1.01})$. The function $2^{-K(x)}$ is an enumerable distribution. Hence there is a positive constant c_3 such that $\mu(x) \geq c_3 2^{-K(x)}$. Hence, for all sufficiently large n, we have $\mu(x) \geq c_3/(2^n n^{1.01})$, $f(\mu(x)) = \mu(x)^{g(\mu(x))} \geq (c_3/(2^n n^{1.01}))^{t_2} = c_3^{t_2}/(2^{t_2 n} n^{1.01 t_2})$ for any $x \in \Sigma^n$, and hence $f(\mu(\Sigma^n)) \geq c_3^{t_2} 2^{(1-t_2)n}/n^{1.01 t_2}$.

Therefore, there are infinitely many n such that $f(\mu(0^n))/f(\mu(\Sigma^n)) \leq (c_2^{t_1} 2^{-c_1 t_1}/c_3^{t_2})(n^{1.01 t_2 - t_1}/2^{(1-t_2)n})$.

Let A be an algorithm whose computation time for an input x of length n is 2^n or n^2 according as $x = 0^n$ or not. Then we have

$$t_A^{\text{av}, f(\mu(x))}(n) = (f(\mu(0^n))/f(\mu(\Sigma^n)))2^n + (1 - (f(\mu(0^n))/f(\mu(\Sigma^n)))n^2$$
$$\leq (f(\mu(0^n))/f(\mu(\Sigma^n)))2^n + n^2.$$

Hence, there are infinitely many n such that

$$t_A^{\text{av}, f(\mu(x))}(n)/t_A^{\text{wo}}(n) \leq (c_2^{t_1} 2^{-c_1 t_1}/c_3^{t_2})(n^{1.01 t_2 - t_1}/2^{(1-t_2)n}) + n^2/2^n,$$

and the righthand side value of this inequality approaches 0 as n becomes large. This means that there cannot be a positive constant c_4 such that $t_A^{\text{av}, f(\mu(x))}(n)/t_A^{\text{wo}}(n) \geq c_4$ for any n, and hence the semi-distribution $f(\mu(x))$ is not malign. $\square$

As a special case, we have the following corollary.

Corollary 5. *If $0 < t < 1$ then $f(x) = x^t$ does not preserve malignness.*

Using the same idea, we can show that $f(x) = x^{g(x)}$ for the two of the four $g(x)$ mentioned at the end of the previous section do not preserve malignness.

Theorem 6. *The functions $f(x) = x^{g(x)}$ for the following two $g(x)$ do not preserve malignness:*

$$g(x) = 1 - 1/(\log^m(1/x))^r, \quad 2 \leq m, 0 < r,$$
$$g(x) = 1 - 1/(\log(1/x))^r, \quad 0 < r < 1.$$

Proof. The proof is almost the same as that of Theorem 4. For the first $g(x)$ we have

$$t_A^{\text{av},f(\mu(x))}(n)/t_A^{\text{wo}}(n) \leq c_1 n^{0.01} n^{(\log n)^{0.01}}/2^{n^{0.99}} + n^2/2^n$$

and for the second $g(x)$ we have

$$t_A^{\text{av},f(\mu(x))}(n)/t_A^{\text{wo}}(n) \leq c_2 n^{0.01} n^{1/(\log n)^{0.99r}}/2^{n^{0.99(1-r)}} + n^2/2^n$$

for infinitely many n (c_1, c_2 are positive constants). Both of the righthand side values approach 0 as n becomes large. The remainder of the proof is completely the same. □

Finally we consider functions $f(x) = x^{g(x)}$ such that $\lim_{x \to 0} g(x) = \infty$.

Theorem 7. *If $g(x)$ is non-increasing in some interval of the form $(0, a]$ and $\lim_{x \to 0} g(x) = \infty$, then $f(x) = x^{g(x)}$ does not preserve malignness.*

Proof. Note that $f(x)$ is non-decreasing in some interval of the form $(0, a']$.

Let $\mu_1(x)$ be an arbitrary universal distribution. There exists a constant c_1 such that $K(1^n) \leq K(0^n) + c_1$ for any n. Hence there exist positive constants c_2, c_3 such that $\mu_1(1^n) \geq c_2 2^{-K(1^n)} \geq 2^{-c_1} c_2 2^{-K(0^n)} \geq 2^{-c_1} c_2 c_3 \mu_1(0^n)$ for any n.

Let $\mu_2(x)$ be defined by

$$\mu_2(x) = \begin{cases} (2/(2^{-c_1} c_2 c_3)) \mu_1(x) & x = 1^n, \\ \mu_1(x) & x \neq 1^n \end{cases}$$

for x of length n. Then $\mu_2(x)$ is a universal distribution because we may assume that $2/(2^{-c_1} c_2 c_3)$ is a rational number, and we have $\mu_2(1^n) = (2/(2^{-c_1} c_2 c_3)) \mu_1(1^n) \geq 2\mu_1(0^n)$. We will show that $\mu_3(x) = f(\mu_2(x)) = \mu_2(x)^{g(\mu_2(x))}$ is not malign.

Let A be an algorithm whose computation time for an input x of length n is 2^n or n^2 according as $x = 0^n$ or not. Then we have $t_A^{\text{wo}}(n) = 2^n$ and

$$\begin{aligned} t_A^{\text{av},\mu_3(x)}(n) &\leq (\mu_3(0^n)/\mu_3(\Sigma^n))2^n + n^2 \\ &\leq (\mu_3(0^n)/\mu_3(1^n))2^n + n^2 \\ &= (f(\mu_2(0^n))/f(\mu_2(1^n)))2^n + n^2 \\ &\leq (f(\mu_1(0^n))/f(2\mu_1(0^n)))2^n + n^2. \end{aligned}$$

Denoting $\mu_1(0^n)$ by s and assuming that n is sufficiently large and hence s is sufficiently small, we have

$$\begin{aligned} t_A^{\text{av},\mu_3(x)}(n) &\leq (s^{g(s)}/(2s)^{g(2s)})2^n + n^2 \\ &= s^{g(s)-g(2s)}(1/2^{g(2s)})2^n + n^2 \\ &\leq (1/2^{g(2s)})2^n + n^2. \end{aligned}$$

Hence, for all sufficiently large n we have

$$t_A^{\text{av},\mu_3(x)}(n)/t_A^{\text{wo}}(n) \leq 1/2^{g(2s)} + n^2/2^n,$$

and the righthand side value of this inequality approaches 0 as n becomes large because $g(2s)$ approaches ∞. Hence, the semi-distribution $\mu_3(x)$ is not malign. $\square$

Note that in the proof of Theorem 7 we did not prove that $f(\mu(x))$ is not malign for any universal distribution $\mu(x)$. The functions $x^{1/x}$, $x^{\log(1/x)}$, $x^{\log\log(1/x)}$, ... are examples of $f(x)$ mentioned in Theorem 7. These functions satisfy the condition (1) of Theorem 1, but not the condition (2). Hence the condition (2) is really necessary in Theorem 1.

References

1. Kobayashi, K.: On malign input distributions for algorithms. IEICE Trans. on Information and Systems **E76-D** (1993) 634–640
2. Li, M., Vitányi, P. M. B.: Worst case complexity is equal to average case complexity under the universal distribution. Inform. Process. Lett. **42** (1992) 145–149
3. Li, M., Vitányi, P. M. B.: An Introduction to Kolmogorov Complexity and Its Applications. Springer-Verlag (1993)
4. Miltersen, P. B.: The complexity of malign ensembles. SIAM J. Computing **22** (1993) 147–156

Intersection Suffices for Boolean Hierarchy Equivalence

Lane A. Hemaspaandra[1]* and Jörg Rothe[2]**

[1] Department of Computer Science, University of Rochester, Rochester, NY 14627, USA
[2] Institut für Informatik, Friedrich-Schiller-Universität Jena, 07743 Jena, Germany

Abstract. It is known that for any class C closed under *union and intersection*, the Boolean closure of C, the Boolean hierarchy over C, and the symmetric difference hierarchy over C all are equal. We prove that these equalities hold for any complexity class closed under *intersection*.

1   Introduction

NP and NP-based hierarchies—such as the polynomial hierarchy [MS72, Sto77] and the Boolean hierarchy over NP [CGH+88, CGH+89, KSW87]—have played such a central role in complexity theory, and have been so thoroughly investigated, that it would be natural to take them as predictors of the behavior of other classes or hierarchies. However, over and over during the past decade it has been shown that NP is a singularly poor predictor of the behavior of other classes.

In light of the many ways in which NP parts company with certain other classes (see [HH88, HJV93, HJ93, BG94, HR92, Reg89]), it is clear that we should not merely assume that results for NP hold for other classes, but, rather, we must carefully check to see to what extent, if any, results for NP suggest results for other classes. In this paper, we study whether the structure of the Boolean hierarchy over NP can be extended to classes that, unlike NP, are not known to be closed under union.

For the Boolean hierarchy over NP, which has generated quite a bit of interest and the collapse of which is known to imply the collapse of the polynomial hierarchy [Kad88, CK90a, BCO93], a large number of definitions are known to be equivalent. For example, for NP, all the following coincide [CGH+88]: the Boolean closure of NP, the Boolean (alternating sums) hierarchy, the nested difference hierarchy, and the Hausdorff hierarchy. The symmetric difference hierarchy also characterizes the Boolean closure of NP [KSW87]. In fact, these equalities are known to hold for all classes that contain Σ^* and $\emptyset$ and are closed under union and intersection [Hau14, CGH+88, KSW87, BBJ+89, Cha91, CK90b]. We prove that both the symmetric difference hierarchy (SDH) and the Boolean hierarchy (CH) remain equal to the Boolean closure (BC) *even in the absence of the assumption of closure under union.* That is, for any class $\mathcal{K}$ containing Σ^* and $\emptyset$ and closed under intersection (e.g.,

* Work done in part while visiting the University of Amsterdam and the Friedrich-Schiller-Universität Jena. Supported in part by grants NSF-CCR-8957604, NSF-INT-9116781/JSPS-ENGR-207, and NSF-CCR-9322513, and by an NAS/NRC COBASE grant.
** Work done in part while visiting the University of Rochester. Supported in part by a grant from the DAAD. Email: rothe@mipool.uni-jena.de.

UP, US, and DP, first defined respectively in [Val76], [BG82], and [PY84], and each of which is not currently known to be closed under union): $\text{SDH}(\mathcal{K}) = \text{CH}(\mathcal{K}) = \text{BC}(\mathcal{K})$. However, for the remaining two hierarchies, we show that not all classes containing Σ^* and $\emptyset$ and closed under intersection robustly (i.e., in every relativized world) display equality. In particular, the Hausdorff hierarchy over UP and the nested difference hierarchy over UP both fail to robustly capture the Boolean closure of UP. In fact, the failure is relatively severe; we show that even low levels of other Boolean hierarchies over UP—the third level of the symmetric difference hierarchy and the fourth level of the Boolean (alternating sums) hierarchy—fail to be robustly captured by either the Hausdorff hierarchy or the nested difference hierarchy.

2 Boolean Hierarchies over Classes Closed Under Intersection

We consider sets of strings over the alphabet $\Sigma = \{0, 1\}$. For each set $L \subseteq \Sigma^*$, $\overline{L} \stackrel{\text{df}}{=} \Sigma^* - L$ denotes the complement of L. For sets A and B, $A \triangle B \stackrel{\text{df}}{=} (A - B) \cup (B - A)$. For any class $\mathcal{C}$ of sets over Σ, define $\text{co}\mathcal{C} \stackrel{\text{df}}{=} \{L \mid \overline{L} \in \mathcal{C}\}$, and let $\text{BC}(\mathcal{C})$ denote the Boolean algebra generated by $\mathcal{C}$. For classes $\mathcal{C}$ and $\mathcal{D}$, define

$$\mathcal{C} \wedge \mathcal{D} \stackrel{\text{df}}{=} \{A \cap B \mid A \in \mathcal{C} \wedge B \in \mathcal{D}\}, \quad \mathcal{C} \triangle \mathcal{D} \stackrel{\text{df}}{=} \{A \triangle B \mid A \in \mathcal{C} \wedge B \in \mathcal{D}\},$$
$$\mathcal{C} \vee \mathcal{D} \stackrel{\text{df}}{=} \{A \cup B \mid A \in \mathcal{C} \wedge B \in \mathcal{D}\}, \quad \mathcal{C} - \mathcal{D} \stackrel{\text{df}}{=} \{A - B \mid A \in \mathcal{C} \wedge B \in \mathcal{D}\}.$$

The Boolean hierarchy is a natural extension of the classes NP [Coo71, Lev73] and $\text{DP} \stackrel{\text{df}}{=} \text{NP} \wedge \text{coNP}$ [PY84]. Both NP and DP contain natural problems, as do the levels of the Boolean hierarchy. For example, graph minimal uncolorability is known to be complete for DP [CM87]. Note that DP clearly is closed under intersection, but is not closed under union unless the polynomial hierarchy collapses (due to [Kad88], see also [CK90a, Cha91]).

Definition 1. [CGH$^+$88, KSW87, Hau14]   Let $\mathcal{K}$ be any class of sets.

1. The *Boolean ("alternating sums") hierarchy over* $\mathcal{K}$: $\text{CH}(\mathcal{K}) \stackrel{\text{df}}{=} \bigcup_{k \geq 1} \text{C}_k(\mathcal{K})$,

$$\text{C}_1(\mathcal{K}) \stackrel{\text{df}}{=} \mathcal{K}, \quad \text{C}_k(\mathcal{K}) \stackrel{\text{df}}{=} \begin{cases} \text{C}_{k-1}(\mathcal{K}) \vee \mathcal{K} & \text{if } k \text{ odd} \\ \text{C}_{k-1}(\mathcal{K}) \wedge \text{co}\mathcal{K} & \text{if } k \text{ even} \end{cases}, \quad k \geq 2.$$

2. The *nested difference hierarchy over* $\mathcal{K}$: $\text{DH}(\mathcal{K}) \stackrel{\text{df}}{=} \bigcup_{k \geq 1} \text{D}_k(\mathcal{K})$, $\text{D}_1(\mathcal{K}) \stackrel{\text{df}}{=} \mathcal{K}$, $\text{D}_k(\mathcal{K}) \stackrel{\text{df}}{=} \mathcal{K} - \text{D}_{k-1}(\mathcal{K})$, $k \geq 2$.

3. The *Hausdorff ("union of differences") hierarchy over* $\mathcal{K}$: $\text{EH}(\mathcal{K}) \stackrel{\text{df}}{=} \bigcup_{k \geq 1} \text{E}_k(\mathcal{K})$, $\text{E}_1(\mathcal{K}) \stackrel{\text{df}}{=} \mathcal{K}$, $\text{E}_2(\mathcal{K}) \stackrel{\text{df}}{=} \mathcal{K} - \mathcal{K}$, $\text{E}_k(\mathcal{K}) \stackrel{\text{df}}{=} \text{E}_2(\mathcal{K}) \vee \text{E}_{k-2}(\mathcal{K})$, $k > 2$.

4. The *symmetric difference hierarchy over* $\mathcal{K}$: $\text{SDH}(\mathcal{K}) \stackrel{\text{df}}{=} \bigcup_{k \geq 1} \text{SD}_k(\mathcal{K})$, $\text{SD}_1(\mathcal{K}) \stackrel{\text{df}}{=} \mathcal{K}$, $\text{SD}_k(\mathcal{K}) \stackrel{\text{df}}{=} \text{SD}_{k-1}(\mathcal{K}) \triangle \mathcal{K}$, $k \geq 2$.

Clearly, for any $\text{X} \in \{\text{C}, \text{D}, \text{E}, \text{SD}\}$, if $\mathcal{K}$ is nontrivial (i.e., contains $\emptyset$ and Σ^*), then for any $k \geq 1$, $\text{X}_k(\mathcal{K}) \cup \text{coX}_k(\mathcal{K}) \subseteq \text{X}_{k+1}(\mathcal{K}) \cap \text{coX}_{k+1}(\mathcal{K})$.

Fact 2 *For every class $\mathcal{K}$ of sets and every $n \geq 1$, $D_{2n-1}(\mathcal{K}) = \mathrm{coC}_{2n-1}(\mathrm{co}\mathcal{K})$ and $D_{2n}(\mathcal{K}) = C_{2n}(\mathrm{co}\mathcal{K})$.*

Corollary 3. $\mathrm{CH}(\mathrm{UP}) = \mathrm{coCH}(\mathrm{UP}) = \mathrm{DH}(\mathrm{coUP})$.

We are interested in the Boolean hierarchies over classes closed under intersection (but perhaps not under union or complementation), such as UP, US, and DP. We state our theorems in terms of the class of primary interest to us, UP. However, many apply to any nontrivial class closed under intersection (see Theorem 7). We first prove that the symmetric difference hierarchy over UP (or any class closed under intersection) equals the Boolean closure. Though Köbler, Schöning, and Wagner [KSW87] proved this for NP, their proof gateways through a class whose proof of equivalence to the Boolean closure uses closure under union, and thus the following result is not implicit in their paper.

Theorem 4 $\mathrm{SDH}(\mathrm{UP}) = \mathrm{BC}(\mathrm{UP})$.

Proof. The inclusion from left to right is clear. For the converse inclusion, it is sufficient to show that SDH(UP) is closed under all Boolean operations, as BC(UP), by definition, is the smallest class of sets that contains UP and is closed under all Boolean operations. Let L and L' be arbitrary sets in SDH(UP). Then, for some $k, \ell \geq 1$, there are sets $A_1, \ldots, A_k, B_1, \ldots, B_\ell$ in UP such that $L = A_1 \Delta \cdots \Delta A_k$ and $L' = B_1 \Delta \cdots \Delta B_\ell$. Thus, $L \cap L' = \left(\Delta_{i=1}^{k} A_i \right) \cap \left(\Delta_{j=1}^{\ell} B_j \right) = \Delta_{i \in \{1, \ldots, k\}, j \in \{1, \ldots, \ell\}} (A_i \cap B_j)$, and since UP is closed under intersection and SDH(UP) is (trivially) closed under symmetric difference, we clearly have that $L \cap L' \in \mathrm{SDH}(\mathrm{UP})$. Furthermore, since $\overline{L} = \Sigma^* \Delta L$ implies that $\overline{L} \in \mathrm{SDH}(\mathrm{UP})$, SDH(UP) is closed under complementation. Since all Boolean operations can be represented in terms of complementation and intersection, our proof is complete. $\square$

Next, we show that for any class closed under intersection, instantiated below to the case of UP, the Boolean (alternating sums) hierarchy over the class equals the Boolean closure of the class. Our proof is inspired by the techniques used to prove equality in the case where closure under union may be assumed.

Theorem 5 $\mathrm{CH}(\mathrm{UP}) = \mathrm{BC}(\mathrm{UP})$.

Proof. We will prove that $\mathrm{SDH}(\mathrm{UP}) \subseteq \mathrm{CH}(\mathrm{UP})$. By Theorem 4, this will suffice. Let L be any set in SDH(UP). Then, $L \in \mathrm{SD}_k(\mathrm{UP})$ for some $k > 1$ (the case $k = 1$ is trivial). Let $U_1, \ldots, U_k$ be the witnessing UP sets; that is, $L = U_1 \Delta U_2 \Delta \cdots \Delta U_k$. By the inclusion-exclusion rule, L satisfies the equalities below. For odd k,

$$L = \left(\cdots \left(\left((U_1 \cup U_2 \cup \cdots \cup U_k) \cap \left(\overline{\bigcup_{j_1 < j_2} (U_{j_1} \cap U_{j_2})} \right) \right) \cup \right. \right.$$
$$\left. \left. \left(\bigcup_{j_1 < j_2 < j_3} (U_{j_1} \cap U_{j_2} \cap U_{j_3}) \right) \right) \cap \cdots \cup \left(\bigcup_{j_1 < \cdots < j_k} (U_{j_1} \cap \cdots \cap U_{j_k}) \right) \right),$$

where each subscripted j term must belong to $\{1, \ldots, k\}$. For even k, the last term in this expression is complemented. For notational convenience, let us use $A_1, \ldots, A_k$ to represent the respective terms in the above expression (ignoring the complementations).

By the closure of UP under intersection, each A_i, $1 \leq i \leq k$, is the union of $\binom{k}{i}$ UP sets $B_{i,1}, \ldots, B_{i,\binom{k}{i}}$. Using the fact that $\emptyset$ is clearly in UP, we can easily turn the union of n arbitrary UP sets (or the intersection of n arbitrary coUP sets) into an alternating sum of $2n - 1$ UP sets. So for instance, $A_1 = U_1 \cup U_2 \cup \cdots \cup U_k$ can be written $(\cdots (((U_1 \cap \bar{\emptyset}) \cup U_2) \cap \bar{\emptyset}) \cup \cdots \cup U_k)$, call this C_1. Clearly, $C_1 \in C_{2k-1}(\text{UP})$. To transform the above representation of L into an alternating sum of UP sets, we need two (trivial) transformations holding for any $m \geq 1$ and arbitrary sets S and $T_1, \ldots, T_m$:

$$S \cap (\overline{T_1 \cup T_2 \cup \cdots \cup T_m}) = (\cdots ((S \cap \overline{T_1}) \cap \overline{T_2}) \cap \cdots) \cap \overline{T_m} \qquad (1)$$
$$S \cup (T_1 \cup T_2 \cup \cdots \cup T_m) = (\cdots ((S \cup T_1) \cup T_2) \cup \cdots) \cup T_m. \qquad (2)$$

Using (1) with $S = C_1$ and $T_1 = B_{2,1}, \ldots, T_m = B_{2,\binom{k}{2}}$ and the fact that $\emptyset$ is in UP, $A_1 \cap \overline{A_2}$ can be transformed into an alternating sum of UP sets, call this C_2. Now apply (2) with $S = C_2$ and $T_1 = B_{3,1}, \ldots, T_m = B_{3,\binom{k}{3}}$ to obtain, again using that $\emptyset$ is in UP, an alternating sum $C_3 = (A_1 \cap \overline{A_2}) \cup A_3$ of UP sets, and so on. Eventually, this procedure of alternately applying (1) and (2) will yield an alternating sum C_k of sets in UP that equals L. Thus, $L \in \text{CH}(\text{UP})$. $\qquad \square$

Note that the proof of Theorem 5 implicitly gives a recurrence yielding an upper bound on the level-wise containments. We find the issue of equality to BC(UP), or lack thereof, to be the central issue, and thus we focus on that. Nonetheless, we point out that losing the assumption of closure under union seems to have exacted a price. Though the hierarchies SDH(UP) and CH(UP) are indeed equal, the above proof embeds $\text{SD}_k(\text{UP})$ in an exponentially higher level of the C hierarchy, namely, $C_{2^{k+1}-k-2}(\text{UP})$. Similarly, it is not hard to see, related to the proof of Theorem 4, that for each $k \geq 1$, $C_k(\text{UP}) \subseteq \text{SD}_{T(k)}(\text{UP})$, where $T(k) = 2^k - 1$ if k is odd, and $T(k) = 2^k - 2$ if k is even. Theorem 6 below shows that the nested difference hierarchy is contained in any of the other hierarchies considered. Surprisingly, it turns out that, relative to a recursive oracle, even the fourth level of CH(UP) and the third level of SDH(UP) are not subsumed by any level $\text{E}_k(\text{UP})$ of the EH(UP) hierarchy. Consequently, neither the D nor the E normal forms of Definition 1 capture the Boolean closure of UP. Theorem 8 (the proof of which is contained in the full version of this paper) is optimal, as clearly $C_3(\text{UP}) \subseteq \text{EH}(\text{UP})$ and $\text{SD}_2(\text{UP}) \subseteq \text{EH}(\text{UP})$, and both these containments relativize.

Theorem 6 *For every $k \geq 1$, $\text{D}_k(\text{UP}) \subseteq \text{C}_k(\text{UP}) \cap \text{E}_k(\text{UP})$.*

Proof. For the first inclusion, by [CH85, Proposition 2.1.2], each set L in $\text{D}_k(\text{UP})$ can be represented as $L = A_1 - (A_2 - (\cdots (A_{k-1} - A_k) \cdots))$, where $A_i = \bigcap_{1 \leq j \leq i} L_j$, $1 \leq i \leq k$, and the L_j's are the original UP sets representing L. Note that since the proof of [CH85, Proposition 2.1.2] only uses intersection, the sets A_i are in UP. A special case of [CH85, Proposition 2.1.3] says that sets in $\text{D}_k(\text{UP})$ via decreasing chains such as the A_i are in $\text{C}_k(\text{UP})$, and so $L \in \text{C}_k(\text{UP})$. The proof of the second inclusion is done by induction on the odd and even levels separately. The induction base follows by definition in either case. For odd levels, assume $\text{D}_{2n-1}(\text{UP}) \subseteq \text{E}_{2n-1}(\text{UP})$ to be valid, and let L be any set in $\text{D}_{2n+1}(\text{UP}) = \text{UP} - (\text{UP} - \text{D}_{2n-1}(\text{UP}))$. By our inductive hypothesis, L

can be represented as $L = A - \left(B - \left(\bigcup_{i=1}^{n-1} \left(C_i \cap \overline{D_i} \right) \cup E \right) \right)$, where A, B, C_i, D_i, and E are sets in UP. Thus,

$$L = A \cap \left(\overline{B \cap \left(\overline{\bigcup_{i=1}^{n-1} \left(C_i \cap \overline{D_i} \right) \cup E} \right)} \right) = A \cap \left(\overline{B} \cup \left(\bigcup_{i=1}^{n-1} \left(C_i \cap \overline{D_i} \right) \cup E \right) \right)$$

$$= (A \cap \overline{B}) \cup \left(\bigcup_{i=1}^{n-1} A \cap C_i \cap \overline{D_i} \right) \cup (A \cap E) = \left(\bigcup_{i=1}^{n} F_i \cap \overline{D_i} \right) \cup G,$$

where $F_i = A \cap C_i$, $1 \le i \le n-1$, $F_n = A$, $D_n = B$, and $G = A \cap E$. Since UP is closed under intersection, each of these sets is in UP. Thus, $L \in E_{2n+1}(\text{UP})$. The proof for the even levels is analogous except that the set E is dropped. $\square$

Theorem 7. *Theorems 4, 5, and 6 apply to all nontrivial classes closed under intersection.*

Remark. Although DP is closed under intersection but seems to lack closure under union (unless the polynomial hierarchy collapses to DP [Kad88, CK90b, Cha91]) we note that the known results about the Boolean hierarchy over NP [CGH+88, KSW87] in fact even for the DP case imply stronger results than those given by our Theorem 7, due to the very special structure of DP. This appears to contrast with the UP case.

Theorem 8 *There are recursive oracles A and D (though we may take $A = D$) such that $C_4(\text{UP}^A) \not\subseteq \text{EH}(\text{UP}^A)$ and $\text{SD}_3(\text{UP}^D) \not\subseteq \text{EH}(\text{UP}^D)$.*

Corollary 9. *There is a recursive oracle A such that $\text{EH}(\text{UP}^A) \neq \text{BC}(\text{UP}^A)$ and $\text{DH}(\text{UP}^A) \neq \text{BC}(\text{UP}^A)$.*

Acknowledgments

We are very grateful to Gerd Wechsung for his help in bringing about this collaboration, and for his kind and insightful advice over many years. We thank Marius Zimand for proofreading, and Nikolai Vereshchagin for helpful discussions during his visit to Rochester.

References

[BBJ+89] A. Bertoni, D. Bruschi, D. Joseph, M. Sitharam, and P. Young. Generalized Boolean hierarchies and Boolean hierarchies over RP. In *Proceedings of the 7th Conference on Fundamentals of Computation Theory*, pages 35–46. Springer-Verlag *Lecture Notes in Computer Science #380*, August 1989.

[BCO93] R. Beigel, R. Chang, and M. Ogiwara. A relationship between difference hierarchies and relativized polynomial hierarchies. *Mathematical Systems Theory*, 26:293–310, 1993.

[BG82] A. Blass and Y. Gurevich. On the unique satisfiability problem. *Information and Control*, 55:80–88, 1982.

[BG94] R. Beigel and J. Goldsmith. Downward separation fails catastrophically for limited nondeterminism classes. In *Proceedings of the 9th Structure in Complexity Theory Conference*, pages 134–138. IEEE Computer Society Press, June/July 1994.

435

[CGH+88] J. Cai, T. Gundermann, J. Hartmanis, L. Hemachandra, V. Sewelson, K. Wagner, and G. Wechsung. The Boolean hierarchy I: Structural properties. *SIAM Journal on Computing*, 17(6):1232–1252, 1988.

[CGH+89] J. Cai, T. Gundermann, J. Hartmanis, L. Hemachandra, V. Sewelson, K. Wagner, and G. Wechsung. The Boolean hierarchy II: Applications. *SIAM Journal on Computing*, 18(1):95–111, 1989.

[CH85] J. Cai and L. Hemachandra. The Boolean hierarchy: Hardware over NP. Technical Report 85-724, Cornell University, Department of Computer Science, Ithaca, NY, December 1985.

[Cha91] R. Chang. *On the Structure of NP Computations under Boolean Operators*. PhD thesis, Cornell University, Ithaca, NY, 1991.

[CK90a] R. Chang and J. Kadin. The Boolean hierarchy and the polynomial hierarchy: A closer connection. In *Proceedings of the 5th Structure in Complexity Theory Conference*, pages 169–178. IEEE Computer Society Press, July 1990.

[CK90b] R. Chang and J. Kadin. On computing Boolean connectives of characteristic functions. Technical Report TR 90-1118, Department of Computer Science, Cornell University, Ithaca, NY, May 1990.

[CM87] J. Cai and G. Meyer. Graph minimal uncolorability is DP-complete. *SIAM Journal on Computing*, 16(2):259–277, 1987.

[Coo71] S. Cook. The complexity of theorem-proving procedures. In *Proceedings of the 3rd ACM Symposium on Theory of Computing*, pages 151–158, 1971.

[Hau14] F. Hausdorff. *Grundzüge der Mengenlehre*. Leipzig, 1914.

[HH88] J. Hartmanis and L. Hemachandra. Complexity classes without machines: On complete languages for UP. *Theoretical Computer Science*, 58:129–142, 1988.

[HJ93] L. Hemachandra and S. Jha. Defying upward and downward separation. In *Proceedings of the 10th Annual Symposium on Theoretical Aspects of Computer Science*, pages 185–195. Springer-Verlag *Lecture Notes in Computer Science #665*, February 1993.

[HJV93] L. Hemaspaandra, S. Jain, and N. Vereshchagin. Banishing robust Turing completeness. *International Journal of Foundations of Computer Science*, 4(3):245–265, 1993.

[HR92] L. Hemachandra and R. Rubinstein. Separating complexity classes with tally oracles. *Theoretical Computer Science*, 92(2):309–318, 1992.

[Kad88] J. Kadin. The polynomial time hierarchy collapses if the Boolean hierarchy collapses. *SIAM Journal on Computing*, 17(6):1263–1282, 1988. Erratum appears in the same journal, 20(2):404.

[KSW87] J. Köbler, U. Schöning, and K. Wagner. The difference and truth-table hierarchies for NP. *R.A.I.R.O. Informatique théorique et Applications*, 21:419–435, 1987.

[Lev73] L. Levin. Universal sorting problems. *Problems of Information Transmission*, 9:265–266, 1973.

[MS72] A. Meyer and L. Stockmeyer. The equivalence problem for regular expressions with squaring requires exponential space. In *Proceedings of the 13th IEEE Symposium on Switching and Automata Theory*, pages 125–129, 1972.

[PY84] C. Papadimitriou and M. Yannakakis. The complexity of facets (and some facets of complexity). *Journal of Computer and System Sciences*, 28(2):244–259, 1984.

[Reg89] K. Regan. Provable complexity properties and constructive reasoning. Manuscript, April 1989.

[Sto77] L. Stockmeyer. The polynomial-time hierarchy. *Theoretical Computer Science*, 3:1–22, 1977.

[Val76] L. Valiant. The relative complexity of checking and evaluating. *Information Processing Letters*, 5:20–23, 1976.

A $\frac{3}{2}$ log 3–Competitive Algorithm for the Counterfeit Coin Problem

Dean Kelley[1]*, Peng-Jun Wan[2], Qifan Yang[3]**

[1] Department of Computer Science, University of Minnesota
Minneapolis, MN 55414 USA, kelley@mail.cs.umn.edu
[2] Department of Computer Science, University of Minnesota
Minneapolis, MN 55414 USA, wan@mail.cs.umn.edu
[3] Department of Mathematics, Zhejiang University
China

Abstract. We study the following *counterfeit coin problem*: Suppose that there is a set of n coins. Each one is either *heavy* or *light*. The goal is to sort them according to weight with a minimum number of weighings on a balance scale. Hu and Hwang gave an algorithm with a competitive ratio of $3 \log 3$ (all logarithms are base-2). Hu, Chen and Hwang also gave an algorithm with a competitive ratio of $2 \log 3$. In this paper we give an improved algorithm whose competitive ratio is $\frac{3}{2} \log 3$.

1 Introduction

Consider a set of n coins C which contains d light coins and $n - d$ heavy coins. The cases where d is known and where d is unknown are considered as different problems. We want to sort the coins by using a balance scale and perform a minimum number of weighings to identify the d light and the $n-d$ heavy coins.

Let $M_A (n : d)$ denote the maximum number of weighings required by algorithm A to sort the n coins when d is *unknown* and let $M_A (n, d)$ denote this maximum when d is *known*. Let $M (n : d) = \min_A M_A (n : d)$ and let $M (n, d) = \min_A M_A (n, d)$. An algorithm A is a *competitive algorithm* if there exist constants c and b such that for all $n > d > 0$ we have

$$M_A (n : d) \leq cM (n, d) + b . \tag{1}$$

The constant c is called the *competitive ratio*.

One can easily find that $M (n : 1) = M (n, 1) = \lceil \log n \rceil$ (all logarithms in this paper are base-2). But when $d \geq 2$ it is surprisingly hard to find $M (n : d)$

---

* Also of Department of Mathematics and Computer Science, Gustavus Adolphus College, St. Peter, MN 56082 USA

** Visiting scholar in the Department of Computer Science, University of Minnesota. Supported by Pao Yu-kong and Pao Zhao-long scholarship and NSF grant CCR-9208913.

and $M(n, d)$. Hu and Hwang [3] gave an algorithm with a competitive ratio of $3 \log 3$. Recently, Hu, Chen, and Hwang [2] gave a new $2 \log 3$–competitive algorithm. In this paper we give an improved algorithm and show that it has a competitive ratio of $\frac{3}{2} \log 3$.

Before presenting the algorithm we give some results needed in the analysis.

Lemma 1. [1] Let $d = d_1 + d_2$ and $n = n_1 + n_2$ where $d_i > 0$ and $n_i > 0$ for $i = 1, 2$. Then

$$d_1 \log \frac{n_1}{d_1} + d_2 \log \frac{n_2}{d_2} \leq d \log \frac{n}{d}.$$

Lemma 2. [3] $M(n, d) \geq \frac{d}{\log 3} \left(\log \frac{n}{d} + \log \frac{e\sqrt{3}}{2} \right) - \frac{\log d}{2 \log 3} - \frac{0.567}{\log 3} - \frac{1}{2}.$

Lemma 3. [3] $M(n : 0) = M(n : 1) = \lceil \log n \rceil.$

2 The Algorithm

Assume that the set of coins to be identified is $C = \{c_1, c_2, \ldots, c_n\}$ for $n \geq 2$. In the algorithm and in the discussion that follows we employ the notation $|A|$ to denote the number of elements in the set A and $\|A\|$ to denote the weight of the elements of A. Weight of a set of elements is relative. That is, we are only interested in comparison weighings as would be done with a balance scale.

An important component of Hu's and Hwang's $3 \log 3$-competitive algorithm is a halving procedure or *binary search*. In searching for a heavy coin this procedure repeatedly splits a set which is know to contain a heavy coin until that coin is identified. Each time the set is split the two halves are compared on the balance scale. If one half weighs more it is selected to be split at the next step. In case the halves are of equal weight (i.e., they balance) the half to be split in the next step is chosen arbitrarily. Note that in this case both halves contain equally many heavy coins.

If any equal weighings occur in such a binary search consider the last one. Suppose that A_1 and A_2 are the two sets involved in that weighing and that A_1 is subsequently chosen to be searched. Then A_1 contains only 1 heavy coin and the search identifies all coins in A_1. Moreover, A_2 contains exactly 1 heavy coin and a subsequent binary search of it will identify all coins in it.

The same holds in case the search is for a light coin.

Algorithm[4] A

1. $V \leftarrow \emptyset, U \leftarrow C$.
2. Take an arbitrary coin c from U, put it into $\overline{V}$, $U \leftarrow U - \{c\}$.
3. Let $V' \subseteq \overline{V}, U' \subseteq U$ with $|V'| = |U'|$, and either $V' = \overline{V}$, or $U' = U$.
4. If $\|V'\| > \|U'\|$ then go to (8). Else If $\|V'\| < \|U'\|$ then go to (8').
5. $\overline{V} \leftarrow \overline{V} \cup U'$, $U \leftarrow U - U'$.
6. If $U \neq \emptyset$ then go to (3).
7. End the algorithm.
8. Binary search U' for a light coin.
9. If there is an equality anywhere in the binary search path then go to (20). Otherwise U' contains 1 light coin and the algorithm has identified the entire set U'.
10. $\overline{V} \leftarrow \overline{V} \cup U'$ $U \leftarrow U - U'$.
11. If $U = \emptyset$ then end algorithm.
12. Let $V' \subseteq \overline{V}, U' \subseteq U$, where $|V'| = |U'|$ and either $V' = \overline{V}$ or $U' = U$ and V' contains one light coin.
13. If $\|V'\| < \|U'\|$ then go to (10) (U' is pure). If $\|V'\| = \|U'\|$ then go to (27) (U' contains 1 light coin).
14. Binary search U' to find one light coin and a set Q which contains exactly one light coin (from the last comparison yielding equality). Let S be the set of coins identified from U'.
15. If $|Q| = 2^k$ for some integer k then go to (19).
16. If $|Q| \leq 3$ then
$$V \leftarrow \emptyset$$
$$U \leftarrow U - (\overline{V} \cup S \cup Q)$$
Go to (2).
17. Binary search Q.
$\overline{Q} \leftarrow 2^{\lfloor \log |Q| \rfloor}$ coins from Q including the 1 light coin from Q.
18. $V \leftarrow V \cup \overline{V} \cup S \cup (Q - \overline{Q})$, $U \leftarrow U - (S \cup Q)$, $\overline{V} \leftarrow \overline{Q}$.
Go to (11).
19. Binary search Q.
$V \leftarrow V \cup \overline{V} \cup S$ $U \leftarrow U - (S \cup Q)$ $\overline{V} \leftarrow Q$
Go to (11).

---

[4] In the algorithm those statements within the scope of an *IF* are indented

20. At the last equality in the binary search path of U' binary search the other child node. Let S be the set of identified coins found from U'.
$Q \leftarrow U' - S$.

21. If $Q = \emptyset$ then go to (28).

22. Perform another test on Q weighing Q against a set containing 1 light coin.
If Q contains no light coins go to (28).
If Q contains 2 or more light coins then
$\quad V \leftarrow \emptyset$
$\quad U \leftarrow U - S$
$\quad$ Go to (2).

23. If $|Q| = 2^k$ for some k then go to (26).

24. If $|Q| \leq 3$ then
$\quad V \leftarrow \emptyset$
$\quad U \leftarrow U - U'$
$\quad$ Go to (2).

25. Binary search Q.
$\overline{Q} \leftarrow 2^{\lfloor \log |Q| \rfloor}$ coins from Q including the 1 light coin from Q.
$V \leftarrow V \cup \overline{V} \cup S \cup (Q - \overline{Q})$
$\overline{V} \leftarrow \overline{Q}$
$U \leftarrow U - U'$.
Go to (11).

26. Binary search Q.
$V \leftarrow V \cup \overline{V} \cup S$
$\overline{V} \leftarrow Q$
$U \leftarrow U - U'$
Go to (11).

27. Binary search U'.

28. $\overline{V} \leftarrow \overline{V} \cup U'$
$U \leftarrow U - U'$

29. If $U = \emptyset$ then end the algorithm.

30. Let $V' \subseteq \overline{V}$ and $U' \subseteq U$ where $|V'| = |U'|$ and $V' = \overline{V}$ or $U' = U$, and V' contains two light coins.

31. If $||V'|| \geq ||U'||$ then
$\quad V \leftarrow \emptyset$
$\quad$ Go to (2).

32. Partition U' into two subsets U_1 and U_2 of the same size (if necessary add a heavy coin).

33. If $||U_1|| \neq ||U_2||$ then go to (35).

34. $\overline{V} \leftarrow \overline{V} \cup U'$
$U \leftarrow U - U'$
Go to (29).

35. Binary search U'
$Q \leftarrow U'$.

36. If $Q \neq 2^k$ for some k then go to (38).

37. $V \leftarrow V \cup \overline{V}$

$\overline{V} \leftarrow Q$
$U \leftarrow U - Q$
Go to (11).

38. If $|Q| \leq 3$ then
    $V \leftarrow \emptyset$
    $U \leftarrow U - U'$
    Go to (2).

39. Let $\overline{Q}$ be $2^{\lfloor \log |Q| \rfloor}$ coins from Q containing one light coin.
    $V \leftarrow V \cup \overline{V} \cup (Q - \overline{Q})$
    $\overline{V} \leftarrow \overline{Q}$
    $U \leftarrow U - U'$.
    Go to (11)

The procedure beginning at (8') is similar to the one in steps 8–39 except that it searches for heavy coins rather than light ones.

3   The Competitive Ratio of Algorithm A

In this section we will discuss the properties of algorithm A and we will show that the competitive ratio is $\frac{3}{2} \log 3$. In steps (3), (12), and (30) we form new sets for comparison. When there are at least as many unidentified coins as there are identified ones these steps double the number of coins under consideration, thus we refer to this process as *doubling*. Beginning with a set V containing no coins at step (2) the algorithm employs doubling together with binary search to find 2 heavy (light) coins and a set V which is either empty or contains 1 heavy (light) coin. We refer to this as a *jump step*. At the completion of a jump step the algorithm begins a new jump step with the set V.

Lemma 4.   *1. Suppose that there are d' light coins and $n' - d'$ heavy coins in the set $\overline{V} \cup U$ at the beginning of a new jump step and set $m' = \min\{d', n' - d'\}$. If two light (heavy) coins are identified in this step then at least two heavy (light) coins are also identified in the same step.*

2. Suppose that there are d'' light coins and $n'' - d''$ heavy coins remaining in the set $\overline{V} \cup U$ at the end of the jump step (or at the beginning of the next doubling step) and set $m'' = \min\{d'', n'' - d''\}$. Then $m'' \leq m' - 2$.

Proof.   1. Follows immediately from the structure of the algorithm.

2. Without loss of generality suppose that $m' = d'$ and two light coins are identified in the step. Then $d'' = d' - 2$ and $n'' - d'' \leq n' - d' - 2$ since at least two heavy coins are also identified. Thus $m'' = \min\{d'', n'' - d''\} \leq m' - 2$.
$\square$

Lemma 5. *If all together $\overline{n}$ coins are identified in a jump step then the number of tests in this step is at most $3 \log \overline{n}$ except in the case of the last jump step.*

Proof. Suppose that algorithm A finds two light coins and at least two heavy coins in a jump step and without loss of generality suppose that the step starts with $\overline{V} = \emptyset$ (or else fewer weighings than the following are needed).

Case 1: If one light coin is found in the ith doubling and the other one in the jth doubling then the number of tests in this step is at most $(2i-1) + (j-i) + (j-1) + 2(k-j) = 2k + i - 2 \le 3k - 3 = 3\log(2^{k-1})$, for $i < j \le k$ and $\overline{n} = 2^{k-1}$.

If $n = 2^j + n_1$ and $n_1 < 2^j$ then the number of tests in this step is at most $(2i-1) + (j-i) + (j-1) + 2 = i + 2j \le 3j - 1 < 3\log(2^j)$ for $i < j$ and $\overline{n} = 2^j$.

If $n = 2^{j-1} + n_1$ and $n_1 < 2^{j-1}$ then the number of tests in this step is at most $(2i-1) + (j-i) + \lceil \log n_1 \rceil \le i + 2j - 2 < 3j - 3 < 3\log(2^{j-1} + n_1)$ where $i < j$ and $\overline{n} = n = 2^{j-1} + n_1$ (the algorithm ends in this case).

Case 2: If one light coin is found in the ith doubling and in the jth doubling it is found that U' contains at least two light coins (there is at least one equality in the binary search path of U') then use the binary search algorithm to identify one of them and also get a set Q which contains one light coin. The number of tests in this step is at most: $(2i-1) + (j-i) + (j-1) = 2j + i - 2 < 3\log(2^{j-1} + |S|)$.

If $n = 2^{j-1} + n_1$ with $n_1 < 2^{j-1}$ then the number of tests is at most $(2i-1) + (j-i) + \lceil \log n_1 \rceil < 2j + i - 2 < 3\log(2^{j-1} + |S|)$.

Case 3: If the algorithm finds two light coins in the ith doubling and $Q = \emptyset$ then the number of tests in this step is at most $(3i-3) + 2(k-i) < 3\log(2^{k-1})$.
If $n = 2^{i-1} + n_1$ with $n_1 < 2^{i-1}$ then the number of tests is at most $i + 2\lceil \log n_1 \rceil - 1 \le 3i - 3 < 3\log(2^{i-1} + n_1)$.

Case 4: If at least two light coins are found in the ith doubling then there are 2 sub-cases.

    case i. If Q contains no light coin and doubling ends with $|V' \cup U'| = 2^k$ (or $2^{k-1} + n_1$ with $n_1 < 2^{k-1}$) the number of tests is at most $(3i-4) + 1 + 2(k-i) = 2k + i - 3 \le 3k - 4 < 3\log \overline{n}$ for $\overline{n} = 2^{k-1}$.
When $n = 2^{i-1} + n_1$ where $n_1 < 2^{i-1}$, the number of tests is at most $i + \lceil \log n_1 \rceil + (\lceil \log n_1 \rceil - 2) + 1 \le 3i - 3 < 3\log n$.

    case ii. If Q contains at least one light coin then the jump step ends. The number of tests is at most $(3i-4) + 1 \le 3\log(2^{i-1} + |S|)$. When $n = 2^{i-1} + n_1$ for $n_1 < 2^{i-1}$ the number of tests is at most $i + \lceil \log n_1 \rceil + (\lceil \log n_1 \rceil - 2) + 1 \le 3i - 3 < 3\log(2^{i-1} + |S|)$.

$\square$

In this analysis the jump step is assumed to begin with $\overline{V} = \emptyset$. If a jump step begins with $\overline{V} = Q$ or $\overline{Q}$ then even fewer tests are needed. In fact, instead of using $2i - 1$ tests to find the first light coin the algorithm begins by binary searching Q. The number of tests used for it is $\lceil \log |Q| \rceil = i + 1$ and $i + 1 \le 2i - 1$ for $i \ge 2$, (in the algorithm $|\overline{Q}| \ge 4$).

Lemma 6. *If* $0 < d_1 < d_2 \le \frac{n}{2}$ *then* $d_1 \log \frac{n}{d_1} + d_1 \le d_2 \log \frac{n}{d_2} + d_2$.

Proof. Let $f(x) = x \log \frac{n}{x} + x$. Since $f'(x) = \frac{2n}{ex} \geq \log \frac{4}{e} > 0$ and so $f(x)$ is increasing for $x \leq \frac{n}{2}$. $\square$ $\qquad$ $\square$

Lemma 7. *Suppose that Algorithm A uses t jump steps to solve an (n, d) problem and in the ith step it finds d_i light coins and $n_i - d_i$ heavy coins $(i = 1, \ldots, t)$. Let $m_i = \min\{d_i, n_i - d_i\}$.*

1. *If $m_t = 2$ then the number of tests used in the last step is at most $3 \log n_t$.*
2. *If $m_t = 1$, consider the last two steps. Let $\bar{n} - n_{t-1} + n_t$. Let $\bar{d}$ and $\bar{n} - \bar{d}$ be the number of light coins and heavy coins respectively and let $\bar{m} = \min\{\bar{d}, \bar{n} - \bar{d}\}$. Then the total number of tests in the last two steps is at most $\frac{3}{2}\left(\bar{m} \log \frac{\bar{n}}{\bar{m}} + \bar{m}\right)$.*

Proof. (1) If $m_t = 2$, there are two possible cases.

Case 1: The algorithm finds one light (or heavy) coin in the ith doubling and then another one in the jth doubling for $j > i$.
1. If $n_t \geq 2^j$ then the number of tests used in this step is at most $(2i - 1) + (j - i) + (j - 1) + 2(\lceil \log n_t \rceil - j) \leq 2 \log n_t + i < 3 \log n_t$.
2. If $n_t = 2^{j-1} + n'_t$, for $n'_t < 2^{j-1}$ then the number of tests used in this step is at most $(2i - 1) + (j - i) + \lceil \log n'_t \rceil \leq 3j - 3 < 3 \log n_t$.

Case 2: The algorithm finds two light (or heavy) coins in the ith doubling.
1. If $n_t \geq 2^i$ then the number of tests used in this step is at most $(3i - 3) + 2(\lceil \log n_t \rceil - i) = 2 \log n_t + i - 1 < 3 \log n_t$.
2. If $n_t = 2^{i-1} + n'_t$ and $n'_t < 2^{i-1}$ then the number of tests used in this step is at most $i + 2\lceil \log n'_t \rceil - 1 \leq 3i - 3 < 3 \log n_t$

(2) If $m_t = 1$, consider the last two jump steps. If $\bar{m} = 3$, for simplicity suppose that $\bar{m} = \bar{d}$. The second to last step does not end until the algorithm finds a set V containing exactly one light coin and the tth step begins with $\bar{V}$.

1. If $|Q| \geq 4$ the total number of tests used in the last two steps is at most

$$3 \log n_{t-1} + (\lceil \log n_t \rceil - \log |\bar{Q}|) + \log |Q| \leq 3 \log n_{t-1} + \log n_t + 2 \qquad (2)$$

$$= \frac{3}{2}\left(2 \log \frac{n_{t-1}}{2} + 2 + \log n_t + 1\right) - \frac{1}{2} \log n_t + \frac{1}{2} \qquad (3)$$

$$< \frac{3}{2}\left(3 \log \frac{\bar{n}}{3} + 3\right) \qquad (4)$$

$$= \frac{3}{2}\left(\bar{m} \log \frac{\bar{n}}{\bar{m}} + \bar{m}\right) . \qquad (5)$$

2. If $|Q| \leq 3$, then the total number of tests used in the last two steps is at most $3 \log n_{t-1} + \lceil \log n_t \rceil + 1 \leq 3 \log n_{t-1} + \log n_t + 2 \leq \frac{3}{2}\left(\bar{m} \log \frac{\bar{n}}{\bar{m}} + \bar{m}\right)$. (*Note:* $n_t \geq 2$).

For $\overline{m} \geq 4$, suppose we seek two light coins in the second to the last step. Since $m_t = 1$ and there are at least two light coins remaining, the last step should have to start beginning with $\overline{V} = \emptyset$ and the algorithm will seek a heavy coin. The total number of tests used in the last two steps is at most $3 \log n_{t-1} + 2 \log n_t + 1$ by Lemma 5.

Case 1: If $n_t \leq 16$ then

$$3 \log n_{t-1} + 2 \log n_t + 1 = \frac{3}{2} \left(2 \log \frac{n_{t-1}}{2} + 2 + \log n_t \right) + \frac{1}{2} \log n_t + 1 \quad (6)$$

$$\leq \frac{3}{2} \left(3 \log \frac{\overline{n}}{3} + 2 \right) + 3 \quad (7)$$

$$= \frac{3}{2} \left(3 \log \frac{\overline{n}}{3} + 4 \right) \quad (8)$$

$$\leq \frac{3}{2} \left(4 \log \frac{\overline{n}}{4} + 4 \right) \quad (9)$$

$$\leq \frac{3}{2} \left(\overline{m} \log \frac{\overline{n}}{\overline{m}} + \overline{m} \right) \quad (10)$$

the last inequality holds since $\overline{m} \geq 4$ and $\overline{n} \geq 2\overline{m}$ (by Lemma 6).

Case 2: If $n_t > 16$ then

$$3 \log n_{t-2} + 2 \log n_t + 1 = \frac{3}{2} \left(2 \log \frac{n_{t-1}}{2} + \log n_t + 2 \right) + \frac{1}{2} \log n_t + 1 \quad (11)$$

$$\leq \frac{3}{2} \left(3 \log \frac{\overline{n}}{3} + \log n_t + 2 \right) - \log n_t + 1 \quad (12)$$

$$\leq \frac{3}{2} \left(4 \log \frac{\overline{n} + n_t}{4} + 2 \right) - \log n_t + 1 \quad (13)$$

$$\leq \frac{3}{2} \left(4 \log \frac{\overline{n}}{4} + 6 \right) - \log n_t + 1 \quad (14)$$

$$\leq \frac{3}{2} \left(4 \log \frac{\overline{n}}{4} + 4 \right) - \log n_t + 4 \quad (15)$$

$$\leq \frac{3}{2} \left(4 \log \frac{\overline{n}}{4} + 4 \right) \quad (16)$$

$$\leq \frac{3}{2} \left(\overline{m} \log \frac{\overline{n}}{\overline{m}} + \overline{m} \right) \quad (17)$$

$\square$

Theorem 8. Let $m = \min\{d, n - d\}$. Then $M_A (n : d) \leq \frac{3}{2} \left(m \log \frac{n}{m} + m \right)$.

Proof. Suppose that Algorithm A takes t jump steps to solve an (n, d) problem and in the ith step it identifies d_i light coins and $n_i - d_i$ heavy coins. Let $m_i =$

$\min\{d_i, n_i - d_i\}$ for $i = 1, 2, \ldots, t$. If $m_t = 2$ then

$$M_A\,(n:d) \leq 3\,(\log n_1 + \log n_2 + \cdots + \log n_t) \quad \text{by Lemmas 6 and 7(1)} \quad (18)$$

$$= \frac{3}{2}\left(2\log\frac{n_1}{2} + 2 + 2\log\frac{n_2}{2} + 2 + \cdots + 2\log\frac{n_t}{2} + 2\right) \quad (19)$$

$$\leq \frac{3}{2}\left(2t\log\frac{n}{2t} + 2t\right) \quad \text{by Lemma 1} \quad (20)$$

$$(21)$$

Since $m_t = 2$, by Lemma 4(2), we have $m \geq 2t$ and $M_A\,(n:d) \leq \frac{3}{2}\left(m\log\frac{n}{m} + m\right)$ by Lemma 6.

If $m_t = 1$ then by Lemmas 6, 7(2), and 1

$$M_A\,(n:d) \leq \frac{3}{2}\left((2\,(t-2) + \overline{m})\log\frac{n}{2\,(t-2) + \overline{m}} + (2\,(t-2) + \overline{m})\right) \quad (22)$$

By Lemma 4(2), $m \geq 2\,(t-2) + \overline{m}$ and we have $M_A\,(n:d) \leq \frac{3}{2}\left(m\log\frac{n}{m} + m\right)$ by Lemma 6 and since $n \geq 2m$. □

Theorem 9. $M_A \leq \frac{3}{2}\,(\log 3)\,M\,(n:d) + 3.$

Proof. Since $M\,(n:0) = M\,(n:1) = \lceil\log n\rceil$ (Lemma 3), we have

$$M_A\,(n:0) = \lceil\log n\rceil = M\,(n:0) \quad (23)$$

$$M_A\,(n:1) \leq 2\lceil\log n\rceil - 1 \leq 2\log n + 1 < \frac{3}{2}\,(\log 3)\,M\,(n:1) + 3 \quad (24)$$

Now, without loss of generality, suppose $m = d$. For $d \geq 2$, by Lemma 2, we have

$$\frac{3}{2}\,(\log 3)\,M\,(n:d) + 3 \geq \frac{3}{2}\,(\log 3)\,M\,(n,d) + 3 \quad (25)$$

$$\geq \frac{3}{2}\left(d\log\frac{n}{d} + d\right) + \frac{3}{2}\left(\log\frac{e\sqrt{3}}{2} - 1\right)d - \frac{3\log d}{4} - \frac{3}{2}\,(0.567) - \frac{3}{4} + 3 \quad (26)$$

Let $h\,(d) = \frac{3}{2}\left(\log\frac{e\sqrt{3}}{2} - 1\right)d - \frac{3\log d}{4} - \frac{3}{2}\,(0.567) - \frac{3}{4} + 3$. Then $h'\,(d) = \frac{3}{2}\left(\log\frac{e\sqrt{3}}{2} - 1\right) - \frac{3}{4d\ln 2}$. Note that $h'\,(d) > 0$ and $h\,(d)$ is increasing when $d \geq 4$. Moreover, $h\,(2)$, $h\,(3)$, and $h\,(4)$ are all positive. Thus we have that $\frac{3}{2}\,(\log 3)\,M\,(n:d) + 3 \geq \frac{3}{2}\,(d\log\frac{n}{d} + d) \geq M_A\,(n:d)$. □

445

References

1. D.Z.Du and H.Park. On competitive group testing. SIAM J. Computation, 23:5 (1994), 1019-1025.
2. X.D.Hu, P.D.Chen, and F.K.Hwang. A new competitive algorithm for the counterfeit coin problem. (preprint, 1993).
3. X.D.Hu and F.K.Hwang. A competitive algorithm for the counterfeit coin problem. (preprint, 1992).
4. D.D.Sleator and R.E.Tarjan. Amortized efficiency of list update and paging rules, *Communications of the ACM*, 28 (1985) 202-208.

Searching Rigid Data Structures

(Extended Abstract)

Svante Carlsson   Jingsen Chen

Department of Computer Science, Luleå University, S-971 87 Luleå, Sweden

Abstract. We study the *exact* complexity of searching for a given element in a rigid data structure (i.e., an implicit data structure consistent with a fixed family of partial orders). In particular, we show how the ordering information available in the structure facilitates the search operation. Some general lower bounds on the search complexity are presented, which apply to concrete rigid data structures as well. Optimal search algorithms for certain rigid structures are also developed. Moreover, we consider a general problem of searching for a number of elements in a given set. Non-trivial lower bounds are derived and efficient search algorithms are constructed.

1 Introduction

Searching for particular elements is a very common operation in data processing. The data is often structured appropriately in order to speed up the search process. Examples of facilitating the search operation include arranging the data in increasing order and imposing tree structures on them. If all the elements are unordered or totally ordered, then the search problem is solved completely in the sense that the upper bounds on the search complexity match exactly the information-theoretic lower bounds [6]. However, if the elements have been preprocessed such that certain ordered structure is built upon them, the *exact* complexity of the search problem has not been so well explored. In this paper, we study the exact complexity of searching general ordered data structures.

In particular, we are interest in investigating the search problem for rigid data structures which are implicit data structures where the only information known about the structure is a fixed partial order on the data locations and the size of the structure. The sorted array and heap are examples of rigid data structures. The study of the search problem for rigid structures can be viewed as the problem of searching partial orders [2,7,9]. More precisely, given a rigid data structure that is modeled by an n-element poset $\mathcal{P}$ and a real number α, we want to locate α in the poset $\mathcal{P}$ (if it is present) by comparing α with $f(p)$ for some elements $p \in \mathcal{P}$, where f is a one-to-one order-preserving function from $\mathcal{P}$ onto $\{1, 2, \cdots, n\}$. We may consider α as the key value of the data to be retrieved and think of the elements of the poset as data locations where real numbers are stored consistent with the poset ordering. A rigid data structure (or a poset) will be represented either by an array so that, for each set size, all sets of that size satisfy some fixed partial order on the array locations or by the Hasse diagram.

The trade-off between the preprocessing cost (of building a data structure), the structure maintenance cost, and the subsequent search cost has been established [2,7-9]. In this paper we show how the relative ordering information discovered during the establishment of a partial order facilitates searching for a given element in the partial order. Some general lower bounds on the complexity of searching an arbitrary partial order are presented. These lower bounds are applied to concrete rigid data structures, especially the implicit priority queue and deque structures, yielding tight lower limits. Optimal search algorithms for the structures are also developed. Moreover, we consider a general search problem. Namely, given a set of n elements, we want to search for k elements in the set. Non-trivial lower bounds are demonstrated and asymptotically optimal search algorithms are constructed.

2 Searching in Posets

Given a partial order, it seems very hard to determine the exact number of operations which are necessary and sufficient to perform a search on the poset in general. Nevertheless, the following observations are straightforward and useful.

Proposition 1. *Let $S(\mathcal{P})$ be the worst number of operations needed to decide if an given item is equal to any of the elements present in an n-element poset $\mathcal{P}$ under the comparison tree model.*

1. *The search costs for isomorphic posets are identical.*
2. *If $\mathcal{P}$ is an antichain, then $S(\mathcal{P}) = n$.*
3. *$\lceil \log(n+1) \rceil \leq S(\mathcal{P}) \leq n$ (All logarithms in this paper are to base 2).*

Intuitively, the search cost will be higher when the search domain is enlarged. However, this conjecture is not true in general, in the sense that there exists some partial order for which the search cost is even cheaper than that for its induced subposet. We say that $\mathcal{P}$ is an *induced subposet* of Q if $\mathcal{P}$ can be obtained from Q by deleting some elements together with their corresponding order relations.

Fact 1 *There exist partial orders $\mathcal{P}$ and Q such that $\mathcal{P}$ is an induced subposet of Q and $S(\mathcal{P}) > S(Q)$.*

Proof. Consider the posets illustrated in the figure below, where the smaller element is placed at a position with a relative higher horizontal coordinate. Clearly, $\mathcal{P}$ is an induced subposet of Q. The search cost $S(\mathcal{P})$ for the poset $\mathcal{P}$ equals n by Proposition 5.

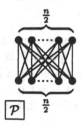
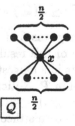

To search for a given element in the poset Q, we first compare the given element with x and then perform a search operation in either the upper-part or the bottom-part of Q according to the outcome of the comparison made. Hence, $S(Q) \leq 1 + \frac{n}{2}$ by Proposition 1 (2).

On the other hand, for partial orders on the same ground set, it is true that the more ordering relations a poset has, the more saving in the search cost there will be. For two partial orders $\mathcal{P}$ and Q on the same ground element-set, we say that Q is an *extension* of $\mathcal{P}$ (or $\mathcal{P}$ is a *suborder* of Q) if one can determine $\mathcal{P}$ from Q with no further comparisons.

Proposition 2. *If $\mathcal{P}$ is an extension of Q, then $S(\mathcal{P}) \leq S(Q)$.*

Proof. Consider a decision tree $\mathcal{T}$ for searching the given element x in Q. We shall now prune $\mathcal{T}$ to obtain a decision tree for the search problem in $\mathcal{P}$. For every internal node v of $\mathcal{T}$ with its comparison between x_i and x_j, if $x_i \prec x_j$ according to both the sequence of comparisons made along the path from the root of $\mathcal{T}$ to the parent of v and the partial order $\mathcal{P}$, then we delete v together with its right branch from $\mathcal{T}$ and connect the left child of v directly to the parent node of v. The resulting decision tree clearly solves the problem of searching x in $\mathcal{P}$, which implies $S(\mathcal{P}) \leq S(Q)$.

For special class of subposets of a partial order, namely the section, the search cost for the section establishes a lower limit on the complexity of searching the partial order. More precisely, $\mathcal{P}$ is a *section* of Q if $\mathcal{P}$ is a subset of Q and if for any $x, y \in \mathcal{P}$ and $z \in Q$, $x \prec_Q z \prec_Q y$ implies $z \in \mathcal{P}$.

Corollary 3. *If $\mathcal{P}$ is a section of Q, then $S(\mathcal{P}) \leq S(Q)$.*

Proof. If $\|\mathcal{P}\| = \|Q\|$, then the result follows immediately from Proposition 2. For the case when $\|\mathcal{P}\| < \|Q\|$, define $Q_1 \triangleq \{q \in Q \backslash \mathcal{P};\ q \prec_Q p \text{ for some } p \in \mathcal{P}\}$ and $Q_2 \triangleq \{q \in Q \backslash \mathcal{P};\ p \prec_Q q \text{ for some } p \in \mathcal{P}\}$. Since $\mathcal{P}$ is a section of Q, the sets Q_1, Q_2, and $\mathcal{P}$ are pairwise disjoint. Now we extend the poset Q to a poset Q' that contains all the ordering relations of Q plus the additional comparability:

$$q_1 \prec p \prec q_2 \text{ in } Q' \text{ whenever } q_1 \in Q_1, p \in \mathcal{P} \text{ and } q_2 \in Q_2$$

Hence, Q is a suborder of Q' and thus $S(Q') \leq S(Q)$ by Proposition 2. Consider the problem of searching x in Q'. If it is known that every element in Q_1 is less than x and the latter is less than every element in Q_2, then the search problem for Q' on this particular input is equivalent to the search problem for $\mathcal{P}$. Therefore, $S(\mathcal{P}) \leq S(Q')$ and the corollary follows.

The above corollary was first observed in [7]. We present here a different and constructive proof of this result. The *height* $h(\mathcal{P})$ of a poset $\mathcal{P}$ is defined as the number of elements in a longest chain in $\mathcal{P}$, and the *width* $w(\mathcal{P})$ of $\mathcal{P}$ is the size of a largest antichain in $\mathcal{P}$. Notice that any antichain of a poset is also a section of the poset. Combining this with Proposition 1 and the fact that $w(\mathcal{P}) \cdot h(\mathcal{P}) \geq n$ (resulted from Dilworth's theorem [5]) we obtain

Corollary 4. *For any n-element poset* $\mathcal{P}$, $S(\mathcal{P}) \geq w(\mathcal{P}) \geq \frac{n}{h(\mathcal{P})}$.

Although the above lower bound is somewhat trivial, it can apply to a number of posets and obtain good lower-bound results. For posets of height 2, we can prove an even better bound.

Proposition 5. *Let* $\mathcal{P}$ *is an n-element poset with* $h(\mathcal{P}) \leq 2$. *Then* $S(\mathcal{P}) = n$.

The above proposition can be verified by an adversary-based argument. If the comparisons between the poset elements are not allowed when searching for a given item in a poset, a result similar to the above proposition was proved in [7]. We have demonstrated in the above that the assertion holds even for the general type of comparisons.

Some special classes of partial orders may admit efficient ways of estimating their search costs. The *sum*, $\mathcal{P} + \mathcal{Q}$, of partial orders $\mathcal{P}$ and $\mathcal{Q}$ on disjoint sets S_1 and S_2, respectively, is a partial order on the union of the sets S_1 and S_2 such that $x \prec y$ in $\mathcal{P} + \mathcal{Q}$ if either $x \prec y$ in $\mathcal{P}$ for $x, y \in S_1$ or $x \prec y$ in $\mathcal{Q}$ for $x, y \in S_2$. The *product*, $\mathcal{P} \times \mathcal{Q}$, of $\mathcal{P}$ and $\mathcal{Q}$ is the poset $\mathcal{P} + \mathcal{Q}$ with the additional comparability relations $x \prec y$ in $\mathcal{P} \times \mathcal{Q}$ whenever $x \in S_1$ and $y \in S_2$.

Fact 2 $S(\mathcal{P} \times \mathcal{Q}) \geq \max\{S(\mathcal{P}), S(\mathcal{Q})\}$.

For a partial order representable as the sum of disjoint posets, namely $\mathcal{P} = \mathcal{P}_1 + \mathcal{P}_2 + \cdots + \mathcal{P}_k$, we know that $S(\mathcal{P}) \leq \sum_{i=1}^{k} S(\mathcal{P}_i)$. Moreover, the equality in the equation above does hold. First, we shall show that

Lemma 6. $S(\mathcal{Q} + \{x\}) = 1 + S(\mathcal{Q})$.

Proof. Let $\mathcal{P} = \mathcal{Q} + \{x\}$. By induction on $\|\mathcal{P}\|$. For any $p \in \mathcal{P}$, $U[p] \triangleq \{y \mid y \preceq_{\mathcal{P}} p\}$ and $D[p] \triangleq \{y \mid p \preceq_{\mathcal{P}} y\}$. From the recursion for $S(\mathcal{P})$ established in [7],

$$S(\mathcal{P}) = 1 + \min_{p \in \mathcal{P}}\{\max\{S(\mathcal{P} \setminus U[p]), S(\mathcal{P} \setminus D[p])\}\}$$
$$= 1 + \min\{S(\mathcal{Q}), \min_{p \in \mathcal{P}-\{x\}}\{\max\{S(\mathcal{P} \setminus U[p]), S(\mathcal{P} \setminus D[p])\}\}\}$$
$$= 1 + \min\{S(\mathcal{Q}), \min_{p \in \mathcal{Q}}\{\max\{S(\mathcal{P} \setminus U[p]), S(\mathcal{P} \setminus D[p])\}\}\}$$
$$= 1 + \min\{S(\mathcal{Q}), \min_{p \in \mathcal{Q}}\{\max\{S(\mathcal{Q} \setminus U[p]) + 1, S(\mathcal{Q} \setminus D[p]) + 1\}\}\}$$
$$= 1 + \min\{S(\mathcal{Q}), 1 + \min_{p \in \mathcal{Q}}\{\max\{S(\mathcal{Q} \setminus U[p]), S(\mathcal{Q} \setminus D[p])\}\}\}$$
$$= 1 + \min\{S(\mathcal{Q}), S(\mathcal{Q})\}$$
$$= 1 + S(\mathcal{Q})$$

Repeating the similar argument yields

Lemma 7. $S(\mathcal{P} + \mathcal{Q}) = S(\mathcal{P}) + S(\mathcal{Q})$.

Therefore, we have

Proposition 8. $S(\mathcal{P}_1 + \mathcal{P}_2 + \cdots + \mathcal{P}_k) = S(\mathcal{P}_1) + S(\mathcal{P}_2) + \cdots + S(\mathcal{P}_k)$.

This is a very interesting property due to the fact that the search complexity can be studied by investigating the complexity of the search problem for simpler and smaller instances.

3 Searching in Concrete Structures

The search problem is now considered for concrete rigid structures, namely the heap [11], the min-max heap [1], the deap [4], the twin-heap [6], and the binomial tree [3,10]. For searching binomial trees, notice that any binomial tree of size four requires 3 comparisons to complete the search process in the worst case. Hence, search for a given element in a binomial tree of size n takes at most $\frac{3}{4}n$ comparisons. The tight lower bound is implied by showing that an n-element binomial tree has a section of size $\frac{3}{4}n$ and of height 2. Therefore,

Theorem 9. *To search for a given element in a binomial tree of size n, $\frac{3}{4}n$ comparisons are necessary and sufficient in the worst case.*

For the exact complexity of searching for a given element in an implicit priority queue/deque structure, the following result can be obtained.

Lemma 10. *Let $\mathcal{H}$ denote the heap, the min-max heap, the deap, or the twin-heap on n elements. Suppose that $\mathcal{P}$ is a collection containing all leaves of $\mathcal{H}$ together with their parents. Then $S(\mathcal{H}) = S(\mathcal{P})$.*

Clearly, the poset $\mathcal{P}$ defined in Lemma 10 is a section of the structure $\mathcal{H}$. Moreover, if $\mathcal{H}$ is either a heap or a min-max heap, then the height of the poset $\mathcal{P}$ is two. On the other hand, we can show that $S(\mathcal{P}) = \frac{2}{3}\|\mathcal{P}\|$ when its corresponding structure is either a deap or a twin-heap. This, combining with Lemma 10 and the fact that $\|\mathcal{P}\| = \frac{3}{4}n$, implies that

Theorem 11. *Let $\mathcal{H}$ denote one of the implicit priority queue/deque structures (the heap, the min-max heap, the deap, and the twin-heap) on n elements. Then*

$$S(\mathcal{H}) = \begin{cases} \frac{3}{4}n, & \text{for heap/min-max-heap} \\ \frac{1}{2}n, & \text{for deap/twin-heap} \end{cases}$$

4 A General Search Problem

In this section, we are interest in the following problem: Given a set X of n elements and a k-element set S, the task is to search for each one of the elements of S in the set X. Assume that $k \leq n$. Let $S(n, k)$ be the worst number of comparisons needed to solve the problem under the comparison tree model.

If we search for just one element s (i.e., $k = 1$), we could simply compare s to every element of X. This would be the preferred method for searching for a small number of elements. On the other hand, if k were large, we could first sort the elements of X, building up a binary search tree for the elements of X.

A lower bound on the complexity of the search problem can be derived from an information-theoretic argument. Consider the case when no element of S presents in X and no pair of elements of S have consecutive ranks among all the elements in both X and S. Hence, any search algorithm establishes the relative ordering between each element of X and each one in S. Moreover, such relative orders partition X into at least $k - 1$ nonempty blocks and the number of all possible partitions is at least $\binom{n}{n_1, n_2, \cdots, n_{k-1}}$, where n_i is the size of the ith partition block $(1 \leq i \leq k - 1)$. Therefore, any decision tree that solves the search problem must have at least $k! \cdot \binom{n}{n_1, n_2, \cdots, n_{k-1}}$ leaves. Notice that

$$\log\left(k! \cdot \binom{n}{n_1, n_2, \cdots, n_{k-1}}\right) = \Omega((n + k) \log k).$$ Thus,

Proposition 12. *The number of operations needed to search for k elements is $\Omega((n + k) \log k)$ in the worst case under the comparison tree model.*

An asymptotically optimal search algorithm can easily be designed by first sorting the element-set S and then searching for each element of X in the sorted list of length k separately.

References

1. M. Atkinson, J.-R. Sack, N. Santoro, and T. Strothotte. Min-max heaps and generalized priority queues. *Commun. ACM*, 29:996–1000, 1986.
2. A. Borodin, L. Guibas, N. Lynch, and A. Yao. Efficient searching using partial ordering. *Inf. Process. Lett.*, 12:71–75, 1981.
3. M. Brown. Implementation and analysis of binomial queue algorithms. *SIAM J. Comput.*, 7:299–319, 1978.
4. S. Carlsson. The deap – A double-ended heap to implement double-ended priority queues. *Inf. Process. Lett.*, 26:33–36, 1987.
5. R. Dilworth. A decomposition theorem for partially ordered sets. *Annals of Mathematics*, 51:161–166, 1950.
6. D. Knuth. *The Art of Computer Programming, Vol. 3: Sorting and Searching.* Addison-Wesley, Reading, Massachusetts, 1973.
7. N. Linial and M. Saks. Searching ordered structures. *J. Algo.*, 6:86–103, 1985.
8. J. Munro and H. Suwanda. Implicit data structures for fast search and update. *J. Comput. Syst. Sci.*, 21:236–250, 1980.
9. H. Noltemeier. On a generalization of heaps. In Proce. Int. Workshop on Graph-Theoretic Concepts in Comput. Sci., pages 127–136, 1980.
10. J. Vuillemin. A data structure for manipulating priority queues. *Commun. ACM*, 21:309–314, 1978.
11. J. Williams. Algorithm 232: Heapsort. *Commun. ACM*, 7:347–348, 1964.

A Better Subgraph of
the Minimum Weight Triangulation

Bo-Ting Yang

Department of Scientific Computing, College of Science,
Xi'an Jiaotong University, Xi'an 710049, China

Abstract. Given a set of n points in the plane, it is shown that the $csc(2\pi/7)$ -skeleton of S is a subgraph of the minimum weight triangulation of S. We improve the results in [2] that the $\sqrt{2}$ -skeleton of S is a subgraph of the minimum weight triangulation of S.

Keywords: Computational geometry; minimum weight triangulation; point sets; Euclidean plane

1 Introduction

Let S be a finite set of points in the Euclidean plane. A triangulation of S is a maximally connected straight-line plane graph whose vertices are the points of S. By maximality, each face is a triangle except for the exterior face, which is the complement of the convex hull of S.

A minimum weight triangulation (MWT) of S is a triangulation that minimizes the sum of the edge lengths. The problem of computing the MWT is included in Garey and Johnson's list of problems that are neither known to be NP-complete nor known to be solvable in polynomial time [1]. Even more annoying is the lack of a constant approximation scheme, that is, an algorithm that in polynomial time constructs a triangulation guaranteed to have total edge length of at most some constant times the optimum. Obviously, the edges which are always in the MWT play a very important role in the structure analysis of the MWT and the approximation algorithm design for computing the MWT. Recently, Keil showed that a polynomially computable Euclidean graph, the $\sqrt{2}$ -skeleton is always a subgraph of the MWT [2]. The $\sqrt{2}$ -skeleton is the β -skeleton of Kirkpatrick and Radke[3] for $\beta = \sqrt{2}$.

In this paper, we improve Keil's result to show that the $csc(2\pi/7)$ -skeleton of S is always a subgraph of the MWT. The $csc(2\pi/7)$ -skeleton is a superset of the usual sense β -skeleton for $\beta = csc(2\pi/7) \approx 1.279$.

2 Results

Let S be a finite set of points in the Euclidean plane, $x, y \in S$. Points x and y are adjacent in the β -skeleton ($\beta > 1$) if and only if there does not exist a point $z \in S$ such that $\angle xzy \geqslant arcsin(1/\beta)$. Thus, points x and y are adjacent in the $\sqrt{2}$ -skeleton if and only if there does not exist a point $z \in S$ such that

$\angle xzy \geqslant \pi/4$ [2]. In this paper, the $\csc(2\pi/7)$ -skeleton is defined as follows.

Definition 1. Given a set S of points in the plane. Points x and y are adjacent in the $\csc(2\pi/7)$ -skeleton if and only if there does not exist a point $z \in S$ such that $\angle xzy > 2\pi/7$.

Obviously, the $\csc(2\pi/7)$ -skeleton is a superset of the usual sense β -skeleton for $\beta = \csc(2\pi/7)$. In the remainder of this paper, we use xy to denote the edge with endpoints x and y and $|xy|$ to denote its length, and $\overset{\frown}{xy}$ to denote the radian of the small arc with endpoints x and y . The following two lemmas bound the length of more remote edges by length of edges that cross $\csc(2\pi/7)$ -skeleton edges.

Lemma 1. *Let x and y be two points in the plane. The planar region R is defined as follows*
$$R = \{p \mid \angle xpy \leqslant 2\pi/7\}.$$
Let a,b,c and d be four points in $R - \{x,y\}$, such that ab passes through x , cd passes through y , ab does not intersects cd , and a and d lie on the same side of the line through xy . Then either $|ab| > |bc|$ or $|cd| > |bc|$.

Proof. Refer to Fig. 1. Obviously, the complement of R is the open region bounded by the two circles through both x and y and with diameter $|xy|\csc(2\pi/7)$. Let o_1 and o_2 be the centers of the two circles such that a,d and o_1 lie on the same side of the line through xy . We have two cases according to the positional relation of ab and cd .

Case 1. If ab does not parallel cd , let z be the intersection point of lines through ab and cd , a_1 be the intersection point of xa and circle o_1 , b_1 be the intersection point of xb and circle o_2 , c_1 be the intersection point of yc and circle o_2 , and d_1 be the intersection point of yd and circle o_1 . If z and o_1 lie on the same side of the line through xy , then $2\angle xzy = \overset{\frown}{xy} - \overset{\frown}{a_1d_1} = \overset{\frown}{b_1c_1} - \overset{\frown}{xy}$. Since $\overset{\frown}{xy} = 4\pi/7$, we have
$$\overset{\frown}{a_1d_1} + \overset{\frown}{b_1c_1} = 8\pi/7 \tag{1}$$
If z and o_2 lie on the same side of the line through xy , then $2\angle xzy = \overset{\frown}{a_1d_1} - \overset{\frown}{xy}$ $= \overset{\frown}{xy} - \overset{\frown}{b_1c_1}$. Hence (1) also holds. Assume without loss of generality that z and o_1 lie on the same side of the line through xy and $\angle abc \leqslant \angle bcd$ in the remainder of Case 1.

(i) Suppose that $\overset{\frown}{a_1d_1} = 0$. In this case, the points a,a_1,d,d_1 and z are co-incident. Since $\angle bac = 2\pi/7$, $\angle abc \leqslant (\pi - \angle bac)/2 = 5\pi/14$, it follows that $\angle acb \geqslant \pi - (2\pi/7) - (5\pi/14) = 5\pi/14 > \angle bac$. Thus $|ab| > |bc|$.

(ii) Suppose that $0 < \overset{\frown}{a_1d_1} \leqslant 2\pi/7$. In this case, $\angle xzy = (\overset{\frown}{xy} - \overset{\frown}{a_1d_1})/2$ $\geqslant \pi/7$, and $\angle abc \leqslant (\pi - \angle xzy)/2 \leqslant 3\pi/7$. Since $\angle bac < 2\pi/7$, we have $\angle acb > \pi - (3\pi/7) - (2\pi/7) = 2\pi/7 > \angle bac$. Moreover, $|ab| > |bc|$.

(iii) Suppose that $2\pi/7 < \overset{\frown}{a_1d_1} < 4\pi/7$. It follows from (1) that $4\pi/7 <$

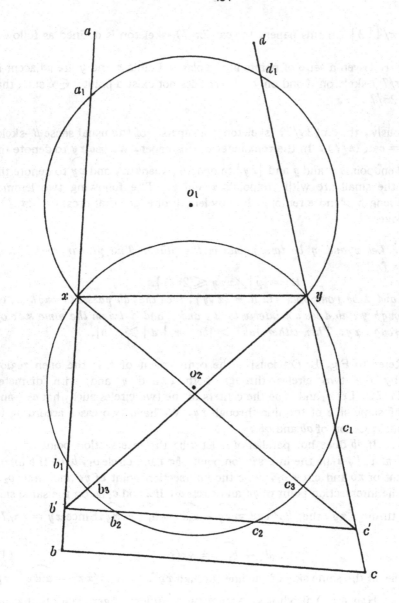

Fig. 1.   Lemma 1.

$\overset{\frown}{b_1 c_1} < 6\pi/7$. We now prove this case by contradiction. Assume that $|bc| > |ab|$ and $|bc| > |cd|$. Since $\angle ayb = \pi - \angle xay - \angle xby \geqslant \pi - (2\pi/7) - (2\pi/7) = 3\pi/7 > \angle bay$, we have $|ab| > |by|$. By assumption, $|bc| > |by|$, thus $\angle byc > \angle bcy$; similarly, $\angle bxc > \angle xbc$. If bc does not intersect circle o_2, then let $b'c'$ be an edge such that $b'c'$ parallels bc, $b' \in ab$, $c' \in cd$, b' and c' lie on outside of circle o_2. Let b_2 and c_2 be the intersection points of $b'c'$ and circle o_2, and circle o_2 intersects $b'y$ and $c'x$ at points b_3 and c_3, respectively. Since $\angle b'yc' > \angle byc > \angle bcy = \angle b'c'y$ and $\angle b'xc' > \angle bxc > \angle xbc = \angle xb'c'$, it follows that

$$b_3c_2c_1 > \overset{\frown}{yxb_1b_2} - \overset{\frown}{c_1c_2} \qquad (2)$$

$$\overset{\frown}{b_1c_2c_3} > \overset{\frown}{c_2c_1yx} - \overset{\frown}{b_1b_2} \qquad (3)$$

(2) and (3) imply that $\overset{\frown}{b_3c_2c_1} + \overset{\frown}{b_1c_2c_3} > \overset{\frown}{c_1yxb_1} + \overset{\frown}{xy}$. Thus $3\overset{\frown}{b_1c_1} \geqslant \overset{\frown}{b_1c_1} + \overset{\frown}{b_3c_2c_1}$ $+ \overset{\frown}{b_1c_2c_3} > \overset{\frown}{b_1c_1} + \overset{\frown}{c_1yxb_1} + \overset{\frown}{xy} = 2\pi + (4\pi/7) = 18\pi/7$. Moreover, $\overset{\frown}{b_1c_1} > 6\pi/7$, a contradiction. Therefore, either $|ab| > |bc|$ or $|cd| > |bc|$.

Case 2. If ab parallels cd, then $\overset{\frown}{a_1d_1} = \overset{\frown}{b_1c_1} = 4\pi/7$. Similar to case (iii), we can show that either $|ab| > |bc|$ or $|cd| > |bc|$. $\quad\square$

Lemma 2. *Let x and y be the end points of an edge in the $\csc(2\pi/7)$-skeleton of a set S of points in the plane. Let a, b, c and d be four other distinct points of S such that ab intersects xy, cd intersects xy, ab does not intersects cd and a and d lie on the same side of the line through xy. Then either $|ab| > |bc|$ or $|cd| > |bc|$.*

Proof. Refer to Fig. 1. Let $a'b$ be an edge such that $a'b$ passes through x and $|a'b| = |ab|$, and cd' be an edge such that cd' passes through y and $|cd'| = |cd|$. If a' or d' lies inside the circle o_1, then, obviously, either $|ab| > |bc|$ or $|cd| > |bc|$; otherwise, it follows from Lemma 1 that either $|a'b| > |bc|$ or $|cd'| > |bc|$, which completes the proof. $\quad\square$

By Lemma 2, we know that the Remote Length Lemma in [2] holds when $\beta = \csc(2\pi/7)$. This enables us to prove the following theorem with exactly the same technique as [2].

Theorem 1. *Let S be a set of points in the plane, the $\csc(2\pi/7)$-skeleton $(S) \subseteq MWT(S)$.*

Acknowledgment

I would like to thank Yin-Feng Xu for many useful discussions on these problems.

References

[1] M. Garey and D. Johnson, Computers and Intractability: A Guide to the Theory of NP-Completeness (Freeman, San Francisco, 1979).

[2] J. M. Keil, Computing a subgraph of the minimum weight triangulation, Computational Geometry 4(1994)13-26.

[3] D. G. Kirkpatrick and J. D. Radke, A framework for computational morphology, in: G. T. Toussaint, ed., Computational Geometry (Elsevier, Amsterdam, 1985) 217-248.

[4] B.-T. Yang, Y.-F. Xu and Z.-Y. You, A chain decomposition algorithm for the proof of a property on minimum weight triangulations, in: D.-Z. Du and X.-S. Zhang, eds., Algorithms and Computation, Lecture Notes in Computer Science 834 (Springer-Verlag, Berlin, 1994) 423-427.

Sequence Decomposition Method for Computing a Gröbner Basis and Its Application to Bivariate Spline

Wen Gao and Baocai Yin

Department of Computer Science, Harbin Institute of Technology,
Harbin 15000,China

Abstract. The purpose of this paper is to discuss basis function of bivariate spline that will be used to express facial surface in facial synthesis. First we present an approach for computing a Gröbner basis of bivariate homogenous polynomials and then applied it to construct a basis for bivariate spline space.

1 INTRODUCTION

The multivariate spline function plays an important role in approximation theory and computer aided geometric design. Many applications, such as geometry modeling and surface interpolation, need to determine a basis of spline. It is difficult to determine a basis even if dimension had been obtained. We denote $S^{\mu}(\Delta, D)$ be bivariate spline function on a given grid partition of a domain $D \in R^2$ with μ-smoothness, $S^{\mu}_k(\Delta, D)$ be the subset of consisting of functions involving only polynomials of degree at most k. The problem of determining a basis of $S^{\mu}_k(\Delta, D)$ can be consider as a problem of linear equation system. The equation system order increase rapidly as k. Ren-Hong Wang and C.K.Chui[3] discuss the dimension and a basis of $S^{\mu}_k(\Delta, D)$ with a basis of solution space of coordinate equation at one point when Δ is a cross-cut partition. This means the problem of computing a basis of spline was put into the problem of computing a basis of solution space of coordinate equation. It needs solving a polynomial equation system or a linear equation system for us to determine a basis for solution space of coordinate equation. To avoid large linear equation system problem, L.J.Billera and L.L.Rose[1] present a Gröbner basis method for multivariate spline. The method transforms polynomial equation problem into that of computing Gröbner basis. Computation and applications of Gröbner basis had been discussed[2][7]. It is difficult to estimate an expression of Gröbner basis from the general ideal. In this paper we present an approach for computing Gröbner basis of bivariate homogenous polynomial system. The algorithm is limit to these ideals that was general by bivariate homogenous polynomial system. A main advantage of the algorithm is that it can be used to estimate an expression of Gröbner basis and computational complex for computing a Gröbner basis. This provide a way to determine a basis of bivariate spline. Applied it and coordinate factor method to bivariate spline, explicit basis of solution space for some coordinate equation was obtained and then basis functions for Type-1,2 triangulation's were given. Gröbner basis is an algorithmic tool for the solution of problems connected with system of multivariate polynomials. We now review the definitions, basic properties and associated concepts. First of all, it is necessary to fix a total order. Let R be the ring of polynomial in two variables over a field K and T be the set of all monomials in R.

Definition 1.1. A admissible order $>$ on R is a total order on T such that
i) $m > 1$ for all $m \in T$.
ii) If $m, n \in T$ and $m > n$, then $um > un$ for all $u \in T$.
For a fixed admissible order $>$, denote by $Lpp(f)$ the Leading Power Product of f in R, which means the largest monomial under the order $>$.

Definition 1.2. Let F be a system of polynomials. A set $G = (G_1, G_2, \cdots, G_k)$ in $ideal(F)$ is called a Gröbner basis for $ideal(F)$ if and only if for any $f \in ideal(F)$, there exist $r_1, r_2, \cdots, r_k \in R$ such that $f = \sum_{i=1}^{k} r_i G_i$, and for all i, $Lpp(r_i G_i) \leq Lpp(f)$.

This definition can be extended to free module R of finite t, provided a noetherian order of the set of movectors, that are monomials in module Rt, Rt has the form $m e_i$ where m is a monomial in R and e_i is the i-th standard basis vector of R^t, is given (cf.[2][7]).

The sequence decomposition method for computing a Gröbner basis will be given in the next section. Then in the third, the basic result is used to discuss basis function of all solutions of coordinate equation.

2 SEQUENCE DECOMPOSITION

In this section, we consider the problem for computing Gröbner basis of bivariate homogeneous polynomials. Our discussion will be stated in graded lexicographic order. We present firstly a concept of sequence triangular decomposition of matrix. In fact it can be consider as a generalized triangular decomposition of matrix.

Definition 2.1. Let A be $m \times n$ matrix, and $A(j)$ denote the matrix of the first j columns of A, the matrix $A^T(j)$ of the first j rows of A. If exist $m \times s$ matrix $L = (l_{ij})$ and $s \times n$ matrix $R = (r_{ij})$ with $l_{ij} = 0$ for $j > rank(A^T(i))$ and $l_{ij} = 1$ for $j = rank(A^T(i))$; $r_{ij} = 0$ for $i = rank(A(j))$ and $r_{ij} \neq 0$ for $i = rank(A(j))$, such that $A = LR$, then it is called sequence triangular decomposition of matrix A. Correspondingly, L is called the sequence down triangular matrix of A, R is called the sequence upper triangular matrix of A.

We denote A_{ij} to be the sub-matrix consisting of all intersection elements of first i rows and first j columns. If matrix A satisfies $rank(A_{ij}) = rank(A(j)) = i$ for $j = 1, 2, \cdots, n$, then the follow algorithm forms a sequence triangular decomposition of matrix A.

Algorithm

1. Let $l_{(1)} = e_1, r_{(1)} = a_{(1)}$. Because of $a_{11} \neq 0$, there exist k such that $r_{1k} \neq 0, r_{1j} = 0$ for $j < k$. Let $l^{(1)} = a^{(k)}/r_{1k}$.
2. Let $r_{(i+1)} = a_{(i+1)} - l_{i+1,s}r_s$, $l^{(i+1)} = (a^{(k)} - r_{s,i+1}l^{(s)})/r_{i+1,k}$. k satisfies $r_{i+1,k} \neq 0$, and $r_{i+1,t} = 0$, $i = 1, 2, \cdots, r-1$, where $a^{(i)}, a_{(i)}$ are the i-th column, i-th row vector of A, respectively.

Let $p = (p_1, p_2, \cdots, p_m)^T = AI_n$ be a bivariate homogeneous polynomial's system of degree n. By exchanging rows of A , we have $rank(A_{ij}) = rank(A^{(i)})$, for $j = 1, 2, \cdots, n$. The exchanging do not change syzygies of p. Without loss generality, we suppose matrix A have sequence triangular decomposition of matrix.

Definition 2.2. Let $A = (a_1, a_2, \cdots, a_n)$ be $m \times n$ matrix , where each a_i is its column vector. The $mj \times (n + j - 1)$ matrix

$$A_j = \begin{pmatrix} a_1 & a_2 & a_3 & \cdots & a_n \\ & a_1 & a_2 & a_3 & \cdots & a_n \\ & & \cdots & & & \cdots \\ & & & a_1 & a_2 & a_3 & \cdots & a_n \end{pmatrix}$$

is called j-th resultant matrix of A.

Theorem 2.1. Let $p = (p_1, p_2, \cdots, p_m)^T = AI_n$ be a bivariate homogeneous polynomial's system of degree n. Suppose that the common factor of p_1, p_2 is a constant . Then there exist an integer number $t \leq n$ such that
$$Gp = (Gp^{(1)}, Gp^{(2)}) = ((I_n R_1^T, I_{n+1}R_2^T, \cdots, I_{n+t-2}R_{t-1}^T), I_{n+t-1})$$
is a Gröbner basis of p, and $rank(A_t) = n + t$. where R_t is sequence upper triangular matrix of A_i, and $I_s = (x^s, x^{s-1}y, \cdots, y^s)$.

Proof. Firstly, we prove that there exists $t \leq n$ such that $rank(A_t) = n + t$. Without loss generality , we suppose that the common factor of p_1, p_2 is constant. Otherwise, we can translate p into $\bar{p}$ by multiplying an $m \times n$ matrix such that $\bar{p}_1, \bar{p}_2$ are irreducible. This translation matrix can be found because of the common factor of $p_1, p_2, \cdots, p_m$ is a constant. Of course , p and $\bar{p}$ have the same ideal. Hence, they have the same Gröbner basis. We now choose the first , second , , (m+1)-th , (m+2)-th , ,((n-1)m+1)-th,((n-1)m+2)-th rows of A_n, respectively to form a new matrix. Determinant of the matrix is the resultant of polynomials p_1, p_2. It is non zero. This show $rank(A_n) = n + n = 2n$. So, there exists $t \leq n$ such that $rank(A_t) = n + t$.

On the other hand, because
$$ideal(Gp^{(2)}) \in ideal(I_{n+t-2}R_{t-1}^T) \in ideal(I_{n+t-3}R_{t-2}^T)$$
$\in \cdots \in ideal(I_n R_1^T) = ideal(p) = ideal(Gp)$. Let $q \in ideal(p)$ be a polynomial, q can be written a linear combination of p. There exist homogeneous polynomials $a = (a_1, a_2, \cdots, a_m)$ such that $q = ap$. Let
$$a = a^{(1)} + a^{(2)}, a^{(j)} = (a_1^{(j)}, a_2^{(j)}, \cdots, a_m^{(j)}),$$
$j = 1, 2$, with $a^{(2)}p \in ideal(Gp^{(2)})$, $deg ree(Lpp(a^{(1)})) < t - 1$.
Hence, $ap = a^{(1)}p + a^{(2)}p, a^{(2)}p \in ideal(Gp^{(2)})$, and $a_i^{(1)}p_i$ is a linear combination of $Gp^{(1)}$ with constant coefficient. So, $a^{(1)}p$ is a linear combination of $Gp^{(1)}$. Let

$a^{(1)}p = r^{(1)}Gp^{(1)}$. Because of $Gp^{(2)}$ is a Gröbner basis, there exists $r^{(2)}$ such that
$a^{(2)}p = r^{(2)}(Gp^{(2)})^T$ and $Lpp(a^{(2)}p \geq r_i^{(2)}x^{n+t-i}y_{i-1}, i = 1,2,\cdots,n+t$. Thus
$ap = \sum_{i=1}^{n+t} r^{(2)}x^{n+t-i}y^{i-1} + \sum_{j=1}^{l} r_j^{(1)}(Gp^{(1)})_j$.
When $r^{(2)} = 0$,
$$Lpp(ap) \geq r_j^{(1)}(Gp^{(1)})_j, j = 1,2,\cdots,l$$
where l is the number of rows of $Gp^{(1)}$.
When $r^{(2)} \neq 0$,
$\deg ree(Lpp(r^{(1)}Gp^{(1)})) < \deg ree(Lpp(r^{(2)}Gp^{(2)}))$,
$Lpp(ap) = Lpp(r^{(1)}Gp^{(1)} + r(2)Gp^{(2)}) =$
$Lpp(r^{(2)}Gp^{(2)}) \geq Lpp(r_i^{(2)}x^{n+t-i}y^{i-1}), i = 1,2,\cdots,n+t$, and
$Lpp(r_j^{(1)}(Gp^{(1)})_j) \leq Lpp(r^{(2)}Gp^{(2)}) = Lpp(ap)$.
This shows that Gp is a Gröbner basis of $ideal(p)$. □

Corollary 2.1. Let p and A be stated as above. Suppose that $rank(A) \geq n+1$ then $(x^n, x^{n-1}y, \cdots, y^n)$.is a Gröbner basis of $ideal(p)$.

3 BASIS OF SOLUTION SPACE OF POLYNOMIAL EQUATION

Purpose of the section is to determine a basis of solution space of some coordinate equation. Suppose Δ is a partition and that $(\alpha_i x + \beta_i y)^{\mu+1}$, $i = 1,2,\cdots,n$ are its interior edges with different slopes through unique interior vertex. The coordinate equation of spline $S_k^\mu(\Delta, D)$ can be express as $\sum_{i=1}^n q_i(\alpha_i x + \beta_i y)^{\mu+1} = 0$, $q_1, q_2, \cdots, q_m \in P_{k-\mu-1}$. The solution space of equation is $V_{k,m}^\mu = \{(q_1, q_2, \cdots, q_m)^T : \sum_{i=1}^n q_i(\alpha_i x + \beta_i y)^{\mu+1} = 0, q_1, q_2, \cdots, q_m \in P_{k-\mu-1}\}$, where $(\alpha_1, \beta_1), (\alpha_2, \beta_2), \cdots, (\alpha_n, \beta_n)$ are linear independent ordered pairs. Schumarker[4][5]had determined the dimension of $V_{k,m}^\mu$. In the section, we compute the basis of $V_{k,m}^\mu$ by Gröbner basis method. To do this, it is enough to get a Gröbner basis of syzygies of $(\alpha_i x + \beta_i y)^{\mu+1}$:$i=1,2,\cdots,n$. Applying theorem 2.1, a Gröbner basis of $(\alpha_i x + \beta_i y)^{\mu+1} : i = 1,2,\cdots,n$can be obtained, by which syzygies can be determined. Without loss generality, we suppose $\alpha_i = 1$. From [2][7], one can find more details about Gröbner basis in module and syzygies. When $m \geq \mu + 2$, we have

Theorem 3.1. Assume $\alpha_i = 1$ and $\beta_i \neq \beta_j$ for $i \neq j$. Let Φ be a basis of $V_{k,\mu+2}^\mu$. Then the set
$$B = \left\{x^a y^b(e_t - \sum_{i=1}^{\mu+2} l_i(\beta_t)e_i), (q,0) : q \in \Phi, 0 \leq a+b \leq k - \mu - 1, \mu + 2 < t \leq m\right\}$$
is a basis of $V_{k,m}^\mu$, where $l_i(x) = \prod_{j=1,j\neq i}^{\mu+2} \frac{x-\beta_j}{\beta_i - \beta_j}$.

Proof. In paper[4], the dimension of $V_{k,m}^\mu$is determined as
$$dim(V_{k,m}^\mu) = \frac{1}{2}m(k+1-\mu)(k-\mu) - \frac{1}{2}((k+1)(k+2) - (\mu+1)(\mu+2)) +$$
$$\sum_{j=1}^{k-m\mu}(m\mu + 1 + j - jm)_+.$$
For$0 \leq s \leq \mu + 1$,we has $x^s = \sum_{i=1}^{\mu+2} \beta_i l_i(x)$. So,

$$(x + \beta_t y)^{\mu+1} = \sum_{i=1}^{\mu+2} l_i(\beta_t)(x + \beta_i y)^{\mu+1} \, .$$

This shows that

$$x^a y^b (et - \sum_{i=1}^{\mu+2} l_i(\beta_t)) \in V_{k,m}^\mu \text{ for } t > \mu + 2.$$

On other hand, $(x^a y^b (et - \sum_{i=1}^{\mu+2} l_i(\beta_t)), (q, 0))$ are linear independent. The number of elements in B is

$$\tfrac{1}{2}(m - \mu - 2)(k - \mu)(k + 1 - \mu) + \dim(V_{k,\mu+2}^\mu) = \dim(V_{k,m}^\mu). \qquad \square$$

Theorem 3.2. Let $f_i = (\alpha_i x + \beta_i y)^n, i = 1, 2, \cdots, n+1$, and $F = (f_1, f_2, \cdots, f_{n+1})$. Assume $\beta_i \neq \beta_j$, then columns of AR is basis of syzygies of F, where $l_i(x) = \prod_{j=1, j \neq i}^{\mu+2} \frac{x - \beta_j}{\beta_i - \beta_j}$,

$$A = \begin{pmatrix} l_1(0) & l_1^{(1)}(0) & \cdots & l_1^{(n)}(0) \\ l_2(0) & l_2^{(1)}(0) & \cdots & l_2^{(n)}(0) \\ \cdots & \cdots & \cdots & \cdots \\ l_{n+1}(0) & l_{n+1}^{(1)}(0) & \cdots & l_{n+1}^{(n)}(0) \end{pmatrix} diag(1, \tfrac{1}{n}, \cdots, \tfrac{(n-k)!}{n!}, \cdots, \tfrac{1}{n!}), \text{ and}$$

$$R = \begin{pmatrix} y & & & \\ -x & y & & \\ 0 & -x & \ddots & \\ \vdots & & & y \\ 0 & 0 & \cdots & -x \end{pmatrix}.$$

Proof. Let $b = (b_1, b_2, \cdots, b_{n+1})$ be polynomials.

$$FARb^T = (f_1, f_2, \cdots, f_{n+1})A^{-1}R(b_1, b_2, \cdots, b_{n+1})^T$$
$$= (x^n, x^{n-1}y, \cdots, y^n)R(b_1, b_2, \cdots, b_{n+1})^T = 0.$$

So, every linear combination of the columns of AR is a syzygy of F.

On the other hand, let $b = (b_1, b_2, \cdots, b_{n+1})$ be a syzygy of F, that is $Fb^T = 0$, Then $(x^n, x^{n-1}y, \cdots, y^n)A^{-1}b^T = 0$. This show that $A^{-1}b^T$ is a syzygy of $(x^n, x^{n-1}y, \cdots, y^n)$. Because of R is a basis of syzygies of $(x^n, x^{n-1}y, \cdots, y^n)$, there exist $\bar{b}$ such that $A^{-1}b^T = R\bar{b}$. Thus $b = AR\bar{b}$, it means that b is a linear combination of the columns of AR. $\qquad \square$

Theorem 3.3. Let R and A be state as above. Then set

$$B = \{x^a y^b AR : 0 \leq a + b \leq k - \mu - 2\}$$

is a basis of $V_{k,\mu+2}^\mu$.

Proof. Theorem 3.2 implies $B \in V_{k,\mu+2}^\mu$. Because of the columns of AR are linear independent. So, the columns of $x^a y^b AR$ are also linear independent. The number of elements in B is $\tfrac{1}{2}(\mu + 1)(k - \mu)(k - \mu - 1) = dim(V_{k,\mu+2}^\mu)$. $\qquad \square$

Using the basis of $V_{k,m}^\mu$, an exact basis of $S_k^\mu(\Delta)$ for $m \geq \mu + 2$ can be gotten when Δ is the cross-cut partition(see[3]).

4 Further Work

We have discussed space construction of spline functions in previous sections. We will use them as surface geometry model in facial image synthesis and hand gesture synthesis. Human face is a complex curved surface. We will choose a bivariate spline that is defined on a triangulation in head contour with one order smoothness as represent of facial surface. The advantages are that spline surface have whole smoothness with lower polynomial function and the partition can be adjusted accounting to facial image. It is beneficial in real-time simulating computation when a basis had been determined.

References

1. L.J.Billera and L.L.Rose, A dimension series for multivariate splines. Discrete Comput Geom 6(1991)107-128.
2. B.Buchberger, Gröbner bases: An algorithmic method in polynomial ideal theory, in Multidimensional system, N.K.Bose(eds), Reidel, Dordrecht, 1985, pp. 184-232.
3. C.K.Chui and R.H.Wang, Multivariate spline space. J.Math.Anal.Appl., 94(1983),197-221.
4. L.L.Schmaker, On the dimension of spaces of piecewise polynomials in two variables. in"Multivariate Approximation Theory" (W.Schempp and K.Zeller. eds), pp. 396-412, Birkhuser, Basel, 1979.
5. L.L.Schumaker, Bounds on the dimension of spaces of multivariate piecewise polynomials, Rocky Mount.J.Math.,14(1984), 251-164.
6. Baocai Yin, Dimension series and basis function of spline, J.math. Research and exposition(to appear).
7. F.Winkler: "A Recursive Method for computing a Gröbner basis of a Module in ",A.AE.C.C-5, Menorca, Spain, 1987.1

A Broadcasting Algorithm on the Arrangement Graph

Leqiang Bai[1], Peter M.Yamakawa[1], Hiroyuki Ebara[2] and Hideo Nakano[3]

[1] Faculty of Engineering, Osaka University, Yamada Oka, Suita, Osaka 565, Japan
[2] Faculty of Engineering, Kansai University, Yamate-cho, Suita, Osaka 564, Japan
[3] Osaka City University, Sugimoto, Sumiyoshi, Osaka 565, Japan

Abstract. In this paper, we propose a distributed algorithm for one-to-all broadcasting on the arrangement graph. The algorithm exploits the rich topological properties of the (n, k)-arrangement graph to constitute the broadcasting binary tree and works recursively. When faulty links are encountered, the concepts of node-disjoint paths and virtual paths are used to deal with the broadcasting procedure. It is shown that the message can be broadcast to all $\frac{n!}{(n-k)!}$ processors in $O(k \lg n)$ steps for fault-free mode, and in $O(k(k + \lg n))$ for less than $k(n - k) - 1$ faulty links.

1 Introduction

A widely studied interconnection network topology is the star graph [AHK87]. It has been proposed as an attractive alternative to the hypercube with many superior characteristics. The arrangement graph [DT92] is a new interconnection network topology that brings a solution to the problem of growth of the number $n!$ of nodes in the n-star graph with respect to the dimension n. It also preserves all the desirable qualities of the star graph topology such as hierarchical construction, vertex and edge symmetry, simple routing and fault tolerance properties.

Broadcasting is one of the fundamental communication problems for a distributed memory multicomputer. The broadcasting problems on the hypercube and the star graph have been investigated intensively in recent years. In [JH89], Johnsson and Ho presented three new communication graphs for hypercube and defined scheduling disciplines. In [MS92], Mendia and Sarker proposed an optimal broadcasting algorithm in the star graph. In [RS88], Ramananathan and Shin proposed a reliable broadcasting algorithm which achieves broadcasting in any n-dimension hypercube, even when as many as $n - 1$ faults of links or nodes occur. In [Fra92], Fraigniaud proposed a reliable broadcasting and gossiping algorithms for the hypercube.

In this paper, we consider one-to-all broadcasting problems on the arrangement graph and propose a broadcasting algorithm for the message passing mode. By using the idea of binary tree construction and concepts of virtual paths and node-disjoint paths, this algorithm performs an optimal broadcasting for the fault-free mode and faster broadcasting for less than $k(n-k) - 1$ faulty links on the (n, k)-arrangement graph.

2 Arrangement Graph

2.1 Definition

Let n and k with $1 \le k \le n$ be two integers, and let us denote $< n >= \{1, 2, \ldots, n\}$ and $< k >= \{1, 2, \ldots, k\}$. Let P_n^k be the set of permutations of the n elements of $< n >$ taken k at a time. The k elements of an arrangement p are denoted $p_1, p_2, \ldots, p_k$; we write $p = p_1 p_2 \ldots p_k$.

Definition 1. The (n, k)–arrangement graph $A_{n,k} = (V, E)$ is an undirected graph given by:

$$V = \{p_1 p_2 \ldots p_k \,|\, p_i \text{ in } < n > \text{ and } p_i \ne p_{i'} \text{ for } i \ne i'\} = P_n^k, \quad \text{and}$$
$$E = \{(p, q) \,|\, p \text{ and } q \text{ in } V,\ p_i \ne q_i \text{ for some } i \text{ in } < k > \text{ and } p_{i'} = q_{i'} \text{ for } i \ne i'\} \ . \tag{1}$$

2.2 Basic Properties

The (n, k)–arrangement graph is regular of degree $k(n - k)$, number of nodes $\frac{n!}{(n-k)!}$, and diameter $\lfloor \frac{3}{2} k \rfloor$. The degree of the arrangement graph $A_{n,k}$ is $k(n - k)$ so it tolerates at most $k(n - k) - 1$ faults. By tuning two parameters n and k, we can make a more suitable choice for the number of nodes and for degree/diameter tradeoff. The arrangement graph $A_{n,k}$ can be considered as a level–n hierarchical graph. For fixed i, the subgraph $1_i, 2_i, \ldots, n_i$ has the disjoint sets of nodes, therefore they form a partitioning of the nodes of $A_{n,k}$. For a more thorough coverage of the arrangement graph, refer to [DT92].

3 Background

Let $p_{in} = p_1 p_2 \ldots p_k$ be an arrangement in P_n^k. We define the set $INT(p_{in}) = \{p_1, p_2, \ldots, p_k\}$ of the k elements of $< n >$ used in the arrangement p_{in}. Similarly we define the set $EXT(p_{in}) = \{< n > -INT(p_{in})\}$ of the $n-k$ elements of $< n >$ not used in the arrangement p_{in}.

Definition 2. The node $p_{n,k}$ of the arrangement graph $A_{n,k}$ is given as follows:

$$p_{n,k} = p_{in} * p_{out} \ . \tag{2}$$

where p_{out} is a dynamic virtual arrangement and meets for $k + 1 \le j \le n$:

$$p_{out} \in \{p_{k+1} \ldots p_j \ldots p_n \,|\, p_j \text{ in } EXT(p_{in}) \text{ and } p_{j_1} \ne p_{j_2} \text{ for } j_1 \ne j_2\} \ . \tag{3}$$

Definition 3. The subgraph $A_{n-l,k-l}$ of the arrangement graph $A_{n,k}$ is the set of the nodes that has l fixed elements in the first l positions of the arrangement p_{in}. For the set of fixed elements $< c >= \{1, 2, \ldots, l\}$, the set of nodes $P_{n-l,k-l}$ is given by:

$$P_{n-l,k-l} = \{c_1 \ldots c_l \, p_{l+1} \ldots p_k * p_{k+1} \ldots p_n\} \in V \ . \tag{4}$$

Definition 4. Generator g_{ij} is the transposition between the element in the position i $(1 \leq i \leq k)$ of the arrangement p_{in} and the element in the position j $(k+1 \leq j \leq n)$ of the virtual arrangement p_{out}, and the special generator g_0 does not act on the arrangement, this is $p_{n,k} = p_{n,k}g_0$.

In this paper, we assume that the source node has the given p_{out} and any node acted by the generator sends its p_{out} ; any node that receives the message revises its p_{out} based on the received message and uses this p_{out} before receiving a new message.

Definition 5. Consider a node p_{xy} and $\coprod_{i=a}^{k} g_{ij}$ that is a sequence of generators in fixed order from g_{aj} to g_{kj}. For any i with $a \leq i \leq k$, $p_{xy}\coprod_{i=a}^{k} g_{ij}$ defines the sequence of the adjacent nodes of the node p_{xy} in the fixed order.

Definition 6. Consider a node p, the node p_d with $p \neq p_d$ and $\prod_{pp_d}$ that is the sequence of fixed generators. $p\prod_{pp_d}$ defines the fixed path between the node p and p_d.

Lemma 7. *Given a node p and its adjacent node $p_d = pg_{xy}$, the set $\prod_{pp_d(j)}$ of the paths for $k+1 \leq j \leq n$ is defined as follows:*

$$\prod_{pp_d(j)} = \{g_{xy}, g_{xj}g_{xy}|j \neq y\} \ . \tag{5}$$

$\prod_{pp_d(j)}$ *forms $n-k$ node-disjoint paths between the node p and p_d.*

Lemma 8. *Given a node p and its adjacent node $p_d = pg_{xy}$, the set $\prod_{pp_d(ij)}$ of the paths for $i \neq x$ and $j \neq y$ is defined as follows:*

$$\prod_{pp_d(ij)} = \{g_{ij}g_{xy}g_{ij}|1 \leq i \leq k \ and \ k+1 \leq j \leq n\} \ . \tag{6}$$

$\prod_{pp_d(ij)}$ *forms $(k-1)(n-k-1)$ node-disjoint paths between the node p and p_d.*

Lemma 9. *Given a node p and its adjacent node $p_d = pg_{xy}$, the set $\prod_{pp_d(i)}$ of the paths for $i \neq x$ is defined as follows:*

$$\prod_{pp_d(i)} = \{g_{iy}g_{xy}g_{iy}g_{xy}g_{iy}|1 \leq i \leq k\} \ . \tag{7}$$

$\prod_{pp_d(i)}$ *forms $k-1$ node-disjoint paths between the node p and p_d.*

Theorem 10. *Given a node p and its adjacent node $p_d = pg_{xy}$, the set $\prod_{pp_d(\Sigma)}$ of the paths:*

$$\prod_{pp_d(\Sigma)} = \prod_{pp_d(j)} \cup \prod_{pp_d(ij)} \cup \prod_{pp_d(i)} \ . \tag{8}$$

$\prod_{pp_d(\Sigma)}$ *forms $k(n-k)$ node-disjoint paths between the node p and p_d.*

Corollary 11. *For a node p and the node $p_d = pg_{xy}$, each of the adjacent nodes of the node p is in one of $k(n-k)$ node-disjoint paths in $\prod_{pp_d(\Sigma)}$. Let the node $p_{ij} = pg_{ij}$ be the adjacent node of the node p, then :*

$$\prod_{p_{ij}p_d} = \begin{cases} g_0 & i = x \ j = y \\ g_{xy} & i = x \ j \neq y \\ g_{xy}g_{ij} & i \neq x \ j \neq y \\ g_{xy}g_{iy}g_{xy}g_{iy} & i \neq x \ j = y \ . \end{cases} \tag{9}$$

4 Broadcasting Algorithm

In this section, we consider one-to-all broadcasting problems in the following two cases:

1. Optimal broadcasting algorithm (OB algorithm) on fault-free arrangement graph $A_{n,k}$
2. Fault-tolerant broadcasting algorithm (FTB algorithm) on faulty arrangement graph $A_{n,k}$ with less than $k(n-k)-1$ faulty links

4.1 Assumptions and Theorem

Here, we make the following assumptions:

A node consists of a processor with full duplex communication links to each of its adjacent nodes. Any node knows the condition of its adjacent links and has enough buffer to preserve the message it sends and the received message. At any given time, a node can communicate with at most one of its adjacent nodes. If there are faulty links on $A_{n,k}$, the number of faulty links is less than $k(n-k)-1$ in order to preserve the connectivity of the network.

Theorem 12. *If an interconnection network consists of N nodes or processors that can communicate with at most one of its adjacent nodes at any given time, then any one-to-all broadcasting algorithm on the network must take at least $\Omega(\lg N)$ steps.*

This means that a one-to-all broadcasting algorithm is optimal on the time complexity based on $O-notation$ if and only if it meets the condition given in Theorem 12.

4.2 OB Algorithm

Our purpose is to develop a broadcasting algorithm that can distribute the message to all nodes in $O(\lg N)$ steps on the arrangement graph with N nodes.

Let $p_{n-l,k-l}$ be the node of the arrangement graph $A_{n-l,k-l}$ in the arrangement graph $A_{n,k}$ and be ready to broadcast a message to all the nodes of $A_{n-l,k-l}$. Notice that $A_{n-l,k-l}$ contains $n-l$ node-disjoint subgraph $A_{n-l-1,k-l-1}$ with the fixed elements $<c>$ from the position 1 to the position l and the different $p_{l+1} \in \{<n>-<c>\}$ in the position $l+1$. Therefore, we can divide all the nodes in $A_{n-l,k-l}$ into $n-l$ parts, and make each of the parts forms a subgraph $A_{n-l-1,k-l-1}$ of $A_{n-l,k-l}$. Let $A_{n-l-1,k-l-1,\text{in}}$ denote the subgraph $A_{n-l-1,k-l-1}$ consisting of the nodes with $p_{l+1} \in \{INT(p_{n-l,k-l})-<c>\}$, and $A_{n-l-1,k-l-1,\text{out}}$ denote the subgraph $A_{n-l-1,k-l-1}$ consisting of the nodes with $p_{l+1} \in OUT(p_{n-l,k-l})$. We deal with the nodes belonging to $n-l$ distinct $A_{n-l-1,k-l-1}$ by considering the following two cases:

(a) The nodes in $A_{n-l-1,k-l-1,\text{in}}$
(b) The nodes in $A_{n-l-1,k-l-1,\text{out}}$

Let $p_{n-l-1,k-l-1}^{\alpha}$ denote the nodes that receive the message in the α-th step. The node $p_{n-l,k-l}$ first sends the message to the intermediate node $p_{n-l-1,k-l-1}^1$ directly connected to the $A_{n-l-1,k-l-1,\text{in}}$, where all nodes have the element p_{l+2} of the node $p_{n-l,k-l}$ in the position $l+1$. The node $p_{n-l-1,k-l-1}^1 = p_{n-l,k-l}g(l+2)n$ has the element p_n of the node $p_{n-l,k-l}$ in the position $l+2$ and the element p_{l+2} of the node $p_{n-l,k-l}$ in the position n. The node $p_{n-l-1,k-l-1}^1$ is also directly connected to $A_{n-l-1,k-l-1,\text{out}}$ with the element $p_{l+1} \in \{\widehat{EXT}(p_{n-l,k-l}) - p_n\}$, where p_n is the element in the position n of the node $p_{n-l,k-l}$. In the next step, the node $p_{n-l,k-l}$ and $p_{n-l-1,k-l-1}^1$ can respectively send the message to two nodes of $p_{n-l-1,k-l-1}^2$ that have the similar property as that of the node $p_{n-l-1,k-l-1}^1$, where $p_{n-l-1,k-l-1}^2 = \{p_{n-l,k-l}g(l+3)n, p_{n-l-1,k-l-1}^1 g(l+4)n\}$. In the same way, an additional step will double the number of the nodes that have received the message. After $k-l-1$ intermediate nodes $p_{n-l-1,k-l-1}$ receive the message, we can make these intermediate nodes and the source node send the message to $k-l$ node-disjoint $A_{n-l-1,k-l-1,\text{out}}$. Similarly, an additional step will double the number of the nodes which receive the message in the node-disjoint $A_{n-l-1,k-l-1,\text{out}}$ until $n-k-1$ distinct $A_{n-l-1,k-l-1,\text{out}}$ receive the message. Then these intermediate nodes send the message to $A_{n-l-1,k-l-1,\text{in}}$, and the source node sends the message to $A_{n-l-1,k-l-1,\text{out}}$ with the element p_n in the position $l+1$ of the node $p_{n-l,k-l}$. It is clear that the broadcasting procedure distributes the message to $n-l$ distinct $A_{n-l-1,k-l-1}$ based on the binary tree construction.

The broadcasting procedure described above can be represented with the sequence of the generators. Let σ_p^{k-l-1} denote the sequence to distribute the message to intermediate nodes in the same $A_{n-l-1,k-l-1}$ as the source node, and let σ_b^{k-l-1} denote the sequence to distribute the message to the other $n-l-1$ node-disjoint $A_{n-l-1,k-l-1}$. Because the sequences of σ_p^{k-l-1} and σ_b^{k-l-1} for the source node and the intermediate nodes are different, we define the notation σ_p^{k-l-1} and σ_b^{k-l-1} as follows:

Definition 13. The sequence σ_p^{k-l-1} (*First phase*);

$$\sigma_p^{k-l-1} = \begin{cases} \sigma_{ps}^{k-l-1} & \text{for } s - \text{node} \\ \sigma_{pm}^{k-l-1} & \text{for } m - \text{node} \\ \sigma_{pf}^{k-l-1} = g_0 & \text{for } f - \text{node} . \end{cases} \tag{10}$$

The sequence σ_b^{k-l-1} (*Second phase*);

$$\sigma_b^{k-l-1} = \begin{cases} \sigma_{bs}^{k-l-1} & \text{for } s - \text{node} \\ \sigma_{bm}^{k-l-1} & \text{for } m - \text{node} \\ \sigma_{bf}^{k-l-1} & \text{for } f - \text{node} . \end{cases} \tag{11}$$

The indices of the sequence σ_p^{k-l-1} and σ_b^{k-l-1} have the following meanings:

p; send the message to the intermediate nodes that are in the same $A_{n-l-1,k-l-1}$ as the source node.

b; send the message to the intermediate nodes that belong to the new node-disjoint $A_{n-l-1,k-l-1}$.

s; the source node that begins to broadcast the message in $A_{n-l,k-l}$.

m; the intermediate nodes that are in the same $A_{n-l-1,k-l-1}$ as the source node.

f; the intermediate nodes that are not in the same $A_{n-l-1,k-l-1}$ as the source node.

Let the notation σ^{k-l-1} denote the broadcasting sequence that distributes the message from the source node to at least one node in each of node-disjoint $A_{n-l-1,k-l-1}$ of $A_{n-l,k-l}$.

Definition 14.

$$\sigma^{k-l-1} = \sigma_{p}^{k-l-1}\sigma_{b}^{k-l-1} = \begin{cases} \sigma_{ps}^{k-l-1}\sigma_{bs}^{k-l-1} = \sigma_{s}^{k-l-1} & \text{for } s-\text{node} \\ \sigma_{pm}^{k-l-1}\sigma_{bm}^{k-l-1} = \sigma_{m}^{k-l-1} & \text{for } m-\text{node} \\ \sigma_{bf}^{k-l-1} = \sigma_{f}^{k-l-1} & \text{for } f-\text{node} . \end{cases} \quad (12)$$

When there is a message to be broadcasted on $A_{n-l,k-l}$, each of the nodes that receive the message in $A_{n-l,k-l}$ sequentially brings about the broadcasting procedure in First and Second phase. The message contains the procedure variables that the algorithm needs. These variables are defined as follows:

$m_1(r)$; the received number of steps in First phase.
$m_1(s)$; the sending number of steps in First phase.
$m_2(r)$; the received number of steps in Second phase.
$m_2(s)$; the sending number of steps in Second phase.
$i(r)$; the received position i.
$i(s)$; the sending position i.
$j(r)$; the received position j.
$j(s)$; the sending position j.
$Z = \{m_1(s), m_2(s), i(s), j(s), l\}$; the set of the procedure variables sent together with the message.

In First phase, $m_2(s) = 0$, $j(s) = n$ and l is constant. M_s^{k-l-1}, M_m^{k-l-1} and M_f^{k-l-1} denote the length of the sequence σ_{ps}^{k-l-1}, σ_{pm}^{k-l-1} and σ_{pf}^{k-l-1} respectively. In Second phase, $m_1(s)$ and l are constant and $i(s) = l+1$. After finishing the corresponding sequence σ_{p}^{k-l-1}, the node executes the corresponding sequence σ_{b}^{k-l-1}. L_{bs}^{k-l-1}, L_m^{k-l-1} and L_f^{k-l-1} denote the length of the sequence σ_{bs}^{k-l-1}, σ_{bm}^{k-l-1} and σ_{bf}^{k-l-1} respectively.

Based on the sequence σ_{p}^{k-l-1} and σ_{b}^{k-l-1}, the broadcasting sequence σ^{k-l-1} can be represented as follows:

i] $0 \le l \le k-2$

$$\sigma^{k-l-1} = \begin{cases} \sigma_{s}^{k-l-1} = (\coprod_{m_1(s)=1}^{M_s^{k-l-1}} g_{i(s)n})(\coprod_{m_2(s)=1}^{L_s^{k-l-1}} g_{(l+1)j(s)})g_{(l+1)n} \\ \sigma_{m}^{k-l-1} = (\coprod_{m_1(s)=m_1(r)+1}^{M_m^{k-l-1}} g_{i(s)n})(\coprod_{m_2(s)=1}^{L_m^{k-l-1}} g_{(l+1)j(s)})g_{(l+1)n} \\ \sigma_{f}^{k-l-1} = \coprod_{m_2(s)=m_2(r)+1}^{L_f^{k-l-1}} g_{(l+1)j(s)} . \end{cases} \quad (13)$$

ii] $l = k - 1$

$$\sigma^0 = \begin{cases} \sigma_{\mathrm{ps}}^0 \sigma_{\mathrm{bs}}^0 = (\coprod_{m_2(s)=1}^{L^0_s} g_{kj(s)}) g_{kn} \\ \sigma_{\mathrm{pm}}^0 \sigma_{\mathrm{bm}}^0 = g_0 \\ \sigma_{\mathrm{bf}}^0 = \coprod_{m_2(s)=m_2(r)+1}^{L^0_f} g_{kj(s)} \end{cases} \tag{14}$$

We denote the notation $\sigma_{\mathrm{A}}^{k-l}$ to the broadcasting sequence that broadcasts the message from the source node to all of the nodes in $A_{n-l,k-l}$. It can be given recursively by:

$$\sigma_{\mathrm{A}}^{k-l} = \begin{cases} \sigma_{\mathrm{ps}}^{k-l-1} \sigma_{\mathrm{bs}}^{k-l-1} \sigma_{\mathrm{A}}^{k-l-1} \\ \sigma_{\mathrm{pm}}^{k-l-1} \sigma_{\mathrm{bm}}^{k-l-1} \\ \sigma_{\mathrm{bf}}^{k-l-1} \sigma_{\mathrm{A}}^{k-l-1} \end{cases} \quad \text{with} \quad \sigma_{\mathrm{A}}^0 = g_0 . \tag{15}$$

Lemma 15. *The sequence $\sigma_{\mathrm{p}}^{k-l-1}$ correctly distributes the message to $k - l - 1$ intermediate nodes, which are in the same $A_{n-l-1,k-l-1,\mathrm{in}}$ as the source node and are adjacent to the other $k - l - 1$ distinct $A_{n-l-1,k-l-1,\mathrm{in}}$ and $n - k - 1$ distinct $A_{n-l-1,k-l-1,\mathrm{out}}$. The length of the sequence $\sigma_{\mathrm{p}}^{k-l-1}$ is $O(\lg(k - l))$.*

Lemma 16. *After finishing the sequence $\sigma_{\mathrm{p}}^{k-l-1}$, the sequence $\sigma_{\mathrm{b}}^{k-l-1}$ correctly distributes the message to at least one node in each of $n-l$ distinct $A_{n-l-1,k-l-1}$ comprising the original $A_{n-l,k-l}$ in $O(\lg \frac{n-l}{k-l})$ steps.*

Lemma 17. *The sequence σ^{n-k-1} correctly distributes the message to at least one node in each of $n - l$ node-disjoint $A_{n-l-1,k-l-1}$ comprising the original $A_{n-l,k-l}$ in $O(\lg(n - l))$ steps.*

Theorem 18. *The sequence $\sigma_{\mathrm{A}}^{k-l}$ forms a one-to-all broadcasting algorithm on $A_{n-l,k-l}$. The length of the sequence $\sigma_{\mathrm{A}}^{k-l}$ is $O((k - l)\lg(n - l))$.*

Corollary 19. *The sequence σ_{A}^k forms a one-to-all broadcasting algorithm on $A_{n,k}$. The length of the sequence σ_{A}^k is $O(k \lg n)$.*

It is clear that the broadcasting algorithm given in Corollary 19 is optimal based on Corollary 12 for fault-free $A_{n,k}$.

4.3 FTB algorithm

When there are some faulty links in $A_{n,k}$, OB algorithm cannot be performed as expected. We confront this problem by using the concept of node-disjoint paths to deal with the faulty links, and develop our fault-tolerant broadcasting algorithm for less than $k(n - k) - 1$ faulty links on $A_{n,k}$.

FTB Procedure. A node in performing broadcasting procedure of the sequence σ^{k-l-1} will send the message to $O(\lg(n - l))$ nodes. When a node encounters some faulty links while executing σ^{k-l-1} of OB algorithm, we make it remember the nodes directly connected through the faulty links and continue to execute its usual broadcasting procedure until the σ^{k-l-1} is ended. Then, it begins to deal with the faulty links.

Procedure FTB_1. We construct the set Q by procedure FTB_1. This set contains the name of every node directly connected through the faulty links that are encountered in executing the sequence σ^{k-l-1}.

Procedure FTB_2. Let p be the node that is ready to send the message through the link g_{xy}, and let g_{xy} be faulty links. The destination node p_d is a node adjacent to the node p that is represented by $p_d = pg_{xy}$. The $k(n-k)$ adjacent nodes of the node p are represented by the set $P_{ij} = \{pg_{ij} | 1 \leq i \leq k, k+1 \leq j \leq n\}$. The set P_{ij} can be divided into k subset $P_{\gamma j}$ for the given $i = \gamma$ with $1 \leq \gamma \leq k$. If there are not faulty links in $A_{n,k}$, the nodes in the set $P_{\gamma j}$ will form a complete graph with $n - k$ nodes. The node p is directly connected to each of the nodes in $P_{\gamma j}$. When there are β faulty links among the node p and the nodes in $P_{\gamma j}$, the node p is directly connected to $n - k - \beta$ nodes in $P_{\gamma j}$. Let $\Gamma_{\gamma j}$ denote the set of $n - k - \beta$ nodes that are directly connected to the node p through non-faulty links in $P_{\gamma j}$, and is named as the non-faulty set of the set $P_{\gamma j}$.

As the faulty links are encountered in executing the sequence σ^{k-l-1}, the node p detects its links and constructs the set $\Gamma_{\gamma j}$ of the nodes. Then the node p sends the message along with the procedure variables to one node $p_{\gamma j_\rho}$ with $k + 1 \leq j_\rho \leq n$ in the set $\Gamma_{\gamma j}$ if $\Gamma_{\gamma j} \neq \emptyset$. The procedure variables contain the set $\Gamma_{\gamma j}$ and $p_{in} * p_{out}$ of the node p.

Lemma 20. *The procedure FTB_2 takes at most $O(k)$ steps to distribute the message to one node in each of k subset $\Gamma_{\gamma j}$ for $1 \leq \gamma \leq k$.*

Procedure FTB_3. As the node $p_{\gamma j_\rho}$ receives the message, it first checks whether there are faulty links that are directly connected to the nodes in the set $\Gamma_{\gamma j}$. If there are β_{j_ρ} faulty links, the node $p_{\gamma j_\rho}$ builds its non-faulty set $\Gamma_{\gamma j_\rho}$, where there are $n - k - \beta - \beta_{j_\rho} - 1$ names of the nodes directly connected to the node $p_{\gamma j_\rho}$ through the non-faulty links and $\Gamma_{\gamma j_\rho} \in \Gamma_{\gamma j}$. Then the node $p_{\gamma j_\rho}$ begins to broadcast the message to the nodes in the set $\Gamma_{\gamma j_\rho}$.

In order to perform fast broadcasting procedure, the set $\Gamma_{\gamma j_\rho}$ is divided into the two node-disjoint parts $\Gamma_{\gamma j_\rho}^1$ and $\Gamma_{\gamma j_\rho}^2$, with $\Gamma_{\gamma j_\rho}^1 \cup \Gamma_{\gamma j_\rho}^2 = \Gamma_{\gamma j_\rho}$, $|\Gamma_{\gamma j_\rho}^1| = \lfloor \frac{|\Gamma_{\gamma j_\rho}|}{2} \rfloor$ and $|\Gamma_{\gamma j_\rho}^2| = \lceil \frac{|\Gamma_{\gamma j_\rho}|}{2} \rceil$. The message is sent to one node in the set $\Gamma_{\gamma j_\rho}^2$ if $\Gamma_{\gamma j_\rho}^2 \neq \emptyset$. At the next step, the set $\Gamma_{\gamma j_\rho}^1$ is divided into the two disjoint parts $\Gamma_{\gamma j_\rho}^{11}$ and $\Gamma_{\gamma j_\rho}^{12}$ as the same as the set $\Gamma_{\gamma j_\rho}$. The message is sent to one node in the set $\Gamma_{\gamma j_\rho}^{12}$ if $\Gamma_{\gamma j_\rho}^{12} \neq \emptyset$. In the same way, the node $p_{\gamma j_\rho}$ will send the message until the set $\Gamma_{\gamma j_\rho}^{11\cdots1}$ is empty. Each of the nodes that receive the message in the above procedure performs the same procedure as the node $p_{\gamma j_\rho}$ does until the message will have been tried to be sent to each of the nodes in the set $\Gamma_{\gamma j_\rho}$. Since an additional step will double the number of the nodes to which the message should have been tried to be sent, this procedure will take at most $O(\lg(n - k))$ steps for the node $p_{\gamma j_\rho}$ to finish the broadcasting procedure in the set $\Gamma_{\gamma j}$.

Lemma 21. *The procedure FTB_3 correctly tries to distribute the message to each node contained in the set $\Gamma_{\gamma j_\rho}$ in $O(\lg(n - k))$.*

Procedure FTB_4. In order to utilize the node-disjoint paths given in Corollary 11, we need to revise p_{out} of the node that receives the message in FTB_3 procedure. We assume that the node p_{ij} has directly received the message through the virtual path g_{ij} through which the message may be not received in fact. The p_{out} of the node p_{ij} can be decided by $p_{ij} = pg_{ij}$. Using this p_{out} and the node-disjoint paths defined in Corollary 11, the message is sent to all the destination nodes contained in the set Q, where the message needs to contain the original procedure variable set Z of OB algorithm. Any destination node p_d continues its OB procedure using the received set Z.

FTB procedure. As shown above, FTB procedure consists of four parts and can be summarized based on FTB_1, FTB_2, FTB_3 and FTB_4 as follows:

procedure FTB
  var Q : *set*;
begin
  if *encounter faulty link in executing* σ^{k-l-1} **then** FTB_1;
    if $Q \neq \emptyset$ *after finishing* σ^{k-l-1} **then** FTB_2;
  if *message received and* $Q \neq \emptyset$ **then** FTB_3 *and* FTB_4
end

Lemma 22. *FTB procedure sends the message to each of the destination nodes contained in the set Q in $O(k + \lg(n - l))$ steps if the number of the faulty links on $A_{n,k}$ is less than $k(n - k) - 1$.*

FTB Algorithm. Combining the sequence σ^{k-l-1} and FTB procedure, the message can be distributed to $n - l$ node-disjoint $A_{n-l-1,k-l-1}$ on $A_{n,k}$ with less than $k(n - k) - 1$ faulty links.

Lemma 23. *The broadcasting sequence σ^{k-l-1} along with FTB procedure on $A_{n,k}$ with less than $k(n - k) - 1$ faulty links correctly distributes the message to at least one node in each of $n - l$ node-disjoint $A_{n-l-1,k-l-1}$ in $O(k + \lg(n - l))$ steps.*

Theorem 24. *The broadcasting sequence σ_A^{k-l} along with FTB procedure on $A_{n,k}$ with less than $k(n - k) - 1$ faulty links forms a one-to-all FTB algorithm with $O((k - l)(k + \lg(n - l)))$ on $A_{n-l,k-l}$.*

Corollary 25. *The broadcasting sequence σ_A^k along with FTB procedure forms a one-to-all FTB algorithm with $O(k(k + \lg n))$ for $A_{n,k}$ with less than $k(n-k)-1$ faulty links.*

FTB algorithm will become OB algorithm in fault-free mode and will broadcast the message to all the nodes on $A_{n,k}$ in $O(k \lg n)$ steps. For the network with $k(n - k) - 1$ faulty links, this algorithm broadcasts the message to all the nodes on $A_{n,k}$ in $O(k(k + \lg n))$. Additionally, using this broadcasting algorithm, we can perform fast broadcasting to all the nodes of some given subgraph $A_{n-l,k-l}$ when the number of the faults is less than $k(n - k) - 1$.

5 Conclusion

In this paper, we have presented a distributed broadcasting algorithm for the arrangement graph. Utilizing the hierarchical construction of the arrangement graph recursively, we have developed the broadcasting sequence σ_A^{k-l} that it can perform an optimal one-to-all broadcasting on fault-free $A_{n-l,k-l}$ in $O((k-l)\lg(n-l))$ steps. For the faulty mode with less than $k(n-k)-1$ faulty links, we have developed the FTB procedure by constructing the set Q and by utilizing the concepts of virtual paths and node-disjoint paths so that the message is sent to at least one node in each of $A_{n-l-1,k-l-1}$ in $O(k+\lg(n-l))$ steps.

Finally, we show that our algorithm broadcasts the message to all $\frac{n!}{(n-k)!}$ nodes in $O(k\lg n)$ steps for fault-free mode and $O(k(k+\lg n))$ for less than $k(n-k)-1$ faults on the arrangement graph $A_{n,k}$.

References

[AHK87] S. B. Akers, D. Harel, and B. Krishnamuthy. "The star graph: An attractive alternative to the n-cube". *Proc. Int. Conf. Parallel Proceeding.*, pages 393–400, 1987.

[CLR90] Thomas H. Cormen, Charles E. Leiserson, and Ronald L Rivest. *"Introduction to Algorithms"*. The MIT Press, Cambridge, 1990.

[DT92] K. Day and A. Tripathi. "Arrangement graph: A class of generalized star graphs". *Information Processing Letters.*, 42:235–241, 1992.

[Fra92] P. Fraigniaud. "Asymptotically Optimal Broadcasting and Gossiping in Faulty Hypercube Multicomputers". *IEEE Trans. Computs.*, 41(11):1410–1419, 1992.

[HZ81] E. Horowitz and A. Zorat. "The binary tree as interconnection network: Application to multiprocessing systems and VLSI". *IEEE Trans. Computs.*, 30(4):245–253, 1981.

[JH89] S. L. Johnsson and C. T. Ho. "Optimal Broadcasting and Presonalized Communcation in Hypercubes". *IEEE Trans. Computs.*, 38(9):1249–1268, 1989.

[MJ94] J. Misic and Z. Jovanovic. "Communication Aspects of the Star Graph Interconnection Network". *IEEE Trans. Parallel. Distrib Systs.*, 7(5):678–687, 1994.

[MS92] V. E. Mendia and D. Sarkar. "Optimal Broadcasting on the Star Graph". *IEEE Trans. Parallel. Distrib Systs.*, 3(4):389–396, 1992.

[RS88] P. Ramanathan and K. G. Shin. "Reliable Broadcast in Hypercube Multicomputers". *IEEE Trans. Computs.*, 37(12):1654–1657, 1988.

[SLC93] J. P. Sheu, W. H. Liaw, and T. S. Chen. "A broadcasting algorithm in star graph interconnection network". *Information Processing Letters .*, 48:237–241, 1993.

[THY93] J. Y. Tien, C. T. Ho, and W. P. Yang. "Broadcasting on Incomplete Hypercubes". *IEEE Trans. Computs.*, 42(11):1393–1398, 1993.

A Fast Maximum Finding Algorithm on Broadcast Communication *

Shyue-Horng Shiau and Chang-Biau Yang

Department of Applied Mathematics
National Sun Yat-sen University
Kaohsiung, . 80424
cbyang@math.nsysu.edu.tw

Abstract. In this paper, we propose a fast algorithm to solve the maximum finding problem under the broadcast communication model, such as an Ethernet. In our maximum finding algorithm, each processor holds one data element initially. The key point of our algorithm is to use broadcast conflicts to build broadcasting layers and then to distribute the data elements into those layers. Only those elements who are on the active layer and still alive can broadcast their values. Thus, the number of broadcast conflicts is reduced. Suppose that there are n input data elements and n available processors. We show that the average time complexity of our algorithm is $\Theta(\log n)$, which is better than any previous result.

1 Introduction

One of the simplest parallel computation models is the *broadcast communication model* [1, 2, 4–9, 11–13]. This model consists of some processors sharing one common channel for communications. Each processor in this model can communicate with others only through this shared channel. Whenever a processor broadcast messages, any other processor can hear the broadcast message via the shared channel. If more than one processor wants to broadcast messages simultaneously, a *broadcast conflict* occurs. When a conflict occurs, a conflict resolution scheme should be invoked to resolve the conflict. This resolution scheme will enable one of the broadcasting processors to broadcast successfully. The Ethernet, one of the famous local area networks, is an implementation of such a model. The study on this model is less than that on other parallel computer structures. However, Ethernet-like networks are very common and popular today. Thus, the distributed or parallel computation on this model is more realistic than that on other parallel computer structures.

The time required for an algorithm to solve a problem under the broadcast communication model includes three parts: (1) resolution time: spent to resolve conflicts, (2) transmission time: spent to transmit data, (3) computation time: spent to solve the problem. We hope that the sum of resolution time , transmission time and computation time, which is the time complexity of the algorithm, is minimized.

* This research work was partially supported by the National Science Council under contract NSC 84-2121-M-110-005 MS.

A straightforward method for finding maximum under the broadcast communication model is to have all processors broadcast their elements one by one. Then the maximum element can be found after n broadcasts. The time required for this method is $O(n)$. Levitan and Foster [5, 6] proposed a maximum finding algorithm, which is nondeterministic and requires $O(\log n)$ successful broadcasts in average. To produce a successful broadcast, we may have to resolve a broadcast conflict. The fastest nondeterministic conflict resolution scheme, which was proposed by Willard [10], requires $O(\log \log n)$ time in average to resolve a conflict. Thus, finding a maximum by Levitan and Foster's algorithm requires $O(\log n \log \log n)$ time in average. Martel [7] showed this problem can be solved in $10.2 \log n$ time slots in average by his algorithm. His basic idea is to estimate a proper broadcasting probability which is held by each data element. Each element decides individually whether it broadcasts in the current time slot or not. We can expect that in average, about one data element wants to broadcast. Thus, his idea can reduce conflict resolution time. Until now, Martel's algorithm is the fastest one for solving the maximum finding problem under the broadcast communication model.

The layer concept proposed by Yang [11] is another idea which can reduce conflict resolution time. In this paper, we shall apply the layer concept to solve the maximum finding problem under the broadcast communication model. Based upon the layer concept, we propose a more efficient algorithm, which reduces the average time slots including conflict slots, empty slots, and slots for successful broadcast, to x, where $4 \ln n - \frac{3}{2} < x < 5 \ln n + 2$.

2 The Algorithm for Maximum-Finding

In our maximum finding algorithm, each processor holds one data element initially. The key point of our algorithm is to use broadcast conflicts to build broadcasting layers and then to distribute the data elements into those layers. Note that for shortening the description in our algorithm, we use the term "data element" to represent "the processor storing the data element" and to represent the data element itself at different times if there is no ambiguity. When a broadcast conflict occurs, a layer is built. All data elements which join the previous conflict randomly choose to continue to broadcast or to abandon broadcasting with equal probabilities. That is , the probability of continuing to broadcast is $\frac{1}{2}$. Each data element which chooses to continue to broadcast will build a new upper layer and bring itself up to the new layer. Each data element which chooses to abandon broadcasting will stay in the current layer.

Algorithm Maximum-Finding

Step 1. Each data element is alive and sets the active layer as the bottom of the stack.

Step 2. Each data alive element on the active layer broadcasts its value.

Step 3. The situations of this time slot can be divided into three cases as follows:

Case 1: A broadcast conflict occurs. Before the next time slot coming, all data elements which cause the conflict randomly, with probability $\frac{1}{2}$, choose to continue to broadcast or to abandon broadcasting. Each data element which chooses to continue to broadcast brings itself into the upper layer of the stack. All alive data elements which choose to abandon broadcasting stay in current layer. The active layer is also changed to the upper layer. Each data element on the active layer broadcasts its value. Then go to step 3.

Case 2: Exactly one data element broadcasts. Each data element which is less or equal to this element sets itself to be dead and never broadcasts hereafter. The active layer is changed to the lower layer. If the active layer is not below the bottom, then go to step 2; otherwise go to step 4.

Case 3: No data element broadcasts. This case results from that either no alive data element chooses to broadcast or each data element in the active layer is dead. There are two subcases for the situation of the last time slot which is not silent as follows:

    Subcase A: It is a conflict. Before the next time slot coming, each alive data element which causes the conflict will randomly, with probability $\frac{1}{2}$, choose to continue or not again. On the current layer, each data element which choose to continue broadcasts its value. Then go to step 3.

    Subcase B: It is a successful broadcast. Change the active layer to the lower layer. If the active layer is not below the bottom, then go to step 2; otherwise go to step 4.

Step 4. Report the last data element which is the last one broadcasts successfully as the maximum.

Figure 1 shows an example for illustrating our algorithm. Initially, all data elements can broadcast. Hence, in time slot 1, all of 1, 2, ..., 8 broadcast and a conflict occurs. Then layer 0 is built. Suppose that in time slot 2, only 3, 5, 6, 8 decide to broadcast. Then layer 1 is built, and 3, 5, 6, 8 bring themselves to layer 1. At the same time, 1, 2, 4, 7 stay in layer 0. In time slot 3, the data elements on the active layer, which have the right to decide if they continue to broadcast or not, are 3, 5, 6, 8. This process will continue. Suppose that one day, in layer 3, only 5 broadcasts successfully. Now 5 is the first winner. Of course, 5 is not the last winner. The elements whose values are less than 5, such as 3 on layer 1 and 1, 2, 4 on layer 0, become dead. The broadcasting right now is passed to the only alive element in layer 2, which is 6. In fact, 6 is the next winner. The broadcasting right is passed to the alive element in layer 1, which is 8. 8 is the next winner. And, 7 will become dead. Thus, 8 is the final winner, which is the maximum.

3 Analysis of the Algorithm

In this section, we shall prove that the average time complexity of our maximum finding algorithm is $\Theta(\log n)$.

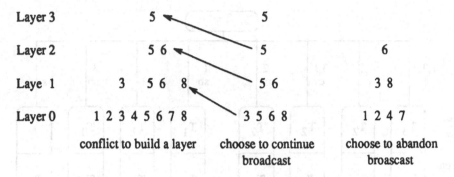

Fig. 1. An example for the layer concept in maximum finding.

Suppose that there are k data elements held by at least k processors in which each processor holds at most one data element. Let T_k denote the average number of time slots, including conflict slots, empty slots and slots for successful broadcasts, required when the algorithm is executed. Let S_k denote the average number of time slots required after step 2 of the algorithm. S_k may include three situations: conflict, successful broadcast and empty slot. Thus, we have

$$T_k = 1 + S_k .$$

When there is zero or one input data element for the algorithm, one empty slot or one successful broadcast is needed. Thus $S_0 = 0$, $T_0 = 1$ and $S_1 = 0$, $T_1 = 1$.

T_2 can be obtained by the following formulas:

$$T_2 = 1 + S_2 \tag{1}$$

$$S_2 = \frac{1}{4}(1 + S_2 + T_0) + \frac{1}{4}(1 + S_1 + T_1) + \frac{1}{4}(1 + S_1 + T_0) + \frac{1}{4}(1 + S_2) . \tag{2}$$

In Eq. (1), the first term, 1, is a conflict in the first time slot. It takes one time slot. The number of time slots needed after the first time slot is included in S_2.

Next, we shall explain Eq. (2) term-by-term. The first term, $\frac{1}{4}(1 + S_2 + T_0)$, is a conflict which was caused by that both elements want to broadcast and bring themselves to the upper layer. The time slots required for the upper layer is expressed by S_2. The conflict is expressed by the subterm 1. The empty lower layer is expressed by the subterm T_0. The second term, $\frac{1}{4}(1 + S_1 + T_1)$, means that the smaller element broadcasts successfully. The subterm, $1 + S_1$, represents this successful broadcast. The other subterm, T_1, implies that the larger element stays in the current layer and is still alive. The fourth term, $\frac{1}{4}(1 + S_1 + T_0)$, means that the larger element broadcasts successfully. The subterm, T_0, implies that the smaller element stays in the current layer and becomes dead. Therefore, the current layer is empty. The final term, $\frac{1}{4}(1 + S_2)$, is a silence which was caused by that no element wants to broadcast. This silence is expressed by 1. The subterm , S_2 , means that there are still two elements in the current layer.

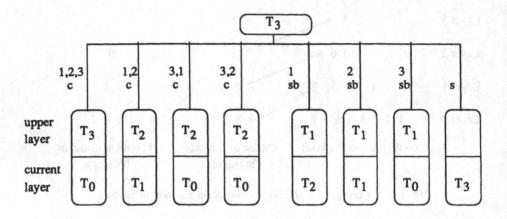

Assume the data elements are 1,2 and 3.   s:silence c:conflict sb:successful broadcast

Fig. 2. The eight cases when the number of input data elements is three.

Applying Eq. (1) into Eq. (2), we have

$$T_2 = 1 + \frac{1}{4}(T_2 + T_0) + \frac{1}{4}(T_1 + T_1) + \frac{1}{4}(T_1 + T_0) + \frac{1}{4}(T_2) . \tag{3}$$

Substituting $T_0 = 1$ and $T_1 = 1$ into Eq. (3) and solving it, we find $T_2 = \frac{9}{2}$.

When $k = 3$, there are eight cases as shown in Figure 2. Therefore, we can obtain T_3 as follows:

$$\begin{aligned}
T_3 = 1 &+ \tfrac{1}{8}(T_3 + T_0) \\
&+ \tfrac{1}{8}(T_2 + T_1) + \tfrac{1}{8}(T_2 + T_0) + \tfrac{1}{8}(T_2 + T_0) \\
&+ \tfrac{1}{8}(T_1 + T_2) + \tfrac{1}{8}(T_1 + T_1) + \tfrac{1}{8}(T_1 + T_0) \\
&+ \tfrac{1}{8}(T_3) .
\end{aligned} \tag{4}$$

The terms in Eq. (4) are one to one correspondence to the cases in Figure 2. For example, the fourth term, $\frac{1}{8}(T_2 + T_0)$, means that the smallest element stays in the current layer (not broadcast). The fifth term, $\frac{1}{8}(T_2 + T_0)$, means that the second smallest element does not choose to broadcast.

Substituting $T_0 = 1$, $T_1 = 1$ and $T_2 = \frac{9}{2}$ into Eq. (4) and solving it, we have $T_3 = \frac{35}{6}$.

Rearranging Eq. (4), we have

$$\begin{aligned}
T_3 - \tfrac{1}{8}(T_3 + T_3) = 1 + \tfrac{1}{8}[\ &T_0 \\
&+ (3T_2 \qquad + T_1 + 2T_0) \\
&+ (3T_1 + T_2 + T_1 + T_0)\ ] .
\end{aligned}$$

Since $T_0 = 1$, we get

$$\begin{aligned}
(1 - \tfrac{1}{2^{3-1}})T_3 = 1 + \tfrac{1}{2^3}[\,&1 \\
&+ (C_2^3 T_2 \qquad\quad + C_1^1 T_1 + C_1^2 T_0) \\
&+ (C_1^3 T_1 + C_0^0 T_2 + C_0^1 T_1 + C_0^2 T_0)\,] .
\end{aligned}$$

Generalizing the above result, for $k \geq 3$, we obtain

$$
\begin{aligned}
(1 - \tfrac{1}{2^{k-1}})T_k \\
= 1 + \tfrac{1}{2^k}[1 \\
+ C_{k-1}^k T_{k-1} && + C_{k-2}^{k-2} T_1 + C_{k-2}^{k-1} T_0 \\
+ C_{k-2}^k T_{k-2} && + C_{k-3}^{k-3} T_2 + C_{k-3}^{k-2} T_1 + C_{k-3}^{k-1} T_0 \\
+ \cdots \\
+ C_i^k T_i && + C_{i-1}^{i-1} T_{k-i} + \cdots + C_{i-1}^{k-3} T_2 + C_{i-1}^{k-2} T_1 + C_{i-1}^{k-1} T_0 \\
+ \cdots \\
+ C_2^k T_2 && C_1^1 T_{k-2} + \cdots + C_1^{i-1} T_{k-i} + \cdots + C_1^{k-3} T_2 + C_1^{k-2} T_1 + C_1^{k-1} T_0 \\
+ C_1^k T_1 && + C_0^1 T_{k-1} + C_0^1 T_{k-2} + \cdots + C_0^{i-1} T_{k-i} + \cdots + C_0^{k-3} T_2 + C_0^{k-2} T_1 + C_0^{k-1} T_0 \;] \; .
\end{aligned}
$$

$$(5)$$

The terms of the above equality can be expressed as follows:

$$
C_i^k T_i + \sum_{j=i-1}^{k-1} C_{i-1}^j T_{k-j-1} \; .
$$

The term , $C_{i-1}^j T_{k-j-1}$, means that the $(j+1)th$ largest element and other $(i-1)$ smaller elements choose to broadcast and bring themselves into the upper layer. After the maximum on the upper layer is found, there are still $k - j - 1$ alive elements on current layer.

The proofs of the following lemmas will appear in the appendix.

Lemma 1.

$$
\begin{aligned}
(1 - \tfrac{1}{2^{n-1}})T_n - (1 - \tfrac{1}{2^{n-2}})T_{n-1} = \tfrac{1}{2^n}[\; & C_{n-2}^{n-1}(T_{n-1} - T_{n-2}) \\
& + C_{n-3}^{n-1}(T_{n-2} - T_{n-3}) \\
& + \cdots \\
& + C_2^{n-1}(T_3 - T_2) \\
& + C_1^{n-1}(T_2 - T_1) \\
& + T_1 \\
& + 2T_{n-1} \qquad \qquad \;] \; , \; for \; n \geq 3 \; .
\end{aligned}
$$

Lemma 2.

$$
\tfrac{4}{n} \leq T_n - T_{n-1} \leq \tfrac{5}{n} \; , \; for \; n \geq 3 \; .
$$

Lemma 3.

$$
4\ln \; n - \frac{3}{2} < T_n < 5\ln \; n \; + \; 2, \; for \; n \geq 3 \; .
$$

Since the computation before each broadcast in each processor requires only constant time, by Lemma 3 , we have the following theorem.

Theorem 4. *The average time complexity of Algorithm-Maximum-Finding is* $\Theta(\log \; n)$.

4 Conclusion

The time required for an algorithm executed under the broadcast communication model is divided into three parts: resolution time, transmission time and computation time. If we want to reduce the required time, we may reduce one of them. However, in many situations, it is hard to reduce either transmission or computation time. The remaining thing we can do is to make every effort to reduce the resolution time. In this paper, we minimize broadcast conflicts to achieve this goal.

The layer concept [11] can help us to reduce conflict resolution when an algorithm is not conflict-free under the broadcast communication model. In this paper, we apply the layer concept to solve maximum finding problem and get good performance. The total number of time slots, including conflict slots, empty slots and slots for successful broadcasts, is x in average, $4 \ln n - \frac{3}{2} < x < 5 \ln n + 2$, which is better than any previous result.

Appendix: This appendix contains all proofs of this paper.

Proof of Lemma 1: Applying $C_n^n + C_{n-1}^n \cdots + C_1^n + C_0^n = 2^n$ into the Eq. (5), we obtain

$(1 - \frac{1}{2^{k-1}})T_k$
$= 1 + \frac{1}{2^k}[1$
$+ C_{k-1}^k T_{k-1}$
$+ C_{k-2}^k T_{k-2}$
$+ \cdots$
$+ C_i^k T_i$
$+ \cdots$
$+ C_2^k T_2$
$+ C_1^k T_1 \qquad + 2^0 T_{k-1} + 2^1 T_{k-2} + \cdots + 2^{k-3}T_2 + 2^{k-2}T_1 + (2^{k-1}T_0 - C_{k-1}^{k-1}T_0)]$

$$= 1 + \frac{1}{2^k}[\, C_{k-1}^k T_{k-1} + C_{k-2}^k T_{k-2} + \cdots + C_3^k T_3 + C_2^k T_2 + C_1^k T_1 \\ + 2^0 T_{k-1} + 2^1 T_{k-2} + \cdots + 2^{k-3}T_2 + 2^{k-2}T_1 + 2^{k-1}T_0 \,] . \qquad (6)$$

Substituting $k = n$ and $k = n - 1$ into Eq. (6), we get the following equations

$$(1 - \frac{1}{2^{n-1}})T_n = 1 + \frac{1}{2^n}[\, C_{n-1}^n T_{n-1} + C_{n-2}^n T_{n-2} + \cdots + C_3^n T_3 + C_2^n T_2 + C_1^n T_1 \\ + 2^0 T_{n-1} + 2^1 T_{n-2} + \cdots + 2^{n-3}T_2 + 2^{n-2}T_1 + 2^{n-1}T_0 \,] \qquad (7)$$

$$(1 - \frac{1}{2^{n-2}})T_{n-1} = 1 + \frac{1}{2^n}[\, 2 \cdot C_{n-2}^{n-1}T_{n-2} + \cdots + 2 \cdot C_2^{n-1}T_2 + 2 \cdot C_1^{n-1}T_1 \\ + 2^1 T_{n-2} + \cdots + 2^{n-3}T_2 + 2^{n-2}T_1 + 2^{n-1}T_0 \,] . \qquad (8)$$

Subtracting Eq. (8) from Eq. (7) , we have

$$(1 - \tfrac{1}{2^{n-1}})T_n - (1 - \tfrac{1}{2^{n-2}})T_{n-1} = \tfrac{1}{2^n}[\ C_{n-1}^n T_{n-1} - C_{n-1}^{n-1} T_{n-1} - C_{n-2}^{n-1} T_{n-2}$$
$$+ C_{n-2}^n T_{n-2} - C_{n-2}^{n-1} T_{n-2} - C_{n-3}^{n-1} T_{n-3}$$
$$+ \cdots$$
$$+ C_3^n T_3 - C_3^{n-1} T_3 - C_2^{n-1} T_2$$
$$+ C_2^n T_2 - C_2^{n-1} T_2 - C_1^{n-1} T_1$$
$$+ C_1^n T_1 - C_1^{n-1} T_1$$
$$+ 2^0 T_{n-1} + C_{n-1}^{n-1} T_{n-1} \qquad]\ .$$

Applying $C_i^n - C_i^{n-1} = C_{i-1}^{n-1}$ into the above equality, we have

$$(1 - \tfrac{1}{2^{n-1}})T_n - (1 - \tfrac{1}{2^{n-2}})T_{n-1} = \tfrac{1}{2^n}[\ C_{n-2}^{n-1} T_{n-1} - C_{n-2}^{n-1} T_{n-2}$$
$$+ C_{n-3}^{n-1} T_{n-2} - C_{n-3}^{n-1} T_{n-3}$$
$$+ \cdots$$
$$+ C_2^{n-1} T_3 - C_2^{n-1} T_2$$
$$+ C_1^{n-1} T_2 - C_1^{n-1} T_1$$
$$+ n T_1 - (n-1) T_1$$
$$+ T_{n-1} + T_{n-1} \qquad]$$

$$(1 - \tfrac{1}{2^{n-1}})T_n - (1 - \tfrac{1}{2^{n-2}})T_{n-1} = \tfrac{1}{2^n}[\ C_{n-2}^{n-1}(T_{n-1} - T_{n-2})$$
$$+ C_{n-3}^{n-1}(T_{n-2} - T_{n-3})$$
$$+ \cdots$$
$$+ C_2^{n-1}(T_3 - T_2)$$
$$+ C_1^{n-1}(T_2 - T_1)$$
$$+ T_1$$
$$+ 2 T_{n-1} \qquad]\ .\square$$

Proof of Lemma 2: We shall prove this lemma by induction on n. The proof is divided into two parts. Now we prove the first part, $T_n - T_{n-1} \leq \tfrac{5}{n}$, for $n \geq 3$. When $n = 3$, it is trivially true since $T_3 - T_2 = \tfrac{35}{6} - \tfrac{9}{2} = \tfrac{4}{3} \leq \tfrac{5}{3}$. By hypothesis, assume that $T_3 - T_2 \leq \tfrac{5}{3}, T_4 - T_3 \leq \tfrac{5}{4}, \cdots, T_{n-1} - T_{n-2} \leq \tfrac{5}{n-1}, n \geq 3$, are all true. We shall verify that $T_n - T_{n-1} \leq \tfrac{5}{n}, n \geq 3$, is also true.

Substituting $T_{n-1} - T_{n-2} \leq \tfrac{5}{n-1}$, for $n \geq 4, T_3 - T_2 = \tfrac{35}{6} - \tfrac{9}{2} = \tfrac{4}{3}$ and $T_2 - T_1 = \tfrac{9}{2} - 1 = \tfrac{7}{2}$, into the inequality of Lemma 1, we obtain the following inequality

$$(1 - \tfrac{1}{2^{n-1}})T_n - (1 - \tfrac{1}{2^{n-2}})T_{n-1} \leq \tfrac{1}{2^n}[\ C_{n-2}^{n-1} \cdot \tfrac{5}{n-1}$$
$$+ C_{n-3}^{n-1} \cdot \tfrac{5}{n-2}$$
$$+ \cdots \qquad \qquad (9)$$
$$+ C_2^{n-1} \cdot \tfrac{4}{3}$$
$$+ C_1^{n-1} \cdot \tfrac{7}{2}$$
$$+ 1 + 2 T_{n-1} \quad]\ .$$

Since

$$C_{n-i-1}^{n-1} \cdot \tfrac{k}{n-i} = \tfrac{k}{n} \cdot C_{n-i}^n ,\ \text{for } i = 1, \cdots, n-2 , \qquad (10)$$

the expression of the right hand of Eq. (9) can be derived as follows:

$$\frac{1}{2^n}[\frac{5}{9}C_{n-1}^n + \frac{5}{n}C_{n-2}^n + \cdots + \frac{4}{n}C_3^n + \frac{7}{n}C_2^n + 1 + 2T_{n-1}]$$
$$= \frac{1}{2^n}[\frac{5}{9}(C_{n-1}^n + C_{n-2}^n + \cdots + C_4^n + C_3^n + C_2^n) - \frac{1}{n}C_3^n + \frac{2}{n}C_2^n + 1 + 2T_{n-1}]$$
$$= \frac{1}{2^n}[\frac{5}{9}(2^n - C_n^n - C_1^n - C_0^n) - \frac{1}{n}C_3^n + \frac{2}{n}C_2^n + 1 + 2T_{n-1}]$$
$$= \frac{1}{2^n}[\frac{5}{n}2^n - \frac{5}{n}(1+1) + 2T_{n-1} + (-\frac{5}{n}\cdot n - \frac{1}{n}C_3^n + \frac{2}{n}C_2^n + 1)] .$$

In the above expression,

$$-\frac{5}{n}\cdot n - \frac{1}{n}C_3^n + \frac{2}{n}C_2^n + 1 = -5 - \frac{1}{n}\cdot\frac{n(n-1)(n-2)}{6} + \frac{2}{n}\cdot\frac{n(n-1)}{2} + 1$$
$$= -\frac{n^2-9n+32}{6} \le 0 , \text{ for } n \ge 3 .$$

We obtain

$$(1 - \frac{1}{2^{n-1}})T_n - (1 - \frac{1}{2^{n-2}})T_{n-1} \qquad \le \frac{1}{2^n}(\frac{5}{n}2^n - \frac{5}{n}\cdot 2 + 2T_{n-1})$$
$$(1 - \frac{1}{2^{n-1}})T_n - (1 - \frac{1}{2^{n-2}})T_{n-1} - \frac{1}{2^n}\cdot 2T_{n-1} \le \frac{1}{2^n}(\frac{5}{n}2^n - \frac{5}{n}\cdot 2)$$
$$(1 - \frac{1}{2^{n-1}})T_n - (1 - \frac{1}{2^{n-1}})T_{n-1} \qquad \le (1 - \frac{1}{2^{n-1}})\cdot\frac{5}{n} .$$

Dividing both sides by $1 - \frac{1}{2^{n-1}}$, we have

$$T_n - T_{n-1} \le \frac{5}{n}, \text{ for } n \ge 3 .$$

The proof of the first part is terminated. The proof of the second part, $T_n - T_{n-1} \ge \frac{4}{n}$ is similar.

This completes the proof. □

Proof of Lemma 3: The proof is separated into two parts. First, we shall show the first part, $T_n < 5\ln n + 2$, for $n \ge 3$.

By Lemma 2, we have

$$T_n - T_{n-1} \quad \le \frac{5}{n} \quad , \text{ for } n \ge 3$$
$$T_{n-1} - T_{n-2} \le \frac{5}{n-1}$$
$$\cdots$$
$$T_3 - T_2 \quad \le \frac{5}{3} .$$

Summing the above inequalities, we obtain

$$T_n - T_2 \le \frac{5}{n} + \frac{5}{n-1} + \cdots + \frac{5}{3}$$
$$T_n \quad \le 5(\frac{1}{n} + \frac{1}{n-1} + \cdots + \frac{1}{3} + \frac{1}{2} + \frac{1}{1}) - 5\cdot\frac{1}{2} - 5\cdot\frac{1}{1} + T_2$$
$$= 5(\frac{1}{n} + \frac{1}{n-1} + \cdots + \frac{1}{3} + \frac{1}{2} + \frac{1}{1}) - 5\cdot\frac{1}{2} - 5\cdot\frac{1}{1} + \frac{9}{2} \qquad (11)$$
$$= 5(\frac{1}{n} + \frac{1}{n-1} + \cdots + \frac{1}{3} + \frac{1}{2} + \frac{1}{1}) - 3 .$$

Let

$$H(n) = \sum_{i=1}^{n}\frac{1}{i} = \frac{1}{1} + \frac{1}{2} + \frac{1}{3} + \cdots + \frac{1}{n-1} + \frac{1}{n} .$$

$H(n)$ is called a harmonic number and it can be expressed as follows [3]:

$$H(n) = \ln n + r + \frac{1}{2n} - \frac{1}{12n^2} + \frac{1}{120n^4} - \epsilon$$

where $0 < \epsilon < \frac{1}{252n^6}$ and r is the Euler's constant, $r = 0.57721 \cdots$. Thus,

$$\ln n < H(n) < \ln n + 1, \text{for } n \geq 3. \tag{12}$$

Applying the above inequality into the Eq. (11), we have

$$T_n < 5 \ln n + 2.$$

The proof of $T_n > 4 \ln n - \frac{3}{2}$ is similar. $\square$

References

1. J. I. Capetanakis, "Tree algorithms for packet broadcast channels," *IEEE Transactions on Information Theory*, Vol. 25, No. 5, pp. 505–515, May 1979.
2. R. Dechter and L. Kleinrock, "Broadcast communications and distributed algorithms," *IEEE Transactions on Computers*, Vol. 35, No. 3, pp. 210–219, Mar. 1986.
3. K. D. . E., *The Art of Computer Programming: Fundamental Algorithms, Vol. 1.* Addison, Wesley Publish Company Inc., 1968.
4. J. H. Huang and L. Kleinrock, "Distributed selectsort sorting algorithm on broadcast communication," *Parallel Computing*, Vol. 16, pp. 183–190, 1990.
5. S. Levitan, "Algorithms for broadcast protocol multiprocessor," *Proc. of 3rd International Conference on Distributed Computing Systems*, pp. 666–671, 1982.
6. S. P. Levitan and C. C. Foster, "Finding an extremum in a network," *Proc. of 1982 International Symposium on Computer Architechure*, pp. 321–325, 1982.
7. C. U. Martel, "Maximum finding on a multi access broadcast network," *Information Processing Letters*, Vol. 52, pp. 7–13, 1994.
8. W. M. Moh, C. U. Martel, and T. S. Moh, "A dynamic solution to prioritized conflict resolution on a multiple access broadcast channel," *Proc. of 1993 International Conference on Parallel and Distributed Systems*, pp. 414–418, 1993.
9. C. Y. Tang and M. J. Chiu, "Distributed sorting on the serially connected local area networks," *Proc. of 1989 Singapore International Conference on Networks*, pp. 458–462, 1989.
10. D. E. Willard, "Log-logarithmic protocols for resolving ethernet and semaphore conflicts," *Proc. of 16th Annual ACM Symposium on Theory of Computing*, pp. 512–521, 1984.
11. C. B. Yang, "Reducing conflict resolution time for solving graph problems in broadcast communications," *Information Processing Letters*, Vol. 40, pp. 295–302, 1991.
12. C. B. Yang, R. C. T. Lee, and W. T. Chen, "Parallel graph algorithms based upon broadcast communications," *IEEE Transactions on Computers*, Vol. 39, No. 12, pp. 1468–1472, Dec. 1990.
13. C. B. Yang, R. C. T. Lee, and W. T. Chen, "Conflict-free sorting algorithm broadcast under single-channel and multi-channel broadcast communication models," *Proc. of 1985 International Conference on Computing and Information*, pp. 350–359, 1991.

Broadcasting in General Networks I : Trees*

Aditya Shastri

Department of Computer Science, Banasthali University
BANASTHALI VIDYAPITH-304022, INDIA.

Abstract. Broadcasting is the process of information dissemination in communication networks whereby a message originated at one vertex becomes known to all members given that at each unit of time a vertex can pass the message to at most one of its neighbours. In this paper we consider the problem of broadcasting in trees which is a step towards studying broadcasting in general graphs, as oppose to the much studied problem of constructing *broadcast graphs* having the smallest number of edges in which message can be broadcast in minimum possible (= $\lceil \log_2 n \rceil$) steps regardless of originator. Trees with the smallest possible broadcast time are exhibited for all $n \leq 326$, and for n sufficiently large existence of trees with broadcast time roughly $\frac{3}{2} \log_2 n$ is shown. It is also shown that broadcast time of a general tree can be computed in $O(n)$ time.

1   Introduction

We represent a communication network by a connected graph G, where the vertices of G represent processors and edges represent bidirectional communication channels. The problem of *broadcasting* is to disseminate a piece of information which originates at one vertex to all the members. This is to be accomplished as quickly as possible by a series of calls under the following constraints:

    1. Each call requires one unit of time,

    2. any member may participate in at most one call per time unit,

    3. a member can only call an adjacent member.

That is if u sends a message to v then neither u nor v can send or receive another message at that time. A *broadcast protocol* for G allows any originator vertex to send messages to all other vertices in the network.

Given G and vertex $v \in G$, let $b(v, G)$ be the minimum time needed to broadcast a message from v. Let $b(G) = \max_v b(v, G)$ be the *broadcast radius* or *broadcast time* of G. Since the number of members knowing the message can at most double at every step, it is clear that $b(G) \geq \lceil \log_2 n \rceil$. Graphs for which broadcst time is equal to $\lceil \log_2 n \rceil$ are called *broadcast graphs*. Broadcast graphs with the fewest number of edges are called *Minimum Broadcast Graphs (MBGs)*

---

 * Part of this work was done while the author was visiting Department D'Informatique, Université Du Maine, Le Mans, France under Marie-Curie Fellowship of CEC (contract No. CI1*-CT-93-0234). Research also supported in part by the Department of Science and Technology, Govt. of India grant no. SR/OY/E-06/93.

or *Minimum Broadcast Networks (MBNs)*, and henceforth we shall use these terms interchangeably. Let $B(n)$ denote the number of edges in a MBN on n vertices.

Determination of $B(n)$, or at least sharp estimation of it, has been the central focal points in most of the papers written thus far on the subject of broadcasting. The progress in this direction has been excruciatingly slow and enormous efforts of last two decades have only resulted in exact determination of $B(n)$ for $n \leq 22$ and some bounds for large n.

It is also of immense theoritical and practical interest to determine broadcast time of general graphs. This problem, though suggested by Farley[1] in 1979, has not been studied at all except for grid graphs[2,4-5], and some attempts to construct broadcast graphs which admit broadcast protocols with slightly more time than the optimal[3,8]. Proskurowski[6] and Slater, Cockayne and Hedetneimi[7] considered what we call *broadcasting in rooted trees*. They constructed trees in which broadcasting was possible from one designated vertex (called root), and not any originator, in $\lceil \log_2 n \rceil$ steps, and designed algorithms to compute the *broadcasting center* of a general tree.

In this paper we study how fast one can broadcast in trees regardless of the originator. Thus, where determination of $B(n)$ amounts to constructing graphs with the fewest number of edges in which one can broadcast in the least possible number of steps, this problem can be thought of as the other extreme where we start with the fewest possible number of edges and ask for what best broadcast time is achievable.

2 Main Results

In one of the earliest papers on broadcasting Farley[1] noted that in order to broadcast we at least require $n - 1$ edges since the underlying network must be connected, and thus raised the question: "What trees produce the best result for broadcast time?" Farley was also quick to note that in the context of trees it is better to consider all three values: the minimum, the average and the maximum number of time units required to complete the broadcasting. However, he only constructed trees in which broadcasting can be accomplished in $\lceil \log_2 n \rceil$ steps from one designated vertex, called the root, and not from any originator. We call such a tree a *Minimum Rooted Broadcast Tree*.

One can immediately realize that broadcasting is not possible from any originator in $\lceil \log_2 n \rceil$ steps. For example, there are only two trees on 4 vertices and both have broadcast radius equal to 3. Thus, we are led to the following definition of what we call the *Tree Broadcast Function* :

$$T(n) \; = \; \min \{ \, b(T) \mid T \text{ is a tree on } n \text{ vertices} \},$$

where the minimum is taken over all trees of size n. Recall that $b(T)$ is the broadcast radius of T which is equals to the maximum of $b(v, T)$, minimum time needed to broadcast a message from v in T, where the maximum is taken over all $v \in T$.

There is a natural way to construct minimum rooted broadcast trees. We start from the root and build the tree in stages where at every step we introduce a new succesor to every vertex which was in the tree at the beginning of the step. More formally, if there are n vertices labelled 1 to n, let 1 be the root and for $2 \leq i \leq n$ connnect i to $i - 2^{\lceil \log_2 i \rceil - 1}$. It is easy to see that broadcasting from root in such trees takes $\lceil \log_2 n \rceil$ steps. Furthermore, since every vertex is at a distance of at most $\lceil \log_2 n \rceil - 1$ apart from the root (or one of its neighbours called first in an optimal protocol), it follows that

$$\lceil \log_2 n \rceil \leq T(n) \leq 2 \lceil \log_2 n \rceil - 1.$$

We shall see that broadcasting takes much fewer steps than the above upper bound for most values of n. In the next theorem we determine $T(n)$ exactly for n upto 326. This is clearly in sharp contrast to determination of $B(n)$ where exact values are known only for $n \leq 22$.

Theorem 1. The minimum number of steps required to broadcast a message in a tree of size n for $n \leq 326$ are as given in Table I.

Table I: Tree Broadcast Time $T(n)$ for $n \leq 326$			
n	$\lceil \log_2 n \rceil$	$T(n)$	$A(n)$
1	0	0	0
2	1	1	0
3	2	2	0
4	2	3	1
5-6	3	4	1
7-8	3	5	2
9	4	5	1
10-14	4	6	2
15-16	4	7	3
17-22	5	7	2
23-32	5	8	3
33	6	8	2
34-52	6	9	3
53-64	6	10	4
65-84	7	10	3
85-128	7	11	4
129-198	8	12	4
199-256	8	13	5
257-326	9	13	4

The constructions which give the sharp values of Table I of $T(n)$ can be generalized to obtain an asymptotic bound which for n sufficiently large shows the existence of trees of order n in which broadcasting takes (roughly) $\frac{3}{2} \log_2 n$ steps.

Theorem 2. For all n sufficiently large, there exist a tree T_n of order n such that

$$b(T_n) = \tfrac{3}{2} \log_2 n (1 + o(1)).$$

However, we believe that the asymptotic bound of Theorem 2, in fact, holds for all n and make the following conjecture.

Conjecture 1. $T(n) \le \lceil \tfrac{3}{2}(\log_2 n + 1) \rceil - 1$ for all n.

In [7] algorithms was presented to compute the broadcasting center of a tree T, denoted by $BC(T)$, which is defined as the set of vertices from where message originated can be broadcast faster as compare to any other vertex not in $BC(T)$. Let T be a tree and (u, v) be an edge in T, then $T(u, v)$ and $T(v, u)$ will denote the subtrees of T consisting of the components of $T - (u, v)$ containing, u and v, respectively. Let $v_1, v_2, ..., v_k$ denote the vertices adjacent to u in T, and assume they are labelled so that $b(v_1, T(v_1, u)) \ge, b(v_2, T(v_2, u)) \ge ... \ge b(v_k, T(v_k, u))$. A simple but important observation of [6,7] is that an optimal calling scheme from u consists of calling $v_1, v_2, ..., v_k$ in that order.

Given a tree T, the algorithm BROADCAST of [7] computes $BC(T)$ by essentially labelling every vertex u with $t(u)$ which is the time required to complete broadcasting in a subtree T_u rooted at u consisting of all the vertices which have been labelled previously. It begins by letting $t(x) = 0$ for every pendent vertex x of T, and at each subsequent step picks a vertex v to be labelled such that the label of v is going to be the smallest among the remaining vertices. Thus, having labelled all the end vertices the algorithm proceeds by moving "inwards" and assigning increased labels to vertices, all but one of whose neighbours have already been labelled, until we label w at the final step which means $w \in BC(T)$.

The algorithm BROADCAST can be easily modified to compute $b(u, T)$ for any $u \in T$. Let $u_1, u_2, ..., u_k$ be the neighbours of u in T. We can start by labelling all the endvertices 0 and proceed as in BROADCAST until we encounter some u_i. At this stage unlike BROADCAST we do not label u and continue to assign increasing labels to remaining vertices until all u_i's are labelled. Suppose π in the symmetric group S_k gives the order of u_i's in the non-increasing order of their labels then we have $b(u, T) = \max\{t(u_{\pi(i)}) + i | 1 \le i \le k\}$. This way, finding $b(u, T)$ for all $u \in T$, and maximizing gives us $b(T)$. The complexity of BROADCAST is $O(n)$, so the running time of this algorithm is going to be $O(n^2)$. However, BROADCAST can be adapted by introducing a *relabelling* step to give $b(T)$ in linear time. Having already labelled the vertices in the broadcast center by their broadcast time t, we can label all their neighbours by $t + 1$. The modified algorithm BROADRAD proceeds by moving "outwards" in a BFS manner labelling neighbours of already labelled vertices by the label of their neighbour plus one. Omitting details we summarize preceding discussion in the following theorem.

Theorem 3. Algorithm BROADRAD computes the broadcast radius of a given tree in $O(n)$ time.

3   General Constructions

Let $S(t)$ be the size of the largest tree in which broadcasting takes no more than t steps regardless of the originator. It turns out that knowing the diameter as well enables us to put sharp bound on the size of the largest tree. So define, $S(t,d)$ be the size of the largest tree of diameter d in which broadcasting takes at most t steps. Also recall that $d(u,v)$ denotes the distance between vertex u and v which equals the length of the shortest path between these two vertices. *Radius* of a vertex v in $G = (V, E)$ denoted by $r(v, G)$ is defined as $r(v, G) = \max\{d(u,v)|u \in V(G)\}$. Let $r(G) = \min r(v, G)$, where the minimum is taken over all $v \in V(G)$, be the *radius of G*, and $d(G) = \max r(v, G)$ be the diameter of G.

Our first lemma determines the size of the largest tree with given diameter d and broadcast time t for $2 \le d \le 9$. It will prove extremely useful in the sequel in exact determination of $T(n)$ for n as high as 300.

Lemma 1. If $S(t,d)$ be the size of the largest tree of diameter d in which broadcasting takes no more than t steps, then
(i) $S(t,2) = t + 1$ for $t \ge 2$,
(ii) $S(t,3) = 2(t-1)$ for $t \ge 3$,
(iii) $S(t,4) = \binom{t-1}{2} + 3$ for $t \ge 4$,
(iv) $S(t,5) = (t-2)(t-3) + 2$ for $t \ge 5$,
(v) $S(t,6) = 3\binom{t-3}{2} + 3$ for $6 \le t \le 9$, and $S(t,6) = \binom{t-2}{3} + t + 1$ for $t \ge 10$,
(vi) $S(t,7) = 2\binom{t-3}{3} + 2(t-3)$ for $t \ge 7$,
(vii) $S(t,8) = 3\binom{t-4}{3} + 3(t-4)$ for $8 \le t \le 13$, $S(t,8) = \binom{t-3}{4} + \binom{t-3}{2} + 9$ for $t \ge 14$,
(viii) $S(t,9) = 2\binom{t-4}{4} + 2\binom{t-6}{2} + 4(t-5)$ for $t \ge 9$.

Proof. Proofs of (i) and (ii) are trivial. To see (iii), let T be a tree of diameter 4, and let $a, b \in T$ such that $d(a,b) = 4$. Let x, y, z be the three vertices encountered, respectively, as we traverse along the shortest path of length 4 from a to b. Vertex x is at a distance 3 from b and therefore can only have some more pendent vertices attached to it. Since message originated at b shall reach x earliest at step 3, the number of such pendent vertices can be at most $(t-4)$ excluding a. In other woeds we can say that we can have a star of *order* (number of vertices) $t-2$ which is rooted at x attached to y. The message originated at b (or any vertex at distance 2 from y) will arrive at y earliest at step 2. Therfore, we can at most have such additional stars of order $t-3$ down to 1 attached to y. The tree so obtained will have total number of vertices equal to

$$\sum_{i=1}^{t-2} i + 2 + 1 = \frac{1}{2}(t-2)(t-1) + 3,$$

and broadcasting in this tree obviously takes t time units. Furthermore, in a tree of diameter 4 the center of a longest path will be informed earliest at step 2 and therefore we cannot possibly have more vertices than argued above. This completes the proof that $S(t,4) = \binom{t-1}{2} + 3$.

Increasingly tedious, but quite similar proofs also yield (iv)-(viii). The case when d is odd differs slightly. Therefore, in what follows we shall outline the proof of $d = 7$ case and leave the rest for the readers.

As before assume T is a tree of diameter equal to 7, and let $a, b \in T$ such that $d(a,b) = 4$. Let $x_1, x_2, ..., x_6$ are the six vertices on the shortest path from a to b in the given order. Messages originating at b shall (at the earliest) arrive at x_2 at step 5. Therefore, as argued above we can (at most) have stars of order $t - 5$ down to 1 attached to x_1. Since the message could arrive at x_3 at step 4, x_3 can continue to broadcast from step 6 onwards and we can have additional subtrees, such as the one rooted at x_2 (of radius 2), of decreasing size attached to x_3. If the originator is a instead, by *symmetry* the same construction can be repeated on x_4. In other words, we can (at most) have a subtree of radius 3 rooted at x_3 in which broadcasting from root x_3 takes $t - 4$ steps, and also an identical copy of this tree rooted at x_4. The size of such a tree can be at most

$$2(\sum_{i=1}^{t-5} (\sum_{j=1}^{i} j + 1) + 2) = 2 \binom{t-3}{3} + 2(t-3). \quad \square$$

Above lemma, though proved using only elementary combinatorial arguments suffices to determine tree broadcast function for n upto 326. The methodology adopted in the proof can certainly be continued to obtain $T(n)$ for higher values of n. In what follows we shall analyze the constructions of Lemma 1, which will help us prove Theorem 2. We begin with a definition. Let $R(t,r)$ be the size of the largest tree, rooted at say v_0, of radius r in which messages from root v_0 can be broadcast in time t. These parameters $R(t,r)$ are more amenable to recursive determination than $S(t,d)$.

Lemma 2. The parameters $R(t,r)$ satisfy the following recurrence relation:

$$R(r,r) = 2^r, R(t,1) = t + 1 \text{ and } R(t,r) = \sum_{i=r}^{t-1} R(i, r-1) + 2^r \text{ for } t > r.$$

Proof. The root v_0 at time units 1 to $t - r$ can pass the message to $v_1, v_2, ..., v_{t-r}$ which are themselves (can be thought of as) roots of subtrees T_{v_i} such that $r(v_i, T_{v_i}) = (r-1)$ and $b(v_i, T_{v_i}) = (t-i)$. In the remaining r steps the root can *multiply* itself to generate 2^r additional vertices. $\quad \square$

Lemma 3. $S(t, 2r+1) = 2R(t-r, r)$ for $t \geq 2r + 1$.

Proof. Follows from a straightforward generalization of the construction given in the proof of Lemma 1 for the $d = 7$ case. Let T be a tree of the largest size of diameter $2r+1$ and broadcast radius t, and let $a, b \in T$ such that $d(a,b) = 2r+1$. The *center* of the induced path of length $2r+1$ from a to b consists of two adjacent vertices, say x and y, each of which is the root of subtrees $T(x,y)$ and $T(y,x)$, respectively, each of them of radius r and broadcast time from the root of $t - r$, and therefore of size at most $R(t-r, r)$. $\quad \square$

4   Proofs of Main Theorems

With the constructions of preceding section we shall now verify all the entries of Table I and also prove Theorem 2.

Proof of Theorem 1. The first few entries for $n \leq 6$ can be directly verified by inspection. For $7 \leq n \leq 9$, consider the values of $S(5, d)$ for $2 \leq d \leq 5$. From Lemma 1 these values are $6, 8, 9, 8$, respectively. This implies that the largest tree of broadcast radius 5 has order 9. We know that the largest tree with broadcast time at most 4 can have no more than 6 vertices. Therefore, $T(n) = 5$ for $7 \leq n \leq 9$. The largest tree in which broadcasting takes 6 steps can also be obtained from Lemma 1 which gives the values $10, 13, 14, 12$, respectively, for $S(6, d)$ for $3 \leq d \leq 6$. Thus, we see that $S(6) = 14$. We have already determined that $S(5) = 9$, and therefore, for n in the given range of $10 - 14$ the tree broadcast time is equal to 6.

Other entries can also be verified similarly. We shall only indicate the proof of one more case, namely $85 \leq n \leq 128$, and leave it for the reader to similarly check the remaining entries. Once again going back to Lemma 1, we see that the largest sizes of trees with broadcast time 10 are 58,80,84,78, respectively, achieved with diameters $5, 6, 7, 8$, respectively. Thus, the smallest tree requiring 11 steps has size 85, and similarly from Lemma 1 we get the largest tree having the same broadcast time cannot have more than 128 vertices.   □

Proof of Theorem 2. From Lemma 3, we have $S(3r + 1, 2r + 1) = 2R(2r, r)$. For large n (and large r) from Lemma 2 it can be seen that $R(2r, r) \sim 2^{2r}$, and the theorem follows.   □

5   Concluding Remarks

Recall that a finite sequence $s_1, s_2, ..., s_n$ is *unimodal* if there exists some k, $1 \leq k \leq n$, such that $s_1 \leq s_2 ... \leq s_k \geq s_{k+1} \geq ... \geq s_n$. Observing the behaviour of $S(t, d)$, first of all we make the following conjecture.

Conjecture 2. For fixed t, the sequence $S(t, d)$, $1 \leq d \leq t$, is unimodal.

If the above conjecture is true, it would be interesting to determine d (in terms of t) where optimality is achieved.

We have said in the beginning that it is of considerable interest to study broadcasting in general graphs. The case of trees considered here is, of course, the first natural choice, and in [9] we deal with the unicyclic graphs, i.e. connected graph on n vertices having exactly n edges. In [8] we have seen how the edges in the MBG drop as we allow additional steps over $\lceil \log_2 n \rceil$. It would be interesting to see the effect on broadcast time as we allow additional edges over n. To begin with, it would be very interesting to determine the extreme cases.

Problem 1. Let $B(n, e)$ and $W(n, e)$ denote the best and the worst broadcast time achievable in a connected graph of order n and e edges. Determine $B(n, e)$ and $W(n, e)$.

To begin with some suitable values of e may be chosen. Results of this paper, of course, imply that $B(n, n-1) \sim \frac{3}{2} \log_2 n$, and quite obviously $W(n, n-1) = n - 1$. The case $e = n$ is dealt with in [9].

We have not assumed any conditions such as some bound on the maximum degree or bound on the message overhead needed to implement the broadcast protocol under local control (in terms of the number of extra bits needed on any message sent), which are often imposed due to physical constraints of a communication networks. It would be interesting to study broadcasting in general graphs with these added constraints. Let us conclude by reiterating a problem whose solution is highly desirable despite the NP-completeness of the problem of determining the broadcast time of a general graph.

Problem 2. Design an efficient sequential (parallel) approximation algorithm to determine (bound) $b(G)$ for general graph G.

References

[1] Farley A.M., Minimum Broadcast Networks, *Networks* 9 (1979), 313-332.

[2] Farley A., Hedetniemi S.T., Broadcasting in Grid graphs, *Congr. Numer.* 21(1978), 275-288.

[3] Grigni M., Peleg D., Tight Bounds on Minimum Broadcast Networks, *SIAM J. of Disc. Math.* 4(1991), 207-222.

[4] Ko C.S., On a Conjecture Concerning Broadcasting in Grid Graphs, *Preliminary Report, AMS Notices* 1979, 196-197.

[5] Peck G.W., Optimal Spreading in an n-dimensional Rectilinear Grid, *Stud. Appl. Math.* 62(1980), 69-74.

[6] Proskurowski A., Minimum Broadcast Trees, *IEEE Transactions on Computers* 30 (1981), 363-366.

[7] Slater P.J., Cockayne E.J. and Hedetniemi S.T., Information Dissemination in Trees, *SIAM J. Computing* 10 (1981), 692-701.

[8] Shastri A., "Time-relaxed Broadcasting in Communication Networks", *Discrete Applied Math.*, submitted.

[9] Shastri A., Broadcasting in General Networks II: Unicyclic graphs, in preparation.

Uni-directional Alternating Group Graphs *

Shyh-Chain Chern[1], Tai-Ching Tuan[1] and Jung-Sing Jwo[2]

[1] Department of Applied Mathematics, CS Program
National Sun Yat-sen University, Kaohsiung,
[2] Department of Computer and Information Sciences
Tunghai University, Taichung,

Abstract. A class of uni-directional Cayley graphs based on alternating groups is proposed in this paper. It is shown that this class of graphs is strongly connected and recursively scalable. The analysis of shortest distance between any pair of nodes in a graph of this class is also given. Based on the analysis, we develop a polynomial time routing algorithm which yields a path distance at most one more than the theoretic lower bound.

1 Introduction

Due to the rapid advances in the large *interconnection networks* such as *massive parallel computers*, the search for large uni-directional graphs with fairly nice topological and symmetric properties has gained much attention in literature recently [2, 4, 5, 6, 8]. Although most graphs studied as interconnection network models for parallel computers are un-directed, an un-directed link in such models is normally realized by two directed links in practice. A generic architecture of a node in parallel computers is shown in Figure 1, where each switch supports direct-in and direct-out channels. Usually a crossbar is used to allow all possible connections between input and output channels within the switch. Thus, the search for uni-directional graphs which possess the properties similar to those of their un-directed counterparts is an immediate and meaningful design consideration.

A class of graphs called *group* graphs or *Cayley* graphs provide a very natural and rich framework for interconnection networks. *Hypercubes* [10], *star* graphs [1] and *alternating group* graphs [9] are examples of Cayley graphs. Recently, Chou and Du [4] propose two different schemes to define the orientations of the edges in hypercube. These two uni-directional schemes are called $UHC1_n$ and $UHC2_n$, respectively. Day and Tripathi [5] propose a scheme called US_n, to define the orientation of edges in a star graph. In this paper, our aim is to propose a uni-directional scheme on alternating group graphs and to explore its properties. In addition, we shall compare some properties of the uni-directional alternating group graph with those of $UHC1_n$, $UHC2_n$ and US_n.

* T. C. Tuan was supported in part by the National Science Council of Taiwan, under Contract 84-2111-M-110-073. Email: t.tuan@ieee.org.
  J. S. Jwo was supported in part by the National Science Council of Taiwan, under Contract 84-0208-M-029-014.

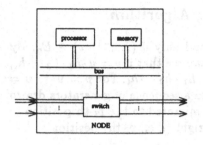

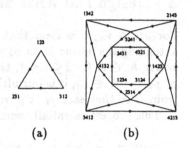

(a)   (b)

Figure 1 A generic node architecture   Figure 2 (a) UAG_3 and (b) UAG_4.

2 The Uni-Directional Model

Let $\langle n \rangle = \{1, 2, \cdots, n\}$ and p be a permutation, $p_1 p_2 \cdots p_n$, of $\langle n \rangle$. Furthermore, let I denote the identity permutation of $\langle n \rangle$. A permutation is said to be *even* (resp. *odd*) if its parity is even (resp. odd). A well known fact regarding permutation representation is that p can be represented by its *cycle* structure, i.e., $p = c_1 c_2 \cdots c_k e_1 e_2 \cdots e_\ell$, where c_i is a nontrivial cycle of length $|c_i| \geq 2$, for $1 \leq i \leq k$ and e_i is *invariant*, i.e., $|e_i| = 1$, for $1 \leq i \leq \ell$.

Let $g_i = (1\ 2\ i)$, $3 \leq i \leq n$. Define $\Omega = \{g_i | 3 \leq i \leq n\}$, and each g_i in Ω is called a *generator*. An *uni-directional* n-dimensional alternating group graph, denoted by $UAG_n(V_n, E_n)$, has the vertex set V_n of all the even permutations of $\langle n \rangle$ and the edge set $E_n = \{(p, q) | p, q \in V_n, q = p \cdot h, \text{ for some } h \in \Omega\}$. Figure 2 gives two examples of UAG_n, where $n = 3$ and $n = 4$ respectively. Clearly each edge in UAG_n is embedded in a circuit of length three. Thus if a packet is mis-routed from vertex a, it can always return to a in three hops.

The following Lemma is given without proof.

Lemma 1. UAG_n *is strongly connected for* $n \geq 3$.

Given a graph $G(V, E)$, an *automorphism* of V is a permutation α of V satisfying the property that if $(u, v) \in E$, then $(\alpha(u), \alpha(v)) \in E$. Observe that the set of all automorphisms of V forms a group under ".", the function composition operator. Furthermore, G is *vertex symmetric* if and only if, for $u, v \in V$, there exists an automorphism α such that $\alpha(u) = v$.

Lemma 2. UAG_n *is vertex symmetric.*

Recall that each vertex in V_n is an even permutation of $\langle n \rangle$. For $1 \leq i \leq n$, let $V_n^i = \{p_1 p_2 \cdots p_n \in V_n | p_n = i\}$. Consider the subgraph of UAG_n induced by V_n^i, denoted by UAG_n^i, whose edge set is represented by E_n^i. The following Lemma guarantees that UAG_n is recursively scalable.

Lemma 3. UAG_n *can be decomposed into* n *subgraphs, namely* UAG_n^1, UAG_n^2, $\ldots$, UAG_n^n, *and each of them is isomorphic to* UAG_{n-1}.

3 Design and Analysis of Routing Algorithm

For $u, v \in V_n$, it is clear that $(u, v) \in E_n$ if and only if $(I, u^{-1} \cdot v) \in E_n$. By extending the notion of an edge in UAG_n, we can see that if $v = u \cdot h_1 \cdot h_2 \cdots h_t$, where $h_i \in \Omega$ for $1 \leq i \leq t$, then $I = v^{-1} \cdot u \cdot h_1 \cdot h_2 \cdots h_t$. For $p = v^{-1} \cdot u = p_1\, p_2 \cdots p_n \in V_n$, it is our ultimate goal to find a sequence of generators drawn from Ω so that each p_i in p will eventually be migrated to the p_i-th position in I. This process essentially sorts p_i in p to the right (i. e., p_i-th) position.

Lemma 4. *Let $p \in V_n$. 1 and 2 in p will automatically be sorted when all the other entries in p are sorted.*

Given $p_1 p_2 \ldots p_n \in V_n$ with some i satisfying $3 \leq i \neq p_i \geq 3$, we can easily see that it takes three steps to move p_i back to the right position. Assume that $(r_0\ r_1 \cdots r_{x-1})$ and $(s_0\ s_1 \cdots s_{y-1})$ are two cycles of p containing neither 1 nor 2. In order to sort each and every element in the first cycle, we first apply the generator $g_{r_{(i-1)\ \mathrm{mod}\ x}}$ to move r_i, where $0 \leq i \leq x - 1$, to the second position. Then, for the second cycle, we apply $g_{s_{(j-1)\ \mathrm{mod}\ y}}$ to move s_j, where $0 \leq j \leq y - 1$, to the second position and r_i is accordingly moved to the first position. Now applying the corresponding generator g_{r_i} will sort r_i, and move s_j and r_{i+1} to the first and second positions respectively. Next, we apply g_{s_j} to sort s_j, and move r_{i+1} and s_{j+1} to the first and second positions respectively. The number of steps needed to sort all the elements in either cycle (of interest) by alternating the above process between the two cycles, is at least one more than the respective cycle length.

Lemma 5. *If cycle $c = (t_1\ t_2\ \cdots\ t_r)$ does not contain 1 nor 2, then at least $r + 1$ steps are needed to sort all the elements in c back to their right positions.*

Suppose each nontrivial cycle is associated with a weight which is one more than its cycle length. Following the above development, the shortest distance between any two nodes can be obtained by first partitioning the nontrivial cycles into 2 groups such that the total weight in one group is as close to the total weight of the other group as possible, and then alternating the sorting process discussed in the previous paragraph between these 2 groups. The former is the weighted set partition problem, which is NP-complete (p. 223, [7]). Before giving a heuristic to efficiently and effectively perform the partition, we shall explore the issue of the shortest distance between two nodes p and I in UAG_n a bit further. Let $p = c_1 \cdots c_k e_1 \cdots e_\ell$, and D_p denote the shortest distance from p to I.

1. $p = 1\ 2\ p_3\ \cdots\ p_n$, and each cycle c_i is associated with a number $l_i = |c_i| + 1$. It suffices to study the following three cases for dealing with the problem of partitioning $\{l_1, l_2, \cdots, l_k\}$ into two sets A and B such that $\sum_{l_i \in A} l_i \geq \sum_{l_i \in B} l_i$.

   (a) $\sum_{l_i \in A} l_i - \sum_{l_i \in B} l_i = 0$: $D_p = \sum_{l_i \in A \cup B} l_i = (n - \ell) + k = n + k - \ell$.

   (b) $\sum_{l_i \in A} l_i - \sum_{l_i \in B} l_i = 1$: Since p is an even permutation, the number of the even-length cycles is even. This case cannot occur.

(c) Otherwise: It takes one step more than case (1)(a) does. That is, $D_p = (n + k - \ell + 1)$.

2. $p = 2\ 1\ p_3 \cdots p_n$, and $c_1 = (1\ 2)$. As before, let $l_i = |c_i| + 1$, for $1 \leq i \leq k$. It suffices to study the following 3 cases for dealing with the problem of partitioning $\{l_2, l_3, \ldots, l_k\}$ into two sets A and B such that $\sum_{l_i \in A} l_i \geq \sum_{l_i \in B} l_i$. Consider the following three cases:

(a) $\sum_{l_i \in A} l_i - \sum_{l_i \in B} l_i = 0$: The number of even length cycles is even. Since cycle c_1 will be automatically sorted, the number of cycles with odd length is odd in A and B. This case will not occur.

(b) $\sum_{l_i \in A} l_i - \sum_{l_i \in B} l_i = 1$: $D_p = (n - \ell - 2) + (k - 1) = n + k - \ell - 2$.

(c) Otherwise: With the same argument as the case (1)(c), $D_p = (n - \ell - 2) + (k - 1) + 1 = n + k - \ell - 2$.

3. $p = 1\ p_2\ p_3 \cdots p_{j-1}\ 2\ p_{j+1} \cdots p_n$ and $c_1 = (2\ p_2 \cdots j)$. Define $l_i = |c_i| + 1$ for $2 \leq i \leq l$ and $l_1 = |c_1| - 1$. It suffices to study the following 3 cases for dealing with the problem of partitioning $\{l_1, l_2, l_3, \ldots, l_k\}$ into two sets A and B such that $\sum_{l_i \in B} l_i \geq \sum_{l_i \in A} l_i$ and l_1 is in A. Consider the following three cases:

(a) $\sum_{l_i \in A} l_i - \sum_{l_i \in B} l_i = 0$: $D_p = \sum_{l_i \in A \cup B} l_i = (n - \ell - 1) + (k - 1) = n + k - \ell - 2$.

(b) $\sum_{l_i \in B} l_i - \sum_{l_i \in A} l_i = 1$: It can be seen that this case can not occur.

(c) otherwise: With the same argument as the case (1)(d), $D_p = (n - \ell - 1) + (k - 1 + 1) = n + k - \ell - 1$.

4. Let $p = p_1\ 2\ p_3 \cdots p_{j-1}\ 1\ p_{j+1} \cdots p_n$ and $c_1 = (1\ p_1\ p_i \cdots j)$. Move p_1 back to the right position. Then, the case can be :

(a) $2\ 1\ p_3 \cdots p_{j-1}\ p_1\ p_{j+1} \cdots p_n$, if $p_1 = j$. Under this case, the analysis is similar to case (2).

(b) $2\ p_i \cdots p_{j-1}\ 1\ p_{j+1} \cdots p_n$, if $p_1 \neq j$. The analysis of this case is similar to case (3).

5. Let $p = 2\ p_2 \cdots p_{j-1}\ 1\ p_{j+1} \cdots p_n$ and $c_1 = (1\ 2\ p_1\ p_i \cdots j)$. Following the same argument as case (3), let $l_1 = |c_1| - 2$. The following discussion can be easily shown.

(a) $\sum_{l_i \in A} l_i - \sum_{l_i \in B} l_i = 0$: This case will not occur.

(b) $\sum_{l_i \in B} l_i - \sum_{l_i \in A} l_i = 1$: $D_p = (n - \ell - 2) + (k - 1) = n + k - \ell - 3$.

(c) otherwise: $D_p = (n - \ell - 2) + (k - 1 + 1) = n + k - \ell - 2$.

6. Let $p = p_1\ p_2 \cdots p_{i-1}\ 1\ p_{i+1} \cdots p_{j-1}\ 2\ p_{j+1} \cdots p_n$ and $c_1 = (1\ \underbrace{p_1 \cdots j}_{m_1}$ $2\ \underbrace{p_2 \cdots i}_{m_2})$. Keep on routing the element in the first position back to its right position until $p_1 \in \{1, 2\}$. Consider the following cases:

(a) $p = 2\ 1\ p_3 \cdots p_n$, if $m_1 = m_2$: The discussion is similar to case (2).

(b) $p = 1\ 2\ p_3 \cdots p_n$, if $m_1 = m_2 + 1$: The discussion is similar to case (1).

(c) $p = 1\ p_2\ p_3 \cdots p_n$, if $m_1 > m_2 + 1$: The discussion is similar to case (3).

(d) $p = 2\ p_2\ p_3 \cdots p_n$, if $m_1 < m_2$: The discussion is similar to case (5).

7. Let $p = p_1 p_2 \cdots p_{i-1} 1 p_{i+1} \cdots p_{j-1} 2 p_{j+1} \cdots p_n$ and $c_1 = (1 \underbrace{p_1 \cdots i}_{m_1})$

and $c_2 = (2 \underbrace{p_2 \cdots j}_{m_2})$. The argument of this case is quite similar to that of case (6). We leave this as an exercise to the readers.

Theorem 6 summarizes the above discussion.

Theorem 6. *If $p(\in V_n)$ is of the form $c_1 c_2 \cdots c_k e_1 e_2 \cdots e_\ell$, then*
$$
\begin{aligned}
D_p &= n + k - \ell \text{ or } n + k - \ell + 1, && \text{if } (p_1 = 1, p_2 = 2); \\
&= n + k - \ell - 2 \text{ or } n + k - \ell - 1, && \text{if } (p_1 \neq 1, p_2 = 2), \text{ or } (p_1 = 1, p_2 \neq 2); \\
&= n + k - \ell - 3 \text{ or } n + k - \ell - 2, && \text{if } (p_1 = 2, p_2 = 1), \text{ or } \\
& && \text{if } 1, 2 \in c_i, 1 \leq i \leq k, 3 \leq |c_i|; \\
&= n + k - \ell - 4 \text{ or } n + k - \ell - 3, && \text{if } 1 \in c_i, 2 \in c_j \text{ and } i \neq j.
\end{aligned}
$$

Since the cycle set partition is NP-complete, we shall develop a polynomial time greedy heuristic to perform routing from $p = p_1 p_2 \cdots p_n$ to I in the following. This heuristic always produces a routing distance at most one more than the theoretical lower bound [9]. Let vertex $r = r_1 r_2 \cdots r_n$ be a node in the routing path from p to I. Its immediate successor r' can be determined by using Algorithm 1 given below.

Algorithm 1 {Let $c_i = (q_1 q_2 \cdots q_m)$ and $q_1 < q_j, 2 \leq j \leq m$.}

1. evaluate the cycle structure of r;
2. if ($r_1 \in \{1, 2\}$) then
    if ($r_2 \in \{1, 2\}$) then
        if ($r = I$) then   { the destination is reached; }
        else {
            Select any nontrivial cycle c_i in r and $c_i \neq (1, 2)$;
            $r' = r \cdot g_{q_m}$; }
    else
        if (there exists a cycle c_i in r and $r_2 \notin c_i$) then $r' = r \cdot g_{q_m}$;
        else {
            There exists only one nontrivial cycle c_i in r and $q_j = r_2$;
            $\ell = |c_i - \{1, 2\}|$;
            if ($\ell > 1$) then $r' = r \cdot g_{q_{(j+\lfloor \frac{m}{2} \rfloor)-1}}$;
            else $r' = r \cdot g_{r_2}$; }
3. else $r' = r \cdot g_{r_1}$.

The worst case analysis of employing Algorithm 1 successively to route vertex p back to vertex I can lead to the following.

Lemma 7. *Let D be the diameter of UAG_n, then*
$$
\begin{aligned}
D &= 3 * \frac{n-2}{2} && , \text{ if } n \text{ is even and } \frac{n-2}{2} \text{ is even.} \\
&= 3 * \frac{n-2}{2} + 1 && , \text{ if } n \text{ is even and } \frac{n-2}{2} \text{ is odd.} \\
&= 3 * \lfloor \frac{n-2}{2} \rfloor + 1 && , \text{ if } n \text{ is odd and } \lfloor \frac{n-2}{2} \rfloor \text{ is even.} \\
&= 3 * \lfloor \frac{n-2}{2} \rfloor + 2 && , \text{ if } n \text{ is odd and } \lfloor \frac{n-2}{2} \rfloor \text{ is odd.}
\end{aligned}
$$

4 Concluding Remarks

Table 1 shows the comparisons among UAG_n, $UHC1_n$ and $UHC2_n$. It can be seen that if we fix the cardinality of the vertex set for the four uni-directional schemes theoretically, then (1) uni-directional alternating group graphs have the smallest diameter, and (2) uni-directional alternating group graphs have far less indegree and outdegree than uni-directional hypercubes (of interest) do. $UHC1_n$, $UHC2_n$ and US_n do not have vertex symmetric property though $UHC1_n$ and $UHC2_n$ can be decomposed into two vertex-symmetric components [3]. These observations show the superiority of uni-directional alternating group graphs.

Networks	No. vertices	Indegree (Outdegree)	vertex Symmetry	Diameter
UAG_n	$\frac{n!}{2}$	$n-2$ $(n-2)$	yes	$3 * \frac{n-2}{2}$, if n is even and $\frac{n-2}{2}$ is even, $3 * \frac{n-2}{2} + 1$, if n is even and $\frac{n-2}{2}$ is odd, $3 * \lfloor \frac{n-2}{2} \rfloor + 1$, if n is odd and $\lfloor \frac{n-2}{2} \rfloor$ is even, $3 * \lfloor \frac{n-2}{2} \rfloor + 2$, if n is odd and $\lfloor \frac{n-2}{2} \rfloor$ is odd.
$UHC1_n$	2^n	$\lceil \frac{n}{2} \rceil$ or $\lfloor \frac{n}{2} \rfloor$ $(\lfloor \frac{n}{2} \rfloor$ or $\lceil \frac{n}{2} \rceil)$	no	$n+1$, if n is even, $n+2$, if n is odd.
$UHC2_n$	2^n	$\lceil \frac{n}{2} \rceil$ or $\lfloor \frac{n}{2} \rfloor$ $(\lfloor \frac{n}{2} \rfloor$ or $\lceil \frac{n}{2} \rceil)$	no	$n+2$, if $n=3$, $n+1$, if $n \neq 3$.
US_n	$n!$	$\lfloor \frac{n-1}{2} \rfloor$ or $\lceil \frac{n-1}{2} \rceil$ $(\lceil \frac{n-1}{2} \rceil$ or $\lfloor \frac{n-1}{2} \rfloor)$	no	$\leq 5 * (n-2) + 1$.

Table 1. A comparison among UAG_n, $UHC1_n$, $UHC2_n$ and US_n.

References

1. Akers, S. B., Harel, D. and Krishnamurthy, B.:The star-graph: an attractive alternative to the n-cube. Proc. the International Conference on Parallel Processing (1987) 393–400
2. Comellas, F., Fiol, M. A.:Vertex symmetric digraphs with small diameter. Technique Report, Department de Mathematica Aplicada i Telematica, Universitat Politecnica de Catalunya, Spain (2/1992).
3. Chern, S. C., Tuan, T. C., Jwo, J. S.:On uni-directional hypercubes. Technical Report, Department of Applied Mathematics, National Sun Yat-sen University, Taiwan (1994).
4. Chou, C. H., Du, David H. C.:Unidirectional hypercubes. Proc. Supercomputing'90 (1990) 254–263
5. Day, K., Tripathi, A.:Unidirectional star graphs. Info. Proc. Lett. 45 (1993) 123–129
6. Faber, V., Moore, W., Chen, W. Y. C.:Cycle prefix digraphs for symmetric interconnection networks. Networks 23 (1993) 641–649.
7. Garey, M. R., Johnson, D. S.:Computers and intractability: a guide to the theory of NP-completeness. W. H. Freeman and Company (1979).
8. Hamidoune, Y.O., Llado, A. S., Serra, O.:The connectivity of hierarchical Cayley digraphs. Disc. Appl. Math. 37/38 (1992) 275–280
9. Jwo, J. S., Lakshmivarahan, S., Dhall, S.K.:A new class of interconnection network based on the alternating group. Networks 23 (1993) 315–326
10. Leighton, F. T.:Introduction to parallel algorithms and architectures: arrays, trees, hypercubes. Morgan Kaufmann (1992).

On separating proofs of knowledge from proofs of membership of languages and its application to secure identification schemes*

(Extended Abstract of COCOON'95)

Kouichi Sakurai

Dept. of Computer Science and Communication Engineering,
Kyushu University, 812-81 Japan.
e-mail:sakurai@csce.kyushu-u.ac.jp

Abstract. A four-move protocol for quadratic residuosity is proposed and the security is discussed. An application of the proposed protocol to a cryptographic identification scheme introduces a new notion of practical soundness. Our basic approach is to separate proofs of knowledge from proofs of membership of languages. Previous works deal with proofs of knowledge as an additional property of proofs of membership.

1 Introduction

The original notion of interactive proof systems (of membership) of languages proposed by Goldwasser, Micali, and Rackoff [GMR85] was defined as two party protocol between a (infinitely powerful) prover and a (probabilistic polynomial time) verifier, in which the prover tries to convince the verifier that a given theorem is true. A variant notion, proofs of *knowledge*, was formulated by Feige-Fiat-Shamir [FFS87] and Tompa and Woll [TW87] from more practical points of view, where practical provers (with only a probabilistic polynomial time computing power) can convince the verifier only when the prover knows a proof of the theorem. Proofs of knowledge have been regarded as an additional property of proofs of languages. Because any proof of knowledge is a proof of languages [TW87, Slo89], and the known concrete zero-knowledge interactive proofs, which require no unproven complexity assumption, for membership of $\mathcal{NP}$-languages [GMR85, GMW86, BMO90] are shown to be zero-knowledge interactive proofs of knowledge.

Recently, Bellare and Goldreich [BG92] gave rigorous definitions of proofs of knowledge and introduced the following weaker version of the knowledge proof: *Proofs of knowledge should be defined in the case of only the correct inputs* (positive-side knowledge). Their motivation was natural; previous definitions were too restrictive when they dealt with some subprotocols in the whole protocol, and they suggested usefulness of the new definition. However, they failed to show a concrete gap of the power between proofs of strong knowledge in the sense of [FFS87, TW87] and proofs of positive-side knowledge in their sense. Feige, Fiat, and Shamir [FFS87] already showed that for any language L in $\mathcal{NP} \cap$ co-$\mathcal{NP}$, under the assumption that secure public key encryption schemes exist, the

---

* This work was inspired by the discussion while the author was writing his thesis [Sak93] under the supervision of Kazuo IWAMA.

prover can show that he knows whether a given x is in L or in its complement $\overline{L}$ without revealing even this single bit of knowledge. This implies that, under such cryptographic assumptions, there are no gap between proofs of strong knowledge in the sense of [FFS87, TW87] and proofs of positive-side knowledge in the sense of [BG92]. However, the idea in [FFS87] is based on a general zero-knowledge interactive proofs for $\mathcal{NP}$-complete languages [GMW86], and requires a complicated transformation from the language L into $\mathcal{NP}$-complete languages, which is not practical in our framework.

This paper investigates more practical and concrete merits on defining proofs of knowledge without relying on proofs of membership of languages. First, a four-move protocol for quadratic residuosity is presented and its properties are explored. In the proposed protocol, provers cannot convince the verifier without a square root of the given quadratic residuosity (proofs of positive-side knowledge). However, powerful provers can cheat the verifier that the given integer which is not quadratic residuosity is quadratic residuosity (not proofs of the language). Nevertheless, we apply this protocol to identification scheme like as the Fiat-Shamir setting by analyzing the cheating prover's knowledge.

The original zero-knowledge Fiat-Shamir scheme [FS89] needs polynomially many round. The direct parallelization decreases the round complexity into 3-move, however, it is no longer zero-knowledge [GK90]. De Santis et al. [DDP94, DiP94] showed a way of obtaining a 4-move perfect ZK interactive proof system for special languages associated with the quadratic residuosity. However, provers in their protocol [DDP94, DiP94] requires not only the exact witness of the problem (e.g. a square root of I module N) but also the factorization of the modulus N as an additional knowledge in order to convince the verifier. So, the protocol by [DDP94, DiP94] cannot be applied to the Fiat-Shamir like ID-based identification schemes. The 5-move protocol based on the technique by Bellare, Micali, and Ostrovsky [BMO90] has the lowest round complexity among the previous known perfect zero-knowledge identification schemes based on the quadratic residuosity problem. Our 4-move protocol further improves the efficiency, and is shown to be optimal with respect to the round complexity.

The properties of the proposed protocol and the application introduce a new definition on practical soundness fitted for the identification systems like as the Fiat-Shamir scheme. Our basic idea is to investigate the knowledge-state of the prover who convinces the wrong input (negative-side knowledge), and we regard the protocol as practically sound if the negative-side knowledge is more powerful than the positive-side knowledge. (Informally speaking, if *cheating is more difficult than proving*, we call the identification system practically sound.) We also discuss comparisons of the new notion to the previous notion of security of identification, and clarify a property of soundness hidden in the notion of "no-transferable" introduced in [FFS87].

2 Notation and Definitions

Our model of computation is the interactive probabilistic Turing machines (both for the prover P and for the verifier V) with an auxiliary input. The common input is denoted by x and, and its length is denoted by $|x| = n$. We use $\nu(n)$ to denote any function vanishing faster than the inverse of any polynomial in n.

More formally,

$$\forall k \in \mathbf{N} \; \exists n_0 \; s.t. \; \forall n > n_0 \; 0 \le \nu(n) < \frac{1}{n^k}.$$

We define *negligible* probability to be the probability behaving as $\nu(n)$, and *overwhelming* probability to be the probability behaving as $1 - \nu(n)$.

Let $A(x)$ denote the output of a probabilistic algorithm A on input x. This is a random variable. When we want to make the coin tosses of A explicit, for any $\rho \in \{0,1\}^*$ we write $A[\rho]$ for the algorithm A with ρ as its random tape. Let $V_P(x)$ denote V's output after interaction with P on common input x, and let $M(x; A)$ (where A may be either P or V) denote the output of the algorithm M on input x, where M may use the algorithm A as a (blackbox) subroutine. Each call M makes to A is counted as a single computation step for M.

Definition 1 [GMR85]. An interactive proof for membership of the language L is a pair of interactive probabilistic Turing machines (P, V) satisfying:

Language Completeness: If x belongs to L, V accepts P's proof with overwhelming probability. Formally:

$$\forall x \in L \; Prob(V_{P(x)}(x) \, accepts) > 1 - \nu(|x|),$$

where the probability is taken over all of the possible coin tosses of P and V.

Language Soundness: If x does not belong to L and P^* may act in any way, V accepts P^*'s proof with negligible probability. Formally:

$$\forall x \notin L \, \forall P^* \; Prob(V_{P^*(x)}(x) \, accepts) < \nu(|x|),$$

where the probability is taken over all of the possible coin tosses of P^* and V.

It should be noted that P(resp. P^*)'s resource is computationally unbounded, while V's resource is bounded by probabilistic polynomial time in $|x|$.

We recall that the *view* of the verifier is everything he sees during an interaction with the prover, that is, his own coin tosses and the conversation between himself and the prover.

Definition 2 [GMR85]. Let (P, V) be an interactive protocol and let $x \in \{0,1\}^*$. The *view* of V' on input x is the probability space

$$VIEW_{(P,V')}(x) = \{(R, C) : R \leftarrow \{0,1\}^{p(|x|)}; \; C \leftarrow (P \leftrightarrow V'[R])(x)\},$$

where p is a polynomial bounding the running time of V', and $(P \leftrightarrow V'[R])(x)$ denotes the probability space of conversations between P and $V'[R]$ on input x (the probability is taken over the all of the possible coin tosses of P).

Definition 3. An interactive proof system (P, V) for the membership of the language L is perfect zero knowledge if there exists a simulator S which runs in expected polynomial time, for every V' and for $\forall x \in L$, $S(x; V'(x)) = VIEW_{(P,V')}(x)$.

Definition 4. A *move* of an interactive proof is a message sent by one of the participants.

Definition 5. Let R be a relation $\{(x, w)\}$ testable in $\mathcal{NP}$. Namely, given x and w, checking whether $(x, w) \in R$ is computed in polynomial time. The language associated with the relation R is defined to be $L_R = \{x : \exists y \text{ such that } (x, y) \in R\}$, and belongs to $\mathcal{NP}$. Conversely, every $\mathcal{NP}$ language L naturally induces a relation R_L, of which checking is done in polynomial time. For any x, its *witness set* $w(x)$ is the set of w such that $(x, w) \in R$.

3 A 4-move protocol for quadratic residuosity

This section presents a 4-move protocol for quadratic residuosity and discusses the properties of the protocol.

3.1 Quadratic Residuosity

A language quadratic residuosity (QR) is defined to be $QR = \{(I, N) | \exists s \in Z_N^* \text{ such that } I \equiv s^2 \pmod{N}\}$, and a language quadratic nonresiduosity (QNR) is defined, as the complement of QR, to be $QNR = \overline{QR}$. Clearly QR is in $\mathcal{NP}$ (the witness is a square root of I module N). Also QNR is in $\mathcal{NP}$ (there is a polynomial time algorithm, which uses the complete factorization of N, to decide if $(I, N) \in QNR$). Especially, for a given modulus N, we denote the set of quadratic residuosity module N by QR_N, i.e. $QR_N = \{I | \exists s \in Z_N^* \text{ such that } I \equiv s^2 \pmod{N}\}$.

3.2 Basic idea

Our protocol is based on the idea proposed by Saito, Kurosawa, and Sakurai [SKS91], who constructed a 4-move perfect ZK proof of knowledge on the certified logarithm problem. In their protocol [SKS91], at the first stage the verifier constructs the basis of the bit commitment and send the basis to the prover. Next the prover commits his random coins using the basis, and send the verifier to these committed values. Then, after receiving the verifier's challenges, the prover uses the EX-OR of the verifier's challenge bits and the prover's previous random bits as the coins which used by the prover in typical 3-move interactive protocols.

We apply this idea to the quadratic residuosity problem. Each element of QR has multiple witnesses. Then the proposed protocol does not require a complicated "one out two" protocol [FS90], and is simpler than one in [SKS91].

Remark. There is another variant of 4-move protocol with the same properties as the following. The final version [Sak95] describes the other protocol with a bitcommitment based on QR/QNR. Although it is less efficient in communication complexity than the following protocol, the other variant clarifies the idea of our construction.

3.3 The protocol

The common inputs of the prover and the verifier is (I, N), where $I = s^2$ (mod N) for some $s \in Z_N^*$. The prover P proves to the verifier V the fact that P knows the witness s. Let $k = |N|$. The proposed protocol consists of the following 3 subprotocols.

Subprotocol A (Construction of the basis of the bit commitment)

A1: V chooses $r \in_R Z_N^*$ and sends $y = r^2$ (mod N) to P.

A2: V proves via a 3-move witness hiding protocol the facts that V knows a square root of y. The following steps is executed k independent times in *parallel*.

Subsubprotocol Aa

Aa1 V chooses independently $u \in_R Z_N^*$ and sends $w = u^2$ (mod N) to P.

Aa2 P independently picks $b \in_R \{0, 1\}$, and sends b to V.

Aa3 V sends $z = r^b \times u$ (mod N) to P, where r is generated in the previous step A1.

Aa4 P verifies V's answer by checking if $z^2 = y^b \times w$ (mod N).

Subprotocol B (Random bits generation by coin flipping)

B1: P chooses $t_i \in_R Z_N^*$ and $e_{P_i} \in_R \{0, 1\}$ for $i = 1, \ldots, k$, then sends $q_i = y^{e_{P_i}} \times t_i^2$ (mod N) to V ($i = 1, \ldots, k$).

B2: V chooses $e_{V_i} \in_R \{0, 1\}$($i = 1, \ldots, k$) and sends $e_{V_1} \ldots, e_{V_k}$ to P.

B3: P sets $E_i = e_{P_i} \oplus e_{V_i}$($i = 1, \ldots, k$). Each E_i is used in the next protocol as V's challenge bits. Then P sends e_{P_i} and t_i for $i = 1, \ldots, k$.

B4: V verifies P^*'s answer, i.e. checks if $q_i = y^{e_{P_i}} \times t_i^2$ (mod N) for $i = 1, \ldots, k$.

Subprotocol C (Basic parallelized protocol for QR)

C1: P chooses $R_i \in_R Z_N^*$ and sends $X_i = R_i^2$ (mod N)($i = 1, \ldots, k$) to V.

C2: P computes $Y_i = s^{E_i} \times R_i$ (mod N) for $i = 1, \ldots, k$, where E_i is obtained in the protocol B, and sends Y_i($i = 1, \ldots, k$) to V.

C3: V verifies if $Y_i^2 = I^{E_i} \times X_i$ (mod N) for $i = 1, \ldots, k$.

We obtain the full protocol Λ by composing these sub(sub)protocols in the following manner;

Full protocol Λ

V1(A1,Aa1), **P1**(Aa2,B1,C1), **V2**(Aa3,B2), **P2**(Aa4,B3,C2), **V3**(B4,C3),

which is 4-move.

3.4 Properties of the protocol

Proposition 6. *If a (probabilistic polynomial time) prover with a $\sqrt{I}$ (mod N) causes the verifier to accept with probability 1.*

Next we consider the soundness of the protocol. However, a powerful cheating prover P^* can convince the verifier even when the input is not in QR.

Proposition 7. *Even if an input $I \notin QR_N$, a (powerful) prover P^* causes the verifier to accept with probability 1.*

Proof: The cheating prover P^* convinces the verifier when the input I is not in QR_N as follows. Suppose that the cheating prover P^*'s power is not restricted. Then P^* can compute $\sqrt{y}$ (mod N). As the prover P^* knows $\sqrt{y}$ (mod N), the commitments q_i is *chameleon* in the sense of [BCY89] for the prover P^*. Namely, P^* can disclose freely both bit 0 and 1 as e_{P_i} after he commit at the stage B1. Then, at the subprotocol C, P^* can choose the value E_i at the stage C1 in advance as his will. This implies the following prover's cheating.

$$P^*\text{'s cheating in subprotocol C}$$

$C1^*$: P chooses $Y_i \in_R Z_N^*$ and $E_i \in_R \{0,1\}$ $(i = 1,\ldots,k)$ and sends $X_i = Y_i^2 / I^{E_i}$ $(i = 1,\ldots,k)$ to V.

$C2^*$: P sends Y_i $(i = 1,\ldots,k)$ to V.

Clearly, the verifier V accepts the prover P^* above. ∎

Corollary 8. *The protocol protocol is not proofs of membership of the language QR (unless $QR \in \mathcal{BPP}$).*

Next we investigate the knowledge of provers who accepts the verifier.

Proposition 9. *For any $I \in QR_N$, if P^* can convince V to accept, then he actually "knows" a witness $\sqrt{I}$ (mod N). Or, for any $I \notin QR_N$, if P^* can convince V to accept, then he actually "knows" the complete factorization of the integer N. A probabilistic polynomial time knowledge extractor M is used in order to demonstrate P^*'s ability to compute a witness (or the factorization). Formally:*

$$\forall a \; \exists M \; \forall P^* \; \forall I \in Z_N^* \; \forall w'$$

$$Prob\Big(V_{P^*(I,w')}(I) \; accepts\Big) > 1/|I|^a \Rightarrow$$

$$Prob\left(M\big(I; P^*(I,w')\big) = \left\{ \begin{array}{l} \text{one of } \sqrt{I} \text{ (mod } N) \quad \text{if } I \in QR_N \\ \text{the complete fact. of } N \text{ otherwise} \end{array} \right. \right) > 1 - \nu(|I|),$$

where the probability is taken over all of the possible coin tosses of M and V.

Remark. P^* is assumed *not* to toss coins, since his favorable coin tosses can be incorporated into the auxiliary input w'. The knowledge extractor M is allowed to use P^* as a blackbox subroutine and runs in expected polynomial time. Each message that P^* sends M costs a single computation step for M. This condition is assumed when we consider the state of knowledge of provers in the following discussion of this paper.

Proof: We describe a knowledge extractor M.

Case 1: common input $x \notin QR_N$

Note that, as the input x is not in QR_N, P^* who honestly in the subprotocol C cannot convince the verifier. If the cheating prover P^* convinces the prover in the case when $x \notin QR_N$ with a non-negligible probability, then the commitment q_i, which is used subprotocol B, have to be chameleon in the sense of [BCY89]. Namely, the prover P^* must reveal both bits 0 and 1 for the committed result q_i.

Based on this observation, the extractor M acts as follows. In the first stage, the extractor M sends random bits $e_{V1}, \ldots, e_{Vk}$ as the (honest) verifier V does. After obtaining P^*'s replies $e_{Pi}, t_i (i = 1, \ldots, k)$, M resets P^* at the state before **V2** (at the step **V1**, M acts as did in the first stage). In the second stage, the extractor M sends different random bits $\tilde{e}_{V1}, \ldots, \tilde{e}_{Vk}$ from the first stage. Suppose that $e_{Vj} \neq \tilde{e}_{Vj}$ for some j. then the P^*'s second answer $\tilde{e}_{Pj}$ has to be changed from the first one e_{Pj}, i.e., $e_{Pj} \neq \tilde{e}_{Pj}$. Thus M obtains

$$q_j = y^{e_{Pj}} \times t_i^2 = y^{\tilde{e}_{Pj}} \times \tilde{t}_i^2 \pmod{N}.$$

M gets a square root $\sqrt{y}$ by computing $t_i / \tilde{t}_i \pmod{N}$ (or its inverse). With probability more than $1/2$, $r \neq \sqrt{y}$. In this case, $GCD(N, a - \sqrt{y})$ gives a non-trivial factor of N. M iterates the above steps to obtain the complete factorization of N. The following lemma, which is formally discussed by Tompa and Woll [TW87], helps us to analyze the algorithm of the extractor M.

> **Lemma 10 [TW87].** *Let N be odd and γ, δ be constants satisfying $0 < \gamma, \delta \leq 1$. Suppose there is a probabilistic algorithm $SQUREROOT(N, x)$ that, for a fraction δ of the quadratic residues x in Z_N^*, outputs a single square root of x modulo N with probability γ in expected time $(\log N)^{O(1)}$. Then there is an algorithm that outputs the complete prime factorization of N in expected time $(\log N)^{O(1)}$.*

As mentioned above, $\gamma \geq 1/2$ in the extractor M. However, nothing is mentioned on the value δ. Then, in parallel with the above steps, M tries to find the complete prime factorization of N by himself using exhaustive research (M runs the procedure 2^k times).

Case 2: common input $x \in QR_N$

Given (a possibly cheating P^*), M first executes the whole protocol (P^*, V) by faithfully simulating V's part. If V rejects, M stop and outputs nothing. Otherwise, when V accepts, next M repeatedly resets P^* to the step V2, with choosing new random challenge bits $\tilde{e}_{V1}, \ldots, \tilde{e}_{Vk}$ until P^* correctly answers to these challenges to obtain two successful executions. In the two successful executions. if M finds an index j such that

$$e_{Vj} \neq \tilde{e}_{Vj} \quad \& \quad e_{Pj} \oplus e_{Vj} \neq \tilde{e}_{Pj} \oplus \tilde{e}_{Vj}$$

in B3 of **P2**, then M can derive a square root $\sqrt{x}$ by computing $Y_i / \tilde{Y}_i \pmod{N}$ (or its inverse) which are P^*'s answers in step C2. Or, if M finds an index j such that

$$e_{Vj} \neq \tilde{e}_{Vj} \quad \& \quad e_{Pj} \oplus e_{Vj} = \tilde{e}_{Pj} \oplus \tilde{e}_{Vj}$$

in B3 of **P2**, then M can M gets a square root $\sqrt{y}$ by computing $t_i / \tilde{t}_i \pmod{N}$ (or its inverse). The same argument as case 1 above is applies to obtain the complete factorization of N. ∎

Thus, in the proposed protocol, provers cannot convince the verifier without $\sqrt{I}$ $\pmod{N}$ or the complete factorization of N. Especially, the cheating prover who convinces the verifier for $I \notin QR_N$ have to know the complete factorization of N, which is regard as a witness of the fact that $I \in QNR_N$.

3.5 A new definition of proofs of knowledge

Previous definitions of proofs of knowledge: (Interactive) Proofs of knowledge was formulated by Feige-Fiat-Shamir [FFS87] and Tompa and Woll [TW87] from practical points of view. The original interactive proof systems [GMR85] is defined to prove membership of the given inputs of the language. In GMR-model, then, the prover's power is assumed to be unbounded. Feige, Fiat, and Shamir [FFS87] observed that, in the proposed protocol for QR [GMR85] (GI [GMW86]), if the polynomial time prover has a knowledge associated with the inputs, then he can convince the verifier that the input belongs the given language. They also gave a precise definition on the state that the prover possesses the knowledge by introducing the extractor, which plays a similar role of the simulator[GMR85] of zero-knowledge.

However, under these definitions [FFS87, TW87], the notion of proofs of knowledge is regards as an additional property of proofs of membership.

Proposition 11 [TW87, Slo89]. *An interactive proof of knowledge (in the sense of [FFS87, TW87]) for the relation R is an interactive proof for membership of the language L_R.*

In fact, known concrete zero-knowledge interactive proofs, which require no unproven complexity assumption, for membership of $\mathcal{NP}$-languages [GMR85, GMW86, BMO90] are shown to be zero-knowledge interactive proofs of knowledge. Thus, there has been no known gap between proofs of languages and proofs of knowledge.

A weaker definition of proofs of knowledge: Bellare and Goldreich [BG92] suggested the following weak definition of proofs of knowledge[2] because the previous definition is too restrictive to deal with "proofs of knowledge" used in a subprotocol inside a larger protocol.

Definition 12. An interactive proof of *positive-side* knowledge for the relation R is a pair of interactive probabilistic Turing machines (P, V) satisfying:

Knowledge Completeness: For any $(x, w) \in R$, V accepts P's proof with overwhelming probability. Formally:

$$\forall(x, w) \in R \ \ Prob(V_{P(x,w)}(x) \ accepts) > 1 - \nu(|x|),$$

where the probability is taken over all of the possible coin tosses of P and V.

Positive-Side Knowledge Validity: For any $x \in L_R$, for any P^*, P^* can convince V to accept only if he actually "knows" a witness for $x \in L_R$. A probabilistic polynomial time knowledge extractor M is used in order to demonstrate P^*'s ability to compute a witness. Formally:

$$\forall a \ \exists M \ \forall P^* \ \forall x \in L_R \ \forall w' \ \forall \rho$$

$$Prob\Big(V_{P^*[\rho](x, w')}(x) \ accepts\Big) > 1/|x|^a \Rightarrow$$

$$Prob\Big(M\big(x; P^*[\rho](x, w')\big) \in w(x)\Big) > 1 - \nu(|x|),$$

where the probability is taken over all of the possible coin tosses of M and V.

---

[2] The original definition of proofs of knowledge [BG92] deals with provers who convince the verifier with probability which is not non-negligible for wider applications, and we shall adopt precisely the idea of Bellare and Goldreich [BG92]. However, in this paper, we consider only provers who convince the verifier with non-negligible probability to clarify our idea of definition and simplify the discussion.

Note that zero-knowledgeness is discussed only for the correct input $x \in L_R$, then zero-knowledge proofs of positive-side knowledge is defined as definition 3.

Proposition 13. *For (P, V') in the protocol Λ, there exists a simulator S which runs in expected polynomial time, for every V' and for $\forall I \in QR_N$, $S(I; V'(I)) = VIEW_{(P,V')}(I)$.*

Proof: We construct a simulator S for any (possibly dishonest) verifier V'. After running V' as the stage V1, the simulator first performs prover P's part of the stage P1 and gets the verifier V''s messages of the stage V2. If V' does not complete this stage successfully, S stops. Otherwise, S repeats the stage P1, each time with different randomly chosen challenges in A2a, until V' again successfully meets S's challenges. From the two successful executions S can find a $\sqrt{y} \pmod{N}$. Once S obtain such a information, S can disclose freely both bit 0 and 1 as e_{P_i} after he committed e_{P_i} at B1. This allows S to carry out P's part P2 without knowing a $\sqrt{I} \pmod{N}$. ■

Thus, we obtain the following.

Theorem 14. *Protocol Λ is a perfect zero-knowledge proof of positive-side knowledge on R_{QR}.*

The known previous protocols with positive-side knowledge are proofs of membership of languages. Our proposed protocol is the first example of proofs of positive-side knowledge but not of membership of languages.

Theorem 15. *There exists a protocol which is a proof of positive-side knowledge on a relation R but not of proofs of membership of the language L_R (unless $QR \in \mathcal{BPP}$).*

3.6 Round-optimality of the proposed scheme

Goldreich and Krawczyk [GK90] showed that the zero-knowledge interactive proofs for a language need at least 4-move unless the language is inside $\mathcal{BPP}$. Although our protocol is not proof of language, the similar argument as one by Itoh and Sakurai [IS91], which extended the result of [GK90] into the case of proofs of knowledge in the sense of [TW87, FFS87], implies the following.

Theorem 16. *If a $\mathcal{NP}$-relation R has a 3-move zero-knowledge proof of positive-side knowledge, then there exists a probabilistic polynomial time algorithm A*

$$A(x) = \begin{cases} y \text{ such that } (x,y) \in R & \text{if } x \in L_R \\ \bot & \text{if } x \notin L_R \end{cases}$$

with overwhelming probability.

Corollary 17. *Protocol Λ is optimal with respect to the round complexity among perfect zero-knowledge proofs of positive-side knowledge on R_{QR} unless $QR \in \mathcal{BPP}$.*

4 An application of the proposed protocol

This section applies our proposed scheme into an identification scheme fitted to identity(ID)-based systems.

4.1   Fiat-Shamir identification scheme

Fiat and Shamir [FiS86] applies the zero-knowledge protocol for QR [GMR85] into an ID-based identification scheme. The scheme assumes the existence of a trusted center which issues users' private/public key as follows:

> The unique trusted center's secret key in the system is (p, q), and the public key is N, where p, q are distinct large primes, $N = p \times q$. The center generates user A's secret key s_A, where $1/s_A = \sqrt{I_A}$ (mod N). I_A is the identity of user A and is published to other users.

At the identification stage between user A and user B, user A sends his identity I_A to the user B, and A shows B that A knows $\sqrt{I_A}$ (mod N) by the GMR-zero-knowledge interactive proof for QR.

4.2   Complexity assumptions in cryptographic setting

Fiat-Shamir identification scheme [FiS86] is based on the difficulty of computing a modular square roots when the factorization of N is unknown. If the factoring assumption fails, the identification system is no longer secure because everybody can convince users. Thus, such intractability assumptions are indispensable to construct identification schemes like as [FiS86]. This is contract to the theoretical result that QR has a perfect zero-knowledge interactive proof of the language, which holds even if QR belongs to $\mathcal{BPP}$.

4.3   Optimal round zero-knowledge identification scheme

Designing optimal-round zero-knowledge interactive proofs without any unproven assumption is an interesting problem from theoretical points of view. On the other hand, constructing optimal-round secure identification scheme with possibly weak assumptions is an important topic from practical points of view.

Note that if QR has a 4-move perfect ZKIP with no assumption has still remained open. The previous known protocols require 5-move interaction [BMO90] or an additional unproven assumption [FS89]. In the 4-move perfect ZK interactive proof system for special languages associated with the quadratic residuosity by De Santis et al. [DDP94, DiP94] provers requires not only the exact witness of the problem (e.g. a square root of I module N) but also the factorization of the modulus N as an additional knowledge in order to convince the verifier.

In our protocol, if we assume the hardness of the factoring, no (polynomial-time powerful) prover convinces the verifier for the inputs $x \notin QR_N$, except the trusted center which generates the modulus N. The soundness of our protocol is the same as of the original Fiat-Shamir scheme[FiS86] in such a practical setting.

Thus, this paper gives a positive answers to the open question that constructing an optimal-round zero-knowledge identification scheme based on QR.

Remark. Feige-Fiat-Shamir [FFS87] characterized the security of the 3-move parallelization of Fiat-Shamir scheme [FiS86], however, it is no longer zero-knowledge [GK90], Sakurai and Itoh [SI93] clarified the essential gap between the zero-knowledge and the security of the direct parallel version, which is important in a practical setting.

5 Proposed new formulation of practical soundness

What is soundness Soundness is a condition on the object which the verifier accepts. In the proofs of membership of languages, soundness implies that a verifier accepts only the inputs which belongs to a language. In the proofs of knowledge, soundness implies that a verifier accepts only the provers who knows the correct knowledge associated with the input. However, previous definition of proofs of knowledge [TW87, FFS87] fails to capture "soundness of knowledge".

5.1 Identification scheme

We first give a definition of identification scheme of Fiat-Shamir like setting. As observed in subsection 4.2, an identification is constructed based on certain NP-relation R, of which hard instances are generated in probabilistic polynomial time [AABFH88].

Definition 18. An identification scheme based on a NP-relation R consists of two stages:

1. Initialization between a center and each user:
   The unique trusted center generates system parameters commonly used among all users as a part of the public key. Furthermore, the center generates user A's secret key SK_A and public key PK_A which satisfy the relation $R(SK_A, PK_A)$, and PK_A is published to other users.
2. Operation between any user A and a verifier B:
   User A demonstrates her identity to the verifier B by proving the fact that "she knows the secret key SK_A for the public key PK_A" via some protocol. At the end of the protocol, B decides if B accepts A or not.

5.2 A new soundness

We define a new soundness fitted for the FSIS-setting by generalizing the observation on the proposed protocol. Our basic idea is that: we accept the cheating prover for $x \notin L_R$ if such a cheating requires much power than (the honest prover's) proving possession of a witness w for any input $x \in L_R$.

In general identification schemes, the trusted center gives a certificate (e.g. via digital signatures) on each user's public key to avoid user from using $x \notin L_R$. However, the ID-based Fiat-Shamir scheme does not have always this mechanism. We have to consider the system, in which only the common public modulus N is certificated but a user A's public information I_A has no certificate. In such a case, a cheating prover would have a chance to use $\tilde{I} \notin QR_N$ as his public information. Thus, it is very important to discuss cheating provers who use $x \notin L_R$.

Definition 19. An identification scheme based on a relation R is called *practically sound* if it satisfying the condition that (1) for any $x \in L_R$, if P^* can convince V to accept, then he actually "knows" a witness of $w(x)$, or (2) for any $x \notin L_R$, if P^* can convince V to accept, then he actually "knows" a witness of $w(y)$ for any $y \in L_R$, A probabilistic polynomial time knowledge extractor M is used in order to demonstrate P^*'s ability to compute such these witnesses. Formally:

$$\forall a \,\exists M \,\forall P^* \,\forall x \in \{0,1\}^* \,\forall w'$$

$$Prob\Big(V_{P^*(x,w')}(x)\, accepts\Big) > 1/|x|^a \Rightarrow$$

$$Prob\left(M\big(x; P^*(x,w')\big) = \left\{ \begin{array}{ll} \text{a witness of } w(x) & \text{if } x \in L_R \\ \kappa \text{ s.t. } \exists D \text{ satisfying} & \\ \forall y \in L_R \;\; D(y;\kappa) \in w(y) & \text{otherwise} \end{array} \right. \right) > 1 - \nu(|x|),$$

where D is a probabilistic polynomial-time algorithm.

Under this definition, Proposition 9 says that the identification scheme based on the relation R_{QR} of which operating stage is of the proposed 4-move protocol is practically sound.

Remark. For a given composite number N and integer g $(0 < g < N)$, consider the following relation $R_{(N,g)}$

$$R_{(N,g)}\,(x,y) \iff y = g^x \pmod{N}.$$

Our 4-move protocol is also described based on this relation $R_{(N,g)}$. We discuss the identification scheme based on the relation $R_{(N,g)}$ of which operating stage is the 4-move protocol. The following property is obtained by the similar argument in Proposition 9.

$$\forall a \; \exists M \; \forall P^* \; \forall y \in Z_N^* \; \forall w'$$
$$Prob\Big(V_{P^*(y,w')}(y)\, accepts\Big) > 1/|y|^a \Rightarrow$$
$$Prob\Big(M\big(y; P^*(y,w')\big) = \Big\{ \begin{array}{c} \text{one of } w(y) \\ \text{the complete factorization of } N \end{array} \Big) > 1 - \nu(|y|).$$

It should be noted that there are no known probabilistic polynomial-time algorithm to compute discrete logarithms over the modulus of the composite integer N even using the complete factorization of N. Then, in the case above, we *cannot* conclude that the identification is practically sound in our sense.

5.3 Comparison to the previous soundness

Though there are some works to propose weaker notions than zero-knowledge [FFS87, FS90], few attention have been paid to weaken the soundness of cryptographic protocols. Recall a notion of the security of the practical identification scheme proposed by [FFS87].

As mentioned in Subsection 4.2, a practical identification scheme needs some intractability problem. We denote the cryptographic assumption used in the identification scheme by $\mathcal{CA}$. Namely, no practical users (in general, probabilistic poly-time power) can break the assumption $\mathcal{CA}$. The previous definitions[FFS87, Oka92] does not include the cryptographic assumptions, however, we give the following definition, which is essentially founded on the cryptographic assumption $\mathcal{CA}$.

Definition 20. A prover A (resp. verifier B) who honestly acts is denoted by $\overline{A}$ (resp. $\overline{B}$). Let $\tilde{A}$ be a dishonest prover who does not complete the Initial stage of Definition 18 and may deviate from the protocols. $\tilde{B}$ is not a dishonest verifier.

An identification scheme (A, B) is no-transferable if

1. $(\overline{A}, \overline{B})$ succeeds with overwhelming probability.
2. If there exists a coalition of $\tilde{A}, \tilde{B}$ with the property that, after a polynomial number of executions of $(\overline{A}, \tilde{B})$ and relaying a transcript of the communication to $\tilde{A}$, it is possible to execute $(\tilde{A}, \overline{B})$ with nonnegligible probability of success, then there exists a probabilistic polynomial time algorithm M which breaks the assumption $\mathcal{CA}$.

Proposition 21. *Suppose that the operating protocol (A,B) of an identification scheme based on R satisfies three conditions that (1) knowledge completeness, (2) practical soundness, and (3) zero-knowledgeness, then (A,B) is no-transferable under the assumption that the relation R is hard, i.e. there are no probabilistic polynomial time algorithm that compute a witness $w(x)$ for an instance x.*

Fiat,Fiat, and Shamir [FFS87] introduced the notion of no-transferable to characterize the security of the 3-move direct parallelization of FS-scheme. A weaken notion of zero-knowledge is proposed as witness hiding [FS90], and a connection to the notion, no-transferable is explored [CD92]. Among these works on how to weaken the security of zero-knowledge, our formulation on practical soundness is the first attempt to weaken soundness of cryptographic protocol. In fact, our protocol gives an evidence that the notion of no-transferable weaken not only the security (e.g. zero-knowledge) but also soundness (e.g. language soundness).

6 Concluding remarks

This paper investigated definitions and properties of convincing possession of the proof of a given theorem in interactive protocols, and proposed 4-move perfect zero-knowledge protocol for ID-based identification scheme in the Fiat-Shamir setting. However, the proposed scheme is not a proof of the language of quadratic residuosity.

Open Problem: Does there exist a 4-move interactive proof system for the general quadratic residuosity problem without any unproven assumption ?

Acknowledgments. The author would like to thank Claude Crépeau for useful discussion at École Normale Supérieure and Tatsuaki Okamoto for his email exchange regarding identification schemes. Finally thanks to anonymous referees for their comments on the submitted version of this paper.

References

[AABFH88] Abadi,A., Allender,E., Broder,A, Feigenbaum,J., and Hemachandra,L.A., "On generating solved instances of computational problems," in Advances in Cryptology – Crypto'88, LNCS 403, *Springer-Verlag*, Berlin (1987).

[BCLL91] Brassard,G., Crepeau, C., Laplante, S., and Leger, C., "Computationally convincing proofs of knowledge," *Proc. of the 8th STACS*, (1991).

[BCY89] Brassard, G., Crépeau, C., and Yung, M., "Everything in $\mathcal{NP}$ Can Be Argued in Perfect Zero-Knowledge in a Bounded Number of Rounds," Proc. of 16th ICALP'89, LNCS 372, *Springer-Verlag*, pp.123-136, Berlin (1989); final version in "Constant-round perfect zero-knowledge computationally convincing protocols," *TCS*, 84, pp. 23-52 (1991).

[BFL89] Boyar, J., Friedl, K., and Lund, C., "Practical zero-knowledge proofs:/ Giving hints and using deficiencies," *J. of Cryptology*, Vol.4, pp.185-206 (1991); preliminary version in *Proc. of Eurocrypt'89*(1989).

[BG92] Bellare, M., and Goldreich,O., "On defining Proofs of Knowledge," in Advances in Cryptology – Crypto'92, LNCS 740, *Springer-Verlag*, Berlin (1993).

[BHZ87] Boppana,R., Hastad,J., and Zachos,S., "Does co-NP have short interactive proofs," *IPL*, Vol.25, No.2, pp.127-132 (1987).

[BM92] Brickell, E. F. and McCurley, K.S "An Interactive Identification Scheme Based on Discrete Logarithms and Factoring," *J. of Cryptology*, Vol.5, pp.29-40 (1992); preliminary version in *Proc. of Eurocrypt'90*(1990).

[BMO90] Bellare, M., Micali, S., and Ostrovsky, R., "Perfect Zero-Knowledge in Constant Rounds," *Proc. of ACM STOC*, pp.482-493 (May 1990).

[CD92] Chen,L., and Damgaard, Y., "Security bounds for parallel versions of identification protocols," in Advances in Cryptology – Eurocrypt'92, LNCS 658, pp.461-466, *Springer-Verlag*, Berlin (1993).

[DDP94] De Santis, A., Di Crescenzo,G. and Persioano G., "The knowledge complexity of quadratic residuosity languages," *TCS*, 132, pp. 291-317 (1991).

509

[DiP94] Di Crescenzo,G. and Persioano G., "Round-optimal perfect zero-knowledge proofs," IPL 50, pp.93-99 (1994).

[FFS87] Feige, U., Fiat, A., and Shamir, A., "Zero-Knowledge Proofs of Identity," *J. of Cryptology*, Vol.1, pp.77-94 (1988); preliminary version in *Proc. of 19th STOC*, pp.210-217 (1987).

[FiS86] Fiat, A. and Shamir, A., "How to Prove Yourself," Advances in Cryptology – Crypto'86, LNCS 263, *Springer-Verlag*, Berlin, pp.186-199 (1987).

[Fo87] Fortnow, L., "The Complexity of Perfect Zero-Knowledge," *Advanced in Computing Research*, Vol.5, *Randomness and Computation*, pp.327-pp.344 (1989); preliminary version in *Proc. of 19th STOC*, pp.204-209 (1987).

[FS89] Feige, U. and Shamir, A., "Zero-Knowledge Proofs of Knowledge in Two Rounds," in Advances in Cryptology – Crypto'89, LNCS 435, pp.526-544, *Springer-Verlag*, Berlin (1990).

[FS90] Feige, U. and Shamir, A., "Witness Indistinguishable and Witness Hiding Protocols," Proc. of STOC, pp.416-426 (May 1990).

[GK90] Goldreich, O. and Krawczyk, H., "On the Composition of Zero-Knowledge Proof Systems," in the Proceedings of ICALP'90, LNCS 443, pp.268-282, *Springer-Verlag*, Berlin (1990).

[GMR85] Goldwasser, S., Micali, S., and Rackoff, C., "The Knowledge Complexity of Interactive Proof Systems," *SIAM J. of Comp.*, Vol.18, No.1, pp.186-208, (1989); preliminary version in *Proc. of 17th STOC*, pp. 291-304 (1985).

[GMW86] Goldreich, O., Micali, S., and Wigderson, A., "Proofs that Yield Nothing But Their Validity or All Languages in NP Have Zero-Knowledge Proofs," *J. of ACM*, Vol.38, No.1, pp.691-729 (July 1991); preliminary version in *Proc. of 27th FOCS*, pp.174-187, (1986).

[IS91] Itoh, T. and Sakurai, K., "On the Complexity of Constant Round ZKIP of Possession of Knowledge," Advances in Cryptology – Asiacrypt'91, LNCS 739, *Springer-Verlag*, Berlin, (1993).

[Oka92] Okamoto,T., "Provably Secure and Practical Identification Schemes and Corresponding Signature Schemes," in Advances in Cryptology – Crypto'92, LNCS 740, pp.31-53, *Springer-Verlag*, Berlin (1993).

[Sak93] Sakurai,K., "Studies on the efficiency ans security of cryptographic protocols based on the zero-knowledge techniques ," *Ph.D thesis*, Kyushu University (June 1993).

[Sak95] Sakurai,K., "Practical proofs of knowledge without relying on theoretical proofs of membership on languages," *manuscript* (1995).

[Slo89] Sloan, R., "All Zero-Knowledge Proofs are Proofs of Language Membership," *Technical Memorandum*, MIT/LCS/TM-385 (February 1989).

[SI93] Sakurai,K. and Itoh, T., "On the discrepancy between the serial and the parallel of zero-knowledge protocols" Advances in Cryptology – Crypto'92, LNCS 740, *Springer-Verlag*, Berlin, (1993).

[SKS91] Saitoh, T., Kurosawa, K., and Sakurai, K., "4-Move Perfect ZKIP of Knowledge with No Assumption," Advances in Cryptology – Asiacrypt'91, LNCS 739, *Springer-Verlag*, Berlin, (1993).

[TW87] Tompa, M. and Woll, H., "Random Self-Reducibility and Zero-Knowledge Interactive Proofs of Possession of Information," *Proc. of 28th FOCS*, pp.472-482 (1987).

Compact Location Problems with Budget and Communication Constraints

S. O. Krumke[1], H. Noltemeier[1], S. S. Ravi[2] and M. V. Marathe[3,*]

[1] University of Würzburg, Am Hubland, 97074 Würzburg, Germany. Email:
{krumke, noltemei}@informatik.uni-wuerzburg.de.
[2] University at Albany - SUNY, Albany, NY 12222, USA. Email:
ravi@cs.albany.edu.
[3] Los Alamos Nat. Lab. P.O. Box 1663, MS M986, Los Alamos, NM 87545, USA.
Email: madhav@c3.lanl.gov.

Abstract. We consider the problem of placing a specified number p of facilities on the nodes of a given network with two nonnegative edge–weight functions so as to minimize the diameter of the placement with respect to the first weight function subject to a diameter– or sum–constraint with respect to the second weight function.

Define an (α, β)–approximation algorithm as a polynomial–time algorithm that produces a solution within α times the optimal value with respect to the first weight function, violating the constraint with respect to the second weight function by a factor of at most β.

We show that in general obtaining an (α, β)–approximation for any fixed $\alpha, \beta \geq 1$ is $\mathcal{NP}$–hard for any of these problems. We also present efficient approximation algorithms for several of the problems studied, when both edge–weight functions obey the triangle inequality.

1 Introduction and Basic Definitions

Several fundamental problems in location theory [HM79, MF90] involve finding a placement obeying certain "covering" constraints. Generally, the goal of such a location problem is to find a placement of minimum cost that satisfies all the specified constraints. The cost of a placement may reflect the price of constructing the network of facilities, or it may reflect the maximum communication cost between any two facilities. Examples of such cost measures are the total edge cost and the diameter respectively.

Finding a placement of sufficient generality minimizing even one of these measures is often $\mathcal{NP}$–hard [GJ79]. In practice, it is usually the case that a facility location problem involves the minimization of a certain cost measure, subject to budget constraints on other cost measures.

The problems considered in this paper can be termed as *compact location* problems, since we will typically be interested in finding a "compact" placement of facilities. The following is a prototypical compact location problem: Given an undirected edge-weighted complete graph $G = (V, E_c)$, place a specified number

---

* Research supported by Department of Energy under contract W-7405-ENG-36.

p of facilities on the nodes of G, with at most one facility per node, so as to minimize some measure of the distances between facilities. This problem has been studied for both diameter and sum objectives [RKM+93]. Some geometric versions of this problem have also been studied [AI+91].

Consider the following extension of the compact location problem. Suppose we are given *two* weight–functions δ_c, δ_d on the edges of the network. Let the first weight function δ_c represent the cost of constructing an edge, and let the second weight function δ_d represent the actual transportation– or communication–cost over an edge (once it has been constructed). Given such a graph, we can define a general bicriteria problem $(\mathcal{A}, \mathcal{B})$ by identifying two minimization objectives of interest from a set of possible objectives. A budget value is specified on the second objective $\mathcal{B}$ and the goal is to find a placement of facilities having minimum possible value for the first objective $\mathcal{A}$ such that this solution obeys the budget constraint on the second objective. For example, consider the *diameter-bounded minimum diameter compact location problem* denoted by DC-MDP: Given an undirected graph $G = (V, E)$ with two different nonnegative integral edge weight functions δ_c (modeling the building cost) and δ_d (modeling the delay or the communication cost), an integer p denoting the number of facilities to be placed, and an integral bound B (on the total delay), find a placement of p facilities with minimum diameter under the δ_c–cost such that the diameter of the placement under the δ_d–costs (the maximum delay between any pair of nodes) is at most B. We term such problems as *bicriteria compact location problems*.

In this paper, we study bicriteria compact location problems motivated by practical problems arising in diverse areas such as statistical clustering, pattern recognition, processor allocation and load–balancing.

2 Preliminaries and Summary of Results

Let $G = (V, E_c)$ be a complete undirected graph with $n = |V|$ nodes and let p $(2 \leq p \leq n)$ be the number of facilities to be placed. We call any subset $P \subseteq V$ of cardinality p a *placement*. Given a nonnegative weight– or cost–function $\delta : E_c \to \mathbb{Q}$, we will use $\mathcal{D}_\delta(P)$ to denote the *diameter* of a placement P with respect to δ; that is

$$\mathcal{D}_\delta(P) = \max_{\substack{u, v \in P \\ u \neq v}} \delta(u, v).$$

Similarly, we will let $\mathcal{S}_\delta(P)$ denote the *sum of the distances* between facilities in the placement P; that is

$$\mathcal{S}_\delta(P) = \sum_{\substack{u, v \in P \\ u \neq v}} \delta(u, v).$$

We note that the average length of an edge in a placement P equals $\frac{2}{p(p-1)}\mathcal{S}_\delta(P)$.

As usual, we say that a nonnegative distance δ on the edges of G satisfies the *triangle inequality*, if we have

$$\delta(v, w) \leq \delta(v, u) + \delta(u, w)$$

for all $v, w, u \in V$,

The *Minimum Diameter Placement Problem* (denoted by MDP) is to find a placement P that minimizes $\mathcal{D}_\delta(P)$. Similarly, the *Minimum Average Placement Problem* (denoted by MAP) is to find a placement P such that $\mathcal{S}_\delta(P)$ is minimized. Both problems are known to be $\mathcal{NP}$-hard, even when the distance δ obeys the triangle inequality [RKM+93]. Moreover, if the distances are not required to satisfy the triangle inequality, then as observed in [RKM+93], there can be no polynomial time relative approximation algorithm for MDP or MAP unless $\mathcal{P} = \mathcal{NP}$.

In the sequel we will restrict ourselves to those instances of the problems where the weights on the edges obey the triangle inequality. Given a problem Π, we use TI-Π to denote the problem Π restricted to graphs with edge weights satisfying the triangle inequality.

Following [HS86], the *bottleneck graph* bottleneck(G, δ, Δ) of $G = (V, E_c)$ with respect to δ and a bound Δ is defined by

$$\text{bottleneck}(G, \delta, \Delta) := (V, E'), \text{ where } E' := \{e \in E_c : \delta(e) \leq \Delta\}.$$

We now formally define the problems studied in this paper.

Definition 1. [*Diameter Constrained Minimum Diameter Placement Problem* (DC-MDP)]

Input: An undirected complete graph $G = (V, E_c)$ with two nonnegative weight functions $\delta_c, \delta_d : E_c \to \mathbb{Q}$, an integer $2 \leq p \leq n$ and a number $\Omega \in \mathbb{Q}$.

Output: A set $P \subseteq V$, with $|P| = p$, minimizing the objective

$$\mathcal{D}_{\delta_c}(P) = \max_{\substack{v, w \in P \\ v \neq w}} \delta_c(v, w)$$

subject to the constraint

$$\mathcal{D}_{\delta_d}(P) = \max_{\substack{v, w \in P \\ v \neq w}} \delta_d(v, w) \leq \Omega.$$

Definition 2. [*Sum Constrained Minimum Diameter Placement Problem* (SC-MDP)]

Input: An undirected complete graph $G = (V, E_c)$ with two nonnegative weight functions $\delta_c, \delta_d : E_c \to \mathbb{Q}$, an integer $2 \leq p \leq n$ and a number $\Omega \in \mathbb{Q}$.

Output: A set $P \subseteq V$, with $|P| = p$, minimizing the objective

$$\mathcal{D}_{\delta_d}(P) = \max_{\substack{v, w \in P \\ v \neq w}} \delta_d(v, w)$$

and satisfying the budget–constraint

$$\mathcal{S}_{\delta_c}(P) = \sum_{\substack{v_i, v_j \in P \\ v_i \neq v_j}} \delta_c(v_i, v_j) \leq \Omega.$$

Let $\Pi \in \{\text{TI-DC-MDP}, \text{TI-SC-MDP}\}$. Define an (α, β)-*approximation algorithm* for Π to be a polynomial–time algorithm, which for any instance I of Π does one of the following:

(a) It produces a solution within α times the optimal value with respect to the first distance function (δ_c), violating the constraint with respect to the second distance function (δ_d) by a factor of at most β.
(b) It returns the information that no feasible placement exists at all.

Notice that if there is no feasible placement but there is a placement violating the constraint by a factor of at most β, an (α, β)-approximation algorithm has the choice of performing either action (a) or (b).

In this paper we study the complexity and approximability of the problems DC-MDP and SC-MDP. We show that, in general, obtaining an (α, β)-approximation for any fixed $\alpha, \beta \geq 1$ is $\mathcal{NP}$-hard for any of these problems. We also present efficient approximation algorithms for several of the problems studied, when both edge–weight functions obey the triangle inequality. For TI-DC-MDP problem, we provide a $(2, 2)$-approximation algorithm. We also show that no polynomial time algorithm can provide an $(\alpha, 2 - \varepsilon)$- or $(2 - \varepsilon, \beta)$-approximation for any fixed $\varepsilon > 0$ and $\alpha, \beta \geq 1$, unless $\mathcal{P} = \mathcal{NP}$. This result is proved to remain true, even if one fixes $\varepsilon' > 0$ and allows the algorithm to place only $2p/|V|^{1/6-\varepsilon'}$ facilities. Our techniques can be extended to devise approximation algorithms for TI-SC-MDP. For this problem, our heuristics provide performance guarantees of $(2 - 2/p, 2)$ and $(2, 2 - 2/p)$ respectively. These techniques can also be used to find efficient approximation algorithms for TI-DC-MDP and TI-SC-MDP when there are node and edge weights. Due to lack of space, the discussion on the node-weighted cases is omitted in this version of the paper.

3 Related Work

While there has been much work on finding minimum-cost networks (see for example [DF85, FG88, Go85, IC+86, LV92, Won80]) for each of the cost measures considered in our bicriteria formulations, there has been relatively little work on approximations for multi-objective network-design. In this direction, Bar-Ilan and Peleg [BP91] considered balanced versions of the problem of assigning network centers, where a bound is imposed on the number of nodes that any center can service. Warburton [Wa87] has considered multi-objective shortest path problems. We refer the reader to [MR+95, RMR+93] for a detailed survey of the work done in the area of algorithms for bicriteria network design and location theory problems. Other researchers have addressed multi–objective approximation algorithms for problems arising in areas other than network design. This includes research in the areas of computational geometry [AF+94], numerical analysis, network design [ABP90, KRY93, Fi93] and scheduling [ST93].

Due to lack of space the rest of the paper consists of selected proof sketches.

4 Diameter Constrained Problems

As shown in [RKM⁺93], TI-MDP is $\mathcal{NP}$-hard. Here we can extend this result to obtain the following non-approximability result.

Proposition 3. *Let $\varepsilon > 0$ and $\varepsilon' > 0$ be arbitrary. Suppose that A is a polynomial time algorithm that, given any instance of TI–DC–MDP, either returns a subset $S \subseteq V$ of at least $\frac{2p}{|V|^{1/6-\varepsilon'}}$ nodes satisfying $\mathcal{D}_{\delta_d}(S) \leq (2-\varepsilon)\Omega$, or provides the information that no placement of p nodes having communication diameter of at most Ω does exist. Then $\mathcal{P} = \mathcal{NP}$.* □

We can interchange the roles of δ_c and δ_d in the proof of the last proposition to show that the optimal value of the problem cannot be approximated by a factor of $(2 - \varepsilon)$. Moreover, replacing 2 by a suitable function $f \in \Theta(2^{\text{poly}(|V|)})$, which given an input length of $\Theta(|V|)$ is polynomial time computable, it is easy to see that, if the triangle inequality is not required to hold, there can be no polynomial time approximation with performance ratio $O(2^{\text{poly}(|V|)})$ for neither the optimal function value nor the constraint (modulo $\mathcal{P} = \mathcal{NP}$). Thus we obtain:

Lemma 4. *Unless $\mathcal{P} = \mathcal{NP}$, for any fixed $\varepsilon > 0$ and $\varepsilon' > 0$ there can be no polynomial time approximation algorithm for TI–DC–MDP that is required to place at least $2p/|V|^{1/6-\varepsilon'}$ facilities and has a performance guarantee of $(\alpha, 2-\varepsilon)$ or $(2-\varepsilon, \beta)$. If the triangle inequality is not required to hold, then the existence of an $(f(|V|), g(|V|))$-approximation algorithm for any $f, g \in O(2^{\text{poly}(|V|)})$ implies that $\mathcal{P} = \mathcal{NP}$.* □

PROCEDURE HEUR-FOR-DIA

1. $G' := \text{bottleneck}(G, \delta_d, \Omega)$
2. $V_{cand} := \{v \in G' : \deg(v) \geq p - 1\}$
3. IF $V_{cand} = \emptyset$ THEN RETURN "certificate of failure"
4. Let $best := +\infty$
5. Let $P_{best} := \emptyset$
6. FOR each $v \in V_{cand}$ DO
   (a) Let $N(v)$ be the set of $p - 1$ nearest neighbors of v in G with respect to δ_c
   (b) Let $P(v) := N(v) \cup \{v\}$
   (c) IF $\mathcal{M}_{\delta_c}(P(v)) < best$ THEN $P_{best} := P(v)$
   $$best := \mathcal{M}_{\delta_c}(P(v))$$
7. OUTPUT P_{best}

Fig. 1. Details of the heuristic for TI–DC–MDP and TI–DC–MAP

Using the results in [RKM⁺93] in conjunction with the results in [MR+95] we can devise an approximation algorithm with a performance guarantee (4, 4)

for TI-DC-MDP. Here we present an improved heuristic HEUR-FOR-DIA for this problem. This heuristic provides a performance guarantee of $(2, 2)$. In view of Lemma 4, this is the best approximation we can expect to obtain in polynomial time. The heuristic is quite simple. The details of the heuristic are shown in Figure 1.

Theorem 5. *Let I be any instance of of TI–DC–MDP such that an optimal solution P^* of diameter cost $OPT(I) = \mathcal{D}_{\delta_c}(P^*)$ exists. Then the algorithm HEUR-FOR-DIA, called with $\mathcal{M}_{\delta_d} := \mathcal{D}_{\delta_d}$, returns a placement P satisfying $\mathcal{D}_{\delta_d}(P) \leq 2\Omega$ and $\mathcal{D}_{\delta_c}(P)/OPT(I) \leq 2$.*

Proof: Consider an optimal solution P^* such that $\mathcal{D}_{\delta_d}(P^*) \leq \Omega$. Then by definition this placement forms a clique of size p in $G' := \text{bottleneck}(G, \delta_d, \Omega)$. Thus in this case V_{cand} is non–empty and the heuristic will not output a "certificate of failure".

Moreover, any placement $P(v)$ considered by the heuristic will form a clique in $(G')^2$. By the definition of G' as a bottleneck graph with respect to δ_d, the bound Ω and by the assumption that edge weights obey triangle inequality, it follows that no edge e in $(G')^2$ has weight $\delta_d(e)$ more than 2Ω. Thus *every* placement $P(v)$ considered by the heuristic has communication diameter $\mathcal{D}_{\delta_d}(P(v))$ no more than 2Ω.

Consider an arbitrary $v \in P^*$. Clearly $v \in V_{cand}$. Consider the step of the algorithm HEU-FOR-DIA in which it considers v. For any $w \in N(v)$ we have $\delta_c(v, w) \leq OPT(I)$, by definition of $N(v)$ as the set of nearest neighbors of v and by the fact that every node from the optimal solution is adjacent to v in G'. Thus for $w, w' \in N(v)$ we have $\delta_c(w, w') \leq \delta_c(v, w) + \delta_c(v, w') \leq 2OPT(I)$ by the triangle inequality. Consequently, $\mathcal{D}_{\delta_c}(P(v)) = \mathcal{D}_{\delta_c}(N(v) \cup \{v\}) \leq 2OPT(I)$.

Now, since the algorithm HEUR-FOR-DIA chooses a placement with minimal diameter among all the placements produced, the claimed performance guarantee with respect to the cost diameter $\mathcal{D}_{\delta_c}$ follows. □

5 Sum Constrained Problems

Next, we study bicriteria compact location problems where the objective is to minimize the diameter $\mathcal{D}_{\delta_d}$ subject to budget–constraints of sum type.

Again, it is not an easy task to find a placement P satisfying the budget–constraint or to determine that no such placement exists. Using a reduction from CLIQUE [GJ79] one obtains the following.

Proposition 6. *If the distances δ_c, δ_d are not required to satisfy the triangle inequality, there can be no polynomial time (α, β)-approximation algorithm for SC–MDP for any fixed $\alpha, \beta \geq 1$, unless $\mathcal{P} = \mathcal{NP}$. Moreover, if there is a polynomial time $(\alpha, 1)$-approximation algorithm for TI–SC–MDP for any fixed $\alpha \geq 1$, then $\mathcal{P} = \mathcal{NP}$.* □

We proceed to present a heuristic for TI-SC-MDP. The main procedure shown in Figure 2 uses the test procedure from Figure 3.

PROCEDURE HEUR-FOR-SUM

1. Sort the edges of G in ascending order with respect to δ_d
2. Assume now that $\delta_d(e_1) \le \delta_d(e_2) \le \cdots \le \delta_d(e_{\binom{n}{2}})$
3. Let $P_{best} :=$ "certificate of failure"
4. $i := 1$
5. Do
   (a) $G_i := \text{bottleneck}(G, \delta_d, \delta_d(e_i))$
   (b) $P_{best} := \text{test}(G_i, \delta_c|_{G_i}, \Omega)$
   (c) $i := i + 1$
6. UNTIL $P_{best} \ne$ "certificate of failure"
7. OUTPUT P_{best}

Fig. 2. Generic bottleneck procedure

PROCEDURE test(G, δ, Ω)

1. $V_{cand} := \{v \in G : \deg(v) \ge p - 1\}$
2. IF $V_{cand} = \emptyset$ THEN RETURN "certificate of failure"
3. Let $best := +\infty$
4. Let $P_{best} := \emptyset$
5. FOR each $v \in V_{cand}$ Do
   (a) Let $N(v)$ be the set of $p - 1$ nearest neighbors of v in G with respect to δ
   (b) Let $P(v) := N(v) \cup \{v\}$
   (c) IF $S_\delta(P(v)) < best$ THEN $P_{best} := P(v)$
   $$best := S_\delta(P(v))$$
6. IF $best > (2 - 2/p)\Omega$ THEN RETURN "certificate of failure"
   ELSE RETURN P_{best}

Fig. 3. Test procedure used for TI–SC–MDP

Lemma 7. *Let I be an instance of* TI-SC-MDP *such than there is an optimal placement P^*. If the test procedure* test(G_i, δ_c, Ω) *returns a "certificate of failure", then we have* $OPT(I) > \delta_d(e_i)$. $\qquad\square$

Now we can establish the result about the performance guarantee of the heuristic:

Theorem 8. *Let I denote any instance of* TI–SC–MDP *and assume that there is an optimal placement P^* of diameter $OPT(I) = \mathcal{R}_{\delta_d}(P^*)$. Then* HEUR-FOR-SUM *with the test procedure* test *returns a placement P with $S_{\delta_c}(P) \le (2-2/p)\Omega$ and $\mathcal{D}_{\delta_1}(I)/OPT(I) \le 2$.*

Proof: Consider the case when $\delta_d(e_i) = OPT(I)$. Since in G_i we have deleted only edges e having weight $\delta_d(e) > OPT(I)$ and we assume that there is a feasible solution satisfying the budget–constraint, it follows that the bottleneck graph G_i must contain a clique C of size p such that $S_{\delta_c}(C) \le \Omega$.

For a node $v \in C$ let

$$S_v := \sum_{\substack{w \in C \\ w \neq v}} \delta_c(v, w).$$

Then we have

$$S_{\delta_c}(C) = \sum_{v \in C} S_v.$$

Now let $v \in C$ be so that S_v is a minimum among all nodes in C. Then clearly

$$S_{\delta_c}(C) \geq pS_v. \tag{1}$$

By definition of the bottleneck graph G_i and the clique C, the node v must have degree at least $p - 1$ in G_i. Thus v is one of the nodes considered by the test procedure. Let $N(v)$ be the set of $p - 1$ nearest neighbors of v in G_i. Then we have

$$\sum_{\substack{w \in N(v) \\ w \neq v}} \delta_c(v, w) \leq S_v, \tag{2}$$

by definition of $N(v)$ as the set of nearest neighbors, $P(v) := N(v) \cup \{v\}$. Let $w \in N(v)$ be arbitrary. Then

$$\sum_{u \in N(v) \cup \{v\} \setminus \{w\}} \delta_c(w, u) = \delta_c(w, v) + \sum_{u \in N(v) \setminus \{w\}} \delta_c(w, u)$$

$$\leq \delta_c(w, v) + \sum_{u \in N(v) \setminus \{w\}} (\delta_c(w, v) + \delta_c(v, u))$$

$$= (p - 1)\delta_c(w, v) + \sum_{u \in N(v) \setminus \{w\}} \delta_c(v, u)$$

$$= (p - 2)\delta_c(v, w) + \sum_{u \in N(v)} \delta_c(v, u)$$

$$\overset{(2)}{\leq} (p - 2)\delta_c(v, w) + S_v. \tag{3}$$

Now using (3) and again (2), we obtain

$$S_{\delta_c}(P(v)) = S_{\delta_c}(N(v) \cup \{v\})$$

$$= \sum_{u \in N(v)} \delta_c(v, u) + \sum_{w \in N(v)} \sum_{u \in N(v) \cup \{v\} \setminus \{w\}} \delta_c(w, u)$$

$$\overset{(2)}{\leq} S_v + \sum_{w \in N(v)} \sum_{u \in N(v) \cup \{v\} \setminus \{w\}} \delta_c(w, u)$$

$$\overset{(3)}{\leq} S_v + \sum_{w \in N(v)} ((p - 2)\delta_c(v, w) + S_v)$$

$$= S_v + (p - 2)S_v + (p - 1)S_v$$

$$= (2p - 2)S_v$$

$$\overset{(1)}{\leq} (2 - 2/p)OPT(I).$$

Thus the placement $P(v)$ violates the budget–constraint by a factor of at most $2 - 2/p$. Consequently, as the algorithm chooses the placement with P_{best} with the least constraint–violation, it follows that the test–procedure called with $G_i =$ bottleneck$(G, \delta_d, OPT(I))$ will not return a "certificate of failure".

The placement P_{best} that is produced by the algorithm turns into a clique in G_i^2. Thus the longest edge in the placement with respect to δ_d is at most $2OPT(I)$. $\qquad\qquad\qquad\qquad\qquad\qquad\qquad\qquad\qquad\qquad\qquad\qquad\qquad\qquad\qquad\quad$ □

References

[ABP90] B. Awerbuch, A. Baratz, and D. Peleg. Cost–sensitive analysis of commu-nication protocols. In *Proceedings of the 9th Symposium on the Principles of Distributed Computing (PODC)*, pages 177–187, 1990.

[AF+94] E.M. Arkin, S.P. Fekete, J.S.B. Mitchell, and C.D. Piatko. Optimal Cov-ering Tour Problems. In *Proceedings of the Fifth International Symposium on Algorithms and Computation*, 1994.

[AI+91] A. Aggarwal, H. Imai, N. Katoh, and S. Suri. Finding k points with Mini-mum Diameter and Related Problems. *J. Algorithms*, 12(1):38–56, March 1991.

[BP91] J. Bar-Ilan and D. Peleg. Approximation Algorithms for Selecting Net-work Centers. In *Proc. 2nd Workshop on Algorithms and Data Structures (WADS)*, pages 343–354, Ottawa, Canada, August 1991. Springer Verlag, LNCS vol. 519.

[BS94] M. Bellare and M. Sudan. Improved Non–Approximability Results. in *Pro-ceedings of the 26th annual ACM Symposium on the Theory of Computing (STOC)*, May 1994.

[DF85] M.E. Dyer and A.M. Frieze. A Simple Heuristic for the p–Center Problem. *Operations Research Letters*, 3(6):285–288, Feb. 1985.

[EN89] E. Erkut and S. Neuman. Analytical Models for Locating Undesirable Fa-cilities. *European J. Operations Research*, 40:275–291, 1989.

[FG88] T. Feder and D. Greene. Optimal Algorithms for Approximate Clustering. In *ACM Symposium on Theory of Computing (STOC)*, pages 434–444, 1988.

[Fi93] T. Fischer. Optimizing the Degree of Minimum Weight Spanning Trees. Technical report, Department of Computer Science, Cornell University, Ithaca, New York, April 1993.

[GJ79] M.R. Garey and D.S. Johnson. *Computers and Intractability.* W.H. Free-man, 1979.

[Go85] T.F. Gonzalez. Clustering to Minimize the Maximum Intercluster Distance. *Theoretical Computer Science*, 38:293–306, 1985.

[Ha92] R. Hassin. Approximation schemes for the restricted shortest path problem. *Mathematics of Operations Research*, 17(1):36–42, 1992.

[HM79] G.Y. Handler and P.B. Mirchandani. *Location on Networks: Theory and Algorithms.* MIT Press, Cambridge, MA, 1979.

[HS86] D. S. Hochbaum and D. B. Shmoys. A Unified Approach to Approximation Algorithms for Bottleneck Problems. *Journal of the ACM*, 33(3):533–550, July 1986.

[IC+86] A. Iwainsky, E. Canuto, O. Taraszow, and A. Villa. Network decomposi-tion for the optimization of connection structures. *Networks*, 16:205–235, 1986.

[KRY93]   S. Khuller, B. Raghavachari, and N. Young. Balancing Minimum Spanning
           and Shortest Path Trees. In *Proceedings of the Fourth Annual ACM–SIAM
           Symposium on Discrete Algorithms (SODA)*, pages 243–250, 1993.

[Lee82]   D.T. Lee. On *k*-nearest neighbor Voronoi diagrams in the plane. *IEEE
           Trans. Comput.*, C-31:478–487, 1982.

[LV92]   J.H. Lin and J. S. Vitter. ϵ-Approximations with Minimum Packing Con-
           straint Violation. In *ACM Symposium on Theory of Computing (STOC)*,
           pages 771–781, May 1992.

[MR+95]   M.V. Marathe, R. Ravi, R. Sundaram, S.S. Ravi, D.J. Rosenkrantz, and
           H.B. Hunt III. Bicriteria Network Design Problems. To appear in *Pro-
           ceedings of the 22nd International Colloquium on Automata Languages and
           Programming (ICALP)*, 1995.

[MF90]   P.B. Mirchandani and R.L. Francis. *Discrete Location Theory*. Wiley-
           Interscience, New York, NY, 1990.

[PS85]   F.P. Preparata and M.I. Shamos. *Computational Geometry: An Introduc-
           tion*. Springer–Verlag Inc., New York, NY, 1985.

[RKM+93]   V. Radhakrishnan, S.O. Krumke, M.V. Marathe, D.J. Rosenkrantz, and
           S.S. Ravi. Compact Location Problems. In *13th Conference on the Foun-
           dations of Software Technology and Theoretical Computer Science (FST-
           TCS)*, volume 761 of *LNCS*, pages 238–247, December 1993.

[RMR+93]   R. Ravi, M.V. Marathe, S.S. Ravi, D.J. Rosenkrantz, and H.B. Hunt III.
           Many birds with one stone: Multi-objective approximation algorithms. In
           *Proceedings of the 25th Annual ACM Symposium on the Theory of Com-
           puting (STOC)*, pages 438–447, 1993.

[RRT91]   S.S. Ravi, D.J. Rosenkrantz, and G.K. Tayi. Facility Dispersion Problems:
           Heuristics and Special Cases. In *Proc. 2nd Workshop on Algorithms and
           Data Structures (WADS)*, pages 355–366, Ottawa, Canada, August 1991.
           Springer Verlag, LNCS vol. 519. (Journal version: *Operations Research*,
           42(2):299-310, March-April 1994.)

[ST93]   D.B. Shmoys and E. Tardos. Scheduling unrelated parallel machines with
           costs. In *Proceedings of the 4th Annual ACM–SIAM Symposium on Discrete
           Algorithms (SODA)*, pages 438–447, 1993.

[Wa87]   A. Warburton. Approximation of Pareto optima in multiple–objective,
           shortest path problems. *Operations Research*, 35:70–79, 1987.

[Won80]   R.T. Wong. Worst case analysis of network design problem heuristics.
           SIAM J. Disc. Math., 1:51–63, 1980.

Minimum Dominating Sets of Intervals on Lines*

(Extended Abstract)

Siu-Wing Cheng**   Michael Kaminski***   Shmuel Zaks[†]

Abstract

We study the problem of computing minimum dominating sets of n intervals on lines in three cases: (1) the lines intersect at a single point, (2) all lines except one are parallel, and (3) one line with t weighted points on it and the minimum dominating set must maximize the weight sum of the weighted points covered. We propose polynomial-time algorithms for the first two problems, which are special cases of the minimum dominating set problem for path graphs which is known to be NP-hard. The third problem requires identifying the structure of minimum dominating sets of intervals on a line so as to be able to select one that maximizes the weight sum of the weighted points covered. Assuming that presorting has been performed, the first problem has an $O(n)$ time solution, while the second and the third problems are solved by dynamic programming algorithms, requiring $O(n \log n)$ and $O(n + t)$ time, respectively.

1 Introduction

Interval graphs play important role in numerous applications, many of which are in scheduling problems. They are a sub-class of perfect graphs (see [Gol80]). An interval graph is defined by its vertex set being given intervals on a line ℓ, and its edges connecting intersecting intervals. A dominating set in such a graph thus corresponds to a set of intervals that intersect all other intervals. A minimum dominating set is a dominating set of minimum cardinality. The minimum dominating set problem has been studied for interval graphs and some special perfect graphs and linear time algorithm is known [BG88, Bra87, RR88]. In this paper, we study the following three problems of computing minimum dominating sets of intervals on lines.

---

* Contact author: Siu-Wing Cheng, email SCHENG@CS.UST.HK

** Research partially supported by RGC CER grant HKUST 190/93E. Department of Computer Science, Hong Kong University of Science and Technology, Clear Water Bay, Hong Kong.

*** Department of Computer Science, Technion - Israel Institute of Technology, Haifa 32000, Israel. Research done when the author was with the Department of Computer Science, Hong Kong University of Science and Technology.

[†] Department of Computer Science, Technion - Israel Institute of Technology, Haifa 32000, Israel. Research done while visiting the Department of Computer Science, Hong Kong University of Science and Technology.

- **Problem 1** : Given a set of lines that have one common intersection point, find a minimum dominating set for a set of given n intervals on these lines.
- **Problem 2** : Given a set of lines such that all except one are parallel, find a minimum dominating set for a set of given n intervals on these lines.
- **Problem 3** : Given n intervals and t positively weighted points on a line, find a minimum dominating set for this set of intervals that maximizes the weight sum of the weighted points covered.

Throughout this paper, we assume that each line contains at least one interval and that given an endpoint, we can tell whether it is the left or right endpoint, and of which interval. We assume that no two intervals have a common endpoint. We also assume that the endpoints of the intervals on each line are given in sorted order. An immediate consequence of these assumptions is that the sorted orderings of the left endpoints alone and the right endpoints alone can be extracted in linear time. For Problem 2, we assume that the parallel lines are sorted from left to right along the line which they intersect. For Problem 3, we assume that the weighted points on the line are also given in sorted order from left to right.

Problems 1 and 2 are special cases of the minimum dominating set problem for path graphs, which are extensions of interval graphs [Gol80] defined as follows. For a given graph and a specified set of paths, build an intersection graph whose vertices correspond to the paths in the graph and two vertices are connected by an edge, if the two corresponding paths intersect. Finding a minimum dominating set for path graphs is known to be NP-hard [BJ82]. In contrast, we propose efficient algorithms for Problem 1 and Problem 2 that run in $O(n)$ time and $O(n \log n)$ time, respectively. The algorithm for Problem 3 uses dynamic programming in a non-trivial fashion and uses appropriate data structures in order to obtain an $O(n + t)$ time algorithm. Note that computing just a minimum dominating set of intervals on a line can be done in linear time using an obvious greedy approach.

Given a line ℓ and a set of intervals on it, we also use ℓ to denote this set of intervals for convenience. We use $m(\ell)$ to denote the cardinality of the minimum dominating set of ℓ. We say that a set of intervals covers x, if one of the intervals covers x. Given any point x on ℓ, we denote the subset of intervals that does not cover x by $\ell(x)$.

The rest of the paper is organized as follows. Problem 1 is discussed in Section 2. In the discussion we make use of detecting the existence of a minimum dominating set of intervals on a line that covers a specified point. This result is later used in solving Problem 2 (Section 3) in a dynamic programming strategy. A straightforward implementation will produce a quadratic time algorithm, and we show how to use priority search trees to reduce the running time. In Section 4 we design a dynamic programming algorithm for Problem 3 and then employ the union-and-find data structure for an efficient implementation. Some of the proofs are only sketched or omitted.

2 Problem 1

In order to solve Problem 1 in linear time, we need the following two lemmas.

Lemma 1. *Given a set of intervals and a point x on a line ℓ, it can be determined in linear time whether there exists a minimum dominating set of ℓ that covers x.*

Proof (sketch): View x as a degenerate interval. Clearly, $m(\ell) \leq m(\ell \cup \{x\}) \leq m(\ell) + 1$. It is easily shown that there is a minimum dominating set of ℓ that covers x if and only if $m(\ell \cup \{x\}) = m(\ell)$. As mentioned before in Section 1, $m(\ell)$ and $m(\ell \cup \{x\})$ can be computed in linear time. □

Lemma 2. *Let L be a set of lines with one common intersection point x. Let L' be the subset of L such that for each $\ell \in L'$, some minimum dominating set of ℓ covers x. Let k be the number of lines ℓ such that $m(\ell(x)) = m(\ell) - 1$. Then*

$$m(L) = \begin{cases} \sum_{\ell \in L} m(\ell) - k & \text{if } L' = \emptyset \text{ or } |L'| > k \\ \sum_{\ell \in L} m(\ell) - k + 1 & \text{otherwise.} \end{cases}$$

Proof (sketch): By Lemma 1, we have $m(\ell \cup \{x\}) = m(\ell)$ for every $\ell \in L'$, and $m(\ell \cup \{x\}) = m(\ell) + 1$ for every $\ell \in L - L'$. Then, $m(\ell(x)) \geq m(\ell)$ for each $\ell \in L - L'$. Otherwise, $m(\ell(x)) \leq m(\ell) - 1$ and adding to the minimum dominating set of $\ell(x)$ any interval that covers x yields a dominating set for $\ell \cup \{x\}$ of size $m(\ell)$. Let L'' be the subset of L' such that for each $\ell \in L''$, $m(\ell(x)) = m(\ell) - 1$. It follows that $|L''| = k$.

Case (1). $L' = \emptyset$ or $|L'| > k$. If $L' = \emptyset$, then clearly $k = 0$ and $m(L) = \sum_{\ell \in L} m(\ell)$. So suppose that $L' \neq \emptyset$. For each $\ell \in L - L'$, select a minimum dominating set of ℓ. For each $\ell \in L' - L''$, select a minimum dominating set of ℓ that covers x. For each $\ell \in L''$, select a minimum dominating set of $\ell(x)$. Let D be the union of these minimum dominating sets. It follows that D is a minimum dominating set.

Case (2). The other possibility is that $L' = L'' \neq \emptyset$ and hence $|L'| = k > 0$. The argument is similar to that used in Case (1). □

A linear time algorithm follows naturally from Lemma 1 and Lemma 2.

Theorem 3. *Given a set of lines with one common intersection point, a minimum dominating set of n intervals on the lines can be computed in $O(n)$ time.*

3 Problem 2

In this section we present an efficient algorithm to solve Problem 2. Given a line ℓ_0 and t parallel lines ℓ_i, $1 \leq i \leq t$, that intersect ℓ_0, our algorithm finds a minimum dominating set for $\bigcup_{i=0}^{t} \ell_i$ in $O(n \log n)$ time, where $n = \sum_{i=0}^{t} |\ell_i|$.

Let the intersection between ℓ_i and ℓ_0 be denoted by x_i, referred to as a *crossing*. Lemma 4 below asserts that if for some k, $1 \leq k \leq t$, no minimum dominating set of ℓ_k covers x_k, then ℓ_k does not play any role in the problem and can be ignored. (Note that all such lines ℓ_k can be identified in linear time by Lemma 1.)

Lemma 4. *Let* k, $1 \leq k \leq t$, *be such that no minimum dominating set of* ℓ_k *covers* x_k. *Then there exists a minimum dominating set of* $\bigcup_{i=0}^{t} \ell_i$ *that contains a minimum dominating set of* ℓ_k *(hence* $m(\bigcup_{i=0}^{t} \ell_i) = m(\bigcup_{i=0}^{t} \ell_i - \{\ell_k\}) + m(\ell_k)$*).*

Proof. Omitted. □

In summary, we can assume that for each ℓ_i, $1 \leq i \leq t$, some minimum dominating set of ℓ_i covers x_i. Lemma 5 below provides the structure of a minimum dominating set of $\bigcup_{i=0}^{t} \ell_i$ and its cardinality. To state the lemma we need the following notation.

We mark all the crossings x_i's for which $m(\ell_i(x_i)) < m(\ell_i)$, i.e., $m(\ell_i(x_i)) = m(\ell_i) - 1$. (Recall that $\ell_i(x_i)$ denotes the subset of intervals on ℓ_i that do not cover x_i.) For a set of intervals $\hat{D}$ on line ℓ_0, we denote by $c(\hat{D})$ the set of marked crossings covered by $\hat{D}$. Let $D_0 \subseteq \ell_0$ be such that it intersects all intervals in ℓ_0 that do not cover a crossing and for which $|D_0| - |c(D_0)|$ is minimum.

Lemma 5. *The cardinality of a minimum dominating set of* $\bigcup_{i=0}^{t} \ell_i$ *is* $|D_0| - |c(D_0)| + \sum_{i=1}^{t} m(\ell_i)$.

Proof. Let D be a dominating set of $\bigcup_{i=0}^{t} \ell_i$. For each marked crossing $x_i \in c(D \cap \ell_0)$, we need at least $m(\ell_i) - 1$ intervals to cover the intervals in $\ell_i(x_i)$. For each crossing $x_i \notin c(D \cap \ell_0)$, we need at least $m(\ell_i)$ intervals to cover the intervals in ℓ_i. Therefore, $|D| \geq |D \cap \ell_0| - |c(D \cap \ell_0)| + \sum_{i=1}^{t} m(\ell_i)$. Clearly, $D \cap \ell_0$ must cover all intervals in ℓ_0 that do not cover a crossing. Therefore, by the definition of D_0, $|D \cap \ell_0| - |c(D \cap \ell_0)| \geq |D_0| - |c(D_0)|$, which implies that $|D| \geq |D_0| - |c(D_0)| + \sum_{i=1}^{t} m(\ell_i)$. Conversely, we can use the above arguments to extend D_0 to a dominating set of $\bigcup_{i=0}^{t} \ell_i$ of size $|D_0| - |c(D_0)| + \sum_{i=1}^{t} m(\ell_i)$.
□

By Lemma 5, a minimum dominating set of $\bigcup_{i=0}^{t} \ell_i$ can be constructed from D_0 in linear time. The rest of this section is devoted to the construction of D_0. For this, in addition to the assumption that for each ℓ_i, $1 \leq i \leq t$, some minimum dominating set of ℓ_i covers x_i, we assume that no crossing is an endpoint of any interval. Next we introduce the following definitions and notation.

- The sequence $p_1, p_2, \ldots, p_f$ consists of all the crossings and the right endpoints of intervals in ℓ_0 sorted from left to right.
- S_i is the set of intervals on ℓ_0 whose right endpoint is at p_i or to the left of p_i. A subset of S_i is a *weak dominating subset* if it intersects all intervals in S_i that do not cover a crossing. A weak dominating subset A of S_i is a *weak dominating subset of the first type*, if p_i is a crossing or p_i is the right endpoint of some interval $I \notin A$. The weak dominating subset of the first type is undefined, if p_i is the right of endpoint of some interval I and every weak dominating subset of S_i contains I. A weak dominating subset A of S_i is a *weak dominating subset of the second type* if p_i is a right endpoint of an interval $I \in A$. A weak dominating subset of the second type is undefined if p_i is a crossing.

- Let $D_{i,1}$ be a weak dominating subset of S_i of the first type such that $|D_{i,1}| - |c(D_{i,1})|$ is minimum. Such a $D_{i,1}$ will be referred to as a *minimum weak dominating set of the first type* and the difference $|D_{i,1}| - |c(D_{i,1})|$ will be denoted by $N_{i,1}$. If $D_{i,1}$ is undefined, then we set $N_{i,1}$ to be ∞.
- Let $D_{i,2}$ be a weak dominating subset of S_i of the second type such that $|D_{i,2}| - |c(D_{i,2})|$ is minimum. Such a $D_{i,2}$ will be referred to as a *minimum weak dominating set of the second type* and the difference $|D_{i,2}| - |c(D_{i,2})|$ will be denoted by $N_{i,2}$. If $D_{i,2}$ is undefined, then we set $N_{i,2}$ to be ∞.

Obviously, the set of intervals D_0 can be taken as $D_{f,1}$, if $N_{f,1} \leq N_{f,2}$, and as $D_{f,2}$ otherwise. Therefore, it suffices to compute $D_{f,1}$, $N_{f,1}$, $D_{f,2}$, and $N_{f,2}$. This can be done by computing $D_{i,1}$, $N_{i,1}$, $D_{i,2}$, and $N_{i,2}$ inductively starting with $i = 1$. If p_1 is a crossing, by definition, $D_{1,1} = \emptyset$, $N_{1,1} = 0$, $D_{1,2}$ is undefined, and $N_{1,2} = \infty$. Otherwise, let p_1 be the right endpoint of interval $I \in \ell_0$. Then $D_{1,1}$ is undefined, $N_{1,1} = \infty$, $D_{1,2} = \{I\}$, and $N_{1,2} = 1$. Lemmas 6 and 7 show how to perform the inductive computation.

The computation of the minimum dominating set is based on the following two lemmas, whose proofs are omitted here.

Lemma 6. *For all $2 \leq i \leq f$, if p_i is a crossing, then $N_{i,1} = N_{i-1,1}$ and $D_{i,1} = D_{i-1,1}$. If p_i is the right endpoint of an interval I, then*

1. *If I covers some crossing, then $N_{i,1} = \min\{N_{i-1,1}, N_{i-1,2}\}$ and $D_{i,1}$ is set to be $D_{i-1,1}$ or $D_{i-1,2}$ correspondingly.*
2. *If I does not cover any crossing, then $N_{i,1} = N_{k,2} = \min\{N_{j,2} : j < i$ and p_j covered by $I\}$ and $D_{i,1}$ is set to be $D_{k,2}$.*

Lemma 7. *For all $2 \leq i \leq f$, if p_i is a crossing, then $D_{i,2}$ is undefined and $N_{i,2} = \infty$. Otherwise, let I be the interval whose right endpoint is p_i and let p_j be the rightmost point not covered by I. Then $N_{i,2}$ is equal to the minimum of:*

1. $1 + \min\{N_{j,1}, N_{j,2}\} - c(\{I\})$.
2. $\min\{1 + N_{k,2} - c(\{[p_k, p_i]\}) : k < i$ and p_k covered by $I\}$.

Also, $D_{i,2}$ is set to be $D_{j,1} \cup \{I\}$ or $D_{j,2} \cup \{I\}$ in (1), corresponding to whether the minimum is attained by $N_{j,1}$ or $N_{j,2}$, and $D_{k,2} \cup \{I\}$ in (2).

Lemmas 6 and 7 pave the way for an efficient dynamic programming algorithm to compute the minimum dominating set for $\bigcup_{i=0}^{t} \ell_i$. We describe below an implementation that runs in $O(n \log n)$ time.

We use the priority search tree introduced in [McC85]. A priority search tree supports the following two operations on a dynamic set of ordered pairs of elements of a linearly ordered set $\mathcal{F} = (x_i, y_i)_{i=1,\ldots,N}$ in $O(\log N)$ time.

- *Insert(x, y)*. Insert a pair (x, y) into $\mathcal{F}$.
- *Query(a, b)*. Among all pairs $(x, y) \in \mathcal{F}$ such that $a \leq x \leq b$, find a pair whose y is minimal.

We shall keep two priority search trees T_1 and T_2.

We scan the points p_i, $1 \le i \le f$, from left to right to determine $c(\{(-\infty, p_i]\})$ for all $1 \le i \le f$. This takes $O(n)$ time. (Recall that $c(\{(-\infty, p_i]\})$ is the number of marked crossing covered by $(-\infty, p_i]$.) This will then enable us to compute in $O(1)$ time $c([p_k, p_i])$ for all $k < i$. Note that this quantity is needed in Lemma 7. For all interval I on ℓ_0, by a similar $O(n)$-time left-to-right scan, we can also compute $c(\{I\})$, decide whether I contains any crossing, and compute $m(I)$ where $p_{m(I)}$ is the rightmost point to the left of I but not covered by I. Note that the decision of whether I contains any crossing is needed in Lemma 6 and $c(\{I\})$ and $p_{m(I)}$ are needed in Lemma 7.

First, we compute $N_{i,1}$ as discussed in Lemma 6. It takes only constant time to handle the cases where p_i is a crossing or p_i is the right endpoint of an interval I but I contains some crossing. To handle the case where I does not contain any crossing, we use a priority search tree T_1 to maintain the set $\{(p_k, N_{k,2}) : k < i\}$. In addition we store with each pair $(p_k, N_{k,2})$ $N_{k,1}$, $D_{k,1}$, and $D_{k,2}$ as auxiliary information. Let a be the left endpoint of I. We call $Query(a, p_i)$ to find the minimum $N_{k,2}$ for some p_k covered by I. Thus, it takes $O(\log n)$ to compute $N_{i,1}$ and $D_{i,1}$.

Next we compute $N_{i,2}$ as discussed in Lemma 7. It takes constant time to handle the case when p_i is a crossing It also takes constant time to compute $1 + \min\{N_{m(I),1}, N_{m(I),2}\} - c(\{I\})$ when p_i is the right endpoint of an interval I. To compute $\min\{1 + N_{k,2} - c(\{[p_k, p_i]\}) : k < i$ and p_k covered by $I\}$, we first observe that since $c(\{[p_k, p_i]\}) = c(\{(-\infty, p_i]\}) - c(\{(-\infty, p_{k-1}]\})$, $\min\{1 + N_{k,2} - c(\{[p_k, p_i]\}) : k < i$ and p_k covered by $I\} = \min\{1 + N_{k,2} + c(\{(-\infty, p_k]\}) : p_k \in I\} - c(\{(-\infty, p_i]\})$. Therefore, we use a priority search tree T_2 to maintain the set $\{(p_k, 1 + N_{k,2} + c(\{(-\infty, p_k]\})) : k < i\}$. In addition we store with each pair $[p_k, N_{k,2}]$ $N_{k,1}$, $D_{k,1}$, and $D_{k,2}$ as auxiliary information. Let a be the left endpoint of I. We call $Query(a, p_i)$ to find the minimum $1 + N_{k,2} + c(\{(-\infty, p_k]\})$ for some p_k covered by I and then subtract $c(\{(-\infty, p_i]\})$ from it. Thus, it takes $O(\log n)$ to compute $N_{i,2}$ and $D_{i,2}$.

Afterwards, we insert $(p_i, N_{i,2})$ into T_1 and $(p_i, 1 + N_{i,2} + c(\{(-\infty, p_i]\}))$ into T_2 and continue in the same manner.

Theorem 8. *Given a line intersected by some other parallel lines, a minimum dominating set of n intervals on the lines can be computed in $O(n \log n)$ time.*

□

4 Problem 3

We are given n intervals and t weighted points on a line. We want to find a minimum dominating set D that maximizes the weight sum of the points covered by intervals in D. Suppose that the endpoints of the intervals and the weighted points are already in sorted order from left to right. We again solve the problem using a dynamic programming approach. The algorithm uses the values $leftmost(I_k)$, defined for each interval I_k, and which have to be computed in advance. We first describe how to compute $leftmost(I_k)$, and then describe our solution to Problem 3.

4.1 Computing $leftmost(I_k)$

We first order the n given intervals so that their right endpoints are in increasing x-coordinates. (This can be done by a linear time scan through the sorted list of endpoints.) Let $I_1, I_2, \cdots, I_n$ denote this ordered list. Let $left(I_k)$ and $right(I_k)$ denote the left and right endpoints of I_k, respectively. We also use $left(I_k)$ or $right(I_k)$ to denote the x-coordinate of these endpoints. For each interval I_k, define $leftmost(I_k)$ to be the interval I_q such that $q < k$, I_q intersects I_k and I_q has the leftmost left endpoint. Let $\mathcal{K}(I_k) = \{I_j : j < k, I_j \cap I_k \neq \emptyset\}$. $leftmost(I_k)$ thus satisfies

$$leftmost(I_k) = \begin{cases} I_q \in \mathcal{K}(I_k) \text{ s.t. } left(I_q) = \min\{left(I_j) : I_j \in \mathcal{K}(I_k)\}, & \text{if } \mathcal{K}(I_k) \neq \emptyset \\ null, & \text{if } \mathcal{K}(I_k) = \emptyset. \end{cases}$$

Consider all the $2n$ interval endpoints in increasing order of x-coordinates: $p_1, p_2, \cdots, p_{2n}$. We assign a value to each endpoint as follows. If p_i is a left endpoint, then $val(p_i) = \infty$. If $p_i = right(I_k)$, then $val(p_i) = left(I_k)$. We also define the value of a set A of endpoints by $val(A) = \min_{p \in A} val(p)$.

Our algorithm ComputeLeftmost works as follows. Initialize each endpoint p_i to be a singleton set; as the algorithm advances, we will maintain a partition of the endpoints into disjoint subsets. To compute $leftmost(I_k)$ for $k \geq 1$, we compute $find(left(I_k))$ which is the subset A' that contains $left(I_k)$. If $val(A') = \infty$, then we set $leftmost(I_k)$ to be null. Otherwise, we set $leftmost(I_k)$ to be I_q, where $right(I_q)$ is the rightmost endpoint in A'. Afterwards, we update the ordered collection of subsets of endpoints as follows. Let $A_1, A_2, \cdots, A_m$ be the disjoint subsets of endpoints that are to the left of $right(I_k)$. We perform repeated merging of subsets as shown below.

```
Procedure MergeA;
    A := {right(I_k)};
    i := m;
    while i ≥ 1 and val(A) ≤ val(A_i) do
        A := union(A, A_i);
        i := i - 1;
    end while
end MergeA
```

The correctness proof (omitted) is based on the following invariants (1) and (2) maintained during the computation (as can be proved by induction) :

1. The collection of subsets is ordered from left to right according to the ordering of the endpoints and the endpoints in each subset form a consecutive sequence.
2. If we are computing $leftmost(I_k)$ and A_j, $1 \leq j \leq m$, are the subsets of endpoints to the left of $right(I_k)$ before performing the repeated merging, then:

(a) There exists $0 \le r \le m$ such that A_j, $1 \le j \le r$, contains at least one right endpoint and A_j, $r < j \le m$, is a singleton set containing a left endpoint.

(b) For $1 \le j \le r$, the rightmost point p in A_j is a right endpoint and $val(A_j) = val(p)$.

(c) For $1 \le j < r$, $val(A_j) < val(A_{j+1}) < \infty$, and for $r < j \le m$, $val(A_j) = \infty$.

Except for the *union* and *find* operations, our algorithm takes $O(1)$ time to process one interval. We need a data structure that can efficiently implement the union of disjoint subsets and finding the subset that contains a given element. There is a total of n *find* operations and at most $2n - 1$ *union* operations.

If we use the union-and-find data structure based on path compression and union by rank [Tar83], then we obtain an algorithm with running time $O(n\alpha(n, n))$, where $\alpha(n, n)$ is the extremely slow growing inverse Ackermann function. However, we can actually employ the special union-and-find data structure in [GT85] to improve the running time to $O(n)$.

In [GT85], the *incremental tree set union* problem is considered. Initially, there is a tree T consisting of a single node, the root, which corresponds to a singleton set containing an element r. The tree will be allowed to grow by one node (leaf), but each node of the tree will always correspond to a unique element. So we shall also use the element name as the node name. The following three operations are allowed.

1. *find(x)*: return the name of the set containing element x.
2. *union(parent(x), x)* : create a new set that is the union of the sets containing x and *parent(x)*, where *parent(x)* is the parent of x in T.
3. *grow(x, y)* : Add a node y to T by making x its parent. The node y is a new node not in T.

If n' denotes the final number of elements in T and m' denotes the total number of unions and finds (note that there are exactly $n' - 1$ *grow* operations), then it is proved in [GT85] that the incremental tree set union problem can be solved in $O(m' + n')$ time. This enables us to implement our algorithm in $O(n)$ time (details are omitted here).

4.2 Maximum weighted minimum dominating set

4.2.1 Definitions

We now turn to the algorithm for Problem 3. For any interval I, define $wt(I)$ to be the weight sum of points covered by I. For any set A of intervals, define $wt(A) = wt(\bigcup_{I \in A} I)$. We want to compute the minimum dominating set with the maximum weight. We first do an $O(n + t)$-time scan of all the intervals to determine $wt(I_k)$ for all intervals I_k.

Given intervals $\{I_1, \cdots, I_k\}$, define $D_1(I_k)$ as follows. Consider the collection of dominating sets of $\{I_1, \cdots, I_k\}$ that contain I_k. (Note that such a dominating set always exists.) First remove those dominating sets with larger than minimum cardinality. Then remove those with less than maximum weight. Define $D_1(I_k)$

to be one of the remaining dominating sets. Note that $D_1(I_k)$ is not necessarily a minimum dominating set for $\{I_1, \cdots, I_k\}$.

Define $D_2(I_k)$ as follows. Consider the collection of dominating sets of $\{I_1, \cdots, I_k\}$ that do not contain I_k. If there is no such dominating set, then define $D_2(I_k)$ to be null. Otherwise, first remove those dominating sets with larger than minimum cardinality. Then remove those with less than maximum weight. Define $D_2(I_k)$ to be one of the remaining dominating sets. Note that $D_2(I_k)$, even when it is not null, is not necessarily a minimum dominating set for $\{I_1, \cdots, I_k\}$.

Define $MD(I_k)$ to be a maximum weighted minimum dominating set for $\{I_1, \cdots, I_k\}$ as follows. If $D_2(I_k) = $ null or $|D_1(I_k)| < |D_2(I_k)|$, then define $MD(I_k)$ to be $D_1(I_k)$. Otherwise, if $|D_1(I_k)| > |D_2(I_k)|$, then define $MD(I_k)$ to be $D_2(I_k)$. Otherwise ($|D_1(I_k)| = |D_2(I_k)|$), define $MD(I_k)$ to be the one between $D_1(I_k)$ and $D_2(I_k)$ with the larger weight (ties broken arbitrarily).

Clearly, $MD(I_n)$ is the desired maximum weighted minimum dominating set. Given $D_1(I_k)$ and $D_2(I_k)$, $MD(I_k)$ can be computed in $O(1)$ time. So we focus on computing $D_1(I_k)$ and $D_2(I_k)$.

Define $neighbor(I_k)$ to be the interval I_q with the largest index such that I_q does not intersect I_k. We can compute $neighbor(I_k)$ for all k in $O(n)$ time as follows. Scan all the $2n$ endpoints from left to right. Keep track of the interval I such that $right(I)$ is the rightmost right endpoint that have been scanned so far. Initially, I is null. Whenever we see a left endpoint of I_k, we set $neighbor(I_k)$ to be I. Whenever we see a right endpoint of I_k, we update I to be I_k. After we have scanned $left(I_n)$, we have computed $neighbor(I_k)$ for all k.

4.2.2 Computing $D_1(I_k)$

We now show how to compute the quantities $D_1(I_k)$ for $k \geq 1$. Let $N_k = MD(neighbor(I_k)) \cup \{I_k\}$ and $L_k = D_1(leftmost(I_k)) \cup \{I_k\}$. $D_1(I_k)$ can be computed by the following algorithm:

Algorithm ComputeD1

1. If $neighbor(I_k) = $ null, then $D_1(I_k) = \{I_k\}$.
2. If $leftmost(I_k) = $ null and $k > 1$, then $D_1(I_k) = N_k$. (Note that it is impossible that $leftmost(I_k) = neighbor(I_k) = $ null for $k > 1$.)
3. Otherwise: if $|L_k| < |N_k|$, then $D_1(I_k) = L_k$; if $|L_k| > |N_k|$, then $D_1(I_k) = N_k$; if $|L_k| = |N_k|$ and $wt(L_k) \geq wt(N_k)$, then $D_1(I_k) = L_k$; if $|L_k| = |N_k|$ and $wt(L_k) < wt(N_k)$, then $D_1(I_k) = N_k$.

Thus, once $MD(I_j)$ and $D_1(I_j)$, $1 \leq j < k$, are known, computing each $D_1(I_k)$ takes $O(1)$ time. The correctness of the computation of $D_1(I_k)$ is omitted.

4.2.3 Computing $D_2(I_k)$

We now show how to compute the quantities $D_2(I_k)$ for $k \geq 1$. For $k > 1$, let $\mathcal{K}(I_k) = \{D_1(I_j) : j < k, I_j \cap I_k \neq \emptyset, |D_1(I_j)| \text{ minimized}\}$, and $\mathcal{L}(I_k) = \{D_1(I_j) \in \mathcal{K}(I_k) : wt(D_1(I_j)) \text{ maximized}\}$. $D_2(I_k)$ can be computed by the following algorithm:

529

Algorithm ComputeD2

1. If $leftmost(I_k) = $ null, then $D_2(I_k) = $ null.
2. Otherwise, $D_2(I_k) = D_1(I_q)$ for any $D_1(I_q) \in \mathcal{L}(I_k)$.

$D_2(I_k)$ can be computed in $O(n)$ time, using a strategy that is a more sophisticated version of the one used for computing $leftmost(I_k)$; the detailed are omitted here.

Theorem 9. *Given n possibly nested intervals and t weighted points on a line, the minimum dominating set that maximizes the weight sum of the weighted points covered can be computed in $O(n + t)$ time.* □

Acknowledgment

We thank Pandu Rangan for a discussion that stimulated the investigation. We also thank Martin Golumbic, Dieter Kratsch, and Ron Shamir for helpful correspondence.

References

[BG88]   A. BERTOSSI AND A. GORI, *Total domination and irredundance in weighted interval graphs*, SIAM J. Discrete Math., 3 (1988), pp. 317–327.

[BJ82]   K. BOOTH AND J. JOHNSON, *Dominating sets in chordal graphs*, SIAM Journal on Computing, 11 (1982), pp. 191–199.

[Bra87]   A. BRANDTSTAEDT, *The computational complexity of feedback vertex set, hamiltonian circuit, dominating set, steiner tree, and bandwidth on special perfect graphs*, J. Inf. Process. Cybern., 23 (1987), pp. 471–477.

[GT85]   H. GABOW AND R. TARJAN, *A linear-time algorithm for a special case of disjoint set union*, Journal of Computer and System Sciences, (1985), pp. 209–221.

[Gol80]   M. C. GOLUMBIC, *Algorithmic graph theory and perfect graphs*, Academic Press, 1980.

[McC85]   E. MCCREIGHT, *Priority search trees*, SIAM Journal on Computing, 14 (1985), pp. 257–276.

[RR88]   G. RAMALINGAM AND P. RANGAN, *A unified approach to domination problems on interval graphs*, Information Processing Letters, 27 (1988), pp. 271–274.

[Tar83]   R. TARJAN, *Data structures and network algorithms*, CBMS-NSF 44, SIAM, 1983.

Two-Dimensional Pattern Matching on a Dynamic Library of Texts

Y. Choi and T.W. Lam

Department of Computer Science
University of Hong Kong
Pokfulam Road, Hong Kong
Email: {ychoi, twlam}@csd.hku.hk

Abstract

Giancarlo's s-trees [9] has been the most efficient data structure for representing a library of text matrices such that we can efficiently search for the occurrences of any pattern matrix in the library, and update the library dynamically through insertion and deletion of texts. This paper presents a new scheme which improves the s-trees and provides faster algorithms for each above-mentioned operation. In particular, this scheme, though allowing the library to change dynamically, can search for a pattern as fast as the currently best algorithm designed for the static case, in which the text matrices are all fixed in advance.

1 Introduction

Two-dimensional pattern matching finds application in areas like image processing. In recent years, significant progress has been made in the design of efficient algorithms for a variety of problems on two-dimensional pattern matching. For instance, consider the most basic problem: Given an $n \times m$ text matrix T and a $p \times d$ pattern matrix P, search for all occurrences of P in T. Galil and Park [7], improving a chain of work [5, 6, 4, 2], have obtained an optimal algorithm that uses $O(n \times m + p \times d)$ time.

In this paper, we consider a different problem. There is a library of texts $L = \{T^1, T^2, \cdots, T^k\}$, where each T^i is a rectangular matrix of arbitrary size and with elements chosen from a fixed alphabet Σ. We are required to devise data structures to store L such that any sequence of the following operations can be executed efficiently.

- Query: Given a pattern matrix P, search for all occurrences of P in L.

- Insertion: Insert a new text matrix T into L.

- Deletion: Delete an existing text matrix T^i from L.

Previous work: This problem has been studied before [10, 8, 9]. The most efficient data structure has been Giancarlo's s-trees [9], which is a generalization of the suffix tree [12, 1] to represent all submatrices of matrices in the library.

To discuss the performance of the s-trees, we need a notation: For any matrix T of size $n \times m$, define $s(T) = nm \min(n, m)$. Let $s(L) = \sum_{T \in L} s(T)$. The s-trees supports a search for a $p \times d$ pattern matrix in $O(pd \log s(L) + tocc)$ time, where $tocc$ is the total number of occurrences of the pattern, and an insertion or deletion of an $n \times m$ text matrix T in $O(s(T)(\log^2 s(L) + \log |\Sigma|))$ time. If the library is fixed in advance, the data structure can be simplified; the library can be preprocessed in $O(s(L)(\log s(L) + \log |\Sigma|))$ time and the search operation is improved to $O(pd \log |\Sigma| + tocc)$ time, which would be optimal if Σ is a bounded alphabet. Giancarlo [9] has also showed a lower bound of $\Omega(s(L))$ on the space requirement of any tree that represents all submatrices of matrices in L in such a way that submatrices having a prefix in common share a path in the tree. This implies that the insertion of a text T requires $\Omega(s(T))$ time.

Our result: In this paper, we present a new scheme to improve the search time to $O(pd \log |\Sigma| + tocc)$ and update time to $O(s(T)(\log s(L) + \log |\Sigma|))$. The preprocessing of a library L can also be done in $O(s(L)(\log s(L) + \log |\Sigma|))$ time. The advantage of our scheme is that it has the same performance as the best static solution, while allowing the library to get updated. Our scheme is still based on s-trees, yet it is simpler from a practical viewpoint and it avoids using those more complicated data structures for finding nearest common ancestors on dynamic trees [13].

Other related work: If the texts and patterns are restricted to squares only, the number of possible submatrices in the library is significantly reduced and the above lower bound is no longer valid. In this case, a much better performance can be attained by using another data structure, the Lsuffix trees [8]. An insertion or deletion of an $n \times n$ square text requires $O(n^2 \log^2(\sum_{T^i \in L} n_i^2))$ time while a search for a $p \times p$ square pattern takes $O(p^2 \log |\Sigma| + tocc)$ time.

Another related problem that has attracted attention is the two-dimensional dictionary matching [8, 3, 11], in which a dictionary of patterns is given in advance and the query is about the occurrences of all the patterns in a certain text.

Notation: Consider any $n \times m$ matrix T. For any integers $i_1 \leq i_2 \leq n$, $j_1 \leq j_2 \leq m$, let $T[i_1{:}i_2, j_1{:}j_2]$ denote the submatrix of T with the upper left corner at (i_1, j_1) and lower right corner (i_2, j_2). The submatrix $T[i_1{:}i_1, j_1{:}j_2]$ is also denoted by $T[i_1, j_1{:}j_2]$.

2 Suffix Trees

The main data structure used in this paper is a two-dimensional analog of the suffix trees [12]. Before proceeding to the details of our scheme, we give a brief review of suffix trees. Readers who are familiar with suffix trees may skip this section.

Suffix trees are data structures for representing strings to allow efficient searching for the occurrences of substrings. Let $x[1{:}n]$ be a string of n characters chosen from a totally ordered alphabet Λ and let \$ be a special symbol that does not match any character, including itself. Note that a suffix of the string $x\$$ cannot be the prefix of any other suffix. A suffix tree of the string $x\$$ is a compacted trie with $n + 1$ leaves satisfying the following properties:

1. The outdegree of an internal node is at least two.

2. Every edge is labeled with a nonempty substring of $x\$$, and the labels of two edges emanating from a node must start with different characters.

3. There is a one-to-one correspondence between the leaves and the suffixes of $x\$$. More importantly, for each suffix $x[i{:}n]$ of x, there is a leaf u such that the concatenation of labels on the path from the root to u is $x[i{:}n]\$$.

4. Each leaf stores the starting position of the corresponding suffix in respect of $x\$$.

A string y is a substring of x if and only if there is a node u in the suffix tree such that the concatenation of the labels on the path from the root to u contains y as a prefix. Moreover, each leaf in the subtree rooted at u stores the starting position of an occurrence of y in x.

The label of an edge can be represented by a pair of integers denoting the starting and ending positions of the label in $x\$$. In some applications where the size of Λ is not fixed, we associate an AVL tree with each internal node u of the suffix tree so that the suffix tree can be traversed easily. For every edge (u, v) in the suffix tree, the AVL tree of u contains a node storing the first character of the string labeling (u, v) as well as the address of the node v. The number of nodes in the AVL tree is equal to the outdegree of u, which is upper bounded by $\min(|\Lambda|, n)$. Searching for a string y of length m takes $O(mc \log(\min(|\Lambda|, n)))$ time, where c is the time required to compare two characters in Λ.[1] It has been shown that a suffix tree occupies only $O(n)$ space and, more interestingly, there is an $O(nc \log(\min(|\Lambda|, n)))$ time algorithm to insert the suffixes of $x\$$, from the

---

[1] In most applications, c would be a constant. However, we will consider in the next section an alphabet Λ whose elements are strings of characters chosen from some fixed alphabet Σ and c is no longer a constant.

longest to the shortest, into an initially empty tree, which then becomes the suffix tree of $x\$$.

To represent a set of strings $S = \{x_1, x_2, \cdots, x_k\}$ so that a substring of any x_i can be searched efficiently, we can build a suffix tree to represent the string $z = x_1\$x_2\$\cdots x_k\$$. The size of this suffix tree is $O(n)$, where $n = \sum_{i=1}^{k} |x_i|$, and it can be built in $O(nc\log(\min(|\Lambda|, n)))$ time. Searching for the occurrence of a string y with m characters can still be done in $O(mc\log(\min(|\Lambda|, n)))$ time. Moreover, this suffix tree can be updated efficiently when a string is inserted into or deleted from S. Using McCreight's algorithms [12], we can convert the suffix tree for z to one representing the string $x_1\$\cdots x_k\$x\$$ for any $x \in \Lambda^*$ in $O(|x|c\log(\min(|\Lambda|, |x|+n)))$ time, or to one representing the string $x_1\$\cdots x_{j-1}\$x_{j+1}\$\cdots x_k\$$ for any $j \in [1, k]$ in $O(|x_j|c\log(\min(|\Lambda|, n)))$ time.

3   Data Structures

We divide the library into two parts: one contains texts with height larger than width and the other contains the rest. Each part is represented by separate data structures. Below, we only show the data structures for those texts with height larger than width. The other case can be handled in a symmetric way.

Let $L = \{T^1, T^2, \cdots, T^k\}$ be a library of texts, where each T^i is of size $n_i \times m_i$ such that $n_i \geq m_i$. Characters in the texts are chosen from a totally ordered alphabet Σ. In order to search for a pattern matrix efficiently, we keep a set of suffix trees to represent all submatrices of matrices in L. The technique is similar to the s-trees [9]. For any integer $d \leq \max_{1 \leq i \leq r}\{m_i\}$, the suffix tree C_d, in a sense, represents all submatrices of width d as follows:

Let $\Lambda = \Sigma^d$. With respect to each T^i in L of size $n \times m$, we consider every $n \times d$ submatrix of the form $T^i[1{:}n, j{:}j+d-1]$ for some $j \leq m-d+1$. Such a submatrix can be regarded as a string with n characters over Λ, i.e., $T^i[1, j{:}j + d - 1]T^i[2, j{:}j + d - 1]\cdots T^i[n, j{:}j + d - 1]$. Denote this string $\langle T^i_{j,d}\rangle$. See Figure 1 for an example.

C_d is a suffix tree for the set of strings $S = \{\langle T^i_{1,d}\rangle, \langle T^i_{2,d}\rangle, \cdots, \langle T^i_{m-d+1,d}\rangle \mid T^i \in L\}$. Note that an edge in C_d is labeled with a string over Λ, which is represented by a pair of integers denoting its starting and ending positions on some string $\langle T^i_{j,d}\rangle \in S$.

Consider a pattern P of size $p \times d$. P also defines a string $\langle P\rangle$ with p characters over Λ $(= \Sigma^d)$. That is, $\langle P\rangle = P[1, 1{:}d]P[2, 1{:}d]\cdots P[p, 1{:}d]$. P is a submatrix of a text in L if and only if $\langle P\rangle$ is a substring of some $\langle T_{j,d}\rangle$, or equivalently, C_d contains a path from the root to a node with edge labels spelling $\langle P\rangle$ as a prefix.

If we search the suffix tree C_d in a straightforward manner for the occurrences of a $p \times d$ pattern matrix P, the time required would be much more than what

$$T$$

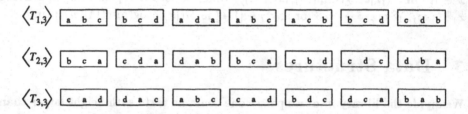

Figure 1: T is a text matrix of size 7×5; its characters are chosen from an alphabet $\Sigma = \{a, b, c, d\}$. $\langle T_{1,3}\rangle$, $\langle T_{2,3}\rangle$ and $\langle T_{3,3}\rangle$ are strings of length 7 over Σ^3.

we have claimed in Section 1. Note that the time required to compare any two characters in Λ (i.e., two strings of length d over Σ) is $O(d)$. Suppose we keep an AVL tree for each node u in C_d to keep track of the first characters of its edge labels. The best upper bound on the outdegree of u is $\min(|\Lambda|, |C_d|)$, and we may need $\Theta(d \log(\min(|\Lambda|, |C_d|)))$ time to search the AVL tree at u in order to pick the right edge to traverse. Thus, in regard to the time for searching C_d for $\langle P\rangle$, we can only guarantee an upper bound of $O(pd \log \min(|\Lambda|, |C_d|))$, i.e., $O(\min(pd^2 \log |\Sigma|, pd \log |C_d|))$. Moreover, when we want to insert a text T of size $n \times m$ into L, it is even more time consuming.

The problem above is that we have spent too much time in repeatedly performing comparison of two characters over Λ (i.e., two strings of length d over Σ) in a brute-force manner. Observe that we are not dealing with arbitrary characters in Λ. We are only interested in those characters, each corresponding to a substring ($\in \Sigma^*$) on a row of a text. Below, we build an auxiliary trie B to capture all possible substrings on a row of a text in L such that each such substring identifies uniquely a node in B. Whenever we want to compare two characters over Λ, we simply check whether the addresses of the corresponding nodes in B are the same. In other words, characters in Λ are transformed into addresses. The latter can be compared in constant time.

Intuitively, we treat each edge in the suffix tree C_d being labeled by a string of addresses instead of characters over Λ. Given a pattern P of size $p \times d$, we first transform $\langle P \rangle$ into a string of p addresses and then search C_d for $\langle P \rangle$. In case P contains a row that cannot be mapped to any node in B, we can conclude, without searching C_d, that $\langle P \rangle$ is not a substring of any string represented by C_d.

Details of the auxiliary data structures are as follows:

1. B: a trie with every edge labeled by a single character in Σ. For each text T in L of size $n \times m$, for all integers i, j, d with $1 \leq i \leq n$, $1 \leq j \leq m$, and $j + d - 1 \leq m$, there is a node u in B such that the path from the root to u consists of d edges spelling $T[i, j:j + d - 1]$. At each node in B, there is a field $Count$ storing the number of substrings in the library that are mapped to this node. $Count$ is for the purpose of text deletion. Also, each node in B maintains an AVL tree to keep track of its edge labels.

2. A_T: an array for mapping a substring on a row of T to an address. Assume T is a text in L with size $n \times m$. A_T is of size $n \times m \times m$. $A_T[i, j, d]$ stores the address of a node u in B such that the path from the root to u spells $T[i, j:j + d - 1]$. In case $j + d - 1 > m$, $A_T[i, j, d]$ is undefined. Obviously, A_T occupies $O(nm^2)$ space.

3. An AVL tree at each node of C_d. As mentioned before, for each internal node u of C_d, we keep an AVL tree to ease the search for the right edge to traverse from u. Each node in the AVL tree has two fields: the key field is an address x to B and the other field points to a child node v of u in C_d. The node x in B corresponds to the first character ($\in \Lambda = \Sigma^d$) of the string labeling the edge (u, v). Note that the number of nodes in the AVL tree is equal to the outdegree of u, which is upper bounded by $\min(|\Sigma|^d, |C_d|)$.

Lemma 1 The total number of nodes in B is at most $\sum_{i=1}^{k} n_i m_i^2$.

Proof: Every node in B defines a distinct substring that appears at least once on a row of some text in L. There are at most $\sum_{i=1}^{k} n_i m_i^2$ such substrings; hence B cannot contain more than $\sum_{i=1}^{k} n_i m_i^2$ nodes. □

Lemma 2 The total number of nodes in all suffix trees C_d is $O(\sum_{i=1}^{k} n_i m_i^2)$.

Proof: C_d is a suffix tree for the set of strings $\{\langle T_{1,d}^i \rangle, \langle T_{2,d}^i \rangle, \cdots, \langle T_{m_i-d+1,d}^i \rangle \mid T^i \in L\}$. The number of nodes in C_d is $O(\sum_{i=1}^{k} n_i(m_i - d + 1))$. Sum over all suffix trees, we obtain a total of $O(\sum_{i=1}^{k} n_i m_i^2)$ nodes. □

Though the AVL tree associated with a particular node of B may contain more than a constant number of nodes, the total number of nodes in all such AVL trees is bounded by the number of nodes in B. Similarly, the number of nodes in all AVL trees associated with C_d is bounded by the number of nodes in C_d. By Lemmas 1 and 2, we can conclude that the space requirement of all data structures mentioned so far is $O(\sum_{i=1}^{k} n_i m_i^2)$, i.e., $s(L)$.

4 Query

To find all occurrences of a $p \times d$ pattern P in texts of L, we search the suffix tree C_d for the first node α with the path from the root to α spelling the string $\langle P \rangle$ as a prefix. Details are given in following procedure **Query**(P). The starting positions of all occurrences of P in L are recorded in the leaves of the subtree rooted at α. These leaves can be found by any kind of traversal on the subtree.

---

procedure Query$(P[1{:}p, 1{:}d])$
(Step 1 transforms each row of P to an address of a node in B; Step 2 searches C_d for the string of addresses defined by P.)

1. **for** $i = 1$ **to** p **do**

   Search B for the node u such that the path from the root to u spells $P[i, 1{:}d]$; **if** such node exists **then** store the address of u in the array entry $P\text{-}Addr[i]$; **else return** "failure" immediately.

2. (a) $i \leftarrow 1$;
   $\alpha \leftarrow$ the root of C_d;

   (b) **while** $i \leq p$ **do**

   i. Search for $P\text{-}Addr[i]$ in the AVL tree associated with the node α;
   if found **then** let α' denote the corresponding child node of α; **else return** "failure" immediately.

   ii. Suppose the edge (α, α') in C_d is labeled with the string
   $T[i_1, j{:}j + d - 1]T[i_1 + 1, j{:}j + d - 1] \cdots T[i_2, j{:}j + d - 1]$
   for some T in L.
   $r \leftarrow \min(i_2 - i_1 + 1, p - i + 1)$;
   for $l = i_1$ **to** $r + i_1 - 1$ **do**
       if $P\text{-}Addr[i] = A_T[l, j, d]$ **then** $i \leftarrow i + 1$;
       else return "failure" immediately.

   iii. $\alpha \leftarrow \alpha'$

   (c) **return** α.

Time Complexity: The time required by Step 1 is $O(pd \log(\min(|\Sigma|, |B|)))$, or simply, $O(pd \log |\Sigma|)$. In one iteration of Step 2b, the index i increases by some integer $r \geq 1$; it takes $O(\log(\min(|\Sigma|^d, |C_d|)))$ time to execute Step (i) and $O(r)$ time for the rest. Note that $\log(\min(|\Sigma|^d, |C_d|))$ is bounded by $d \log |\Sigma|$. Step 2b requires in total $O(pd \log |\Sigma|)$ time.

After **Query**(P) returns a node α, we traverse the subtree rooted at α and visit all the leaves. The time required is determined by the number of leaves in the subtree, which is bounded by the number of occurrences of P in L.

5 Insertion and Deletion

In order to insert a matrix T of size $n \times m$ into L, we update the data structures as follows.

- For all integers $i \leq n$ and $j \leq m$, the string $T[i, j : m]$ is inserted into the trie B. Every time we visit or create a node in B, the field $Count$ of the node is incremented by one.

- The array A_T is created. While the string $T[i, j : m]$ is inserted into B, we encounter $m - j + 1$ nodes (excluding the root) in B, whose addresses are stored in the entries $A_T[i, j, 1], A_T[i, j, 2], \cdots, A_T[i, j, m - j + 1]$.

- For $1 \leq d \leq m$, we update the suffix tree C_d to accommodate the strings $\langle T_{1,d} \rangle, \langle T_{2,d} \rangle, \cdots, \langle T_{m-d+1,d} \rangle$. Note that when we traverse or update C_d, we actually make use of the addresses precomputed in the arrays A_T instead of dealing with characters in Λ (i.e., Σ^d).

Time Complexity: Let $L' = L \cup \{T\}$. Step 1 inserts at most nm strings, each of length no more than m, into B. This can be done in $O(nm^2 \log |\Sigma|)$ time. Step 2 obviously requires $O(nm^2)$ time. As mentioned in Section 2, updating C_d to accommodate a string $\langle T_{i,d} \rangle$ takes $O(nc \log \min(|\Lambda|, n + |C_d|))$ time, where $\Lambda = \Sigma^d$ and c is the time required to compare two characters in Λ. With the help of the arrays A_T, we can compare two characters (in the form of their addresses) in constant time. Thus, updating C_d with the strings $\langle T_{1,d} \rangle, \langle T_{2,d} \rangle, \cdots, \langle T_{m-d+1,d} \rangle$ requires $O(n(m - d + 1) \log \min(|\Lambda|, |C_d'|))$ time, where C_d' denotes the updated suffix tree afterward. In respect of all suffix trees, the time required to insert an $n \times m$ matrix into L is $O(\sum_{d=1}^{k} n(m - d + 1) \log \min(|\Sigma|^d, |C_d'|)) = O(s(T) \log s(L'))$ time.

The deletion of an $n \times m$ matrix T from L can also be done in a straightforward manner. Similar to insertion, we can update C_d to remove the strings $\langle T_{1,d} \rangle, \langle T_{2,d} \rangle, \cdots, \langle T_{m-d+1,d} \rangle$ from C_d in $O(n(m - d + 1) \log \min(|\Lambda|, |C_d|))$ time. Then we remove the array A_T. To update B, we search for $T[i, j:m]$ for all $i \leq n$ and $j \leq m$. The field $Count$ of every node encountered is reduced by one. When $Count$ becomes zero, the node is deleted from B.

References

[1] A. Apostolico and Z. Galil, editors, *Combinatorial Algorithms on Words*, Springer-Verlag, New York, 1985.

[2] A. Amir, G. Benson, M. Frach, Alphabet independent two dimensional matching, *Proc. 24th Annual ACM Symposium on the Theory of Computing*, 1992, 59-68.

[3] A. Amir, M. Farach, R. Idury, J.L. Poutre, and A. Schäffer, Improved dynamic dictionary matching, *Proc. 4th Annual ACM-SIAM Symposium on Discrete Algorithms*, 1993, 392-401.

[4] A. Amir, G.M. Landau, and U. Vishkin, Efficient pattern matching with scaling, *Proc. 1st Annual ACM-SIAM Symposium on Discrete Algorithms*, 1990, 344-357.

[5] T.J. Baker, A technique for extending rapid exact-match string matching to arrays of more than one dimension, *SIAM J. of Computing*, 7 (1978), 533-541.

[6] R.S. Bird, Two dimensional pattern matching, *Information Processing Letters*, 6 (1978), 168-179.

[7] Z. Galil and K. Park, A truly alphabet independent two-dimensional pattern matching algorithm, *Proc. 33th Annual IEEE Symposium on Foundations of Computer Science*, 1992, 247-256.

[8] R. Giancarlo, The suffix of a square matrix, with applications, *Proc. 4th Annual ACM-SIAM Symposium on Discrete Algorithms*, 1993, 402-411.

[9] R. Giancarlo, An index data structure for matrices, with applications to fast two-dimensional pattern matching, *Proc. 3rd Workshop on Algorithms and Data Structures*, 1993, 337-348.

[10] G.H. Gonnet, Efficient searching of text and pictures- extended abstract, Technical Report, University of Waterloo, OED-88-02, 1988.

[11] R.M. Idury and A.A. Schäffer, Multiple matching of rectangular patterns, *Proc. 25th Annual ACM Symposium on Theory of Computing*, 1993, 81-89.

[12] E.M. McCreight, A space-economical suffix tree construction algorithm, *J. of ACM*, 23 (1976), 262-272.

[13] D.D. Sleator and R.E. Tarjan, A data structure for dynamic trees, *Journal of Computer and System Sciences*, 26 (1983), 362-391.

Structure in Approximation Classes
(Extended abstract)

P. Crescenzi[1], V. Kann[2], R. Silvestri[1], and L. Trevisan[1]

[1] Dipartimento di Scienze dell'Informazione
Università degli Studi di Roma "La Sapienza"
Via Salaria 113, 00198 Rome, Italy
E-mail: {piluc,silver,trevisan}@dsi.uniroma1.it***
[2] Department of Numerical Analysis and Computing Science
Royal Institute of Technology
S-100 44 Stockholm, Sweden
E-mail: viggo@nada.kth.se

1 Introduction

In his pioneering paper on the approximation of combinatorial optimization problems [11], David Johnson formally introduced the notion of approximable problem, proposed approximation algorithms for several problems, and suggested a possible classification of optimization problems on grounds of their approximability properties. Since then it was clear that, even though all NP-hard optimization problems are many-one polynomial-time reducible to each other, they do not share the same approximability properties. The main reason of this fact is that many-one reductions not always preserve the objective function and, even if this happens, they rarely preserve the quality of the solutions. It is then clear that a stronger kind of reducibility has to be used. Indeed, an approximation preserving reduction not only has to map instances of a problem A to instances of a problem B, but it also has to be able to come back from "good" solutions in B to "good" solutions in A. Surprisingly, the first definition of this kind of reducibility was given as long as 13 years after Johnson's paper [23] and, after that, at least seven different definitions of approximation preserving reducibilities appeared in the literature. These definitions are identical with respect to the overall scheme but differ essentially in the way they preserve approximability: they range from the Strict reducibility [23] in which the error cannot increase to the PTAS-reducibility [8] in which there are basically no restrictions (see also Chapter 3 of [12]). By means of these reducibilities, several notions of completeness in approximation classes have been introduced and, basically, two different approaches were followed. On the one hand, the attention was focused on computationally defined classes of problems whose approximability properties were well understood, such as NPO and APX: along this line of research, however, almost all completeness results dealt either with artificial optimization problems

---

*** Research partially supported by the MURST project *Algoritmi, Modelli di Calcolo, Strutture Informative*.

or with problems for which lower bounds on the quality of the approximation were easily obtainable [23, 7]. On the other hand, researchers focused on the logical definability of optimization problems and introduced several syntactically defined classes for which natural completeness results were obtained [25, 24, 17]: unfortunately, the approximability properties of the problems in these latter classes were not related to standard complexity-theoretic conjectures. A first step towards the reconciling of these two approaches consisted of proving lower bounds (modulo P $\neq$ NP or some other likely condition) on the approximability of complete problems for syntactically defined classes [1, 20]. More recently, another step has been performed since the closure of syntactically defined classes with respect to approximation preserving reducibility has been proved to be equal to the more familiar computationally defined classes [16].

In spite of this important achievement, beyond APX we are still forced to distinguish between maximization and minimization problems as long as we are interested in completeness proofs. Indeed, a result of [17] states that it is not possible to rewrite every NP maximization problem as an NP minimization problem unless NP=co-NP. A natural question is thus whether this duality extends to approximation preserving reductions. Moreover, even though the existence of "intermediate" artificial problems, that is, problems for which lower bounds on their approximation are not obtainable by completeness results was proved in [7], a natural question arises: do natural intermediate problems exist? Observe that this question is also open in the field of decision problems even though the existence of artificial NP-intermediate problems has been already proved [19]. For example, it is known that the graph isomorphism problem cannot be NP-complete unless the polynomial-time hierarchy collapses [26], but no result has ever been obtained giving evidence that the problem does not belong to P.

The first goal of this paper is to define an approximation preserving reducibility that can be used for as many approximation classes as possible and such that all reductions appeared in the literature still hold. Notwithstanding the L-reducibility [25] has been the most widely used so far, we will give strong evidence that it cannot be used to obtain completeness results in "computationally defined" classes such as APX, log-APX (i.e., the class of problems approximable within a logarithmic factor), and poly-APX (i.e., the class of problems approximable within a polynomial factor). Indeed, on the one hand the L-reducibility is too weak and is not approximation preserving (unless P $=$ NP $\cap$ co-NP), on the other it is too strict and does not allow to reduce problems, which are known to be easy to approximate, to problems which are known to be hard to approximate. The weakness of the L-reducibility is, essentially, shared by all reducibilities appeared in the literature but the Strict reducibility [23] and the E-reducibility [16], while the strictness of the L-reducibility is shared by all of them (unless $P^{NP} = P^{NP[O(\log n)]}$) but the PTAS-reducibility [8]. The reducibility we propose is *a combination of the E-reducibility and of the PTAS-reducibility* and, as far as we know, it is the strictest reducibility that allows to obtain all approximation completeness results that have appeared in the literature, such as, for example, the APX-completeness of MAXIMUM SATISFIABILITY [8, 16] and the poly-APX-completeness of MAXIMUM CLIQUE [16].

The second group of results refers to the existence of natural complete problems for NPO. Indeed, both [23] and [7] provide examples of natural complete problems for the class of minimization and maximization NP problems, respectively. In Sect. 3 we will show *the existence of both maximization and minimization* NPO-*complete natural problems*. In particular, we prove that MAXIMUM 0 − 1 PROGRAMMING, MINIMUM 0 − 1 PROGRAMMING, and MINIMUM WEIGHTED INDEPENDENT DOMINATING SET are NPO-complete. This result shows that making use of a natural approximation preserving reducibility is powerful enough to encompass the "duality" problem raised in [17]. Moreover, the same result can also be obtained when restricting ourselves to the class NPO PB (i.e., the class of polynomially bounded NPO problems). In particular, we prove that MAXIMUM PB 0 − 1 PROGRAMMING, MINIMUM PB 0 − 1 PROGRAMMING and MINIMUM INDEPENDENT DOMINATING SET, are NPO PB-complete. Indeed, this result can also be obtained as a consequence of Theorem 6(a) of [16]. However, *our proof does not make use of the PCP model.*

The third group of results refers to the existence of natural APX-intermediate problems. In particular, in Sect. 4, we will prove that MINIMUM BIN PACKING (and other natural NPO problems) cannot be APX-complete unless the polynomial-time hierarchy collapses. Since it is well-known [22] that this problem belongs to APX and that it does not belong to PTAS (i.e., the class of NPO problems with polynomial-time approximation schemes) unless P=NP, our result thus yields *the first example of a natural* APX-*intermediate problem* (under a natural complexity-theoretic conjecture). Roughly speaking, the proof of our result is structured into two main steps. In the first step, we show that if MINIMUM BIN PACKING is APX-complete then the problem of answering any set of k non-adaptive queries to an NP-complete problem can be reduced to the problem of approximating an instance of MINIMUM BIN PACKING within a ratio depending on k. In the second step, we show that the problem of approximating an instance of MINIMUM BIN PACKING within a given ratio can be solved in polynomial-time by means of a constant number of non-adaptive queries to an NP-complete problem. These two steps will imply the collapse of the query hierarchy which in turn implies the collapse of the polynomial-time hierarchy. As a side effect of our proof, we will show that *if a problem is* APX-*complete, then it does not admit an asymptotic approximation scheme*: as far as we know, no general technique to obtain this kind of results was previously known.

In the last group of results, we state new connections between the approximability properties and the query complexity of NP-hard optimization problems. In several recent papers the notion of query complexity (that is, the number of queries to an NP oracle needed to solve a given problem) has been shown to be a very useful tool for understanding the complexity of approximation problems. In [6, 4] upper and lower bounds have been proved on the number of queries needed to approximate certain optimization problems (such as MAXIMUM SATISFIABILITY and MAXIMUM CLIQUE): these results dealt with the complexity of approximating the value of the optimum solution and not with the complexity of computing approximate solutions. In this paper, instead, the complexity of "constructive" approximation will be addressed by considering the languages that can be recognized by polynomial-time machines which have a function oracle that solves the approximation problem. In particular, we will be able to solve an open question of [4] proving that *finding the vertices of the largest clique is more difficult than merely finding the vertices of a 2-approximate clique* (that is, a clique with at least half the size of the largest clique) unless the polynomial-time hierarchy collapses. The results of [6, 4] show that the query complexity is a good measure of complexity to study approximability properties of optimization problems. Our results show that completeness in approximation classes implies lower bounds on the query complexity. In Sect. 5 we finally show that the two approaches are basically equivalent by giving *sufficient and necessary conditions for approximation completeness in terms of query-complexity hardness and combinatorial properties.* The importance of these results is twofold: they give new insights into the structure of complete problems in approximation classes and they reconcile the approach based on standard computation models with that based on the computation model for approximation proposed by [5]. As a final observation, our result can be seen as an extension of a result of [16] in which general sufficient conditions for APX-completeness are proved.

Due to the lack of space, the proofs of our results are not contained in this extended abstract. We now give some preliminary definitions and results.

A language L belongs to the *class* $P^{NP[f(n)]}$ if it is decidable by a polynomial-time oracle Turing machine which asks at most $f(n)$ queries to an NP-complete oracle on input of size n. The *class* QH is equal to the union $\bigcup_{k>1} P^{NP[k]}$.

Theorem 1 [27]. *For any function $f(n) \in O(\log n)$, if $P^{NP[f(n)+1]} \subseteq P^{NP[f(n)]}$ then the polynomial-time hierarchy collapses.*

The basic ingredients of an NP optimization problem are the set of *instances*, the set of *feasible solutions* associated to any instance, and the *measure* $m(x, y)$ defined for any instance x and for any feasible solution y. The problem is specified as a maximization problem or a minimization problem depending whether its goal is to find a solution whose measure is maximum or minimum (in the following *opt* will denote the function mapping an instance x to the measure of an optimum solution). The *class* NPO is the set of all NP optimization problems (for a formal definition of this class see [3]). Moreover, Max NPO is the set of maximization NPO problems and Min NPO is the set of minimization NPO problems.

An NPO problem is said to be *polynomially bounded* if a polynomial q exists such that, for any instance x and for any solution y of x, $m(x, y) \leq q(|x|)$. The *class* NPO PB is the set of all polynomially bounded NPO problems. NPO PB = Max PB $\cup$ Min PB where Max PB is the set of all maximization problems in NPO PB and Min PB is the set of all minimization problems in NPO PB.

Let A be an NPO problem. Given an instance x and a feasible solution y of x, we define the *performance ratio of y with respect to x* as

$$R(x,y) = \max\left\{\frac{m(x,y)}{opt(x)}, \frac{opt(x)}{m(x,y)}\right\}.$$

Let T be an algorithm that, for any instance x of A, returns a feasible solution $T(x)$. Given an arbitrary function $r : N \to (1,\infty)$, we say that T is an $r(n)$-*approximate algorithm for* A if, for any instance x, the performance ratio of the feasible solution $T(x)$ with respect to x verifies the following inequality:

$$R(x, T(x)) \leq r(|x|).$$

Given a class of functions F, an NPO problem A belongs to the *class* F-*APX* if an $r(n)$-approximate polynomial-time algorithm T for A exists, for some function $r \in F$. In particular, APX, log-APX, and poly-APX will denote the classes F-APX with F equal to the set of constant functions, to the set $O(\log n)$, and to the set of polynomials, respectively.

An NPO problem A belongs to the *class* PTAS if an algorithm T exists such that, for any fixed rational $r > 1$, $T(\cdot, r)$ is a polynomial-time r-approximate algorithm for A.

2   A new approximation preserving reducibility

We will justify our definition by emphasizing the disadvantages of previously known reducibilities. The first reducibility we shall consider is the L-reducibility (for *linear* reducibility) [25] which is often most practical to use in order to show that a problem is at least as hard to approximate as another. This reducibility preserves membership in PTAS. However, the next result gives a strong evidence that, in general, it is not approximation preserving. It also shows that the behavior of L-reductions depends on the type (that is, maximization or minimization) of the problems involved.

Theorem 2. *The following hold:*

1. *L-reductions from minimization problems to optimization problems are approximation preserving.*
2. *L-reductions from maximization problems to optimization problems are not approximation preserving if and only if the γ-reducibility is different from the many-one reducibility.*

The E-reducibility (for *error* reducibility) has been defined in [16] and imposes a linear relation between the performance ratios. For any function r, an E-reduction maps $r(n)$-approximate solutions into $(1+\alpha(r(n)-1))$-approximate solutions so that it not only preserves membership in PTAS but also membership in any F-APX class where F is closed with respect to linear applications, such as poly-APX, log-APX, and APX. As a consequence of this observation and of the results in [8], we have that NPO problems should exist which are L-reducible to each other but not E-reducible. However, the following result shows that within the class APX the E-reducibility is just a generalization of the L-reducibility.

Proposition 3. *For any two* NPO *problems A and B, if $A \leq_L B$ and $A \in$ APX, then $A \leq_E B$.*

The converse of the above result does not hold since no problem in NPO − NPO PB can be L-reduced to a problem in NPO PB while any problem in PO can be E-reduced to any NPO problem.

The main drawback of the E-reducibility consists of preserving optimum solutions. This is due to the fact that the linear relation between the performance ratios is too restrictive. The final step thus consists in letting the two functions mapping instances to instances and solutions to solutions, respectively, depend on the performance ratio.

Let A and B be two NPO problems. A is said to be *AP-reducible* to B, in symbols $A \leq_{AP} B$, if two functions f and g and a positive constant α exist such that:

1. For any $x \in I_A$ and for any $r > 1$, $f(x, r) \in I_B$.
2. For any $x \in I_A$, for any $r > 1$, and for any $y \in sol_B(f(x, r))$, $g(x, y, r) \in sol_A(x)$.
3. f and g are computable by two algorithms T_f and T_g, respectively, whose running time is polynomial for any fixed r.
4. For any $x \in I_A$, for any $r > 1$, and for any $y \in sol_B(f(x, r))$,

$$R_B(f(x, r), y) \leq r \text{ implies } R_A(x, g(x, y, r)) \leq 1 + \alpha(r - 1).$$

The AP-reducibility is a generalization of the E-reducibility. Moreover, it is easy to see that any PTAS problem is AP-reducible to any NPO problem. As far as we know, this reducibility is the strictest one appearing in the literature that allows to obtain natural APX-completeness results (for instance, the APX-completeness of MAXIMUM SATISFIABILITY [8, 16]).

3 NPO-complete problems

We will in this section prove that there are natural problems that are complete for the classes NPO and NPO PB. Previously, completeness results have been obtained just for Max NPO, Min NPO, Max PB, and Min PB [7, 23, 2, 13]. One example of such a result is the following theorem.

Theorem 4 [23, 7]. MINIMUM WEIGHTED SATISFIABILITY *is* Min NPO-*complete and* MAXIMUM WEIGHTED SATISFIABILITY *is* Max NPO-*complete*.

Using AP-reductions from maximization problems to minimization problems and vice versa, we can prove the following results showing that any problem that is Max NPO-complete or Min NPO-complete in fact is complete for the whole of NPO, and that a problem that is Max PB-complete or Min PB-complete is complete for the whole of NPO PB.

Theorem 5. MINIMUM WEIGHTED SATISFIABILITY *and* MAXIMUM WEIGHTED SATISFIABILITY *are* NPO-*complete.*

Corollary 6. *If* A *is a* Min NPO-*complete (respectively,* Max NPO-*complete) problem, then* A *is* NPO-*complete.*

We can also show that there are natural complete problems for the class of polynomially bounded NPO problems.

Theorem 7. MAXIMUM PB $0 - 1$ PROGRAMMING *and* MINIMUM PB $0 - 1$ PROGRAMMING *are* NPO PB-*complete.*

Corollary 8. *If* A *is a* Min PB-*complete (respectively,* Max PB-*complete) problem, then* A *is* NPO PB-*complete.*

By using similar techniques together with the result that LONGEST INDUCED PATH is not approximable within $|V|^{1-\varepsilon}$ for any $\varepsilon > 0$ unless P = NP [21], one can also show new hardness results for some NPO PB-complete problems [14].

4 Query complexity and APX-intermediate problems

Let A be an NPO problem and r be a function, then $A_{r(n)}$ is the following multi-valued partial function: given an instance x of A, $A_{r(n)}(x)$ is the set of feasible solutions y of x such that $R(x, y) \leq r(|x|)$.

Given an NPO problem A and a rational $r \geq 1$, a language L belongs to P^{A_r} if two polynomial-time computable functions f and g exist such that, for any x, $f(x)$ is an instance of A, and, for any $y \in A_r(f(x))$, $g(x, y) = 1$ if and only if $x \in L$. The *class* AQH(A) is equal to the union $\bigcup_{r>1} P^{A_r}$.

Using techniques similar to those of [4, 6], we can prove the following result.

Proposition 9. *For any problem* A *in* APX, AQH(A) $\subseteq$ QH.

Recall that an NPO problem admits an *asymptotic polynomial-time approximation scheme* if an algorithm T exists such that, for any x and for any $r > 1$, $R(x, T(x, r)) \leq r + k/opt(x)$ with k constant and the time complexity of $T(x, r)$ is polynomial with respect to $|x|$. The class of problems that admit an asymptotic polynomial-time approximation scheme is usually denoted as PTAS$^\infty$. The following result shows that, for this class, the previous fact can be strengthened.

Proposition 10. *For any* $A \in$ PTAS$^\infty$, AQH(A) $\subseteq$ P$^{\text{NP}[h]}$ *for some* h.

The following fact, instead, states that any language L in the query hierarchy can be decided using just one query to A_r where A is APX-complete and r depends on the level of the query hierarchy L belongs to.

Proposition 11. *For any* APX-*complete problem* A, QH $\subseteq$ AQH(A).

By combining Propositions 11 and 9, we thus have the following characterization of the approximation query hierarchy of the hardest problems in APX.

Theorem 12. *For any* APX-*complete problem* A, $\text{AQH}(A) = \text{QH}$.

Finally, as a consequence of this theorem, Proposition 10, Theorem 1, and the results of [10, 15, 9] we have the following.

Corollary 13. *If the polynomial-time hierarchy does not collapse, then* MINIMUM DEGREE SPANNING TREE, MINIMUM BIN PACKING, *and* MINIMUM EDGE COLORING *are* APX-*intermediate*.

By proving the analogue of Proposition 9 within NPO PB, using the fact that $\text{P}^{\text{NP}[\log n]}$ is contained in P^{MC_1} where MC stands for MAXIMUM CLIQUE [18], and from Theorem 1, it thus follows the next result that solves an open question posed in [4].

Theorem 14. *If* $\text{P}^{\text{MC}_1} \subseteq \text{P}^{\text{MC}_2}$ *then the polynomial-time hierarchy collapses.*

5 Query complexity and completeness in approximation classes

In the following $\text{NPF}^{\text{NP}[q(n)]}$ will denote the class of partial multi-valued functions computable by nondeterministic polynomial-time Turing machines which ask at most $q(n)$ queries to an NP oracle in the *entire* computation tree.

Let F and G be two partial multi-valued functions. We say that F many-one reduces to G (in symbols, $F \leq_{\text{mv}} G$) if two polynomial-time algorithms t_1 and t_2 exist such that, for any x in the domain of F, $t_1(x)$ is in the domain of G and, for any $y \in G(t_1(x))$, $t_2(x, y) \in F(x)$. We shall say that a function F is hard for $\text{NPF}^{\text{NP}[q(n)]}$ if, for any $G \in \text{NPF}^{\text{NP}[q(n)]}$, $G \leq_{\text{mv}} F$.

A problem A is *self-improvable* if two algorithms t_1 and t_2 exist such that, for any instance x of A and for any two rationals $r_1, r_2 > 1$, $x' = t_1(x, r_1, r_2)$ is an instance of A and, for any $y' \in A_{r_2}(x')$, $t_2(x, y', r_1, r_2) \in A_{r_1}(x)$. Moreover, for any fixed r_1 and r_2, the running time of t_1 and t_2 is polynomial. From [24] it follows that the equivalence with respect to the AP-reducibility preserves the self-improvability property. We are now ready to state the main result of this section.

Theorem 15. *A* poly-APX *problem* A *is* poly-APX-*complete if and only if it is self-improvable and* A_{r_0} *is* $\text{NPF}^{\text{NP}[\log \log n + O(1)]}$-*hard for some* $r_0 > 1$.

An NPO problem A is *linearly additive* if a constant β and two algorithms t_1 and t_2 exist such that, for any rational $\varepsilon > 0$ and for any sequence $x_1, \ldots, x_k$ of instances of A, $x' = t_1(x_1, \ldots, x_k, \varepsilon)$ is an instance of A and, for any $y' \in A_{1+\varepsilon\beta/k}(x')$, $t_2(x_1, \ldots, x_k, y', \varepsilon) = y_1, \ldots, y_k$ where each y_i is a $(1+\varepsilon)$-approximate solution of x_i. Moreover, the running time of t_1 and t_2 is polynomial for every fixed $\varepsilon > 0$.

Theorem 16. *An* APX *problem A is* APX-*complete if and only if it is linearly additive and a constant r_0 exists such that A_{r_0} is* $\text{NPF}^{\text{NP}[1]}$-*hard.*

It is also possible to establish query complexity results for log-APX-complete problems. In particular, even though we have not been able to establish a full characterization of log-APX-complete problems, we can prove the following result.

Theorem 17. *No* log-APX-*complete problem can be self-improvable unless the polynomial time-hierarchy collapses.*

It is then an interesting open question to find a characterizing combinatorial property for log-APX-complete problems. Moreover, as a consequence of the above theorem and of the results of [16], we conjecture that MINIMUM SET COVER is not self-improvable.

References

1. Arora, S., Lund, C., Motwani, R., Sudan, M., and Szegedy, M. (1992), "Proof verification and hardness of approximation problems", *Proc. of 33rd Ann. IEEE Symp. on Foundations of Comput. Sci.*, IEEE Computer Society, 14–23.
2. Berman, P., and Schnitger, G. (1992), "On the complexity of approximating the independent set problem", *Inform. and Comput.* **96**, 77–94.
3. Bovet, D.P., and Crescenzi, P. (1993), *Introduction to the theory of complexity.* Prentice Hall.
4. Chang, R. (1994), "On the query complexity of clique size and maximum satisfiability", *Proc. 9th Ann. Structure in Complexity Theory Conf.*, IEEE Computer Society, 31–42 (an extended version is available as Technical Report TR-CS-95-01, Department of Computer Science, University of Maryland Baltimore County, April 1995).
5. Chang, R. (1994), "A machine model for NP-approximation problems and the revenge of the Boolean hierarchy", *EATCS Bulletin* **54**, 166–182.
6. Chang, R., Gasarch, W.I., and Lund, C. (1994), "On bounded queries and approximation", Technical Report TR CS-94-05, Department of Computer Science, University of Maryland Baltimore County.
7. Crescenzi, P., and Panconesi, A. (1991), "Completeness in approximation classes", *Inform. and Comput.* **93**, 241–262.
8. Crescenzi, P., and Trevisan, L. (1994), "On approximation scheme preserving reducibility and its applications", *Proc. 14th FSTTCS*, Lecture Notes in Comput. Sci. 880, Springer-Verlag, 330–341.
9. Fürer, M., and Raghavachari, B. (1992), "Approximating the minimum degree spanning tree to within one from the optimal degree", *Proc. Third Ann. ACM-SIAM Symp. on Discrete Algorithms*, ACM-SIAM, 317–324.
10. Holyer, I. (1981), "The NP-completeness of edge-coloring", *SIAM J. Computing* **10**, 718–720.
11. Johnson, D.S. (1974), "Approximation algorithms for combinatorial problems", *J. Comput. System Sci.* **9**, 256–278.

12. Kann, V. (1992), *On the approximability of NP-complete optimization problems*, PhD thesis, Department of Numerical Analysis and Computing Science, Royal Institute of Technology, Stockholm.

13. Kann, V. (1994), "Polynomially bounded minimization problems that are hard to approximate", *Nordic J. Computing* 1, 317–331.

14. Kann, V. (1995), "Strong lower bounds of the approximability of some NPO PB-complete maximization problems", *Proc. MFCS*, to appear.

15. Karmarkar, N., and Karp, R. M. (1982), "An efficient approximation scheme for the one-dimensional bin packing problem", *Proc. of 23rd Ann. IEEE Symp. on Foundations of Comput. Sci.*, IEEE Computer Society, 312–320.

16. Khanna, S., Motwani, R., Sudan, M., and Vazirani, U. (1994), "On syntactic versus computational views of approximability", *Proc. of 35th Ann. IEEE Symp. on Foundations of Comput. Sci.*, IEEE Computer Society, 819–830.

17. Kolaitis, P. G., and Thakur, M. N. (1991), "Approximation properties of NP minimization classes", *Proc. Sixth Ann. Structure in Complexity Theory Conf.*, IEEE Computer Society, 353–366.

18. Krentel, M.W. (1988), "The complexity of optimization problems", *J. Comput. System Sci.* 36, 490–509.

19. Ladner, R.E. (1975), "On the structure of polynomial-time reducibility", *J. ACM* 22, 155–171.

20. Lund, C., and Yannakakis, M. (1994), "On the hardness of approximating minimization problems", *J. ACM* 41, 960–981.

21. Lund, C., and Yannakakis, M. (1993), "The approximation of maximum subgraph problems", *Proc. of 20th International Colloquium on Automata, Languages and Programming*, Lecture Notes in Comput. Sci. 700, Springer-Verlag, 40–51.

22. Motwani, R. (1992), "Lecture notes on approximation algorithms", Technical Report STAN-CS-92-1435, Department of Computer Science, Stanford University, 1992.

23. Orponen, P., and Mannila, H. (1987), "On approximation preserving reductions: Complete problems and robust measures", Technical Report C-1987-28, Department of Computer Science, University of Helsinki.

24. Panconesi, A., and Ranjan, D. (1993), "Quantifiers and approximation", *Theoretical Computer Science* 107, 145–163.

25. Papadimitriou, C. H., and Yannakakis, M. (1991), "Optimization, approximation, and complexity classes", *J. Comput. System Sci.* 43, 425–440.

26. Schöning, U. (1986), "Graph isomorphism is in the low hierarchy", *Proc. 4th Ann. Symp. on Theoretical Aspects of Comput. Sci.*, Lecture Notes in Comput. Sci. 247, Springer-Verlag, 114–124.

27. Wagner, K. (1988), "Bounded query computations", *Proc. 3rd Ann. Structure in Complexity Theory Conf.*, IEEE Computer Society, 260–277.

Improved Lower Bounds for the Randomized Boppana-Halldórsson Algorithm for MAXCLIQUE

Marcus Peinado

Department of Computer Science, Boston University,
Boston, MA 02215, USA, e-mail: mpe@cs.bu.edu

Abstract. It is shown that the randomized version of the MAXCLIQUE approximation algorithm by Boppana and Halldórsson analyzed in [5] does not come to within a factor of $n/e^{3\sqrt{\ln n}\ln\ln n}$ of the maximum clique. The lower bound derived in [5] was $\sqrt{n}$. Furthermore. we show that the randomized greedy algorithm for MAXCLIQUE does not come to within a factor of $n/\log^{5+\epsilon} n$ of the maximum clique. The lower bounds derived in this paper come close to the known upper bounds.

1 Introduction

The best known P-time approximation algorithm for the MAXCLIQUE problem was published by Boppana and Halldórsson [2] in 1990. Its *performance guarantee*, i.e. the worst-case ratio of the size of the largest clique in the input graph and the clique size found by the algorithm, is $O(n/\log^2 n)$. Although CLIQUE, being one of Karp's original NP-complete problems, has been studied extensively for a long time, no P-time approximation algorithm with a non-trivial performance guarantee had been published before that result. In this sense, the algorithm of Boppana and Halldórsson is unique and deserves special consideration.

Boppana and Halldórsson show that the performance guarantee of their algorithm is tight. The result applies to the deterministic version of the algorithm studied in [2]. It has been suggested that a randomized version of the algorithm might have improved performance. This idea has been investigated in [5], where a randomized version of the algorithm of Boppana and Halldórsson is studied. The main result is a lower bound on the performance guarantee of the randomized algorithm. It is shown that with high probability, the randomized algorithm will not come to within a factor of $\sqrt{n}$ of the largest clique if the input graph has certain properties. This bound has been criticized as even a P-time performance guarantee of $\sqrt{n}$ would be considered a breakthrough for clique approximation algorithms.

This paper improves the lower bound of [5] substantially. It is shown that the randomized version of the algorithm of Boppana and Halldórsson with high probability will not approximate MAXCLIQUE to within a factor of $n/e^{3\ln\ln n/\sqrt{\ln n}} > n^{1-\epsilon}$ for all $\epsilon > 0$. Furthermore, it is shown that the randomized greedy algorithm will not approximate MAXCLIQUE to within a factor of $\Omega(n/\log^{5+\epsilon} n)$ for all $\epsilon > 0$. The graphs used to prove these lower bounds are simpler and more natural than those used in [5] and similar to those used by Jerrum [4] to show a lower bound of $\Omega(\sqrt{n})$ for the Metropolis process for MAXCLIQUE. The lower bounds proved in this paper come very close to the known upper bounds.

The rest of the introduction explains the algorithm and some basic concepts in its analysis. Section 2 outlines the proof of the lower bound for the randomized greedy algorithm and introduces the reader to the basic strategy used in the proof of the main result (the lower bound for the randomized version of the algorithm of Boppana and Halldórsson) which is proved in Section 3. Due to space limitations, most of the proofs had to be omitted or shortened. The full version of this paper can be requested from the author.

1.1 Notation

Throughout this paper, let $\alpha > 1$ be a constant. We will say that an event E holds with high probability (w.h.p.) if $\mathbf{P}(E) = 1 - n^{-\omega(1)}$ where $n^{-\omega(1)}$ denotes a function which approaches zero faster than any polynomial. Given a graph $G = (V, E)$, $n = |V|$ denotes the number of its vertices. For $v \in V$ let the neighborhood of v in G be $\mathcal{N}_G(v) = \{u \in V | \{v, u\} \in E\}$ and let the non-neighborhood of v be $\bar{\mathcal{N}}_G(v) = \{u \in V | u \neq v, \{v, u\} \notin E\}$. If the graph is clear from the context, the subscript G may be dropped. The base 2 logarithm is denoted by $\log n$ and the natural logarithm is denoted by $\ln n$. Given a distribution $\mathcal{D}$, the expression '$X \in \mathcal{D}$' is used as a shorthand for 'generate X according to distribution $\mathcal{D}$'.

1.2 The Algorithm

The algorithm consists of a subgraph exclusion procedure and a recursive subprocedure (RAMSEY) which is motivated by Ramsey theory and which, given an input graph, returns a clique and an independent set. The subgraph exclusion procedure calls RAMSEY, stores the clique returned, and removes the independent set from the graph. This is repeated until the graph has become empty. The RAMSEY subprocedure is a generalization of the greedy method:

greedy(G):   IF G is empty THEN return $\emptyset$
                ELSE choose a vertex v and return $\{v\} \cup$ greedy($\mathcal{N}(v)$)

The selected vertex is called a **pivot vertex**. The vertex set returned is a clique because when a vertex is selected, its non-neighborhood is no longer considered.

RAMSEY improves and generalizes the greedy method by making an additional call to search the non-neighborhood of the pivot vertex. Thus, each recursive call has two cliques to choose from: the clique found in the neighborhood of the pivot together with the pivot and the clique found in the non-neighborhood. RAMSEY returns the larger one. Clearly, the same idea can be used to find an independent set by interchanging the terms neighborhood and non-neighborhood. RAMSEY returns both an independent set and a clique in the input graph.

RAMSEY$((V, E))$:
```
1    IF (V, E) is empty THEN return (∅, ∅)
2    ELSE choose a vertex v ∈ V
3        (C₁, I₁) := RAMSEY(𝒩(v))
4        (C₂, I₂) := RAMSEY(𝒩̄(v))
5        return (larger of (C₁ ∪ {v}, C₂), larger of(I₁, I₂ ∪ {v}))
```

It can be shown that the clique or the independent set returned by RAMSEY has a minimum size. The idea behind the subgraph exclusion algorithm is to modify the graph such that, eventually, there will be no large independent sets in the graph and the returned clique has to be large. This is done by repeatedly calling RAMSEY and excluding (removing) the returned independent sets:

```
IS Exclusion(G):    DO   (C, I) := RAMSEY(G)
                         G := G \ I
                    WHILE G ≠ ∅
                    return largest C
```

A clique in G can lose at most one vertex per iteration because a clique and an independent set can share at most one vertex. If the graph has a large enough clique, a constant fraction of the graph will be left even if all independent sets of a certain minimum size k are excluded.

An important concept in the analysis of RAMSEY which was used in [2] and [5] and which is fundamental to the analysis in this paper is the *computation tree* of recursive calls made by RAMSEY. Each node in the computation tree corresponds to a recursive call made by RAMSEY. If the input graph to the recursive call is empty, RAMSEY returns in line 1 and the corresponding node has no children. Otherwise, the node has two children corresponding to the two recursive calls of lines 3 and 4. We adopt the convention of identifying the recursive call of line 3 which searches the neighborhood with the left child and of identifying the call in line 4 (non-neighborhood) with the right child. We will ignore all nodes corresponding to recursive calls with an empty input graph. Each node can be labeled with the pivot vertex chosen in line 2 or with the input graph to the corresponding recursive call. We will use both kinds of labels to identify nodes.

A third piece of information we associate with each node is an abstraction of the path that leads to it. The input graph of a recursive call depends on the pivot vertices in the path from the root of the computation tree to the node representing the recursive call. However, the exact sequence of the vertices in the path is not important. If RAMSEY is run on input G, the vertex set of the input graph to a recursive call is

$$\mathcal{N}_{CD} = \bigcap_{v \in C} \mathcal{N}(v) \cap \bigcap_{v \in D} \bar{\mathcal{N}}(v), \tag{1}$$

where $C \subseteq V$ is the set of vertices in the path which are parents of left edges (neighborhood) and $D \subseteq V$ is the set of vertices in the path which are parents of right edges (non-neighborhood). If G is not clear from the context, we may write $\mathcal{N}_{CD}^G$ for $\mathcal{N}_{CD}$. Each node determines such a pair of sets (C, D), and given any node, we will often refer to its corresponding pair (C, D). The pairs (C, D) also determine the clique and independent set found by RAMSEY: C is a clique, D is an independent set and the clique (independent set) returned by RAMSEY is the largest C (D respectively) corresponding to any node in the computation tree. Furthermore, the size of the largest clique plus the size of the largest independent set in the input graph limits the maximum path length.

This paper analyzes a randomized version of the algorithm in which RAMSEY chooses the pivots at random, i.e. in each recursive call the pivot is chosen uniformly at random from the vertex set of the input graph to the recursive call. We call the randomized RAMSEY subprocedure R-RAMSEY. Furthermore, we allow polynomial amplification, i.e. we analyze a procedure PAR-RAMSEY, which calls R-RAMSEY $n^{O(1)}$ times and returns the largest clique and the largest independent set found in all runs. Finally, let PAR-IS-EXCLUSION denote the subgraph exclusion procedure which calls PAR-RAMSEY instead of RAMSEY. The main result of this paper can now be stated:

Theorem 1. *There are graphs on which with high probability* PAR-IS-EXCLUSION *does not come to within a factor of $\Omega(n/e^{3\sqrt{\ln n}\ln\ln n})$ of the maximum clique.*

Furthermore,

Theorem 2. *For every $\epsilon > 0$, there are graphs on which with high probability* PAR-RAMSEY *does not come to within a factor of $\Omega(n/\ln^{5+\epsilon} n)$ of the maximum clique.*

2 Hard Graphs for PAR-RAMSEY

The objective of this section is to show that the performance of PAR-RAMSEY is worse than $\Omega(n/\log^{5+\epsilon} n)$ (for any $\epsilon > 0$) on almost every graph generated according to the following distribution.

Definition 3. Given $n, l \in \mathbb{N}$ ($l < n$) and $p \in (0, 1)$, generate pairs (G, L) as follows, where $G = (V, E)$ is a graph. Let $V = \{1, \ldots, n\}$ and let $L \subseteq V$ be a randomly chosen subset of V of size $|L| = l$. Force L to be a clique by setting $\{u, v\} \in E$ for all $u, v \in L, u \neq v$. Determine all other edges of G by independent random coin flips with probability p. Let $\mathcal{G}(n, p, l)$ be the distribution defined by this procedure. Let $\mathcal{G}_n = \mathcal{G}(n, 1 - 1/\ln^\alpha n, \lfloor 5n/\ln^{\alpha+1} n \rfloor)$.

$\mathcal{G}_n$ is a distribution of pairs (G, L), where G is a randomly generated graph of size n with an embedded clique of size $l = \lfloor 5n/\ln^{\alpha+1} n \rfloor$. The intuition behind our proof strategy is as follows. The graph must contain a large clique – which is L. However, there are no large cliques outside L as the remaining parts of G are random and w.h.p. no clique or independent set will be larger than $4\log^{\alpha+1} n$ which is much smaller than L. Thus, it is sufficient to show that the algorithm will not find many vertices from L. Because R-RAMSEY chooses the pivots uniformly at random, the focus of the analysis is to show that in any recursive call, the fraction of vertices from L in the input graph is small. The first lemma shows that for $(G, L) \in \mathcal{G}_n$, the number Y_{CD} of $v \notin L$ is much larger than the number Z_{CD} of $v \in L$ for all potential input graphs to recursive calls with sufficiently large vertex set $\mathcal{N}_{CD}$. To be more precise, for $C, D \subseteq V$ let

$$Y_{CD} = |\mathcal{N}_{CD} \setminus L| \quad \text{and} \quad Z_{CD} = |\mathcal{N}_{CD} \cap L| \tag{2}$$

and

$$\mathcal{C} = \{(C, D) \subseteq V^2 : C \cap D = \emptyset \text{ and } lp^{|C|}(1-p)^{|D|} > (|C| + |D|) \ln n\} \tag{3}$$

The last condition in (3) limits the size of $C \cup D$ because Lemma 4 is not true for larger sets. As with $\mathcal{N}_{CD}$, we may write Y_{CD}^G and Z_{CD}^G for Y_{CD} and Z_{CD} if G is not clear from the context.

Lemma 4. *Let $(G, L) \in \mathcal{G}_n$ and let $g \in (0, 1/2)$ be a constant. Then w.h.p.*

$$\forall (C, D) \in \mathcal{C} : Y_{CD} > (1 - g)np^{|C|}(1-p)^{|D|} \quad \text{and} \tag{4}$$
$$\forall (C, D) \in \mathcal{C} : Z_{CD} < (1 + g)lp^{|C \setminus L|}(1-p)^{|D|} \tag{5}$$

Proof. (sketch): Use Chernoff bounds to estimate the probability that for any given (C, D), $Y_{CD} > (1 - g)np^{|C|}(1-p)^{|D|}$. $\square$

The next lemma puts these 'raw' random graph properties into a form which can be directly used in the following analysis of the algorithm itself.

Lemma 5. *With high probability, $(G, L) \in \mathcal{G}_n$ has the following property: for all $(C, D) \in V \times V$ such that $C \cap D = \emptyset$, C is a clique, D is an independent set, and $|L \cap C| < \ln^\alpha n$:*

$$Z_{CD} < 24 \ln^{\alpha+2} n \quad \text{or} \quad \frac{1}{3} Z_{CD} z / e^{2k/\ln^\alpha n} < Y_{CD} \tag{6}$$

where $k = |L \cap (C \cup D)|$

The lemma states that either there are negligibly few (less than $24 \ln^{\alpha+2} n$) vertices from L or there are many more vertices not from L than there are from L in the subgraph induced by $\mathcal{N}_{CD}$. Furthermore, k denotes the number of vertices from L in the path.

Lemma 6. *IF $(G, L) \in \mathcal{G}_n$ then*

$$\mathbf{P}(\text{PAR-RAMSEY}(G) \text{ finds more than } 25 \ln^{\alpha+2} n \text{ vertices } v \in L) < n^{-\omega(1)} \tag{7}$$

Theorem 2 follows readily from this.

3 Hard Graphs for PaR-IS-EXCLUSION

The analysis of PaR-RAMSEY has been simplified by making use of basic random graph properties. Lemma 4 relies on the independence of the random edges in the input graph. The situation becomes more complicated as subgraphs are excluded. Even if the edges in the input graph are independent, they might not be independent in the graph which is obtained after several iterations of the subgraph exclusion loop of PaR-IS-EXCLUSION.

In spite of this difficulty we show that graphs similar to those from $\mathcal{G}(n, p, l)$ retain most of the properties which were necessary to show the result of the previous section for a limited number of subgraph exclusions. We show that in each exclusion, the size of the largest clique in L is reduced by one. Thus, only the first l exclusions have to be considered since the graph contains no large clique after that. Our strategy is based on the observation that even for quite large $l = o(n)$, the total number of excluded vertices makes up only a tiny fraction of the original (random) graph. The main random graph property required in the previous sections was a bound on the sizes of the neighborhoods $\mathcal{N}_{CD}$. The hope is that the absence of a tiny fraction of the vertices of the original graph should not change the neighborhood sizes by too much. However, even though the number of excluded vertices is much smaller than n, it is larger than the neighborhood sizes in the recursive calls at a certain depth. We solve this problem by showing that, as recursive calls are made, not only the input graph is split into neighborhood and non-neighborhood of the pivot, but the set of excluded vertices is also. This is complicated by the fact that the set of excluded vertices cannot be assumed to be random. However, we show that *every* sufficiently large set of vertices is split in this way by all but very few pivots.

We begin by modifying the graphs slightly in order to insure that the size of the largest clique in L shrinks by one in each of the first $l/\ln n$ exclusions.

Definition 7. Given $n, l \in \mathbb{N}$ ($l < n$) and $p \in (0, 1)$, generate $(G_1 = (V_1, E_1), L_1)$ according to $\mathcal{G}(n - l + \lfloor l/\ln n \rfloor, p, \lfloor l/\ln n \rfloor)$. Let the graph G be the same graph as G_1 except that every vertex $v_i \in L_1$ of G_1 is replaced by an independent set L_i of size $\ln n$ in G. For any two such sets L_i and L_j and any $u \in L_i$ and $w \in L_j$, let $\{u, w\} \in E$ if and only if $i \neq j$. Furthermore, for each edge $\{v_i, u\}$ ($u \in V_1 \setminus L_1$) in G_1, each vertex in L_i is connected to u. Let $L = \bigcup_{i \leq \lfloor l/\ln n \rfloor} L_i$. Let $\mathcal{H}(n, p, l)$ be the distribution on (G, L) defined by this procedure and let $\mathcal{H}_n = \mathcal{H}(n, 1 - 1/\ln^\alpha n, n/e^{\sqrt{\ln n} \ln \ln n})$.

L induces a complete $l/\ln n$-partite subgraph. G is essentially a random graph of size n with built-in cliques of size $l/\ln n$. For the rest of this paper, we will be considering graphs generated according to $\mathcal{H}_n$. Furthermore, let $p = 1 - 1/\ln^\alpha n$ and $l = n/e^{\sqrt{\ln n} \ln \ln n}$. Even though $\mathcal{H}_n$ is formally a distribution on pairs (G, L) we may abuse notation and treat it as a distribution on graphs G.

Given G, let $G^j = (V^j, E^j)$ be what remains of G in the j-th iteration of the while-loop of PaR-IS-EXCLUSION(G), i.e. G^j is the input to PaR-RAMSEY after the first $j - 1$ independent sets found by PaR-Ramsey have been removed from G.

Lemma 8. *If $(G, L) \in \mathcal{H}_n$ then the largest clique in $L \cap V^j$ has at most $l/\ln n - j$ vertices for $i \in \{0, \dots, l/\ln n\}$ w.h.p. (over $\mathcal{H}$).*

Proof. Similar to the proof of Lemma 6 in [5]. $\qquad\qquad\qquad\qquad\qquad\qquad\qquad$ □

After $l/\ln n$ exclusions, L has been completely removed and no large clique can be found by any algorithm because there is none in the graph. Let B be the set of vertices from $V \setminus L$ which is removed in the first $l/\ln n$ exclusions, i.e. the union of the independent sets returned in the first $l/\ln n$ calls by PaR-Ramsey. With high probability (over $\mathcal{H}_n$), the largest independent set in the subgraph induced by $V \setminus L$ is smaller than $\ln n$. Hence, $|B| < l$. This is the only property of B used in the rest of this section. B is only a tiny fraction of the input graph. It remains to show that the same holds for the inputs $\mathcal{N}_{CD}$ to the recursive calls, i.e. we have to show that with high probability $|\mathcal{N}_{CD} \cap B| \ll |\mathcal{N}_{CD}|$. The following lemmas show that with high probability most pivots split $\mathcal{N}_{CD} \cap B$ between neighborhood and non-neighborhood by the same ratio as $\mathcal{N}_{CD}$.

Given a set V and $k \in \mathbb{N}$, let $V_k^V = \{S \subseteq V : |S| \geq k\}$ be the set of all subsets of V of size at least k. Furthermore, given a graph $G = (V, E)$, a vertex set $S \subseteq V$ and $v \in V \setminus S$, let $Y_v^S = |\mathcal{N}(v) \cap S|$. The next lemma says that with high probability over $\mathcal{H}_n$, all sufficiently large sets S of vertices are split by approximately the same ratio by all but a small number of pivots $A \subseteq V$.

Lemma 9. *Let $G = (V, E) \in \mathcal{H}_n$. Then for $s \geq 3\ln s/(g^2 p)$ and $k \geq 3\ln n/(pg^2 (1 - 1/\ln s))$ and $g \in (0, 1)$:*

$$\mathbf{P}(\exists S \in V_s^{V \setminus L}, A \in V_k^V, \forall v \in A : Y_v^S > (1 + g)|S|p) < n^2 e^{s \ln n + k \ln s - skpg^2/3} \quad (8)$$

The lemma leaves open the possible existence of small numbers of pivots which may split the sets S by different ratios. The next step is to bound the number of such 'bad' pivots on any path of the computation tree.

Bounding the number of 'bad' pivots on any path: Consider any path of recursive calls in the computation tree of the algorithm. Let v_i be the pivot vertex and (C_i, D_i) the pair of vertex sets corresponding to the i-th node in the path. Furthermore, let $G_i = (\mathcal{N}_{C_i D_i}, E_i)$ be input graph of the i-th node, let $B_{C_i D_i} = \mathcal{N}_{C_i D_i} \cap B$ be the set of B-vertices of G_i, and let

$$A_i^C = \{v \in \mathcal{N}_{C_i D_i} : |\mathcal{N}_{G_i}(v) \cap B_i| > (1 + g_C)|B_i|p\} \quad (9)$$

and

$$A_i^D = \{v \in \mathcal{N}_{C_i D_i} : |\bar{\mathcal{N}}_{G_i}(v) \cap B_i| > (1 + g_D)|B_i|(1 - p)\} \quad (10)$$

where $g_C = 1/\ln^\alpha n$ and $g_D = 1/\ln n$. A_i^C is the set of pivots at the i-th node which we consider 'bad' because their neighborhood contains too many vertices from B. Similarly, A_i^D are the 'bad' pivots whose nonneighborhood contains too many B-vertices. It is an immediate consequence of the previous lemma that w.h.p. (over the distribution $\mathcal{H}_n$):

$$|A_i^C| < 4\ln^{2(\alpha+1)+1} n \text{ as long as } |B_i| > 15\ln^{2(\alpha+1)} n \ln\ln n \text{ and} \tag{11}$$

$$|A_i^D| < 4\ln^{\alpha+3} n \text{ as long as } |B_i| > 15\ln^{\alpha+2} n \ln\ln n \tag{12}$$

In the following, assume that G has these properties.

Fact 1 *Let* $g \in (0,1)$ *be constant. If* $Y_{C_i,D_i} \geq e^{\sqrt{\ln n}+8\alpha\ln\ln n}$ *then w.h.p.*

$$Y_{C_i,D_i} \geq (1-g)p^{|C|}(1-p)^{|D|}n \tag{13}$$

Furthermore, $|C| \leq \ln^{\alpha+1} n$ *and* $|D| \leq \ln n$.

Proof. The proof is similar to the proof of Lemma 4. □

Consider node i and for $j < i$, let X_j be the indicator variable for the event $v_j \in (A_j^C \cap C_i) \cup (A_j^D \cap D_i)$. Then $\sum_{j=1}^{i-1} X_j$ is the number of 'bad' pivots in the path to node i which do not reduce the number of vertices from B sufficiently.

Lemma 10. *Let* k *be the number of 'bad' pivots on the path to node* i. *Then*

$$|B_{C_i,D_i}| \leq 2e^4 \frac{|B|}{n} Y_{C_i,D_i} \ln^{\alpha k} n \tag{14}$$

w.h.p. over $\mathcal{H}_n$ *for all nodes of the computation tree at which* $|T_i| \geq e^{\sqrt{\ln n}+8\alpha\ln\ln n}$

Proof. We use Fact 1 to bound Y_{C_i,D_i}. Let $\bar{C} = \{v_j \in A_j^C \cap C_i : 0 \leq j < i\}$ be the set of 'bad' pivots in the path to node i at positions where the path turns left (neighborhood). Similarly, let $\bar{D}$ be the set of 'bad' pivots at positions where the path turns right. Note that $k = |\bar{C}| + |\bar{D}|$. Then

$$\frac{|B_{C_i,D_i}|}{Y_{C_i,D_i}} \leq \frac{|B|((1+g^C)p)^{|C\setminus\bar{C}|}((1+g^D)(1-p))^{|D\setminus\bar{D}|}}{np^{|C|}(1-p)^{|D|}/2}$$

$$\leq 2\frac{|B|}{n}p^{-|\bar{C}|}(1-p)^{-|\bar{D}|}(1+1/\ln^{\alpha+1} n)^{|C|}(1+1/\ln n)^{|D|}$$

$$\leq 2e^2\frac{|B|}{n}e^{2|\bar{C}|/\ln^\alpha n}\ln^{\alpha|\bar{D}|}n(1+1/\ln n)^{|D|}$$

$$\leq 2e^4\frac{|B|}{n}\ln^{\alpha k} n$$

□

The next lemma bounds the number $\sum_{j=1}^{i-1} X_j$ of bad vertices in a path.

Lemma 11. *W.h.p. over the coin flips of the algorithm and* H_n

$$\sum_{j=1}^{i-1} X_j < \sqrt{\ln n}/(2\alpha) \tag{15}$$

for all nodes of the computation tree at which $Y_{C_i,D_i} \geq e^{\sqrt{\ln n}\ln\ln n+8\alpha\ln\ln n}$ *and* $|B_{C_i,D_i}| > 15\ln^{2(\alpha+1)} n \ln\ln n$.

Proof. We begin by estimating $\mathbf{P}(X_i = 1)$:

$$\mathbf{P}(X_i = 1 | \sum_{j=1}^{i-1} X_j = k) \leq q_k = \begin{cases} \frac{|A_i^C|}{Y_{C,D_i} - |B_{C,D_i}|} & \text{if } k \leq \sqrt{\ln n}/(2\alpha) \\ 1 & \text{otherwise} \end{cases}$$

The q_k are used to define a Markov chain $(U_i)_{i \in \mathbb{N}}$ on $\mathbb{N}$ with transition probabilities $p_{i\,i+1} = q$, $p_{i\,i} = 1 - q$, and $p_{i\,j} = 0$ for $i, j \in \mathbb{N}$ and $i \neq j \neq i + 1$. Let $U_0 = 0$. It can be shown by induction on i that for all $x \in \mathbb{R}$ and $i \in \mathbb{N}$ $\mathbf{P}(\sum_{j=1}^i X_j > x) \leq \mathbf{P}(U_i > x)$. Therefore, it is sufficient to bound $\mathbf{P}(U_Z \geq k)$ for $k = \sqrt{\ln n}/(2\alpha)$ and $Z = \ln^{\alpha+1} n$ (Fact 1). Let T_i ($i \in \mathbb{N}$) be the number of steps the Markov chain spends in state i provided that it reaches that state. Then, for sufficiently large n

$$\mathbf{P}(U_Z \geq k) = \prod_{i=1}^{k-1} \mathbf{P}(T_i < Z) = \prod_{i=1}^{k-1}(1 - p_{ii}^Z) \leq \prod_{i=1}^{k-1}\left(1 - \left(1 - \frac{|A_i^C|}{Y_{C,D_i} - |B_i|}\right)^Z\right)$$

$$\leq \left(1 - \exp\left(-e^{-\sqrt{\ln n}\ln\ln n}\right)\right)^k \leq \left(2e^{-\sqrt{\ln n}\ln\ln n}\right)^k \leq n^{-\omega(1)}.$$

The step from the first to the second line is based on (11) and (12) and uses the bound $x \geq \ln(1 + x)$ (for $|x| < 1$). The step from $1 - \exp\left(-e^{-\sqrt{\ln n}\ln\ln n}\right)$ to $2e^{-\sqrt{\ln n}\ln\ln n}$ follows by considering the limit $n \to \infty$ of the quotient of the two functions. $\square$

Lemma 12.
$$\mathbf{P}(|B_{C,D_i}|/Y_{C,D_i} = o(1)) \geq 1 - n^{-\omega(1)} \qquad (16)$$
for all nodes i of the computation tree at which $Y_{C,D_i} \geq e^{\sqrt{\ln n}\ln\ln n + 8\alpha\ln\ln n}$.

Proof. If $B_{C,D_i} \leq 15\ln^{2(\alpha+1)} n \ln\ln n$, the statement is trivially true. Otherwise, it is an immediate consequence of the previous two lemmas. $\square$

The result of this subsection is that as long as the input graphs to the recursive calls are not too small ($\geq e^{\sqrt{\ln n}\ln\ln n + 8\alpha\ln\ln n}$), w.h.p. only a negligible fraction of their vertices will have been excluded. This means that the strategy used to prove the lower bound for PAR-RAMSEY can also be used to prove a lower bound for PAR-IS-EXCLUSION.

3.1 Putting it all Together

Recall that G^j is the input graph to the j-th call to PAR-RAMSEY made by PAR-IS-EXCLUSION and that only $j \leq l/\ln n$ has to be considered.

Lemma 13. *Let $G \in \mathcal{H}_n$ be the input of PAR-IS-EXCLUSION. The following statement holds w.h.p. (over $\mathcal{H}_n$ and the coin flips of the algorithm): In the first $l/\ln n$ calls to PAR-RAMSEY(G^j), in all recursive calls for which $Y_{CD}^G \geq e^{\sqrt{\ln n}\ln\ln n + 8\alpha\ln\ln n}$:*

$$\frac{1}{3}Z_{CD}^{G^j} z e^{-2k/\ln^\alpha n} < Y_{CD}^{G^j} \qquad (17)$$

where $k = |C \cap L|$

Lemma 14. *W.h.p. (over the coin flips of the algorithm and over $\mathcal{H}_n$) the clique returned by* PAR-IS-EXCLUSION *is not larger than* $O(e^{\sqrt{\ln n}\,\ln\ln n + 8\alpha\ln\ln n})$ *if the input distribution is* $\mathcal{H}_n$.

Proof. It has to be shown that for $G \in \mathcal{H}_n$ w.h.p. all calls to PAR-RAMSEY made by PAR-IS-EXCLUSION(G) return only small cliques. As w.h.p. there are no big cliques outside L and, by Lemma 8, L is completely removed from the graph after the first $l/\ln n$ calls to PAR-RAMSEY, only these first $l/\ln n$ calls have to be considered. Let G^j be what remains of G after the first $j \leq l/\ln n$ exclusions and consider any path in the computation tree of PAR-RAMSEY(G^j). Let (C_i, D_i) be the pair of sets associated with the i-th node in the path and let the random variable S be the depth at which Z_{C,D_i} falls below $e^{\sqrt{\ln n}\,\ln\ln n + 8\alpha\ln\ln n}$. The path can contain at most $e^{\sqrt{\ln n}\,\ln\ln n + 8\alpha\ln\ln n}$ vertices from L after this point.

The main part of the proof is to show that with high probability the algorithm will not find more than $\ln^\alpha n$ vertices $v \in L$ before S. It follows from Lemma 13 that w.h.p.

$$\frac{1}{3} Z_{C_i,D_i}^{G^j} z e^{-2k_i/\ln^\alpha n} < Y_{C_i,D_i}^{G^j}, \quad \forall i < S \text{ such that } k_i < \ln^\alpha n \qquad (18)$$

Given (18), it can be shown that

$$\mathbf{P}(\sum_{i=1}^{S} X_i > \ln^\alpha n) \leq 2\,\mathbf{P}(\sum_{i=1}^{5\ln^{\alpha+1} n} X_i > \ln^\alpha n) < n^{-\omega(1)}. \qquad (19)$$

Thus, w.h.p. $\sum_{i=1}^{S} X_i \leq \ln^\alpha n$.

On any given path, the algorithm finds not more than $\ln^\alpha n$ vertices $v \in L$ before S and not more than $e^{\sqrt{\ln n}\,\ln\ln n + 8\alpha\ln\ln n}$ $v \in L$ after S. Considering all polynomially many paths, all $l/\ln n$ calls by PAR-IS-EXCLUSION and polynomial amplification can increase the probability of finding more $v \in L$ only by a polynomial factor. $\qquad\Box$

Theorem 1 is just a restatement of this lemma.

References

1. N. Alon, J. Spencer, and P. Erdös. *The Probabilistic Method.* Wiley, 1992.
2. R. Boppana and M. M. Halldórsson. Approximating maximum independent sets by excluding subgraphs. *BIT*, 32:180–196, 1992. (see also SWAT'90).
3. T. Hagerup and C. Rüb. A guided tour of Chernoff bounds. *Information Processing Letters*, 33:305 – 308, 1989.
4. M. Jerrum. Large cliques elude the metropolis process. *Random Structures and Algorithms*, 3(4):347–360, 1992.
5. M. Peinado. Hard graphs for the randomized Boppana-Halldórsson algorithm for maxclique. *Nordic Journal of Computing*, 1:493–515, 1994. (see also SWAT'94).

MNP: A Class of NP Optimization Problems *

(Extended Abstract)

Qi Cheng   Hong Zhu

Dept. of Computer Science, Fudan University,
Shanghai 200437, China.
E-mail:hzhu@solaris.fudan.edu.cn,

Abstract. We investigate a large class of NP optimization problems which we call **MNP**. We show that **Rmax(2)** [PR] are in our class and some problems which are not likely in **Rmax(2)** are in our class. We also define a new kind of reductions, WL-reductions, to preserve approximability and unapproximability, so it is more general version of L-reductions[PY] and A-reductions [PR]. Then we show some complete problems of this class under WL-reductions and prove that the max-clique problem is one of them. So all complete problems in this class are as difficult to approximate as the max-clique problem.

1 Introduction

For last two decades, many optimization problems were shown to be NP-equivalent [Ka, GJ]. Solving them precisely in polynomial time seems impossible. So finding their approximation algorithms has long been an attractive area in theorical computer science. Due to the recent development of the theory of interactive proof systems, probabilistically checkable proofs and their surprising connection to approximation, many open problems about the hardness of approximating optimatization problems were settled during the past three years.

Definition 1.1 *We say that function $f(x)$ approximates a maximum (or minimum) problem $g(x)$ within a factor $h(n)$ ($n = |x|$) if $1 \leq \frac{g(x)}{f(x)}$ (or $\frac{f(x)}{g(x)}$) < $h(n)$.*

Papadimitriou and Yannakakis[PY] initiated a classification of **NP** optimization problems based on their logic characterization. They defined the class **MAX NP** and **MAX SNP** as extensions of syntactic definition of **NP**[Fa].

Many **NP** problems on input structure G can be expressed as $\exists S \forall \bar{x} \exists \bar{y} \psi(x, y, G, S)$, where S is a structure and ψ is quantifier free. For such a problem $\Pi \in$ **NP**, they defined maxΠ (the maximization version of Π) as

$$\max_S |\{\bar{x} : \exists \bar{y} \psi(\bar{x}, \bar{y}, G, S)\}| \tag{1}$$

MAX NP is the class of these maximization problems. Problems such as **3SAT** can be expressed as $\exists S \forall x \psi(x, G, S)$. They form a subclass of **NP** denoted

---

* This work is supported under the National Science Fundation of China grant 69073303 and National High Technology Projection of China grant 863-306-05-03-04.

as SNP(*strict* NP). Similarly, **MAX SNP** is the class of optimization versions of such problems (no $\exists y$ quantifier).

Many results about the hardness of approximating **MAX SNP** problems have been obtained. [PY] showed that every problem in **MAX SNP** can be approximated within some constant factor, and [ALMSS] proved that for every **MAX SNP**-complete problem there is a constant $c > 0$ such that approximating it within the constant c means P=NP. These results almost fully characterize the approximability of **MAX SNP**.

The clique problem also caught a lot of attention in this area. It was shown by [FGLSS, AS, ALMSS, BGLR] that this problem is very hard to be approximated. This paper will show that this problem is complete in a kind of class.

Panconesi and Ranjan [PR] introduced a class **Rmax(2)** in which max-clique problem is a complete problem. A problem in **Rmax(2)** if its optimization function can be expressed as

$$opt_F(I) = max_S\{\| S \|: \forall y\Phi(\bar{y}, I, S)\}$$

where Φ is a quantifier-free CNF formula with all occurences of S in Φ being negative, S a single predicate appearing twice in each clause, and $\| S \|$ denotes $\| \{\bar{x} : S(\bar{x})\} \|$. But **Rmax(2)** relates loosely to **MAX NP**, for example, **MAX SAT** is not likely in **Rmax(2)**. In addition, the reductions they defined are relatively weak since recent result of unapproximability [FGLSS, AS, ALMSS, BGLR] are not preserved under it.

2 WL-reductions and their properties

Papadimitriou and Yannakakis defined a strict form of transformation in their paper which was called "L-reduction", Panconesi and Ranjan defined a stronger one. We give a more general version of it.

Definition 2.1 *Let Π and Π' be two maximum (minimum) problems, we say that Π WL-reduces to Π' if there are two polynomial-time algorithms f, g, for each instance I of Π, algorithm f produces an instance $I' = f(I)$ of Π', and for any feasible solution of I' with cost c', algorithm g produces a feasible solution of I with cost c, such that there is a constant $\alpha > 0$,*

$$\frac{OPT(I)}{c}(\text{or } \frac{c}{OPT(I)}) \leq \alpha \frac{OPT(I')}{c'}(\text{or } \frac{c'}{OPT(I')})$$

where $OPT(I)$ and $OPT(I')$ are the optima of I and I' respectively.

Proposition 1 WL-reductions compose.

Proposition 2 Let **MPro** be a class of optimization problems, Π be a complete problem under WL-reductions of this class, and $h(n), g(n)$ be two monotone increasing functions whose range domains are $\mathbf{R}^+$. If Π has a polynomial-time approximation algorithm within a factor $h(n)$(n is size of instance), then for every problem in **MPro** there are constants p and β such that it has one within factor $\beta h(n^p)$. If Π don't have a polynomial-time approximation algorithm within a factor $e(n)$, then for every **MPro**-complete problem there are constant q and β such that it can't have one within factor $\beta e(n^q)$

It is easy to see that proposition 2 is a property that L-reductions and A-reductions don't hold. But actually most reductions used in [PY] and [PR] are WL-reductions.

Definition 2.2 Π *WL-equals to* Π' *if* Π *can WL-reduce to* Π' *and* Π' *can WL-reduce to* Π.

3 Relationship between MAX NP and MAX Π_1

Since the predicate in (1) is fixed, independent of the input G, there are a polynomial number of possible values of x and y. If we change $\exists y$ to $\forall y$,

$$\max_S |\{\bar{x} : \forall \bar{y}\psi(\bar{x}, \bar{y}, G, S)\}| \qquad (2)$$

the problem is also **NP**-easy. This class of problems is called **MAX** Π_1[PR].

Next we will show **MAX NP** is in **MAX** Π_1.

Definition 3.1 *Given a set of boolean formulas each of which is a conjunction of literals,* **MSAT** *is to find the maximum formulas set which can be satisfied under the same assignment.*

In fact, MSAT is very similar to MAX SAT except that each of the formulas is a conjunction of literals while in MAX SAT it is a disjunction.

Definition 3.2 *For a logic formula A, we say a set of boolean formulas is its a.m.o.s.(at most one satisfied) set if*

(a) Every member of this set is a conjunction of literals.

(b) An assignment satisfies A if and only if it satisfies one member of the set.

(c) For any assignment at most one member of this set is satisfied.

The a.m.o.s. set of $x_1 \vee x_2 \vee \cdots x_n$ is

$$\{x_1, \neg x_1 \wedge x_2, \neg x_1 \wedge \neg x_2 \wedge x_3, \cdots, \neg x_1 \wedge \neg x_2 \wedge \cdots \neg x_{n-1} \wedge x_n\}$$

This indicates the following Lemma.

Lemma 3.1 MAX SAT can WL-reduce to MSAT.

Theorem 1 Every problem in **MAX NP** WL-equals to a problem in **MAX** Π_1.

Proof: For a problem Π in MAX NP,

$$\max_S |\{\bar{x} : \exists y \psi(\bar{x}, y, G, S)\}|$$

Let "$<$" be a arbitrary linear order on the set of all possible y (universe of G), consider following **MAX** Π_1 problem Π':

$$\max_S |\{< \bar{x}, y' >: \forall y((\neg(y < y') \vee \neg\psi(\bar{x}, y, G, S)) \wedge \psi(\bar{x}, y', G, S))\}|$$

The WL-equivalence of Π and Π' follows similarly to Lemma 1. $\qquad \square$

4 MNP and the Motivation of Introducing It

We define the subclass **MNP** of **MAX** Π_1.

Definition 4.1 *A problem belongs to* **MNP** *if it can be expressed as:*

$$\max_S|\{\bar{x} : \forall \bar{y}\psi(\bar{x}, \bar{y}, G, S)\}| \tag{3}$$

where ψ is a quantifier-free CNF formula, in each clause only one component of $\bar{y}$ occures and occures only once.

A typical problem of this class is MSAT. We can see that all problems in **MAX NP** can WL-reduce to a problem in **MNP**. Every problem in **Rmax(2)** is also in **MNP**. The verification of this assertion is left to our full paper.

In regard to language problems, **NPC** problems are all polynomial transformable. but as to optimization problems, NPC class has different complexity. The character may help us estalish approximability or unapproximability of some problems. More importantly, they may lead us to find the structure inside **NPC** class

5 The Completeness of MSAT

Theorem 2(Main) MSAT is MNP-complete.

Proof(skeleton) W.l.o.g, we consider a problem in MNP whose logic formula expression is the same as (3), but $\bar{y}$ has only one component and S have only one unary predicate:

$$\max_S|\{\bar{x} : \forall y \psi(\bar{x}, y, G, S)\}|$$

ψ is a boolean formula with leaves of the form $S(z)$ and $G(\bar{w})$, where $\bar{w}$ is the projections of $<\bar{x}, y>$ and z is one projection of it.

Every boolean formula can be converted to a CNF.Assume

$$\psi(\bar{x}, y, G, S) = \tau^1_{\bar{x},y} \wedge \tau^2_{\bar{x},y} \wedge \cdots \wedge \tau^p_{\bar{x},y}$$

where $\tau^i_{\bar{x},y}(1 \le i \le p)$ are disjunctions of some of $S(z), G(\bar{w})$. Assume that the universe of G is $\{1, 2, \cdots, n\}$,

$$\begin{aligned}
\forall y \psi(\bar{x}, y, G, S) &= \psi(\bar{x}, 1, G, S) \wedge \psi(\bar{x}, 2, G, S) \wedge \cdots \wedge \psi(\bar{x}, n, G, S) \\
&= \tau^1_{\bar{x},1} \wedge \cdots \wedge \tau^p_{\bar{x},1} \\
&\quad \wedge \tau^1_{\bar{x},2} \wedge \cdots \wedge \tau^p_{\bar{x},2} \\
&\quad \vdots \\
&\quad \wedge \tau^1_{\bar{x},n} \wedge \cdots \wedge \tau^p_{\bar{x},n} \\
&= \tau^1_{\bar{x},1} \wedge \cdots \wedge \tau^1_{\bar{x},n} \\
&\quad \wedge \tau^2_{\bar{x},1} \wedge \cdots \wedge \tau^2_{\bar{x},n} \\
&\quad \vdots \\
&\quad \wedge \tau^p_{\bar{x},1} \wedge \cdots \wedge \tau^p_{\bar{x},n}
\end{aligned}$$

Let $C_i = \tau_{\bar{x},1}^i \wedge \cdots \wedge \tau_{\bar{x},n}^i$. For each value of $\bar{x}$, $\tau_{\bar{x},j}^i$ is a disjunction of the leaves of the form $S(z)$—since G is given. The literals are the form of $S(z)$ or $\neg S(z)$. Ignoring these $\tau_{\bar{x},j}^i$ which are tautology, we have

$$C_i = \tau_{\bar{x},y_1}^i \wedge \cdots \wedge \tau_{\bar{x},y_m}^i$$

If $\tau_{\bar{x},y}^i$ has no leaf of $S(y)$, w.l.o.g, assume $\tau_{\bar{x},y}^i = S(x_1) \vee S(x_2) \vee \cdots \vee S(x_q)$, then $C_i = \tau_{\bar{x},y_1}^i = \cdots = \tau_{\bar{x},y_m}^i$. Its a.m.o.s. set is $S_{\bar{x}}^i = \{S(x_1), \neg S(x_1) \wedge S(x_2), \cdots, \neg S(x_1) \wedge \neg S(x_2) \wedge \cdots \wedge \neg S(x_{q-1}) \wedge S(x_q)\}$. If $\tau_{\bar{x},y}^i$ has a leaf of $S(y)$, w.l.o.g, assume $\tau_{\bar{x},y}^i = S(x_1) \vee S(x_2) \vee \cdots \vee S(x_q) \vee S(y)$, then

$$C_i = (S(x_1) \vee S(x_2) \vee \cdots \vee S(x_q) \vee S(y_1))$$
$$\wedge (S(x_1) \vee S(x_2) \vee \cdots \vee S(x_q) \vee S(y_2))$$
$$\vdots$$
$$\wedge (S(x_1) \vee S(x_2) \vee \cdots \vee S(x_q) \vee S(y_m))$$

Its a.m.o.s. set is $S_{\bar{x}}^i = \{S(x_1), \neg S(x_1) \wedge S(x_2), \cdots, \neg S(x_1) \wedge \neg S(x_2) \wedge \cdots \wedge \neg S(x_{q-1}) \wedge S(x_q), \neg S(x_1) \wedge \neg S(x_2) \wedge \cdots \wedge \neg S(x_{q-1}) \wedge \neg S(x_q) \wedge S(y_1) \wedge \cdots \wedge S(y_m)\}$.

For each value of $\bar{x}$, $\forall y \psi(\bar{x}, y, G, S)$'s a.m.o.s. set is

$$S_{\bar{x}} = \{h_1 \wedge h_2 \wedge \cdots \wedge h_p | h_1 \in S_{\bar{x}}^1, \cdots, h_p \in S_{\bar{x}}^p\}$$

So

$$\bigcup_{\bar{x}} S_{\bar{x}}$$

forms an instance of MSAT. it is easy to check that this transformation is a WL-reduction. $\square$

6 The Completeness of Max-clique Problem

Lemma 6.1 The max-clique problem is in MNP.
Proof The max-clique problem can be expressed as following:

$$\max{}_S |\{x : \forall y (x \in S \wedge (E(x,y) \vee y \notin S))\}|$$

where S is a set of vertices and $E(x,y)$ means that there is an edge between x and y. $\square$

The max-clique problem can easily WL-reduces to MSAT, for every vertex v_1, assume that $v_2, \cdots, v_i$ are all vertices not adjacent to v_1, the instance of MSAT contains a formula such that

$$S(v_1) \wedge \neg S(v_2) \cdots \wedge \neg S(v_i)$$

where $S(v)$ means $v \in S$. Therefore, if the same assignment satisfies $S(v_1) \wedge \neg S(v_2) \cdots \wedge \neg S(v_i)$ and $S(v_1') \wedge \neg S(v_2') \cdots \wedge \neg S(v_j')$, v_1 must be adjacent to v_1'.

Theorem 3 The max-clique problem is MNP-complete.

Proof We only need to prove that MSAT can WL-reduce to the max-clique problem. For an instance of MSAT, construct a graph: for every formula in the instance, there is a vertex in the graph, and there is an edge between the two vertices iff there is an assignment that can satisfy both the corresponding two formulas. It is very easy to check such a reduction is WL-reduction. □

It was shown by [FGLSS, AS, ALMSS, BGLR] that for some constant $c > 0$ approximating the size of max-clique within factor n^c implies **P=NP**. And the best known factor of approximating max-clique problem is $\frac{n}{\log^2 n}$ [BH].

Corollary 6.1 For any MNP-complete problem, there is an $\epsilon > 0$ such that to approximate it within a factor of n^ϵ means **P= NP**.

Corollary 6.2 If **P≠NP**, **MAX SAT** isn't MNP-complete.

7 Conclusion and Acknowledgement

We define a class of NP optimization problems **MNP** and show that there are a large number of NP problems in it, **Rmax(2)** [PR] is the subclass of **MNP**. We also introduce a new kind of reductions, WL-reductions, which has better property than L-reductions and A-reductions. Under WL-reductions, **MSAT** is a complete problem in **MNP**, and so is max-clique problem.

A lot of work need doing. For example, many people believe that the longest path in a graph is as hard to approximate as max-clique problem. Is this problem in **MNP**? If so, is it **MNP**-hard? In the definition of **MNP**, can the restriction on components be released?

Acknowledgement

We are very grateful to Mr. Y.-M. Zhao and Miss F. Fang for their helpful discussions.

References

[ALMSS] S. Arora, C. Lund, R. Motwani, M. Sudan and M. Szegedy, Proof Verification and Hardness of Approximation Problems. *Proc. 33 FOCS*, 1992.

[AS] S. Arora and S. Safra, Probabilistic Checking of Proofs: A New Characterization of **NP**. *Proc. 33 FOCS*, 1992.

[BGLR] M. Bellare, S. Goldwasser, C. Lund and A. Russell, Efficient Probabilistically Checkable Proofs and Applications to Approximation, *25th STOC*, 1993

[BH] R. B. Boppana and M. M. Halldórsson, Approximating maximum independent set by excluding subgraphs, In *Proc. of 2nd Scand. Workshop on Algorithm Theory, Springer-Verlag, Lecture Notes in Computer Science 447*, 1990.

[Fa] R. Fagin, Generalized first-order spectra and polynomial-time recognizable sets, in "Complexity of Computations", AMS, 1974.

[FGLSS] U. Feige, S. Goldwasser, L. Lovász, S. Safra and M. Szegezy, Approximating clique is almost **NP**-hard. *Proc. 32nd FOCS*, 1991.

[GJ] M. R. Garey and D. S. Johnson, *Computers and Intractability*, W. H. Freeman and Company,1979.

[Ka]   R. M. Karp, Reducibility among combinatorial problems, In R. E. Miller and J. W. Thatcher, editors,*Complexity of Computer Computations*, Plenum Press, 1972.

[PR]   A. Panconesi and D. Ranjan, Quantifiers and Approximation, *22nd STOC*, 446-456, 1990.

[PY]   C. H. Papadimitriou and M. Yannakakis. Optimization, Approximation, and Complexity Classes. *Journal of Computer and System Sciences, vol. 43, 1991*

Semidefinite Programming and its Applications to NP Problems

Roman Bačík[1] and Sanjeev Mahajan[2]

[1] School of Computing Science, Simon Fraser University, Canada
[2] Max Planck Institut für Informatik, Im Stadtwald, Saarbrücken, Germany

Abstract. The graph homomorphism problem is a canonical NP-complete problem in a sense that it generalizes various other well-studied problems such as graph coloring and finding cliques. To get a better insight into a combinatorial problem, one often studies relaxations of the problem. We define *fractional homomorphisms* and *pseudo-homomorphisms* as natural relaxations of graph homomorphisms. In their paper [4], Feige and Lovász defined a semidefinite relaxation of the homomorphism problem, which allowed them to obtain polynomial time algorithms for certain special cases of the problem. Their relaxation is defined in terms of the solution to a semidefinite program. Hence a characterization of their relaxation in terms of known combinatorial notions is desirable. We show that our pseudo-homomorphism is equivalent to the relaxation defined by Feige and Lovász [4]. Although general graph homomorphism does not admit a simple forbidden subgraph characterization, surprisingly we can show that there is a simple forbidden subgraph characterization of the fractional homomorphism (the forbidden subgraph is a clique in this case). As a byproduct, we obtain a simpler proof of the NP hardness of the fractional chromatic number, first proved by Grötschel, Lovász and Schrijver using the ellipsoid method [6] Finally, we briefly discuss how to apply these techniques to general NP problems and describe a unified setting in which a wide variety of seemingly disparate polynomial time problems can be decided.

1 Introduction

A *homomorphism* between two graphs G and H is a function from the vertex set of G to the vertex set of H, which maps adjacent vertices to adjacent vertices. We say that G is *homomorphic* to H if such a homomorphism exists, and in this case we will write $G \to H$. Determining whether $G \to H$ for input graphs G and H is known to be NP-complete (cf. [10]). This problem is a generalization of a large number of well-studied problems. For example, determining if a graph G is k-colorable is equivalent to setting H to be a complete graph on k vertices and determining if $G \to H$. Determining if a graph H has a k-clique is equivalent to setting G to a complete graph on k vertices and determining if $G \to H$. Therefore it is desirable to identify families of instances for which a polynomial algorithm exists. Recent research in semidefinite optimization shows that it is a powerful tool towards obtaining approximation algorithms for NP hard problems (see

[5, 12]). We use semidefinite programming to study problems in NP such as the graph homomorphism problem. We define the notions *pseudo-homomorphism* and *fractional homomorphism* which are, respectively, semidefinite and linear relaxations of a certain integer program corresponding to the graph homomorphism problem. In fact, the pseudo-homomorphism is defined in terms of the classical ϑ number of [14]. As semidefinite programs can be solved in polynomial time, pseudo-homomorphism is a polynomial time notion. It turns out, as we will show in a later section, that our notion of pseudo-homomorphism coincides with another semidefinite relaxation defined by Feige and Lovász in [4]. Feige and Lovász used this relaxation to show polynomial time solvability of various special cases of graph homomorphism.

Although the general graph homomorphism does not admit of a simple forbidden subgraph characterization, we show a surprisingly simple such characterization for the fractional homomorphism which also yields several other interesting consequences. First, we obtain a much simpler proof of the NP-hardness of fractional clique number, a result which was first proven by [6] using the ellipsoid algorithm. A longstanding conjecture in graph theory is the graph product conjecture [7]. Using our characterization of the fractional homomorphism, we show that the above conjecture is true if we replace homomorphism by fractional homomorphism.

The notions of pseudo-homomorphism and fractional homomorphism can be generalized to all sets in NP. Observing that the natural generalization of pseudo-homomorphism to other NP sets can be computed in polynomial time, we obtain polynomial time algorithms for certain NP problems when the instances are drawn from certain families.

1.1 Definitions

Let G be a finite undirected graph without loops. The set of the vertices, and edges of the graph G is denoted by $V(G)$, and $E(G)$, respectively. For $(u, v) \in E(G)$, we also use $u \sim v$ and say that they are *adjacent*. We say that two vertices of G are *incident* if they are either adjacent or are equal. The *complementary* graph $\overline{G}$ of the graph G be the graph with $V(\overline{G}) = V(G)$ and $(u, v) \in E(\overline{G})$ iff $(u, v) \notin E(G)$. All vectors will be row vectors.

Definition 1. Let G and H be two graphs. The *hom-product* $G \circ H$ is defined as the graph with $V(G \circ H) = V(G) \times V(H)$, in which $((s, u), (t, w)) \in E(G \circ H)$ iff $(s \neq t)$ and $((s, t) \in E(G)$ implies $(u, w) \in E(H))$.

We first formulate the clique number, $\omega(G)$, and the chromatic number, $\chi(G)$, of a graph G, as integer linear programs. In these formulations we take v to range over the vertex set of G and I to range over all maximal independent sets in G.

$$\omega(G) = \max \sum_v x_v \qquad \chi(G) = \min \sum_I y_I$$
$$\sum_{v \in I} x_v \leq 1 \text{ for each } I \qquad \sum_{I \ni v} y_I \geq 1 \text{ for each } v$$
$$x_v \in \{0, 1\} \qquad y_I \in \{0, 1\}$$

If we replace the integrality conditions $x_v \in \{0, 1\}$, $y_{\mathcal{I}} \in \{0, 1\}$ by $x_v \geq 0$, $y_{\mathcal{I}} \geq 0$ respectively, we obtain linear programs with optima $\omega_f(G)$, the fractional clique number of G, and $\chi_f(G)$, the fractional chromatic number of G. Note that these linear programs are dual of each other and hence $\omega_f(G) = \chi_f(G)$ by the duality theorem of linear programming (see [2]). Therefore we have the inequalities:

$$\omega(G) \leq \omega_f(G) = \chi_f(G) \leq \chi(G).$$

To compute any of these numbers is NP hard. The NP hardness of the fractional chromatic number was proved by Grötschel, Lovász and Schrijver in [6]. They also proved that one can compute in polynomial time a real number ϑ (see also [13]) which satisfies

$$\omega(G) \leq \vartheta \leq \omega_f(G).$$

One possible definition for ϑ is as follows [16]:

Let $\mathcal{S}$ be the set of all $|V(G) \times V(G)|$ positive semidefinite matrices, let I denote the unit matrix and J denote the matrix of all 1's. The $*$ product of two matrices $A = (a_{ij})$ and $B = (b_{ij})$ is the number $A * B = \sum_{i,j} a_{ij} b_{ij}$. Then

$$\vartheta = \max\{B * J | B \in \mathcal{S}; B * I = 1; \forall (s, t) \notin E(G) : b_{st} = 0\}.$$

Another semidefinite programming upper bound for ω is the real number $\vartheta_{1/2}$ of Schrijver [15], which can be expressed as:

$$\vartheta_{1/2} = \max\{B * J | B \in \mathcal{S}; B * I = 1; \forall (s, t) \notin E(G) : b_{st} = 0; \forall s, t \in V(G) : b_{st} \geq 0\}.$$

We need a notation that explicitly expresses the dependence of $\vartheta, \vartheta_{1/2}$ on the graph G. For historical reasons [16, 14, 13] we write $\vartheta = \vartheta(\overline{G})$ and $\vartheta_{1/2} = \vartheta_{1/2}(\overline{G})$.

2 Overview

In section 4, we define the notions of fractional homomorphism and pseudo-homomorphism, and establish certain relationships between them. In section 5 we give a forbidden subgraph characterization of the fractional homomorphism. This characterization yields a much simpler proof of the NP-hardness of fractional clique number, which was first proved using the ellipsoid method [6]. Using this characterization, we also show a fractional version of the well known graph product conjecture [7]. In section 6 we show that the notion of pseudo-homomorphism is equivalent to the notion of hoax as defined in [4] and give a necessary condition for existence of a pseudo-homomorphism. In section 7, we discuss the applications of this theory to other problems in NP. Some proofs are omitted for insufficient space and can be found in [1].

3 Relaxations of Homomorphisms

Theorem 2. *For any graphs G, H, the following inequalities hold:*

$$\omega(G \circ H) \leq \vartheta_{1/2}(\overline{G \circ H}) \leq \vartheta(\overline{G \circ H}) \leq \omega_f(G \circ H) = \chi_f(G \circ H) \leq \chi(G \circ H) \leq |V(G)|,$$

and $\omega(G \circ H) = |V(G)|$ iff there is a homomorphism from G to H.

Proof. Omitted. ☐

The previous theorem suggests the following relaxations of homomorphisms:
Let G, H be two graphs.

Definition 3. We say that G is *fractionally homomorphic* to H
(denote by $G \to_f H$) if $\omega_f(G \circ H) = |V(G)|$.

Definition 4. We say that G is *pseudo-homomorphic* to H
(denote by $G \to_p H$) if $\vartheta(\overline{G \circ H}) = |V(G)|$.

Definition 5. We say that G is *pseudo$_{1/2}$-homomorphic* to H
(denote by $G \to_{p/2} H$) if $\vartheta_{1/2}(\overline{G \circ H}) = |V(G)|$.

Corollary 6. $(G \to H) \Rightarrow (G \to_{p/2} H) \Rightarrow (G \to_p H) \Rightarrow (G \to_f H)$.

Later we will show that we cannot reverse the first and the last implications in
general. This is not surprising because $G \to_p H$, and $G \to_{p/2} H$ are polynomial
while both $G \to H$ and $G \to_f H$ are NP hard. However for certain families
of graphs we can prove the reverse implications, and so obtain a polynomial
algorithm for these families.

4 Fractional Homomorphisms

In this section we will give a forbidden subgraph characterization of fractional
homomorphism, which will enable us to prove a fractional version of the graph
product conjecture [7]. This also gives an alternative proof of the NP hardness
of computing the fractional clique number of a graph.

Theorem 7. $G \to_f H$ iff $\omega(G) \leq \omega_f(H)$.

Proof. Omitted. ☐

This theorem shows that although the graph homomorphism problem is NP-
complete, for any fixed non-bipartite graph H [10], the fractional graph homomorphism problem is polynomial, for every fixed graph H. The theorem also
shows that if H is part of the input, then fractional homomorphism is co-NP-
complete. Another interesting consequence of this theorem is that for a given k
and a graph G, it is NP-hard to determine if G has fractional chromatic number
at least k (or equal to k). This result was originally proved in [6] by the ellipsoid
method.

Corollary 8. *The following problem is NP hard:*
Instance:A graph G and a number n
Question:Is $\omega_f(G)$ at least (equal to) n?

Proof. We give a reduction from the clique problem, i.e., given a graph G and an integer n, it is known to be NP complete to decide whether $\omega(G) > n$. But $n = \omega_f(K_n)$ and therefore by our theorem this is equivalent to the question whether $G \not\to_f K_n$ which is equivalent to $\omega_f(G \circ K_n) < |V(G)|$. ☐

A longstanding conjecture in graph theory states that whenever $G \not\to K_n$ and $H \not\to K_n$, then also $G \times H \not\to K_n$, where $G \times H$ is the categorical product of G and H, defined by $V(G \times H) = V(G) \times V(H)$ and $((s,u),(t,w)) \in E(G \times H)$ iff $(s,t) \in E(G) \wedge (u,w) \in E(H)$ c.f. [7].

Corollary 9. $G \not\to_f W$ and $H \not\to_f W$ imply $G \times H \not\to_f W$

Proof. By our theorem $G \not\to_f W$ and $H \not\to_f W$ imply $\omega(G) > \omega_f(W)$ and $\omega(H) > \omega_f(W)$. One can easily observe that $\omega(G \times H) = \min\{\omega(G), \omega(H)\}$ and therefore also $\omega(G \times H) > \omega_f(W)$ which means $G \times H \not\to_f W$. ☐

The fractional version of the product conjecture follows by taking $W = K_n$.

5 Pseudo-homomorphisms

In this section we will relate pseudo-homomorphism to the concept of a hoax introduced by Feige and Lovász [4].First we describe the notion of a hoax, as defined by Feige and Lovász. It is based on the following two-prover interactive proof system:

A verifier is trying to determine if a graph G is homomorphic to a graph H. The verifier chooses randomly and independently vertices s and t from G and sends them, respectively to the provers P_1 and P_2. P_1 replies with a vertex u from H as the claimed image of s, and P_2 with a vertex w from H as the claimed image of t. The verifier accepts just in the following situations:
1) If $s = t$ then $u = w$
2) If s is adjacent to t in G, then u is adjacent to w in H.
For each input (G, H), consider the 0-1 matrix V (with rows and columns labeled by su where the s's run through the vertices of G and the u's runs through the vertices of H) where $V_{su,tw} = 1$ iff the verifier accepts the answer u and w to the requests s and t respectively. Note that V is the incidence matrix of $G \circ H$.

Let p_{su} be the probability that P_1 or P_2 (we can assume that P_1 and P_2 use the same strategies as the game is symmetric) answers u on s. The provers P_1 and P_2 want to maximize the probability that the verifier accepts. This probability is given by $|V(G)|^{-2} \sum_{su,tw} V_{su,tw} p_{su} p_{tw}$ subject to the condition that for all s : $\sum_u p_{su} = 1$, and for all s, u: $p_{su} \geq 0$. The optimum of this quadratic program is 1 iff G is homomorphic to H because only in this case is the probability equal to 1. By rewriting $Q_{su,tw} = p_{su} p_{tw}$, and $C = |V(G)|^{-2} V$, we can convert this quadratic program to the following form:

maximize $\sum_{s,u,t,w} C_{su,tw} Q_{su,tw}$

s.t.

Q is a rank 1 matrix

Q is symmetric

$\forall s, t : \sum_{u,w} Q_{su,tw} = 1$

$\forall s, t, u, w : Q_{su,tw} \geq 0.$

To make the above program convex, Feige and Lovász replace the rank 1 constraint with the condition 'Q is positive semidefinite'. Let us denote this modified program by $(*)_{G,H}$. The ellipsoid algorithm can be used to solve the new program c.f. [4].

The optimal solution of the modified program with objective value 1 is called a *hoax* (c.f. [4]). If we replace both the rank 1 constraint and the nonnegativity constraint with the condition 'Q is positive semidefinite', we obtain a modified program $(**)_{G,H}$; any optimum solution with objective value 1 of this modified program is called a *semi-hoax*. Feige and Lovász [4] proved the following necessary and sufficient conditions for a given instance to admit a hoax.

Lemma 10. *[4] The system $(*)_{G,H}$ has a hoax iff there exists a system of vectors $v_{su}, s \in V(G), u \in V(H)$ satisfying the following conditions:*

$$\tilde{v} = \sum_u v_{su} \tag{1}$$

is independent of s,

$$\tilde{v} v_{su}^T = |v_{su}|^2 \tag{2}$$

$$|\tilde{v}| = 1 \tag{3}$$

$$\forall s, u, t, w : v_{su} v_{tw}^T \geq 0 \tag{4}$$

and

$$v_{su} v_{tw}^T = 0 \text{ whenever } V_{su,tw} = 0. \tag{5}$$

One can similarly prove the following necessary and sufficient conditions for a semi-hoax.

Lemma 11. *The system $(**)_{G,H}$ has a semi-hoax iff there exists a system of vectors $v_{su}, s \in V(G), u \in V(H)$ satisfying the conditions (1),(2),(3), and (5).*

We now prove that our notions of pseudo$_{1/2}$-homomorphism and pseudo-homomorphism coincide with the notions of hoax and semi-hoax respectively.

Theorem 12. *The system $(*)_{G,H}$ has a hoax iff $G \to_{p/2} H$.*

Proof. Omitted. □

We can similarly prove the following.

Theorem 13. *The system $(**)_{G,H}$ has a semi-hoax iff $G \to_p H$.*

Although we do not have a complete characterization of pseudo-homomorphisms, we can prove a necessary condition for existence of a pseudo-homomorphism which will be enough for our applications. First let us recall some definitions and results from [14].

If G and H are two graphs, then their *strong product* $G \cdot H$ is defined as the graph with $V(G \cdot H) = V(G) \times V(H)$, in which (s, u) is incident with (t, w) iff s is incident to t in G and u is incident to w in H.

The *tensor product* of two vectors $u = (u_1, \ldots, u_n)$ and $w = (w_1, \ldots, w_m)$ is the vector

$$u \otimes w = (u_1 w_1, \ldots, u_1 w_m, u_2 w_1, \ldots, u_n w_m)$$

of length nm. One can easily verify that

$$(s \otimes t)(u \otimes w)^T = (su^T)(tw^T). \tag{6}$$

Let G be a graph with $V(G) = \{1, \ldots, n\}$. An *orthonormal representation* of G is a system $\{v_1, \ldots, v_n\}$ of unit vectors in a vector space $\mathcal{R}^r$ such that if $(i \neq j) \wedge (i, j) \notin E(G)$ then $v_i v_j^T = 0$. Every graph has an orthonormal representation, for instance an orthonormal basis of $\mathcal{R}^n$.

Lemma 14. *[14] Let $\{u_1, \ldots, u_n\}$ and $\{v_1, \ldots, v_m\}$ be orthonormal representations of G and H respectively. Then the vectors $u_i \otimes v_j$ form an orthonormal representation of $G \cdot H$.*

Lemma 15. *Let $\{u_1, \ldots, u_n\}$ and $\{v_1, \ldots, v_m\}$ be orthonormal representations of G and $\overline{H}$ respectively. Then the vectors $u_i \otimes v_j$ form an orthonormal representation of $\overline{G \circ H}$.*

Proof. The proof follows from the previous Lemma 14 and the fact that $G \cdot \overline{H} \subseteq \overline{G \circ H}$. □

Theorem 16. *[14] Let $(u_1, \ldots, u_n)$ range over all orthonormal representations of G and c over all unit vectors. Then*

$$\vartheta(G) = \min \max_{1 \leq i \leq n} \frac{1}{(c u_i^T)^2}.$$

Theorem 17.

$$\vartheta(\overline{G \circ H}) \leq \vartheta(G) \vartheta(\overline{H}).$$

Proof. Omitted. □

Corollary 18. *If $G \to_p H$ then $\vartheta(G) \vartheta(\overline{H}) \geq |V(G)|$.*

In the last part of this section we prove that in general we cannot reverse implications in the Corollary 6. We need the following results (see [14, 13]): for odd n

$$\vartheta(C_n) = \frac{n \cos(\pi/n)}{1 + \cos(\pi/n)},$$

$$\vartheta(\overline{C_n}) = \frac{1 + \cos(\pi/n)}{\cos(\pi/n)},$$

and for every n

$$\vartheta(K_n) = 1,$$

$$\vartheta(\overline{K_n}) = n.$$

Theorem 19. *For odd n, m such that $3 \leq n < m$, $C_m \not\rightarrow_p K_2$ and $C_n \not\rightarrow_p C_m$.*

Proof. Omitted. □

As $C_5 \rightarrow_f K_2$, and the above theorem shows that $C_5 \not\rightarrow_p K_2$, pseudo-homomorphsim is strictly stronger than fractional homomorphism.

The next Theorem shows that homomorphism is strictly stronger than pseudo$_{1/2}$-homomorphism. Let G, H be the graphs from the next figure.

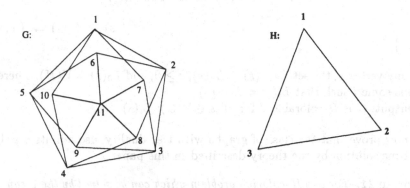

Fig. 1. $G \not\rightarrow H$, $G \rightarrow_{p/2} H$.

Theorem 20. $G \not\rightarrow H$ but $G \rightarrow_{p/2} H$.

Proof. Omitted. □

6 Discussion

As alluded to earlier, the theory presented here applies to any NP problem, although in this paper we have concentrated on graph homomorphism. We now briefly describe how the theory applies to other problems. The *hoax set* is a set of instances for which there is a hoax according to [4].

Hell, Nešetřil, and Zhu found polynomial algorithms for classes of graphs which have a property referred to as *tree-duality* in [8, 11, 9]. They showed that the following labeling algorithm works for H-coloring (the problem whether $G \to H$ for a fixed graph H) for H with tree-duality. We will later refer to this algorithm as the 1-*consistency test* (cf. [8]).

> Instance: A digraph G;
> Question: Is $G \to H$?
> Define labels $L_k : V(G) \to 2^{V(H)}, k \geq 0$ by induction as follows:
> $L_0(s) = V(H)$ for all $s \in V(G)$
> $L_{k+1}(s) = \{u \in L_k(s) \mid \forall t \in V(G) \, \exists w \in L_k(t) : ((s,t) \in E(G) \Rightarrow (u,w) \in E(H)) \wedge$
>
> $\qquad\qquad\qquad\qquad\qquad\qquad\qquad\qquad\qquad\qquad \wedge ((t,s) \in E(G) \Rightarrow (w,u) \in$
>
> $E(H))\}$

For any vertex s, the set $L_{k+1}(s) \subseteq L_k(s), k \geq 0$, and $L_0(s) = V(H)$. Therefore there is some i such that $L_i(s) = L_{i+1}(s)$.

> Output: G is H-colorable if for all $s \in V(G)$: $L_i(s) \neq \emptyset$.

We now prove that the class of graphs with tree duality also admits a polynomial time solution by the theory described in this paper.

Theorem 21. *For any H-coloring problem which can be solved by the 1-consistency test we have $G \to H$ iff $G \to_f H$.*

Proof. Omitted. $\qquad\qquad\qquad\qquad\qquad\qquad\qquad\qquad\qquad\qquad\qquad\qquad\qquad$ □

According to the latest results of Hell, Zhu, and Nešetřil [8], and independently of Feder and Vardi [3], the case of digraphs H with tree duality is extended to digraphs H with bounded tree width duality, where 1-consistency check is replaced by a more general k-consistency check. We can also modify our game so that the verifier gives each prover k vertices. It then shows out that as in Theorem 21, any H-coloring problem which can be solved by a k-consistency test has $G \to H$ iff $\omega_f(G \circ_k H) = |V(G)|$ where $G \circ_k H$ is the graph whose incidence matrix equals to the verifier's matrix. This provides another proof of the polynomiality of these problems.

6.2 SAT

For the satisfiability problem, the two-prover game is as follows. The two provers claim a formula in the conjunctive normal form is satisfiable. The verifier randomly picks two clauses and gives one to prover P_1 and the other to P_2. The provers are supposed to give truth assignments to the variables occurring in their respective clauses. The verifier accepts iff the corresponding truth assignments make the respective clauses true, the two truth assignments are compatible and whenever the verifier gives the same clause to both the provers, the provers return the same truth assignment. We can then show that if all clauses are of size

575

at most 2, then the hoax set contains exactly the satisfiable clauses, and hence gives a polynomial time algorithm for 2-SAT.

Acknowledgments: We would like to thank Pavol Hell, Arvind Gupta and Graham Finlayson for valuable discussions regarding this paper. We would also like to thank NSERC for financial support.

References

1. Bačík R. and Mahajan S.: Semidefinite Programming and its Applications to NP Problems. Electronic Colloquium on Computational Complexity (1995) TR95-011
2. Chvátal V.: Linear Programming. W.H. Freeman and Company (1983), New York.
3. Feder T. and Vardi M.Y.: Monotone Monadic SNP and constraint Satisfaction. 25th ACM Symposium on Theory of Computing (1993) 612-622
4. Feige U. and Lovász L.: Two-Prover One-Round Proof Systems: Their Power and Their Problems. 24th Annual ACM Symposium on Theory of Computing (1992) 733-744
5. Goemans M.X. and Williamson D.P.: .878-Approximation Algorithms for MAX CUT and MAX 2SAT. 26th Annual ACM Symposium on Theory of Computing (1994) 422-431
6. Grötschel M., Lovász L. and Schrijver A.: The Ellipsoid Method and Its Consequences in Combinatorial Optimization. Combinatorica 1(2) (1981) 169-197
7. Häggkvist R., Hell P., Miller D.J., and Neumann Lara V.: On Multiplicative Graphs and the Product Conjecture. Combibatorica 8(1) (1988) 63-74
8. Hell P., Nešetřil J. and Zhu X.: Duality and Polynomial Testing of Tree Homomorphism. Manuscript (1994)
9. Hell P., Nešetřil J. and Zhu X.: Duality of graph homomorphism. Combinatorics Paul Erdös is Eighty 2, Bolyai Society Mathematical Studies, Budapest (to appear)
10. Hell P. and Nešetřil J.: On the complexity of H-colouring. J. Combin. Th. (B) 48 (1990) 92-110
11. Hell P. and Zhu X.: Homomorphism to oriented paths. Discrete Math. (to appear)
12. Karger D., Motwani R. and Sudan M.: Approximate Graph Coloring by Semidefinite Programming. 35th Annual IEEE Symposium on Foundations of Computer Science (1994) 2-13
13. Knuth D.E.: The Sandwich Theorem. The Electronic Journal of Combinatorics 1 (1994) #A1.
14. Lovász L.: On the Shannon Capacity of a Graph. Transactions on Information Theory II-25 No.1 (1979) 1-6
15. Schrijver A.: A comparison of the Delsarte and Lovász Bounds. IEEE Transactions on Information Theory 25 No.4 (1979)
16. Szegedy M.: A note on the θ number of Lovász and the generalized Delsarte bound. 35th Annual IEEE Symposium on Foundations of Computer Science (1994) 36-39

Analysis and Experimentation on List Update Algorithms

Lucas Chi-Kwong Hui and Kwok-Yan Lam

Department of Information Systems and Computer Science
National University of Singapore
Republic of Singapore
lhui@iscs.nus.sg

Abstract. This paper addresses the problem of efficient successful and unsuccessful search in deterministic self-adjusting linear lists. We describe and analyze a new list update algorithm which is 2-competitive to a large class of optimal offline static adversaries when considering both successful and unsuccessful searches. Analysis of the new scheme shows that it has a lower bookkeeping cost as fewer data movement is required. Experiments are conducted to compare the performance of this new algorithm and other algorithms.

1 Introduction

One fundamental issue in algorithm design is the fast implementation of the abstract data type *Dictionary*, which supports three operations: *insert*, *delete*, and *find*. An important issue is the design of dictionaries which have faster lookup times for more frequently accessed elements. There exist a large volume of work on the use *self-adjusting* data structures which dynamically change itself to reflect the past access pattern [1, 2, 3, 4, 5].

This paper describes and analyzes a new competitive self-adjusting linear list algorithm. This algorithm is 2-competitive to a large class of optimal offline static adversaries when considering both successful and unsuccessful searches. Note that if unsuccessful searches are not considered, the cost of the Move-To-Front heuristic proposed by [5] is within a factor of 2 of a large class of self-adjusting offline strategies.

In this paper an improved version of the online self-adjusting algorithm MP, denoted M*, is introduced and analyzed. MP supports fast *successful* and *unsuccessful* searches in a linear list. The cost of MP is within a factor of 3 of a large class of static or self-adjusting offline strategies. Our analysis shows that the cost of M* is within a factor of 2 of a large class of static offline strategies, and is within a factor of 3 of self-adjusting offline strategies. In addition, the self-adjusting strategy of M* is simpler than that of MP. When handling an unsuccessful search, MP moves two keys to the front, while M* moves only one key to the front.

We start with a description of the linear list setting in §1.1. In §2 we describe and analyze M*, in §3 we give experimental study of the performance of M*. §4 concludes this paper.

1.1 Linear List Setting

The most classical self-adjusting dictionary implementation is an unsorted linear list. In this implementation, we process a $find(x)$ request by scanning the list until we hit x or until we determine that x is not in the list. For any key value x, let x_p denote the *predecessor* of x, the largest key in the list smaller than x, and let x_s denote the *successor* of x, the smallest key in the list greater than x. The predecessor of the smallest key in the list, and any smaller value is $-\infty$. Similarly the successor of the largest key in the list, and any larger value is $+\infty$.

We only consider algorithms (and adversaries) that have the following behavior: During a $find(x)$ operation, if x is *not* in the list, then the algorithm will stop after finding both x_p and x_s (conceptually we treat $+\infty$ and $-\infty$ are "found" when we access the first element in the list). After x_p and x_s are found the algorithm is free to move x_p and x_s forward to any desired position. The action of moving x_p (or x_s) one position forward is called a *free* exchange. If the searched key, x, is in the list, then the algorithm will stop after finding x. After that the algorithm can move x forward by free exchanges. Besides free exchanges, the algorithm can pay one unit cost to swap two neighboring keys. This is called a *paid* exchange.

To insert key x, we first perform a $find(x)$ and then add x to the head of the list if *find* is unsuccessful. Otherwise it is treated as a successful $find(x)$ operation. The deletion of x also starts with $find(x)$ and key x is deleted if *find* is successful. Otherwise it is treated as an unsuccessful find operation.

The access cost of *find* is the number of elements in the list traversed. The access cost of *insert* is the cost of the associated *find* plus one. The cost of *delete* is just that of the associated *find*. Finally, the cost of an access request is the sum of all costs due to paid exchanges, and traversal of the list.

In this paper, $find(x)$ is written as $find([x_p, x_s])$ when x is not in the list. Note that, x_p and x_s are as defined before and are located by the search algorithm.

Competitive analysis is used to study the performance of these algorithms. For a list update algorithm A and an access request sequence σ, let $A(\sigma)$ denote the cost of A in serving all requests in σ, and $OPT(\sigma)$ be the minimum cost to serve a request sequence σ. A deterministic algorithm, M (like MP), is *c-competitive* if there is a constant b such that for list of any size and all request sequences σ, $M(\sigma) \leq c \cdot OPT(\sigma) + b$.

2 The M* List Update Algorithm

Our new algorithm M* is an improvement of the MP algorithm described in [2]. Each key x in the list of M* is associated with two "ps" bits. The function of these two bits are the same as the ps bits used in MP. Properties of these bits are given by Lemma 1 [2].

Lemma 1. *With the ps bits, for an unsuccessful search $find([x_p, x_s])$, we conclude that x is not in the list once we have found both x_p and x_s.*

The self-adjusting actions of M* are characterized by the following. After a successful search for the key x, M* moves x to the front. After an unsuccessful search $find([x_p, x_s])$, if x_p is in front of x_s, M* moves x_s to the front; if x_s is in front of x_p, M* moves x_p to the front. To insert a key x, we first perform a search to make sure that x is not in the list. If x is not in the list then we add x to the front of the list. To delete a key x, we first perform a search to make sure that x is in the list, then we delete x. The ps bits are maintained with constant extra time using the techniques described in [2].

Theorem 2 below compares the performance of M* and OPT, the optimal offline static adversary which finishes an unsuccessful search $find([x_p, x_s])$ by finding the two boundary keys x_p and x_s. The lists of M* and OPT are assumed to be identical initially.

Theorem 2. *For a request sequence σ which consists of successful and unsuccessful searches, we have $M^*(\sigma) \leq 2 \cdot OPT(\sigma)$.*

Proof. We assume that for each request in the request sequence σ, M* serves the request, performs its move-to-front actions, and then OPT serves the request.

We use $x \prec^{M^*} y$ to denote the fact that x is in front of y in the list of M*, and use $x \prec^{OPT} y$ to denote the fact that x is in front of y in OPT's list. Define a pair of keys (a, b) to be an *inversion* if b occurs before a in M*'s list while b occurs after a in OPT's list. That is $b \prec^{M^*} a$ and $a \prec^{OPT} b$. Define $\Phi =$ the total number of inversions to be the *potential function* for the competitive analysis. The definition assures that Φ is always non-negative. As we assume that both M* and OPT begin with the same list, there are no inversions, Φ is initially zero.

For the request i, let $M^*(i)$ be the amortized cost of M*, and $OPT(i)$ be the cost of OPT in accessing the list. Note that given an access sequence σ, $OPT(i)$ is fixed for every i. $M^*(i)$ consists of two parts: (1) the cost of M* in accessing the list, denoted as $A_i^{M^*}$; and (2) the change of potential function by M*'s move-to-front self-adjusting actions, denoted as $\overline{\Delta\Phi_i}^{mtf}$.

A case analysis is used to compare $M^*(i)$ to $OPT(i)$, for request i in σ.

(Case I) Request i is a successful search $find(x)$

Let Y be the set of keys that are in front of x in both M*'s list and OPT's list, that is $Y = \{y | y \prec^{M^*} x \text{ and } y \prec^{OPT} x\}$. Similarly, define Z to be the set of keys that are in front of x in M*'s list and behind x in OPT's list, that is $Z = \{z | z \prec^{M^*} x \text{ and } x \prec^{OPT} z\}$, and define W to be the set of keys that are behind x in M*'s list and in front of x in OPT's list, that is $W = \{w | x \prec^{M^*} w \text{ and } w \prec^{OPT} x\}$.

M* has to search all keys in Y, all keys in Z, and x itself to find x, therefore $A_i^{M^*} = |Y| + |Z| + 1$. Similarly, OPT has to search all keys in Y, all keys in W, and x itself to find x, therefore $OPT(i) = |Y| + |W| + 1$.

When M* moves x to the front of its list, $|Y|$ new inversions are created and $|Z|$ inversions are destroyed, so $\overline{\Delta\Phi_i}^{mtf} = |Y| - |Z|$. Therefore

$$A_i^{M^*} + \overline{\Delta\Phi_i}^{mtf} \leq |Y| + |Z| + 1 + |Y| - |Z|$$
$$< 2 \cdot (|Y| + |W| + 1) = 2 \cdot OPT(i). \tag{1}$$

(Case II) Request i is an unsuccessful search $find([x_p, x_s])$

We only analyze the sub-case $x_p \prec^{M*} x_s$, and $x_p \prec^{OPT} x_s$, as other sub-cases ($x_s \prec^{M*} x_p$ or $x_s \prec^{OPT} x_p$) can be handled similarly. With reference to the two lists before the access i, we define the followings:

$Y = \{y | y \prec^{M*} x_s, y \prec^{OPT} x_s, \text{ and } y \notin \{x_p, x_s\}\};$
$Z = \{z | z \prec^{M*} x_s, x_s \prec^{OPT} z, \text{ and } z \notin \{x_p, x_s\}\};$
$W = \{w | x_s \prec^{M*} w, w \prec^{OPT} x_s, \text{ and } w \notin \{x_p, x_s\}\}.$

We have $A_i^{M*} = |Y| + |Z| + 2$, $OPT(i) = |Y| + |W| + 2$, and $\overline{\Delta\Phi_i}^{mtf} = |Y| + 1 - |Z|$, so

$$A_i^{M*} + \overline{\Delta\Phi_i}^{mtf} \leq 2 \cdot |Y| + 3 \leq 2 \cdot (|Y| + |W| + 2) = 2 \cdot OPT(i). \qquad (2)$$

For the other sub-cases we still get equation (2).

Therefore, for request i, no matter it is a successful search or an unsuccessful search, equations (1) and (2) imply that

$$M^*(i) = A_i^{M*} + \overline{\Delta\Phi_i}^{mtf} \leq 2 \cdot OPT(i). \qquad (3)$$

Finally, summing up eq. (3) for all requests in σ, the theorem is proved. $\qquad\square$

We also show that M* is 3-competitive to ADJ, the optimal self-adjusting offline adversary which finishes an unsuccessful search $find([x_p, x_s])$ by finding the two boundary keys x_p and x_s. We start by considering an access sequence which consists of successful searches and unsuccessful searches. The only difference between ADJ and OPT is that ADJ can perform paid exchanges and free exchanges, which can in turn change the potential function Φ. In the analysis we assume that for request i, ADJ first performs paid exchanges, then M* serves the request and performs move-to-front actions, then ADJ serves the request, and finally ADJ performs free exchanges.

For request i, let $M^*(i)$ be the amortized cost of M*, and $ADJ(i)$ denotes the cost of ADJ. We have $M^*(i) = A_i^{M*} + \overline{\Delta\Phi_i}^{mtf} + \overline{\Delta\Phi_i}^{fe} + \overline{\Delta\Phi_i}^{pe}$, where $\overline{\Delta\Phi_i}^{fe}$ and $\overline{\Delta\Phi_i}^{pe}$ are the change of Φ due to free exchanges and paid exchanges, respectively. Also $ADJ(i) = A_i^{adj} + \mathcal{P}^{adj}{}_i$ where A_i^{adj} is the cost of ADJ in accessing the list, and $\mathcal{P}^{adj}{}_i$ is the cost of ADJ is performing paid exchanges.

Using similar arguments as that in the proof of theorem 2, we have

$$A_i^{M*} + \overline{\Delta\Phi_i}^{mtf} \leq 2 \cdot A_i^{adj}. \qquad (4)$$

Since ADJ has to paid one unit cost to perform a paid exchange, which can increase Φ by at most 1, so we have

$$\overline{\Delta\Phi_i}^{pe} \leq \mathcal{P}^{adj}{}_i. \qquad (5)$$

In addition, we consider free exchanges. If request i is a successful search, a free exchange will only remove one inversion, so $\overline{\Delta\Phi_i}^{fe} \leq 0$. If request i is an unsuccessful search $find([x_p, x_s])$ where (without loss of generality) $x_p \prec^{M*} x_s$, then a free exchange involving x_s will remove one inversion, and a free exchange involving x_p will at most increase Φ by one. Note that there are at most $A_i^{adj}-1$ free exchanges involving x_p, so $\overline{\Delta\Phi_i^{fe}} \leq A_i^{adj}$. Therefore we have

$$\overline{\Delta\Phi_i^{fe}} \leq A_i^{adj}. \tag{6}$$

Summing up equations (4), (5), and (6), we have $M^*(i) \leq 3 \cdot \text{ADJ}(i)$.

We now extend the analysis to include insertions. For an insertion request, if x is in the list, it is the same as a successful search; if x is not in the list, it is the same as an unsuccessful search, except that both M^* and ADJ have to pay 1 more to add x in the front of its list. In any case we still get $M^*(i) \leq 3 \cdot \text{ADJ}(i)$.

Finally, by summing up $M^*(i) \leq 3 \cdot \text{ADJ}(i)$ for every request i, we have the following theorem:

Theorem 3. *For an access sequence σ which consists of finds (successful and unsuccessful) and insertions, we have $M^*(\sigma) \leq 3 \cdot \text{ADJ}(\sigma)$.*

3 Experimental Work

This section describes experiments that compare the performance of M^* to that of MP and MTF (the original move-to-front heuristic proposed in [5]) in serving a request sequence of successful and unsuccessful searches. When comparing MP and M^*, we are interested in the total number of **data movement**, which is the number of swaps required to move an item to the front of the list, as well as the **total cost**, which is defined as the total number of nodes traversed by the algorithm to serve a request sequence.

Each experiment has several parameters: (i) the length of the list n; (ii) the length of the request sequence m; (iii) the percentage of unsuccessful searches in the request sequence, $u\%$; and (iv) the method of generating the requests.

The generated request sequence consists of $m \times (1 - u\%)$ successful searches and $m \times u\%$ unsuccessful searches. Individual requests are generated either by (a) the **uniform random** method or (b) the **skewed random** method. When using (a), a successful search is generated by randomly picking a key. Unsuccessful searches are generated in a similar manner. When using (b), we first generate a random permutation of the keys $\mathcal{P}_K$. Let $\mathcal{P}_K[i]$ denotes the ith element in $\mathcal{P}_K$. A successful search is generated by picking $\mathcal{P}_K[i]$ with probability $(1/i)/\sum_{j=1}^{n}(1/j)$ for $i = 1, ..., n$. The unsuccessful searches are generated similarly.

For each combination of n, m, and request generation method, several experiments are performed and the average values of the total cost and data movement are recorded. It is observed that the performance of MP or M^* is quite independent of n, m, and the request generation method used. Therefore only results of combinations with $n = 20$, $m = 5000$, and skewed random method are presented.

The following table shows the (average) total number of nodes traversed by M^*, MP and MTF for different values of the $u\%$ parameter. When $u\% = 0$, all the m accesses in the request sequence are all successful searches, so the three algorithms are identical. When $u\% > 0$, both MP and M^* outperform MTF. Also, MP and M^* have similar total cost, with MP performs slightly better. The last two columns show that when counting the total number of data movement, M* has a much lower cost than MP.

u%	Total cost (nodes traversed)			Data movement	
	MTF	MP	M*	MP	M*
0	36326	36326	36326	31326	31326
25	52345	43905	44336	44868	39336
50	68248	47581	49173	54417	44173
75	84204	49164	51722	62289	46722
100	100000	48465	51845	67985	46845

4 Conclusions

This paper described and analyzed M*, a new list update algorithm which is 2-competitive to optimal offline static adversaries, when considering both successful and unsuccessful searches. Analysis shown that this new algorithm has a lower bookkeeping cost as it has fewer data movement. Experiments were also conducted to compare the performance of M* to MP and MTF.

Experimental results shown that when there are unsuccessful searches, both M* and MP outperform MTF. If data movement is essentially free, it is better to use MP as it has a slightly smaller total cost than M*. Otherwise, M* will be a better choice.

In the analysis, M* outperforms MP against the optimal offline static adversaries. This is achieved by simplifying the self-adjusting actions. The major difference of M* and MP is that for every unsuccessful search $find([x_p, x_s])$, MP moves both endpoints x_p and x_s to the front, while M* only moves the endpoint which occurs later in the list. This idea is also applicable to the k-forest data structures [2] where the use of ps bits in Frederickson's data structures [1] is also addressed. One future research work is to test whether this simplification scheme is applicable to Frederickson's data structures or not.

References

1. G. Frederickson. "Self-organizing heuristics for implicit data structures", *SIAM Journal on Computing*, 13, 2 (1984) 277–291.
2. L. Hui and C. Martel. "Unsuccessful search in self-adjusting data structures", *Journal of Algorithm*, 15 (1993) 447–481.
3. L.C.K. Hui and C. Martel. Randomized Competitive Algorithms for Successful and Unsuccessful Search. In *Proc. of the Fourth Annual International Symposium on Algorithms and Computation*, 426–435, December 1993, Hong Kong.
4. L.C.K. Hui and C. Martel. Analysing Deletions in Competitive Self-adjusting Linear List Algorithms. In *Proc. of the Fifth Annual International Symposium on Algorithms and Computation*, 433–441, August 1994, Beijing, P.R. China.
5. D. Sleator and R. Tarjan. "Amortized efficiency of list update and paging rules", *Communications of the A.C.M.*, 28, 2 (1985) 202–208.

An Exact Branch and Bound Algorithm for the Steiner Problem in Graphs

B. N. Khoury and P. M. Pardalos

Center for Applied Optimization and ISE Department
303 Weil Hall, University of Florida, Gainesville, Florida 32611

Abstract. An exact branch and bound algorithm for the Steiner Problem in Graphs (SPG) is presented, based on a new integer programming formulation. The algorithm proves to be robust and relatively efficient based on preliminary results obtained for average-size problems.

1 Introduction

Given a graph $G = (\mathcal{N}, \mathcal{A}, \mathcal{C})$ represented by a set of nodes $\mathcal{N}$ and a set of arcs $\mathcal{A}$ associated with a set of arc costs $\mathcal{C}$, the Steiner problem in graphs (SPG), consists of connecting a subset of given nodes on a graph with the minimum cost tree [6]. It is well known that the SPG decision problem is NP-complete in general. Even for some restrictions like grid graphs, equal-edge-weight graphs, bipartite graphs, chordal and split graphs, the problem remains NP-complete. Moreover, on directed graphs, it is known that the variant of the rooted Steiner arborescence of a directed graph is NP-complete. However, there are some instances, for which polynomial time algorithms exist. See the books by Hwang et al. [7], Voss [15], and MacGregor Smith and Winter [11] for excellent sources of information on the subject.

All known exact SPG algorithms for general graphs are in some way enumerative algorithms. However, they differ in how the enumeration is done and how clever their strategies are to avoid total enumeration. In order to avoid total enumeration, topological and other properties of general graphs can be used. Another alternative is to develop good lower and upper bounds in order to reduce the possible search space. Two simple approaches are proposed by Hakimi [6]. While one of them is based on spanning tree enumeration, the other approach is a topology enumeration recursive algorithm. Another algorithm, which generates spanning trees for a derived problem in order of increasing total length until a solution to the original problem can be inferred, is proposed by Balakrishnan and Patel [1]. Also, implicit enumeration approaches are given by Shore et al. [13], and Yang and Wing [16], where the set of all trees spanning the set of given points is, in a systematic way, separated into smaller subsets to be analyzed, using upper and lower bounds, in order to determine whether or not they contain the optimal feasible solution. A branch and bound approach that uses heuristics to provide good lower bounds and to choose the next edge for consideration in the backtracking process, is proposed by Shore et al. [13]. Other approaches, including branch and bound are discussed in [7]. Exact algorithms for the SPG, however, can benefit from preprocessing the graph so that the size of the problem to solve is reduced. Many types of reduction techniques are discussed in Chopra et al. [5]. While some of those algorithms mentioned above are better then others, they still run in exponential time in the worst case.

All the known exact approaches to the Steiner problem in graphs (SPG) are based on a kind of enumeration technique. Some of these approaches are better than others in terms of avoiding total enumeration by using properties of optimality or by using good lower and upper bounds to reduce the search space in size. One of those approaches is the branch and bound procedure where nodes in the binary search tree correspond to binary branching variables. For any branch and bound procedure, the choices of preprocessing strategies, lower bound generators and upper bounds are crucial in order to reduce the total number of nodes in the search tree. In this paper we propose a branch and bound algorithm and we discuss reduction techniques in order to preprocess the original graph, upper and lower bounds, and branching strategies. In addition, some preliminary computational experimental results are presented.

2 An Exact Branch and Bound Algorithm for SPG

2.1 Graph Preprocessing

As experience has shown, preprocessing an integer program reduces the computational burden on an exact-solution procedure. Given a graph $G = (\mathcal{N}, \mathcal{A}, \mathcal{C})$ and a subset of nodes $\mathcal{R}$, we present in the following three reduction rules for the SPG which are easy to implement.

Edge Contraction: Edge contraction can be performed in two cases. The first case corresponds to the situation where the degree of a regular node i is one. In such a case, the only arc (i, j) incident to i has to be in any solution tree. As a result, (i, j) can be contracted. In the second case, let i and j be two regular nodes of $\mathcal{R}$. If $c_{ij} \leq c_{ik}$ for all nodes k adjacent to i, then it is easy to see that arc (i, j) is a part of an optimal solution. Therefore, edge (i, j) can be contracted. After the contraction, if parallel edges or arcs occur, only the edge with the smallest cost is retained.

Edge Deletion: Suppose there exist $k \in \mathcal{R}$ and $i, j \in \mathcal{N}$ such that $c_{ij} \geq \max\{c_{ik}, c_{jk}\}$. Then edge (i, j) can be deleted from the graph for the simple reason that it will never be a part of an optimal solution. In fact, suppose that an optimal solution contains arc (i, j), then we can make an improvement by replacing (i, j) by either (i, k) or (j, k), a contradiction.

Node Deletion: There are two situations where nodes not in $\mathcal{R}$ can be deleted. The first case is when the degree of the node is equal to one, an obvious case since this node will never be in a Steiner tree. In the second situation, the degree of the node is equal to two. In such a case, if the node is to be in a Steiner tree, both incident arcs have to be used. Therefore, in a reduced equivalent graph, this node can be deleted and its two incident arcs can be replaced by a single arc of cost equal to the total of their costs and with incident nodes the two adjacent nodes of the node in question; if parallel edges occur in this process, only the edge with the smallest cost is kept.

2.2 Upper and Lower Bounds

A good starting feasible solution provides a good upper bound which combined with good lower bounds causes early node pruning in the search tree, an essential objective. In the course of the branch and bound algorithm, the upper bound is updated whenever a better integer feasible solution is identified. Usually, the starting upper bound is provided by a heuristic. For this purpose, we use the polynomial-time heuristic WPSPGH proposed in [8].

A good technique to generate lower bounds is necessary in order to reduce the possible search space. In this section, we propose a lower bound obtained by relaxing the integrality constraints in one of the integer formulations of the (SPG) proposed in [10]. Given a graph $G = (\mathcal{N}, \mathcal{A}, \mathcal{C})$, construct its corresponding symmetric directed graph $G^d = (\mathcal{N}, \mathcal{A}^d, \mathcal{C}^d)$ (every undirected arc $(i,j) \in \mathcal{A}$ corresponds to two oppositely directed arcs (i,j) and $(j,i) \in \mathcal{A}^d$ associated with the same cost $(c_{ij}^d = c_{ji}^d = c_{ij})$, $\mathcal{R} \subseteq \mathcal{N}$ and $r \in \mathcal{R}$ as mentioned in [10]. Consider the following equivalent formulation $\mathcal{IF}$:

$$\text{minimize} \quad \sum_{(i,j) \in \mathcal{A}^d} c_{ij} x_{ij} \tag{1}$$

$$\text{subject to:} \quad \sum_{(i,j) \in \mathcal{I}(j)} x_{ij} = 1 \quad (\forall j \in \mathcal{R} \backslash \{r\}) \tag{2}$$

$$\sum_{(k,i) \in \mathcal{I}(i) \backslash \{(j,i)\}} x_{ki} \geq x_{ij} \quad (\forall (i,j) \in \mathcal{A}^d \backslash \mathcal{O}(r)) \tag{3}$$

$$y_i - y_j + |\mathcal{N}| x_{ij} \leq |\mathcal{N}| - 1 \ (\forall (i,j) \in \mathcal{A}^d) \tag{4}$$

$$x_{ij} \in \{0, 1\} \quad (\forall (i,j) \in \mathcal{A}^d) \tag{5}$$

$$y_i \geq 0 \quad (\forall i \in \mathcal{N}). \tag{6}$$

Involving additional nonnegative variables corresponding to nodes, constraint (4) eliminates directed cycles, which correspond to tours and subtours; this type of constraint is used in formulating the traveling salesman problem. Let $\mathcal{LP}$ be the linear program obtained form $\mathcal{IF}$ by relaxing the integrality constraint. In the process of generating lower bounds for the branch and bound algorithm, every node in the search tree is associated with a pair of arc subsets $(\mathcal{A}_0, \mathcal{A}_1)$ and a lower bound $\mathcal{LP}(\mathcal{A}_0, \mathcal{A}_1)$ which is the optimal objective value of $\mathcal{LP}$ where the binary variables corresponding to arcs in $\mathcal{A}_0$ and $\mathcal{A}_1$ are fixed to zero and one respectively.

The choice of the Steiner tree root node r used in the mathematical model above has an effect on the quality of the lower bound. One way to determine the root node is to choose among the $|\mathcal{R}|$ regular nodes the root node that has the largest linear programming relaxation objective value to provide the largest starting lower bound. Another way to determine the root node is to choose among the regular graph nodes the one that has the largest average preference weights found by dividing the sum of preference (the preference weights in question are mentioned in [8]) for arcs emanating from the node by the number of these arcs. For the purpose of efficiency, the second alternative to choose the root node is adopted for the branch and bound algorithm.

2.3 Branching Strategies

In a branch-and-bound scheme, there are two different aspects of the branching strategy related to constructing and exploring the search tree.

Choosing the Branching Variable: In constructing the search tree, choosing the branching variable at a parent node has to be done in a systematic way. One alternative to choose the branching variable is to choose the variable with value above a threshold value close to one and whose corresponding arc has the smallest depth. The depth of an arc is defined as the smallest number of arcs it takes to reach the beginning node of that arc from the root node. With this rule, good solutions are found relatively fast. However, to prove optimality, it takes a very long time. The other way to choose the branching variable is to take the smallest-depth arc whose value is within an interval centered at 0.5 and length equal to twice a threshold value, a value that is close to 0.1 and could be increased if needed. With this rule, it takes time to find good solutions; however, it takes less time, relatively speaking, to prove optimality. Due to the good quality of the starting upper bound solution, the second alternative is adopted for the sake of efficiency in proving optimality.

Exploring the Search Tree: There are several alternatives to deal with the second aspect of the branching strategy, i.e, exploring the search tree. Examples of such alternatives are breadth-first search, best-first search and depth-first search. The breadth-first search and best-first search require more storage space. However, they explore the search tree in a parallel fashion making the chance of finding good solutions better which in turn eliminates exploring bad branches in the search tree. Depth-First search, on the other hand, requires much less memory storage space, at the expense of increasing the chance of exploring bad branches in the search tree. With a good starting upper bound like the one we proposed, this chance of exploring bad branches in the search tree is reduced, and the depth-first search becomes an effective strategy for the branch-and-bound algorithm proposed in the following section.

2.4 A Branch and Bound Algorithm

In this section, we propose a branch and bound algorithm for the SPG which maintains a branching binary tree. Consider a graph $G = (\mathcal{N}, \mathcal{A}, \mathcal{C})$ and its corresponding symmetric directed graph $G^d = (\mathcal{N}, \mathcal{A}^d, \mathcal{C}^d)$ obtained as mentioned in [10]. Every node i in the branching tree is associated with a set of directed arcs $\mathcal{A}_0^i \in \mathcal{A}^d$ whose variables are fixed to zero, a set of directed arcs $\mathcal{A}_1^i \in \mathcal{A}^d$ whose variables are fixed to one, a lower bound $\mathcal{LP}(\mathcal{A}_0^i, \mathcal{A}_1^i)$ obtained as mentioned in the fourth section and a branching arc variable obtained as mentioned in the fifth section. The algorithm explores the set of active nodes, leaf nodes in the branching tree whose lower bounds are less than the best known upper bound, in a depth-first-search manner. In this fashion, the upbranch at a node is completely exhausted before the downbranch at the same node is considered. We chose this way to explore the search tree because it is very efficient in terms of memory space requirements and because we start with a very good upper bound which reduces the need for exploring the search tree in parallel fashion.

The active nodes are sorted in a stack according to a selection rule, which is the depth-first-search rule in this case. The state of the branching tree is represented by six variables: BBN, $BBAN$, $BBCN$, $BBBV(i)$, $BBLB(i)$ and $BBP(i)$. BBN is the total number of nodes in the search tree. $BBAN$ is the total number of active nodes in the stack. $BBCN$ is the current node to be analyzed. $BBBV(i)$ contains the index of the branching variable associated with node i in the branching tree. $BBLB(i)$ contains the lower bound value associated with node i in the search tree. $BBP(i)$ points out to the parent node of node i in the branching tree; if $BBP(i) = j > 0$ or $BBP(i) = -j < 0$, then i is an upbranch or downbranch, respectively, of j. In what follows, we propose an algorithm for the SPG.

SPG Branch and Bound Algorithm

Given: A graph $G = (\mathcal{N}, \mathcal{A}, \mathcal{C})$ and a subset of regular nodes $\mathcal{R}$.
Output: A Steiner minimum tree.
Step 0

- Do graph preprocessing.
- Find a starting upper bound ub .
- Find a starting lower bound $lb = \mathcal{LP}(\emptyset, \emptyset)$.
- If lb is equal to ub or if the starting lower bound solution is integer, then stop an optimal solution with value lb is at hand.
- Create the first active node as follows: $BBN = 1, BBAN = 1$; $\mathcal{A}_0^1 = \mathcal{A}_1^1 = \emptyset$; $BBLB(1) = lb$; $BBP(1) = 0$; choose the branching variable bv, $BBBV(1) = bv$; $BBCN = 1$.
- Go to step 1.

Step 1: Upbranch

- Let $k = BBCN$, and let $(i, j) = BBBV(k)$.
- Let $\mathcal{A}_0 = \mathcal{A}_0^k \cup \{\forall(l, j) \in \mathcal{A}^d, \text{s.t. } l \neq i\}$. Let $\mathcal{A}_1 = \mathcal{A}_1^k \cup \{(i, j)\}$, and find $lb = \mathcal{LP}(\mathcal{A}_0, \mathcal{A}_1)$.
- If lb is larger than or equal to ub then go to step 2.
- If the lower bound solution is integer, then $ub = lb$, and go to step 2.
- Create an active node as follows: $BBN = BBN + 1$; $BBAN = BBAN + 1$; $\mathcal{A}_0^{BBN} = \mathcal{A}_0$, $\mathcal{A}_1^{BBN} = \mathcal{A}_1$; $BBLB(BBN) = lb$; $BBP(BBN) = +k$; choose the branching variable bv, $BBBV(BBN) = bv$;
- $BBCN = BBN$.
- Repeat step 1.

Step 2: Downbranch

- Let $k = BBCN$, and let $(i, j) = BBBV(k)$.

- Let $\mathcal{A}_0 = \mathcal{A}_0^k \cup \{(i,j)\}$. Let $\mathcal{A}_1 = \mathcal{A}_1^k$, and find $lb = \mathcal{LP}(\mathcal{A}_0, \mathcal{A}_1)$.
- If lb is larger than or equal to ub then go to step 3.
- If the lower bound solution is integer, then $ub = lb$, and go to step 3.
- Create an active node as follows: $BBN = BBN + 1$; $BBAN = BBAN + 1$; $\mathcal{A}_0^{BBN} = \mathcal{A}_0, \mathcal{A}_1^{BBN} = \mathcal{A}_1$; $BBLB(BBN) = lb$; $BBP(BBN) = -k$; choose the branching variable bv, $BBBV(BBN) = bv$;
- $BBCN = BBN$.
- Go to step 1.

Step 3

- Let $k = BBP(BBCN)$.
- If k is positive, then set $BBCN$ to k, and go to step 2.
- If k is negative, then decrease $BBAN$ by one.
- If $BBAN = 0$, stop. There are no more active nodes, and the best known integer solution is the optimal solution with value ub.
- Repeat step 3.

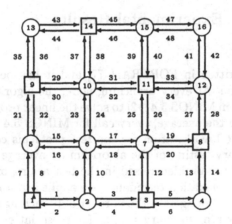

Fig. 1. A directed grid graph with unit-cost edges.

Example Here, we illustrate how the branch and bound algorithm works. Consider a unit-edge-cost grid graph where the arcs of the corresponding directed graph are numbered as shown in Figure 1, and the regular nodes are the nodes marked by squares. If applied to this grid graph, WPSPGH, proposed in [8], yields a Steiner tree of length 9. Also, node 11 is chosen as the Steiner tree root node according to the rule discussed earlier. As shown in Figure 2, the branch and bound algorithm given above considered one node in the branching tree before finding an optimal solution of value 8. The operations performed at the branching node are presented below.

node 1: The optimal solution of $\mathcal{LP}$ at this node is $x_2 = 0.354, x_4 = 0.354, x_8 = 0.646, x_{19} = 1, x_{12} = 1, x_{22} = 0.646, x_{37} = 1, x_{30} = 1, x_{26} = 1, x_{32} = 1$. The branching variable associated with this node is x_{22}. The lower bound at node 1 is equal to 8.

The upbranch at node 1 yields an integer solution of value 8, and the downbranch yields a solution with an objective value of 8. Therefore both branches are pruned (not added to the set of active nodes), and the search tree contains only one node.

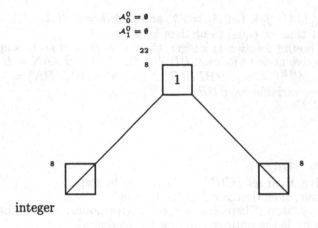

Fig. 2. A branch and bound tree with a starting upper bound of 9.

3 Preliminary Experimental Results

The algorithm is written in FORTRAN 77, and it has been implemented on a SPARC workstation. At every branching node, the algorithm interfaces with subroutine MINOSS in MINOS 5.4 [12] to solve the linear program that generates the lower bound. On the average, every call for MINOS 5.4 takes about 1 CPU second on the SPARC workstation for the test problems considered. We have done some preliminary testing of the algorithm on average size problems. The solutions of some of the problems used in the test are reported by Beasley [2], where a Lagrangean relaxation techniques are applied to a minimum-spanning-tree based formulation, along with the number of branching nodes and the CPU time. To compare our preliminary results for these latter problems with the previous ones is a difficult matter. Although our algorithm considered a smaller number of branching nodes as compared to Beasley's algorithm, the number of branching nodes is not a decisive measurement since the lower bound techniques are different. The CPU time is not relevant either because the previous results were obtained on a Cray X-MP/48 with vector processing capability. However, we may conclude that our algorithm is robust, and it finds the optimal solution in a reasonable amount of time. Furthermore, one problem from the set of grid-graph test problems given by Soukup and Chow [14], problem 15 in Table 1, for which Beasley's algorithm did not find an optimal solution [4], can be solved using our algorithm with 250357 branching nodes.

In Table 2, we give the results corresponding to some problems from sets steinc and steinb obtained from Beasley's OR-library [3]. As we can see the problem size is dramatically reduced after applying reduction techniques. And the reduced-size problems can be solved with a small number of branching nodes. In Table 3 the optimal solutions of problems spgtp generated by the test problem generator given in [9] are verified with a small number of iterations.

4 Conclusions

In this paper, we have presented a branch-and-bound algorithm for the SPG. This algorithm uses an integer formulation presented in [10] to develop lower bounds, and it uses a heuristic presented in [8] to obtain a starting feasible solution which provides an upper bound. From the preliminary results we have obtained, the algorithm is robust and compares favorably to the state of the art.

Table 1. Performance on SPRG problems.

Prob. #	$\|\mathcal{R}\|$	opt.	bran. node #
1	5	1.87	3
2	6	1.64	1
9	7	1.64	0
15	14	1.48	250357
17	10	2.0	3206

Table 2. Performance on SPG problems from sets **steinb** and **steinc**.

Prob. #	bef. red.			aft. red.			opt.	bran. node #
	$\|\mathcal{N}\|$	$\|\mathcal{A}\|$	$\|\mathcal{R}\|$	$\|\mathcal{N}\|$	$\|\mathcal{A}\|$	$\|\mathcal{R}\|$		
b1	50	63	9	10	16	4	82	0
b2	50	63	13	14	25	7	83	0
b3	50	63	25	8	12	4	138	0
c1	500	625	5	150	274	5	85	0
c4	500	625	125	149	272	57	1079	8

References

1. Balakrishnan, A., and N. R. Patel. Problem reduction methods and a tree generation algorithm for the Steiner network problem. *Networks*, 17:65–85, 1987.
2. Beasley, J. E. An SST-based algorithm for the Steiner problem on graphs. *Networks*, 19:1–16, 1989.
3. J. E. Beasley, OR-Library: Distributing test problems by electronic mail. *J. of Oper. Res. Soc. 41(11)* (1990) 1069-1072.
4. Beasley, J. E. A heuristic for the Euclidean and rectilinear Steiner problems. *Europ. J. of Oper. Res.*, 58:284–292, 1992.

Table 3. Performance on problems of set spgtp.

Prob. #	$\mathcal{N}$	$\mathcal{A}$	$\mathcal{R}$	opt.	bran. node #
spgtp1	20	40	5	986	0
spgtp2	20	40	5	3562	0
spgtp3	20	40	7	2213	0
spgtp4	20	40	8	2864	0
spgtp5	20	40	10	2776	0

5. Chopra, S., E. R. Gorres, and M. R. Rao. Solving the Steiner tree problem on graphs using branch and cut. *ORSA Journal on Computing*, 4(3):320–335, 1992.

6. S. L. Hakimi, Steiner's problem in graphs and its implications. *Networks 1* (1971) 113-133.

7. F. K. Hwang, D. S. Richards and P. Winter, *The Steiner Tree Problem*, Elsevier, Amsterdam (1992).

8. B. N. Khoury and P. M. Pardalos, A Heuristic for the Steiner Problem in Graphs. *Computational Optimization and Applications*, (1995).

9. B. N. Khoury, P. M. Pardalos and D.-Z. Du, A test problem generator for the Steiner problem in graphs. *ACM Transactions on Mathematical Software*, Vol. 19, No. 4 (1993), pp. 509-522.

10. B. N. Khoury, P. M. Pardalos and D. W. Hearn, Equivalent formulations for the Steiner problem in graphs. In *Network Optimization Problems*, 111-124, P.M. Pardalos and D.-Z. Du (Eds.), World Scientific (1993).

11. J. MacGregor Smith and P. Winter (Eds.), *Topological Network Design Vol. 31(1-4) of Annals of Operations Research* (1991)

12. Murtagh, B. A., and M. A. Saunders. MINOS 5.4 user's guide(preliminary). Technical Report SOL 83-20R, Department of Operations Research, Stanford University, Stanford, CA, 1983.

13. Shore, M. L., L. R. Foulds, and P. B. Gibbons. An algorithm for the Steiner problem in graphs. *Networks*, 12:323–333, 1982.

14. J. Soukup and W. F. Chow, Set of test problems for the minimum length connection networks. *ACM/SIGMAP Newsletter 15* (1973) 48-51.

15. S. Voss, *Steiner-Probleme in Graphen* [in German], Hain, Frankfurt (1990).

16. Yang, Y. Y., and O. Wing. An algorithm for the wiring problem. *Digest Ieee Int. Symp. Electrical Networks*, 14–15, 1971.

A Physical Model for the Satisfiability Problem

Huang Wenqi[1]   Li Wei[2]   Lu Weifeng[2] and Zhang Yuping[2]

[1] Dept. of Computer Science,
Huazhong University of Science and Technology, 430074, Wu Han, P.R.China
[2] Dept. of Computer Science,
Beijing University of Aeronautics and Astronautics, 100083, Beijing, P.R.China

Abstract. An one to one and onto mapping between the set of conjunctive normal forms and a subset of the potential functions of static electricity fields is constructed; and it has been further proved that a conjunctive normal form is satisfiable if and only if the minimum of the corresponding potential function is zero. It is also shown that the local search method has the same physical model as the gradient method given in this paper.

1 Transformation of the SAT Problem

The satisfiability problem, or the SAT problem for short, is perhaps one of the most important problems in computer science. It can be informally described as below:

Definition 1. Conjunctive normal form and SAT problem. Let L be a propositional language, and $\mathcal{P}$ be the set of proposition names, consisting of P_1, ..., and P_m. A clause is a proposition having the form $Q_1 \vee Q_2 \vee \cdots \vee Q_l$, where Q_i is a literal. A literal is either a proposition name or the negation of a proposition name. A conjunction normal form, called a CNF for short, is a proposition having the form $C_1 \wedge C_2 \wedge \cdots C_n$, where C_i is a clause. The set of conjunction normal forms is denoted by $\mathcal{F}_m$.

Let A be a given CNF. The SAT problem is the problem to determine whether there exists a truth assignment from $\mathcal{P}$ to $\{T, F\}$, which makes A true.

Using the optimization methods to solve the SAT problem has attracted great attention in the computer society. It usually consists of two steps: transforming the SAT problem into some minimization (or maximization) problem of certain object functions, and then solving the minimization problem by applying the optimization techniques to the object function. This method was introduced to solve SAT problem first by Gu [1, 2]. In [1, 2], Gu found a simple discrete object function and proposed a family of local search algorithms. Selman et al. discovered and studied the hard distributions of some similar local search algorithms for the same object function [6]. In [2, 3], Gu further proposed a family of continuous object functions and studied a family of algorithms called global optimization.

In this paper, we propose a class of continuous functions as the object functions for the SAT problem. It is similar to but different from the family of continuous object functions mentioned as above. In fact, our object functions are motivated from a physical interpretion of the problem. They are potentials of a certain kind of static electric fields. Searching a truth assignment of a CNF is equivalent to calculating the minimum of the corresponding potential. This interpretation enlightens us to find efficient algorithms. For example, a particle always moves along the direction of gradient decent of the electric field, it is the fastest decreasing direction of potential of the particle. If a CNF is satisfiable, then the gradient method for its corresponding potential function becomes one of fast algorithms to solve the SAT problem.

Transformation T

$$T : \mathcal{F}_m \to (R^m \to R)$$

$$T(C_1 \wedge \cdots \wedge C_n) = T(C_1) + \cdots + T(C_n)$$
$$T(Q_1 \vee \cdots \vee Q_l) = \begin{cases} T(Q_1) \cdot \cdots \cdot T(Q_l), & \text{if } T(Q_i) > 0, i = 1, 2, \cdots, l; \\ 0, & \text{otherwise.} \end{cases}$$
$$T(P_i) = 1 - x_i, \quad i = 1, \cdots, m$$
$$T(\neg P_i) = x_i, \quad i = 1, \cdots, m$$

Let us see an example which will be used later:

Example 1 Let the CNF A be $(P_1 \vee P_2) \wedge (\neg P_1 \vee P_2) \wedge (\neg P_1 \vee \neg P_2)$. According to the definition, $T(P_i) = 1 - x_i$ and $T(\neg P_i) = x_i$ for $i = 1, 2$. By using the transformation T, we have:

$$T(A) = T(P_1 \vee P_2) + T(\neg P_1 \vee P_2) + T(\neg P_1 \vee \neg P_2), \quad \text{where}$$

$$T(P_1 \vee P_2) = \begin{cases} (1 - x_1)(1 - x_2), & \text{if } x_1 < 1 \text{ and } x_2 < 1; \\ 0, & \text{otherwise} . \end{cases}$$

$$T(\neg P_1 \vee P_2) = \begin{cases} x_1(1 - x_2), & \text{if } x_1 > 0 \text{ and } x_2 < 1; \\ 0, & \text{otherwise} . \end{cases}$$

$$T(\neg P_1 \vee \neg P_2) = \begin{cases} x_1 x_2, & \text{if } x_1 > 0 \text{ and } x_2 > 0; \\ 0, & \text{otherwise} \end{cases}$$

Theorem 1 *Let A be a CNF. A is satisfiable if and only if that there exists a minimal value of $T(A)$ which is zero.*

Proof: The proof can be found from [4].

The object function $T(A)$ has a clear physical interpretation. It will be shown that $T(A)$ is just the potential function a particle with negative electricity in some m-dimensional static electric field.

2   Static electric fields of metal plates

Consider the following three regions in the two dimensional space:

$$\text{I}: x_1 \geq 1 \text{ or } x_2 \geq 1, \text{ II}: x_1 \leq 0 \text{ or } x_2 \geq 1, \text{ III}: x_1 \leq 0 \text{ or } x_2 \leq 0,$$

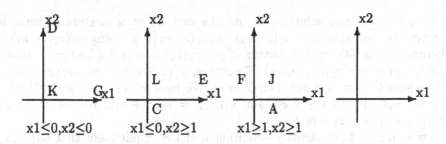

$$x1 \leq 0, x2 \leq 0 \qquad x1 \leq 0, x2 \geq 1 \qquad x1 \geq 1, x2 \geq 1$$

Suppose that the regions I, II and III are the metal plates with the positive electric charge, and there exists an isolating layer between every two metal plates. Assume that the metal plates are conductors, and so in each metal plate, the static electric potential is constant. Let the coordinate of a particle $\mathcal{P}$ with a unite negative charge be (x_1, x_2). The particle $\mathcal{P}$ is attracted by the forces of static electricity of the plates I, II and III respectively. It is moving under the action of the three forces. Every plate wants the particle to be on itself. If the particle is not on the plate, it is attracted by the force of static electricity of the plate. When it is on, the force disappears, the particle is attracted by other plates, and moves until it is also on the other plates. In brief, the particle moves under the action of the force of static electricity and the action of metal shield phenomenon.

This moving picture of the particle can be described mathematically as below:

Let the potential function of the static electric fields of the metal plates I, II and III be: $u_1(x_1, x_2)$, $u_2(x_1, x_2)$ and $u_3(x_1, x_2)$ respectively. Thus, the potential function of the static electric field of the total space is:

$$U(x_1, x_2) = u_1(x_1, x_2) + u_2(x_1, x_2) + u_3(x_1, x_2) \tag{1}$$

The force of static electric potential attracting the particle $\mathcal{P}$ is:

$$\vec{F} = \vec{F_1} + \vec{F_2} + \vec{F_3} = -\text{grad}U = (-\text{grad}u_1, -\text{grad}u_2, -\text{grad}u_3) \tag{2}$$

where grad is the gradient operator, the gradient vector of U is:

$$\text{grad}U(x_1, x_2) = \left(\frac{\partial U(x_1, x_2)}{\partial x_1}, \frac{\partial U(x_1, x_2)}{\partial x_2} \right) \tag{3}$$

Let us derive the analytic expression of $u_3(x_1, x_2)$ in (1). From the theory of static electric fields, we know that $u_3(x_1, x_2)$ is zero in the second, third and fourth quadrants, in other word, the potential of static electricity is zero in the metal plate III. In the first quadrant, $u_3(x_1, x_2)$ is in vacuum, it satisfies the

Laplace equation. Therefore, the condition of definiting solution for $u_3(x_1, x_2)$ is:

$$\begin{cases} \frac{\partial^2 u_3(x_1,x_2)}{\partial x_1^2} + \frac{\partial^2 u_3(x_1,x_2)}{\partial x_2^2} = 0, & \text{if } (x_1, x_2) \text{ is in the first quadrant;} \\ u_3(x_1, x_2) = 0, & \text{if } (x_1, x_2) \text{ is on the boundary of the first quadrant.} \end{cases} \quad (4)$$

To get a non-zero solution, we use the method of separation of variables to solve the equation. Let $u_3(x_1, x_2) = v_1(x_1)v_2(x_2)$. Using $v_1(x_1)v_2(x_2)$ to substitute u_3 in (4), we can obtain $v_1''(x_1)v_2(x_2) + v_1(x_1)v_2''(x_2) = 0$, that is $v_1''(x_1)/v_1(x_1) = -v_2''(x_2)/v_2(x_2)$. Therefore, $v_1''(x_1)/v_1(x_1) = -v_2''(x_2)/v_2(x_2) = C$ where C is a constant. Let $C = 0$. We have $v_1''(x_1) = 0$ and $v_2''(x_2) = 0$. Thus, $v_1(x_1) = c_{11}x_1 + c_{12}$ and $v_2(x_2) = c_{21}x_2 + c_{22}$. We get $u_3(x_1, x_2) = (c_{11}x_1 + c_{12})(c_{21}x_2 + c_{22})$.

According to the boundary condition and the requirement that $u_3(x_1, x_2)$ is a non-zero solution, we know $c_{12} = c_{22} = 0$ and $u_3(x_1, x_2) = c_{11}c_{21}x_1x_2$. Thus, $u_3(x_1, x_2) = x_1x_2$ is a non-zero solution, and it satisfies the condition (4). Finally, the analytic expression of the potential function $u_3(x_1, x_2)$ of static electricity on the total space is:

$$u_3(x_1, x_2) = \begin{cases} x_1x_2, & \text{if } x_1 > 0 \text{ and } x_2 > 0; \\ 0, & \text{if } x_1 \le 0 \text{ or } x_2 \le 0. \end{cases} \quad (5)$$

It is obvious that $u_3(x_1, x_2)$ cannot be negative. $u_3(x_1, x_2)$ denotes the potential of the particle $\mathcal{P}$ having coordinate (x_1, x_2) with respect to the metal plate III. Similarly, the analytic expression of $u_2(x_1, x_2)$ in the total space is:

$$u_2(x_1, x_2) = \begin{cases} x_1(1 - x_2), & \text{if } x_1 > 0 \text{ and } x_2 < 1; \\ 0, & \text{if } x_1 \le 0 \text{ or } x_2 \ge 1. \end{cases} \quad (6)$$

The analytic expression of $u_1(x_1, x_2)$ in the total space is:

$$u_1(x_1, x_2) = \begin{cases} (1 - x_1)(1 - x_2), & \text{if } x_1 < 1 \text{ and } x_2 < 1; \\ 0, & \text{if } x_1 \ge 1 \text{ or } x_2 \ge 1. \end{cases} \quad (7)$$

By summing up (5), (6) and (7), we can obtain the analytic expression of the total potential of the system $U(x_1, x_2)$.

It is easy to see that the potential functions u_1, u_2, u_3 and $U(x_1, x_2)$ are exactly the same as the transformations of $T(P_1 \vee P_2)$, $T(\neg P_1 \vee P_2)$, $T(\neg P_1 \vee \neg P_2)$, and $T(A)$ derived in example 1. Thus, to say "at the point (x_1^*, x_2^*), the potential of static electricity of the particle $\mathcal{P}$ is zero, or $U(x_1^*, x_2^*) = 0$" is equivalent to say " the conjunctive normal form A is satisfiable for the corresponding truth assignment V^*." In fact, this conclusion is true not only for this particular A given in example 1, but also for any conjunctive normal form.

The physical model can inspire us to find fast algorithms for solving the SAT problem. For example, since a free particle always moves along the direction in which the potential of the static electricity decreases most quickly. According to the electric field theory, this direction is the direction of gradient descent of the corresponding static electricity field. The particle will eventually reach a position with a local minimum potential. This is to say that we can obtain a fast algorithm to find a local minimum by applying the gradient method to the object function U.

Finally, as we have seen: the solution of the Laplace equation is $u_3(x_1, x_2) = (c_{11}x_1 + c_{12})(c_{21}x_2 + c_{22})$. It is the boundary condition which determines the form of u_3 given in (5). The boundary condition comes from the model of metal plate. This is why we say that the static electric fields of metal plates are the physical model for the SAT problem. Without the boundary condition, the reason to take $u_3(x_1, x_2) = x_1 x_2$ and especially $u_2(x_1, x_2) = x_1(1 - x_2)$, where $(x_1, x_2) \in R^2$, as in [3], is by the mathematical intuition.

3　Physical model for the local search

Let $A = C_1 \wedge \cdots \wedge C_n$ be a CNF. If we restrict the domain R^m of $T(A) = T(C_1) + \cdots + T(C_n)$ to the discrete space $\{0,1\}^m$, then the value of $T(A)$ in $\{0,1\}^m$ is a natural number, which is exact the number of unsatisfiable clauses of the corresponding CNF. In other word, the interpretation of the object function of local search methods given in [1, 6] is also the potential function of some static electric field, but for discrete space $\{0,1\}^m$. This fact explains why local search method is fast.

There are many methods to calculate the minimum of continuous function with domain R^m, the m - dimensional Euclid space. We outline one of the typical methods below. By this algorithm, we provide a mathematical framework for the family of algorithms of local search given in [1, 6].

Algorithm 3.1 Let the object function be $U(x_1, \cdots, x_m) = T(A)$ defined by the *Transformation* T, we use notation $x_j^{(i)}$ to define a sequence convergent to a local minimum of $U(x_1, \cdots, x_m)$, where j means the j-th variable, the supperscript (i) means the ith loop.

1) Let $(x_1^{(0)}, \cdots, x_m^{(0)})$ be an initial point;
2) Let $i = 0$.
Do the following loop for i until find a local minimum of $U(x_1, \cdots, x_m)$.
　　For $j = 1$ to m do the loop:
　　　　Find the point $(x_1^{(i+1)}, \cdots, x_j^{(i+1)}, x_{j+1}^{(i)}, \cdots, x_m^{(i)})$,
　　　　which is a minimal point of the unary function
　　　　$U(x_1^{(i+1)}, \cdots, x_{j-1}^{(i+1)}, x, x_{j+1}^{(i)}, \cdots, x_m^{(i)})$ in the subspace
　　　　$\{(x_1^{(i+1)}, \cdots, x_{j-1}^{(i+1)}, x, x_{j+1}^{(i)}, \cdots, x_m^{(i)}) | \ x \in (-\infty, +\infty)\}$.

For providing a mathematical framework for the family of algorithms of local search given in [1, 6], we first prove the following lemmas. Here $U(x_1, \cdots, x_m) = u_0(x_1, \cdots, x_m) + \cdots + u_m(x_1, \cdots, x_m)$ is the function defined above, which is the potential function of a static electricity field corresponding to a conjunctive normal form.

Lemma 1 *If* $x \geq 1$, *then* $U(x_1, \cdots, x_{i-1}, x, x_{i+1}, x_m) \geq U(x_1, \cdots, x_{i-1}, 1, x_{i+1}, x_m)$, *where* $i = 1, ..., m$.

Proof We have the inequality in this lemma because that, for every $u_l(x_1, \cdots, x_m)$, if $x \geq 1$, then $u_l(x_1, \cdots, x_{i-1}, x, x_{i+1}, x_m) \geq u_l(x_1, \cdots, x_{i-1}, 1, x_{i+1}, x_m)$, $\qquad\square$

Lemma 2 *If $x \leq 0$, then $U(x_1, \cdots, x_{i-1}, x, x_{i+1}, x_m) \geq U(x_1, \cdots, x_{i-1}, 0, x_{i+1}, x_m)$, where $i = 1, ..., m$.*

Proof This lemma can be proved as lemma 1. $\qquad\square$

Lemma 3 *The potential function $U(x_1^{(0)}, \cdots, x_{i-1}^{(0)}, x, x_{i+1}^{(0)}, \cdots x_m^{(0)})$ is a linear function in x with domain $[0, 1]$, where $x_1^{(0)}, \cdots, x_{i-1}^{(0)}, x_{i+1}^{(0)}, \cdots$, and $x_m^{(0)}$ are real numbers.*

Proof It is easy to see that, in every $u_l(x_1, \cdots, x_m)$, each variable x_i does not occur or only occurs once. This implies the linearity of the function U. $\qquad\square$

Lemma 4 *Let the object function U in the above example be $U(x_1^{(0)}, \cdots, x_{i-1}^{(0)}, x, x_{i+1}^{(0)}, \cdots x_m^{(0)})$. Then the sequence, $\{(x_1^{(n)}, \cdots, x_m^{(n)}) | n$ is a natural number $\}$, in the above Algorithm can be defined as following.*

If $U(x_1^{(n)}, \cdots, x_i^{(n)}, 1, x_{i+2}^{(n-1)}, \cdots, x_m^{(n-1)}) \geq U(x_1^{(n)}, \cdots, x_i^{(n)}, 0, x_{i+2}^{(n-1)}, \cdots, x_m^{(n-1)})$, then $x_i^{(n)} = 0$, otherwise $x_i^{(n)} = 1$.

Proof From lemma 1 and lemma 2, we know that the minimal point must be in the closed interval $[0, 1]$. lemma 3 implies that point is either 0 or 1. $\qquad\square$

Having proved the above lemmas, we can reach the following conclusion: Let the object function be the potential function $U(x_1, \cdots, x_m)$ of a static electricity field corresponding to a conjunctive normal form, and let the domain of U be $\{0, 1\}^m$. Then the object function of local search is the same as the object function given above, and *Algorithm 3.1* provides the mathematical framework for the family of local search given in [1, 6].

Acknowledgement

Authors wish to thank Dr. Jun Gu for sending his publications to us.

References

1. Gu, J., *Efficient Local Search for Very Large-Scale Satisfiability Problem*, SIGART Bulletin, Vol. 3, No. 1, Jan. 1992, pp.8-12.
2. Gu, J., *Local Search For Satisfiability (SAT) Problem*, IEEE Transactions on Systems, Man, and Cybernetics, Vol. 23, No. 4, July/August 1993.
3. Gu, J., *Global Optimization For Satisfiability (SAT) Problem*, IEEE Transactions on Knowledge and Data Engineering, Vol.6 , No. 3, June 1994.
4. Li, W., Huang, W., A mathematical-physical approach to the satisfiability problem, Science in China, Vol.38, No.1, 1995.
5. Mitchell,D.,Selman,B.,and Levesque,H.J.(1992),Hard and easy distributions of SAT problems. Proceedings AAAI-92,San Jose,CA,459-465.
6. Selman,B.and Levesque,H.J.,and Mitchell,D.G.(1992), A New Method for Solving Hard Satisfiability Problems, Proc. AAAI-92,San Jose,CA,440-446.

An Efficient Algorithm for Local Testability Problem of Finite State Automata*

Sam Myo Kim[1] and Robert McNaughton[2]

[1] Department of Computer Engineering
Kyungpook National University
Taegu, Korea
[2] Computer Science Department,
Rensselaer Polytechnic Institute
Troy, New York 12180

Abstract. A locally testable language is a language with the property that, for some positive integer j, whether or not a string x is in the language depends on the prefix and suffix of x of length $j - 1$, and the set of substrings of x of length j, without regard to the order in which these substrings occur or the number of times each substring occurs. For any j for which this is true we say the language is j-testable. In previous papers we have proved that (1), given a deterministic finite automaton, it is decidable in $O(n^2)$ time if the language of the automaton is locally testable, (2), for a given k, to decide whether or not the language is k-testable is NP-hard. This paper introduces the concepts of "partially locally testable" and "locally testable with respect to a string x." An $O(k^2n^2)$ time algorithm is introduced which, given a deterministic finite automaton and a string x of length k, decides whether or not the language of the automaton is locally testable with respect to x. When k is a small fixed integer, we can used this algorithm and determine practically whether the language is k-testable.

1 Introduction

The concept of local testability is rooted in the study of pattern recognition. It is best understood in terms of a kind of computational procedure used to classify a two-dimensional image: A window of relatively small size is moved around on the image and a record is made of the various attributes of the image that are detected by what is observed through the window. No record is kept of the order in which the attributes are observed, the positions of the image where each attribute occurs, or how many times it occurs. We say that a classification on the possible images is *locally testable* if a decision about how the image is classified can be made simply on the basis of the set of attributes that occur. Certainly, some patterns involve global constraints and therefore cannot be recognized by local testing. Nevertheless, for many patterns, local testing is sufficient. In [6],

* Partial support for this research was provided by NSF grants CDA-8805910 and CCR-9114725.

local testability is discussed in terms of diameter-limited perceptrons. The one-dimensional analogy to this concept has been well studied and is the subject of this paper. If we think of such an image as a character string, then the classification yields a language. Formally, a locally testable language is defined as follows [1, 7].

Definition 1.1 Let Σ be a finite alphabet. For an integer $k \geq 0$ and $x \in \Sigma^*$ such that $|x| \geq k + 1$, define $f_k(x)$ as the prefix ("front end") of x of length k, $t_k(x)$ as the suffix ("tail") of x of length k, and $I_{k+1}(x)$ as the set of ("intermediate") substrings of x of length $k + 1$, i.e., $\{v | v \in \Sigma^{k+1}$ and $x = uvw$, for some $u, w \in \Sigma^*\}$. For $|x| \leq k$, $f_k(x) = t_k(x) = x$ and $I_{k+1} = \emptyset$ (the empty set). Let $SV_{k+1}(x)$ denote the triple $(f_k(x), I_{k+1}(x), t_k(x))$, called the substring vector of x. For $L \subseteq \Sigma^*$,

(a) L is 0-testable if and only if it is Σ^* or $\emptyset$.
(b) For an integer $k \geq 0$, L is $(k + 1)$-testable if and only if, for all $x, y \in \Sigma^*$, $SV_{k+1}(x) = SV_{k+1}(y)$ implies that either both x and y are in L or neither is in L.
(c) An automaton is j-testable if it accepts a j-testable language.
(d) A language or an automaton is locally testable, if it is j-testable for some $j \geq 0$.

It is well known that all locally testable languages are regular. The only automata considered in this paper are deterministic finite automata which are reduced. Throughout this paper we let $M = (Q, \Sigma, \delta, q_{st}, F)$ be a deterministic finite automaton, where Q is the set of states, Σ the input alphabet, δ the state-transition function, $q_{st} \in Q$ the initial state and $F \subseteq Q$ the set of final (or accepting) states. For any $q \in Q$ and $w \in \Sigma^*$, by $\delta(q, w)$ we mean the state that results when input w is applied to M in state q. For notation we follow [2]. By SCC we mean a maximal strongly connected component. *Transient state* is a state of an automaton that does not belong to any SCC.

Clearly, if a language or an automaton is j-testable, it is j'-testable, for any $j' > j$. We say a locally testable language or automaton has *testability order* j (*order*, for brevity) if it is j-testable, but not $(j-1)$-testable. In the early 1970's, these languages and their finite state automata were extensively investigated [1, 7, 8, 10]. However, there were several open problems concerning locally testable automata that did not yield to practical solutions: (1) Is a given deterministic finite automaton locally testable? (2) If it is locally testable, what is the order? (3) What is the largest order that a locally testable finite automaton with n states can have?

To solve these problems, three characterizations for locally testable languages were introduced in the literature[1, 8, 3]. Brzozowski and Simon [1] introduced the following characterization theorem for locally testable automata. (Much of the theoretical work in this paper is based on part (b) of this theorem.)

Theorem 1.1 For any M:

(a) M is locally testable if and only if, for all $q \in Q$, $x \in \Sigma^+$, $y, z \in \Sigma^*$, and $n \geq |Q|$,

      (i) $\delta(q, x^n y x^n) = \delta(q, x^n y x^n y x^n)$, and

      (ii) $\delta(q, x^n y x^n z x^n) = \delta(q, x^n z x^n y x^n)$.

(b) M is $(k+1)$-testable, if and only if, for all $q \in Q$, $y, z \in \Sigma^*$ and $x \in \Sigma^k$,

      (i) $xy = zx$ implies $\delta(q, xy) = \delta(q, xyy)$, and

      (ii) $\delta(q, xyxzx) = \delta(q, xzxyx)$.

In [8], a characterization of locally testable automata was introduced in terms of algebraic properties of the semigroups of locally testable automata. The papers in [1, 8] suggested algorithms for problems (1) and (2), which simply check whether the semigroup of the given automaton satisfies the appropriate algebraic properties. However, both algorithms take exponential time in the worst case, since the size of the semigroup of a deterministic automaton can be exponential in the number of states of the automaton. The papers left it open whether it was possible to find better algorithms for problems (1) and (2). There was an answer to problem (3) in [1], which was a bound equal to the size of the semigroup of the automaton, which is also exponential in the number of state in the worst case. In [3] we gave a characterization of locally testable deterministic finite automata and introduced an $O(n^2)$ time algorithm for problem (1), the local testability problem, where n is the number of states of the automaton. Recently, we investigated the following problem to study problem (2), i.e., computing the order.

    The j-testability problem: Given a deterministic finite automaton M, and a nonnegative integer j, is M j-testable?

In [4], we showed that this problem is co-NP-complete and presented a polynomial-time ϵ- approximation algorithm for computing the order. The algorithm, given a constant $\epsilon > 0$ and a locally testable deterministic automaton, computes an approximate order $\hat{j}$ of the automaton such that if j is the order then $\hat{j} \geq j$ and $(\hat{j} - j)/j \leq \epsilon$. For any constant ϵ the algorithm runs in polynomial time, the degree of the polynomial depending on ϵ. With this approximate algorithm it is possible to solve the j-testability problem in polynomial time if j is a fixed integer. However, the algorithm is not practical even for small j, since its time complexity is $O(n^{max(10,6+2j)})$.

This paper has two major objectives: One is to investigate a practical algorithm for the j-testability problem for a small fixed integer j, and the other is to investigate automata whose languages are partially locally testable in the sense defined as follows.

Definition 1.2 An automaton M is locally testable with respect to a string $x \in \Sigma^*$ if and only if, for all $y, z \in \Sigma^*$ and $q \in Q$, it satisfies

(i) xy = zx implies $\delta(q, xy) = \delta(q, xyy)$, and

(ii) $\delta(q, xyxzx) = \delta(q, xzxyx)$.

Clearly, from this definition and Theorem 1.1-(b), we have

Lemma 1.1 A deterministic finite automaton is $(k+1)$-testable if it is locally

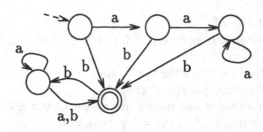

Fig. 1. Partially locally testable automaton.

testable with respect to all $x \in \Sigma^k$. Notice that M can be locally testable with respect to string x, while it is not with respect to others, i.e., it is partially locally testable. Figure 1 shows an automaton which is locally testable w.r.t. $x = a$, but not w.r.t. $x = b$. (Actually, it is not locally testable w.r.t. b^i, for all $i \geq 1$.) The question is; given x, how fast can we identify such a property? More specifically, we are interested in the following problem.

> **Partial testability problem:** Given a deterministic finite automaton M and a string x, is M locally testable with respect to x?

Following some preliminaries in Section 2, Section 3 introduces a new characterization of locally testable deterministic finite automata. Based on this characterization, Section 4 presents an efficient algorithm for the partial testability problem, which can be used to solve the k-testability problem efficiently for a small fixed integer k.

2 Preliminaries

Throughout this paper with no further specification we assume that a problem instance is given in terms of the state transition graph of a deterministic automaton M which is reduced.

Definition 2.1 (reaching component) Let G be the state transition graph of a finite state automaton. For a pair of states s_i and s_j, the **reaching component** from s_i to s_j, written m_{ij}, is the component of G that consists of s_i and all of its descendants from which s_j is reachable. Note that if s_j is in an SCC m, then m is also included in m_{ij}. In particular, when s_i is the start state, we shall write the reaching component as m_{0j} and simply call the reaching component to s_j.

Definition 2.2 (input span) Let $M = (Q, \Sigma, \delta, q_0, F)$ be finite state automaton. A state $q \in Q$ has an input span x if $\delta(p, x) = q$, for some $p \in Q$. We call state p an **origin** of input span x of q.

Definition 2.3 (repeated span, repeating span) Let w be a span of a path such that $w = xy = zx$. The prefix x of w is called **repeated span** and the suffix x **repeating span**.

Note that repeated span and repeating span belong to a common path of the state transition graph, possibly overlapped.

Definition 2.4 (s-local) A component m of a state transition graph of an automaton is **s-local** with respect to a string x if and only if no two distinct states exist in m that have the same repeating input span x. The component m is s-local of order k if and only if m is s-local with respect to every string x of length k. Figure 2 illustrates typical examples which violates s-locality condition with respect to a string x, where circles denote SCC's. For clearity the figure shows only the cases of repeated and repeating spans that do not overlap.

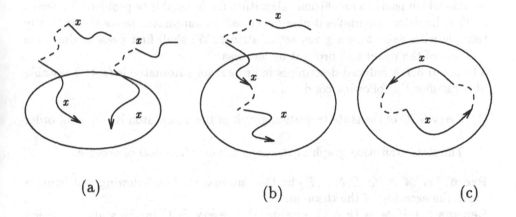

(a)   (b)   (c)

Fig. 2. Typical transition patterns violating s-locality.

Definition 2.5 (TS-local) A finite state automaton is **TS-local** with respect to a string x if and only if for every pair of states p and q, either

(1) x is not an input span of p that is repeated by an input span of q, or
(2) if the input span x of p is repeated by an input span of q, then, for all $w \in \Sigma^*$, state q is reachable from state $\delta(p, w)$ if and only if q is reachable from $\delta(q, w)$.

A finite state automaton is TS-local of order k if and only if it is TS-local with respect to every string x of length k.

Definition 2.6 (pair-graph) Let $Q_1, Q_2 \subseteq Q$ which are not necessarily disjoint, and $*$ be a distinct symbol not in Q. The **pair-graph** on $Q_1 \times Q_2$ is the edge-labeled directed graph $G(V, E)$, where $V = (Q_1 \cup \{*\}) \times (Q_2 \cup \{*\}) - \{(*, *)\}$, and edges in E are defined as follows: Let $a \in \Sigma$ and $(p, q) \in V$ such that $p \in Q_1$, $q \in Q_2$ and $p \neq q$.

(1) G has an edge from (p, q) to (r, s) with label a if $\delta(p, a) = r$ and $\delta(q, a) = s$, for some $r \in Q_1$ and $s \in Q_2$.

(2) G has an edge from (p, q) to $(r, *)$ with label a if $\delta(p, a) = r$, for some $r \in Q_1$, and $\delta(q, a)$ is not in Q_2.

(3) G has an edge from (p, q) to $(*, s)$ with label a if $\delta(p, a)$ is not in Q_1 and $\delta(q, a) = s$, for some $s \in Q_2$.

Notice that there are no outgoing edges from a node (p, q), if either $p = q$, $q = *$ or $p = *$. Such nodes are terminal nodes.

3 A Characterization of $(k + 1)$-testable Automata

This section introduces a new characterization theorem for locally testable automata which leads to an efficient algorithm for k-testability problem, for small k. The algorithm also makes it possible to test an automaton to see if it is locally testable with respect to a given set of strings. We shall first present the main theorem of the paper and prove it by lemmas.

Theorem 3.1 A reduced deterministic finite state automaton is $(k+1)$-testable iff it satisfies the following conditions:

(1) Every SCC of the state transition graph of the automaton is s-local of order k.

(2) The state transition graph of the automaton is TS-local of order k.

Proof. Let $M = (Q, \Sigma, \delta, q_0, F)$ be the automaton. The following two lemmas prove the necessity of the theorem.

Lemma 3.1 If M is $(k + 1)$-testable, then every SCC in the state transition graph of M is s-local of order k.

Proof. Suppose that the lemma is not true, and there is an SCC m which is not s-local of order k. Then m contains two distinct states p and q that have the same repeated input span $x \in \Sigma^k$. This implies that there are two spans w_1 and w_2 such that, for some states r and s,

(a) $w_1 = xy_1 = z_1 x$ and $\delta(r, xy_1) = p$, and
(b) $w_2 = xy_2 = z_2 x$ and $\delta(s, xy_2) = q$.

Since M is $(k + 1)$-testable, by Theorem 1.1-(b) $\delta(r, xy_1) = p = \delta(r, xy_1 y_1)$. State p has a loop on y_1, which must be in SCC m. Clearly, $xy_1{}^j = z_1{}^j x$, for all $j \geq 0$, which implies that there is a state t in m (actually, on the loop corresponding to $\delta(p, y_1) = p$) such that $\delta(t, x) = p$. By the same argument with part (b) above, we can show that m has a state t' such that $\delta(t', x) = q$.

Now, let u be a span on a path from p to t' and v be a span on a path from q to t. Since M is $(k + 1)$-testable, by Theorem 1.1-(b) we have $\delta(t, xux) = q = \delta(t, xuxux)$ and $\delta(t', xvx) = p = \delta(t', xvxvx)$, which implies that q has a loop on ux and p has a loop on vx. It follows that $\delta(t, xuxvx) = p \neq \delta(t, xvxux) = q$. It violates Theorem 1.1-(b). Automaton M is not $(k+1)$-testable, a contradiction. $\square$

Lemma 3.2 If M is $(k + 1)$-testable, then the state transition graph of M is TS-local of order k.

Proof. Suppose the lemma is not true. Then there exist two states p and q such that p has an input span $x \in \Sigma^k$ which is repeated by an input span of q and, for some string w, either

(a) there is a path from $\delta(p, w)$ to q, while no path exists from $\delta(q, w)$ to q, or

(b) there is no path from $\delta(p, w)$ to q, while there is a path from $\delta(q, w)$ to q.

Since p's span x is repeated by q's span, there is a span w such that $w = xy = zx$ and $\delta(r, x) = p$ and $\delta(r, xy) = q$, for some state r. Since M is $(k+1)$-testable, by Theorem 1.1-(b) $\delta(r, xyy) = q$, which implies that q has a loop on y. State q is a state in an SCC. Let m be this SCC. Since $xy^j = z^j x$, for all $j \geq 0$, there is a state s in m such that $\delta(s, x) = q$.

Suppose part (a) above is true. Find a path from $\delta(p, w)$ to s and let u be a span on this path. Then $\delta(r, xwux) = q \neq \delta(r, xwuxwux)$ because $\delta(r, xwuxwux) = \delta(q, wux)$ and there is no path from $\delta(q, w)$ to q, i.e., $\delta(q, wux)$ is out side of m. The automaton does not satisfy Theorem 1.1-(b) and, hence, is not $(k+1)$-testable, a contradiction.

Now, suppose part (b) above is true. Let y be a span on a path from p to q and let u be a span on a path from $\delta(q, w)$ to s. Then $\delta(r, xywuxwux) = q \neq \delta(r, xwuxywux)$ because $\delta(r, xwuxywux)$ is not in m. Again the automaton does not satisfy Theorem 1.1-(b) and, hence, is not $(k+1)$-testable, a contradiction.

Now, the following lemma proves the sufficiency of the theorem.

Lemma 3.3 Let $M = (Q, \Sigma, \delta, q_{st}, F)$ be a reduced deterministic automaton. If M satisfies the following conditions, then it is $(k+1)$-testable.

(1) Every SCC of the state transition graph of the automaton is s-local of order k.

(2) The state transition graph of the automaton is TS-local of order k.

Proof. Suppose M is not $(k+1)$-testable. Then by Theorem 1.1-(b) there exist $p \in Q$, $x \in \Sigma^k$ and $y, z \in \Sigma^*$ such that either

(a) $xy = zx$ and $\delta(p, xy) \neq \delta(p, xyy)$, or

(b) $\delta(p, xyxzx) \neq \delta(p, xzxyx)$.

Case 1. $xy = zx$ and $\delta(p, xy) \neq \delta(p, xyy)$. Let $q = \delta(p, z)$ and $r = \delta(p, x)$ and $s = \delta(q, x)$. States r has an input span x which is repeated by an input span of s. Obviously, there is a path from $\delta(r, y) = s$ to s. By condition (2) (i.e., TS-locality property) of the lemma, there should be a path from $\delta(s, y)$ to s. It follows that s and $\delta(s, y)$ are in the same SCC. Let m be this SCC. Let $t = \delta(p, xyy) = \delta(s, y)$. Since $xyy = zzx$, state t has repeating input span x. We have two distinct states s and t in SCC m that have the same repeating input span x whose length is k. The automaton is not s-local of order k violating condition (1) of the lemma.

Case 2. $\delta(p, xyxzx) \neq \delta(p, xzxyx)$. Let $r = \delta(p, x)$, $s = \delta(r, yxzx)$ and $s' = \delta(r, zxyx)$. Clearly, state r has an input span x which is repeated by an input span of s. Since there is a path from $\delta(r, yxzx) = s$ to s, by condition (2) of the lemma there exists a path from $\delta(s, yxzx)$ to s, i.e., states s and $\delta(s, yxzx)$

are in the same SCC. By condition (1) of the lemma, we have $\delta(s, yx) = s$ and $\delta(s, zx) = s$. It follows that $\delta(s, zxyx) = s$, which, by condition (2) of the lemma, implies that there is a path from state $\delta(r, zxyx) = s'$ to s.

By the same argument with the role of s and s' interchanged, we can show that s' is in an SCC and there is a path from s to s'. Since s and s' are in an SCC and there is a path from the one to the other, they must belong to the same SCC. Since s and s' have identical input spans x which are repeating, the automaton violates condition (1) of the lemma, a contradiction. $\square$.

If an automaton has a transient state s_j which has repeating input span x, it is not TS-local w.r.t. x since state s_j is not reachable from $\delta(s_j, w)$, for any $w \in \Sigma^+$. Hence, we have the following corollary.

Corollary. If M has a transient state which has a repeating input span x of length $k \geq 0$, then the automaton is not $(k+1)$-testable.

Using Theorem 3.1 we can easily prove the following theorem which was presented in [4] with a different proof.

Theorem 3.2 A reduced deterministic automaton M is 1-testable if and only if, for all $q \in Q$ and $a, b \in \Sigma$,

(1) $\delta(q, a) = \delta(q, aa)$ and
(2) $\delta(q, ab) = \delta(q, ba)$.

Before we go to Section 3 which introduces an algorithm for the j-testability problem, for arbitrary j, we present the following lemma which will be conveniently used by the algorithm.

Lemma 3.4. Let G be the state transition graph of a deterministic finite automaton M. For a state $s_j \in Q$ which has repeating input span x, let Q_{0j} be the set of states in the reaching component to s_j, and let Q_{jj} be the set of states in the reaching component from s_j to s_j. Let T_j be the set of states in Q_{0j} that have input span x which is repeated by an input span of s_j. Automaton M is not TS-local w.r.t. x if and only if the pair-graph on $Q_{0j} \times Q_{jj}$ has a path from a node in $T_j \times \{s_j\}$ to a node $(t, *)$ or $(*, t)$, for some $t \in Q_{0j}$.

Proof. Suppose M is not TS-local w.r.t. x. Then, by Definition 2.3 there is a pair of states s_i and s_j such that s_i's input span x is repeated by s_j's input span and, for a string $w \in \Sigma^*$, either

(a) s_j is reachable from $\delta(s_i, w)$, but not from state $\delta(s_j, w)$, or
(b) s_j is reachable from $\delta(s_j, w)$, but not from $\delta(s_i, w)$.

Let $t = \delta(s_i, w)$ for case (a), and $t = \delta(s_j, w)$ for case (b). Clearly, $s_i \in T_j$ and the pair-graph will have a path from (s_i, s_j) to either $(t, *)$ or $(*, t)$, depending on case (a) or (b), with span w. Now, suppose that the pair-graph has a path from a node $(s_i, s_j) \in T_j \times \{s_j\}$ to a node $(t, *)$, for some state $t \in Q_{0j}$. Obviously s_j is reachable from $\delta(s_i, w) = t$ because t is in the reaching component to s_j. However, s_j is not reachable from $\delta(s_j, w)$ by the implication of flag $*$ (recall Definition 2.6). The automaton violates condition (2) of Definition 2.5, and, by Lemma 3.3 it is not TS-local w.r.t. x. When there is a path from a node $(s_i, s_j) \in T_j \times \{s_j\}$ to a node $(*, t)$, the argument is similar. $\square$

4 The Algorithm

Now, we present our main theorem together with an algorithm for the proof. The algorithm, given an automaton M and a string $x \in \Sigma^k$, decides whether or not M is locally testable w.r.t. x in $O(\alpha n^2 + k^2 n^2)$ time, where n is the number of states and α is the alphabet size.

Theorem 4.1 The partial testability problem is solvable in $O(\alpha n^2 + k^2 n^2)$ time, where α, k and n are, respectively, the alphabet size, the length of string x and the number of states of the automaton.

Proof. Because of limited space, we omit detailed algorithm which is presented in [5]. Suppose that the partial testability problems is w.r.t. string x. The algorithm takes the following four major steps and tests conditions (1) and (2) of Theorem 3.1. Let α be the size of the alphabet and n be the number of states of the automaton.

(1) Identify SCC's: We use well know Tarjan's algorithm[9] which can mark all SCC's of the state transition graph in time linear to the size of the graph.

(2) Test input span: The algorithm marks all states which have input span x and the origins of the span. This step takes no more than $O(kn)$ steps.

(3) Test s-locality: For every state the algorithm first tests if it has an repeating input span x. If a string has a repeating string x, then it has substring w which has the property $w = xy = zx$, for some nonnull strings y and z. Thus, if a state p's input span x is repeated by an input span of a state q, then there is a string y such that $xy = zx$ and $\delta(p, y) = q$. If w has nonoverlapped repeating x, then $y = ux$ and using the marks put in step (2), we can easily test if a state has nonoverlapping repeating span x in time $O(\alpha^2 n^2 + k^2 n^2)$. Let $x = a_1 a_2 \dots a_k$, and let $y_r = a_{k-r+2} a_{r+3} \dots a_k$ and $z_r = a_1 a_2 \dots a_{r-1}$, for $1 < r \le k$. If $a_{r+i} = a_{1+i}$, for all i, $0 \le i \le k-r$, then we have $xy_r = z_r x$ with two x's overlapped. For each state p which has input span x, we test if there exist y_r, $1 < r \le k$, such that $\delta(p, y_r) = q$, for some q which is marked as an origin of span x. This can be carried out in time $O(\alpha^2 n^2 + k^2 n^2)$. Once we identify all states which have repeating input span x, we can easily chech if the automaton has two such states. If it does, the automaton is not s-local.

(4) Test TS-locality: Suppose that the automaton has passed step (3) for s-locality test. By Corollary of Theorem 3.1, if a transient state has repeating input span x, the automaton is not TS-local w.r.t. x. Suppose that the automaton has no such state. Having passed this test and step (3), every state q which has repeating input span x must be in an SCC and no other state in that SCC will have repeating input span x. Let m_q be the SCC. Let Q_{0q} and Q_q be, respectively, the set of states in the reaching component to q and the set of states in m_q. Using the pairgraph on $Q_{0q} \times Q_q$, we can test if, for each node (p, q) in the pairgraph, states p and q satisfy the condition of Definition 2.5, in time $O(\alpha^2 n^2)$.

$\square$

5 Conclusion

We have presented an $O(\alpha n^2 + k^2 n^2)$ time algorithm for the partial testability problem. Using the algorithm, the j-testability problem can be practically solvable for small j, which partially answers an open question raised in [4]. This algorithm has a potential application in the field of pattern recognition, since oftentimes we may only interested in the local testability of a language with respect to a relatively small set of test patterns.

References

1. Brzozowski, J, Simon, I.: Characterizations of locally testable events, Discrete Mathematics, 4 (1973) 243-271.
2. Hopcroft, J. E., Ullman, J. D.: Introduction to Automata Theory, Languages, and Computation, Addison Wesley, 1979.
3. Kim, S. M., McNaughton, R., McCloskey, R.: A polynomial time algorithm for the local testability problem of deterministic finite automata, IEEE Trans. Computers, 40 (1991) 1087-1093.
4. Kim, S. M., McNaughton, R.: Computing the Order of a Locally Testable Automaton, SIAM J. Computing 23 (1994) 1193-1215.
5. Kim, S. M., McNaughton, R.: An Efficient Algorithm for Local Testability Problem of Finite State Automata, Computer Science Department, Rensselaer Polytechnic Institute. Tech. Report 93-8.
6. Minsky, M., Papert, S.: Perceptrons, M.I.T. Press, 1969.
7. McNaughton, R., Papert, S.: Counter-free Automata, M.I.T. Press, 1971.
8. McNaughton, R.: Algebraic decision procedures for local testability, Mathematical Systems Theory, 8 (1974) 60-76.
9. Tarjan, R. E.: Depth first search and linear graph algorithms, SIAM J. Computing 1 (1972) 146-160.
10. Zalcstein, Y.: Locally testable languages, Journal of Computer and System Sciences, 6 (1972) 151-167.

Scheduling Task-Tree with Additive Scales on Parallel/ Distributed Machines

Xiangdong Yu[1][2] and Moti Yung[3]

[1] Department of Computer Science, Columbia University, New York, NY 10027
[2] Partially supported by NSF grant CCR-93-16209 and CISE Institutional Infrastructure Grant CDA-90-24735
[3] IBM Research Division, T.J. Watson Research Center, Yorktown Heights, NY 10598

Abstract. Scheduling interdependent tasks on parallel/distributed architectures is a hard problem. We consider jobs consisting of tasks whose interdependency relationships form a tree, such as QuickSort, Brute-Force Search, and other Divide-and-Conquer jobs. Tree nodes (tasks) have a scale which satisfies an "additive condition": a leaf's scale is 1 and a non-leaf's is the sum of its children's. The execution time L of a task is a function of its scale S, like $aS^\alpha + b$ for some known constants a, b, α. The task set and the shape of the tree, however, may or may not be known in advance. We provide a general algorithm that assigns these tasks to processors in a large set of architectures (including meshes, linear arrays, and rings). We also discuss the relationship between certain complexity parameters and the tree shape. We show that for almost all cases considered, the scheduling achieves optimal or nearly optimal time.

1 Introduction

1.1 The Problem

Definition 1. A task tree job $J = <T, D>$ consists of a task set T and dependency relation D such that the graph J is a rooted tree with maximum outdegree d and minimum outdegree 2 for non-leaf nodes. Each task $t \in T$ has a scale $S(t)$ and a length $L(t)$. L is a function of S and it takes time $L(t)$ for a processor to finish task t. The scale of the job $S = S(root)$.

For any node $t \in T$, the *subtree rooted at t*, or simply subtree t, is the subtree consisting of t and all its descendants. On the other hand, t's subtrees refers to the subtrees rooted at t's children. The work $W(t)$ is defined as the sum of the lengths of all the tasks in subtree t. The work of the job $W = W(root)$.

The most well-known task tree job is divide-and-conquer (D&C) where the execution starts from the root node. In an *expanding* stage, starting from the root, each non-leaf node spawns out children. In a *shrinking* stage, starting from leaf nodes, each task is finished and the result is reported to the parent. The job is done when the root reports result. For this kind of problem, the function S may satisfy a so-called additive condition.

Definition 2. A job $J = < T, D >$ satisfies additive condition if

1. $S(v) = 1$ if v is a leaf node.
2. Otherwise $S(v) = \sum_{u \text{ is } v's \text{ child}} S(u)$.

Intuitively, $S(v)$ approximates the data size or the search space size handled at node v. The length may depend on S in different ways. In QuickSort, $L = aS + b$ for some constants a and b. In a brute force checking of the Satisfiability of a boolean expression, dividing is simply done by fixing the value of a variable to split the search space $\{0, 1\}^k$ into two $\{0, 1\}^{k-1}$. Therefore L is a constant, written as $aS^0 + b$. In this paper we consider functions L of the form $aS^\alpha + b$ for some constant $a, b, \alpha \geq 0$.

The underlying computing architecture can be represented as a graph $G = < V, E >$ where V is the set of processors and E is the set of communication links. We assume the computer is synchronous. A processor can perform a computation step and all-link communication at the same time unit. We do not explicitly specify the topology of the computing architecture. It turns out that a tree job has very flexible scheduling strategy that is easily applicable to many topologies. The performance of the algorithm depends on what we define latter as the "dimension of the underlying graph".

As we defined, a processor takes time $L(t)$ to finish task t. We also assume that similar amount of time is needed to transmit the task (or its result) from one processor to its neighbor, although we do not require this time to be continuous, i.e. the transmission is allowed to be interrupted and be finished within a few time units. Our scheduling is non-preemptive. The expanding stage defines a mapping M from J to G. More specifically, task T_i is assigned to a processor $M(T_i)$, and tree edge $< X, Y >$ is mapped to a path $M(X, Y)$.

Strictly speaking, the task lengths in the expanding stage and the shrinking stage are not necessarily the same, normally one stage is dominant. The scheduling is based on the dominant task lengths. We assume shrinking is the dominant stage and the goal is to schedule the tasks so that the shrinking can be done efficiently. We stress that the same results hold when the expanding is dominant or when the two stages are balanced.

1.2 Related Work

Task scheduling is a well-studied problem. Its general version is NP-hard [9][1]. Various assumptions and simplifications are typically made. Regarding the job, one may assume restrictions on task length, dependency, arrival time, etc. Regarding the computing architecture, one may choose to ignore or account for computation time, communication time, and topology constraints. The goal of the scheduling can be: time efficiency, processor economy, load balance, on-line competitive ratio, fault tolerance, and various tradeoffs. Different versions reflect different characteristics of the investigated problems.

Some models focus on the unpredictability of task length and arrival time, ignore interdependency among tasks and communication costs. Such results include the competitive scheduling appearing in[8] and [19]. One such related result is the optimal/near optimal scheduling algorithm presented in [6] which also deals with task trees with unknown shapes and task lengths. However, [6] ignores the topology of the machine and measures the complexity under a uniform (universal) communication delay proposed in [15] and [16]; it is thus somewhat over-pessimistic when compared with our results. In [17] one can find an on-line protocol for load balancing. In other models, tasks are processors in another architecture (referred to as task graph) that is to be simulated by the current architecture (referred to as system graph)[3],[5],[14],[20], scheduling then becomes graph mapping. In general, the objective is to obtain a mapping which minimizes the *simulation overhead*, characterized by dilation, congestion, expansion, or communication time. The interconnection among tasks is more for communication purpose than for logical dependency. For models considering both logical dependency and communication cost, there are randomized algorithms [2][4] that optimize load balance, dilation, or congestion.

Hu[12] studied the parallel processing of precedence graph in early 60's. More algorithms can be found in [10]. Wu and Kung [21] found a dynamic algorithm achieving optimal load communication tradeoff in solving the D&C, their algorithm ignores the architecture topology. In [22] any tree job with unit task length and arbitrary degree and height was considered and it was shown how to schedule it optimally in time on typical architectures like meshes and arrays. Using the language of this paper, the job in [22] is one with $\alpha = 0$, which is a limited case we generalize here. Dynamic allocation of trees with uniform costs are also considered in [13, 18, 4].

1.3 Organization of This Paper

In this work we give task scheduling algorithms that make use of partial knowledge on S and W and achieve different levels of optimality.

The paper is organized as follows: First we give a lower bound, and some properties of scale, length, height and work. Then we describe a scheduling strategy that results in optimal or nearly optimal job execution time in static, semi-dynamic and dynamic cases.

2 The Lower Bound

Definition 3. Given job $J =< T, D >$, for each $t \in T$, define height $H(t) = MAX_{p \text{ is a } t \text{ to leaf path}}\{\sum_{u \in p} L(u)\}$. A path p that achieves the maximum is called a leg path of $H(t)$. The height of the job tree, $\mathcal{H}$, is $H(root)$.

Definition 4. Given job $J =< T, D >$, the co-length of task t $C(t)$ is defined as $\sum_{u \text{ is } t's \text{ child}} L(u)$, which represents the communication time of transmitting all the child results into t's processor through a single link. The co-length of the job tree $\mathcal{C}$ is $MAX_{t \in T}C(t)$.

Definition 5. The dimension of a family of graphs $F = \{G_1, G_2, \ldots\}$ is the minimum number m satisfying the following condition: for any G_i and any node $v \in G_i$, the number of nodes in v's x-distance neighborhood is $O(x^m)$ for all $x \leq R_{G_i}(v)$ where $R_{G_i}(v)$ is the radius of G_i around v.

According to this definition, the family of linear arrays has dimension 1 and the family of meshes has dimension 2.

Theorem 6. *Given tree job* $J = < T, D >$, *with height* $\mathcal{H}$ *and work* $\mathcal{W}$,
(1) Let $B = max\{\mathcal{H}, \mathcal{W}^{1/(m+1)}\}$, $\Omega(B)$ *time is needed to execute all tasks in* J *on any architecture of dimension* m, *under any task mapping scheme.*
(2) If the architecture graph has constant degree, the above is true for $B = max\{\mathcal{C}, \mathcal{H}, \mathcal{W}^{1/(m+1)}\}$.

3 Scale, Length, Height, and Work

In the scheduling of a job tree, we always assume $S(u)$ is known before task u is scheduled. But knowing only $S(u)$ is much different from knowing also $W(u)$. As we will see, the latter results in optimal scheduling, while the former not necessarily. Given a job tree $J = < T, D >$, the lemmas below describe a few properties about S, L, H, W and $\mathcal{S}, \mathcal{H}, \mathcal{W}$.

Lemma 7. *Given tree job* J. *if* $\alpha > 0$, *we have,*

- $\mathcal{H} = O(\mathcal{S}^{1+\alpha})$ *and* $\mathcal{H} = \Omega(\mathcal{S}^\alpha)$;
- $\mathcal{W} = O(\mathcal{S}^{1+\alpha})$; $\mathcal{W} = \Omega(\mathcal{S})$ *if* $0 < \alpha < 1$ *and* $\mathcal{W} = \Omega(\mathcal{S}^\alpha)$ *if* $\alpha > 1$.

The most intrinsic factor that affects the parallelizability is the height of a job tree. In the design and analysis of the algorithm, we will use this intuition: we may not care that an algorithm does not do well for jobs with large work, because the height would also be large in those cases and a bad performance would not look so bad as compared to the best parallel execution of the job. To quantify this, we give the following definition and lemma.

Definition 8. The serial exponent function (SEF) $u_\alpha(c)$, or $u(c)$ when α is clear, is defined on domain $[1, 1 + \alpha]$ when $0 < \alpha < 1$ and on $[\alpha, 1 + \alpha]$ when $\alpha \geq 1$, as the maximum number x satisfying the following condition: Any job tree $J = < T, D >$ with $W = \Omega(\mathcal{S}^c)$ has height $\mathcal{H} = \Omega(\mathcal{S}^x)$. This function characterizes the unparallelizability of jobs with a given work.

By this definition, a job with work $\mathcal{S}^c$ can expect a factor of at most $\mathcal{S}^{c-u(c)}$ speedup through parallelization. As we show, the scheduling algorithm always gives optimal solution when $\alpha \geq 1$, we are, therefore, more concerned with the case of $0 < \alpha < 1$ in the lemma below.

Lemma 9. (Serial Exponent Function).
If $0 < \alpha < 1$, *then* $u(c) = max\{(c-1)(1+\alpha)/\alpha, \alpha\}$.
If $\alpha \geq 1$, *then* $u(c) \leq max\{(c-1)(1+\alpha)/\alpha, \alpha\}$.

4 A Generic Scheduling Algorithm

We will have a two stage design. We first map the tasks into a "virtual" architecture which is a linear array. Then we map the linear array into real architecture by a "snaking" technique as was proposed in [22].

The virtual machine is viewed as a half number axis $[0, \infty)$, interval $[x, x+1)$ is treated as a processor for integer x. $[0, 1)$ is the starting processor. The scheduling algorithm decides on a number B called *bin size*, roughly corresponding to the amount of work to be loaded in a processor. Each node (or subtree) with length (work) X is viewed as a line segments of length X/B. The mapping of tasks to processors is carried out by embedding line segments into the number axis. Note that there may be inclusion relations between line segments, to reflect the inclusion relation between subtrees and between nodes and subtrees.

Since the root task has length $L(root) = aS^\alpha + b$, we will always choose $B \geq L(root)$. On the other hand, before a subtree t is scheduled, the algorithm must have some estimate or bound on its work $W(t)$. We denote this estimate or bound by $\tilde{W}$. As we mentioned before, we assume $S(t)$, thus also $L(t)$, is always known before t is scheduled (this is natural for D&C, so we do not consider it as a serious restriction).

Definition 10. If the shape of the tree is fully known beforehand, we call the scheduling static. If $\tilde{W}(u) = W(u)$ for each child u of v after v is processed, we call the scheduling semi-dynamic. If only $S(t)$ is known, we call the scheduling dynamic.

The execution starts at processor P_0 on subtree *root*. The working interval is $I = [0, \infty)$. In general, as a branch of the parallel computation, processor P_i is scheduling subtree t, whose working interval is $[x, x + N(t))$ with $i \leq x < i+1$. Here $N(t) = MAX\{L(t), \tilde{W}(t) - L(t)\}$, is an upper bound on the resources subtree t can use at a time. P_i first processes task t, in time $L(t)$, finds out it has children $r_1, r_2, \ldots, r_k$ and makes estimate $\tilde{W}(r_i)$ for $1 \leq i \leq k$. Then it spawns out the children by simply embedding line segments of length $N(r_1)/B, N(r_2)/B, \ldots, N(r_k)/B$ sequentially side by side into $[x, x + N(t))$, starting from point x. Then P_i transmits r_i's task data to the left most processor in r_i's working interval, for $i = k, (k - 1), \ldots, 1$. Then the above procedure repeats in each of the subintervals.

One subtlety is: what if $\tilde{W}$ under-estimates W? The parent's line segment will not be able to hold all the children's segment without overlapping. We solve this problem by proportionally contracting children's segments so that they fit into the parent's segment exactly. If $\tilde{W} = S^\beta$, then an estimated work A may actually be $A^{(1+\alpha)/\beta}$. Below we use $\Delta(A)$ to denote the possible actual work of an estimated amount A. So,

$$\Delta(A) = \begin{cases} A & \tilde{W} = W \\ A^{(1+\alpha)/\beta} & \tilde{W} = S^\beta \end{cases}$$

There are a few details we want to explain:

(1) If $N(t) \leq B$ for some t, the subtree will not be scheduled any more and will be treated as a single task (i.e., put into a single processor).

(2) If the line segment of a single task is not fully within a processor, it will be assigned to the left processor it touches. Since any original single task has length $\leq L(root) \leq B$, and the pseudo-single task obtained in (1) also has work $\leq B$, the total amount of work associated to the *unprocessed* tasks in any single processor is no more than $\Delta(2B)$ (but, considering those already processed parents, the work may be more than $\Delta(2B)$ but less than $\Delta(2B) + \mathcal{H}$).

(3) There are totally $O(\tilde{W}(root)/B)$ number of processors involved.

(4) Before mapping out children $r_1, r_2, \ldots, r_k$, the algorithm sorts them in a scale increasing order and maps larger children to farther area. The idea is to let the computation and communication time of smaller tasks "hide" in that of the larger ones.

In the shrinking stage, each processor first finishes all the tasks that do not depend on outside results. This action is called *in-bin* computation and takes time $\leq \Delta(2B) + \mathcal{H}$.

Lemma 11. *After in-bin computation, all the unprocessed tasks in a processor form a dependency chain.*

After in-bin computation, the results of tasks are reported to parents where further computation and report is carried out. Assume task t has the children $r_1, r_2, \ldots, r_k$. Then the results from children come to t's processor in batches, there is a *time gap* between any two consecutive batches, and the last batch determines the time when t can start to be processed. Therefore, if we consider a new scenario where all children not in the last batch are removed while leaving the mapping of other children unchanged, the time to start t will be the same.

In the transmissions of results back to a parent, contention may occur on certain communication links. When "snaking" the virtual machine onto a real machine, adjacent nodes in the former are mapped to adjacent nodes in the latter, nodes closer to the starting node on the virtual machine are mapped to nodes closer to the starting node on the real machine, and the result of any task is routed to a parent through areas occupied by t and t's left brothers only, where left brothers are those mapped to t's left on the virtual machine. In our method, any left brother's scale is no larger than $S(t)$. This mapping is called a **monotone mapping**. Whenever contention occurs at a link, priority is always determined based on (1) which is mapped more to the left; (2) whose parent is mapped more to the left on the virtual machine.

Given the above mapping and routing method M, we can tell that results in the last batch coming to t, say $v_1, v_2, \ldots, v_l$, have increasing order in scale. The time when t is finished will be $\tau_t = L(t) + \tau_{v_1} + route\_length(v_1, t) + \sum_{i=1}^{l} S(v_i)$. Repeating this analysis bottom up, we will find a sequence of tasks $root = t_0, t_1, \ldots, t_z$, such that for $1 \leq i \leq z$, t_{i-1} has chlidren $t_i = v_1^i, v_2^i, v_3^i, \ldots, v_{d_i}^i$ in a last batch and

$$\tau_{t_{i-1}} = L(t_{i-1}) + \tau_{t_i} + route\_length(t_i, t_{i-1}) + same\_bin(t_i, t_{i-1}) \sum_{j=1}^{d_i} S(v_j^i)$$

where $same\_bin(t_i, t_{i-1}) = 1 - \delta(route\_length(t_i, t_{i-1}))$. Let $Q_i = \sum_{j=1}^{d_i} S(v_j^i)$, we get the finishing time of the entire job:

$$\tau_{root} = \sum_{i=1}^{z} L(t_{i-1}) + \tau_{t_z} + route\_length(t_z, root) + same\_bin(t_i, t_{i-1}) \sum_{i=1}^{z} Q_i.$$

$$\leq \mathcal{H} + (\Delta(2B) + \mathcal{H}) + span(M) + \sum_{i=1}^{z} Q_i.$$

where $span(M)$ is the largest $route\_length(leaf, root)$ among all *leaves* under mapping M.

Lemma 12. $\sum_{i=1}^{z} Q_i = O(\mathcal{H} + C\phi(\alpha, d))$, *where*

$$\phi(\alpha, d) = \begin{cases} 1 & \alpha \geq 1 \text{ or } d = O(1) \\ \log \log d & d \geq \beta \end{cases}$$

Theorem 13. *Given a task tree J, an architecture G. The above mapping method M and contention resolving protocol give a processing time of $O(\Delta(2B) + \mathcal{H} + C\phi(\alpha, d) + span(M))$.*

There are $O(\tilde{W}(root)/B)$ processors used in the virtual machine. For many architectures it is easy to show that a monotone mapping can be implemented by snaking, i.e., mapping the virtual machine node one by one following a traversal on the real machine that covers processors closer to the starting processor first, such that $span = O((\tilde{W}(root)/B)^{1/m})$. For example, on linear arrays the snaking is straight, on 2D mesh the snaking can be done by cycling around the starting processor from near to far. One needs to keep two things in mind: 1. It is fine to map a constant number of virtual machine nodes to a real machine node; 2. The traversal may go through a non-existing edge if the next virtual node belongs to another subtree, as long as there is a good alternative route to the parent. Also, splitting a virtual node into two parts that belong to different subtrees may be useful sometimes.

5 Optimality

5.1 Static and Semi-Dynamic Scheduling

In these cases $\tilde{W} = W$.

Theorem 14. *In an m-dimensional architecture, if we take bin size $B = W^{1/(m+1)}$ and span is made $O((\tilde{W}(root)/B)^{1/m}) = O(W^{1/(m+1)})$, we have execution time $O(\mathcal{H} + C\phi(\alpha, d) + W^{1/(m+1)})$ which is*
(1) optimal within a constant factor if the tree has constant degree (in which case $C = O(\mathcal{H})$),
(2) optimal within a constant factor if $\alpha \geq 1$, and
(3) optimal within $O(\log \log d)$ factor if the architecture has constant degree.

5.2 Dynamic Scheduling

In this situation we can not choose $B = \mathcal{W}^{1/(m+1)}$ anymore because we do not know $\mathcal{W}$. We must represent B and $\tilde{\mathcal{W}}$ solely by $\mathcal{S}$.

If $\alpha \geq 1, \mathcal{H} \geq \mathcal{S}^{\alpha}, \mathcal{C} \leq \mathcal{H}$, and $\mathcal{W} = O(\mathcal{S}^{1+\alpha})$, if we always choose $B = \mathcal{S}^{\alpha}$ and always assume worst case work $\tilde{\mathcal{W}} = \mathcal{S}^{1+\alpha}$, then both upper and lower bounds become $\Theta(\mathcal{H})$ since $\tilde{\mathcal{W}}^{1/(m+1)} \leq \sqrt{\tilde{\mathcal{W}}} \leq \mathcal{S}^{(1+\alpha)/2} \leq \mathcal{S}^{\alpha} \leq \mathcal{H}$. So the scheduling is always optimal. There are jobs with $\mathcal{W} = \Theta(\mathcal{S}^c)$ and $\mathcal{H} = O(\mathcal{S}^{(c-1)(1+\alpha)/\alpha} + \mathcal{S}^{\alpha})$, when $\alpha < c < 1+\alpha$, parallelization provides certain speedup.

For the case of $0 < \alpha < 1$. We can not get optimal time by the approach presented in this paper. Our goal is changed to bounding a quantity called slow down factor.

Definition 15. On an m-dimensional architecture, the geometric processing time g of job J by scheduling $< \tilde{\mathcal{W}}, B >$ is defined as $\mathcal{H} + \Delta(B) + span$, where $\Delta(B)$ is the maximum amount of work "squeezed" into a single processor in the scheduling. The slow down factor r of the scheduling is defined as the ratio of its geometric time to $g_o = \mathcal{H} + \mathcal{W}^{1/(m+1)}$. In the discussion below, when $r = O(1)$, we say the scheduling is (geometrically) optimal.

Slow down factor is a simplified performance measure. It takes the shape of the job tree and the topology of the architecture into account, but ignores the queuing delay (which is fine, at least in the case where the job tree has constant outdegrees). The dimension of the architecture is a major factor concerning the finishing time of the job. If unlimited number of processors can be used and no communication delay is incurred, one can always schedule the job to finish in time $\mathcal{H}$ by using parallelism wherever possible. However, as shown in the lower bound theorem, for a fixed dimensional architecture, larger number of processors implies larger span (also larger processor time product). Our approach of using estimated work to decide on the length of the line segment on the virtual machine is both natural and unnatural. The latter is due to the fact that when the work goes larger, the scheduler would get more processors involved while the height also becomes larger and the parallelizability does not necessarily increase. We can classify jobs into two categories: height dominant for those satisfying $\mathcal{H} \geq \mathcal{W}^{1/(m+1)}$ and span dominant for the rest. Height dominant jobs' performance is less sensitive to the scheduling strategy.

Below we evaluate the performance of the scheduling algorithm on fairly balanced trees; there is some evidence showing that less balanced trees tend to have better (smaller) slow down ratio.

Definition 16. A job J is q-balanced for some $0 \leq q \leq 1$ if it is balanced from the root until a level where $\mathcal{S}^q$ branches (of approximately the same scale) appear. This level is called the cut level and each task in the cut level has scale $\mathcal{S}^{1-q}$. Below this level each branch goes in the extremely unbalanced way, resulting in a work of $\mathcal{S}^{(1-q)(1+\alpha)}$ in that part (called a trunk).

Lemma 17. A q-balanced job J has total work $\mathcal{W} = \mathcal{S}^{1+(1-q)\alpha}$ and height $\mathcal{H} = \mathcal{S}^{max\{(1-q)(1+\alpha),\alpha\}}$. A $((1+\alpha-c)/\alpha)$-balanced job has $\mathcal{W} = \mathcal{S}^c$ and $\mathcal{H} = \mathcal{S}^{u(c)}$.

Our dynamic scheduling algorithm takes $\tilde{W} = S^x$ and $B = S^y$ for some constant $1 + \alpha \geq x \geq 1$ and $1 + \alpha \geq y \geq \alpha$. The number of processors will then be S^{x-y} and the $span = S^{(x-y)/m}$.

Theorem 18. *Let J be a q-balanced tree job.*

- *If $m \geq 1/(\alpha(1+\alpha))$, J can be scheduled with optimal geometric time for all $0 \leq q \leq 1$.*
- *else if $\alpha(m+1) \geq 1$, J can be scheduled with slow down factor bounded by $S^{(q_2-\alpha m)/(m+1)}$.*
- *Otherwise, the slow down factor is bounded by $S^{(q_2-m/(m+1))/(m+1)}$.*

The table below gives a more accurate feeling about the the above theorem.

α	0.2	0.5	0.8
m Slow Down			
1	0.025	0.05	0.00
2	0.02	0.00	0.00

6 Conclusions

We studied the scheduling of task trees on parallel/distributed machines. The major contribution is modeling the lengths of tasks as a function of their additive scales. We investigated relationships between different parameters of the tree, and used this knowledge to design and analyze a general scheduling algorithm that almost always achieves optimal or near optimal time. We suspect that the lower bound result can be improved and consequently the algorithm may turn out to have more optimal cases.

References

1. B. Berger and L. Cowen, "Complexity Results and Algorithms for the {<≤=}-Constrained Scheduling", *SODA*, 1991, pp. 137-147.
2. S. Bhatt and J. Cai, "Take a walk, Grow a Tree", *FOCS*, 1988, pp. 469-478.
3. S. Bhatt, F. Chung, T.Leighton, and A. Rosenberg, Optimal Simulations of Tree Machines, *FOCS*, 1986, pp. 274-282.
4. S. Bhatt, D. Greenberg, T.Leighton, and P. Liu, Tight Bounds for On-line Tree Embeddings, *SODA*, 1990, pp. 344-350.
5. M.Y. Chan, Embedding of Grids into Optimal Hypercubes, *SIAM Journal on Computing*, Vol 20, No 5, Oct., 1991, pp. 834-864.
6. X. Deng and E. Koutsoupias, "Competitive Implementation of Parallel Programs", Proceedings of the 4th ACM-SIAM Symposium on Discrete Algorithms, 1993, pp. 455-461.
7. A.L.Fisher and H.T.Kung, Synchronizing Large VLSI Array Processor, Proc. of *10th Annual IEEE/ACM Symposium on Computer Architecture*,1983, pp.54-58.
8. A. Feldmann, J. Sgall, and S. Teng, "Dynamic Scheduling on Parallel Machines", *FOCS 91*, pp. 111-120.

9. M. R. Garey and D. S. Johnson, *Computers and Intractability : A Guide to Theory of NP-Completeness.* San Francisco, CA, Freeman Publishing Co., 1979.
10. K. Hwang and F. A. Briggs, Computer Architecture and Parallel Processing, *McGRAW-Hill International Editions*, 1987.
11. J. Hong, K. Melhorn, and A.L. Rosenberg, Cost trade-offs in graph embeddings, with applications, *JACM*, 30, 1983, pp. 709-728.
12. T. C. Hu. Parallel Sequencing and Assembly Line Problem, *Operations Research*, Vol. 9, Nov. 1961, pp. 841-848.
13. C. Kaklamanis and G. Persiano, "Branch-and-Bound and Backtrack Search on Mesh-Connected Arrays and Processors", 4th Annual ACM Symposium on Parallel Algorithms and Architectures, 1992, pp. 118-128.
14. R.Koch, T. Leighton, B.Maggs, S.Rao, and A.Rosenberg, Work-Preserving Emulations of Fixed-Connection Networks, *STOC*, 1989, pp. 227-240.
15. C.H. Papadimitriou and J.D. Ullman, A communication time tradeoff, *SIAM J. of Computing*, 16 (4), 1987, pp. 639-646.
16. C.H. Papadimitriou and M. Yannakakis, "Towards An Architecture-Independent Analysis of Parallel Algorithms", 20th ACM STOC, 1988, pp 510-513.
17. S. Phillips and J. Westbrook, "Online Load Balancing and Network Flow", 25th ACM STOC, 1993, pp 402-411.
18. A. Ranade, "Optimal Speed-up for Backtrack Search on a Butterfly Network", 3-d Annual ACM Symposium on Parallel Algorithms and Architectures, 1991, pp. 40-48.
19. D. Shmoys, J. Wein, and D. Williamson, "Scheduling Parallel Machine Online", *FOCS 91*, pp. 131-140.
20. A. Wagner, Embedding arbitrary binary trees in a Hypercube, *J. of Parallel and Distributed Computing*, 7, 1989, pp. 503-520.
21. I.-C. Wu and H.T. Kung, "Communication Complexity for Parallel Divide-and-Conquer", *FOCS 91*, pp. 151-162.
22. X.Yu and D.Ghosal, "Optimal Dynamic Scheduling of Task Tree on Constant-dimensional Architectures", 4th Annual ACM Symposium on Parallel Algorithms and Architectures, 1992, pp. 138-146.

Single-vehicle Scheduling Problem on a Straight Line with Time Window Constraints *

Chi-lok Chan and Gilbert H. Young

High Performance Computing Laboratory
Department of Computer Science
The Chinese University of Hong Kong
Shatin, N.T., Hong Kong
Email: clchan@cs.cuhk.hk, young@cs.cuhk.hk

Abstract. The problem of single-vehicle scheduling with ready time and deadline constraints can be applicable to many practical transportation problems. But the general case is a generalization of classical Traveling Salesman Problem and it is NP-hard. Psaraftis *et al.* [5] presented an $O(n^2)$ algorithm for vehicle scheduling problems where the topological structure of sites are restricted to straight line and there is no deadline for each site. They conjectured that this problem is NP-hard if each site has an arbitrary ready time and deadline (called *time window*). In this paper, we prove the conjecture and show that the decision versions of vehicle scheduling problems on a straight line topology with time window are NP-complete for both path and tour versions. We also give an $O(n^2)$ algorithm to solve a special case of this problem in which all sites have a common ready time. This algorithm is applicable to both path and tour versions.

Keywords: deadline, NP-complete, optimal schedule, ready time, vehicle scheduling problem, time windows

1 Introduction

Single-vehicle scheduling problems (VSP) are of great practical importance. One vehicle is going to visit a set of sites which are connected according to some arbitrary graph structures. The practical goals of scrutinizing these problems are minimization of total time taken for the vehicle to visit all sites. In practice, each site has its earliest (*ready time*) and latest (*deadline*) visiting times. The problem become constrained by time windows so that the vehicle should visit every site after the ready time but no later than the deadline. This is called the *path version* of the VSP. If the vehicle has to return to the starting site at the end of the trip, it is called a *tour version*. The literature on the problems with time windows has been growing at an explosive rate over the last few years[1][4][6]. Unfortunately, many of the interesting routing problems are NP-complete[1]. Karuno proved the VSP on a tree topology with arbitrary degree and ready

* This research was supported in part by CUHK GRANT NO.220500610 and CUHK GRANT NO.220501090.

time only is NP-hard [3]. Psaraftis *et al.* [5] proposed a new model for routing of ships for a shoreline topology. They gave an $O(n^2)$ algorithm for solving the path version of the VSP on a convex shoreline with ready time only, where n denotes the number of sites. This topology is equivalent to a straight line structure. The situation of one common deadline can be solved in $O(n^2)$ time by applying Psaraftis *et al.*'s algorithm as if no deadline were present and then checking whether the total time taken is no greater than the common deadline. The trivial case of non-overlapping time windows can be solved by visiting sites with increasing order of ready time in $O(n \log n)$ time[5]. Psaraftis *et al.* conjectured that the generalized problem with both ready time and deadline is NP-hard. In this paper, we show the Psaraftis *et al.*'s conjecture[5] and prove that the decision versions of the VSP on a straight line with time windows for both path and tour versions are NP-complete. In addition, we present an $O(n^2)$ algorithm based on dynamic programming approach for solving the optimization problem of a special case, in which all sites have a common ready time.

The remainder of this paper is organized as follow: section 2 gives the formulation and some notations of the problem. In section 3, we show that the decision version of the VSP on a straight line topology with time windows is NP-complete. Section 4 gives an $O(n^2)$ dynamic programming solution for solving a special case in which all sites have a common ready time for both path and tour versions. This algorithm can be generalized to tackle some variants of this special case and they are discussed in this section as well. Finally, we draw some concluding remarks in the last section.

2   Problem Formulation and Notations

Given a set of n sites, $S = \{s_1, s_2, \ldots, s_n\}$, in which each site s_i has a *ready time* $r(s_i)$ and *deadline* $d(s_i)$. The distance between any two sites, s_i and s_j, is denoted as $Dist(s_i, s_j)(= Dist(s_j, s_i))$, $1 \leq i, j \leq n$. A vehicle with unit speed starts at site s_1 is going to visit all sites in S.

A *schedule*, $\pi = (\pi_1, \pi_2, \ldots, \pi_n) \in \mathcal{P}_s = \{$set of permutations of $S\}$, is a sequence of sites so that the vehicle visits sites according to the order in π. $t_\pi(\pi_i)$ denotes the time instance at which site π_i is visited, $\forall \pi_i \in S$,

$$t_\pi(\pi_1) = \begin{cases} \max\{r(\pi_1), Dist(s_1, \pi_1)\} & \text{if } Dist(s_1, \pi_1) \leq d(\pi_1) \\ \infty & \text{otherwise} \end{cases}$$

$$t_\pi(\pi_i) = \begin{cases} \max\{r(\pi_i), t_\pi(\pi_{i-1}) + Dist(\pi_{i-1}, \pi_i)\} & \text{if } t_\pi(\pi_{i-1}) + Dist(\pi_{i-1}, \pi_i) \leq d(\pi_i) \\ \infty & \text{otherwise} \end{cases}$$

The *completion time* of a schedule π is the total time taken to visit all sites, i.e., $C_\pi = t_\pi(\pi_n)$.

A schedule π is called *feasible* if all sites can be visited in between its time window (*time window constraint*). This means that $t_\pi(\pi_i) < \infty$, $\forall \pi_i \in S$. In other words, π is feasible iff $C_\pi < \infty$ and $\pi \in \mathcal{P}_s$. Let $\mathcal{F}$ denotes the set of all feasible schedules.

An *optimal schedule* π^o is a feasible schedule with the minimal completion time, i.e., $\exists \pi^o \in \mathcal{F}$ such that

$$C_{min} = C_{\pi^o} \leq C_\pi \qquad , \forall \pi \in \mathcal{F}.$$

The *optimization problem* of the *path version* (VSP-WINDOW-SLP) is to find a π^o if $\mathcal{F} \neq \emptyset$.

Another problem restricts the vehicle return to site s_1 at the end of the trip and called a *tour version* (VSP-WINDOW-SLT). The problem formulation is exactly the same except the definition of the completion time of a schedule π is defined as

$$C_\pi^t = t_\pi(\pi_n) + Dist(s_1, \pi_n) = C_\pi + Dist(s_1, \pi_n).$$

In addition, the following notations are used in this paper,

- $i \mapsto j$ denotes site i must be visited just before j. $i \rightsquigarrow j$ if $i \mapsto s_1 \mapsto s_2 \mapsto \cdots \mapsto s_x \mapsto j$ where $s_i \in S$, or $i \mapsto j$.
- $i \xrightarrow{x} j$ denotes $i \mapsto j$ or $i \mapsto x \mapsto j$ where $i, j, x \in S$.
- $T(x_1 \mapsto \cdots \mapsto x_p)$ denotes the total time taken to travel along the path $x_1, x_2, \ldots, x_p$ where $x_i \in S$, i.e.,

$$T(x_1 \mapsto \cdots \mapsto x_p) = \sum_{i=1}^{p-1} Dist(x_i, x_{i+1}).$$

3 NP-hardness of VSP-WINDOW-SLP

In this section, we show the following decision question is NP-complete,

VSP-WINDOW-SLP(decision)

INSTANCE: Given a set of n sites, $S = \{s_1, s_2, \ldots, s_n\}$, $r(s_i)$, $d(s_i)$ and $Dist(s_j, s_{j+1})$, $1 \leq i \leq n, 1 \leq j < n$.

QUESTION: Is there a feasible schedule for a vehicle to visit all sites and satisfy the time window constraint?

This decision problem is easier than the decision problem of the optimization problem which asks for a feasible schedule with schedule length no greater than a given constant. As a result, the NP-completeness of VSP-WINDOW-SLP(decision) is a stronger result. We are going to show the reduction from PARTITION which is known to be NP-complete[2].

PARTITION

INSTANCE: A finite set $I = \{1, 2, \ldots, N\}$, a "size" $a_i \geq 0$ for each $i \in I$.

QUESTION: Is there a subset $A \subseteq I$ such that

$$\sum_{i \in A} a_i = \sum_{i \in A' = I - A} a_i = \frac{1}{2} \sum_{i \in I} a_i = B?$$

3.1 A Transformation from PARTITION

Given an instance of PARTITION, we can transform it into an instance of VSP-WINDOW-SLP(decision) as follows,

Let $S = \{s, E_i, F_i, P_i, Q_i, 1 \leq i \leq N\}$ be the set of sites where s is the starting site, P_i and Q_i are the partition sites, E_i and F_i are the enforcer sites, and $n = 4N + 1$, $B = \frac{1}{2} \sum_{i=1}^{N} a_i$, $C = 2B + 1$, $X = NC + 2C \sum_{i=1}^{N} i + B = N^2 C + 2NC + B$.

Distance The topological structure (Fig. 1) of the VSP-WINDOW-SLP (decision) instance is

$$(Q_1, Q_2, \ldots, Q_N, s, E_1, E_2, \ldots, E_N, F_N, F_{N-1}, \ldots, F_1, P_N, P_{N-1}, \ldots, P_1)$$

and

$$
\begin{aligned}
Dist(Q_i, Q_{i+1}) &= Dist(Q_N, s) = C &&, 1 \leq i < N \\
Dist(s, E_1) &= Dist(E_i, E_{i+1}) = 0 &&, 1 \leq i < N \\
Dist(E_N, F_N) &= Dist(F_i, F_{i-1}) = 0 &&, 1 < i \leq N \\
Dist(F_1, P_N) &= C + a_N/2 && \\
Dist(P_i, P_{i-1}) &= C + (a_i - a_{i-1})/2 &&, 1 < i \leq N \\
i.e., \quad Dist(F_1, P_i) &= (N - i + 1)C + a_i/2 &&, 1 \leq i \leq N.
\end{aligned}
$$

Ready Time and Deadline

$$
\begin{aligned}
r(s) &= d(s) = NC \\
r(P_i) &= r(Q_i) = NC + 2C \sum_{N-i+2}^{N} k + (N - i + 1)C \\
d(P_i) &= d(Q_i) = X + 2C \sum_{1}^{N-i} k + (N - i + 1)C + B \\
r(E_i) &= \begin{cases} NC + 2C \sum_{1}^{N} k + B = X & \text{if } i = N \\ NC + 2C \sum_{N-i+1}^{N} k & \text{otherwise} \end{cases} \\
d(E_i) &= NC + 2C \sum_{N-i+1}^{N} k + B \\
r(F_i) &= \begin{cases} X + 2C \sum_{1}^{N} k + B & \text{if } i = 1 \\ X + 2C \sum_{1}^{N-i+1} k & \text{otherwise} \end{cases} \\
d(F_i) &= X + 2C \sum_{1}^{N-i+1} k + B.
\end{aligned}
$$

Fig. 2 shows the time windows of the constructed instance.

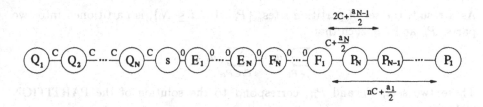

Fig. 1. The topological structure of the VSP-WINDOW-SLP (decision) instance for the reduction.

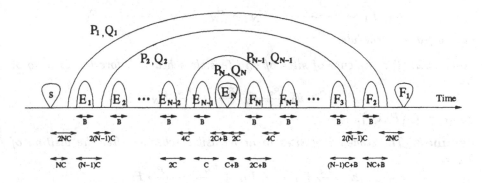

Fig. 2. The ready times and deadline of the VSP-WINDOW-SLP (decision) instance for the reduction.

3.2 Intuitive Idea of the Reduction

The intuitive idea of the reduction can be explained as follows, the starting site (s) and enforcer sites (E_i and F_i, $1 \le i \le N$) have non-overlapping time windows (Fig. 2) so that there is a fixed order of visiting of these sites in any feasible schedule (proved in Lemma 1). The vehicle can only choose either P_i or Q_i in between the visiting of E_{i-1} and E_i (proved in Lemma 2). It is because the partition sites Q_j and P_j ($1 \le j < i$) are too far away from the enforcer sites and the other partition sites, Q_k and P_k ($i < k \le N$) have not been ready yet.

Similarly, the vehicle can only choose either P_i or Q_i in between the visiting of F_{i+1} and F_i (proved in Lemma 3). The vehicle is now constrained by the deadline of the partition sites.

Lastly, sites s, E_n and F_1 should be visited at their deadline, i.e., $t = NC$, X and $2X - NC$ respectively. The elapsed times between the visiting times of these sites are

$$X - NC = 2C \sum_{1}^{N} i + B.$$

As a result, the set of partition sites, $\{P_i : 1 \leq i \leq N\}$, is partitioned into two parts, P_L and P_R, such that

$$\sum_{P_i \in P_L} a_i = \sum_{P_i \in P_R} a_i = B.$$

These two sets, P_L and P_R, correspond to the solution of the PARTITION instance.

3.3 NP-completeness Proof

Lemma 1. *s must be the first site to be visited and*

$$s \rightsquigarrow E_1 \rightsquigarrow E_2 \rightsquigarrow \cdots \rightsquigarrow E_N \rightsquigarrow F_N \rightsquigarrow F_{N-1} \rightsquigarrow \cdots \rightsquigarrow F_1$$

in any feasible schedule.

Lemma 2. *The sequence of sites of any feasible schedule before the visiting of E_N is*

$$s \xrightarrow{R_1} E_1 \xrightarrow{R_2} E_2 \xrightarrow{R_3} \cdots \xrightarrow{R_N} E_N$$

where $R_i \in \{P_i, Q_i\}$.

Lemma 3. *The sequence of sites in any feasible schedule after the visiting of E_N is*

$$E_N \xrightarrow{R_N} F_N \xrightarrow{R_{N-1}} F_{N-1} \xrightarrow{R_{N-2}} \cdots \xrightarrow{R_1} F_1$$

where $R_i \in \{P_i, Q_i\}$.

Lemma 4. *If π is a feasible schedule, then*

$$\pi = (s, R_1, E_1, R_2, E_2, \ldots, R_N, E_N, R'_N, F_N, R'_{N-1}, F_{N-1}, \ldots, R'_1, F_1)$$

where

$$R_i = P_i \text{ or } Q_i,$$
$$R'_i = \begin{cases} P_i & \text{if } R_i = Q_i \\ Q_i & \text{if } R_i = P_i. \end{cases}$$

Lemma 5. *PARTITION has a solution iff the corresponding VSP-WINDOW-SLP(decision) instance has a feasible schedule.*

It is easy to see that VSP-WINDOW-SLP(decision) is NP and the transformation in section 3.1 can be done in polynomial time. By Lemma 5, we have the following result,

Theorem 6. *VSP-WINDOW-SLP(decision) is NP-complete.*

It is obvious that π is a feasible schedule for the path version iff it is a feasible schedule for the tour version. Thus, the reduction is applicable to the tour version, VSP-WINDOW-SLT, so that the decision problem of the tour version is NP-complete by Theorem 6.

Theorem 7. *The decision problem of asking for a feasible schedule for the tour version (VSP-WINDOW-SLT) is NP-complete.*

4 Polynomial Time Algorithm for the VSP-WINDOW on a Straight Line with Common Ready Time

The result in the last section implies that there is no hope for finding an efficient algorithm to solve the general VSP-WINDOW-SLP problem unless $P = NP$. In this section, we try to develop an efficient algorithm that can solve a special case of VSP-WINDOW-SLP in which all sites have the same ready time, i.e., $r(i) = r(j), \forall i, j \in S$. For the rest of this section, the sites are labelled as $1, 2, \ldots, n$ so that sites i and $i + 1$ are connected adjacently, $\forall i, 1 \leq i < n$ and the vehicle is initially located at site 1. If the common ready time is zero, a trivial solution for an optimal schedule is $\pi = (1, 2, \ldots, n)$ for both path and tour versions provided that the problem has a feasible solution. The time complexity is $O(n)$. For the general case, the feasible schedule is characterized as follows,

Lemma 8. *In the path version VSP-WINDOW-SLP with common ready time, for all feasible schedules, π, there exists a feasible schedule π' s.t. $\forall 1 < i \leq n$,*

$$\pi_i' = \min_{1 \leq j < i}\{\pi_j'\} - 1 \ or \ \max_{1 \leq j < i}\{\pi_j'\} + 1.$$

i.e., at any instant, the visited sites form a consecutive region on the straight line and the last visited site is either the leftmost or rightmost site in this region (Fig. 3(a)). In addition,

$$C_{\pi'} + Dist(\pi_n', \pi_n) \leq C_\pi.$$

i.e., the completion time of π' plus the time taken for the vehicle returns to the last visited site in π is no greater than the completion time of π.

Fig. 3 show the main idea of Lemma 8. For any schedule π that does not satisfy Lemma 8, we can apply the transformation shown in Fig. 3(b)-(c). This transformation does not introduce any additional waiting time because π_{j-1} has been visited and all sites have the same ready time.

By Lemma 8, we can solve the VSP-WINDOW-SLP problem with common ready time by an algorithm based on dynamic programming as follows,

Algorithm A (Dynamic Programming Approach)
Input: A set of n sites, $\{1, 2, \ldots, n\}$, their ready time, $r(i)$, deadline, $d(i)$ and the distance between adjacent sites, $Dist(j, j + 1)$, $1 \leq i \leq n$ and $1 \leq j < n$.
Output: The minimal completion time if $\mathcal{F} \neq \emptyset$; otherwise ∞.

(1) for $i = 1$ to n //boundary conditions
(2) $V(i, i, i) = \delta_i(\max\{r(i), Dist(1, i)\})$
(3) $V(n + 1, n + 1, i) = V(i, n + 1, i) = \infty$
(4) for $l = 2$ to n //l is the number of visited sites
(5)   for $i = 1$ to $n - l + 1$ //i is the visited site with the smallest index
(6) $j = i + l - 1$ // j is the visited site with the largest index

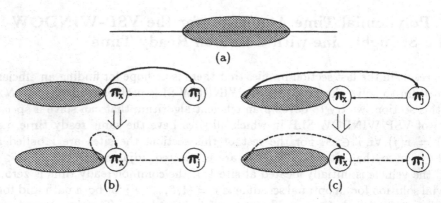

Fig. 3. (a) The visited sites (shaded) form a consecutive region on the straight line. The last visited site is either the leftmost or rightmost site (the two dots) in this region. **(b)** and **(c)** The two possible cases of the transformation. The shaded region represents the visited sites and the dot is the last visited site, π_{j-1}. The thickened line is replaced by the dotted line without increasing the time taken from π_{j-1} to π_j.

(7) $V(i,i,j) = \min\{\delta_i(Dist(i+1,i) + V(i+1,i+1,j)),$
$\delta_i(Dist(j,i) + V(j,i+1,j))\}$

(8) $V(j,i,j) = \min\{\delta_j(Dist(j-1,j) + V(j-1,i,j-1)),$
$\delta_j(Dist(i,j) + V(i,i,j-1))\}$

(9) $C_{min} = \min\{V(1,1,n), V(n,1,n)\}$

(10) return C_{min}

In Algorithm A, $V(k,i,j)$ denotes the time instance that sites $i, i+1, \ldots, j$ are visited and the last visited site is k, $i \le k \le j$. The function $\delta_i(x)$ is defined as follows,

$$\delta_i(x) = \begin{cases} x & \text{if } x \le d(i) \\ \infty & \text{otherwise.} \end{cases}$$

There are altogether $O(n^2)$ iterations. $V(i,i,j)$ and $V(j,i,j)$ can be calculated in constant time by using an $O(n)$ array (line 7 and 8). The time complexity of Algorithm A is $O(n^2)$. In addition, we can record the optimal schedule by some simple modifications.

Theorem 9. *Algorithm A returns the minimal completion time for the VSP-WINDOW-SLP with common ready time in $O(n^2)$ time.*

Algorithm A is also applicable to the tour version by modifying line 9 and 10 as follows,

line 9: $C_{min}^t = \min\{V(1,1,n), V(n,1,n) + Dist(1,n)\}$
line 10: return C_{min}^t

Theorem 10. *We can find the optimal completion time for VSP-WINDOW-SLT with common ready time in $O(n^2)$ time.*

In addition, this algorithm can be extended to cases in which the vehicle starts at time $t = 0$ not at site 1, but at any interior site s between 1 and n. The boundary condition (line 2) is changed as follows, $\forall i, 1 \le i \le n$,

$$V(i, i, i) = \delta_i(\max\{r(i), Dist(s, i)\}).$$

For the tour version, the additional modification is line 9,

$$C^t_{min} = \min\{V(1, 1, n), V(n, 1, n) + Dist(s, n)\}.$$

Algorithm A is only applicable to the cases that all sites have a common ready time, since Lemma 8 will not be valid if sites have different ready times. The reason is that the thickened route in Fig. 3(b)-(c) cannot be replaced by the dotted route, or else waiting time may be introduced at site π_x (due to the different ready time) and thus the deadline of π_j may be missed.

5 Conclusion

In this paper, we study the computational complexity of the single-vehicle scheduling problems (VSP) on a straight line topology. In section 3, we prove the conjecture given by Psaraftis *et al.* [5] and show that the decision problems of the vehicle scheduling problem (VSP) on a straight line with time window constraints for both path (VSP-WINDOW-SLP) and tour (VSP-WINDOW-SLT) versions are NP-complete (Theorem 6 and Theorem 7). An extension to the case in which the vehicle is required to spend a known "service" time (called *handling time*) at each site is also an interesting problem. It is easily observed that the SEQUENCING WITHIN INTERVALS[2], which is a strongly NP-complete problem, is a restricted case of the VSP on a straight line with time windows and handling time. Thus the extension of VSP-WINDOW-SLP and VSP-WINDOW-SLT with handling time are NP-hard in the strong sense. If the problem has only ready time, deadline or handling time, it can be solved in polynomial time. The path and tour versions of the VSP on a straight line with ready time only can be solved in $O(n^2)$ and $O(n)$ times respectively, where n is the number of sites[5]. The trivial cases of the VSP on a straight line with deadline or handling time only can be solved in $O(n)$ time for both path and tour versions. The optimal schedule is a trip visiting all sites successively from the starting site to the other end of the straight line topology. However, the status of the VSP on a straight line with both ready and handling times (or both deadline and handling time) is still open. Psaraftis *et al.* conjectured they are NP-hard[5]. Besides, it remains to examine whether the VSP-WINDOW-SLP and VSP-WINDOW-SLT are strongly NP-hard or not. The status of different single-vehicle scheduling problems on a straight line topological structure is summarized in Table 1.

A special case of the VSP on a straight line with time windows, in which all sites have a common ready time, is investigated in section 4. We present a dynamic programming algorithm, Algorithm A, for solving both the path and tour versions. The time and space complexities of Algorithm A are $O(n^2)$. Furthermore, our algorithm can be extended to tackle problems in which the vehicle is initially located at a given interior site on the straight line topology.

	Path version	Tour version
ready time (r)	$O(n^2)^\dagger$	$O(n)^\dagger$
deadline (d)	$O(n)$	$O(n)$
handling time (h)	$O(n)$	$O(n)$
r and d	NP-hard	NP-hard
r and h	?‡	?
d and h	?	?
r,d and h	NP-hard in the strong sense	NP-hard in the strong sense

†: By Psaraftis *et al.*'s result[5].

‡: A ? denotes an unknown status.

Table 1. The computation complexities of the different variants of the VSP on a straight line topology.

References

1. L. Bodin, B. Golden, A. Assad, and M. Ball. "Routing and Scheduling of Vehicles and Crews: The State of the Art". *Computers and Operations Research*, vol 10:pp. 62–212, 1983.
2. M.R. Garey and D.S. Johnson. *"Computers and Intractability: A Guide to the Theory of NP-Completeness"*. Freeman, 1979.
3. Y. Karuno, H. Nagamochi, and T. Ibaraki. "Vehicle Scheduling on a Tree with Release and Handling Times". In *"Lecture Notes in Computer Science(762)"*, pages 486–495, 1993.
4. H. Psaraftis. "An Exact Algorithm for the Single Vehicle Dial-A-Ride Problem with Time Windows". *Transportation Science*, vol 17:pp. 351–357, 1983.
5. H. Psaraftis, M. Solomon, L. Magnanti, and Tai-Up Kim. "Routing and Scheduling on a Shoreline with Release Times". *Management Science*, vol 36:pp. 212–223, February 1990.
6. Marius M. Solomon. "Algorithms for the vehicle routing and Scheduling problems with Time Window Constraints". *Operations Research*, vol 35:pp. 254–265, 1987.

An On-Line Algorithm for Some Uniform Processor Scheduling

RONGHENG LI & LIJIE SHI[†]

[†]Institute of Applied Mathematics, Academia Sinica, 100080 Beijing China

Abstract. This paper considers the problem of on-line scheduling a set of independent jobs on m uniform machines $(M_1, M_2, \cdots, M_m)$ in which machine $M_i's$ processing speed is $s_i = 1 (i = 1, \cdots, m-1)$ and $s_m = s > 1$. List Scheduling [Yookum Cho and Sartaj Sahni. *Bounds for list schedules on uniform processors*. SIAM J. Compute. 9(1980), pp91-103.] guarantees a worst case performance of $\frac{3m-1}{m+1} (m \geq 3)$ and $\frac{1+\sqrt{5}}{2} (m = 2)$ for this problem. We prove that this worst case bound cannot be imporved for $m = 2$ and $m = 3$ and for every $m \geq 4$, an algorithm with worst case performance at most $\frac{3m-1}{m+1} - \varepsilon$ is presented when $s_m = 2$, where ε is a fixed positive number, and then we improve the bound for general $s_m = s > 1$.

1 Introduction

A uniform machine system consists of m, $m \geq 1$, machines $(M_1, M_2, \cdots, M_m)$. A speed $s_i, s_i \geq 1$, is associated with each machine. In one unit of time M_i can carry out s_i units of processing. A list $(t_1, t_2, \cdots, t_n)$ of n jobs is given to us on-line, that means we get the jobs one by one. Both the total number of jobs that need to be scheduled and the size of the jobs are not known previously. The processing time of job t_i becomes known only when t_{i-1} has already been scheduled. As soon as job t_i appears, it must irrevocably be scheduled. Our goal is to minimize the makespan, i.e., the maximum completion time over all jobs in a schedule. The quality of an algorithm H is measured by its worst case ratio

$$R^H(m) = \sup_L \{C^H(L)/C^*(L) : L \text{ is a list of jobs}\}$$

where $C^H(L)$ denotes the makespan produced by the heuristic H on the machines and the list L of jobs and $C^*(L)$ denotes the corresponding makespan in some optimal schedule. List Scheduling (LS for short) which always assigns the current job to the machine that will complete it first is a simple example of non-preemptive on-line algorithm and is always used. When $s_i = 1 (i = 1, 2, \cdots, m)$, the machine system is well known as identical machine system. For identical machine system, Graham showed that $R^{LS}(m) = 2 - 1/m$ in 1969. For $m \geq 4$, this bound was improved by Gabor Galambos and Gerhard J. Woeginger in 1993 [3]. They designed an algorithm called Refined List Scheduling (RLS) and showed that $R^{RLS}(m) \leq 2 - 1/m - \eta_m$ where $\eta_m > 0$ for $m \geq 4$. Bartal et al. [4] made some progress for large numbers of machines by divising an algorithm whose worst-case guarantee is $2 - \frac{1}{7}$ for large m. When $s_i = 1 (i = 1, \cdots, m)$ and $s_m = s \geq 1$, Yookun et al. in 1980 showed that $R^{LS}(m, s) \leq 1 + \frac{m-1}{m+s-1} min\{2, s\} \leq 3 - \frac{4}{m+1}$ and the bound $3 - \frac{4}{m+1}$ is achieved when s=2.

In this paper, for $m \geq 4$ machines we present a heuristic that has a significantly better worst case performance guarantee than LS when $s_i = 1(i = 1, \cdots, m-1)$ and $s_m = 2$, and then show that the bound $3 - \frac{4}{m+1}$ can be improved when $s_i = 1(i = 1, \cdots, m-1)$ and $s_m = s > 1$. For $m = 2$ and 3, we will show that the worst case performance guarantee of LS can not be improved by any heuristic.

2 Lower bounds for on-line scheduling

For a heuristic H, let

$$
R^H(m,s) = \sup_L \{C^H(L)/C^*(L) \mid L \text{ is a list of jobs and} \tag{2.1}
$$
$$
s_i = 1(i = 1, 2, \cdots, m-1) \text{ and } s_m = s \geq 1 \text{ is fixed}\},
$$

$$
R^H(m) = \sup_{s \geq 1} R^H(m,s) \tag{2.2}
$$

In the following we always assume that the speed $s_i = 1(i = 1, \cdots, m-1)$ and $s_m = s \geq 1$ if there is no any special notation.

Theorem 2.1. The following inequalities hold

   (i) $R^{LS}(2, \frac{1+\sqrt{5}}{2}) = R^{LS}(2) = \frac{1+\sqrt{5}}{2}$

   (ii) $R^{LS}(m, s) \leq 1 + \frac{m-1}{m+s-1} \min\{2, s\}$

   (iii) $R^{LS}(m) = R^{LS}(m,2) = 3 - \frac{4}{m+1}, (m \geq 3)$.

Proof. To see [1]. ∎

Theorem 2.2. For any heuristic H, the following inequalities hold

   (i) $R^H(2) \geq \frac{1+\sqrt{5}}{2}$

   (ii) $R^H(m) \geq 2$ $(m \geq 3)$.

Proof. Claim (i) is easily proved by using list $L_1 = \{1, \frac{1+\sqrt{5}}{2}\}$ when $s_1 = 1$ and $s_2 = \frac{1+\sqrt{5}}{2}$. Claim (ii) can be verified by considering lists $l_k = \{1, 2, \cdots, 2^k\}$ ($k = 0, 1, 2, \cdots$) for $s = 2$. ∎

From Theorem 2.1 and 2.2, we know $R^{LS}(2)$ and $R^{LS}(3)$ can not be improved any more.

3 The algorithm

In this section we will give a heuristic. As the algorithm gets the jobs one by one, the values of C^* and C^H vary during the algorithm. To simplify notation, we will identify each job with its length. The load L_i of a machine M_i is the sum of processing times over all jobs assigned to it. In the following, we always assume that the speed $s_i = 1(i = 1, 2, \cdots, m-1)$ and $s_m = s \geq 1$ if there is no special notation.

We are ready to present our heuristic A. In the algorithm, two real mumbers α_m and β_m are used that satisfy $0 < \alpha_m < 1$ and $\beta_m > 1$. In order to keep the presentation simple, we drop the indices and write α and β instead. The exact values of α and β will be specified later.

Algorithm A

Step 1. Reorder the machines such that $L_1 \leq L_2 \leq \cdots \leq L_{m-1}$ holds. Let x be a new job given to the algorithm.

Step 2. Let $L = \sum_{i=1}^{m-1} L_i + sL_m$ be the total length of jobs given before x. If $L_m \geq \frac{(3m-5)\alpha}{s(m+1)(m-2)}L$ and $x \leq \frac{\beta}{m-2}L$, then put x on L_1.

Step 3. Assign x by LS, that is to say, assign x to the machine that will complete it first.

Throughout the following analysis, we always denote by x current job.

Lemma 3.1. If x is put on L_1 in step 2, then $(L_1 + x)/C^* \leq \frac{m+s-1}{m-1}[1 + \frac{(m-2)\beta}{m+\beta-2} - \frac{(3m-5)\alpha}{(m+1)(m+\beta-2)}]$.

Proof : Because $x \leq \frac{\beta}{m-2}L$, we have

$$x \leq \frac{\beta}{m+\beta-2}(L+x). \tag{3.1}$$

Since

$$L+x = \sum_{1}^{m-1} L_i + sL_m + x$$

$$\geq (m-1)(L_1+x) + \frac{(3m-5)\alpha}{(m+1)(m-2)}(L+x) - [m + \frac{(3m-5)\alpha}{(m+1)(m-2)} - 2]x.$$

From the above and (3.1) we get

$$[1 + \frac{(m-2)\beta}{m+\beta-2} - \frac{(3m-5)\alpha}{(m+1)(m+\beta-2)}](L+x) \geq (m-1)(L_1+s). \tag{3.2}$$

By (3.2), Lemma 3.1 is proved. ∎

Lemma 3.2. If $x \leq \frac{\beta}{m-2}L$ and x is assigned to M_i in step 3, then $L_i^x/C^* \leq \frac{m+m\beta-2}{m+\beta-2}$, where L_i^x represents the load of the machine M_i after x has been assigned to M_i.

Proof : The proof is similar to that of Lemma 3.1. ∎

In the following, the real number $\bar{\alpha}$ satisfies that $\bar{\alpha} > \alpha$.

Lemma 3.3. If $\frac{\beta}{m-2}L < x \leq \frac{\bar{\beta}}{m-1}L$ and $L_m < \frac{(3m-5)\bar{\alpha}}{s(m+1)(m-2)}L$ then

$$L_i^x/C^* \leq \frac{(m+s-1)(m-1)}{(m+\bar{\beta}-1)s}[\frac{(3m-5)\bar{\alpha}}{(m+1)(m-2)} + \frac{\bar{\beta}}{(m-1)}]$$

Proof : The proof is similar to that of Lemma 3.1. ∎

Lemma 3.4. If x is assigned to machine M_i and

$$L_m \geq \frac{(3m-5)\bar{\alpha}}{s(m+1)(m-2)}L, \quad \frac{\beta}{m-2}L \leq x < \frac{\bar{\beta}}{m-1}L$$

and

$$\frac{(3m-5)\alpha}{(m+1)(m+\beta-2)} + \frac{\beta}{m+\beta-2} \leq \frac{(3m-5)\bar{\alpha}}{(m+1)(m-2)}, \tag{3.3}$$

then

$$L_i^x/C^* \leq max\{\frac{s(m-2)(m\bar{\beta}+m-1)(m\beta+m-2)(m+1)}{(m-1)(m+s-1)^2(3m-5)\bar{\alpha}\beta},$$

$$1 + \frac{m-1}{m+s-1}max\{1, s-1\}\}.$$

The proof appears in the full paper.

Lemma 3.5. If $x > \frac{\bar{\beta}}{m-1}L$ and x is assigned to M_i in step 3 then

$$L_i^x/C^* \leq \frac{sm}{m+s+1} + \frac{s(m-1)}{\bar{\beta}(m+s-1)}$$

where L_i^x represents the load of machine M_i after x has been assigned to M_i.
Proof : The proof is similar to that of Lemma 3.1. ∎

Lemma 3.6. Let $R_m = \min R$
Subject to

$$
\begin{cases}
(m+1)R^2 - 2(m-1)R - 2m > 0 \\
2 < R < \dfrac{3m-1}{m+1} \\
\dfrac{(m+1)R^2 - 2(m-1)R - m - 1}{(m+1)^2 R - 2m(m+1)} \\
\geq \dfrac{1}{2}\left[\dfrac{(m+1)R^2 - 2(m-1)R - m - 1}{(m+1)R^2 - 2(m-1)R - 2m} - \dfrac{(m-1)R}{m+1}\right]
\end{cases}
\tag{3.4}
$$

then R_m is the only real root of equation

$$(m+1)^3 R^4 - (m+1)(5m^2 + 4m - 5)R^3 + (6m^3 - 18m + 4)R^2$$
$$+ (m^3 + 11m^2 - m - 3)R - 2m(m^2 - 1) = 0 \tag{3.5}$$

in interval $(2, 3 - \frac{4}{m+1})$ and R_m is an increasing function of m. $\lim\limits_{m \to \infty} R_m$ is the only real root of equation $R^3 - 3R^2 + 1 = 0$ between 2 and 3.

Theorem 3.7. For algorithm A there exists $\varepsilon_m > 0$ such that

$$R^A(m, 2) \leq \frac{3m - 1}{m + 1} - \varepsilon_m \qquad (3.6)$$

and there exists a positive ε such that $\varepsilon_m \geq \varepsilon$ for every $m \geq 4$.

proof: Lemma 3.1, 3.2, 3.3, 3.4 and 3.5 give upper bounds on the worst case ratios in the five scenarios. If $\alpha < 1$ and $\beta > 1$, we can easily verify that

$$\frac{m + 1}{m - 1}[1 + \frac{(m - 2)\beta}{m + \beta - 2} - \frac{(3m - 5)\alpha}{(m + 1)(m + \beta - 2)}] > \frac{m + m\beta - 2}{m + \beta - 2}. \qquad (3.7)$$

Let R be the real root of equation (3.3), and

$$\alpha_1 =$$
$$\frac{2(m + 1)^2 R^4 - 8(m^2 - 1)R^3 + 2(3m^2 - 10m + 3)R^2 + (3m^2 - 2m - 5)R + 2m(m + 1)}{[(m + 1)R - 2m][(m + 1)R^2 - 2(m - 1)R - 2m - 1]}$$

$$\alpha = \frac{m - 2}{3m - 5}\alpha_1$$

$$\beta = \frac{m - 2}{(m + 1)R^2 - 2(m - 1)R - 2m - 1}$$

$$\bar{\alpha} = \frac{2(m - 2)[(m + 1)R^2 - 2(m - 1)R - m - 1]}{(3m - 5)[(m + 1)R - 2m]}$$

$$\bar{\beta} = \frac{2(m - 1)}{(m + 1)R - 2m}$$

then the inequality (3.3) becomes equality and the upper bounds given in Lemma 3.1, 3.3, 3.4 and 3.5 are all equal to R. From Lemma 3.6, let $\varepsilon_m = \frac{3m - 1}{m + 1} - R$, then we get

$$R^A(m, 2) \leq R = \frac{3m - 1}{m + 1} - \varepsilon_m.$$

Because ε_m tends to a positive number when m tends to infinity, there exists a positive number ε such that $\varepsilon_m \geq \varepsilon$ for every $m \geq 4$. ∎

The comparison for some m between the two algorithms is showed in Table 1.

m	α_m	β_m	$R^{LS}(m, 2)$	$R^A(m, 2)$
4	0.9482	1.1512	2.2	2.1835
5	0.8951	1.2557	2.3333	2.3025
9	0.7634	1.4854	2.6	2.5353
∞	0.5357	1.8779	3	2.8795

Table 1

In the following we devise an algorithm $\bar{A}$ for general $s_m = s > 1$.

Algorithm $\overline{A}$:

Let ε_1 and ε_2 be two positive numbers. If $2 - \varepsilon_1 \leq s \leq 2 + \varepsilon_2$, then we use algorithm A to schedule, otherwise we use algorithm LS.

Theorem 3.8. There exist suitable $\varepsilon_1 > 0$ and $\varepsilon_2 > 0$ and $\varepsilon_m > 0$ such that

$$R^{\overline{A}}(m) \leq \frac{3m-1}{m+1} - \varepsilon_m$$

for every $m \geq 4$.

Proof : Because the upper bounds given in Lemma 3.1, 3.2, 3.4 and 3.5 are continuous functions of s, from **Theorem 2.1** and **Theorem 3.7**, **Theorem 3.8** is proved. ∎

4 Conclusion

In this paper we derived two on-line algorithms that beat **List Scheduling** in the measure of worst case performance (for $m \geq 4$) on two conditions respectively. However, the following question may be very intresting. First, asymptotically our analysis did not improve the heuristic LS in Theorem 3.8, since ε_m may tend to zero as m tends to infinity. Second, in Theorem 3.8 $R^{\overline{A}}(m.s) = R^{LS}(m,s)$ for most $s > 1$. Are there on-line scheduling algorithms with worst case better then $R^{LS}(m,s)$ for any fixed $s \geq 1$? Third, for $m \geq 4$ machines, we gave a lower bound of 2. How can we get a greater lower bound ?

Acknowledgement We are grateful to our supervisor professor Minyi Yue for his encouragement on this problem.

References

1. Yookum Cho and Sartaj Sahni. *Bounds for list schedules on uniform processors*. SIAM J. Compute. 9(1980), pp91-103.

2. R.L. Graham. *Bounds on multiprocessing timing anomalies*. SIAM J. Appl. Math. 17(1969), pp416-429.

3. G. Galambos and G.J. Woeginger. *An on-line scheduling heuristic with better worst case ratio than Graham's List Scheduling*. SIAM J. Comput. 22(1993), pp349-355.

4. Y. Bartal, A. Fiat, H.Karloff and R. Vohra. *New algorithms for an ancient scheduling problem*. In Proceedings of 24th ACM Symposium on Theory of Computing. 1992, pp51-58.

An Algebraic Characterization of Tractable Constraints

Peter Jeavons and David Cohen

Department of Computer Science, Royal Holloway, University of London, UK
email: p.jeavons@dcs.rhbnc.ac.uk

Abstract. Many combinatorial search problems may be expressed as 'constraint satisfaction problems', and this class of problems is known to be NP-complete. In this paper we investigate what restrictions must be imposed on the allowed constraints in order to ensure tractability. We describe a simple algebraic closure condition, and show that this is both necessary and sufficient to ensure tractability in Boolean valued problems. We also demonstrate that this condition is necessary for problems with arbitrary finite domains.

1 Introduction

Many combinatorial search problems may be expressed as constraint satisfaction problems, in which values must be found for a set of variables which satisfy a number of constraints.

Deciding whether a constraint satisfaction problem has any solutions is known to be an NP-complete problem in general [14] even when the constraints are restricted to binary constraints. However, many of the problems which arise in practice have special properties which allow them to be solved efficiently. The question of identifying restrictions to the general problem which are sufficient to ensure tractability is important from both a practical and a theoretical viewpoint, and has been extensively studied.

Such restrictions may either involve the structure of the constraints, in other words which variables may be constrained by which other variables, or they may involve the nature of the constraints, in other words which combinations of values may be allowed for variables which are mutually constrained. Examples of the first approach may be found in [6, 7, 10, 16, 17] and examples of the second approach may be found in [1, 2, 12, 16, 21, 22]. In this paper we take the second approach, and investigate those families of constraints which ensure tractability in whatever way they are combined.

One of the earliest papers on this question was written by Schaefer in 1978 [20]. He considered the special case of constraint satisfaction problems in which all the variables are Boolean. In this special case, each constraint represents a logical relation between some subset of the variables, and this class of problems is usually referred to as the GENERALIZED SATISFIABILITY problem [8].

Schaefer proved that this GENERALIZED SATISFIABILITY problem remains NP-complete unless the set, Γ, of logical relations allowed as constraints satisfies one of the following six conditions:

1. Every relation in Γ is satisfied when all variables are False.
2. Every relation in Γ is satisfied when all variables are True.
3. Every relation in Γ is definable by a formula in conjunctive normal form in which each conjunct has at most one negated variable.
4. Every relation in Γ is definable by a formula in conjunctive normal form in which each conjunct has at most one unnegated variable.
5. Every relation in Γ is definable by a formula in conjunctive normal form in which each conjunct contains at most 2 literals.
6. Every relation in Γ is the set of solutions of a system of linear equations over the finite field GF(2).

In this paper, we show that these 6 criteria for tractability may be reduced to one simple criterion, which may be defined in terms of an algebraic closure property of the relations in Γ.

Furthermore, we consider how this simple algebraic criterion may be applied to more general constraint satisfaction problems in which there are more than two possible values for each variable. We show that any family of constraints which does not satisfy this criterion gives rise to a class of constraint satisfaction problems which is NP-complete.

The paper is organised as follows. In Section 2 we give the basic definitions, and describe the algebraic closure condition for constraints which will be used to characterize tractability. In Section 3 we show that this closure condition provides a necessary and sufficient condition for tractability in the GENERALIZED SATISFIABILITY problem. In Section 4 we examine the more general case of larger domain sizes, and show that the same closure condition is a necessary condition for tractability in all cases.

2 Definitions

2.1 The Constraint Satisfaction Problem

We now describe the general constraint satisfaction problem which has been widely studied in the Artificial Intelligence community [16, 14, 13]

Definition 1. An instance of a *constraint satisfaction problem* consists of:

- A finite set of variables, N, identified by the natural numbers $1, 2, \ldots, n$.
- A domain of values, D
- A set of constraints $\{C(S_1), C(S_2), \ldots, C(S_c)\}$.
  Each S_i is an ordered subset of the variables, and each constraint $C(S_i)$ is a set of tuples indicating the mutually consistent values for the variables in S_i.

The length of the tuples in a given constraint will be called the 'arity' of that constraint[1]. In particular, unary constraints specify the allowed values for a single

---

[1] Strictly speaking we also need to associate an arity with an empty set of tuples, but we shall neglect this special case, in order to simplify the presentation.

variable, and binary constraints specify the allowed combinations of values for a pair of variables.

A *solution* to a constraint satisfaction problem is an assignment of values to the variables which is consistent with all of the constraints. Deciding whether or not a given problem instance has a solution is NP-complete in general [14] even when the constraints are restricted to binary constraints. In this paper we shall consider how restricting the allowed constraints to some fixed subset of all the possible constraints affects the complexity of this decision problem. We therefore make the following definition, where Γ is a set of sets of tuples.

Definition 2. $\mathrm{CSP}(\Gamma)$ is the class of decision problems with

INSTANCE: A constraint satisfaction problem P in which all constraints are elements of Γ.

QUESTION: Does P have a solution?

Example 1. Consider the following set of 4-tuples:

$$C = \{ (0, 1, 1, 1),$$
$$(1, 0, 0, 1),$$
$$(1, 0, 1, 0),$$
$$(1, 0, 1, 1),$$
$$(0, 1, 1, 0) \}$$

$\mathrm{CSP}(\{C\})$ contains all constraint satisfaction problems in which the constraints are all equal to C.

Example 2. When the domain of values is $\{\text{True}, \text{False}\}$, then the decision problem for constraint satisfaction problems is equivalent to the SATISFIABILITY problem for propositional formulae [8].

When Γ is a set of logical relations, then the class of problems $\mathrm{CSP}(\Gamma)$ is equivalent to the GENERALIZED SATISFIABILITY problem [20, 8].

If every problem in $\mathrm{CSP}(\Gamma)$ may be solved in polynomial time, then we shall say that Γ is a 'tractable' set of constraints.

2.2 Combining Constraints

By combining constraints in various ways it is possible to generate new constraints. We therefore define the following set of constraints:

Definition 3. The set of constraints which is 'generated' by a set of constraints, Γ over a domain D, will be denoted Γ^*. It is defined to be the smallest set of constraints such that:

1. $\Gamma \subseteq \Gamma^*$;
2. For any $C \in \Gamma^*$ of arity r, and any permutation π on r points,
   $\{(x_{\pi(1)}, x_{\pi(2)}, \ldots, x_{\pi(r)}) \mid (x_1, x_2, \ldots, x_r) \in C\} \in \Gamma^*$;

3. For any $C \in \Gamma^*$ of arity r,
$$\{(x_1, x_2, \ldots, x_r, x_{r+1}) \mid (x_1, x_2, \ldots, x_r) \in C, x_{r+1} \in D\} \in \Gamma^*;$$

4. For any $C \in \Gamma^*$ of arity r
$$\{(x_1, x_2, \ldots, x_{r-1}) \mid \exists (x_1, x_2, \ldots, x_r) \in C\} \in \Gamma^*;$$

5. For any arity r, and any set $\Gamma' \subseteq \Gamma^*$ of constraints with arity r, the intersection $\bigcap_{C \in \Gamma'} C \in \Gamma^*$.
(If Γ' is empty, this intersection is defined to be D^r).

Note that Γ^* contains the set of solutions to any constraint satisfaction problem in CSP(Γ).

2.3 Operations on Tuples

Any operation on the elements of a domain may be extended to an operation on tuples over that domain by applying the operation in each coordinate position separately.

Hence, any operation defined on a domain of values may be used to define an operation on the elements of a constraint over that domain, as follows:

Definition 4. Let C be a constraint over domain D, and let $\otimes : D^k \rightarrow D$ be a k-ary operation on D.

For any collection of k tuples, $t_1, t_2, \ldots, t_k \in C$, (not necessarily all distinct) where $t_i = (x_{i1}, x_{i2}, \ldots, x_{ir})$, define $\otimes(t_1, t_2, \ldots, t_k)$ as follows:
$$\otimes(t_1, t_2, \ldots, t_k) =$$
$$(\otimes(x_{11}, x_{21}, \ldots, x_{k1}), \otimes(x_{12}, x_{22}, \ldots, x_{k2}), \ldots, \otimes(x_{1r}, x_{2r}, \ldots, x_{kr}))$$

Using this definition, we now define the following closure property of constraints.

Definition 5. Let C be a constraint over domain D, and let $\otimes : D^k \rightarrow D$ be a k-ary operation on D.

C is said to be $\otimes$-closed if, for all $t_1, t_2, \ldots, t_k \in C$ (not necessarily all distinct),
$$\otimes(t_1, t_2, \ldots, t_k) \in C$$

Example 3. Let $\triangle$ denote the ternary operation which returns the first repeated value of its three arguments, or the first value if they are all distinct.

The constraint C given in Example 1 is $\triangle$-closed, since applying the $\triangle$ operation to any 3 elements of C yields an element of C. For example
$$\triangle((0, 1, 1, 1), (1, 0, 0, 1), (1, 0, 1, 0)) = (1, 0, 1, 1)$$

If Γ is a set of constraints, and $\otimes$ is an operation such that every $C \in \Gamma$ is $\otimes$-closed, then we shall say that Γ is $\otimes$-closed.

Lemma 6. *If a set of constraints Γ is $\otimes$-closed, then Γ^* is also $\otimes$-closed.*

Proof. Follows immediately from Definitions 5 and 3.

Note that this result implies that if all the constraints in some problem instance are ⊗-closed, then the set of solutions is also ⊗-closed.

We shall be primarily interested in operations which do not simply rename the values in D. Hence we make the following definition.

Definition 7. Any operation $\otimes : D^k \to D$ will be called *essentially unary* if there exists an index $i \in \{1, 2, \ldots, k\}$ such that for all $d_1, d_2, \ldots, d_k \in D$ we have

$$\otimes(d_1, d_2, \ldots, d_k) = f(d_i)$$

where f is some non-constant function on D.

Any operation which is either constant, or depends on more than one of its arguments will be called *essentially non-unary*.

3 Tractability in Binary Domains

Using the result obtained by Schaefer [20] described in Section 1, we now establish the very close connection between tractability and closure for Boolean valued problems.

Theorem 8. *Let Γ be a set of constraints over some domain D with $|D| = 2$.*

If Γ is $\otimes$-closed for some essentially non-unary operation $\otimes$, then $CSP(\Gamma)$ is solvable in polynomial time. Otherwise $CSP(\Gamma)$ is NP-complete.

To prove this result we first show that if C is $\otimes$-closed, then this implies that it is also closed under at least one of a set of special operations:

Lemma 9. *Let C be a constraint over the domain $D = \{True, False\}$.*

If C is $\otimes$-closed for some essentially non-unary operation $\otimes$, then C is closed under one of the following operations:

$$
\begin{array}{lll}
\otimes_0 : D \to D, & \text{where } \otimes_0(x) = False & \text{for all } x \in D. \\
\otimes_1 : D \to D, & \text{where } \otimes_0(x) = True & \text{for all } x \in D. \\
\otimes_2 : D^2 \to D, & \text{where } \otimes_2(x, y) = x \vee y & \text{for all } x, y \in D. \\
\otimes_3 : D^2 \to D, & \text{where } \otimes_3(x, y) = x \wedge y & \text{for all } x, y \in D. \\
\otimes_4 : D^3 \to D, & \text{where } \otimes_4(x, y, z) = (x \wedge y) \vee (y \wedge z) \vee (x \wedge z) & \text{for all } x, y, z \in D. \\
\otimes_5 : D^3 \to D, & \text{where } \otimes_5(x, y, z) = x \oplus y \oplus z & \text{for all } x, y, z \in D,
\end{array}
$$
(where $\oplus$ denotes the exclusive-or operator).

Proof. If $\otimes$ is constant, then C must be closed under $\otimes_0$ or $\otimes_1$, so the result is immediate.

For the other cases, note that if C is $\otimes$-closed, then it must also be closed under all operations which can be obtained by composing and iterating the operation $\otimes$ in arbitrary ways. The operations which can be obtained in this way constitute an 'iterated function system'[2]. Such systems were studied by

---

[2] In the terminology of universal algebra, this set of functions is called the 'clone generated by $\otimes$' [15].

Emil Post, who published a complete description of all possible iterated function systems over Boolean variables in 1941 [18].

When the possible systems which contain non-constant essentially non-unary functions are ordered by inclusion there are just four minimal elements, which Post denotes by S_1, P_1, D_2 and L_4. Hence, if C is closed under any non-constant essentially non-unary operation then there exists $\mathcal{F} \in \{S_1, P_1, D_2, L_4\}$ such that C is $\otimes$-closed for all $\otimes \in \mathcal{F}$.

Finally, we observe from [18] that

- $\otimes_2 \in S_1$;
- $\otimes_3 \in P_1$;
- $\otimes_4 \in D_2$;
- $\otimes_5 \in L_4$.

We now show that closure under one of these special operations corresponds to membership in one of Schaefer's six tractable classes.

Lemma 10. *Let C be a non-empty logical relation.*

1. *C is $\otimes_0$-closed if and only if C contains the tuple in which all variables are False.*
2. *C is $\otimes_1$-closed if and only if C contains the tuple in which all variables are True.*
3. *C is $\otimes_2$-closed if and only if C is definable by a formula in conjunctive normal form in which each conjunct has at most one negated variable.*
4. *C is $\otimes_3$-closed if and only if C is definable by a formula in conjunctive normal form in which each conjunct has at most one unnegated variable.*
5. *C is $\otimes_4$-closed if and only if C is definable by a formula in conjunctive normal form in which each conjunct contains at most 2 literals.*
6. *C is $\otimes_5$-closed if and only if C is the set of solutions of a system of linear equations over the finite field GF(2).*

Proof. 1. Immediate.
2. Immediate.
3. A simple corollary of Lemma 4.5 in [5]. (see also Corollary 4.3 in [2].)
4. Lemma 4.5 in [5].
5. Footnote to Lemma 3.1B in [20].
6. Lemma 3.1A in [20].

Proof. (**Theorem 8**) Assume, without loss of generality, that $D = \{True, False\}$.

If Γ is $\otimes$-closed, for some essentially non-unary operation $\otimes$, then by Lemma 9 Γ must be closed under one of the six operations listed in Lemma 9.

Hence, by Lemma 10, the subset Γ' of Γ consisting of all non-empty elements of Γ must be contained in one of the six tractable classes described in [20], which implies that any problem in CSP(Γ) may be solved in polynomial time.

Conversely, if there is no essentially non-unary operation $\otimes$ such that Γ is $\otimes$-closed, then by Lemma 10, Γ is not contained in any of the tractable classes described in [20]. Hence by Theorem 2.1 of [20], CSP(Γ) is NP-complete.

We have shown that the algebraic closure condition described in Theorem 8 is both necessary and sufficient to ensure tractability (assuming P≠NP).

4 Tractability in Larger Domains

It is natural to ask whether the strikingly simple result given in Theorem 8 may be extended to larger domain sizes.

In this Section we establish the following result for constraints with arbitrary finite domain sizes:

Theorem 11. *For any set of constraints Γ over a finite domain D at least one of the following conditions must hold:*

1. *$CSP(\Gamma)$ is NP-complete.*
2. *Γ is $\otimes$-closed for some essentially non-unary operation $\otimes$.*

For the proof of this result it is convenient to extend the notion of closure to partial operations.

Definition 12. Let C be a constraint over domain D, and let $\otimes : D' \rightarrow D$, where $D' \subseteq D^k$, be a partial k-ary operation on D.

C is said to be *$\otimes$-closed* if, for all $t_1, t_2, \ldots, t_k \in C$ (not necessarily all distinct), such that $\otimes(t_1, t_2, \ldots, t_k)$ is defined

$$\otimes(t_1, t_2, \ldots, t_k) \in C.$$

We now show that a set of constraints must either generate a binary 'not-equal' constraint, or else be closed under a non-trivial partial operation.

Lemma 13. *For any set of constraints Γ over domain D, at least one of the following conditions must hold:*

1. *The 'not-equal' constraint $Q_D = \{(u, v) \mid u, v \in D,\ u \neq v\}$ is an element of Γ^*.*
2. *Γ^* is $\otimes$-closed for some non-trivial partial operation $\otimes$.*

Proof. Consider the constraint $C_0 = \bigcap \{C' \in \Gamma^* \mid Q_D \subseteq C'\}$. Clearly, $C_0 \in \Gamma^*$ and $Q_D \subseteq C_0$.

If $Q_D = C_0$ then $Q_D \in \Gamma^*$ and we are done, so assume that $Q_D \neq C_0$, and let $t \in C_0 \setminus Q_D$. By the definition of Q_D, we have $t = (d, d)$ for some $d \in D$.

Now label the elements of Q_D, so that $Q_D = \{(u_1, v_1), (u_2, v_2), \ldots, (u_m, v_m)\}$ and let $D' = \{(u_1, u_2, \ldots, u_m), (v_1, v_2, \ldots, v_m)\}$. We claim that every $C \in \Gamma^*$ must be $\otimes$-closed for the partial operation $\otimes : D' \rightarrow D$ such that $\otimes(u_1, u_2, \ldots, u_m) = d$ and $\otimes(v_1, v_2, \ldots, v_m) = d$.

To establish this claim, assume for contradiction that there exists some $C \in \Gamma^*$ such that C is not $\otimes$-closed. This means that C contains m tuples, $t_1, t_2, \ldots, t_m$, such that at each coordinate position, j, either $t_i|_j = u_i$ for

$i = 1, 2, \ldots, m$, or $t_i|_j = v_i$ for $i = 1, 2, \ldots, m$, but C does not contain the tuple $(d, d, \ldots, d)$.

But Γ^* contains all projections of C, so this implies that Γ^* contains a binary constraint C' such that $Q_D \subseteq C'$ but $(d, d) \notin C'$. This contradicts the definition of C_0, so the claim is established.

We also make use of the fact that closure under a (non-trivial) partial operation implies closure under some extension of that partial operation to a total operation.

Lemma 14. *Let Γ be a set of constraints.*

If Γ is $\otimes$-closed for some partial operation $\otimes' : D' \rightarrow D$, where $\emptyset \neq D' \subseteq D^k$, then Γ^* is $\otimes$-closed for some total operation $\otimes : D^k \rightarrow D$ which agrees with $\otimes'$ on D'.

Proof. Omitted. (Similar to Theorem 3 in [9].)

Proof. (**Theorem 11**) First note that if Γ is $\otimes$-closed for some unary operation $\otimes$, then for any problem $P \in \mathrm{CSP}(\Gamma)$ and any solution t of P, we know that $\otimes(t)$ is also a solution to P. Hence we may replace each $C \in \Gamma$ with the constraint $\otimes(C) = \{\otimes(t) \mid t \in C\}$ without affecting the existence of solutions.

Hence, if Γ is $\otimes$-closed for some unary non-injective operation $\otimes$, then we may replace each $C \in \Gamma$ with $\otimes(C)$ to obtain a set of constraints over a smaller domain. By repeating this process, we may obtain a set of constraints $\bar{\Gamma}$ over domain $\bar{D} \subseteq D$, such that $\bar{\Gamma}$ is not $\otimes$-closed for any non-injective unary function $\otimes$.

If $|\bar{D}| \leq 2$ then the result holds by Theorem 8.

Otherwise, we have $|\bar{D}| \geq 3$. If the 'not-equal' constraint Q_D, defined in Lemma 13, is an element of $\bar{\Gamma}^*$, then the $|\bar{D}|$-COLORING problem [8] may be reduced to $\mathrm{CSP}(\bar{\Gamma})$, so $\mathrm{CSP}(\bar{\Gamma})$ is NP-complete. Hence $\mathrm{CSP}(\Gamma)$ is NP-complete, and we are done.

Otherwise, by Lemma 13 and Lemma 14, $\bar{\Gamma}$ must be $\bar{\otimes}$-closed for some operation $\bar{\otimes}$. Furthermore, by the proof of Lemma 13, if $\bar{\otimes}$ is essentially unary, then it is non-injective. Hence, by the construction of $\bar{\Gamma}$, $\bar{\otimes}$ must be essentially non-unary, and hence Γ is $\otimes$-closed for some essentially non-unary operation $\otimes$.

If we assume that P is not equal to NP, then Theorem 11 establishes that closure under some essentially non-unary operation is a necessary condition for tractability.

In the previous Section we were able to show that this condition is also sufficient to ensure tractability when $|D| = 2$, by considering all minimal 'iterated function systems' [18] or, in other words, 'minimal clones' [15]. The number of possible minimal clones rises rapidly with the size of the domain, and a complete description is not known for $|D| > 3$ [3, 19].

However, considerable progress has been made in the study of minimal clones during the last ten years. In particular, it has been shown [19] that every minimal clone containing essentially non-unary operations is generated by an operation which is either

- a binary operation of a restricted kind (see [4] for details); or
- a ternary 'majority' operation (that is, an operation $\otimes : D^3 \to D$ such that $\otimes(x,x,y) = \otimes(x,y,x) = \otimes(y,x,x) = x$); or
- a semiprojection (that is, an operation $\otimes : D^k \to D$ such that $\otimes(x_1,x_2,\ldots,x_k) = x_i$ for some fixed i, whenever $|\{x_1,x_2,\ldots,x_k\}| < k$); or
- the ternary operation $x + y + z$, where $+$ corresponds to the operation of an elementary Abelian 2-group.

Some of these types of operations may easily be shown to give rise to tractable constraints, (see [11] for examples).

A complete study of each possible type of minimal clone, and the complexity of the corresponding constraint classes, is currently being completed, but is beyond the scope of this paper.

5 Conclusion

The results presented in this paper lay the foundation for an algebraic theory of tractability in combinatorial problems.

In a separate paper [11], we show how the currently known classes of tractable constraints may be characterized by the property of being $\otimes$-closed for a specific operation $\otimes$. The results presented above establish the strong link between the study of such algebraic operations and the study of computational complexity. We expect this link to lead to considerable further progress in understanding the boundary between tractable and intractable combinatorial problems. For example, this work may provide a useful approach to combinatorial problems, such as GRAPH ISOMORPHISM, whose complexity is not yet established.

References

1. Cooper, M.C., Cohen, D.A., Jeavons, P.G., "Characterizing tractable constraints", *Artificial Intelligence 65*, (1994), pp. 347–361.
2. Cooper, M.C., & Jeavons, P.G., "Tractable constraints on ordered domains", Technical Report, Dept of Computer Science, Royal Holloway, University of London, (1994) and submitted to *Artificial Intelligence*.
3. Csakany, B., "All minimal clones on the three-element set", *Acta Cybernetica 6*, (1983), pp. 227–238.
4. Csakany, B., "On conservative minimal operations", in *Lectures in Universal Algebra (Proc. Conf. Szeged 1983)*, Colloq. Math. Soc. Janos Bolyai 43, North-Holland, (1986), pp. 49–60.
5. Dechter, R., & Pearl, J., "Structure identification in relational data", *Artificial Intelligence 58* (1992) pp. 237–270.
6. Dechter, R. & Pearl J. "Network-based heuristics for constraint-satisfaction problems", *Artificial Intelligence 34* (1988), pp. 1–38.
7. Freuder, E. C. "A sufficient condition for backtrack-bounded search", *Journal of the ACM 32* (1985) pp. 755–761.
8. Garey, M.R., & Johnson, D.S., *Computers and intractability: a guide to NP-completeness*, Freeman, San Francisco, California, (1979).

9. Geiger, D., "Closed systems of functions and predicates" *Pacific Journal of Mathematics 27* (1968) pp. 95–100.

10. Gyssens, M., Jeavons, P., Cohen, D., "Decomposing constraint satisfaction problems using database techniques", *Artificial Intelligence 66*, (1994), pp. 57–89.

11. Jeavons, P.G., Cohen, D.A., Gyssens, M., "A unifying framework for tractable constraints", Technical Report, Dept of Computer Science, Royal Holloway, University of London, (1995) to appear in Proceedings of *Constraint Programming '95*.

12. Kirousis, L., "Fast parallel constraint satisfaction", *Artificial Intelligence 64*, (1993), pp. 147–160.

13. Ladkin, P.B., & Maddux, R.D., "On binary constraint problems", *Journal of the ACM 41* (1994), pp. 435–469.

14. Mackworth, A. K. "Consistency in networks of relations", *Artificial Intelligence 8* (1977) pp. 99–118.

15. McKenzie, R.N., McNulty, G.F., Taylor, W.F., *Algebras, lattices and varieties. Volume I*, Wadsworth and Brooks, California (1987).

16. Montanari, U., "Networks of constraints: fundamental properties and applications to picture processing", *Information Sciences 7* (1974), pp. 95–132.

17. Montanari, U., & Rossi, F., "Constraint relaxation may be perfect", *Artificial Intelligence 48* (1991), pp. 143–170.

18. Post, E.L., *"The two-valued iterative systems of mathematical logic"*, Annals of Mathematical Studies 5, Princeton University Press, (1941).

19. Rosenberg, I.G., "Minimal clones I: the five types", in *Lectures in Universal Algebra (Proc. Conf. Szeged 1983)*, Colloq. Math. Soc. Janos Bolyai 43, North-Holland, (1986), pp. 405–427.

20. Schaefer, T.J., "The complexity of satisfiability problems", *Proc 10th ACM Symposium on Theory of Computing (STOC)* , (1978) pp. 216–226.

21. van Beek, P., "On the minimality and decomposability of row-convex constraint networks", *Proceedings of the Tenth National Conference on Artificial Intelligence, AAAI-92*, MIT Press, (1992) pp. 447–452.

22. Van Hentenryck, P., Deville, Y., Teng, C-M., "A generic arc-consistency algorithm and its specializations", *Artificial Intelligence 57* (1992), pp. 291–321.

Limit Property of Unbalanced Development in Economic Network

Jiyu Ding[1], Chengxiang Qing[2] and Guodong Song[3]

[1]*Dept. of Math. ,Qiqihar Teachers' College,China*
[2]*Qiqihar economic information centre,China*
[3]*Dept. of Basis Science,Qiqihar Light Industry Institute,China*

Abstract. In this paper,the limit property of output vector of unbalanced development in economic network are discussed. We obtain a main result: this limit vector has relation to initial input vector in general case.

1. Introduction

Suppose that an economic system S is cosisted of p sectors $v_1, v_2, \cdots, v_p$. If economic connections form v_i to v_j is measured by a real number $a_{ij}(\geq 0)$, then a_{ij} is called intensity from v_i to v_j . If every a_{ij} is constant for every i and j ,the system S is called linear multi—sectors economic system. If the weight matrix $A(G)$ of graph G is $A(G) = A = [a_{ij}]$,the graph G is called network of the economic system S .

The algebraic multiplicity of the eigenvalue λ_i of $A(G)$ is denoted by $m(\lambda_i)$,the eigenvector associated with the eigenvalue λ_i is $X_i, i = 1,2,\cdots,r$, and we have $\sum_i m(\lambda_i) = p$. We can postulate $|\lambda_1| \geq |\lambda_2| \geq \cdots \geq |\lambda_r|$.

From theory of input—output model we know[1] that if input vector is $X(t)$ at t period,then the output vector is $X(t+1)$ and the growth model of this economic network can be represented by

$$A(G)X(t+1) = X(t) \tag{1}$$

When $X(0)$ is given,the limit properties of $X(t+1)$ as $t \to \infty$ are very important for an economic network, because the equilibrium stability and growth trend of this economic system are shown by the various cases of the limit.

2. Balanced and unbalanced growth of economic network

Lemma 1[2] Suppose that economic network $G = (V,E), p \geq 2, A(G) = [a_{ij}]$,then

    (1) If G is not connected,then the spectrum of G is the union of spectrums of every components of G .

    (2) If G is strongly connected, $|\lambda_1| = \lambda_1, m(\lambda_1) = 1$ and $X_1 > 0$.

Supported by the Heilongjiang Foundation of Natural Science.

(3) $|\lambda_i| < \lambda_1$, for $i = 2,3,\cdots,r$.

Lemma 2[3] If G is strongly connected and $A = A(G) = [a_{ij}]$, then

$$\lim_{t\to\infty}(A^t/\lambda_1^t) = L = [a_1X_1,a_2X_1,\cdots,a_pX_1] \qquad (2)$$

By above—mentioned results, when initial vector $X(0)$ and vector X_1 is propotional $(X(0) = \alpha X_1)$ we have

$$X(t) = \frac{1}{\lambda_1}X(t-1) = \frac{\alpha}{\lambda_1^t}X_1$$

Since $X(t)$ and X_1 are always propotional in growth process, so the network is always in balance development.

Next we study unbalanced development system. We extended the results about simple matrix case in [2] into general full rank matrix.

Lemma 3 Suppose that A is a $p \times p$ full rank matrix and its eigenvector corresponding to eigenvalue λ_i is $X_1^{(i)}$, $m(\lambda_i) \geqslant 1$, $i = 1,2,\cdots,r$ and $\sum_i m(\lambda_i) = p$. then there exist a group of linear independent vectors

$$\{X_1^{(1)},\cdots,X_{m_1}^{(1)},\cdots,X_1^{(r)},\cdots,X_{m_r}^{(r)}\} \qquad (3)$$

and satisfy

$$AX_j^{(i)} = \lambda_i X_j^{(i)} + X_{j-1}^{(i)}, \quad i = 1,2,\cdots,r; \quad j = 1,2,\cdots,m_i \quad (4)$$

where when $j = 0$ we define $X_j^{(i)} = 0$.

Proof: Since A is full rank, so there are an invertible matrix H that

$$H^{-1}AH = J = \text{diag}(J_1,\cdots,J_r) \qquad (5)$$

where J_i is i th $m_i \times m_i$ submatrix (Jordan partitioned matrices) when $m_i = 1, J_i = [\lambda_i]$, where that $m_i = m(\lambda_i)$ is not always true, because corresponding to one eigenvalue can be many Jordan partitioned matrices.
Let

$$H = [H_1,\cdots,H_r], \quad H_i = [X_1^{(i)},\cdots,X_{m_i}^{(i)}],$$

where $X_j^{(i)}$ $(i = 1,2,\cdots,r; \quad j = 1,2,\cdots,m_i)$ is m_i th column vector. By (5), we have $AH_i = H_iJ_i$, therefore

$$A[X_1^{(i)},\cdots,X_{m_i}^{(i)}] = [X_1^{(i)},\cdots,X_{m_i}^{(i)}]J_i \qquad (6)$$

so that (4) is true.

Theorem 1. Suppose that G is strongly connected, eigenvector corresponding to the largest eigenvalue λ_1 of economic network G is X_1 and $A = A(G)$,and input vector $X(t)$ of G is not propotional with $X_1^{(1)}$ at t period, then there exactly are a positive number K ,so that $X(t+k) \geqslant 0$ is not true, for $k > K$.

Theorem 2. Suppose that $A = A(G)$ and λ_u is smallest eigenvalue of A . Under conditions of theorem 1 when $k \to \infty$,the limit of $X(t+k)$ depend on $X(t)$ if $|\lambda_u| < |\lambda_i|$,for $i < u$;The limit does not exist if $|\lambda_u| = |\lambda_j|$ for some j is true.

645

3. Proof of main results

Proof of theorem 1

Suppose $\alpha(>0)$ is a constant and costant vector $X = X(t) \neq \alpha X_1^{(1)}$. Because A and A^T is cospectral so λ_1 is their the largest eigenvalue. Suppose corresponding eigenvector of A^T is y, then $y^T A^k = \lambda_1^k y^T$

By formula (1)

$$A^k X(t+k) = X \tag{7}$$

$$\lambda_1^k y^T X(t+k) = y^T X \tag{8}$$

By lemma 1, $\lambda_1 > 0$ and $y^T > 0$, so right of equality (8) is a postive constant number.

If $k > K > 0, X(t+k) \geqslant 0$, since right of equality (8) is positive, so three factors of left of equality (8) all are positive ($\lambda_1 > 0, y^T > 0, X(t+k) \geqslant 0$ and $X(t+k) \neq 0$) that the left of equlity (8) is bounded for any k. By Weierstrass theorem exactly there is a subsequence $\{l\}$ so that

$$\lim_{l \to 0} \lambda_1^l X(t+l) = X \tag{9}$$

By formula (2),(7) and (9),

$$X = \lim_{l \to \infty} A^l X(t+l) = \lim_{l \to \infty} \frac{A^l}{\lambda_1^l} \lambda_1^l X(t+l)$$
$$= LX = X_1^{(1)} [\alpha_1, \cdots, \alpha_p] X$$

Let $Q^T = [\alpha_1, \cdots, \alpha_p], \alpha = Q^T X$ is constant, so $X = \alpha X_1^{(1)}$ is contradictory with postulation, the theorem is true.

Proof of theorem 2

Suppose the group of generalized eigenvector of A is (3), we first prove
$$A^{-k} X_j^{(i)} = \lambda_i^{-k}(X_j^{(i)} + \alpha_1^i \lambda_i^{-1} X_{j-1}^{(i)} + \cdots + \alpha_{j-1}^i \lambda_i^{-j+1} X_1^{(i)}) \tag{10}$$
where $\alpha_1^i, \cdots, \alpha_{j-1}^i$ are all constants independenting to λ_i. By definition of generalized eigenvector and (4), we have recurrence formula.
$$A^{-1} X_j^{(i)} = \lambda_i^{-1}(X_j^{(i)} - A^{-1} X_{j-1}^{(i)}), j = 1,2,\cdots, m_i \tag{11}$$
so
$$A^{-1} X_j^{(i)} = \lambda_i^{-1}(X_j^{(i)} - \lambda_i^{-1} X_{j-1}^{(i)} + \lambda_i^{-2} X_{j-2}^{(i)} - \cdots + (-1)^{-j+1} \lambda_i^{-j+1} X_1^{(i)})$$
then by mathematical induction, formula (10) is proved.

Assume linear represantation of $X(t)$ of A on basis(3) is
$$X(t) = \sum_{i=1}^r (C_1^{(i)} X_1^{(i)} + \cdots + C_{m_i}^{(i)} X_{m_i}^{(i)}) \tag{12}$$
where $C_1^{(1)}, \cdots, C_{m_1}^{(1)}, \cdots, C_1^{(r)}, \cdots, C_{m_r}^{(r)}$ are linear representation coefficients on basis (3). Suppose that the first nonzero coefficient from right in (12) is $C = C_v^{(u)}, 1 \leqslant u \leqslant r, 1 \leqslant v \leqslant m_u$, so
$$X(t+k) = A^{-k} X(t) = \sum_{i=1}^u A^{-k}(C_1^{(i)} X_1^{(i)} + \cdots + C_{m_i}^{(i)} X_{m_i}^{(i)})$$
By (11) in above formula and let $\phi_j^{(i)}(\lambda_i^{-1})$ be polynomial of degree j of λ_i^{-1} (independent k), we have

$$X(t+k) = \sum_{i=1}^{u} \lambda_i^{-k}[\varphi_{m_i-1}^{(i)}(\lambda_i^{-1})X_1^{(i)} + \varphi_{m_i-2}^{(i)}(\lambda_i^{-1})X_2^{(i)} + \cdots$$
$$+ \varphi_0^{(i)}(\lambda_i^{-1})X_{m_i}^{(i)}]$$

Suppose that $|\lambda_u| < |\lambda_i|, \lambda_u \neq \lambda_i$ and $i < u$

$$X(t+k) = \lambda_u^{-k}\sum_{i=1}^{u}\left(\frac{\lambda_u}{\lambda_i}\right)^k [\varphi_{m_i-1}^{(i)}(\lambda_i^{-1})X_1^{(i)} + \varphi_{m_i-2}^{(i)}(\lambda_i^{-1})X_2^{(i)}$$
$$+ \cdots + \varphi_0^{(i)}(\lambda_i^{-1})X_{m_i}^{(i)}]$$

Now consider two cases:

1. Since $|\frac{\lambda_u}{\lambda_i}| < 1$, when $k \to \infty$, except terms corresponding Jordan partitioned matrices including eigenvalue λ_u, other terms all converge to zero, so that

$$\lim_{k \to \infty} \frac{X(t+k)}{\|X(t+k)\|} = \beta \sum_{\lambda_i = \lambda_u}[\varphi_{m_i-1}^{(i)}(\lambda_i^{-1})X_1^{(i)} + \cdots + \varphi_0^{(i)}(\lambda_i^{-1})X_{m_i}^{(i)}] \quad (13)$$

where β is a constant. Because $\varphi_j^{(u)}(\lambda_u^{-1})$ is dependent on the coefficients of $X(t)$, in general case, $X(t+k)$ is dependent on value of $X(t)$.

2. Suppose that $|\lambda_u| = |\lambda_j|, \lambda_u \neq \lambda_j$, for some j and $j < u$. Since $|\frac{\lambda_u}{\lambda_j}| = 1$, we know that $|\frac{\lambda_u}{\lambda_j}|^k$ are the points on the unit circumference, so that the sum of these terms in (13)

$$\sum_{|\lambda_j| = |\lambda_u|}\left(\frac{\lambda_u}{\lambda_j}\right)^k [\varphi_{m_j-1}^{(j)}(\lambda_j^{-1})X_1^{(j)} + \cdots + \varphi_0^{(j)}(\lambda_j^{-1})X_{m_j}^{(j)}]$$

canont be convergent. That is to say the limit $X(t+k)/\|X(t+k)\|$ does not exist.

Corollary 1. When $u = r$ and $m(\lambda_u) = 1$ in linear representation of $X(t)$ on basis(3), then

$$X(t+k) \to aX_1^{(r)}, (k \to \infty)$$

where a is constant.

Corollary 2. Suppose that A is simple matrix we have
$$X(t+k) \to aX_1^{(u)}, (k \to \infty)$$

where a is constant, $X_1^{(u)}$ is the minimal eigenvector of A.

Reference

[1] R. G. Hwa, Mathematical theory of large scale optimazation of planed economy, Chinese Science Bulletin 18(1984), 20−25.

[2] Guodoug Song and Jiyu Ding, Characteristic root of nonnegative weighted network and its applications in linear multi−sectors economic system, OR and Decision making, Chengdu University of Science and Technology Prees, Vol. 2(1992), 1824−1829.

[3] P. Lancaster and M. Tismeneisky, The Theory of Matrix Second Edition with Application, Academic Press, London, 1985.

Document Processing, Theory, and Practice *

Derick Wood

Department of Computer Science, Hong Kong University of Science & Technology,
Clear Water Bay, Kowloon, Hong Kong

Abstract: Document processing has a very wide scope; therefore, I will concentrate on a few aspects of document processing (to the exclusion of others), namely, the communication, manipulation, querying, and typesetting of textual documents. I aim to persuade you that document processing is a fertile topic in that it raises questions in areas as diverse as: programming languages, databases, systems, algorithms, and formal language theory. In addition, the kinds of questions are also diverse: they range from the theoretical to the practical with stops in between.

* This work was supported under Natural Sciences and Engineering Research Council of Canada grants.

Matching and Comparing Sequences in Molecular Biology

Tao Jiang*

Dept of Computer Science, McMaster Univ., Hamilton, Ontario L8S 4K1, Canada.
Email: jiang@maccs.mcmaster.ca

ABSTRACT

The primary structure of a deoxyribonucleic acid (DNA) molecule is a sequence consisting of four types of letters, A, C, G, and T, each stands for a nucleotide. The length of such a DNA sequence ranges from several thousand letters for a simple virus to three billion letters for a human. We all know that these long and mysterious sequences encode Life as well as genetic diseases, but decoding the sequences is perhaps one of the most challenging tasks in the world. The ultimate goal of molecular biology is to understand what segments of a DNA are responsible for a biological function such as the color of eyes or a genetic disease such as cancer, and how these segments are formed and work. These functionally meaningful segments of a DNA are usually called genes. To find the genes which are responsible for some biological function, a biologist often compares a set of DNA sequences that share the same function and tries to identify regions which are "conserved" in all of these sequences. On the other hand, a biologist may also infer the "closeness" of two organisms by comparing their DNA sequences and computing the degree of similarity of the sequences. Such "closeness" information is useful in the reconstruction of evolutionary histories.

In this talk, we survey various frameworks for the problem of comparing a pair or a set of sequences, including approximate string matching, string edit, (pairwise) sequence alignment, local sequence alignment, and multiple sequence alignment. For the comparison of a pair of sequences, we will describe some standard algorithms and several techniques to improve the time and space efficiency such as preprocessing and divide-and-conquer. Multiple sequence alignment has been identified as one of the most challenging problems in computational molecular biology. In the rest of the talk, we will concentrate on two important variants of multiple sequence alignment: multiple alignment with sum-of-all-pairs (SP) score and multiple alignment with tree score. Since both are NP-hard, we discuss polynomial-time approximation algorithms for these problems with a guaranteed performance bound. If time allows, we will also mention some popular heuristics for performing multiple alignment that seem to work well in practice but do not have a guaranteed performance.

Computational molecular biology is emerging as a fast growing interdisciplinary field involving biology, computer science, statistics, applied mathematics, etc. and is providing a lot of interesting algorithmic questions for theoretical

---

* Research supported in part by NSERC Research Grant OGP0046613 and MRC/NSERC CGAT Grant GO-12278.

computer scientists to solve. We hope that this brief survey will serve as an introduction to one (important) aspect of the field. Most of the results mentioned above or a pointer to them can be found in the following literature.

References

1. V. Bafna, E. Lawler and P. Pevzner, Approximation algorithms for multiple sequence alignment, *Proc. 5th Combinatorial Pattern Matching Conference*, 1994, Asilomar, California.
2. S. Chan, A. Wong and D. Chiu, A survey of multiple sequence comparison methods, *Bulletin of Mathematical Biology* 54(4), 563-598, 1992.
3. T. Jiang and M. Li, Optimization problems in molecular biology, in *Advances in Optimization and Approximation*, D.Z. Du and J. Sun (eds.), Kluwer Academic Publishers, MA, 195-216, 1994.
4. T. Jiang, E. Lawler and L. Wang, Aligning sequences via an evolutionary tree: complexity and approximation, *Proc. 26th ACM Symposium on Theory of Computing*, 1994, Montreal, Canada; final version to appear in *Algorithmica*.
5. E. Lander, R. Langridge and D. Saccocio, Mapping and interpreting biological information, *Communications of the ACM* 34(11), 33-39, 1991.
6. R. Lipton, T. Mar and J. Welsh, Computational approaches to discovering semantics in molecular biology, *Proceedings of the IEEE* 77(7), 1056-60, 1989.
7. E. Myers, An overview of sequence comparison algorithms, *Technical Report* 91-29, Dept of Computer Science, University of Arizona, 1991.
8. D. Sankoff and J. Kruskal (Eds), *Time Warps, String Edits, and Macromolecules: the Theory and Practice of Sequence Comparison*, Addison-Wesley, Reading, MA, 1983.
9. M. Waterman, Sequence alignments, in *Mathematical Methods for DNA Sequences*, M.S. Waterman (ed.), CRC, Boca Raton, FL, 53-92, 1989.

Primal-Dual Schema Based Approximation Algorithms

Vijay V. Vazirani

College of Computing
Georgia Institute of Technology

Abstract] One runs into the following dilemma while designing an approximation algorithm for an NP-hard optimization problem: for establishing the performance guarantee of the algorithm, the cost of the solution found needs to be compared with that of the optimal; however, computing the cost of the optimal is NP-hard as well. Hence a key consideration is establishing a good lower bound on the cost of the optimal solution, assuming we have a minimization problem. For optimization problems that can be expressed as integer programming problems, the following general methodology has been quite successful: use the cost of the optimal solution to the LP-relaxation as the lower bound. In fact, LP-duality not only provides a method of lower bounding the cost of the optimal solution, but also a general schema for designing the algorithm itself: the primal-dual schema. This schema has been applied to several problems including set cover and its generalizations [Jo, Lo, Ch, RV], the generalized Steiner network problem [AKR, GW, KR, WGMV, G+], and finding integral multicommodity flow and multicut in trees [GVY].

The primal-dual schema enables one to find special solutions to an LP, although so far it has only been used for obtaining good integral solutions to an LP-relaxation. In the past, the primal-dual schema has yielded the most efficient known algorithms to some cornerstone problems in P, including matching, network flow and shortest paths. These problems have the property that their LP-relaxations have optimal solutions that are integral, and so the primal-dual schema is able to find an optimal solution to the original integer program. Since numerous NP-hard optimization problems can be expressed as integer programs, this schema holds even more promise in the area of approximation algorithms.

The main idea behind this schema is the following: Consider the LP-relaxation, the primal LP, and obtain its dual; we are assuming that the primal is a minimization problem, and the dual is a maximization problem. We start with a primal infeasible solution and a dual feasible solution. Next, we iteratively improve the feasibility of the primal solution, and the optimality of the dual solution, always ensuring that the primal solution is extended integrally. During these iterations, the current primal solution is used to determine the improvement to the dual, and vice versa. We terminate when a primal feasible solution is obtained; clearly, this solution will be integral. Furthermore, the cost of the dual solution will be a lower bound on the cost of the optimal primal solution, and hence also a lower bound on the cost of the optimal solution to the original integer program. Hence, by comparing the cost of the primal solution with that of the dual solution, we obtain the approximation guarantee. Despite being so general, this schema leaves sufficient scope for exploiting the special combinatorial structure of the specific problem: in designing the solution improving algorithms, and in comparing the final solutions, thereby yielding very good approximation guarantees.

Solutions found using this schema have been used in practice. In particular, the algorithm for the generalized Steiner nework problem given in [WGMV] was implemented at Bellcore [MDMS], and forms the algorithmic core of their product, the CCSN Toolkit, a network design package.

References

[AKR]   A. Agrawal, P. Klein, and R. Ravi, "When trees collide: An approximation algorithm for the generalized Steiner problem in networks," *Proceedings, 23rd Annual ACM Symposium on Theory of Computing*, 1991.

[Ch]   V. Chvatal, "A greedy heuristic for the set covering prob lem," *Math. Oper. Res.* 4, 233-235, 1979.

[GVY]   N. Garg and V.V. Vazirani and M. Yannakakis, "Primal-dual approximation algorithms for integral flow in trees, with applications to matching and set cover," *Proceedings, 20th International Colloquium on Automata, Languages and Programming*, 1993.

[GW]   M.X. Goemans and D.P. Williamson, "A general approximation technique for constrained forest problems," *Proceedings, 3rd Annual ACM-SIAM Symposium on Discrete Algorithms, 307-316*, 1992.

[G+]   M.X. Goemans, A. Goldberg, S. Plotkin, D. Shmoys, E. Tardos, and D.P. Williamson, "Improved approximation algorithms for network design problems," *Proceedings, 5th Annual ACM-SIAM Symposium on Discrete Algorithms*, 1994.

[Jo]   D. S. Johnson, "Approximation Algorithms for Combinatorial Problems," *J. Comput. Sys. Sci.* 9, 256-278, 1974.

[KR]   P.N. Klein and R. Ravi "When cycles collapse: A general approximation technique for constrained two connectivity problems", *Proceedings, 3rd Integer Programming and Combinatorial Optimization Conference, 39-56*, 1993.

[Lo]   L. Lovasz, "On the ratio of Optimal Integral and Fractional Covers," *Discrete Math.* 13, 383-390, 1975.

[MDMS] M. Mihail, N. Dean, M. Mostrel, and D. Shallcross "Algorithm Specifications for the Common Channel Signaling Network Topology Analyzer Feature of Integrated Technology Planning," August 1994 release, Issue 3, Bellcore TM-ARA-24103, 1994.

[RV]   S. Rajagopalan and V. V. Vazirani, "Primal-Dual RNC Approximation Algorithms for (multi)Set (multi)Cover and Covering Integer Programs," *Pro-*

ceedings, 34th Annual Conference on the Foundations of Computer Science,
1993.

[WGMV] D. P. Williamson, M. X. Goemans, M. Mihail, and V. V. Vazirani, "A Primal-Dual Approximation Algorithm for Generalized Steiner Network Problems," *Proceedings, 25th ACM Symposium on Theory of Computing,* 1993.

Author Index

Springer-Verlag
and the Environment

We at Springer-Verlag firmly believe that an international science publisher has a special obligation to the environment, and our corporate policies consistently reflect this conviction.

We also expect our business partners – paper mills, printers, packaging manufacturers, etc. – to commit themselves to using environmentally friendly materials and production processes.

The paper in this book is made from low- or no-chlorine pulp and is acid free, in conformance with international standards for paper permanency.

Lecture Notes in Computer Science

For information about Vols. 1–886

please contact your bookseller or Springer-Verlag